This graduate text introduces relativistic quantum theory, emphasizing its important applications in condensed matter physics.

Basic theory, including special relativity, angular momentum and particles of spin zero are first reprised. The text then goes on to discuss the Dirac equation, symmetries and operators, and free particles. Physical consequences of solutions including hole theory and Klein's paradox are considered. Several model problems are solved. Important applications of quantum theory to condensed matter physics then follow. Relevant theory for the one-electron atom is explored. The theory is then developed to describe the quantum mechanics of many electron systems, including Hartree–Fock and density functional methods. Scattering theory, band structures, magneto-optical effects and superconductivity are amongst other significant topics discussed. Many exercises and an extensive reference list are included.

This clear account of relativistic quantum theory will be valuable to graduate students and researchers working in condensed matter physics and quantum physics.

Relativistic Quantum Mechanics

RELATIVISTIC
QUANTUM MECHANICS

WITH APPLICATIONS IN CONDENSED MATTER
AND ATOMIC PHYSICS

Paul Strange

Keele University

CAMBRIDGE
UNIVERSITY PRESS

PUBLISHED BY THE PRESS SYNDICATE OF THE UNIVERSITY OF CAMBRIDGE
The Pitt Building, Trumpington Street, Cambridge CB2 1RP, United Kingdom

CAMBRIDGE UNIVERSITY PRESS
The Edinburgh Building, Cambridge CB2 2RU, UK http://www.cup.cam.ac.uk
40 West 20th Street, New York, NY 10011-4211, USA http:/www.cup.org
10 Stamford Road, Oakleigh, Melbourne 3166, Australia

First published 1998

Printed in the United Kingdom at the University Press, Cambridge

Typeset in Monotype Times 11/13, in TEX

A catalogue record for this book is available from the British Library

Library of Congress cataloguing in Publication data

Strange, Paul, 1956–
Relativistic quantum mechanics / Paul Strange.
p. cm.
Includes bibliographical references and index.
ISBN 0 521 56271 6
1. Relativistic quantum theory. I. Title.
QC174.24.R4S87 1998
530.12–dc21 97-18019 CIP

ISBN 0 521 56271 6 hardback
ISBN 0 521 56583 9 paperback

Dedication

Unlike so many other supportive families, mine did not proof-read, or type this manuscript, or anything else. In fact they played absolutely no part in the preparation of this book, and distracted me from it at every opportunity. They are not the slightest bit interested in physics and know nothing of relativity and quantum theory. Their lack of knowledge in these areas does not worry them at all. Furthermore they undermine one of the central tenets of the theory of relativity, by providing me with a unique frame of reference. Nonetheless, I would like to dedicate this book to them, Jo, Jessica, Susanna and Elizabeth.

Contents

Preface

I always thought I would write a book and this is it. In the end, though, I hardly wrote it at all, it evolved from my research notes, from essays I wrote for postgraduates starting work with me, and from lecture handouts I distribute to students taking the relativistic quantum mechanics option in the Physics department at Keele University. Therefore the early chapters of this book discuss pure relativistic quantum mechanics and the later chapters discuss applications of relevance in condensed matter physics. This book, then, is written with an audience ranging from advanced students to professional researchers in mind. I wrote it because anyone aiming to do research in relativistic quantum theory applied to condensed matter has to pull together information from a wide range of sources using different conventions, notation and units, which can lead to a lot of confusion (I speak from experience). Most relativistic quantum mechanics books, it seems to me, are directed towards quantum field theory and particle physics, not condensed matter physics, and many start off at too advanced a level for present day physics graduates from a British university. Therefore, I have tried to start at a sufficiently elementary level, and have used the SI system of units throughout.

When I started preparing this book I thought I might be able to write everything I knew in around fifty pages. It soon became apparent that that was not the case. Indeed it now appears to me that the principal decisions to be taken in writing a book are about what to omit. I have written this much quantum mechanics and not used the word Lagrangian. This saddens me, but surely must make me unique in the history of relativistic quantum theory. I have not discussed the very interesting quantum mechanics describing the neutrino and its helicity, another topic that invariably appears in other relativistic quantum mechanics texts. However, as we are leaning towards condensed matter physics in this book, there are sections on topics such as magneto-optical effects and magnetic anisotropy which don't appear in other books despite being intrinsically relativistic and quantum mechanical in nature. In the end, what is included and what is omitted is just a question of taste, and it is up to the reader to decide whether such decisions were good or bad.

This book is very mathematical, containing something like two thousand equations. I make no apology for that. I think the way the mathematics works is the great beauty of the subject. Throughout the book I try to make the mathematics clear, but I do not try to avoid it. Paraphrasing Niels Bohr I believe that "If you can't do the maths, you don't understand it." If you don't like maths, you are reading the wrong book.

There are a lot of people I would like to thank for their help with, and influence on, my understanding of quantum mechanics, particularly the relativistic version of the theory. They are Dr E. Arola, Dr P.J. Durham, Professor H. Ebert, Professor W.M. Fairbairn, Professor J.M.F. Gunn, Prof B.L. Györffy, Dr R.B. Jones, Dr P.M. Lee, Dr J.B. Staunton, Professor J.G. Valatin, and Dr W. Yeung.

Several of the examples and problems in this book stem from projects done by undergraduate students during their time at Keele, and from the work of my Ph.D students. Thanks are also due to them, C. Blewitt, H.J. Gotsis, O. Gratton, A.C. Jenkins, P.M. Mobit, and E. Pugh, and to the funding agencies who supported them (Keele University physics department, the EPSRC, and the Nuffield foundation).

There are several other people I would like to thank for their general influence, encouragement and friendship. They are Dr T. Ellis, Professor M.J. Gillan, Dr M.E. Hagen, Dr P.W. Haycock, Mr J. Hodgeson and Mr B.G. Locke-Scobie. I would also like to thank R. Neal and L. Nightingale of Cambridge University Press for their encouragement of, and patience with, me. Finally, my parents do not have a scientific background, nonetheless they have always supported me in my education and have taken a keen interest in the writing of this book. Thanks are also due to them, R.J. and V.A. Strange.

I hope you enjoy this book, although I am not sure 'enjoy' is the right word to describe the feeling one has when reading a quantum mechanics textbook. Perhaps it would be better to say that I hope you find this book informative and instructive. What I would really like would be for you to be inspired to look deeper into the subject, as I was by my undergraduate lectures many years ago. Many people think quantum mechanics is not relevant to everyday life, but it has certainly influenced my life for the better! I hope it will do the same for you.

1

The Theory of Special Relativity

Relativistic quantum mechanics is the unification into a consistent theory of Einstein's theory of relativity and the quantum mechanics of physicists such as Bohr, Schrödinger, and Heisenberg. Evidently, to appreciate relativistic quantum theory it is necessary to have a good understanding of these component theories. Apart from this chapter we assume the reader has this understanding. However, here we are going to recall some of the important points of the classical theory of special relativity. There is good reason for doing this. As you will discover all too soon, relativistic quantum mechanics is a very mathematical subject and my experience has been that the complexity of the mathematics often obscures the physics being described. To facilitate the interpretation of the mathematics here, appropriate limits are taken wherever possible, to obtain expressions with which the reader should be familiar. Clearly, when this is done it is useful to have the limiting expressions handy. Presenting them in this chapter means they can be referred to easily.

Taking the above argument to its logical conclusion means we should include a chapter on non-relativistic quantum mechanics as well. However, that is too vast a subject to include in a single chapter. Furthermore, there already exists a plethora of good books on the subject. Therefore, where it is appropriate, the reader will be referred to one of these (Baym 1967, Dicke and Wittke 1974, Gasiorowicz 1974, Landau and Lifschitz 1977, Merzbacher 1970, and McMurry 1993).

This chapter is included for revision purposes and for reference later on, therefore some topics are included without much justification and without proof. The reader should either accept these statements or refer to books on the classical theory of special relativity. In the first section of this chapter we state the fundamental assumptions of the special theory of relativity. Then we discuss the Lorentz transformations of time and space. Next we come to discuss velocities, momentum and energy. Then we go on to think about relativity and the electromagnetic field. Finally, we look at the Compton effect where relativity and quantum theory are brought together for the first time in most physics courses.

1.1 The Lorentz Transformations

Newton's laws are known to be invariant under a Galilean transformation from one reference frame to another. However, Maxwell's equations are not invariant under such a transformation. This led Michelson and Morley (1887) to attempt their famous experiment which tried to exploit the non-invariance of Maxwell's equations to determine the absolute velocity of the earth. Here, I do not propose to go through the Michelson–Morley experiment (Shankland *et al.* 1955). However, its failure to detect the movement of the earth through the ether is the experimental foundation of the theory of relativity and led to a revolution in our view of time and space. Within the theory of relativity both Newton's laws and the Maxwell equations remain the same when we transform from one frame to another. This theory can be encapsulated in two well-known postulates, the first of which can be written down simply as

(1) All inertial frames are equivalent.

By this we mean that in an isolated system (e.g. a spaceship with no windows moving at a constant velocity **v** (with respect to distant stars or something)) there is no experiment that can be done that will determine **v**. According to Feynman (1962) this principle has been verified experimentally (although a bunch of scientists standing around in a spaceship not knowing how to measure their own velocity is not a sufficient verification). Here, we are implicitly assuming that space is isotropic and uniform. The second postulate is

(2) There exists a maximum speed, c. If a particle is measured to have speed c in one inertial frame, a measurement in any other inertial frame will also give the value c (provided the measurement is done correctly). That is, the speed of light is independent of the speed of the source and the observer.

The whole vast consequences of the theory of relativity follow directly from these two statements (French 1968, Kittel *et al.* 1973). It is necessary to find transformation laws from one frame of reference to another that are consistent with these postulates (Einstein 1905). Consider a Cartesian frame S in which there is a source of light at the origin. At time $t = 0$ a spherical wavefront of light is emitted. The distance of the wavefront

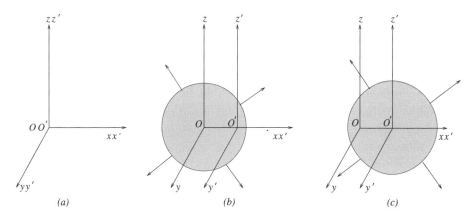

Fig. 1.1. (a) At $t = t' = 0$ the two frames are coincident and observers O and O' are at the origin. At this time a spherical wavefront is emitted. (b) At a time $t > 0$ as viewed by an observer stationary in the unprimed frame. Observer O is at the centre of the wavefront. (c) At a time $t' > 0$ as viewed by an observer stationary in the primed frame. Observer O' is at the centre of the wavefront. Note that in (b) and (c) it is not possible for the observer not at the centre of the wavefront to be outside the wavefront.

from the origin at any subsequent time t is given by

$$x^2 + y^2 + z^2 = c^2 t^2 \tag{1.1}$$

Now consider a second frame S' moving in the x-direction with velocity v relative to S. Let us set up S' such that its origin coincides with the origin of S at $t' = t = 0$ when the wavefront is emitted. Now the equation giving the distance of the wavefront from the origin of S' at a subsequent time t' as measured in S' is

$$x'^2 + y'^2 + z'^2 = c^2 t'^2 \tag{1.2}$$

So, at all times $t, t' > 0$, observers at the origin of both frames would believe themselves to be at the centre of the wavefront. However, each observer would see the other as being displaced from the centre. This is illustrated in figure 1.1. It can easily be seen that a Galilean transformation relating the coordinates in equations (1.1) and (1.2) does not give consistent results. A set of coordinate transformations that are consistent with (1.1) and (1.2) is

$$x' = \frac{x - vt}{\sqrt{1 - v^2/c^2}}, \quad y' = y, \quad z' = z, \quad t' = \frac{t - (xv/c^2)}{\sqrt{1 - v^2/c^2}} \tag{1.3a}$$

and the inverse transformations are

$$x = \frac{x' + vt'}{\sqrt{1 - v^2/c^2}}, \quad y = y', \quad z = z', \quad t = \frac{t' + (x'v/c^2)}{\sqrt{1 - v^2/c^2}} \tag{1.3b}$$

These equations are known as the Lorentz transformations. Under these transformations the interval s defined by

$$s^2 = (ct')^2 - x'^2 - y'^2 - z'^2 = (ct)^2 - x^2 - y^2 - z^2 \tag{1.4}$$

is a constant in all frames.

It is conventional to adopt the notation

$$\gamma = \frac{1}{\sqrt{1 - v^2/c^2}}, \qquad \beta = v/c \tag{1.5}$$

Equations (1.3) lead to some startling conclusions. Firstly consider measurements of length. If we measure the length of a rod by looking at the position of its ends relative to a ruler, then if in the S frame the rod is at rest we can measure the ends at x_1 and x_2 and infer that its length is $L = x_2 - x_1$. Now consider the situation in the primed frame. The observer will measure the ends as being at points x'_1 and x'_2 and hence $L' = x'_2 - x'_1$. We want to know the relation between these two lengths. The rod is moving at velocity $-v$ in the x-direction relative to the observer in S'. To find the length this observer must have measured the position of the ends simultaneously (at t'_0) in his frame. So, considering the first of equations (1.3b) we have

$$x_1 = \frac{x'_1 + vt'_0}{\sqrt{1 - v^2/c^2}}, \qquad x_2 = \frac{x'_2 + vt'_0}{\sqrt{1 - v^2/c^2}} \tag{1.6}$$

Subtracting these equations leads directly to

$$L' = x'_2 - x'_1 = \sqrt{(1 - v^2/c^2)}(x_2 - x_1) = \sqrt{(1 - v^2/c^2)}L \tag{1.7}$$

This is the famous Lorentz–Fitzgerald contraction and is illustrated in figure 1.2. It shows that observers in different inertial frames of reference will measure lengths differently. The length of any object takes on its maximum value in its rest frame. Let us emphasize that nothing physical has happened to the rod. Measuring the length of the rod from one reference frame is a different experiment to measuring the length from another reference frame, and the different experiments give different answers. The process of measuring correctly gives a different result in different inertial frames of reference.

The above description of Lorentz–Fitzgerald contraction depended crucially on the fact that the observer in S' performed his measurements of

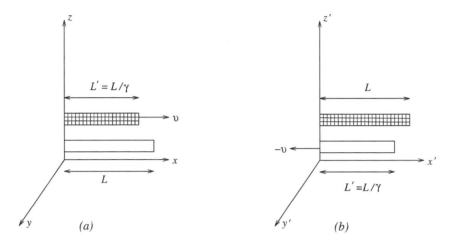

Fig. 1.2. Here are two rods that are identical in their rest frames with length L. In (a) we are in the rest frame of the lower rod. The upper rod is moving in the positive x-direction with velocity v and is Lorentz–Fitzgerald contracted so that its length is measured as L/γ. In (b) we are in the rest frame of the upper rod and the lower rod is moving in the negative x-direction with velocity $-v$. In this frame of reference it is the lower rod that appears to be Lorentz–Fitzgerald contracted.

the position of the end points simultaneously. It is important to note that simultaneous in S' does not mean simultaneous in S. So the fact that the light from the ends of the rod arrived at the observer in S' at the same time does not mean it left the ends of the rod at the same time. This is trivial to verify from the time transformations in equations (1.3).

Next we consider intervals of time. Imagine a clock and an observer in frame S at rest with respect to the clock. The observer can measure a time interval easily enough as the time between two readings on the clock

$$\tau = t_2 - t_1 \tag{1.8}$$

Now we can use the Lorentz time transformations to find the times t_2' and t_1' as measured by an observer in S' again moving with velocity v in the x-direction relative to the observer in S:

$$t_1' = \frac{t_1 - (x_1 v/c^2)}{\sqrt{1 - v^2/c^2}}, \qquad t_2' = \frac{t_2 - (x_2 v/c^2)}{\sqrt{1 - v^2/c^2}} \tag{1.9}$$

We can subtract one of these from the other to discover how to transform time intervals from one frame to another:

$$t_2' - t_1' = \frac{t_2 - t_1}{\sqrt{1 - v^2/c^2}} = \gamma(t_2 - t_1) \tag{1.10}$$

where we have set $x_2 - x_1 = 0$. This is obviously true as the clock is defined as staying at the same coordinate in S. What we have found here is the time dilation formula. The time interval measured in S' is longer than the time interval measured in S. Another way of stating the same thing is to say that moving clocks appear to run more slowly than stationary clocks. This, of course, is completely counter-intuitive and takes some getting used to. However, it has been well established experimentally, particularly from measurements of the lifetime of elementary particles. It is also responsible for one of the most famous of all problems in physics, the twin paradox.

Next, let me describe a thought experiment that one can do, which de-mystifies time dilation to some extent, and shows explicitly that it arises from the constancy of the speed of light. Consider a train in its rest frame S as shown in the top diagram in figure 1.3 (with a rather idealized train). Light is emitted from a transmitter/receiver on the floor of the train in a vertical direction at time zero. It is reflected from a mirror on the ceiling and the time of its arrival back at the receiver is noted. The ceiling is at a height L above the floor, so the time taken for the light to make the return journey is

$$t = \frac{2L}{c} \tag{1.11}$$

Now suppose there is an observer in frame S', i.e. sitting by the track as the train goes past while the experiment is being done, and there is a series of synchronized clocks in this frame. This is shown in the lower part of figure 1.3. The observer in S' can also time the light pulse. Using Pythagoras's theorem it is easy to see from the figure that when the light travels a distance L in S, it travels a distance $(L^2 + (\frac{1}{2}vt')^2)^{1/2}$ in S', and it goes the same distance for the reflected path. So the total distance travelled as viewed by the observer in S' is

$$d = 2(L^2 + (\tfrac{1}{2}vt')^2)^{1/2} \tag{1.12}$$

But the velocity of light is the same in all frames. So

$$d^2 = c^2 t'^2 = 4L^2 + v^2 t'^2 \tag{1.13}$$

Rearranging this

$$t' = \frac{2L}{(c^2 - v^2)^{1/2}} = \frac{2L}{c}\gamma = \gamma t \tag{1.14}$$

Thus if the clock in the train tells us the light's journey time was t, the clocks by the side of the track tell us it was $\gamma t > t$. So, to the observer at the side of the track, the clock in S will appear to be running slowly. Equation (1.14) is exactly the same as equation (1.10) which was obtained directly from the Lorentz transformations.

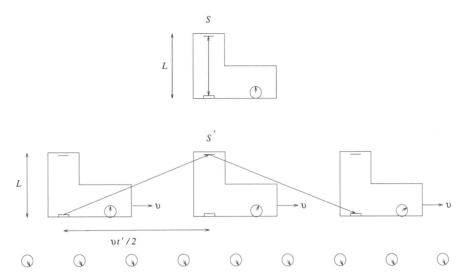

Fig. 1.3. Thought experiment illustrating time dilation, as discussed in the text. The upper figure shows the experiment in the rest frame of the train and the lower figure shows it in the frame of an observer by the side of the track.

Equations (1.3) are easy to derive from the postulates, and easy to apply. However, their meaning is not so clear. In fact they can be interpreted in several ways. Depending on the circumstances, I tend to think of them in two ways. Firstly, a rather woolly and obvious statement. At low velocities non-relativistic mechanics is OK because the time taken for light to get from the object to the detector (your eye) is infinitesimal compared with the time taken for the object to move, so the velocity of light does not affect your perception. However, when the object is moving at an appreciable fraction of the speed of light, the time taken for the light to reach your eye does have an appreciable effect on your perception. Secondly, a rather grander statement. Let us consider space and time as different components of the same thing, as is implied by equations (1.1) and (1.3). Any observer (Observer 1) can split space–time into space and time unambiguously, and will know what he or she means by space and time separately. Any observer (Observer 2) moving with a non-zero velocity with respect to Observer 1 will be able to do the same. However, Observer 2 will not split up time and space in the same way as Observer 1. Observers in different inertial frames separate time and space in different ways!

1.2 Relativistic Velocities

Once we have the Lorentz transformations for position and time, it is an easy matter to construct the velocity transformation equations. As before, we have a frame S in which we measure the velocity of a particle to have three components u_x, u_y and u_z. Now let the frame S' be moving with velocity v relative to S in the x-direction.

We can write the Lorentz transformations (1.3) in differential form:

$$\Delta x' = \frac{\Delta x - v\Delta t}{\sqrt{1 - v^2/c^2}}, \qquad \Delta y' = \Delta y,$$

$$\Delta z' = \Delta z, \qquad \Delta t' = \frac{\Delta t - (\Delta x v/c^2)}{\sqrt{1 - v^2/c^2}} \qquad (1.15)$$

Now velocity in the x-direction in the S frame is given by $\Delta x/\Delta t$ and in the S' frame by $\Delta x'/\Delta t'$, and similarly for the other components. So we simply divide each of the space transformations in (1.15) by the time transformation, and divide top and bottom of the resulting fraction by Δt to obtain

$$u'_x = \frac{u_x - v}{1 - vu_x/c^2} \qquad (1.16a)$$

$$u'_y = \frac{u_y(1 - v^2/c^2)^{1/2}}{1 - vu_x/c^2}, \qquad u'_z = \frac{u_z(1 - v^2/c^2)^{1/2}}{1 - vu_x/c^2} \qquad (1.16b)$$

Equations (1.16) enable us to find the velocity of an object in any other Lorentz frame given its velocity in one such frame (see figure 1.4). These equations are certainly consistent with the postulate that c is the ultimate speed. If, for example, we substitute $u_x = c$ in (1.16a) it is trivial to see that $u'_x = c$ as well for any value of v, the relative velocity of the frames. If, instead of choosing our photon velocity parallel to the relative motion of the frames, we choose it in an arbitrary direction, the magnitude of the velocity as measured in the S' frame also works out as c. However, the angle the photon makes to the axes of S as measured in S is, in general, different to the angle it makes to the axes of S' as measured in S'.

1.3 Mass, Momentum and Energy

Perhaps the most famous equation in the whole of physics, and certainly one of the most important and fundamental in the theory of relativity, is the equivalence of mass and energy described by

$$E = mc^2 \qquad (1.17)$$

where m is the mass of a particle as measured in its rest frame. It is also known that, for photons with zero rest mass, the energy E and frequency

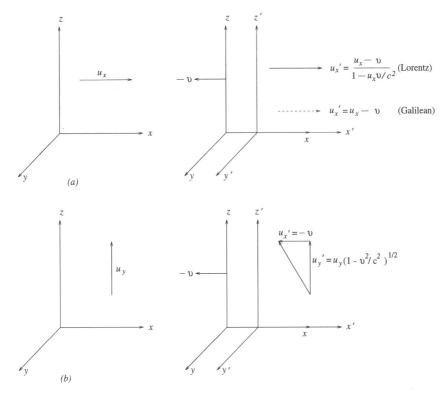

Fig. 1.4. Velocity transformation between different frames. (*a*) The upper left frame contains a particle moving at velocity v_x. In the upper right figure the dashed line indicates the velocity in the primed frame found from a Galilean transformation and the full line indicates the velocity from a Lorentz transformation using (1.16*a*). (*b*) In the lower left figure we have a particle moving parallel to the *y*-axis of an unprimed frame. Its velocity as viewed from a frame moving in the *x*-direction relative to the unprimed frame has components in both the x' and y' directions.

v are related by

$$E = hv = \frac{hc}{\lambda} = pc \tag{1.18}$$

where p and λ are the photon momentum and wavelength respectively. Putting these two equations together, the total energy of any free particle is given by

$$E^2 = p^2c^2 + m^2c^4 \tag{1.19}$$

Obviously these statements do not constitute a derivation of (1.19). For a full discussion of the origin of this equation the reader should refer to standard texts on relativity. Equation (1.19) is usually developed for a single particle. However, if we have a collection of particles and observe them in a frame S, we can measure their energy E and the magnitude of the vector sum of their momenta p and we can form the quantity $E^2 - p^2c^2$. The same thing can be done again in a frame S', and we find

$$E^2 - p^2c^2 = E'^2 - p'^2c^2 \tag{1.20}$$

This is not just a statement about the rest mass of the particles because there is no necessity for them all to be at rest in the same frame. The quantity on both sides of equation (1.20) is described as being relativistically invariant, i.e. it doesn't change under a Lorentz transformation. Compare equation (1.20) with the fundamental definition of an interval given by (1.4) which is also a Lorentz invariant. From equation (1.19) it can be shown that both the energy and mass as measured in a frame moving with velocity $\mathbf{v}$ are given in terms of their rest frame values by

$$E(v) = \frac{mc^2}{\sqrt{1 - v^2/c^2}} = \gamma mc^2 \tag{1.21}$$

and

$$m(v) = \frac{m}{\sqrt{1 - v^2/c^2}} = \gamma m \tag{1.22}$$

where $v = |\mathbf{v}|$. Equation (1.22) is the inertial property of a body moving with velocity $\mathbf{v}$ such that the momentum is given by

$$\mathbf{p} = m(v)\mathbf{v} = \gamma m\mathbf{v} \tag{1.23}$$

and

$$E = m(v)c^2 \tag{1.24}$$

It is not immediately clear that equations (1.21 − 1.23) are consistent with (1.19). The easiest way to prove this is to substitute (1.22) and (1.23) into (1.19) and we find

$$E^2 = m(v)^2\mathbf{v}^2c^2 + m^2c^4 = \frac{m^2v^2c^2}{1 - v^2/c^2} + \frac{m^2c^4(1 - v^2/c^2)}{1 - v^2/c^2}$$

$$= \frac{m^2c^4}{1 - v^2/c^2} \tag{1.25}$$

Taking the square root of this gives (1.21) directly.

Note that now velocity and momentum are no longer proportional to each other as they are in classical mechanics. The velocity is bounded by $-c < v < c$, but the momentum can take on any value $-\infty < p < \infty$. In figure 1.5 we show the relativistic momentum and the classical momentum

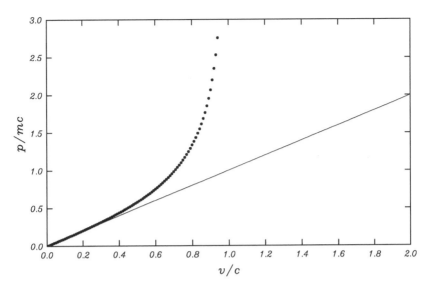

Fig. 1.5. Graph showing the classical *mv* momentum (full line) and the relativistic momentum (dotted line) given by equation (1.18). The momentum is divided by *mc*, where *m* is the rest mass, so the plotted numbers are independent of the rest mass of the particle. Note the agreement between the two curves at low velocities.

as a function of velocity. Clearly they agree at low velocities but diverge strongly when the velocity becomes an appreciable fraction of *c*. The velocity of a particle in terms of its energy and momentum is

$$\mathbf{v} = \frac{\mathbf{p}c^2}{E} \tag{1.26}$$

It is easy to see that this has the correct non-relativistic limit. As $c \to \infty$ we have

$$\mathbf{v} = \frac{\mathbf{p}c^2}{\sqrt{p^2c^2 + m^2c^4}} = \frac{\mathbf{p}c^2}{mc^2\sqrt{1 + p^2/m^2c^4}} \approx \frac{\mathbf{p}}{m(1 + p^2/2m^2c^2)}$$

$$\approx \frac{\mathbf{p}}{m} = \mathbf{v} \tag{1.27}$$

Newton's laws are still valid within a relativistic framework. By direct substitution of equation (1.23) we can define a relativistic force

$$\mathbf{F} = \frac{d\mathbf{p}}{dt} = \frac{d}{dt}\frac{m\mathbf{v}}{\sqrt{1 - v^2/c^2}} \tag{1.28}$$

and by integrating this over a path we can define work as

$$K = mc^2(\gamma - 1) \tag{1.29}$$

Now to ensure that we are on the right track it is useful to take the non-relativistic limit of some of these equations to make sure they correspond to the quantity we think they do. Firstly, let us look at the total energy, as given by equation (1.21):

$$E(v) = \frac{mc^2}{\sqrt{1 - v^2/c^2}} = mc^2(1 - v^2/c^2)^{-1/2}$$

$$\approx mc^2(1 + v^2/2c^2 - 3v^4/8c^4) = mc^2 + \frac{1}{2}mv^2 - 3mv^4/8c^2 \qquad (1.30)$$

This is just the rest mass energy plus the kinetic energy in the non-relativistic limit. Rest mass energy does not appear in non-relativistic physics and only corresponds to a redefinition of the zero of energy. So, we have correctly found the non-relativistic limit of the total energy. In a similar way we find the non-relativistic limit of K as

$$K = \frac{1}{2}mv^2 \qquad (1.31a)$$

This, of course, is the kinetic energy as we expect. Let us consider the first correction to this due to relativistic effects. It is

$$K^{\mathrm{corr}} = -3mv^4/8c^2 = -3m^4v^4/8m^3c^2 = -3p^4/8m^3c^2 \qquad (1.31b)$$

This expression will be useful in interpreting the non-relativistic limit of our relativistic wave equations in later chapters. In figure 1.6 we can see the classical (full line) and relativistic (lower dotted line) expressions for kinetic energy plotted as a function of velocity. This illustrates rather clearly the agreement between the two up to velocities of order $0.5c$, and the continually increasing divergence between them as $v \to c$. We also include on this figure the total energy of equation (1.21) (upper dotted line) directly. Clearly, from equation (1.19), this must take on the value mc^2 when the particle is at rest.

Energy and momentum become united in special relativity in the same way as space and time. The energy–momentum Lorentz transformations in one dimension are

$$p'_x = \frac{p_x - vE/c^2}{\sqrt{1 - v^2/c^2}}, \quad p'_y = p_y, \quad p'_z = p_z, \quad E' = \frac{E - vp_x}{\sqrt{1 - v^2/c^2}} \qquad (1.32)$$

The conservation of momentum and energy unite into one law, which is the conservation of the four-component energy–momentum vector. At this stage we should recall that mass and energy also are no longer separate concepts. A photon moving in the x-direction has

$$E = h\nu, \qquad p_x = h/\lambda = h\nu/c \qquad (1.33)$$

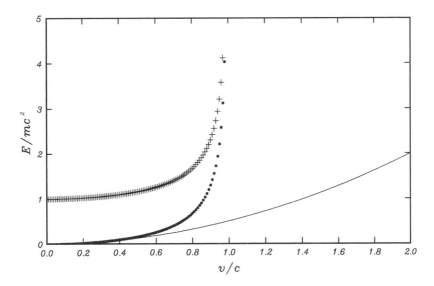

Fig. 1.6. Graph showing the classical kinetic energy (full line) and the relativistic kinetic energy (or work) (dotted line) given by equation (1.29). Note the agreement between the two at low velocities. Also shown is the total relativistic energy including the rest mass energy (crosses). All energies in this figure are divided by mc^2, where m is the rest mass, thus making the plotted numbers independent of the rest mass of the particle.

and substituting these into the last equation of (1.32) gives

$$hv' = \frac{hv - vhv/c}{\sqrt{1 - v^2/c^2}} = hv \frac{1 - v/c}{\sqrt{(1 - v/c)(1 + v/c)}} \qquad (1.34a)$$

i.e.

$$v' = v \left(\frac{1 - v/c}{1 + v/c} \right)^{1/2} \qquad (1.34b)$$

This is the well-known formula for the Doppler effect. If a source of radiation is emitting at frequency v we can use (1.34) to tell us the frequency detected by an observer travelling at velocity v relative to the source. For a relativistic effect this has a surprising number of mundane applications. It is the physics underlying the change in pitch in the sound of a train passing you as you stand on a railway station platform. It is also used to measure the speed of motor vehicles by the police. In physics perhaps its most important uses are astrophysical where the famous red shift enables us to estimate stellar and galactic relative velocities.

Finally in this section we mention briefly the problem of defining a relativistic centre of mass. We will not do any mathematics associated with this, but just point out the problem. Consider two particles (*A* and *B*) of equal mass (in their rest frames), moving uniformly towards each other. We can certainly imagine sitting in a frame of reference in which they move with equal and opposite velocities, and the point at which they collide will be the centre of mass. That will surely be the centre of mass frame. Now consider the same situation from the rest frame of particle *A*. In that frame *A* has mass *m*, but particle *B* has mass γm, from (1.22), and so the centre of mass is closer to *B* than to *A*. An observer in the rest frame of *B* will see *A* as having the greater mass and so will think that the centre of mass is nearer to *A*. Obviously these two points of view cannot be reconciled and we are left with the conclusion that the centre of mass depends upon the inertial frame in which observations are made.

1.4 Four-Vectors

Next we will discuss four-vectors (see Greiner 1990 for example). They are a familiar concept in relativity, and do provide a very neat way of writing down equations. However, they are used sparingly in this book. There are two reasons for this. They are only a mathematical nicety, and for people inexperienced in using them they can obscure the physical meaning of the equations they represent. Furthermore, they are not used much in the condensed matter physics literature. On the other hand, 4-vectors are very much used in many applications of relativity and an understanding of them is undoubtedly useful. Therefore they are introduced here and used occasionally in this book. To really understand 4-vectors and the mathematical advantage gained from them, the reader should go through the work in this book that uses 4-vectors and then redo it without 4-vector notation.

A 4-vector is defined as a set of four quantities that can be written in four component form such as $\mathbf{r} = (x, y, z, ct)$. Other familiar examples of 4-vectors are the energy–momentum 4-vector $p^{\mu} = (p_x, p_y, p_z, E/c)$, the 4-current $J^{\mu} = (J_x, J_y, J_z, c\rho)$ with ρ the charge density, and the vector potential $A^{\mu} = (A_x, A_y, A_z, \Phi/c)$ where Φ is the scalar potential. Using 4-vectors is easier if we simplify the notation. We write the space–time 4-vector as

$$x^1 = x, \quad x^2 = y, \quad x^3 = z, \quad x^4 = ct \qquad (1.35)$$

A general component of the space–time 4-vector can be written x^{μ} with $\mu = 1, 2, 3, 4$. Any vector with four components that transforms under Lorentz transformations like x^{μ} is written with an upper greek suffix, e.g. a^{μ}. Now, we may lower the suffix according to the rule

$$x_1 = -x^1, \quad x_2 = -x^2, \quad x_3 = -x^3, \quad x_4 = x^4 \qquad (1.36)$$

and similarly for any other 4-vector. The x^μ are called the contravariant components of x, and the x_μ are called the covariant components of x. For two arbitrary 4-vectors a and b we may define the scalar products

$$a_\mu b^\mu = a_4 b^4 + a_1 b^1 + a_2 b^2 + a_3 b^3$$
$$= a^4 b^4 - a^1 b^1 - a^2 b^2 - a^3 b^3 \tag{1.37}$$

This illustrates another important convention. If on one side of an equation the same greek letter appears as a lower and upper suffix it is automatically summed over. This is known as the summation convention. Let us define the fundamental tensor $g^{\mu\nu}$ such that

$$g^{44} = 1, \quad g^{11} = g^{22} = g^{33} = -1, \quad g^{\mu\nu} = 0 \ \ \mu \neq \nu \tag{1.38}$$

then we can use this to relate the covariant and contravariant elements of a 4-vector

$$x^\mu = g^{\mu\nu} x_\nu \tag{1.39}$$

where the summation convention is used. These definitions of 4-vectors and scalar product mean we can often write equations in a very neat and succinct way. For example, the definition of the space–time interval, equation (1.4), can be written

$$s^2 = x_\mu x^\mu \tag{1.40}$$

and, using the definition of the momentum 4-vector above, we can write equation (1.20) in 4-vector form as

$$c^2 p_\mu p^\mu = m^2 c^4 \tag{1.41}$$

In this book, 4-vectors will only be used a few times, and then only when they really do simplify the mathematics, without unduly obscuring the physics of what they describe.

1.5 Relativity and Electromagnetism

The whole of classical electromagnetism is described by the Maxwell equations together with the conservation of charge (Reitz and Milford 1972, Jackson 1962):

$$\nabla \cdot \mathbf{E} = \rho/\epsilon_0 \tag{1.42a}$$

$$\nabla \times \mathbf{E} = -\frac{\partial \mathbf{B}}{\partial t} \tag{1.42b}$$

$$\nabla \cdot \mathbf{B} = 0 \tag{1.42c}$$

$$\nabla \times \mathbf{B} = \mu_0 \mathbf{J} + \mu_0 \epsilon_0 \frac{\partial \mathbf{E}}{\partial t} \tag{1.42d}$$

and

$$\nabla \cdot \mathbf{J} = -\frac{\partial \rho}{\partial t} \tag{1.43}$$

Here, μ_0 is known as the permeability of free space and ϵ_0 is the permittivity of free space. If we use the vector identity

$$\nabla \times \nabla \times \mathbf{E} = \nabla \nabla \cdot \mathbf{E} - \nabla^2 \mathbf{E} \tag{1.44}$$

and substitute from equations (1.42) repeatedly for a region of space containing no charges or currents (i.e. ρ and $\mathbf{J}$ are set equal to zero), we end up with

$$\mu_0 \epsilon_0 \frac{\partial^2 \mathbf{E}}{\partial t^2} = \frac{1}{c^2} \frac{\partial^2 \mathbf{E}}{\partial t^2} = -\nabla^2 \mathbf{E} \tag{1.45a}$$

If we replace $\mathbf{E}$ in (1.44) with $\mathbf{B}$ we can derive

$$\mu_0 \epsilon_0 \frac{\partial^2 \mathbf{B}}{\partial t^2} = \frac{1}{c^2} \frac{\partial^2 \mathbf{B}}{\partial t^2} = -\nabla^2 \mathbf{B} \tag{1.45b}$$

where we have

$$c^2 = \frac{1}{\mu_0 \epsilon_0} \tag{1.46}$$

The speed of light is directly related to the electromagnetic quantities ϵ_0 and μ_0. Obviously (1.45) is a wave equation. This shows a direct relation between Maxwell's equations of electromagnetism and the wave theory of light. Although this is a triumph for Maxwell's equations, it does relate only to the wave theory and not to the photon picture of light. This is a shortcoming that we will touch upon in chapter 12.

A particle with charge e and velocity $\mathbf{v}$ in an electromagnetic field feels the Lorentz force

$$\mathbf{F} = e\mathbf{E} + e\mathbf{v} \times \mathbf{B} \tag{1.47}$$

The current density $\mathbf{J}$ is related to the electric field by Ohm's law

$$\mathbf{J} = \sigma \mathbf{E} \tag{1.48}$$

The electric and magnetic fields $\mathbf{E}$ and $\mathbf{B}$ are often written in terms of the scalar and vector potentials

$$\mathbf{B} = \nabla \times \mathbf{A} \tag{1.49}$$

$$\mathbf{E} = -\frac{\partial \mathbf{A}}{\partial t} - \nabla \Phi \tag{1.50}$$

It is clear, though, that these definitions of the fields in terms of potentials $\mathbf{A}$ and Φ are not unique. In electromagnetism, observables depend upon the forces which are written in terms of the electric and magnetic fields in equation (1.47). Any alternative expressions for the potentials that leave the fields unchanged are equally acceptable. In fact we can define equivalent potentials to **A** and Φ using any function θ such that

$$\mathbf{A}' = \mathbf{A} - \nabla\theta, \quad \Phi' = \Phi + \frac{\partial\theta}{\partial t} \tag{1.51}$$

Substituting these back into (1.49) and (1.50) gives us back (1.49) and (1.50) because $\nabla \times \nabla\theta = 0$, which follows directly from the properties of the ∇ operator, and the terms including θ we find when we substitute into (1.50) cancel immediately. This freedom to choose appropriate forms for **A** and Φ is known as gauge freedom. It is often possible to make calculations significantly easier by choosing to work in a suitable gauge. We will make use of this in chapter 9.

Maxwell's equations can provide us with our first non-trivial use of the 4-vector formalism. Let us define the electromagnetic field tensor. Written out explicitly this is

$$F^{\mu\nu} = \begin{pmatrix} 0 & -cB_z & cB_y & E_x \\ cB_z & 0 & -cB_x & E_y \\ -cB_y & cB_x & 0 & E_z \\ -E_x & -E_y & -E_z & 0 \end{pmatrix} = c\left(\frac{\partial A^\nu}{\partial x_\mu} - \frac{\partial A^\mu}{\partial x_\nu}\right) \tag{1.52}$$

We have already defined the current 4-vector, and we also define the 4-vector operator:

$$\nabla^\mu = \frac{\partial}{\partial x^\mu} = \left(\frac{\partial}{\partial x}, \frac{\partial}{\partial y}, \frac{\partial}{\partial z}, \frac{\partial}{\partial ct}\right) \tag{1.53}$$

Next we operate with $\partial/\partial x^\mu$ on $F^{\mu\nu}$, and two of Maxwell's equations can be written

$$\frac{\partial F^{\mu\nu}}{\partial x^\mu} = \frac{J^\nu}{c\epsilon_0} \tag{1.54}$$

To see this explicitly, operate with (1.53) on $F^{\mu\nu}$ and perform the required multiplications. It is then only necessary to equate the 4-vector elements on each side of (1.54) to obtain the familiar form of Maxwell's equations (1.42).

The other two Maxwell equations depend on the covariant form of the field tensor:

$$F_{\mu\nu} = \begin{pmatrix} 0 & -cB_z & cB_y & -E_x \\ cB_z & 0 & -cB_x & -E_y \\ -cB_y & cB_x & 0 & -E_z \\ E_x & E_y & E_z & 0 \end{pmatrix} = c\left(\frac{\partial A_\nu}{\partial x^\mu} - \frac{\partial A_\mu}{\partial x^\nu}\right) \tag{1.55}$$

and can be written

$$\frac{\partial F_{\mu\nu}}{\partial x^\rho} + \frac{\partial F_{\nu\rho}}{\partial x^\mu} + \frac{\partial F_{\rho\mu}}{\partial x^\nu} = 0 \tag{1.56}$$

A natural question to ask is how the electromagnetic fields behave under a Lorentz transformation. We answer that here. Firstly we write the space–time Lorentz transformations for relative motion in the x-direction in 4-vector form

$$\begin{pmatrix} x' \\ y' \\ z' \\ ct' \end{pmatrix} = \begin{pmatrix} \gamma & 0 & 0 & -\beta\gamma \\ 0 & 1 & 0 & 0 \\ 0 & 0 & 1 & 0 \\ -\beta\gamma & 0 & 0 & \gamma \end{pmatrix} \begin{pmatrix} x \\ y \\ z \\ ct \end{pmatrix} \tag{1.57a}$$

or equivalently:

$$x^\mu = L^\mu_\nu x^\nu \tag{1.57b}$$

which just defines the Lorentz tensor L^ν_μ. We can use the Lorentz tensor to write the transformations of the electromagnetic field as

$$F^{\mu\nu} = L^\mu_\rho F^{\rho\alpha} L^\nu_\alpha \tag{1.58a}$$

In component form this amounts to

$$E'_x = E_x \qquad\qquad\qquad B'_x = B_x$$

$$E'_y = \gamma(E_y - vB_z) \qquad\qquad B'_y = \gamma\left(B_y + \frac{v}{c^2}E_z\right)$$

$$E'_z = \gamma(E_z + vB_y) \qquad\qquad B'_z = \gamma\left(B_z - \frac{v}{c^2}E_y\right)$$

$$\tag{1.58b}$$

Clearly the electromagnetic fields transform in a way somewhat reminiscent of the space–time transformations. In actual fact, though, these transformations are quite different to (1.3). In particular, note that it is the fields in the direction of the relative motion that are unaffected by the transformation and the fields perpendicular to the motion that are transformed. Equation (1.58a) is a neat way to write the Lorentz transformations, and is more general than (1.58b) because it does not require us to choose a particular axis for the relative motion of the reference frames. However, it is also rather abstract, and when we implement the transformations the form of equations (1.58b) is simpler to use.

1.6 The Compton Effect

The Compton effect is an illustration of the application of relativity theory to the properties of fundamental physics. It is a very elegant and convincing illustration of the particle nature of electromagnetic radiation. Furthermore, it is useful to include it here in its elementary form as we shall be considering a more sophisticated and complete (but less physically transparent) description of it in chapter 12.

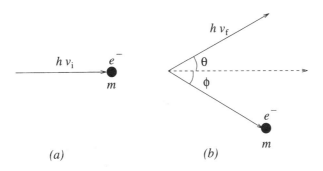

Fig. 1.7. The kinematics of the Compton effect, in which a photon is scattered from an electron (in its rest frame), (a) before and (b) after the collision. The conservation of energy and momentum yields a unique wavelength for the scattered photon as a function of scattering angle θ.

The physics of the Compton effect is shown in figure 1.7. An electromagnetic wave is incident upon a stationary free electron. The photon has frequency ν_i (in the X-ray region of the spectrum). After the collision the photon has been scattered through an angle θ and has a reduced frequency ν_f. The electron recoils making an angle ϕ with the direction of the initial photon. Let us write down the conservation of energy and momentum for this collision. The initial energy of the electron is mc^2 and the final energy γmc^2, so

$$h\nu_i + mc^2 = h\nu_f + \gamma mc^2 \tag{1.59}$$

We must conserve momentum both parallel and perpendicular to the direction of the initial photon, so

$$\frac{h\nu_i}{c} = \frac{h\nu_f}{c}\cos\theta + \gamma m\nu \cos\phi \tag{1.60a}$$

$$0 = \frac{h\nu_f}{c}\sin\theta - \gamma m\nu \sin\phi \tag{1.60b}$$

Now we want to eliminate ϕ between (1.60a) and (1.60b). The best way to do this is to rearrange them so that the term in ϕ is on its own on one side of the equations. The equations can be squared and added. Using $\cos^2\phi + \sin^2\phi = 1$ then leads to

$$\frac{h^2\nu_i^2}{m^2c^4} + \frac{h^2\nu_f^2}{m^2c^4} - \frac{2h^2\nu_i\nu_f}{m^2c^4}\cos\theta = \gamma^2\beta^2 \tag{1.61}$$

The conservation of energy (1.59) can be manipulated to give a rather cumbersome expression for $\beta^2\gamma^2$ which can be substituted into (1.61) to

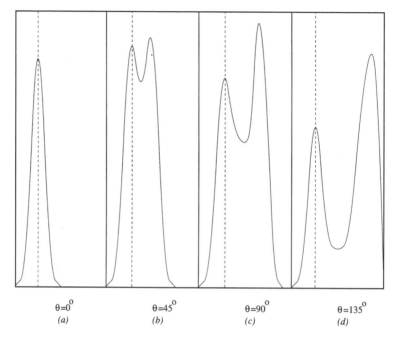

$$\theta=0^\circ \qquad \theta=45^\circ \qquad \theta=90^\circ \qquad \theta=135^\circ$$
(a)   (b)   (c)   (d)

Fig. 1.8. Schematic results of Compton scattering experiments at different angles. (a) The line of the incident radiation. In (b), (c) and (d) the left hand peak shows the incident photon wavelength and is indicated by the vertical dashed lines. The right hand peaks are the Compton scattered radiation. The angles of scatter are 45° in (b), 90° in (c) and 135° in (d). The shift in wavelength of the scattered radiation obeys equation (1.63).

give

$$\frac{1}{v_f} - \frac{1}{v_i} = \frac{h}{mc^2}(1 - \cos\theta) \qquad (1.62)$$

or, as is more familiar, in terms of wavelength

$$\lambda_f - \lambda_i = \frac{h}{mc}(1 - \cos\theta) \qquad (1.63)$$

The quantity h/mc has dimensions of length and is known as the Compton wavelength λ_C. Numerically $\lambda_C = 2.43 \times 10^{-12}$ m.

This scattering was first observed by Compton (1923a and b). The scattered radiation consists of two components. Firstly there is radiation scattered with the same frequency as the incident X-ray. This occurs because the radiation sets the electron oscillating at its own frequency. This radiation is emitted in all directions. There is also radiation scattered at a lower energy which corresponds to that which is Compton scattered according to the formula above (Wichman 1971). This is illustrated in figure 1.8 which is a schematic picture of Compton scattering results at several different scattering angles. Note that the scattering will be largest for $\theta = 180°$ as (1.63) requires and is zero for $\theta = 0°$. Equation (1.63) contains both h and c and so depends on both quantum mechanics and relativity for its derivation. The fact that the angular distribution of the scattered radiation was found to obey (1.63) lent strong support to the photon picture of light. Furthermore, it was necessary to use the relativistic equations of motion (1.59) and (1.60) to arrive at (1.63), so this experiment also helps to validate relativistic mechanics on a microscopic scale. Unfortunately this is not the end of the story of Compton scattering and we will return to it in chapter 12.

1.7 Problems

(1) In a certain inertial frame two events occur a distance 5 kilometres apart and separated by 5 μs. An observer, who is travelling at velocity v parallel to the line joining the two points where the events occurred, notes that the events are simultaneous. Find v.

(2) Observer A sees two events occurring at the same place and separated in time by 10^{-6} s. A second observer B sees them to be separated by a time interval 3×10^{-6} s. What is the separation in space of the two events according to B? What is the speed of B relative to A?

(3) In a scattering experiment an electron with speed $0.85c$ and a proton of speed $0.7c$ are moving towards each other. (a) What is the velocity of the electron as seen by an observer in the rest frame of the proton? (b) What is the velocity of the proton as seen by an observer in the rest frame of the electron?

(4) In frame A two particles of equal mass move at equal velocity towards each other. They collide and stick together. Clearly momentum is conserved in this frame. (a) Perform a Lorentz velocity transformation to the rest frame of one of the particles (frame B) and show that momentum is not conserved in such a frame. (b) Use the Lorentz energy–momentum transformations to show that conservation of momentum in frame B depends upon conservation of energy in frame A. (c) How would an observer in frame A explain that energy is conserved in the collision?

(5) (a) Show that the potentials felt by a charged particle in electric and magnetic fields $\mathbf{E} = (0, E, 0)$ and $\mathbf{B} = (0, 0, B)$ respectively are

$$\mathbf{A} = \frac{1}{2}(-yB, xB, 0) \qquad\qquad \Phi = -yE$$

(b) Show that the potentials in (a) describe the same electromagnetic fields as the potentials

$$A' = (-yB, 0, 0) \qquad \Phi = -yE$$

(6) Starting from the electrostatic field due to a point particle at rest, use the Lorentz transformations for the electromagnetic field to find the electric and magnetic fields due to a point particle moving with velocity **v**.

(7) Use equations (1.54) and (1.56) and the definition of the electromagnetic field tensors to write down Maxwell's equations in their more familiar form of equation (1.42).

2

Aspects of Angular Momentum

Spin is well known to be an intrinsically relativistic property of particles. Nonetheless its effects are seen in many physical situations which are not obviously relativistic, perhaps the most obvious examples being magnets and the quantum mechanically permitted electronic configurations of the elements in the periodic table. The view of spin adopted in these and other problems is that the electron has a quantized spin ($s = 1/2$) and that is a fundamental tenet of the theory, rather than something that has to be explained. In this chapter we are going to discuss the behaviour of spin-1/2 particles (electrons, protons and neutrons for example) without much direct reference to relativistic quantum theory and the origins of spin. This chapter should be instructive in its own right and as a guide to understanding spin when we come to discuss it in a fully relativistic context at various stages throughout this book. Unless otherwise specified, we will refer to electrons, but it should be borne in mind that the theory is equally applicable to any spin-1/2 particle.

Students of quantum theory cannot delve very deeply into the subject without coming across the quantization of angular momentum. The orbital angular momentum usually surfaces in the theory leading up to the quantum description of the hydrogen atom. Spin is often introduced through the Stern–Gerlach experiment and the Goudsmit–Uhlenbeck hypothesis, and the anomalous Zeeman effect (Eisberg and Resnick 1985). One of the things that makes non-relativistic quantum mechanics simpler than its relativistic counterpart is that, whereas only total angular momentum is conserved in relativistic quantum mechanics, in the non-relativistic theory both orbital angular momentum and spin angular momentum are conserved separately. The simplest way to appreciate this is to observe that spin does not appear in the Schrödinger equation and therefore there is no mechanism within non-relativistic quantum theory to change it. Non-relativistic quantum theory does conserve angular momentum, but as orbital angular momentum is all that appears in the Schrödinger equation, that is all that it is able to conserve. Spin is simply not a non-relativistic phenomenon. It is possible, however, to 'fix up' non-relativistic quantum theory by including spin *ad hoc*. Non-relativistic quantum theory including spin is a theory based on the Pauli equation, to which we shall be devoting some of this chapter.

Firstly we will revise the properties of general angular momentum operators. To gain further insight into the nature of quantum mechanical angular momentum we go on to show that the operator necessary to generate rotations can be written in terms of the angular momentum operators. Next we look at the operators and eigenvectors that describe spin-1/2 particles, and through this we introduce the Pauli matrices. As an aside, in section 2.4 we examine a topic not often covered in quantum theory textbooks, the matrix representation of higher spin operators. Then we go on to look at the interaction of the orbital angular momentum of an electron in an atom with an external magnetic field, leading to Larmor precession. Next we write down the Pauli equation using a plausibility argument to justify it, and see that this leads directly to the concept of spin–orbit coupling. We see how the relativistic velocity transformations lead us to postulate Thomas precession of the electron rest frame relative to the nucleus, and how this affects the spin–orbit coupling term in the Pauli Hamiltonian. Next we look at the solutions of the Pauli Hamiltonian in a central field. The eigenfunctions can be separated into spin-angular functions and radial parts, and we show this explicitly. This leads to the coupling of angular momentum using the Clebsch–Gordan coefficients, and we examine some of their properties. It is convenient to introduce some new quantum numbers when considering the solution of the Pauli equation in a central field, and we explain this and show their relation with the more familiar angular momentum quantum numbers. Also in this section we look at the properties of the spin-angular functions. Finally we discuss how to expand a plane wave in terms of the spin-angular functions, as this will be useful when we come to discuss scattering theory.

This chapter provides a grounding in the quantum theory of angular momentum. An understanding of the material covered here should enable the reader to cope easily with angular momentum operators where they appear later in the book, in particular when we come to discuss atoms.

2.1 Various Angular Momenta

In this chapter we will assume that the reader is familiar with angular momentum up to the level required to solve the hydrogen atom in non-relativistic quantum theory. The application of quantum mechanics to the description of atoms cannot proceed much beyond the hydrogen atom without a discussion of spin.

The name *spin* is very suggestive, but we do not necessarily mean by it that an electron spins about its own axis. We know, however, that the spin operator obeys the commutation relations

$$[\hat{S}_x, \hat{S}_y] = i\hbar \hat{S}_z, \quad [\hat{S}_y, \hat{S}_z] = i\hbar \hat{S}_x, \quad [\hat{S}_z, \hat{S}_x] = i\hbar \hat{S}_y, \quad [\hat{S}^2, \hat{S}_i] = 0 \qquad (2.1a)$$

which are written in analogy to the orbital angular momentum operator commutation relations

$$[\hat{L}_x, \hat{L}_y] = i\hbar \hat{L}_z, \quad [\hat{L}_y, \hat{L}_z] = i\hbar \hat{L}_x, \quad [\hat{L}_z, \hat{L}_x] = i\hbar \hat{L}_y, \quad [\hat{L}^2, \hat{L}_i] = 0 \qquad (2.1b)$$

In (2.1) subscript $i = x, y, z$. The orbital angular momentum operator is usually defined in terms of the momentum and position operators as

$$\hat{\mathbf{L}} = \hat{\mathbf{r}} \times \hat{\mathbf{p}} = \frac{\hbar}{i} \hat{\mathbf{r}} \times \nabla \qquad (2.2)$$

with the usual definition of the momentum operator. The definition of the spin and total angular momentum in terms of position and momentum is usually not discussed. In non-relativistic quantum theory we think of the electron (for example) as a point particle, so it is not possible to define spin in an analogous way to equation (2.2) as $\mathbf{r} = 0$ and what would be meant by the linear momentum at the surface of a spinning point particle is anybody's guess. Hence we can define spin only as a degree of freedom that obeys the same commutation relations as orbital angular momentum. The orbital angular momentum operators have the following commutation relations with the position and momentum operators:

$$[\hat{L}_x, \hat{y}] = -[\hat{L}_y, \hat{x}] = i\hbar \hat{z}, \qquad\qquad [\hat{L}_x, \hat{x}] = [\hat{L}_x, \hat{p}_x] = 0 \qquad (2.3a)$$

$$[\hat{L}_x, \hat{p}_y] = -[\hat{L}_y, \hat{p}_x] = i\hbar \hat{p}_z \qquad (2.3b)$$

and all cyclic permutations of (2.3) are also valid. Now that we have the orbital and spin angular momenta, the total angular momentum operator is

$$\hat{\mathbf{J}} = \hat{\mathbf{L}} + \hat{\mathbf{S}} \qquad (2.4a)$$

and

$$[\hat{J}_x, \hat{J}_y] = i\hbar \hat{J}_z, \quad [\hat{J}_y, \hat{J}_z] = i\hbar \hat{J}_x, \quad [\hat{J}_z, \hat{J}_x] = i\hbar \hat{J}_y, \quad [\hat{\mathbf{J}}^2, \hat{J}_i] = 0 \qquad (2.4b)$$

The operators in (2.4a) and (2.4b) have eigenvalues which we denote

$$\hat{\mathbf{J}}^2 \psi_j^{m_j} = j(j+1)\hbar^2 \psi_j^{m_j} \qquad (2.5a)$$

$$\hat{J}_z \psi_j^{m_j} = m_j \hbar \psi_j^{m_j} \qquad (2.5b)$$

where $\psi_j^{m_j}$ are eigenfunctions of both $\hat{\mathbf{J}}^2$ and $\hat{J}_z$, and we have arbitrarily chosen the z-direction as the quantization direction. Let us define the raising and lowering operators

$$\hat{J}_\pm = J_x \pm iJ_y \qquad (2.6)$$

The reason for this notation will soon become obvious and the reader may well anticipate it. Let us state that

$$\psi_j^{m_j \pm 1} = a\hat{J}_\pm \psi_j^{m_j} \qquad (2.7)$$

where a is a constant to be determined. Now

$$\begin{aligned}
\hat{J}_z \psi_j^{m_j \pm 1} &= a\hat{J}_z \hat{J}_\pm \psi_j^{m_j} = a(\hat{J}_\pm \hat{J}_z + \hat{J}_z \hat{J}_\pm - \hat{J}_\pm \hat{J}_z)\psi_j^{m_j} \\
&= a(\hat{J}_\pm \hat{J}_z + [\hat{J}_z, \hat{J}_\pm])\psi_j^{m_j} = a\hat{J}_\pm(\hat{J}_z \pm \hbar)\psi_j^{m_j} \qquad (2.8) \\
&= a\hat{J}_\pm(m_j \pm 1)\hbar\psi_j^{m_j} = a(m_j \pm 1)\hbar\hat{J}_\pm\psi_j^{m_j} = (m_j \pm 1)\hbar\psi_j^{m_j \pm 1}
\end{aligned}$$

where we have made use of the commutation rules (2.4b). This proves that if $\psi_j^{m_j}$ is an eigenfunction of $\hat{J}_z$ and $\hat{\mathbf{J}}^2$ then with the definitions (2.6) and (2.7) $\hat{J}_\pm\psi_j^{m_j}$ is also an eigenfunction of these operators with the same value of j but with m_j raised or lowered by one unit of $\hbar$. We have not proved that (2.7) is valid in any absolute way because we used it in (2.8). Equation (2.8) only shows that equation (2.7) is consistent with the more familiar properties of angular momentum operators. Let us assume all eigenfunctions of $\hat{\mathbf{J}}^2$ are normalized and try to find the value of the constant a. The functions $\psi_j^{m_j}$ form an orthonormal set and we note that $(\hat{J}_\pm)^\dagger = \hat{J}_\mp$ form a pair of hermitian conjugate operators. The strategy is to write the normalization of $\psi_j^{m_j \pm 1}$ in terms of the normalization of $\psi_j^{m_j}$. This will then tell us the value of the constant a.

$$\begin{aligned}
\int (\psi_j^{m_j + 1})^*(\psi_j^{m_j + 1})d\mathbf{r} &= a^2 \int (\hat{J}_+\psi_j^{m_j})^* \hat{J}_+\psi_j^{m_j} d\mathbf{r} \\
&= a^2 \int \psi_j^{m_j *} \hat{J}_- \hat{J}_+\psi_j^{m_j} d\mathbf{r} \\
&= a^2 \int \psi_j^{m_j *} (\hat{\mathbf{J}}^2 - \hat{J}_z(\hat{J}_z + \hbar))\psi_j^{m_j} d\mathbf{r} \\
&= (j(j+1) - m_j(m_j + 1))\hbar^2 a^2 \int \psi_j^{m_j *} \psi_j^{m_j} d\mathbf{r} \\
&= (j - m_j)(j + m_j + 1)\hbar^2 a^2 = 1 \qquad (2.9)
\end{aligned}$$

where we have set the final integral equal to one. So, the right normalization means that we can replace a in equation (2.7) using (2.9) and we get

$$\hat{J}_+\psi_j^{m_j} = (j - m_j)^{1/2}(j + m_j + 1)^{1/2}\hbar\psi_j^{m_j + 1} \qquad (2.10a)$$

and similarly

$$\hat{J}_{-}\psi_j^{m_j} = (j + m_j)^{1/2}(j - m_j + 1)^{1/2}\hbar\psi_j^{m_j-1} \qquad (2.10b)$$

Clearly the raising and lowering is a process that must have both an upper and a lower bound. It would be absurd to allow a situation where the magnitude of the z-component of angular momentum exceeded that of the total angular momentum. Hence

$$\hat{J}_{+}\psi_j^{m_j^{\max}} = 0 \qquad (2.11a)$$

$$\hat{J}_{-}\psi_j^{m_j^{\min}} = 0 \qquad (2.11b)$$

We can operate on (2.11a) with $\hat{J}_{-}$ and on (2.11b) with $\hat{J}_{+}$ and, again making use of the commutation relations, we find

$$\hat{J}_{-}\hat{J}_{+}\psi_j^{m_j^{\max}} = (\hat{\mathbf{J}}^2 - \hat{J}_z^2 - \hbar\hat{J}_z)\psi_j^{m_j^{\max}}$$
$$= (j(j + 1) - m_j^{\max}(m_j^{\max} + 1))\hbar^2\psi_j^{m_j^{\max}} = 0 \qquad (2.12a)$$

$$\hat{J}_{+}\hat{J}_{-}\psi_j^{m_j^{\min}} = (\hat{\mathbf{J}}^2 - \hat{J}_z^2 + \hbar\hat{J}_z)\psi_j^{m_j^{\min}}$$
$$= (j(j + 1) - m_j^{\min}(m_j^{\min} - 1))\hbar^2\psi_j^{m_j^{\min}} = 0 \qquad (2.12b)$$

On the right hand side of these two equations the wavefunction is just some function of position and time. It is not zero, in general. Therefore the terms in brackets must be zero. This can only be true if

$$m_j^{\max} = j \qquad (2.13a)$$

$$m_j^{\min} = -j \qquad (2.13b)$$

In figure 2.1 we illustrate these properties of the angular momentum operators for $j = 3/2$. Note that, as for all angular momenta, the maximum value of the square of the z-component is $m_j^2 = j^2$ whereas the total angular momentum is $j^2 + j$ so the angular momentum can never point precisely along the z-axis. This is an example of the uncertainty principle in action. In the classical (correspondence) limit $j \to \infty$ the eigenvalues of $\hat{\mathbf{J}}^2 \to j^2$ and thus the maximum value of $\hat{J}_z$ gets asymptotically closer to the total angular momentum. Thus this quantized theory of angular momentum is consistent with the classical limit.

Our results so far can be summarized

$$-j \le m_j \le j \qquad (2.14)$$

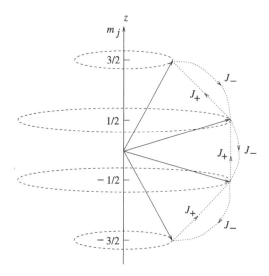

Fig. 2.1. This figure shows the angular momentum vectors for the possible values of m_j for $j = 3/2$ (the full lines). These vectors can actually lie anywhere on the appropriate dashed circle because of the uncertainty in J_x and J_y. The action of the raising $(\hat{J}_+)$ and lowering $(\hat{J}_-)$ operators is illustrated by the dotted arrowed lines.

and m_j takes on values with an interval of unity between these extremes. Clearly there are $2j + 1$ values of m_j for each value of j which implies that there are $2j + 1$ different eigenfunctions with the same value of j.

From equations (2.5) to here all total angular momentum operators and eigenvalues could be replaced by their counterparts for the orbital or spin angular momentum and all the equations would follow through identically. Therefore we can draw the same conclusions about their behaviour

$$\hat{S}^2\psi_s^{m_s} = s(s+1)\hbar^2\psi_s^{m_s} \qquad (2.15a)$$

$$\hat{S}_z\psi_s^{m_s} = m_s\hbar\psi_s^{m_s} \qquad (2.15b)$$

with

$$-s \le m_s \le s \qquad (2.16)$$

For the most part we shall be interested in $s = 1/2$. There are two good reasons for this. As stated earlier, all the common elementary particles have $s = 1/2$ and these are the ones we are primarily interested in describing. Also, the correspondence principle tells us that the classical limit is the limit of high values of the quantum numbers, so $s = 1/2$ particles should exhibit the most extreme forms of non-classical behaviour associated with spin. One further point on terminology is that spin-1/2 particles with

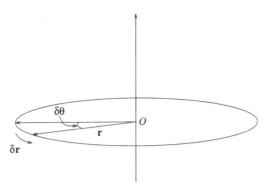

Fig. 2.2. The infinitesimal translation $\delta\mathbf{r}$ corresponds to a rotation $\delta\boldsymbol{\theta}$ about the origin.

$m_s = 1/2$ are known as spin-up particles and those with $m_s = -1/2$ are said to have spin down. Next, let us delve a bit deeper into the general properties of angular momentum operators.

2.2 Angular Momentum and Rotations

Further insight into the nature of the angular momentum operators can be gained by considering them as operators that generate rotations. To see this consider a well behaved arbitrary function in space $F(\mathbf{r})$. We can create a second function $G(\mathbf{r})$ which is defined as

$$G(\mathbf{r} + \delta\mathbf{r}) = F(\mathbf{r}) \tag{2.17}$$

This describes some sort of general translation, but it can be taken to represent a rotation where the new function is just the old function with everything rigidly shifted by $\delta\mathbf{r}$. Next we let $\delta\mathbf{r}$ be an infinitesimal quantity, and then we can write down a Taylor expansion for $G(\mathbf{r})$ retaining terms to first order (for the present) to obtain

$$G(\mathbf{r}) = F(\mathbf{r} - \delta\mathbf{r}) = F(\mathbf{r}) - \delta\mathbf{r} \cdot \nabla F(\mathbf{r}) \tag{2.18}$$

Now let our displacement $\delta\mathbf{r}$ be due to an infinitesimal rotation through angle $\delta\boldsymbol{\theta}$ about the origin, as shown in figure 2.2. Then we can write

$$\delta\mathbf{r} = \delta\boldsymbol{\theta} \times \mathbf{r} \tag{2.19}$$

where the vector quantity $\delta\boldsymbol{\theta}$ points along the axis of rotation. Then

$$G(\mathbf{r}) - F(\mathbf{r}) = -\delta\boldsymbol{\theta} \times \mathbf{r} \cdot \nabla F(\mathbf{r}) = -\delta\boldsymbol{\theta} \cdot (\mathbf{r} \times \nabla F(\mathbf{r})) \tag{2.20}$$

So,

$$\delta F(\mathbf{r}) = G(\mathbf{r}) - F(\mathbf{r}) = -\frac{i}{\hbar}\delta\boldsymbol{\theta} \cdot \hat{\mathbf{L}}F(\mathbf{r}) \tag{2.21}$$

where we have used the definition of the momentum operator $\hat{\mathbf{L}} = \hat{\mathbf{r}} \times \hat{\mathbf{p}}$. What this shows us is that we can operate with the angular momentum operator on a function and that will give us the change in that function due to an infinitesimal rotation according to (2.21).

Now note that we only retained the first order terms in the Taylor expansion to obtain (2.21). If we had kept higher order terms we would have obtained a series of terms for δF. It is easy to convince oneself that this results in an exponential series, and the rotation operator operating on the function $F(\mathbf{r})$ is

$$\mathbf{R}F(\mathbf{r}) = e^{-i\theta \cdot \hat{\mathbf{L}}/\hbar} F(\mathbf{r}) \tag{2.22}$$

Equation (2.22) defines a rotation operator in terms of an angular momentum operator. It is easy to see that it has the right properties. Clearly two rotations about the same axis must commute. From (2.22) we can see that this will be so. However, two rotations about different axes do not commute in general. A rotation about the z-axis followed by a rotation about the y-axis does not bring a vector to the same point as the two rotations performed in reverse order. In fact the difference between the two sets of rotations is equal to a rotation around the third perpendicular axis. A little thought should enable you to convince yourself of this. The commutation relations (2.4b) put into the exponent in (2.22) gives exactly this behaviour. So angular momentum operators obey the same commutation relations as straightforward rotations.

2.3 Operators and Eigenvectors for Spin 1/2

In non-relativistic quantum theory we know how to solve problems. In principle it is easy. The model we are trying to describe is represented by some scalar and/or vector potential, and the Schrödinger equation can be solved for it. We obtain the eigenfunctions from which observable quantities can be determined. The Schrödinger equation we solve is

$$i\hbar \frac{\partial \psi_r(\mathbf{r}, t)}{\partial t} = \frac{1}{2m} \left(\frac{\hbar}{i} \nabla - \frac{e}{c} \mathbf{A}(\mathbf{r}, t) \right)^2 \psi_r(\mathbf{r}, t) + e\Phi(\mathbf{r}, t)\psi_r(\mathbf{r}, t) \tag{2.23}$$

where $\mathbf{A}(\mathbf{r}, t)$ is the vector potential and $e\Phi(\mathbf{r}, t)$ is a scalar potential. To include spin in this theory we require that the Schrödinger equation remains valid, but that we can superimpose on top of it the spin degrees of freedom. One could imagine that this should be done in the simplest way. The wavefunction can be assumed to be a product of the space- and time-dependent part described by equation (2.23) and a spin-dependent part described by the function $\chi_s^{m_s}$.

$$\psi_s^{m_s}(\mathbf{r}, t) = \psi_r(\mathbf{r}, t)\chi_s^{m_s} \tag{2.24}$$

Equation (2.23) contains no dependence on s or m_s so we can write two Schrödinger equations down, one for $m_s = 1/2$ and one for $m_s = -1/2$. One of the $\chi_s^{m_s}$ will appear in every term in each equation and hence will cancel. Then we will just be left with two identical Schrödinger equations again.

Equation (2.24) is rigorously valid in non-relativistic quantum theory, but, looking forward, we want to introduce spin–orbit coupling later on in this chapter and so we require a form for the wavefunction which is easy to generalize to the relativistic case. Therefore we write

$$\psi(\mathbf{r}, t) = c(1/2)\psi_s^{1/2}(\mathbf{r}, t) + c(-1/2)\psi_s^{-1/2}(\mathbf{r}, t) \tag{2.25}$$

where the coefficients $c(1/2)$ and $c(-1/2)$ are the probability amplitudes for the particle to have spin up or down respectively. They are given by

$$c(1/2) = \int d\mathbf{r} \psi_s^{1/2*}(\mathbf{r}, t)\psi(\mathbf{r}, t) \tag{2.26a}$$

$$c(-1/2) = \int d\mathbf{r} \psi_s^{-1/2*}(\mathbf{r}, t)\psi(\mathbf{r}, t) \tag{2.26b}$$

All of this suggests that we use a two-component wavefunction to describe the quantum mechanics of particles with spin. To do this we just define the spin eigenvectors

$$\chi_{1/2}^{1/2} = \begin{pmatrix} 1 \\ 0 \end{pmatrix}, \quad \chi_{1/2}^{-1/2} = \begin{pmatrix} 0 \\ 1 \end{pmatrix} \tag{2.27}$$

With this definition

$$\psi(\mathbf{r}, t) = \begin{pmatrix} c(1/2)\psi_r(\mathbf{r}, t) \\ c(-1/2)\psi_r(\mathbf{r}, t) \end{pmatrix} \tag{2.28}$$

The definitions (2.27) are chosen because the χs are clearly orthogonal to one another and have a normalization of unity. Furthermore they form a complete set: that is, any arbitrary two-component wavefunction can be written as a linear combination of them. Such a two-component wavefunction, where the components are differentiated by the direction of the spin they represent, is often called a *spinor*. Now that we have the spin-dependent part of the wavefunction, we can write down the operators that correspond to the spin operators of the previous section in matrix form. These can be written in terms of the Pauli spin matrices

$$\tilde{\sigma}_x = \begin{pmatrix} 0 & 1 \\ 1 & 0 \end{pmatrix}, \quad \tilde{\sigma}_y = \begin{pmatrix} 0 & -i \\ i & 0 \end{pmatrix}, \quad \tilde{\sigma}_z = \begin{pmatrix} 1 & 0 \\ 0 & -1 \end{pmatrix} \tag{2.29}$$

(a tilde indicates a matrix). These three plus the identity matrix form a complete set of 2×2 matrices. The spin operators are written as

$$\hat{S}_x = \frac{\hbar}{2}\tilde{\sigma}_x, \quad \hat{S}_y = \frac{\hbar}{2}\tilde{\sigma}_y, \quad \hat{S}_z = \frac{\hbar}{2}\tilde{\sigma}_z \tag{2.30}$$

It is straightforward to verify that these definitions give us operators that
obey (2.5) to (2.13) with $\hat{J}$ replaced by $\hat{S}$. To gain familiarity with matrix
operators, we will do this shortly. Before that, it is necessary to become
acquainted with the properties of the Pauli matrices in equation (2.29).
We will just state these here, as the proofs are trivial:

$$\tilde{\sigma}_x^2 = \tilde{\sigma}_y^2 = \tilde{\sigma}_z^2 = \tilde{I}_2 \tag{2.31}$$

where $\tilde{I}_2$ is the 2×2 identity matrix. Furthermore

$$\tilde{\sigma}_x\tilde{\sigma}_y + \tilde{\sigma}_y\tilde{\sigma}_x = \tilde{\sigma}_y\tilde{\sigma}_z + \tilde{\sigma}_z\tilde{\sigma}_y = \tilde{\sigma}_z\tilde{\sigma}_x + \tilde{\sigma}_x\tilde{\sigma}_z = 0 \tag{2.32}$$

and

$$\tilde{\sigma}_x\tilde{\sigma}_y - \tilde{\sigma}_y\tilde{\sigma}_x = 2i\tilde{\sigma}_z \tag{2.33a}$$

$$\tilde{\sigma}_y\tilde{\sigma}_z - \tilde{\sigma}_z\tilde{\sigma}_y = 2i\tilde{\sigma}_x \tag{2.33b}$$

$$\tilde{\sigma}_z\tilde{\sigma}_x - \tilde{\sigma}_x\tilde{\sigma}_z = 2i\tilde{\sigma}_y \tag{2.33c}$$

Finally, a very useful relation obeyed by the Pauli matrices is as follows.
If $\mathbf{A}$ and $\mathbf{B}$ are any arbitrary vectors, or vector operators, then

$$\tilde{\sigma} \cdot \mathbf{A}\tilde{\sigma} \cdot \mathbf{B} = \mathbf{A}.\mathbf{B} + i\tilde{\sigma} \cdot (\mathbf{A} \times \mathbf{B}) \tag{2.34}$$

Now, we can use these properties to show that equation (2.30) describes
an acceptable representation of the spin operators. Let us consider (2.15a)
with the wavefunction given by (2.24). Using (2.31) we have

$$\hat{S}^2\psi_r(\mathbf{r}, t)\chi_s^{m_s} = (\hat{S}_x^2 + \hat{S}_y^2 + \hat{S}_z^2)\psi_r(\mathbf{r}, t)\chi_s^{m_s} = \frac{3\hbar^2}{4}\begin{pmatrix} 1 & 0 \\ 0 & 1 \end{pmatrix}\psi_r(\mathbf{r}, t)\chi_s^{m_s}$$

$$= \frac{3\hbar^2}{4}\psi_r(\mathbf{r}, t)\chi_s^{m_s} = s(s+1)\hbar^2\psi_r(\mathbf{r}, t)\chi_s^{m_s} \tag{2.35}$$

The last equality is true if $s = 1/2$. Next consider (2.15b) for $m_s = +1/2$:

$$\hat{S}_z\psi_r(\mathbf{r}, t)\chi_s^{1/2} = \frac{\hbar}{2}\begin{pmatrix} 1 & 0 \\ 0 & -1 \end{pmatrix}\psi_r(\mathbf{r}, t)\begin{pmatrix} 1 \\ 0 \end{pmatrix} = \frac{\hbar}{2}\psi_r(\mathbf{r}, t)\chi_s^{1/2} \tag{2.36a}$$

and for $m_s = -1/2$:

$$\hat{S}_z\psi_r(\mathbf{r}, t)\chi_s^{-1/2} = \frac{\hbar}{2}\begin{pmatrix} 1 & 0 \\ 0 & -1 \end{pmatrix}\psi_r(\mathbf{r}, t)\begin{pmatrix} 0 \\ 1 \end{pmatrix} = -\frac{\hbar}{2}\psi_r(\mathbf{r}, t)\chi_s^{-1/2} \tag{2.36b}$$

Note that we can drop the part of the wavefunction that depends on
position and time as that is not affected by the matrix multiplications,
being only a one-component function. Let us consider the spin raising

and lowering operators analogous to those defined in equation (2.6):

$$\hat{S}_+ = \hat{S}_x + i\hat{S}_y = \hbar \begin{pmatrix} 0 & 1 \\ 0 & 0 \end{pmatrix} \tag{2.37a}$$

$$\hat{S}_- = \hat{S}_x - i\hat{S}_y = \hbar \begin{pmatrix} 0 & 0 \\ 1 & 0 \end{pmatrix} \tag{2.37b}$$

It is easily seen that these satisfy (2.11):

$$\hat{S}_+ \chi_s^{1/2} = \hbar \begin{pmatrix} 0 & 1 \\ 0 & 0 \end{pmatrix} \begin{pmatrix} 1 \\ 0 \end{pmatrix} = 0 \tag{2.38a}$$

$$\hat{S}_- \chi_s^{-1/2} = \hbar \begin{pmatrix} 0 & 0 \\ 1 & 0 \end{pmatrix} \begin{pmatrix} 0 \\ 1 \end{pmatrix} = 0 \tag{2.38b}$$

and (2.10):

$$\hat{S}_+ \chi_s^{-1/2} = \hbar \begin{pmatrix} 0 & 1 \\ 0 & 0 \end{pmatrix} \begin{pmatrix} 0 \\ 1 \end{pmatrix} = \hbar \begin{pmatrix} 1 \\ 0 \end{pmatrix} = \hbar \chi_s^{1/2} \tag{2.39a}$$

$$\hat{S}_- \chi_s^{+1/2} = \hbar \begin{pmatrix} 0 & 0 \\ 1 & 0 \end{pmatrix} \begin{pmatrix} 1 \\ 0 \end{pmatrix} = \hbar \begin{pmatrix} 0 \\ 1 \end{pmatrix} = \hbar \chi_s^{-1/2} \tag{2.39b}$$

The $\hbar$ in these last two equations comes from the normalization prefactor found in equation (2.10). Finally the definitions (2.30) must satisfy the commutation relations (2.1). This can be seen immediately by multiplying both sides of equation (2.33) by $(\hbar/2)^2$ and substituting (2.30) into the result.

The $\chi_s^{m_s}$ will appear several times in this book. Henceforth we shall drop the suffix s as being redundant; we shall always be referring to spin-1/2 particles when we use these functions.

2.4 Operators for Higher Spins

There are many particles whose spin quantum number is greater than 1/2. The photon has spin one, for example, although that is a rather special case, being massless, and being the *particle* that carries the electromagnetic interaction. We will discuss the properties of the photon in chapter 12. Other examples of spin-1 particles are the **W** and **Z** particles. Of course even higher spin particles are also possible (atoms with unfilled shells, and the graviton, for example). The spin of such particles can also be represented in matrix form. Indeed any angular momentum can be represented in this way. Below we illustrate this for spin-1 and spin-3/2. Interested readers can try to derive the equivalent matrices for states with higher angular momentum if they have some time to waste. For spin 1

$$S_x = \hbar \begin{pmatrix} 0 & 0 & 0 \\ 0 & 0 & -i \\ 0 & i & 0 \end{pmatrix} \qquad (2.40a)$$

$$S_y = \hbar \begin{pmatrix} 0 & 0 & i \\ 0 & 0 & 0 \\ -i & 0 & 0 \end{pmatrix} \qquad (2.40b)$$

$$S_z = \hbar \begin{pmatrix} 0 & -i & 0 \\ i & 0 & 0 \\ 0 & 0 & 0 \end{pmatrix} \qquad (2.40c)$$

Using a similar analysis to equations (2.35) to (2.39), it is easy to see that these matrices have all the right properties. The equivalent eigenvectors are what one might expect:

$$\psi_1^{-1}(\mathbf{r}, t) = \psi(\mathbf{r}, t)\frac{1}{\sqrt{2}} \begin{pmatrix} 1 \\ -i \\ 0 \end{pmatrix} \qquad (2.41a)$$

$$\psi_1^0(\mathbf{r}, t) = \psi(\mathbf{r}, t) \begin{pmatrix} 0 \\ 0 \\ 1 \end{pmatrix} \qquad (2.41b)$$

$$\psi_1^{+1}(\mathbf{r}, t) = \psi(\mathbf{r}, t)\frac{1}{\sqrt{2}} \begin{pmatrix} 1 \\ i \\ 0 \end{pmatrix} \qquad (2.41c)$$

The reader should check that these eigenvectors satisfy the requirements of the raising and lowering operators.

For spin-3/2

$$S_x = \frac{\hbar}{2} \begin{pmatrix} 0 & \sqrt{3} & 0 & 0 \\ \sqrt{3} & 0 & 2 & 0 \\ 0 & 2 & 0 & \sqrt{3} \\ 0 & 0 & \sqrt{3} & 0 \end{pmatrix} \qquad (2.42a)$$

$$S_y = \frac{\hbar}{2} \begin{pmatrix} 0 & -\sqrt{3}i & 0 & 0 \\ \sqrt{3}i & 0 & -2i & 0 \\ 0 & 2i & 0 & -\sqrt{3}i \\ 0 & 0 & \sqrt{3}i & 0 \end{pmatrix} \qquad (2.42b)$$

$$S_z = \frac{\hbar}{2} \begin{pmatrix} 3 & 0 & 0 & 0 \\ 0 & 1 & 0 & 0 \\ 0 & 0 & -1 & 0 \\ 0 & 0 & 0 & -3 \end{pmatrix} \qquad (2.42c)$$

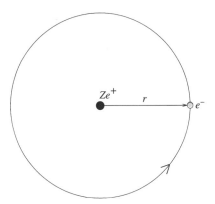

Fig. 2.3. In the classical model of the atom the nucleus is stationary and the electron is in orbit around it. The electron carries a charge and hence forms a current loop.

Equations (2.42) are illustrative only; we shall not be referring to them again. Equations (2.40) and (2.41) will be of use when we come to discuss the angular momentum of the photon.

2.5 Orbital Magnetic Moments

Here we are going to present a simple classical description of the orbital magnetic moment of a one-electron atom. Quantum theory will be included only in a rather arbitrary manner at the end.

Our atom consists of a stationary nucleus with an electron of mass m, charge e and speed v in a circular orbit of radius r. This can be regarded as a current loop as shown in figure 2.3. The current is just the charge divided by the time taken for the electron to orbit the nucleus once.

$$I = \frac{ev}{2\pi r} \tag{2.43}$$

As we learn from electromagnetism, the field at large distances away from a current loop is the same as the field due to a dipole at the centre of the current loop with its axis perpendicular to the plane of the loop. The magnetic moment due to a current loop of area A is given by IA and in the present case this becomes

$$\mu_l = IA = \frac{ev}{2\pi r} \times \pi r^2 = \frac{evr}{2} = \frac{emvr}{2m} = \frac{e}{2m}\mathbf{L} \tag{2.44}$$

Here we have used the fact that for a circular Bohr orbit the orbital velocity and the radial position vector of the electron are always perpendicular. Now we do the usual trick when we want to transform a classical expression into a quantum mechanical one. We put hats on the relevant quantities and call them operators, so

$$\hat{\boldsymbol{\mu}}_l = -\frac{e}{2m}\hat{\mathbf{L}} \qquad (2.45)$$

The minus sign arises here because, by convention, the magnetic moment and the angular momentum are taken to point in opposite directions.

   This is an interesting relation. It tells us that for any orbital the ratio of the magnetic moment to the orbital angular momentum that causes it is equal to a universal constant. Now we know that both the orbital angular momentum and its z-component are quantized and their eigenvalues are given by equations analogous to (2.5), so

$$|\mu_l| = -\frac{e\hbar}{2m}\sqrt{l(l+1)} = -\mu_B\sqrt{l(l+1)} \qquad (2.46a)$$

$$\mu_{lz} = -\frac{e\hbar}{2m}m_l = -g_l\mu_B m_l \qquad (2.46b)$$

where $g_l = 1$ is the orbital g-factor and we have defined the Bohr magneton, which is the magnetic moment of the electron

$$\mu_B = \frac{e\hbar}{2m} = 9.27 \times 10^{-24} \text{ amp m}^2 \qquad (2.47)$$

Equations (2.46) show that the magnetic moment of an atom is quantized in direction and magnitude as a direct consequence of the quantization of orbital angular momentum.

   Now let us ask what will happen if a one-electron atom with orbital magnetic moment given by (2.44) is placed in a magnetic field. Elementary electromagnetism tell us that a dipole in such a field will experience a torque given by

$$\boldsymbol{\tau} = \frac{d\mathbf{L}}{dt} = -\boldsymbol{\mu}_l \times \mathbf{B} \qquad (2.48)$$

and will have an orientational potential energy

$$E = -\boldsymbol{\mu}_l \cdot \mathbf{B} \qquad (2.49)$$

There is no mechanism for an atom in its ground state to get rid of this energy, so the magnetic moment precesses around the direction of the magnetic field, keeping a constant angle to it. Substituting (2.45) into

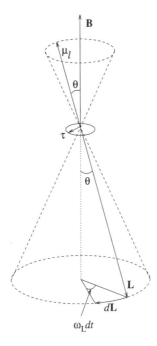

Fig. 2.4. A schematic diagram illustrating Larmor precession. The atom has a magnetic moment μ_L due to its orbital angular momentum **L**. The magnetic moment feels the torque τ and will precess about the applied field **B** with angular frequency ω_L.

(2.48) gives

$$\frac{d\mathbf{L}}{dt} = \frac{e}{2m}\mathbf{L} \times \mathbf{B} \tag{2.50}$$

The torque described by (2.50) causes the direction of the angular momentum (or the magnetic moment) vector to move through an angle in time dt (although its magnitude remains constant, of course). This is shown schematically in figure 2.4. The angle can be written in terms of an angular frequency ω_L as $\omega_L dt$. Geometrically, we can see from the figure that if $d\mathbf{L}$ is infinitesimal we can make a small angle approximation $\sin(\omega_L dt) \approx \omega_L dt$ and so

$$d\mathbf{L} = L \sin\theta(\omega_L dt) \tag{2.51}$$

where θ is the angle between the direction of the applied field and the magnetic moment. Dividing this through by dt and equating it to equation (2.50) gives

$$\omega_L L \sin\theta = \frac{e}{2m}\mathbf{L} \times \mathbf{B} = \frac{e}{2m}LB \sin\theta \tag{2.52}$$

where we have assumed in the final step that we are only interested in the magnitude of the angular frequency ω_L. Trivial cancellation gives

$$\omega_L = \frac{eB}{2m} \tag{2.53}$$

The direction of the precession is in the direction of **B**. The frequency ω_L is known as the Larmor frequency. This precession will split the $2l + 1$ degenerate energy levels (for each value of the quantum numbers n and l) of the one-electron atom. Let us assume that the applied field points along the z-direction. Then the shift in the energy levels due to the field is given by (2.49)

$$\Delta E = \boldsymbol{\mu} \cdot \mathbf{B} = \mu_z \mathbf{B}_z = -\frac{e\hbar}{2m} m_l B = \hbar \omega_L m \tag{2.54}$$

The splitting of the levels is proportional to the m_l quantum number and so most of the degeneracies of non-relativistic quantum theory are removed. This is one way of looking at the normal Zeeman effect. If in addition an oscillating magnetic field with the same frequency is applied to an atom, the system may absorb or emit energy, and hence change its precessional motion. This forms the basis of magnetic resonance techniques used in atomic physics.

2.6 Spin Without Relativity

It is well known that spin does not appear in the Schrödinger equation, but as we shall see it is inherent in relativistic quantum theory. However, spin can be included within a non-relativistic framework, and this was done initially by Pauli (1927). The central philosophy of this approach is to use a standard non-relativistic Hamiltonian for spin-up and spin-down particles separately and to couple these together with a term that represents the interaction of the spin with the magnetic field **B(r)** felt by the particle (Baym 1967).

$$i\hbar \frac{\partial \psi(\mathbf{r}, t)}{\partial t} = \frac{1}{2m} \left(\frac{\hbar}{i} \nabla - e\mathbf{A}(\mathbf{r}) \right)^2 \psi(\mathbf{r}, t) - \hat{\boldsymbol{\mu}} \cdot \mathbf{B}(\mathbf{r}) \psi(\mathbf{r}, t) + V(\mathbf{r}) \psi(\mathbf{r}, t) \tag{2.55}$$

Equation (2.55) is known as the Pauli equation. To investigate the consequences of the term in the magnetic field let us examine the relatively simple case of the one-electron atom. To find the magnetic field felt by the electron, the easiest way is to sit in its rest frame (although what is meant by that may be difficult to define). If we sit in the electron's rest frame we see the nucleus moving round it. This is illustrated in figure 2.5. It is the same sort of transformation as moving from the sun's frame of reference

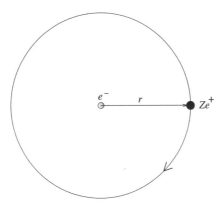

Fig. 2.5. This figure is the same as figure 2.3, but is drawn in the rest frame of the electron rather than the rest frame of the nucleus. This is a convenient frame when we want to calculate the magnetic field felt by the electron due to the relative motion of the nucleus.

where the earth goes round the sun to the earth's rest frame where the sun is seen to go round the earth.

The electron is now sitting at the centre of a circulating charged nucleus which we regard as a current loop. The current in this loop is given by

$$\mathbf{I} = -Ze\mathbf{v} \tag{2.56}$$

where the nucleus has charge Ze and is moving with velocity $\mathbf{v}$ relative to the electron.

At any instant Coulomb's electrostatic law can be used to give us the electric field felt by the electron due to the nucleus.

$$\mathbf{E} = \frac{Ze}{4\pi\epsilon_0 r^3}\mathbf{r} \tag{2.57}$$

The magnetic field due to this current can be found from the Biot–Savart Law

$$\mathbf{B} = \frac{\mu_0}{4\pi}\frac{\mathbf{I} \times \mathbf{r}}{r^3} \tag{2.58}$$

We can substitute from (2.56) into (2.58) for $\mathbf{I}$ and then substitute for $\mathbf{r}$ from (2.57) to obtain

$$\mathbf{B} = -\mu_0\epsilon_0\mathbf{v} \times \mathbf{E} = -\frac{1}{c^2}\mathbf{v} \times \mathbf{E} \tag{2.59}$$

where we have used the well-known relationship $\mu_0\epsilon_0 = 1/c^2$. Equation (2.59) is a measure of the size of $\mathbf{B}$ relative to $\mathbf{E}$. Clearly when the charge is moving at non-relativistic velocities $\mathbf{B}$ is very small.

Now if our electron has a spin associated with it there will be an orientational potential energy of the spin with this magnetic field. This is exactly what the magnetic field term in (2.55) represents. However, it is measured in the rest frame of the nucleus. Our expression (2.59) for the magnetic field is in the rest frame of the electron. We need to transform back to the nuclear frame. Calling the magnetic field term in (2.55) $\hat{H}_1$, we can introduce the well-known expression for the electron magnetic moment operator in terms of the spin angular momentum operator (written in analogy with (2.45)):

$$\hat{H}_1 = -\hat{\boldsymbol{\mu}} \cdot \mathbf{B} = \frac{e}{m} \hat{\mathbf{S}} \cdot \mathbf{B} = \frac{g_s \mu_B}{\hbar} \hat{\mathbf{S}} \cdot \mathbf{B} \qquad (2.60)$$

where $g_s = 2$ is the spin g-factor. Now, when we transform (2.60) back into the nuclear frame a factor of 2 is introduced which is due to the Thomas precession. This is a rather subtle effect which we describe in the next section. Here it should just be accepted. So, in the nuclear rest frame

$$\hat{H}_1 = \frac{e}{2m} \hat{\mathbf{S}} \cdot \mathbf{B} = -\frac{e}{2mc^2} \mathbf{S} \cdot \mathbf{v} \times \mathbf{E} \qquad (2.61a)$$

Note that if $\mathbf{B}$ is parallel to the z-axis we can use (2.5) to give

$$\hat{H}_{1z} = \frac{e\hbar B}{2m} m_s = \hbar \omega_L m_s \qquad (2.61b)$$

where ω_L is the Larmor frequency. Equation (2.61a) is a perfectly valid expression for $\hat{H}_1$, but it turns out to be more convenient for a lot of applications to express it in terms of angular momentum operators. To do this we proceed as follows. From equation (1.47) we can substitute for $\mathbf{E}$ in (2.61) and, of course, the force felt by the electron due to the potential it experiences is

$$\mathbf{F} = -\frac{dV(r)}{dr} \hat{\mathbf{r}} \qquad (2.62)$$

where we have used the fact that for our single-electron atom the potential is spherically symmetric. So (2.61) becomes

$$\hat{H}_1 = -\frac{1}{2mc^2} \mathbf{S} \cdot \mathbf{v} \times \mathbf{F} = \frac{1}{2mc^2} \frac{1}{r} \frac{dV(r)}{dr} \mathbf{S} \cdot \mathbf{v} \times \mathbf{r} \qquad (2.63)$$

Now we can multiply and divide by m and then substitute $\mathbf{L} = m\mathbf{v} \times \mathbf{r}$. Next we replace the classical angular momentum vector with the quantum mechanical angular momentum operator. Finally we have

$$\hat{H}_1 = -\frac{1}{2m^2c^2} \frac{1}{r} \frac{dV(r)}{dr} \hat{\mathbf{S}} \cdot \hat{\mathbf{L}} \qquad (2.64)$$

This is the well-known expression for spin–orbit coupling and is the least favourite equation of so many physics students. It is a direct consequence of the existence of the electron spin and can be derived directly from relativistic considerations. Equation (2.64) is of order v^2/c^2 and as a first approximation can be regarded as a perturbation on non-relativistic calculations. It is worthwhile to look at its magnitude, and we do that in section 2.11. The importance of spin–orbit coupling, especially in a condensed matter context, is not the magnitude of the energy corrections, but the fact that it breaks symmetries that hold in non-relativistic quantum theory. In non-relativistic quantum mechanics both orbital and spin angular momentum are conserved separately. However, because these two angular momenta are able to 'talk to one another' through the spin–orbit coupling, when this is included, only total angular momentum is conserved. So the angular momentum conservation laws are less strict in a relativistic theory than in a non-relativistic theory.

2.7 Thomas Precession

The derivation of the spin–orbit coupling term in the Hamiltonian of a one-electron atom described above is incomplete because when we transformed from the electron's rest frame to the nuclear rest frame we introduced an apparently spurious factor of 2. In this section we will investigate the origin of this factor, and see that it is not spurious after all. This is actually a very subtle effect which is a consequence of the Lorentz velocity transformations. Our model is of a point nucleus and a point electron in a circular Bohr orbit around it. This has already been illustrated in figure 2.3.

Initially let us sit in the nuclear rest frame and observe the orbiting electron. The nuclear frame of reference has the origin at the nucleus and the (x_n, y_n) axes as shown in figure 2.6. Now, when the electron is instantaneously at the lowest point of its orbit (see figure 2.7) a set of axes (x_1, y_1) is associated with it which are parallel to the (x_n, y_n) axes. Another set of axes (x_2, y_2) which are also on the electron's orbital path can be drawn. These are displaced from the (x_1, y_1) set by a very small amount, but have still been drawn parallel to (x_n, y_n). At time t_1 the electron is at the origin of (x_1, y_1) and at time $t_2 = t_1 + \Delta t$ it is at the origin of frame (x_2, y_2). The way we are going to approach this problem is to perform two Lorentz transformations of the electronic motion. Firstly we will transform from (x_n, y_n) into (x_2, y_2) and secondly we will transform from (x_2, y_2) into (x_n, y_n). The results of these two sets of operations must be equivalent. We make use of the central tenet of relativity that if frame A is moving with velocity $\mathbf{v}$ with respect to frame B then frame B must be moving with velocity $-\mathbf{v}$ with respect to frame A. We will see that this can only be the case if the frames are precessing with respect to one another.

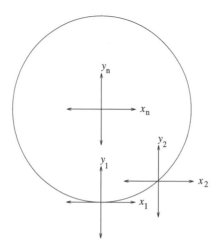

Fig. 2.6. The frames of reference used in our discussion of Thomas precession. (x_n, y_n) is the rest frame of the atomic nucleus, (x_1, y_1) is a frame on the electron's orbital path which we have chosen for convenience at the lowest point of the orbit. (x_2, y_2) is infinitesimally displaced from (x_1, y_1). In the figure this displacement is greatly exaggerated. Note that all axes have been drawn parallel.

We could have drawn (x_1, y_1) anywhere on the electron's orbital path, but now we see the reason for choosing the lowest point in the figure. All the velocities and accelerations are parallel to the axes we have set up. From the point of view of the observer in (x_n, y_n), when the electron is at the origin of (x_1, y_1) it is moving with velocity v_x in the $+x$-direction. Considering now an observer at rest in the (x_1, y_1) frame: the (x_n, y_n) frame is moving with velocity $-v_x$ in the x-direction and $v_y = 0$. The electron acceleration $\mathbf{a}$ is in the $+y_1$-direction. So when the electron has got to the origin of (x_2, y_2) its component of velocity in the y-direction is $\Delta v_y = a\Delta t$, while the change in the x-component of its velocity is $\Delta v_x = 0$. Therefore Δv_y is the velocity of the (x_2, y_2) frame as seen by an observer in (x_1, y_1). Now we can use the relativistic velocity transformations (equations (1.16)) to calculate the velocity of frame (x_2, y_2) as observed from frame (x_n, y_n). The x and y components are

$$u_{2x} = \frac{\Delta v_x + v_x}{1 + v_x \Delta v_x / c^2} = \frac{0 + v_x}{1 + 0} = v_x \qquad (2.65a)$$

$$u_{2y} = \frac{\Delta v_y \sqrt{1 - v_x^2/c^2}}{1 + v_x \Delta v_x / c^2} = \Delta v_y \sqrt{1 - v_x^2/c^2} \qquad (2.65b)$$

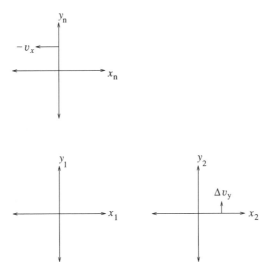

Fig. 2.7. To an observer at rest in the (x_1, y_1) frame the nuclear rest frame appears to be moving in the negative x-direction with velocity v_x. The frame (x_2, y_2) appears to be moving in the positive y-direction with velocity Δv_y.

Now we can do the reverse transformation and find the velocity of the (x_n, y_n) frame as observed from the (x_2, y_2) frame. In this case we have to recall that the relative velocity between the (x_1, y_1) and the (x_2, y_2) frames is in the y-direction so the velocity transformations are permuted. The y-component is

$$u_{ny} = \frac{v_y - \Delta v_y}{1 - v_y \Delta v_y / c^2} = -\Delta v_y \qquad (2.66a)$$

and the x-component

$$u_{nx} = \frac{-v_x \sqrt{1 - (\Delta v_y^2 / c^2)}}{1 - v_y \Delta v_y / c^2} = -v_x \sqrt{1 - \Delta v_y^2 / c^2} \qquad (2.66b)$$

Let us calculate the angle between $\mathbf{u}_2$ and the x-axis of the (x_n, y_n) frame.

$$\theta_2 = \tan^{-1} \left(\frac{\Delta v_y \sqrt{1 - v_x^2 / c^2}}{v_x} \right) \approx \frac{\Delta v_y \sqrt{1 - v_x^2 / c^2}}{v_x} \qquad (2.67a)$$

Next we calculate the angle between $\mathbf{u}_n$ and the x-axis of the (x_2, y_2) frame

$$\theta_n = \tan^{-1} \left(\frac{-\Delta v_y}{-v_x \sqrt{1 - \Delta v_y^2 / c^2}} \right) \approx \frac{-\Delta v_y}{-v_x \sqrt{1 - \Delta v_y^2 / c^2}} \qquad (2.67b)$$

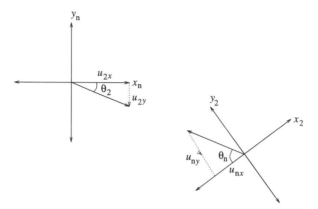

Fig. 2.8. Figure illustrating Thomas precession, the precession of the electron rest frame as it orbits around the nucleus.

Now, because of the central postulate of relativity that all inertial frames are equivalent, $\mathbf{u}_n$ and $\mathbf{u}_2$ must be equal in magnitude and opposite in direction. However, (2.67a) and (2.67b) are at different angles relative to their respective x-axes. The only way these two statements can be reconciled is if the sets of axes are rotated relative to each other, as illustrated in figure 2.8. We could go on with this argument and show that a fourth set of axes must be rotated with respect to (x_2, y_2) in the same way, but such a calculation would be pointless and messy. It is clear now that an observer in (x_1, y_1) sees both the (x_n, y_n) and (x_2, y_2) axes parallel to his axes. However, an observer in (x_n, y_n) sees the axes of (x_2, y_2) rotated with respect to his own and *vice versa*. Let us work out the angular frequency of this precession. We define the vector angle of rotation as being of magnitude $\theta_n - \theta_2$ and having a direction perpendicular to the plane of rotation. The rotation angle is given by

$$v_x \Delta\theta \approx \left(1 - \sqrt{\left(1 - \frac{v_x^2}{c^2} \right)} \right) \Delta v_y \tag{2.68}$$

where we have ignored terms in Δv^2.

From the standard non-relativistic Bohr theory of the atom we know that the velocity of atomic electrons is much less than the speed of light, so to a very good approximation we can perform a binomial expansion on the square root in (2.68) to obtain

$$v_x \Delta\theta \approx \Delta v_y - \left(1 - \frac{v_x^2}{2c^2} \right) \Delta v_y = \frac{v_x^2}{2c^2} \Delta v_y = \frac{v_x^2}{2c^2} a \Delta t \tag{2.69}$$

Including the direction of rotation we have

$$\Delta\theta \approx \frac{1}{2c^2}\mathbf{v} \times \mathbf{a}\Delta t \qquad (2.70)$$

Finally we can divide by Δt and take the limit as $\Delta t \to 0$, giving

$$\omega_T = \frac{d\theta}{dt} = \frac{1}{2c^2}\mathbf{v} \times \mathbf{a} = \frac{1}{2mc^2}\mathbf{v} \times \mathbf{F} = \frac{e}{2mc^2}\mathbf{v} \times \mathbf{E} \qquad (2.71)$$

This precessional effect is known as Thomas precession and ω_T is the Thomas frequency. It is the frequency at which the rest frame of the electron precesses as it moves round the nucleus.

Now we have to bring together several threads from earlier in this chapter. Firstly recall the Larmor precession frequency of equations (2.53) and (2.54). This is the frequency at which a dipole precesses about the direction of a magnetic field. In that case the dipole was due to the electron's orbital angular momentum in an external magnetic field. Note that (2.53) does not depend on the angular momentum. Now in equation (2.60) we have an identical physical situation. In this case it is the electron's spin angular momentum interacting with the magnetic field due to the nucleus which is circling it (in the electron rest frame). In this case we could follow through an argument very similar to that in section 2.3 to find the frequency of precession. Comparison of (2.60) with (2.54) shows that the spin Larmor precession is a factor of 2 larger than the orbital Larmor frequency. The spin Larmor precessional frequency is

$$\omega_{L_s} = \frac{e}{m}\mathbf{B} = \frac{e}{mc^2}\mathbf{v} \times \mathbf{E} \qquad (2.72)$$

Now both (2.71) and (2.72) are measured in the rest frame of the electron and are taken around the same axis. However, we subtract rather than add them because equation (2.72) is really the precessional frequency of the spin angular momentum vector which points in the opposite direction to the spin magnetic moment vector. Hence the total precessional frequency is

$$\omega = \omega_T - \omega_{L_s} = -\frac{e}{2mc^2}\mathbf{v} \times \mathbf{E} \qquad (2.73)$$

where we have included the negative sign which is indicative of the direction of the precession.

So we see that a straight transformation of the electron precessional frequency from the electron rest frame to the nuclear rest frame does not give the correct answer because the electron frame is itself precessing around the nuclear rest frame at the Thomas frequency. Equations (2.71) to (2.73) show that the Thomas frequency is half the magnitude of the Larmor frequency and in the opposite direction. So when they are added they bring the magnitude of the frequency down to half its original value.

This is the origin of the factor of 2 introduced into equation (2.61) without explanation at the time.

The question of why this frequency is a factor of 2 larger than ω_L is not easy to answer at this level. It is due to the relative sizes of the spin and orbital dipole moments, or g-factors. The calculation of the orbital magnetic moment can be taken as correct here. The spin magnetic moment falls out of relativistic quantum theory, as will be seen in chapter 4.

2.8 The Pauli Equation in a Central Potential

The Pauli equation (2.55) can be written in matrix form as in equation (2.74). There is a separate Schrödinger-like equation for each electron spin, and these equations are only coupled through the spin–orbit term.

$$i\hbar\frac{\partial}{\partial t}\begin{pmatrix}\psi_\uparrow(\mathbf{r},t)\\\psi_\downarrow(\mathbf{r},t)\end{pmatrix} =$$

$$\left(\frac{1}{2m}(\hat{\mathbf{p}} - e\mathbf{A}(\mathbf{r}))^2 + V(\mathbf{r}) + \frac{\hbar}{4m^2c^2}\frac{1}{r}\frac{dV(r)}{dr}\tilde{\boldsymbol{\sigma}}\cdot\hat{\mathbf{L}}\right)\begin{pmatrix}\psi_\uparrow(\mathbf{r},t)\\\psi_\downarrow(\mathbf{r},t)\end{pmatrix} \qquad (2.74)$$

where all terms in the Hamiltonian except the spin–orbit one are understood to be multiplied by the unit matrix (in fact we can multiply the spin–orbit term by it as well; it makes no difference). In the spin–orbit term we have introduced the 2×2 nature of the operator by using the Pauli matrix representation of the spin operator from equation (2.30).

Equation (2.74) is an equation of the form

$$i\hbar\frac{\partial}{\partial t}\psi(\mathbf{r},t) = \hat{H}\psi(\mathbf{r},t) \qquad (2.75)$$

and so the variables can be separated in the usual way, giving us

$$\psi(\mathbf{r},t) = \begin{pmatrix}\psi_\uparrow(\mathbf{r},t)\\\psi_\downarrow(\mathbf{r},t)\end{pmatrix} = \begin{pmatrix}\psi_\uparrow(\mathbf{r})\\\psi_\downarrow(\mathbf{r})\end{pmatrix}e^{-iEt/\hbar} = \psi(\mathbf{r})e^{-iEt/\hbar} \qquad (2.76)$$

where E is an eigenvalue of the Hamiltonian operator, and

$$\left(\frac{1}{2m}\left(\frac{\hbar}{i}\nabla - e\mathbf{A}(\mathbf{r})\right)^2 + V(\mathbf{r}) + \frac{\hbar}{4m^2c^2}\frac{1}{r}\frac{dV(r)}{dr}\tilde{\boldsymbol{\sigma}}\cdot\hat{\mathbf{L}} - E\right)\begin{pmatrix}\psi_\uparrow(\mathbf{r})\\\psi_\downarrow(\mathbf{r})\end{pmatrix} = 0 \qquad (2.77)$$

This is the Hamiltonian form of the Pauli equation.

For applications in both atomic and condensed matter physics it is useful to know the form of the solutions to equation (2.77) in a central potential. We know from the standard separation of the variables of the Schrödinger equation that the solutions in the absence of the spin–orbit coupling term take the form

$$\psi_{nlm}(\mathbf{r}) = R_{nl}(r, E)Y_l^m(\hat{\mathbf{r}}) \qquad (2.78)$$

where $Y_l^m(\hat{\mathbf{r}})$ are the usual spherical harmonics (see appendix C) and $R_{nl}(r, E)$ are radial functions. Clearly it would be desirable to have solutions to (2.77) that are close to this form and reduce to (2.78) when the spin–orbit term tends to zero (as $c \to \infty$). Some hint of this has already been given in equation (2.28). We know that if the spin–orbit term in (2.77) is omitted the Hamiltonian will commute with $\hat{S}^2$, $\hat{S}_z$, $\hat{L}^2$, and $\hat{L}_z$. It commutes with the spin operators because there would be no spin dependence in the Hamiltonian at all, and it commutes with the orbital angular momentum operators as we know it does from the non-relativistic theory (Gasiorowicz 1974).

Now we can ask how $\hat{\mathbf{L}}$ and $\hat{\mathbf{S}}$ commute with the Hamiltonian in equation (2.77). This is a very important question and the answer is responsible for many of the differences between a relativistic and a non-relativistic description of the properties of condensed matter. Therefore, we will examine how to evaluate the commutators in detail. We start by evaluating the commutator of $\hat{S}_z$ with the spin–orbit coupling term in (2.77). Making use of equations (2.1) we have

$$
\begin{aligned}
[\hat{S}_z, \hat{\mathbf{L}} \cdot \hat{\mathbf{S}}] &= [\hat{S}_z, \hat{L}_z\hat{S}_z + \hat{L}_y\hat{S}_y + \hat{L}_x\hat{S}_x] \\
&= [\hat{S}_z, \hat{L}_y\hat{S}_y] + [\hat{S}_z, \hat{L}_x\hat{S}_x] = [\hat{S}_z, \hat{S}_y]\hat{L}_y + [\hat{S}_z, \hat{S}_x]\hat{L}_y \\
&= -i\hbar(\hat{S}_x\hat{L}_y - \hat{S}_y\hat{L}_x)
\end{aligned}
\tag{2.79}
$$

We can follow a similar procedure to calculate the commutator of $\hat{L}_z$ with the spin–orbit term. Again we will write out the mathematical steps in detail

$$
\begin{aligned}
[\hat{L}_z, \hat{\mathbf{L}} \cdot \hat{\mathbf{S}}] &= [\hat{L}_z, \hat{L}_z\hat{S}_z + \hat{L}_y\hat{S}_y + \hat{L}_x\hat{S}_x] \\
&= [\hat{L}_z, \hat{L}_y\hat{S}_y] + [\hat{L}_z, \hat{L}_x\hat{S}_x] = [\hat{L}_z, \hat{L}_y]\hat{S}_y + [\hat{L}_z, \hat{L}_x]\hat{S}_x \\
&= -i\hbar(\hat{L}_x\hat{S}_y - \hat{L}_y\hat{S}_x)
\end{aligned}
\tag{2.80}
$$

Now look what happens if we add (2.79) and (2.80) together

$$
\begin{aligned}
[\hat{S}_z, \hat{\mathbf{L}} \cdot \hat{\mathbf{S}}] + [\hat{L}_z, \hat{\mathbf{L}} \cdot \hat{\mathbf{S}}] &= [\hat{S}_z + \hat{L}_z, \hat{\mathbf{L}} \cdot \hat{\mathbf{S}}] = [\hat{J}_z, \hat{\mathbf{L}} \cdot \hat{\mathbf{S}}] \\
&= -i\hbar(\hat{S}_x\hat{L}_y - \hat{S}_y\hat{L}_x) - i\hbar(\hat{L}_x\hat{S}_y - \hat{L}_y\hat{S}_x) = 0
\end{aligned}
\tag{2.81}
$$

Equations (2.79) and (2.80) show that neither $\hat{S}_z$ nor $\hat{L}_z$ commutes with the Pauli Hamiltonian. However, equation (2.81) shows that, despite this, $\hat{J}_z$ does commute with the Pauli Hamiltonian. Clearly we could have gone through the above mathematics for the x- and y-components of $\hat{\mathbf{J}}$ and found that they also commute with the Hamiltonian. We know that if a quantity commutes with the Hamiltonian it is conserved and may have simultaneous eigenfunctions with the Hamiltonian. Hence we expect our solutions to (2.77) to be simultaneous eigenfunctions of $\hat{H}$, $\hat{\mathbf{J}}^2$ and $\hat{J}_z$. Furthermore, equations (2.79) to (2.81) (and their cyclic permutations) are proofs that in relativistic quantum mechanics the spin angular momentum and orbital angular momentum are not conserved separately, but that total angular momentum is conserved.

In matrix form the $\hat{J}_z$ operator is

$$\hat{J}_z = \tilde{I}\hat{L}_z + \frac{\hbar}{2}\tilde{\sigma}_z \tag{2.82}$$

and so

$$\begin{aligned}
\hat{J}_z\psi(\mathbf{r}) &= \begin{pmatrix} \hat{L}_z + \hbar/2 & 0 \\ 0 & \hat{L}_z - \hbar/2 \end{pmatrix} \begin{pmatrix} \psi_\uparrow(\mathbf{r}) \\ \psi_\downarrow(\mathbf{r}) \end{pmatrix} \\
&= \begin{pmatrix} (\hat{L}_z + \hbar/2)\psi_\uparrow(\mathbf{r}) \\ (\hat{L}_z - \hbar/2)\psi_\downarrow(\mathbf{r}) \end{pmatrix} = m_j\hbar \begin{pmatrix} \psi_\uparrow(\mathbf{r}) \\ \psi_\downarrow(\mathbf{r}) \end{pmatrix}
\end{aligned} \tag{2.83}$$

So

$$\begin{pmatrix} (\hat{L}_z - (m_j - 1/2)\hbar)\psi_\uparrow(\mathbf{r}) \\ (\hat{L}_z - (m_j + 1/2)\hbar)\psi_\downarrow(\mathbf{r}) \end{pmatrix} = 0 \tag{2.84}$$

A solution of this is that $\psi_\uparrow$ and $\psi_\downarrow$ both have an angular dependence which is defined by a spherical harmonic with the same value of the l quantum number and with values of the m_l quantum number differing by one. If this is the case then the eigenfunctions of the Hamiltonian are also eigenfunctions of $\hat{L}^2$. So we have

$$\psi_\uparrow(\mathbf{r}) = g(r)Y_l^{m_l}(\hat{\mathbf{r}}), \qquad \psi_\downarrow(\mathbf{r}) = f(r)Y_l^{m_l+1}(\hat{\mathbf{r}}) \tag{2.85}$$

Next we are going to define the operator $\hat{K}$.

$$\hat{K} = \hat{\mathbf{L}} \cdot \tilde{\sigma} + \hbar \tag{2.86}$$

This turns out to be a useful operator, although here it has been defined rather arbitrarily. The $\hbar$ in (2.86) is understood to be multiplied by the unit matrix.

Let us see what the eigenvalues of $\hat{K}$ are. To discover this we will operate with it on ψ, and use the square of equation (2.4a).

$$\begin{aligned}
\hat{K}\psi &= (\hat{\mathbf{L}} \cdot \tilde{\sigma} + \hbar)\psi = \left(\tfrac{2}{\hbar}\hat{\mathbf{L}} \cdot \hat{\mathbf{S}} + \hbar\right)\psi = \left(\tfrac{2}{\hbar}\tfrac{1}{2}(\hat{\mathbf{J}}^2 - \hat{\mathbf{L}}^2 - \hat{\mathbf{S}}^2) + \hbar\right)\psi \\
&= (j(j+1) - l(l+1) - s(s+1)) + 1)\hbar\psi \\
&= (j(j+1) - l(l+1) + \tfrac{1}{4})\hbar\psi = -\hbar\kappa\psi
\end{aligned} \tag{2.87}$$

where we have explicitly used the fact that $s = 1/2$ for all the particles in which we are interested. In (2.87) we have defined the eigenvalues of $\hat{K}$ as $-\kappa\hbar$, but the expression for them can be simplified considerably. There are two possibilities, $j = l + 1/2$ and $j = l - 1/2$. In each case it is simple to use (2.87) to show that

$$j = l + 1/2 \rightarrow \kappa = -l - 1 \qquad (2.88a)$$

$$j = l - 1/2 \rightarrow \kappa = l \qquad (2.88b)$$

So we see that the eigenvalues of this operator are integers and are an equivalent representation of the total angular momentum to $\hat{\mathbf{J}}$. If $\kappa < 0$ we have $j = l + 1/2$ and if $\kappa > 0$ we have $j = l - 1/2$. There are two values of κ for every value of l. Finally note that any integer value of κ is permissible except $\kappa = 0$. Let us operate with $\hat{K}$ on $\psi(\mathbf{r})$

$$\hat{K}\psi(\mathbf{r}) = \begin{pmatrix} \hat{L}_z + \hbar & \hat{L}_x - i\hat{L}_y \\ \hat{L}_x + i\hat{L}_y & -\hat{L}_z + \hbar \end{pmatrix} \begin{pmatrix} \psi_\uparrow(\mathbf{r}) \\ \psi_\downarrow(\mathbf{r}) \end{pmatrix} = -\hbar\kappa \begin{pmatrix} \psi_\uparrow(\mathbf{r}) \\ \psi_\downarrow(\mathbf{r}) \end{pmatrix} \qquad (2.89)$$

Writing (2.89) out in component form

$$(\hat{L}_z + \hbar)\psi_\uparrow + (\hat{L}_x - i\hat{L}_y)\psi_\downarrow = -\hbar\kappa\psi_\uparrow \qquad (2.90a)$$

$$(-\hat{L}_z + \hbar)\psi_\downarrow + (\hat{L}_x + i\hat{L}_y)\psi_\uparrow = -\hbar\kappa\psi_\downarrow \qquad (2.90b)$$

It turns out that (2.90b) is just a multiple of (2.90a), so we only need to work with one of them. We can use equations analogous to (2.10) for orbital angular momentum to solve these, and we obtain

$$f(r)[(l + m_l + 1)(l - m_l)]^{1/2} Y_l^{m_l} = g(r)(-\kappa - m_l - 1) Y_l^{m_l} \qquad (2.91)$$

Here the spherical harmonics cancel, and this relation shows us that the radial parts of $\psi_\uparrow$ and $\psi_\downarrow$ only differ from one another by a constant. For $\kappa < 0$ equation (2.91) gives

$$(l + m_l + 1)^{1/2} f(r) = (l - m_l)^{1/2} g(r) \qquad (\kappa = -l - 1) \qquad (2.92a)$$

For $\kappa > 0$ we have

$$(l - m_l)^{1/2} f(r) = -(l + m_l + 1)^{1/2} g(r) \qquad (\kappa = l) \qquad (2.92b)$$

Equations (2.92) show that for a particular set of quantum numbers, the radial parts of the eigenfunctions defined in (2.85) are the same except for a simple multiplicative factor. So we can write

$$\psi(\mathbf{r}) = f(r) \begin{pmatrix} c_1 Y_l^{m_l}(\hat{\mathbf{r}}) \\ c_2 Y_l^{m_l+1}(\hat{\mathbf{r}}) \end{pmatrix} \qquad (2.93)$$

We have the relative sizes of c_1 and c_2 from (2.92), but their absolute values depend upon the normalization. Let us define the normalization by

$$\int \psi_\uparrow^*(\mathbf{r})\psi_\uparrow(\mathbf{r}) + \psi_\downarrow^*(\mathbf{r})\psi_\downarrow(\mathbf{r})d\mathbf{r} = 1 \qquad (2.94)$$

There is still some arbitrariness in this integral because we can choose the values of the constants c and the radial part of the wavefunction in equation (2.93) in an infinite number of ways and still satisfy (2.94), but the usual and the most intuitively sensible choice is

$$\int r^2 |f(r)|^2 dr = 1 \tag{2.95}$$

and hence

$$|c_1|^2 + |c_2|^2 = 1 \tag{2.96}$$

Comparison of (2.96) and (2.92) gives

$$c_1 = \left(\frac{l - m_l}{2l + 1}\right)^{1/2}, \quad c_2 = \left(\frac{l + m_l + 1}{2l + 1}\right)^{1/2} \qquad (\kappa < 0) \tag{2.97a}$$

$$c_1 = -\left(\frac{l + m_l + 1}{2l + 1}\right)^{1/2}, \quad c_2 = \left(\frac{l - m_l}{2l + 1}\right)^{1/2} \qquad (\kappa > 0) \tag{2.97b}$$

These are more symmetrically written in terms of m_j rather than m_l and are shown in that form in table 2.2, later in this chapter.

2.9 Dirac Notation

In this section we introduce a shorthand method of writing down quite complicated mathematical formulae. We have not used Dirac notation so far in this book, but in fact it is very intuitive and you can probably guess what the shorthand means, even if you haven't met it before. Dirac notation has several advantages. It is both concise and precise, and also rather general. For example one can write down a matrix element without stating what space it is written in. Hence it tends to take us from the particular to the general. We will use Dirac notation in the following section and in several subsequent ones, so let us define it now. A wavefunction is a four-component column and is written

$$|\psi\rangle = \begin{pmatrix} \psi_1 \\ \psi_2 \\ \psi_3 \\ \psi_4 \end{pmatrix} \tag{2.98a}$$

and

$$\langle \phi| = (\phi_1^*, \phi_2^*, \phi_3^*, \phi_4^*) \tag{2.98b}$$

$|\psi\rangle$ is sometimes called a state vector. Another label is that $\langle \psi|$ is known as a bra, and $|\psi\rangle$ is known as a ket. We can take a scalar product of (2.98a and b):

$$\langle \phi|\psi\rangle = \int (\phi_1^* \psi_1 + \phi_2^* \psi_2 + \phi_3^* \psi_3 + \phi_4^* \psi_4) dz \tag{2.99}$$

where z just represents the space in which we are working – it could be position or momentum, for example. The normalization integral is

$$\langle \psi | \psi \rangle = 1 \tag{2.100}$$

and an expectation value for an operator $\hat{O}$ is written

$$\langle \hat{O} \rangle = \langle \psi | \hat{O} | \psi \rangle = \int \psi^\dagger \hat{O} \psi \, dz \tag{2.101}$$

Finally, if ψ_n represents one of a complete set of eigenfunctions, we can write

$$\sum_n |\psi_n\rangle\langle\psi_n| = 1 \tag{2.102}$$

This is about all we need to know about Dirac notation. The ket $|\psi\rangle$ can represent any type of wavefunction. It could be a very complicated many-body wavefunction or one of the simple one-electron type.

2.10 Clebsch–Gordan and Racah Coefficients

The coefficients c_1 and c_2 in the above section are examples of the Clebsch–Gordan coefficients. These are coefficients that we come across when trying to add angular momenta. This is not a book about angular momentum, so this section contains only a short description of the Clebsch–Gordan coefficients and the relations between them that we will require in this book. This section follows the work of Rose (1957) and readers are referred there or to one of the many other good books on the quantum theory of angular momentum (e.g. Edmonds 1957, Tinkham 1964) for a more thorough discussion.

Suppose we have two particles, each with an angular momentum that obeys commutation relations like (2.4b). Another possibility for which the same formalism is applicable is when we have a single particle whose angular momentum is defined as a sum of two parts, spin and orbital for example. In each case both angular momenta obey eigenvalue equations like equations (2.5). The eigenfunctions are defined by

$$\hat{J}_1^2 \psi_{j_1}^{m_{j_1}} = j_1(j_1 + 1)\hbar^2 \psi_{j_1}^{m_{j_1}} \tag{2.103a}$$

$$\hat{J}_{1z} \psi_{j_1}^{m_{j_1}} = m_{j1}\hbar \psi_{j_1}^{m_{j_1}} \tag{2.103b}$$

$$\hat{J}_2^2 \psi_{j_2}^{m_{j_2}} = j_2(j_2 + 1)\hbar^2 \psi_{j_2}^{m_{j_2}} \tag{2.104a}$$

$$\hat{J}_{2z} \psi_{j_2}^{m_{j_2}} = m_{j_2}\hbar \psi_{j_2}^{m_{j_2}} \tag{2.104b}$$

Now we are going to define the total angular momentum

$$\hat{\mathbf{J}} = \hat{\mathbf{J}}_1 + \hat{\mathbf{J}}_2 \tag{2.105}$$

and we want to add the angular momenta in such a way that the total angular momentum operator $\hat{\mathbf{J}}$ also obeys angular momentum commutation relations and eigenvalue equations like (2.103) and (2.104). Firstly let us point out some simple properties that $\hat{\mathbf{J}}$ must have. If we add together two angular momenta their z-components must just behave additively, so $m_j = m_{j_1} + m_{j_2}$. Now we know that

$$-j_1 \leq m_{j_1} \leq j_1 \qquad\qquad -j_2 \leq m_{j_2} \leq j_2 \qquad\qquad (2.106)$$

So, the maximum value that m_j can take on is $j_1 + j_2$, and this must also be the maximum possible value of j. That is,

$$j_{\max} = j_1 + j_2 \qquad\qquad (2.107)$$

This is consistent with our understanding of the addition of vectors. If we just add the vectors $\hat{\mathbf{J}}_1$ and $\hat{\mathbf{J}}_2$ together it is easy to see from drawing simple vector diagrams that the values j can take on are

$$j = |j_1 - j_2|, |j_1 - j_2| + 1, |j_1 - j_2| + 2, \cdots, j_1 + j_2 - 2, j_1 + j_2 - 1, j_1 + j_2 \quad (2.108a)$$

This is known as the triangle condition. We also have

$$-m_j \leq j \leq m_j \qquad\qquad (2.108b)$$

Equations (2.103) and (2.104) describe the two-component system in a way in which the wavefunctions are completely uncoupled. Let us define a new representation in which the two angular momenta are coupled via

$$\psi_j^{m_j} = \sum_{m_{j_1}, m_{j_2}} C(j_1 j_2 j; m_{j_1} m_{j_2} m_j) \psi_{j_1}^{m_{j_1}} \psi_{j_2}^{m_{j_2}} \qquad\qquad (2.109)$$

The coefficients C are the Clebsch–Gordan coefficients. Clearly they require six numbers in the brackets to be fully defined, although we shall see below that not all these numbers are independent. There is a bewildering number of different symbols and notations used for the Clebsch–Gordan coefficients. We stick with the one used by Rose (1961). There is also a variety of names for them in the literature, including 'vector addition coefficients' and 'Wigner coefficients'.

Let us apply the z-component of the operator in (2.105) to both sides of (2.109). This immediately yields

$$\sum_{m_{j_1}, m_{j_2}} (m_j - m_{j_1} - m_{j_2}) C(j_1 j_2 j; m_{j_1} m_{j_2} m_j) \psi_{j_1}^{m_{j_1}} \psi_{j_2}^{m_{j_2}} = 0 \qquad (2.110)$$

The wavefunctions in equation (2.110) are linearly independent quantities, and therefore it is the coefficients that must vanish. Indeed the

$C(j_1 j_2 j; m_{j_1} m_{j_2} m_j)$ must vanish unless $(m_j - m_{j_1} - m_{j_2})$ vanishes. So

$$m_j = m_{j_1} + m_{j_2} \tag{2.111}$$

and we can suppress one of the quantum numbers in (2.109) to give

$$\begin{aligned} C(j_1 j_2 j; m_{j_1} m_{j_2} m_j) &= C(j_1 j_2 j; m_{j_1} m_{j_2}(m_{j_1} + m_{j_2})) \\ &= C(j_1 j_2 j; m_j - m_{j_2}, m_{j_2}) \end{aligned} \tag{2.112}$$

In the last step in (2.112) the sixth quantum number is understood to be the sum of the fourth and fifth quantum numbers. Hence we can write

$$\psi_j^{m_j} = \sum_{m_{j_2}} C(j_1 j_2 j; m_j - m_{j_2}, m_{j_2}) \psi_{j_1}^{m_j - m_{j_2}} \psi_{j_2}^{m_{j_2}} \tag{2.113}$$

The eigenfunctions in (2.103) to (2.113) are all orthonormal to one another. If we assume that both j and j' are found from the same initial values of j_1 and j_2 we have

$$\left\langle \psi_j^{m_j} \middle| \psi_{j'}^{m'_j} \right\rangle = \sum_{m_{j_2}} \sum_{m'_{j_2}} C(j_1 j_2 j; m_j - m_{j_2}, m_{j_2}) C(j_1 j_2 j'; m'_j - m'_{j_2}, m'_{j_2})$$

$$\times \left\langle \psi_{j_1}^{m_j - m_{j_2}} \middle| \psi_{j_1}^{m'_j - m'_{j_2}} \right\rangle \left\langle \psi_{j_2}^{m_{j_2}} \middle| \psi_{j_2}^{m'_{j_2}} \right\rangle =$$

$$\sum_{m_{j_2}} \sum_{m'_{j_2}} C(j_1 j_2 j; m_j - m_{j_2}, m_{j_2}) C(j_1 j_2 j'; m'_j - m'_{j_2}, m'_{j_2}) \delta_{m_{j_2}, m'_{j_2}} \delta_{m_j m'_j}$$

$$= \sum_{m_{j_2}} C(j_1 j_2 j; m_j - m_{j_2}, m_{j_2}) C(j_1 j_2 j'; m'_j - m_{j_2}, m_{j_2}) \delta_{m_j m'_j}$$

$$= \delta_{jj'} \delta_{m_j, m'_j} \tag{2.114}$$

i.e.

$$\sum_{m_{j_2}} C(j_1 j_2 j; m_j - m_{j_2}, m_{j_2}) C(j_1 j_2 j'; m_j - m_{j_2}, m_{j_2}) = \delta_{jj'} \tag{2.115}$$

This is a useful orthogonality relation obeyed by the Clebsch–Gordan coefficients. There exists an inverse transformation to (2.113). The easiest way to prove this is to write it down and then proceed to show that it leads to (2.113). The inverse is

$$\psi_{j_1}^{m_j - m_{j_2}} \psi_{j_2}^{m_{j_2}} = \sum_j C(j_1 j_2 j; m_j - m_{j_2}, m_{j_2}) \psi_j^{m_j} \tag{2.116}$$

Next we multiply (2.116) by $C(j_1 j_2 j'; m_j - m_{j_2}, m_{j_2})$ and sum over m_{j_2}.

$$\sum_{m_{j_2}} C(j_1 j_2 j'; m_j - m_{j_2}, m_{j_2}) \psi_{j_1}^{m_j - m_{j_2}} \psi_{j_2}^{m_{j_2}}$$

$$= \sum_{j} \sum_{m_{j_2}} C(j_1 j_2 j'; m_j - m_{j_2}, m_{j_2}) C(j_1 j_2 j; m_j - m_{j_2}, m_{j_2}) \psi_{j}^{m_j}$$

$$= \sum_{j} \psi_{j}^{m_j} \left(\sum_{m_{j_2}} C(j_1 j_2 j'; m_j - m_{j_2}, m_{j_2}) C(j_1 j_2 j; m_j - m_{j_2}, m_{j_2}) \right)$$

$$= \sum_{j} \psi_{j}^{m_j} \delta_{jj'} = \psi_{j'}^{m_j} \tag{2.117}$$

We see that (2.116) leads to (2.117) which is identical to (2.113). The reason for writing (2.116) is that it leads to another useful orthogonality relation. Multiplying (2.116) by itself and taking matrix elements we obtain

$$\left\langle \psi_{j_1}^{m_j - m_{j_2}} | \psi_{j_1'}^{m_j' - m_{j_2}'} \right\rangle \left\langle \psi_{j_2}^{m_{j_2}} | \psi_{j_2'}^{m_{j_2}'} \right\rangle =$$

$$\sum_{j'} \sum_{j} C(j_1 j_2 j; m_j - m_{j_2}, m_{j_2}) C(j_1' j_2' j'; m_j' - m_{j_2}', m_{j_2}') \left\langle \psi_{j}^{m_j} | \psi_{j'}^{m_j'} \right\rangle \tag{2.118}$$

Using the orthonormality of the eigenfunctions, we can reduce this to

$$\sum_{j} C(j_1 j_2 j; m_j - m_{j_2}, m_{j_2}) C(j_1 j_2 j; m_j' - m_{j_2}', m_{j_2}') = \delta_{m_j m_j'} \delta_{m_{j_2} m_{j_2}'} \tag{2.119}$$

Equations (2.115) and (2.119) are useful orthogonality relations obeyed by Clebsch–Gordan coefficients. There also exist formulae for finding the coefficients directly. These are unwieldy and only an extremely devoted fan of Clebsch–Gordan algebra would evaluate them by hand, but they can be used to calculate the coefficients on a computer, which may be useful for large values of the quantum numbers. The first of these was inflicted on us by Wigner (1931), as described by Rose (1957):

$$C(j_1 j_2 j; m_1 m_2 m) = (2j + 1)^{1/2} \delta_{m(m_1 + m_2)}$$

$$\times \left(\frac{(j - m)!(j + m)!(j_1 + j_2 - j)!(j + j_1 - j_2)!(j_2 + j - j_1)!}{(j_1 - m_1)!(j_1 + m_1)!(j_2 - m_2)!(j_2 + m_2)!(j_1 + j_2 + j + 1)!} \right)^{\frac{1}{2}} \tag{2.120}$$

$$\times \sum_{\mu} \left(\frac{(-1)^{\mu + j_2 + m_2}(j + j_2 + m_1 - \mu)!(j_1 - m_1 + \mu)!}{\mu!(j + m - \mu)!(j + j_2 - j_1 - \mu)!(\mu + j_1 - j_2 - m)!} \right)$$

and the second by Racah (1942):

$$
C(j_1 j_2 j; m_1 m_2 m) = \delta_{m(m_1+m_2)}
$$
$$
\times \left((2j+1)(j-m)!(j+m)!(j_1-m_1)!(j_1+m_1)!(j_2-m_2)!(j_2+m_2)! \right)^{\frac{1}{2}}
$$
$$
\times \left(\frac{(j_1+j_2-j)!(j+j_1-j_2)!(j_2+j-j_1)!}{(j_1+j_2+j+1)!} \right)^{1/2}
$$
$$
\times \sum_{\mu} \frac{-1^{\mu}}{\mu!} ((j_1-m_1-\mu)!(j_2+m_2-\mu)!
$$
$$
\times (j_1+j_2-j-\mu)!(j-j_2+m_1+\mu)!(j-j_1-m_2+\mu)!)^{-1} \qquad (2.121)
$$

Clearly, these two expressions are very complicated indeed. In both equations (2.120) and (2.121) the summations go over only those integral values for which none of the factorials are negative. We define $1/(-n)! = 0$ where n is a positive integer.

There also exist some useful symmetry relations that relate different Clebsch–Gordan coefficients to one another. Here we will simply list these without any derivation:

$$
C(j_1 j_2 j; m_1 m_2 m) = (-1)^{j_2+m_2} \left(\frac{2j+1}{2j_1+1} \right)^{1/2} C(j j_2 j_1; -m\, m_2\, -m_1)
$$
$$
= (-1)^{j_1-m_1} \left(\frac{2j+1}{2j_2+1} \right)^{1/2} C(j j_1 j_2; m\, -m_1\, m_2)
$$
$$
= (-1)^{j_2+m_2} \left(\frac{2j+1}{2j_1+1} \right)^{1/2} C(j_2 j j_1; -m_2\, m\, m_1)
$$
$$
= (-1)^{j_1-m_1} \left(\frac{2j+1}{2j_2+1} \right)^{1/2} C(j_1 j j_2; m_1\, -m\, -m_2)
$$
$$
= (-1)^{j_1+j_2-j} C(j_2 j_1 j; m_2 m_1 m)
$$
$$
= (-1)^{j_1+j_2-j} C(j_1 j_2 j; -m_1\, -m_2\, -m) \qquad (2.122)
$$

One final point to note, which we have glossed over until now, is that the Clebsch–Gordan coefficients originally defined in equation (2.109) are actually only defined there to within an arbitrary phase factor. The phase convention chosen in this book is such that the Clebsch–Gordan coefficients are always real.

The Clebsch–Gordan coefficients most commonly found in theoretical condensed matter physics are those for which $j_2 = 1$ and $j_2 = 1/2$. Therefore algebraic expressions for these are written out explicitly in tables 2.1 and 2.2. As you can see the expressions are very simple and it is a matter of straightforward arithmetic to find the numerical values.

Table 2.1. The Clebsch–Gordan coefficients $C(j_1 1 j; m - m_2, m_2)$.

j	$m_2 = -1$	$m_2 = 0$	$m_2 = 1$
$j_1 - 1$	$\left(\frac{(j_1+m+1)(j_1+m)}{2j_1(2j_1+1)}\right)^{1/2}$	$-\left(\frac{(j_1-m)(j_1+m)}{j_1(2j_1+1)}\right)^{1/2}$	$\left(\frac{(j_1-m+1)(j_1-m)}{2j_1(2j_1+1)}\right)^{1/2}$
j_1	$\left(\frac{(j_1+m+1)(j_1-m)}{2j_1(j_1+1)}\right)^{1/2}$	$\frac{m}{(j_1(j_1+1))^{1/2}}$	$-\left(\frac{(j_1-m+1)(j_1+m)}{2j_1(j_1+1)}\right)^{1/2}$
$j_1 + 1$	$\left(\frac{(j_1-m+1)(j_1-m)}{(2j_1+2)(2j_1+1)}\right)^{1/2}$	$\left(\frac{(j_1-m+1)(j_1+m+1)}{(j_1+1)(2j_1+1)}\right)^{1/2}$	$\left(\frac{(j_1+m+1)(j_1+m)}{(2j_1+2)(2j_1+1)}\right)^{1/2}$

Higher values of the quantum numbers require use of one of the formulae (2.120) or (2.121).

So far, we have been discussing the coupling of two angular momenta to make a third. Of course, physics doesn't stop there, any number of angular momenta can be coupled together. Therefore, in the final part of this section we will mention the coefficients that occur when we couple together three angular momenta. These are known as Racah coefficients. In fact we will not really discuss their underlying physics. We will simply present their relationship with the Clebsch–Gordan coefficients and their elementary properties, as that is all that is necessary for an understanding of the rest of this book.

Following Rose (1957) and Tinkham (1964) we see that the Racah coefficients contain six angular momenta in their argument. It is conventional to use the letters a to f to represent these. In the relations that follow, we also use the greek symbols α to δ to represent angular momenta. A Racah coefficient is written in the form $W(abcd; ef)$ and can be expressed in terms of the Clebsch–Gordan coefficients using one of several different versions of a relation often referred to as the recoupling formula:

$$\sum_f ((2e + 1)(2f + 1))^{1/2} W(abcd; ef) C(bdf; \beta, \delta) C(afc; \alpha, \beta + \delta)$$

$$= C(abe; \alpha, \beta) C(edc; \alpha + \beta, \delta) \qquad (2.123a)$$

$$((2e + 1)(2f + 1))^{1/2} W(abcd; ef) C(afc; \alpha, \beta + \delta)$$
$$= \sum_\beta C(abe; \alpha, \beta) C(edc; \alpha + \beta, \delta) C(bdf; \beta, \delta) \qquad (2.123b)$$

$$W(abcd; ef) = ((2e + 1)(2f + 1))^{-1/2} \times$$
$$\sum_\alpha \sum_\beta C(abe; \alpha, \beta) C(edc; \alpha + \beta, \delta) C(bdf; \beta, \delta) C(afc; \alpha, \beta + \delta) \qquad (2.123c)$$

Table 2.2. The Clebsch–Gordan
coefficients $C(l\frac{1}{2}j; m - m_2, m_2)$.

j	$m_2 = -1/2$	$m_2 = 1/2$
$l - 1/2$	$\left(\frac{l+m+1/2}{2l+1}\right)^{1/2}$	$-\left(\frac{l-m+1/2}{2l+1}\right)^{1/2}$
$l + 1/2$	$\left(\frac{l-m+1/2}{2l+1}\right)^{1/2}$	$\left(\frac{l+m+1/2}{2l+1}\right)^{1/2}$

The Clebsch–Gordan coefficients here can be evaluated using (2.120) or (2.121). We can use these expressions to find an equation for the Racah coefficients

$$W(abcd; ef) = \triangle_{abe}\triangle_{cde}\triangle_{acf}\triangle_{bdf}$$
$$\times \sum_k \frac{(-1)^{k+a+b+c+d}(k+1)!}{(k-a-b-e)!(k-c-d-e)!(k-a-c-f)!(k-b-d-f)!}$$
$$\times \frac{1}{(a+b+c+d-k)!(a+d+e+f-k)!(b+c+e+f-k)!}$$

(2.124a)

where

$$\triangle_{abc} = \left(\frac{(a+b-c)!(a-b+c)!(b+c-a)!}{(a+b+c+1)!}\right)^{1/2}$$

(2.124b)

The value of $\triangle_{abc} = 0$ unless the triangle condition mentioned in the discussion of Clebsch–Gordan coefficients is obeyed. The six angular momenta in $W(abcd; ef)$ may be interchanged provided the four triangle relations above are maintained. This leads to a set of Racah coefficient symmetry relations:

$$W(abcd; ef) = W(badc; ef) = W(cdab; ef) = W(dcba; ef)$$
$$= W(acbd; fe) = W(cadb; fe) = W(bdac; fe) = W(dbca; fe) \quad (2.125a)$$

$$(-1)^{b+c-e-f}W(abcd; ef) = W(aefd; bc) = W(eadf; bc)$$
$$= W(fdae; bc) = W(dfea; bc) = W(afed; cb)$$
$$= W(fade; cb) = W(defa; cb) = W(edaf; cb) \quad (2.125b)$$

$$(-1)^{a+d-e-f}W(abcd; ef) = W(befc; ad) = W(fcbe; ad)$$
$$= W(cfeb; ad) = W(ebcf; ad) = W(fbce; da)$$
$$= W(bfec; da) = W(ecbf; da) = W(cefb; da) \quad (2.125c)$$

It often occurs that formulae involving Clebsch–Gordan coefficients become complicated and tedious, and the Racah coefficients can be used to simplify them, particularly via the recoupling formula. An example of this occurs in chapter 10 in the derivation of the Hartree–Fock equations.

2.11 Relativistic Quantum Numbers and Spin-Angular Functions

We have introduced the quantum number κ in section 2.8. Here we are going to introduce two more numbers which turn out to be useful, and we will also summarize all the relations between quantum numbers.

$$\kappa = -l - 1 = -(j + 1/2) \qquad\qquad (j = l + 1/2) \qquad (2.126a)$$

$$\kappa = l = (j + 1/2) \qquad\qquad (j = l - 1/2) \qquad (2.126b)$$

It will also be worthwhile to define the quantum number $\bar{l}$, which is the value of the l quantum number associated with $-\kappa$

$$\begin{aligned}\bar{l} = l + 1 = -\kappa \qquad\qquad (\kappa < 0)\\ \bar{l} = l - 1 = \kappa - 1 \qquad\qquad (\kappa > 0)\end{aligned} \qquad (2.127)$$

We also define

$$S_\kappa = \frac{\kappa}{|\kappa|} \qquad (2.128)$$

so that

$$S_\kappa = -1 \qquad\qquad (j = l + 1/2) \qquad (2.129a)$$

$$S_\kappa = +1 \qquad\qquad (j = l - 1/2) \qquad (2.129b)$$

and

$$S_\kappa = l - \bar{l} = 2(l - j) \qquad (2.130)$$

We are now in a position to examine the properties of the angular part of our solution to the Pauli equation in a central field. The symbol $\chi_\kappa^{m_j}(\hat{\mathbf{r}})$ is usually used to describe these quantities and they are commonly known as spin-angular functions. Using equations (2.27) and (2.93) and the notation of table 2.1 we have

$$\psi_j^{m_j}(\mathbf{r}) = f(r)\chi_\kappa^{m_j}(\hat{\mathbf{r}}) = f(r)\sum_{m_s} C(l\tfrac{1}{2}j; m_j - m_s, m_s)Y_l^{m_j - m_s}(\hat{\mathbf{r}})\chi^{m_s} \quad (2.131a)$$

where we have replaced the subscripts 1 and 2 with the quantum numbers j and m_j in the wavefunctions. With this definition it is easy to see that

$$\tilde{I}_2 \sum_m Y_l^{m*}(\hat{\mathbf{r}})Y_l^m(\hat{\mathbf{r}}') = \sum_{j,m_j} \chi_\kappa^{m_j}(\hat{\mathbf{r}})\chi_\kappa^{m_j\dagger}(\hat{\mathbf{r}}') \quad (2.131b)$$

Equations (2.131) define the spin-angular functions. It can be shown that

$$\int \chi_\kappa^{m_j\dagger}(\hat{\mathbf{r}}) \chi_{\kappa'}^{m_j'}(\hat{\mathbf{r}}) d\hat{\mathbf{r}} = \delta_{\kappa\kappa'} \delta_{m_j m_j'} \qquad (2.132)$$

where the dagger indicates that the complex conjugate and transpose have to be taken. There is a further useful relation obeyed by the spin-angular functions. Let us define the operator

$$\tilde{\sigma}_r = \frac{1}{r} \mathbf{r} \cdot \tilde{\sigma} \qquad (2.133)$$

This operator commutes with $\hat{\mathbf{J}}$, as is easy to prove. Specifically it commutes with $\hat{J}_z$ and $\hat{\mathbf{J}}^2$ and so $\tilde{\sigma}_r \chi_\kappa^{m_j}(\hat{\mathbf{r}})$ must have the same values of j and m_j as $\chi_\kappa^{m_j}(\hat{\mathbf{r}})$. Furthermore $\tilde{\sigma}_r$ anticommutes with $\hat{K}$. It is not entirely trivial to see this, so we show it explicitly. Making use of equation (2.34) and the fact that $\mathbf{r}$ and $\mathbf{L}$ are by definition at right angles to one another, the anticommutator becomes

$$\hat{K}\tilde{\sigma} \cdot \hat{\mathbf{r}} + \tilde{\sigma} \cdot \hat{\mathbf{r}}\hat{K} = (\tilde{\sigma} \cdot \hat{\mathbf{L}} + \hbar)\tilde{\sigma} \cdot \hat{\mathbf{r}} + \tilde{\sigma} \cdot \hat{\mathbf{r}}(\tilde{\sigma} \cdot \hat{\mathbf{L}} + \hbar)$$
$$= \tilde{\sigma} \cdot \hat{\mathbf{L}}\tilde{\sigma} \cdot \hat{\mathbf{r}} + \tilde{\sigma} \cdot \hat{\mathbf{r}}\tilde{\sigma} \cdot \hat{\mathbf{L}} + 2\hbar\tilde{\sigma} \cdot \hat{\mathbf{r}}$$
$$= \hat{\mathbf{L}} \cdot \hat{\mathbf{r}} + i\tilde{\sigma} \cdot \hat{\mathbf{L}} \times \hat{\mathbf{r}} + \hat{\mathbf{r}} \cdot \hat{\mathbf{L}} + i\tilde{\sigma} \cdot \hat{\mathbf{r}} \times \hat{\mathbf{L}} + 2\hbar\tilde{\sigma} \cdot \hat{\mathbf{r}}$$
$$= i\tilde{\sigma} \cdot (\hat{\mathbf{L}} \times \hat{\mathbf{r}} + \hat{\mathbf{r}} \times \hat{\mathbf{L}}) + 2\hbar\tilde{\sigma} \cdot \hat{\mathbf{r}} \qquad (2.134a)$$

Now the simplest thing to do is to take a single component of this equation and evaluate it. It will be easy to see the corresponding expressions for the other components. So, selecting the z-component, and using (2.3):

$$(\hat{K}\tilde{\sigma} \cdot \hat{\mathbf{r}} + \tilde{\sigma} \cdot \hat{\mathbf{r}}\hat{K})_z = i\tilde{\sigma}_z(\hat{L}_x y - \hat{L}_y x + x\hat{L}_y - y\hat{L}_x) + 2\hbar\tilde{\sigma}_z \cdot z$$
$$= i\tilde{\sigma}_z([\hat{L}_x, y] - [\hat{L}_y, x]) + 2\hbar\tilde{\sigma}_z z$$
$$= i\tilde{\sigma}_z(i\hbar z + i\hbar z) + 2\hbar\tilde{\sigma}_z z$$
$$= -2\hbar\tilde{\sigma}_z z + 2\hbar\tilde{\sigma}_z z = 0 \qquad (2.134b)$$

and similarly for the x- and y-components. We can use this in

$$\hat{K}\tilde{\sigma}_r \chi_\kappa^{m_j}(\hat{\mathbf{r}}) = -\tilde{\sigma}_r \hat{K} \chi_\kappa^{m_j}(\hat{\mathbf{r}}) = -\tilde{\sigma}_r(-\hbar\kappa)\chi_\kappa^{m_j}(\hat{\mathbf{r}}) = \hbar\kappa\tilde{\sigma}_r \chi_\kappa^{m_j}(\hat{\mathbf{r}}) \qquad (2.135)$$

and comparison with (2.87) shows that this can only be true if

$$\tilde{\sigma}_r \chi_\kappa^{m_j}(\hat{\mathbf{r}}) = -\chi_{-\kappa}^{m_j}(\hat{\mathbf{r}}) \qquad (2.136)$$

This is a relation that will be crucial in our discussion of the relativistic one-electron atom.

2.12 Energy Levels of the One-Electron Atom

In this section we are going to use results derived earlier in this chapter to find the effect of the spin–orbit interaction on the energy levels of the one-electron atom using first order perturbation theory. Our total Hamiltonian is the Pauli Hamiltonian of equation (2.74) with the vector potential set equal to zero. The first two terms then form the Schrödinger Hamiltonian and can be solved for the energy levels of the hydrogen atom in the usual way, although it is now a two-component equation. These two terms form our unperturbed Hamiltonian. The spin–orbit coupling term is the perturbation. We know from the standard theory of the one-electron atom that the radial part of the unperturbed wavefunction is a Laguerre polynomial. So we need to evaluate

$$\Delta E = \int \psi_j^{m_j\dagger}(\mathbf{r})\hat{H}_1\psi_j^{m_j}(\mathbf{r})d\mathbf{r} = \frac{1}{2m^2c^2}\int \psi_j^{m_j\dagger}(\mathbf{r})\frac{1}{r}\frac{dV}{dr}\hat{\mathbf{S}}\cdot\hat{\mathbf{L}}\psi_j^{m_j}(\mathbf{r})d\mathbf{r} \quad (2.137)$$

Now $V(r)$ is the usual Coulomb electrostatic potential, so

$$\frac{1}{r}\frac{dV}{dr} = +\frac{Ze^2}{4\pi\epsilon_0 r^3} \quad (2.138)$$

The spin-angular functions defined in (2.131) have been set up because they are eigenfunctions of $\hat{\mathbf{J}}^2$, $\hat{J}_z$ and the Pauli Hamiltonian, as described below equation (2.81). It is easy to see by inspection that they are also eigenfunctions of $\hat{\mathbf{L}}^2$ because any particular spin-angular function contains spherical harmonics which depend on a particular value of the l quantum number. They are not eigenfunctions of $\hat{L}_z$ as the spherical harmonics in the spin-angular functions belong to differing values of the m quantum number. Finally they are also eigenfunctions of $\hat{\mathbf{S}}$, because they contain the spinors χ^{m_s}. These spin-angular functions are simply the angular part of the wavefunction for a particle in a spherical potential, the analogues of the simple spherical harmonics in the Schrödinger theory of the one-electron atom.

Now that we have these wavefunctions we can operate with the spin–orbit term in the Hamiltonian on them. For $j = l + \frac{1}{2}$ we have

$$\begin{aligned}
\hat{\mathbf{S}}\cdot\hat{\mathbf{L}}\psi_j^{m_j}(\mathbf{r}) &= \tfrac{1}{2}((\hat{\mathbf{L}}+\hat{\mathbf{S}})^2 - \hat{\mathbf{L}}^2 - \hat{\mathbf{S}}^2)\psi_j^{m_j}(\mathbf{r}) \\
&= \tfrac{1}{2}(j(j+1) - l(l+1) - s(s+1))\hbar^2\psi_j^{m_j}(\mathbf{r}) \\
&= \tfrac{1}{2}((l+\tfrac{1}{2})(l+\tfrac{3}{2}) - l^2 - l - \tfrac{1}{2}\times\tfrac{3}{2})\hbar^2\psi_j^{m_j}(\mathbf{r}) \\
&= \tfrac{1}{2}l\hbar^2\psi_j^{m_j}(\mathbf{r}) \quad (2.139a)
\end{aligned}$$

and for $j = l - \frac{1}{2}$

$$\hat{\mathbf{S}} \cdot \hat{\mathbf{L}} \psi_j^{m_j}(\mathbf{r}) = \frac{1}{2}((\hat{\mathbf{L}} + \hat{\mathbf{S}})^2 - \hat{\mathbf{L}}^2 - \hat{\mathbf{S}}^2)\psi_j^{m_j}(\mathbf{r})$$

$$= \frac{1}{2}(j(j+1) - l(l+1) - s(s+1))\hbar^2 \psi_j^{m_j}(\mathbf{r})$$

$$= \frac{1}{2}((l - \frac{1}{2})(l + \frac{1}{2}) - l^2 - l - \frac{3}{4})\hbar^2 \psi_j^{m_j}(\mathbf{r})$$

$$= \frac{1}{2}\hbar^2(-l - 1)\psi_j^{m_j}(\mathbf{r}) \qquad (2.139b)$$

Now we are able to substitute (2.138) and (2.139) into (2.137) for ΔE, and make use of (2.78) to obtain

$$\Delta E = \frac{Ze^2}{8\pi\epsilon_0 m^2 c^2} \frac{\hbar^2}{2} \left\{ \begin{matrix} l \\ -l-1 \end{matrix} \right\} \int r^2 R_{nl}^2(r) \frac{1}{r^3} dr \quad j = \left\{ \begin{matrix} l + \frac{1}{2} \\ l - \frac{1}{2} \end{matrix} \right\} \qquad (2.140)$$

Now, the radial integral here is standard

$$\int r^2 R_{nl}^2(r) \frac{1}{r^3} dr = \frac{1}{n^3 l(l + \frac{1}{2})(l+1)} \frac{Z^3}{a_0^3} \qquad (2.141)$$

where a_0 is the first Bohr radius. Let us substitute this into (2.140)

$$\Delta E = \frac{Ze^2}{8\pi\epsilon_0 m^2 c^2} \frac{\hbar^2}{2} \left\{ \begin{matrix} l \\ -l-1 \end{matrix} \right\} \frac{Z^3}{a_0^3} \frac{1}{n^3 l(l + \frac{1}{2})(l+1)}$$

$$= \frac{Z^4 e^8 m}{(4\pi\epsilon_0)^4 4\hbar^4 c^2} \frac{1}{n^3 (l + \frac{1}{2})(l+1)} \qquad j = l + \frac{1}{2} \qquad (2.142)$$

$$= -\frac{Z^4 e^8 m}{(4\pi\epsilon_0)^4 4\hbar^4 c^2} \frac{1}{n^3 l(l + \frac{1}{2})} \qquad j = l - \frac{1}{2}$$

This tells us about the magnitude and sign of the corrections to the energy of the one-electron atom due to the existence of spin–orbit coupling. Firstly, we see that the prefactor is just $1/mc^2$ times the square of the usual Rydberg energy which can be worked out trivially. For hydrogen we are talking about shifts of order tenths of microelectronvolts, so this is a very small correction. For atoms with larger values of Z this quickly increases to a significant size as it goes as Z^4. For neon with $Z = 10$ we are already up to the millielectronvolt range and for molybdenum with $Z = 42$ we will be looking at shifts of order one electronvolt. This splitting is shown schematically in figure 2.9. Secondly, we can comment that some degeneracy is lifted by spin–orbit coupling. When the Schrödinger equation is solved for the energy levels of the one-electron atom they come out dependent on the n quantum number only. Here we see that if the spin–orbit coupling is included the energy levels depend on the l quantum number as well. For $j = l + \frac{1}{2}$ we can see that all quantities in the middle line of equation (2.142) are positive and so the energy is raised. For $j = l - \frac{1}{2}$ we have a negative sign, so the energy is lowered. Although

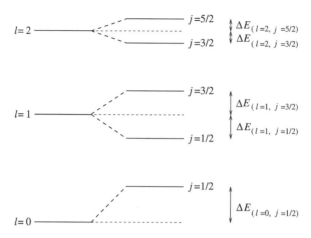

Fig. 2.9. Schematic, exaggerated figure showing spin–orbit splitting of energy
levels for the quantum numbers $l = 0, 1, 2$ for the same value of n. The $l = 0$
level is shifted but not split. Beyond that the splitting decreases with increasing
l. Equation (2.142) says that ΔE differs for $j = l + 1/2$ and $j = l - 1/2$ although
that difference is not shown on the figure.

we have only considered the one-electron atom here, this is a common
feature. The $j = l - \frac{1}{2}$ states generally occur at a lower energy than the
$j = l + \frac{1}{2}$ states.

2.13 Plane Wave Expansions

In scattering theory it is extremely important to be able to expand a plane
wave in terms of spherical waves. We illustrate how this is done in terms
of the spin-angular functions here. Our wavefunction for a plane wave
with a spin part to its wavefunction is

$$\psi(\mathbf{k}, \mathbf{r}) = \chi^{m_s} e^{i\mathbf{k}\cdot\mathbf{r}} \tag{2.143}$$

and we are going to set this equal to a sum over radial functions and
spin-angular functions. For a free particle the radial function is a spherical
Bessel function. So

$$\chi^{m_s} e^{i\mathbf{k}\cdot\mathbf{r}} = \sum_{\kappa m_j} c_{\kappa m_j} j_l(kr) \chi_\kappa^{m_j}(\hat{\mathbf{r}}) \tag{2.144}$$

and our task is to determine the coefficients $c_{\kappa m_j}$. Let us premultiply each
side of (2.144) by the conjugate transpose of a spin-angular function with
a particular value of κ and m_j, and then integrate over angles. From the

orthonormality of the spin-angular functions (equation (2.132)) we have

$$j_l(kr)c_{\kappa m_j} = \int \chi_\kappa^{m_j\dagger}(\hat{\mathbf{r}})\chi^{m_s}e^{i\mathbf{k}\cdot\mathbf{r}}d\hat{\mathbf{r}}$$

$$= \int \chi_\kappa^{m_j\dagger}(\hat{\mathbf{r}})\chi^{m_s}4\pi\sum_{l',m'}i^{l'}j_{l'}(kr)Y_{l'}^{m'*}(\hat{\mathbf{k}})Y_{l'}^{m'}(\hat{\mathbf{r}})d\hat{\mathbf{r}}$$

$$= \int \sum_{m'_s}C(l\tfrac{1}{2}j;m_j-m'_s,m'_s)Y_l^{m_j-m'_s}(\hat{\mathbf{r}})\chi^{m'_s\dagger}\chi^{m_s}$$

$$\times 4\pi\sum_{l',m'}i^{l'}j_{l'}(kr)Y_{l'}^{m'*}(\hat{\mathbf{k}})Y_{l'}^{m'}(\hat{\mathbf{r}})d\hat{\mathbf{r}}$$

$$= 4\pi\sum_{m'_s}C(l\tfrac{1}{2}j;m_j-m'_s,m'_s)\sum_{l',m'}i^{l'}j_{l'}(kr)Y_{l'}^{m'*}(\hat{\mathbf{k}})$$

$$\times \delta_{l,l'}\delta_{m_j-m'_s,m'}\delta_{m'_s m_s} \tag{2.145}$$

where we have used the well-known expansion of the exponential function given in equation (C.21). From (2.145) we see immediately that

$$c_{\kappa m_j} = 4\pi i^l C(l\tfrac{1}{2}j;m_j-m_s,m_s)Y_l^{m_j-m_s*}(\hat{\mathbf{k}}) \tag{2.146}$$

and so finally we have

$$\chi^{m_s}e^{i\mathbf{k}\cdot\mathbf{r}} = 4\pi\sum_{\kappa m_j}i^l C(l\tfrac{1}{2}j;m_j-m_s,m_s)Y_l^{m_j-m_s*}(\hat{\mathbf{k}})j_l(kr)\chi_\kappa^{m_j} \tag{2.147}$$

This is the required result: it is a plane wave on the left hand side expanded in spherical waves on the right hand side. Note that the wave is spin-polarized. For the unpolarized case we have to sum both spin directions and halve the result.

2.14 Problems

(1) Verify that the operators defined in equations (2.40) do have the required effect when applied to the eigenvectors of equation (2.41). If the photon is described using these eigenfunctions what is the role of the three-component nature of the wavefunction?
(2) The spin operators for spin-3/2 particles are given in equations (2.42). Find the corresponding eigenvectors and show that they have all the right properties.
(3) Verify equation (2.34).
(4) From the fact that an arbitrary two-component spinor is an eigenfunction of $\hat{\mathbf{S}}^2$, show that $\hat{\mathbf{S}}^2$ must be diagonal.
(5) Operate with $\tilde{\sigma}_y$ on the complex conjugate of a spin-angular function, and comment on your answer.

3

Particles of Spin Zero

Perhaps a natural reaction to the title of this chapter is that it should be rather short. In condensed matter physics we are interested in the particles that make up the world. They are the protons, neutrons and electrons, of course, and they all have spin 1/2. A few particles, such as the pi-meson, which we come across in particle physics do have spin zero, but they only exist for a very brief time before decaying, so how much can be said that is relevant to condensed matter physics?

This point can actually be answered rather easily. The generalization of quantum mechanics to include relativity is, to say the least, a non-trivial problem. As we shall see, spin-1/2 particles are well described by the Dirac equation. That equation describes both the relativistic nature of the particles and their spin (although the two are not really divisible). Treating spin-zero particles first means we can understand many aspects of the relativistic nature of quantum theory without the added complication of spin. Furthermore, formulae derived in this chapter can be compared with those in later chapters to give added insight into the nature of spin. This is particularly true when we look at the properties of a mythical spin-zero electron in a central Coulomb potential. Spin plays a key role in determining the properties of atoms, and comparison of the results of this chapter with those of chapter 8 lends considerable insight into atomic properties.

In this chapter we first develop the Klein–Gordon equation which describes relativistic spin-zero particles and look at its properties. In particular we concentrate on the free particle probability density and show that it takes on a very different, but nonetheless physically reasonable, form which we are able to interpret in terms of Lorentz–Fitzgerald contraction. For a particle that is not free we see that it becomes impossible to define a probability density rigorously as the expression one obtains for it can become negative. The Klein–Gordon equation as originally derived is a differential equation which is second order in time. It turns out to be more convenient to write it as two equations which are first order in time and to separate the wavefunction into two components. In section five we show how and why this is done and then go on to interpret the two components in terms of particles and antiparticles. We then discuss the position, momentum and velocity of a Klein–Gordon particle, and this forces us to reassess what we mean by a point particle. Next we solve the Klein–Gordon equation for a particle incident upon a potential barrier, which leads to some surprising results and to the idea of spontaneous creation of particle/antiparticle pairs. Finally we derive the radial form of the Klein–Gordon equation and solve it for the Coulomb potential. This is the direct generalization of the non-relativistic hydrogen atom theory so familiar from countless quantum mechanics courses. This theory models an atom where the orbiting electron has no spin. It enables us to see that relativity causes the angular momentum quantum number to enter the expression for the energy eigenvalues of the spinless-electron atom. To aid understanding as much as possible, wherever it is useful in this chapter we take the non-relativistic limit of our expressions to obtain equations that should be familiar from non-relativistic quantum theory.

3.1 The Klein–Gordon Equation

The equation that describes relativistic spin-zero particles is called the Klein–Gordon equation. This can be 'derived' using a method similar to the derivation of the Schrödinger equation. Recall the dispersion relation for a relativistic free particle, given by equation (1.19):

$$W = \sqrt{p^2c^2 + m^2c^4} \tag{3.1}$$

Just as in the Schrödinger case, we substitute into (3.1) the operators

$$\hat{W} = i\hbar\frac{\partial}{\partial t}, \qquad \hat{\mathbf{p}} = \frac{\hbar}{i}\nabla \tag{3.2}$$

Doing this, we find

$$i\hbar\frac{\partial\psi(\mathbf{r},t)}{\partial t} = (-\hbar^2c^2\nabla^2 + m^2c^4)^{1/2}\psi(\mathbf{r},t) \tag{3.3}$$

This could be the basic equation of relativistic quantum mechanics, but it is not because it is in such an impractical form. What is meant by taking the square root of the operator in brackets in (3.3) before doing the operation is anybody's guess.* To remedy this problem we pull all the

---

* In fact equation (3.3) can be solved, but the solutions are plagued by non-locality problems, i.e. the wavefunction at **r** depends directly on the wavefunction at other points in space.

terms in (3.3) onto the left hand side:

$$\left(i\frac{\partial}{\partial t} - c \left(-\nabla^2 + \frac{m^2 c^2}{\hbar^2} \right)^{1/2} \right) \psi(\mathbf{r}, t) = 0 \tag{3.4}$$

and premultiply both sides by $\left(i\partial/\partial t + c \left(-\nabla^2 + m^2 c^2/\hbar^2 \right)^{1/2} \right)$. This gives

$$\left(-\frac{1}{c^2}\frac{\partial^2}{\partial t^2} + \nabla^2 - \frac{m^2 c^2}{\hbar^2} \right) \psi(\mathbf{r}, t) = 0 \tag{3.5}$$

Equation (3.5) is the free-particle Klein–Gordon equation. We could have obtained it by direct substitution of the operators into the square of equation (3.1). Note that all solutions of (3.3) are also solutions of (3.5) although the reverse is not true. Inclusion of electromagnetic potentials into the Klein–Gordon formalism is straightforward. We make the usual substitution of $\mathbf{p} \to \mathbf{p} - e\mathbf{A}(\mathbf{r})$ in equation (1.19) where $\mathbf{A}(\mathbf{r})$ is the vector potential. We simply add the scalar potential to the left hand side to give

$$(W - V(\mathbf{r}))^2 = c^2 (\mathbf{p} - e\mathbf{A}(\mathbf{r}))^2 + m^2 c^4 \tag{3.6}$$

Substituting in the operators again

$$\left(i\hbar\frac{\partial}{\partial t} - V(\mathbf{r}) \right)^2 \psi(\mathbf{r}, t) - c^2 \left(\frac{\hbar}{i}\nabla - e\mathbf{A}(\mathbf{r}) \right)^2 \psi(\mathbf{r}, t) = m^2 c^4 \psi(\mathbf{r}, t) \tag{3.7}$$

We can write this in 4-vector form using the definitions of chapter 1

$$c^2 \left(\hat{p}^\mu - e A^\mu \right) \left(\hat{p}_\mu - e A_\mu \right) \psi(\mathbf{r}, t) = m^2 c^4 \psi(\mathbf{r}, t) \tag{3.8}$$

Equations (3.7) and (3.8) are the full Klein–Gordon equation (Gordon 1926, Schrödinger 1926, Klein 1927).

Finally, consider the non-relativistic limit of the free-particle Klein–Gordon equation. Classically, the non-relativistic limit of the energy is

$$W \to E + mc^2 \tag{3.9}$$

where E is the non-relativistic energy. So let us write our relativistic wavefunction as

$$\psi(\mathbf{r}, t) = \psi(\mathbf{r}) e^{-iWt/\hbar} = \psi_0(\mathbf{r}, t) e^{-imc^2 t/\hbar} \tag{3.10}$$

Now we substitute the right hand side of (3.10) into (3.5). The second derivative of (3.10) with respect to time is trivial, and we are left with

$$\left(-\frac{1}{c^2}\frac{\partial^2}{\partial t^2} + \frac{2im}{\hbar}\frac{\partial}{\partial t} + \frac{m^2 c^2}{\hbar^2} + \nabla^2 - \frac{m^2 c^2}{\hbar^2} \right) \psi_0(\mathbf{r}, t) = 0 \tag{3.11}$$

Now we are trying to take the limit $c \to \infty$, so the first term in the brackets in (3.11) disappears. Obviously, the third and fifth terms cancel. Then, if we multiply (3.11) by $\hbar^2/2m$ and rearrange we have

$$i\hbar \frac{\partial \psi_0(\mathbf{r}, t)}{\partial t} = -\frac{\hbar^2}{2m} \nabla^2 \psi_0(\mathbf{r}, t) \qquad (3.12)$$

If you don't recognize this as the free-particle Schrödinger equation, go to the bottom of the class! Indeed, this is what the Klein–Gordon equation had to reduce to in the non-relativistic limit for it to make sense. This proves that the Klein–Gordon equation (rather than the Dirac equation which we shall come across later) is the direct relativistic generalization of the Schrödinger equation.

3.2 Relativistic Wavefunctions, Probabilities and Currents

Now that we have the relativistic equation for the wavefunctions, we must consider what is meant by a relativistic wavefunction. It is important to maintain the same general philosophy of quantum theory as in the non-relativistic case. The wavefunction for a particle must contain all the information that can be known about the particle. In non-relativistic quantum theory, the probability of finding a particle in the region between x and $x + dx$ is given by $P(x)dx = \psi^*(x)\psi(x)dx$, and hence the probability density is $\rho(x) = |\psi(x)|^2$. We can ask what the equivalent expression is in the relativistic theory. As we shall see, the interpretation of the wavefunction is not so straightforward in this case. Let us rewrite (3.5) and define the Compton wavevector $k_C = mc/\hbar$

$$-\frac{1}{c^2} \frac{\partial^2 \psi(\mathbf{r}, t)}{\partial t^2} = (-\nabla^2 + k_C^2)\psi(\mathbf{r}, t) \qquad (3.13)$$

Next we take (3.13) and premultiply it by $\psi^*(\mathbf{r}, t)$, then we take the complex conjugate of (3.13) and premultiply it by $\psi(\mathbf{r}, t)$. This gives

$$-\psi^*(\mathbf{r}, t) \frac{\partial^2 \psi(\mathbf{r}, t)}{\partial (ct)^2} = \psi^*(\mathbf{r}, t)(-\nabla^2 + k_C^2)\psi(\mathbf{r}, t) \qquad (3.14a)$$

$$-\psi(\mathbf{r}, t) \frac{\partial^2 \psi^*(\mathbf{r}, t)}{\partial (ct)^2} = \psi(\mathbf{r}, t)(-\nabla^2 + k_C^2)\psi^*(\mathbf{r}, t) \qquad (3.14b)$$

Now we can subtract (3.14b) from (3.14a), and a little manipulation yields

$$\frac{\partial}{\partial (ct)} \left(-\psi^*(\mathbf{r}, t) \frac{\partial \psi(\mathbf{r}, t)}{\partial (ct)} + \psi(\mathbf{r}, t) \frac{\partial \psi^*(\mathbf{r}, t)}{\partial (ct)} \right)$$
$$= -\nabla \cdot (\psi^*(\mathbf{r}, t)\nabla \psi(\mathbf{r}, t) - \psi(\mathbf{r}, t)\nabla \psi^*(\mathbf{r}, t)) \qquad (3.15)$$

Multiplying both sides of (3.15) by $i\hbar/2m$ to make it as similar as possible to the non-relativistic case leaves a continuity equation

$$\frac{\partial \rho}{\partial t} = -\nabla \cdot \mathbf{j} \tag{3.16}$$

with

$$\rho = \frac{i\hbar}{2mc}\left(\psi^*(\mathbf{r},t)\frac{\partial \psi(\mathbf{r},t)}{\partial(ct)} - \psi(\mathbf{r},t)\frac{\partial \psi^*(\mathbf{r},t)}{\partial(ct)}\right) \tag{3.17a}$$

$$\mathbf{j} = \frac{i\hbar}{2m}(\psi(\mathbf{r},t)\nabla\psi^*(\mathbf{r},t) - \psi^*(\mathbf{r},t)\nabla\psi(\mathbf{r},t)) \tag{3.17b}$$

Equation (3.16) is the usual conservation law, stating that the rate of change of density ρ in an infinitesimal volume is balanced by the current leaving that volume. If we use the 4-vectors discussed in chapter 1 this can be written rather tidily as

$$\nabla^\mu J^\mu = 0 \tag{3.18}$$

Initially it appears rather hard to devise a physical interpretation of (3.17a) for the probability density, particularly as, in contrast to the non-relativistic case, it contains time derivatives. However, it turns out that there are sensible underlying reasons for ρ and $\mathbf{j}$ to take on this form. We examine these now.

Equations (3.17) have been derived for a free particle. The eigenstates of the free-particle Schrödinger equation are written

$$\psi(\mathbf{r},t) = A\,\exp(-i\omega t) \tag{3.19}$$

with $\omega = W/\hbar$. We know that a free particle must always be described by a wavefunction that looks something like (3.19). Let us try inserting (3.19) into the Klein–Gordon equation and see what restrictions are placed upon A and ω for it to be a solution. If we make the identification $W = i\hbar\partial/\partial t$ the Klein–Gordon equation gives us

$$W^2 = \hbar^2\omega^2 = c^2\hbar^2(k^2 + k_C^2) \tag{3.20}$$

which is just another way of writing equation (1.19). Now let us substitute (3.19) into (3.17a) The differentiation is trivial and we find

$$\rho = \frac{i\hbar}{2mc^2}(A^*e^{i\omega t}(-i\omega A)e^{-i\omega t} - Ae^{-i\omega t}(i\omega A^*)e^{i\omega t}) = \frac{\hbar\omega}{mc^2}|A|^2 \tag{3.21}$$

Now A just defines the normalization. Let us set this equal to one. Then we can write ρ as a function of total energy W which is proportional to the momentum $\mathbf{p}$:

$$\rho(\mathbf{p}) = \frac{\hbar\omega(\mathbf{p})}{mc^2} = \frac{W(\mathbf{p})}{mc^2} \tag{3.22}$$

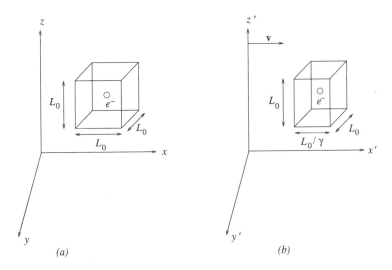

Fig. 3.1. (a) A single particle in a cubic box with sides of length L_0 as measured in frame S, the rest frame of the box. (b) The same box as in (a), but here we measure the lengths of the sides of the box from a reference frame S' moving with velocity **v** in the x-direction as measured in S.

Equation (3.22) is rather unexpected. In the non-relativistic case we would have $\rho = |\psi|^2 = 1$. Obviously, in the relativistic theory ρ has a very different meaning. Note that if $\mathbf{p} = 0$ in (3.22) the total energy $W(\mathbf{p}) = mc^2$ and we find the correct non-relativistic limit. To better interpret (3.22) let us write it in more succinct form:

$$\rho = \frac{W}{mc^2} = \frac{\gamma mc^2}{mc^2} = \gamma \tag{3.23}$$

where γ is the usual relativistic factor.

So we see that in relativistic quantum theory there is an amplification of the probability density proportional to γ. This clearly makes sense in a relativistic context. Consider the meaning behind equation (3.16). It says that the rate of change of probability in a unit volume of space is equal to the negative of the amount of probability leaving the volume per unit time – that is, it just states the conservation of probability. Consider a single particle in a box of volume L_0^3 in figure 3.1a. If the observer O sits in the rest frame of the box, the density of particles in the box is one per volume L_0^3. Now consider figure 3.1b. Here we have the same box containing a single particle, but the observer O' sits in an inertial frame moving with velocity **v** in the x-direction. From classical relativity we know that O' will see the dimensions of the box differently. The y- and z-dimensions of the box will be measured as equal in both frames, but the observer O' in the moving frame will see the x-dimension Lorentz–Fitzgerald contracted by a fractional amount γ^{-1} compared with observer O. Hence O' measures the volume of the box as a fraction γ^{-1} of that measured by O. However, there is still one particle in the box, so the density of particles has increased from $1/L_0^3$ to $\gamma/L_0^3 = 1/(L_0^3(1 - v^2/c^2)^{1/2})$. Hence we should not be surprised by the velocity dependence of the probability density. For (3.16) to be fulfilled, the current density must also have a velocity dependence, and a similar calculation to that above yields

$$\mathbf{j} = \mathbf{p}/m = \gamma\mathbf{v} = \mathbf{v}\rho \qquad (3.24)$$

So there is an analogous relativistic magnification of the probability current density. Equation (3.24) shows explicitly that this is directly due to the magnification of the probability density.

We have seen that the energy of a relativistic free particle is given by (3.20). This is not surprising given the way we set up the Klein–Gordon equation in the first place. However, we have always taken the positive square root to obtain the energy. Mathematically, there is an equally valid negative square root which we have been ignoring. If we had taken the energy as negative in equations (3.22) and (3.23) we would have found

$$\rho = -(1 - v^2/c^2)^{-1/2} = -\gamma \qquad (3.25)$$

The probability density would have been negative! This is the first hint we have of the existence of antiparticles. The current is still given by (3.24) and the momentum is in the *opposite* direction to the velocity. The way we have to interpret this is that the flow of negative particle probability in one direction is equivalent to the flow of positive antiparticle probability in the other direction. We will have more to say about antiparticles later.

So far, our discussion of probabilities and currents has focussed on free-particle solutions of the Klein–Gordon equation. Now we will consider the probability density in the presence of a scalar potential. This is not a simple generalization of the free-particle case, and will lead us to some new and strange physics. Firstly, it is convenient to write the Klein–Gordon equation in a modified form. Equation (3.7) in this case is

$$\left(\left(i\hbar\frac{\partial}{\partial t} - V(\mathbf{r})\right)^2 + \hbar^2 c^2 \nabla^2\right)\psi(\mathbf{r}, t) = m^2 c^4 \psi(\mathbf{r}, t) \qquad (3.26)$$

We can make this equation look a bit neater by dividing through by $\hbar^2 c^2$ and defining $\Phi = 4\pi\epsilon_0 V(\mathbf{r})/e^2$. We then find

$$\left(\left(i\frac{\partial}{\partial(ct)} - \alpha\Phi\right)^2 + \nabla^2\right)\psi(\mathbf{r}, t) = k_C^2 \psi(\mathbf{r}, t) \qquad (3.27)$$

and α is a fundamental quantity. It is a dimensionless constant written in terms of several fundamental constants as

$$\alpha = \frac{e^2}{4\pi\epsilon_0\hbar c} \approx 1/137.037 \tag{3.28}$$

and is known as the *fine structure constant*. We will discuss its role in relativistic quantum theory in the following section. Multiplying out the brackets in (3.27) gives

$$\left(-\frac{\partial^2}{\partial(ct)^2} - 2i\alpha\Phi\frac{\partial}{\partial(ct)} + \alpha^2\Phi^2\right)\psi(\mathbf{r},t) = (k_C^2 - \nabla^2)\psi(\mathbf{r},t) \tag{3.29}$$

Now to find the probability density we use the same procedure as in the free-particle case. We premultiply (3.29) by $\psi^*(\mathbf{r},t)$ and premultiply the complex conjugate of (3.29) by $\psi(\mathbf{r},t)$, and then subtract one from the other. This leaves us with

$$\frac{\partial}{\partial(ct)}\left(\psi^*\frac{\partial\psi}{\partial(ct)} - \psi\frac{\partial\psi^*}{\partial(ct)}\right) + 2i\alpha\Phi\frac{\partial}{\partial(ct)}|\psi|^2 = -\nabla\cdot(\psi\nabla\psi^* - \psi^*\nabla\psi) \tag{3.30}$$

where we have dropped the $\mathbf{r}$ and t dependence of ψ for clarity. If we multiply by $i\hbar/2m$, this is as close to equation (3.17a) as we can get and is in the form of equation (3.16), so we can write

$$\rho = \frac{i\hbar}{2mc}\left(\psi^*\frac{\partial\psi}{\partial(ct)} - \psi\frac{\partial\psi^*}{\partial(ct)} + 2i\alpha\Phi|\psi|^2\right) \tag{3.31}$$

As before, we take the time dependence of the wavefunction to be

$$\psi \propto Ae^{-iWt/\hbar} \tag{3.32}$$

The differentiation is simple, and (3.31) reduces to

$$\rho = \frac{i\hbar}{2mc^2}\left(-\frac{2iW}{\hbar} + 2ic\alpha\Phi\right) = \frac{W - c\alpha\hbar\Phi}{mc^2} = \frac{W - V(\mathbf{r})}{mc^2} \tag{3.33}$$

where we have again set the normalization to one.

This is an odd result. It appears that if the potential is strong enough the probability density can become negative. This is nonsense, of course; a negative probability just indicates that the theory is breaking down. However, it is breaking down in a very instructive way. It turns out to be more sensible to multiply the right hand side of (3.33) by e and interpret ρ as a charge, rather than a probability, density. The consequences of (3.33) will be discussed when we come to the Klein paradox, which is treated in detail in section 3.6.

3.3 The Fine Structure Constant

Here we stop our development of relativistic quantum theory for a while
to discuss briefly the significance of the fine structure constant α defined
by equation (3.28). This constant is a surprising quantity. It contains e^2
from electromagnetism, $\hbar$ from quantum theory and c from relativity, and
it is dimensionless. It seems to derive from several areas of physics.

Let us consider the limits of α. In the non-relativistic limit $c \to \infty$ and
$\alpha \to 0$, so it does not appear in non-relativistic quantum mechanics. In
the non-quantum limit $\hbar \to 0$ and $\alpha \to \infty$, and again α can play no part.
In the classical non-relativistic limit α is undefined.

In relativistic quantum theory α tends to be the ratio of two natural
quantities. We give three examples of this here, although the latter two
are not independent.

The energy of a photon of wavelength λ is

$$E = h\nu = hc/\lambda = \hbar ck \qquad (3.34)$$

where k is the photon wavevector. The potential energy of two electrons
separated by a distance k^{-1} is

$$E = \frac{e^2 k}{4\pi\epsilon_0} \qquad (3.35)$$

The ratio of these two energies is

$$\alpha = \frac{e^2 k}{4\pi\epsilon_0}/(\hbar ck) = \frac{e^2}{4\pi\epsilon_0 \hbar c} \qquad (3.36)$$

Another natural way to obtain the fine structure constant is to determine
the ratio of the two natural lengths that appear in relativistic quantum
theory, firstly the Bohr radius (we shall see in chapter 8 that the Bohr
radius is a natural length in relativistic quantum theory, just as it is in the
non-relativistic theory)

$$a_0 = \frac{4\pi\epsilon_0 \hbar^2}{e^2 m} \qquad (3.37)$$

and secondly the Compton wavelength

$$\lambda_C = \frac{\hbar}{mc} \qquad (3.38)$$

Taking the ratio of the Compton wavelength to the Bohr radius again
gives us the fine structure constant.

There is yet another ratio we can take that yields α. Consider the
hydrogen atom. The energy of the lowest energy level is given (in non-

relativistic theory) by the Rydberg energy

$$E = -\frac{e^4 m}{(4\pi\epsilon_0)^2 2\hbar^2} \tag{3.39}$$

Now consider the electrostatic potential between the electron and proton if they were separated by two Compton wavelengths

$$V = -\frac{e^2 mc}{4\pi\epsilon_0 2\hbar} \tag{3.40}$$

Clearly, taking the ratio of these two gives α yet again. This case is closely related to the previous one, as the first electron in hydrogen is one Bohr radius from the proton. Nonetheless we see that α plays a rather profound role in relativistic theory. It will arise naturally several times in subsequent chapters.

3.4 The Two-Component Klein–Gordon Equation

In this section we are going to divide the Klein–Gordon equation into two equations, both of which are first order in time (Feshbach and Villars 1958). We do this in a fairly standard mathematical way by defining a second wavefunction in terms of the derivative of the first, then finding the equations that connect them, and finally looking for a symmetric form of the equations (Baym 1967). The Klein–Gordon equation is written down in its full glory in equation (3.7). Let us define a new wavefunction

$$\psi_0(\mathbf{r}, t) = \left(\frac{\partial}{\partial t} + \frac{i}{\hbar} V(\mathbf{r}) \right) \psi(\mathbf{r}, t) \tag{3.41}$$

Simply by multiplying out the necessary brackets it is straightforward to show that

$$\left(\frac{\partial}{\partial t} + \frac{i}{\hbar} V(\mathbf{r}) \right) \psi_0(\mathbf{r}, t) = -\frac{1}{\hbar^2} \left(i\hbar \frac{\partial}{\partial t} - V(\mathbf{r}) \right)^2 \psi(\mathbf{r}, t) \tag{3.42}$$

Now this can be substituted back into the Klein–Gordon equation, and a little fiddling of the constants gives

$$\left(\frac{\partial}{\partial t} + \frac{i}{\hbar} V(\mathbf{r}) \right) \psi_0(\mathbf{r}, t) = c^2 \left(\nabla - \frac{ie}{\hbar} \mathbf{A}(\mathbf{r}) \right)^2 \psi(\mathbf{r}, t) - \frac{m^2 c^4}{\hbar^2} \psi(\mathbf{r}, t) \tag{3.43}$$

Equations (3.41) and (3.43) are two equations for $\psi(\mathbf{r}, t)$ and $\psi_0(\mathbf{r}, t)$ that together are equivalent to the Klein–Gordon equation. These two are both first order in time. They can be made to look more tidy if we define two other functions in terms of $\psi(\mathbf{r}, t)$ and $\psi_0(\mathbf{r}, t)$:

$$\phi = \frac{1}{2} \left(\psi(\mathbf{r}, t) + \frac{i\hbar}{mc^2} \psi_0(\mathbf{r}, t) \right) \tag{3.44a}$$

$$\Xi = \frac{1}{2} \left(\psi(\mathbf{r}, t) - \frac{i\hbar}{mc^2} \psi_0(\mathbf{r}, t) \right) \tag{3.44b}$$

Substituting these two into (3.41) and (3.43) enables us to manipulate them into a symmetric form

$$\left(i\hbar \frac{\partial}{\partial t} - V(\mathbf{r}) - mc^2 \right) \phi = \frac{1}{2m} \left(\frac{\hbar}{i} \nabla - e\mathbf{A}(\mathbf{r}) \right)^2 (\phi + \Xi) \tag{3.45a}$$

$$\left(i\hbar \frac{\partial}{\partial t} - V(\mathbf{r}) + mc^2 \right) \Xi = -\frac{1}{2m} \left(\frac{\hbar}{i} \nabla - e\mathbf{A}(\mathbf{r}) \right)^2 (\phi + \Xi) \tag{3.45b}$$

These two equations are fully equivalent to the Klein–Gordon equation. If we write our wavefunction as a two-component quantity

$$\Psi = \begin{pmatrix} \phi \\ \Xi \end{pmatrix} \tag{3.46}$$

then (3.45a) and (3.45b) can be combined into a single matrix equation:

$$\left(i\hbar \frac{\partial}{\partial t} - V(\mathbf{r}) - mc^2 \tilde{\sigma}_z \right) \Psi = \frac{1}{2m} \left(\frac{\hbar}{i} \nabla - e\mathbf{A}(\mathbf{r}) \right)^2 (\tilde{\sigma}_z + i\tilde{\sigma}_y) \Psi \tag{3.47}$$

where the $\tilde{\sigma}$s are the 2×2 Pauli matrices whose properties are discussed in chapter 2. Here and throughout the book, if no matrix is shown as multiplying a term in a matrix equation, the identity matrix is assumed to multiply it. Equation (3.47) is now in the form

$$i\hbar \frac{\partial \Psi}{\partial t} = \hat{H} \Psi = W \Psi \tag{3.48}$$

with

$$\hat{H} = \frac{1}{2m} \left(\frac{\hbar}{i} \nabla - e\mathbf{A}(\mathbf{r}) \right)^2 (\tilde{\sigma}_z + i\tilde{\sigma}_y) + mc^2 \tilde{\sigma}_z + V(\mathbf{r}) \tag{3.49}$$

This looks non-hermitian as $(\tilde{\sigma}_z + i\tilde{\sigma}_y)^\dagger = (\tilde{\sigma}_z - i\tilde{\sigma}_y)$. However, expectation values contain a $\tilde{\sigma}_z$ (see equation (3.54b)) and $\tilde{\sigma}_z \hat{H}^\dagger \tilde{\sigma}_z = \hat{H}$ so there is no problem with energies coming out complex. In this representation the probability density takes on a particularly simple form. To see this, add equations (3.45a and b):

$$\left(i\hbar \frac{\partial}{\partial t} - V(\mathbf{r}) \right) (\phi + \Xi) = mc^2 (\phi - \Xi) \tag{3.50}$$

The probability density comes from (3.31). This can be simply rewritten as

$$\rho(\mathbf{r}, t) = \frac{1}{2mc^2} \left(\psi^* \left(i\hbar \frac{\partial}{\partial t} - V(\mathbf{r}) \right) \psi + \psi \left(-i\hbar \frac{\partial}{\partial t} - V(\mathbf{r}) \right) \psi^* \right) \quad (3.51)$$

We can see from the definitions of ϕ and Ξ that

$$\psi = \phi + \Xi \quad (3.52)$$

Substituting (3.50) and its complex conjugate into (3.51) and using the definition (3.52) leads to

$$\rho(\mathbf{r}, t) = |\phi|^2 - |\Xi|^2 = \Psi^\dagger \tilde{\sigma}_z \Psi \quad (3.53)$$

where $\Psi^\dagger$ has the form of a row vector (Φ^*, Ξ^*). Hence the normalization condition is

$$\int \Psi^\dagger(\mathbf{r}) \tilde{\sigma}_z \Psi(\mathbf{r}) d\mathbf{r} = 1 \quad (3.54a)$$

The expectation value of an arbitrary operator $\hat{A}$ is given by

$$\langle \hat{A} \rangle = \int \Psi^\dagger(\mathbf{r}) \tilde{\sigma}_z \hat{A} \Psi(\mathbf{r}) d\mathbf{r} \quad (3.54b)$$

which leads to the correct result in the correspondence limit. An example of this can easily be seen for the example of the Hamiltonian operator by premultiplying (3.48) by $\tilde{\sigma}_z$ and $\Psi^\dagger$ and then integrating over all space. If we transform (3.54b) back into the representation of the original wavefunction it becomes

$$\langle \hat{A} \rangle = \frac{i\hbar}{2mc^2} \int \left(\psi^* \hat{A} \frac{\partial \psi}{\partial t} - \frac{\partial \psi^*}{\partial t} \hat{A} \psi + \frac{2i}{\hbar} \psi^* \hat{A} V(\mathbf{r}) \psi \right) d\mathbf{r} \quad (3.54c)$$

The matrix element between two-component wavefunctions is

$$\langle \Psi_1 | \Psi_2 \rangle = \int \Psi_1^\dagger(\mathbf{r}) \tilde{\sigma}_z \Psi_2(\mathbf{r}) d\mathbf{r} \quad (3.54d)$$

Equations (3.54a, b, d) are a set of definitions which enable us to calculate observables in a way that is analogous to the non-relativistic procedure. It is also clear, by comparison with (3.54c), that this representation in terms of a two-component wavefunction gives a comparatively simple way of calculating the observables, which avoids the use of time derivatives. However, given that this is the case, it is necessary to interpret the two-component nature of the wavefunction in a physical way. We proceed in this direction in the following section.

Now, let us consider a very simple argument which has some very profound consequences. Look again at equations (3.45). One should remember that the scalar potential felt by the particle in (3.45) is proportional to the particle's own charge. So we write $V(\mathbf{r}) = e\Phi(\mathbf{r})$, where the charge of

the particle under consideration is e. Let us rewrite equations (3.45) for a
particle that has the same mass, but opposite charge:

$$\left(i\hbar \frac{\partial}{\partial t} + e\Phi(\mathbf{r}) - mc^2 \right) \phi = \frac{1}{2m} \left(\frac{\hbar}{i} \nabla + e\mathbf{A}(\mathbf{r}) \right)^2 (\phi + \Xi) \tag{3.55a}$$

$$\left(i\hbar \frac{\partial}{\partial t} + e\Phi(\mathbf{r}) + mc^2 \right) \Xi = -\frac{1}{2m} \left(\frac{\hbar}{i} \nabla + e\mathbf{A}(\mathbf{r}) \right)^2 (\phi + \Xi) \tag{3.55b}$$

Forget these two equations for just a moment. Next we write down the
complex conjugate of (3.45) and multiply it through by -1:

$$\left(i\hbar \frac{\partial}{\partial t} + e\Phi(\mathbf{r}) + mc^2 \right) \phi^* = -\frac{1}{2m} \left(\frac{\hbar}{i} \nabla + e\mathbf{A}(\mathbf{r}) \right)^2 (\phi^* + \Xi^*) \tag{3.56a}$$

$$\left(i\hbar \frac{\partial}{\partial t} + e\Phi(\mathbf{r}) - mc^2 \right) \Xi^* = \frac{1}{2m} \left(\frac{\hbar}{i} \nabla + e\mathbf{A}(\mathbf{r}) \right)^2 (\phi^* + \Xi^*) \tag{3.56b}$$

Now compare (3.55) with (3.56). Clearly equation (3.56a) is identical to
(3.55b) with the wavefunctions ϕ and Ξ replaced by Ξ^* and ϕ^* respectively.
The same replacements can be made in (3.55a) to give us (3.56b). We
obtained (3.56) without reference to any particles of charge $-e$. What this
tells us is that for a given scalar and vector potential, if we have a solution
to the Klein–Gordon equation (3.45) for a particle of mass m and charge
e, we can immediately determine the solution for a particle of mass m and
charge $-e$. This second solution is given in terms of the first by

$$\Psi_A = \tilde{\sigma}_x \Psi^* \tag{3.57}$$

We have said nothing about the energy of this particle. The energy of this
charge conjugate state can be found from the matrix form of the Klein–
Gordon equation (3.48). Let us take the complex conjugate of (3.48) and
premultiply by $\tilde{\sigma}_x$. Recalling that the σ-matrices are their own inverses

$$-i\hbar \frac{\partial}{\partial t} \tilde{\sigma}_x \Psi^* = \tilde{\sigma}_x \hat{H}^* \tilde{\sigma}_x \tilde{\sigma}_x \Psi^* = W \tilde{\sigma}_x \Psi^* \tag{3.58a}$$

so $\tilde{\sigma}_x \Psi^*$ is verified as being an eigenfunction of the Klein–Gordon equation with eigenvalue $-W$. Now let us look at the explicit form of the
Hamiltonian. From (3.58a) we see that we have to pre- and postmultiply the complex conjugate of the Hamiltonian (3.49) by $\tilde{\sigma}_x$. Owing to
the anticommutation relations of the Pauli spin matrices we find that
$\hat{H}^- = \tilde{\sigma}_x \hat{H}^* \tilde{\sigma}_x$ is guaranteed to have energy eigenvalue $-W$ provided the
particle being described has charge $-e$. Then (3.58a) becomes

$$i\hbar \frac{\partial}{\partial t} \Psi_A = \hat{H}^- \Psi_A = -W \Psi_A \tag{3.58b}$$

So the complex conjugate of the Klein–Gordon equation for a particle
of charge e, mass m and energy W is the same as the Klein–Gordon

equation for a particle of charge $-e$, mass m and energy $-W$. Comparison with (3.55) shows that there is a one-to-one correspondence between the negative energy solutions of the Klein–Gordon equation (which arise because of the quadratic nature of $W^2 = p^2c^2 + m^2c^4$) and the positive energy solutions with opposite charge. This leads us to postulate that the negative energy solutions represent a particle of opposite charge to, and the same mass as, the original particle, i.e. to postulate the existence of antiparticles. This is a very definite prediction of the theory, that to every particle there is a corresponding antiparticle with very specific properties relative to the particle. The existence of antiparticles has, of course, long been an experimentally established fact, and their discovery was an outstanding confirmation of relativistic quantum theory.

3.5 Free Klein–Gordon Particles/Antiparticles

In this section we are going to examine, in detail, some of the properties of the solutions of the Klein–Gordon equation in the absence of any potentials. This will enable us to gain considerable insight into the nature and behaviour of relativistic particles. We know from non-relativistic considerations that the wavefunction describing a free particle must be of the form

$$\psi^+(\mathbf{r}, t) = A \, \exp(i(\mathbf{p}.\mathbf{r} - W^+t)/\hbar) \tag{3.59}$$

where the $+$ superscript anticipates the fact that we are going to differentiate between positive energy (superscript $+$) and negative energy (superscript $-$) solutions of the Klein–Gordon equation. Substitution of equation (3.59) into the free-particle Klein–Gordon equation shows that it is a solution provided $W^{+2} = p^2c^2 + m^2c^4$, where W^+ is a positive number.

We saw in the previous section that if a particle has energy W then the theory suggests a particle of equal mass and energy $-W$. The theory implies this symmetry. This is not surprising as it was set up from an expression for W^2, not W. The minimum energy of a positive energy particle is mc^2, whereas the maximum energy of a negative energy particle is $-mc^2$. Both positive and negative energy particles can take on any value of momentum between $+\infty$ and $-\infty$, so for a specific particle there exist two energy continua, one of which is accessible to the particle at positive energies and the other at negative energies. These are separated by a forbidden region of energy defined by $-mc^2 < W_{\text{forbidden}} < mc^2$. Figure 3.2 illustrates this energy separation. The implication of this is that an excitation with energy greater than $2mc^2$ may excite a negative energy particle up into the positive energy states.

You may well ask why particles described by the Klein–Gordon equation do not continually decay to lower energy states all the way down to

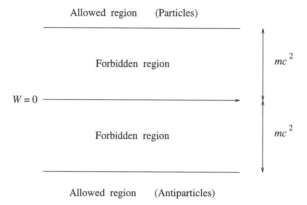

Fig. 3.2.   Illustration of the two energy continua available to a particle and antiparticle, and the forbidden region of width $2mc^2$ separating them.

$-\infty$, and hence disappear up their own backsides. The implication of the theory presented here is that they should do so, but they do not. This is a dilemma which is not solved within the theory presented here, and whose resolution had to wait for quantum field theory. Here we interpret a particle at rest with energy $-mc^2$ as an antiparticle at rest with energy mc^2 and do not worry about the negative energy states. This is not terrifically satisfactory, but is the best we can do, and the experimentally verified existence of antiparticles does support such a viewpoint.

We can ask what is the two-component form of the wavefunction of equation (3.59). This can be derived straightforwardly. In the absence of any potentials, (3.41) simplifies considerably to

$$\psi_0^+(\mathbf{r}, t) = \frac{\partial \psi^+(\mathbf{r}, t)}{\partial t} \tag{3.60}$$

and doing the differentiation therefore gives us

$$\psi_0^+(\mathbf{r}, t) = -\frac{iW^+}{\hbar} A \, \exp(i(\mathbf{p} \cdot \mathbf{r} - W^+ t)/\hbar) \tag{3.61}$$

Now we can substitute (3.59) and (3.61) into (3.44) to find the two components of the free-particle wavefunction. This leads to

$$\phi = \frac{1}{2}\left(1 + \frac{W^+}{mc^2}\right)\psi^+(\mathbf{r}, t) \tag{3.62a}$$

$$\Xi = \frac{1}{2}\left(1 - \frac{W^+}{mc^2}\right)\psi^+(\mathbf{r}, t) \tag{3.62b}$$

and so

$$\Psi^+(\mathbf{r}, t) = A \left(\frac{mc^2 + W^+}{mc^2 - W^+} \right) \exp(i(\mathbf{p} \cdot \mathbf{r} - W^+ t)/\hbar) \qquad (3.63)$$

and from (3.57) the antiparticle wavefunction is

$$\Psi^-(\mathbf{r}, t) = \tilde{\sigma}_x \Psi^{+*}(\mathbf{r}, t)$$
$$= A \left(\frac{mc^2 - W^+}{mc^2 + W^+} \right) \exp(-i(\mathbf{p} \cdot \mathbf{r} - W^+ t)/\hbar)$$
$$= A \left(\frac{mc^2 + W^-}{mc^2 - W^-} \right) \exp(-i(\mathbf{p} \cdot \mathbf{r} + W^- t)/\hbar) \qquad (3.64)$$

The next step is to normalize these functions to a unit volume according to equation (3.54a). This leaves us with

$$A = A^* = \frac{1}{2\sqrt{mc^2 W^+}} \qquad (3.65)$$

It is also easy to see that the matrix element (3.54d) between the two solutions is zero:

$$\langle \Psi^+ | \Psi^- \rangle = \int \Psi^{\dagger+}(\mathbf{r}) \tilde{\sigma}_z \Psi^-(\mathbf{r}) d\mathbf{r}$$
$$= \frac{1}{4mc^2 W^+} (mc^2 + W^+, mc^2 - W^+) \begin{pmatrix} 1 & 0 \\ 0 & -1 \end{pmatrix} \begin{pmatrix} mc^2 - W^+ \\ mc^2 + W^+ \end{pmatrix} = 0$$
$$(3.66)$$

So the positive energy and negative energy states are orthogonal to one another.

To aid us in our interpretation of the Klein–Gordon wavefunction and to help to ensure that the manipulations we have done make sense, it is helpful to examine the non-relativistic limit. The wavefunction for the positive energy particle is given by (3.63), and to take the non-relativistic limit we make the replacement $W^+ = +\sqrt{p^2 c^2 + m^2 c^4} \approx mc^2 \left(1 + p^2 c^2 / 2m^2 c^4\right)$. With the normalization (3.65), Ψ then reduces to

$$\lim_{c \to \infty} \Psi^+(\mathbf{r}, t) = \left(\frac{1}{-v^2/4c^2} \right) \exp(i(\mathbf{p} \cdot \mathbf{r} - W^+ t)/\hbar)) \qquad (3.67a)$$

For a non-relativistic antiparticle we can use the same procedure on (3.64) or take the complex conjugate of (3.67a) and premultiply by $\tilde{\sigma}_x$ to get

$$\lim_{c \to \infty} \Psi^-(\mathbf{r}, t) = \left(\frac{-v^2/4c^2}{1} \right) \exp(-i(\mathbf{p} \cdot \mathbf{r} - W^+ t)/\hbar)) \qquad (3.67b)$$

We see that for particles travelling at non-relativistic velocities, the upper component of the wavefunction is much larger than the lower component, and the other way round for the antiparticle. In the non-relativistic limit, Ξ

therefore becomes vanishingly small. In this limit equation (3.45a) becomes

$$\left(i\hbar\frac{\partial}{\partial t} - V(\mathbf{r}) - mc^2\right)\phi = \frac{1}{2m}\left(\frac{\hbar}{i}\nabla - e\mathbf{A}(\mathbf{r})\right)^2\phi \qquad (3.68)$$

This is just the Schrödinger equation for ϕ with an extra rest energy term. We can use the same argument in (3.45b) and we obtain the Schrödinger equation for the antiparticle.

The next topic we are going to discuss is the behaviour of Klein–Gordon particles under the influence of the position operator. One might expect this not to be terrifically interesting, but it yields a surprising result which forces us to review our concept of a *point particle*. The strategy we adopt is to write down a general form for a wavefunction and then calculate a value for the expectation value of the mean velocity. Following this we will try to set up a wavefunction that is completely localized in space. By calculating overlap integrals we can see that such a wavefunction contains both particle and antiparticle components. Combining these results will enable us to draw some surprising conclusions about what relativistic quantum theory means by a point particle.

The general expression for the expectation value of an operator is given by (3.54b). Inserting the position operator in this gives

$$\langle \hat{\mathbf{r}} \rangle = \int \Psi^\dagger(\mathbf{r})\tilde{\sigma}_z\hat{\mathbf{r}}\Psi(\mathbf{r})d\mathbf{r} \qquad (3.69)$$

Recalling the definition of Ψ from (3.46) we can see that this is

$$\langle \hat{\mathbf{r}} \rangle = \int (\phi^*\hat{\mathbf{r}}\phi - \Xi^*\hat{\mathbf{r}}\Xi)d\mathbf{r} \qquad (3.70)$$

and now we work backwards again using equation (3.60) and the definitions of ϕ and Ξ in equations (3.44) for free particles. We find

$$\langle \hat{\mathbf{r}} \rangle = \frac{i\hbar}{2mc^2}\int\left(\psi^*\hat{\mathbf{r}}\frac{\partial\psi}{\partial t} - \frac{\partial\psi^*}{\partial t}\hat{\mathbf{r}}\psi\right)d\mathbf{r} \qquad (3.71)$$

Now let's differentiate both sides of (3.71) with respect to time to get an expression for something like the expectation value of the velocity. The position operator has no explicit time dependence and so

$$\frac{\partial\langle \hat{\mathbf{r}} \rangle}{\partial t} = \frac{i\hbar}{2mc^2}\int\mathbf{r}\left(\psi^*\frac{\partial^2\psi}{\partial t^2} - \frac{\partial^2\psi^*}{\partial t^2}\psi\right)d\mathbf{r} \qquad (3.72)$$

where we have written the position operator in its actual form *multiply by* **r** to enable us to commute it through the wavefunctions. Next we use equations (3.14) to replace the part of the integrand in brackets in (3.72) as follows

$$\frac{\partial \langle \hat{\mathbf{r}} \rangle}{\partial t} = \frac{i\hbar}{2m} \int \mathbf{r} \left(\psi^* \nabla^2 \psi - \psi \nabla^2 \psi^* \right) d\mathbf{r}$$

$$= -\frac{i\hbar}{2m} \int \mathbf{r} \nabla \cdot \left(\psi^* \nabla \psi - \psi \nabla \psi^* \right) d\mathbf{r} \qquad (3.73)$$

This can easily be integrated by parts to leave

$$\frac{\partial \langle \hat{\mathbf{r}} \rangle}{\partial t} = -\frac{i\hbar}{2m} \int \left(\psi^* \nabla \psi - \psi \nabla \psi^* \right) d\mathbf{r} \qquad (3.74)$$

This is a relatively simple expression for the expectation value of the average velocity. The wavefunction is chosen as a wavepacket made up of the free particle and antiparticle states in (3.59) and (3.64) with parallel momenta.

$$\psi(\mathbf{r}, t) = \int \left(A(\mathbf{p}) \exp[i(\mathbf{p} \cdot \mathbf{r} - Wt)/\hbar] + B(\mathbf{p}) \exp[i(\mathbf{p} \cdot \mathbf{r} + Wt)/\hbar] \right) d\mathbf{p} \qquad (3.75)$$

where W is a positive number, the energy of the particle as opposed to the antiparticle. The first term with the coefficient $A(\mathbf{p})$ represents the particle part of the wavepacket and the term with coefficient $B(\mathbf{p})$ is the antiparticle part. If $\mathbf{p}$ were a discrete variable we would have a summation sign here, but as it is continuous we have an integral, so (3.75) is just a wavepacket made up of a linear combination of plane wave particle and antiparticle states. Now we have to substitute this into (3.74) and do all the differentiations. After some tedious algebra we arrive at

$$\frac{\partial \langle \hat{\mathbf{r}} \rangle}{\partial t} = \frac{1}{m} \int \left(|A(\mathbf{p})|^2 + |B(\mathbf{p})|^2 \right) \mathbf{p} d\mathbf{p} \qquad (3.76)$$

$$+ \frac{1}{m} \int \left(A^*(\mathbf{p})B(\mathbf{p})e^{2iWt/\hbar} + B^*(\mathbf{p})A(\mathbf{p})e^{-2iWt/\hbar} \right) \mathbf{p} d\mathbf{p}$$

This is an extraordinary result. We have taken a general wavefunction for a free particle and calculated the average velocity. Clearly, the first term on the right hand side can be identified with the group velocity of the wavepacket. However, we find that there are in general two oscillatory terms in addition.[*] The oscillatory terms can only be set equal to zero if our wavepacket is either purely particle-like or purely antiparticle-like. Clearly the implication of this result is that if we have a general wavepacket its motion contains this strange oscillatory behaviour. To start investigating this further we create a wavefunction that is completely localized in space.

---

[*] This oscillatory behaviour is known as *Zitterbewegung* and is discussed more fully for Dirac particles in chapter 7.

The way we do this is to set it up as

$$\Psi_L = \begin{pmatrix} \phi \\ \Xi \end{pmatrix} \delta(\mathbf{r}) \tag{3.77}$$

This is a completely general localized Klein–Gordon wavefunction. There are no restrictions on ϕ and Ξ except that they are not both equal to zero (Feshbach and Villars 1958).

The strategy now is to evaluate the overlap integral between our localized wavefunction and the particle and antiparticle wavefunctions (3.63) and (3.64), and then to consider the implications of the result. This is done according to the prescription (3.54d). We make things easy for ourselves by localizing the particle at the origin in its rest frame. A Lorentz transformation can be used to take us to any other inertial frame. Then for the particle part the overlap integral is

$$\langle \Psi^+ | \Psi_L \rangle = \frac{1}{2\sqrt{W mc^2}} (mc^2 + W, mc^2 - W) \begin{pmatrix} 1 & 0 \\ 0 & -1 \end{pmatrix} \begin{pmatrix} \phi \\ \Xi \end{pmatrix}$$

$$= \frac{W(\phi + \Xi) + mc^2(\phi - \Xi)}{2\sqrt{W mc^2}} \tag{3.78a}$$

and for the antiparticle part it is

$$\langle \Psi^- | \Psi_L \rangle = \frac{1}{2\sqrt{W mc^2}} (mc^2 - W, mc^2 + W) \begin{pmatrix} 1 & 0 \\ 0 & -1 \end{pmatrix} \begin{pmatrix} \phi \\ \Xi \end{pmatrix}$$

$$= \frac{-W(\phi + \Xi) + mc^2(\phi - \Xi)}{2\sqrt{W mc^2}} \tag{3.78b}$$

Next, consider what (3.78) implies. Suppose first that our plane wave particle/antiparticle states have $\mathbf{p} = 0$, then $W = mc^2$. If $\phi \neq 0$ and $\Xi \neq 0$ both of equations (3.78) are non-zero. If $\Xi = 0$ then (3.78b) is equal to zero, and (3.78a) is not. Therefore in this case there is no overlap between our localized particle and the zero-momentum free-antiparticle state. However, if we make this choice for Ξ, then for every other free-antiparticle state $\langle \Psi^- | \Psi_L \rangle \neq 0$, so there is overlap. Clearly there is no way of choosing ϕ and Ξ so that either (3.78a) or (3.78b) vanishes for all values of W. We have already seen that Ψ^+ and Ψ^- do not overlap, but (3.78) tells us it is impossible for a localized particle to be purely particle-like or purely antiparticle-like. Compare this with our inference from (3.76) that the oscillatory behaviour of the velocity is only quenched when the particle is purely particle- or antiparticle-like. We can take this discussion further. Consider an experiment in which we require a beam of spin-zero particles, e.g. pions. The pions are created in a nuclear reaction and then collimated. Collimation means confining the pions to a smaller volume – in other words, localizing them. The above argument implies that this can only be done if the antipion nature of the wavefunction is increased, by creating antipions in our beam!

Now let us confuse the issue about relativistic velocities still further. From the Heisenberg picture of quantum mechanics we know that if a particle obeys an equation like

$$i\hbar \frac{\partial \phi}{\partial t} = \hat{H}\phi \tag{3.79}$$

then an operator for the rate of change of an observable with time is given in terms of the commutator of that operator and the Hamiltonian by

$$\frac{\widehat{\partial O}}{\partial t} = \frac{\partial \hat{O}}{\partial t} + \frac{i}{\hbar}[\hat{H}, \hat{O}] \tag{3.80}$$

The Klein–Gordon equation has been written in the form (3.79) where the Hamiltonian is given by equation (3.49). For simplicity we consider the free-particle version of (3.49), and evaluate (3.80) for $\hat{O}$ being the position operator. With trivial application of the position/momentum commutator we find

$$\frac{\widehat{\partial \mathbf{r}}}{\partial t} = \frac{\hat{\mathbf{p}}}{m}(\tilde{\sigma}_z + i\tilde{\sigma}_y) = \frac{\hat{\mathbf{p}}}{m}\begin{pmatrix} 1 & 1 \\ -1 & -1 \end{pmatrix} \tag{3.81}$$

At first glance this looks all right. It has the right sort of structure and order of magnitude. Classically, $\mathbf{p}/m$ is the velocity, and we are multiplying this by a matrix of order unity. The antiparticle states, represented by the lower components of the wavefunction, seem at first sight to have a velocity that is the negative of that of the particle states. This is not the case because the $\tilde{\sigma}_z$ that appears in the expression for an expectation value (3.54*b*) changes the sign of the lower components of the matrix in (3.81). What is meant by a matrix representing velocity is not entirely clear, but (3.81) looks as if it would give a reasonable value for the expectation of the velocity.

Now consider the operator we would use to find the speed. To do this we would have to find the operator that is the square of (3.81), evaluate its expectation value, and take the square root of that. However, if we try to take the square of (3.81) we find the square of the matrix is zero. Hence the velocity-squared operator is *multiply by zero*, clearly a ludicrous result. It is easy to see that this comes about from the existence of the antiparticle states. The particle and antiparticle states interfere with one another to give an expectation value of the speed always equal to zero. This phenomenon will surface again in our discussion of Dirac particles.

Let us collect together our thoughts about the motion of Klein–Gordon particles and try to interpret them. What do these mind-boggling results tell us about the nature of a point particle?

Any point particle described by the Klein–Gordon equation necessarily has both particle and antiparticle character. According to equation (3.76), this particle also has a velocity which can be divided into two components: the average velocity of the particle and an oscillatory component. The particle can still be a point particle, but the oscillatory velocity implies that it exists within a small volume, rather than at a point. In this small volume the existence of the particle necessarily means there is some antiparticle character there as well. If we calculate the speed of the total particle, as given by the square of equation (3.81), we are not calculating the speed of the point particle, but rather that of the total particle/antiparticle. The velocity of the particle component is always in the opposite direction to that of the antiparticle component. Hence, the average speed within the whole little volume is zero.

There are a lot of questions that can be asked about the reasoning in the above paragraphs. For example, you may well ask what happens to the charge of our particle in this small volume. If we don't have a particle, but rather some particle/antiparticle composite, why does it still have a charge? Don't the charge of the particle and antiparticle cancel each other out? Our mathematical manipulations have given us some results which are correct within the single-particle Klein–Gordon theory. What is really true is that we are at the limits of what single-particle Klein–Gordon theory can tell us, and we are trying to put an interpretation on top of mathematics that really is insufficient to describe the physics involved. We are finding it necessary to invoke the (infinite number of) negative energy states to maintain the single-particle theory.

There is one special case of the free-particle Klein–Gordon equation which we have not examined and which really corresponds to no physical reality. Nonetheless we mention it here for completeness. If we consider a particle with zero mass, the free-particle Klein–Gordon equation (3.5) becomes

$$\left(-\frac{1}{c^2}\frac{\partial^2}{\partial t^2} + \nabla^2\right)\psi(\mathbf{r}, t) = 0 \qquad (3.82)$$

But this is just the wave equation for a massless particle travelling at the speed of light. Such an equation can be derived from Maxwell's equations of electromagnetism. It is tempting to think that Maxwell's equations and the Klein–Gordon equation are somehow related. This is not the case. The Klein–Gordon equation describes particles of spin zero only, whereas photons have spin 1. The $m \to 0$ limit of the Klein–Gordon equation would be valid only for massless spin-zero particles of which I am aware of no examples.

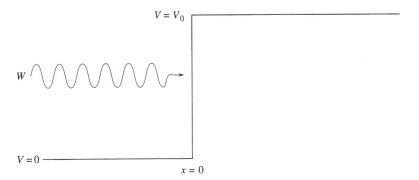

Fig. 3.3. The model problem we solve to illustrate the Klein paradox is a particle incident from the left upon a one-dimensional potential step.

3.6 The Klein Paradox

An expression for the probability density of a Klein–Gordon particle in a potential $V(\mathbf{r})$ is given in equation (3.33). There we commented on how unreasonable the expression looked and left it at that. In this section we discover the physics underlying that result. For completeness we rewrite (3.33) including the normalization which we previously set equal to one:

$$\rho = \frac{W - V(\mathbf{r})}{mc^2}|\psi|^2 \tag{3.83}$$

The model we are going to use to illustrate the Klein paradox (Klein 1929) is a familiar one from non-relativistic quantum mechanics, a particle incident upon a one-dimensional potential step (see figure 3.3). The wavefunction for the particles can be obtained from the one-dimensional Klein–Gordon equation. In the region of zero potential ($x < 0$), this is

$$\left(-\frac{1}{c^2}\frac{\partial^2}{\partial t^2} + \frac{\partial^2}{\partial x^2} - \frac{m^2c^2}{\hbar^2}\right)\psi(x,t) = 0 \tag{3.84}$$

or, in the region where $V = V_0$ (i.e, $x > 0$)

$$\left(i\hbar\frac{\partial}{\partial t} - V_0\right)^2\psi(x,t) + c^2\hbar^2\frac{\partial^2}{\partial x^2}\psi(x,t) = m^2c^4\psi(x,t) \tag{3.85}$$

The solution for $x < 0$ is easy to write down:

$$\psi(x,t) = (e^{ikx} + Re^{-ikx})e^{-iWt/\hbar} \tag{3.86}$$

The first term here represents the incident wave (with an arbitrarily chosen normalization coefficient of unity) and the second term represents the reflected wave with amplitude reflection coefficient R. It is equally easy to see that this is a satisfactory solution by substituting it back into (3.86). If one does this, the usual expression

$$W^2 = \hbar^2 k^2 c^2 + m^2 c^4 \tag{3.87}$$

emerges for the energy. For particles at $x > 0$ we expect non-propagating solutions if the kinetic energy of the particles is less than V_0. We take as our solution

$$\psi(x,t) = T e^{-\kappa x} e^{-iWt/\hbar} \tag{3.88}$$

where T is an amplitude transmission coefficient and κ must be positive for the wavefunction to be normalizable. The variable κ should become complex when the kinetic energy is greater than V_0. Substituting (3.88) into (3.85) gives

$$\hbar c \kappa = (m^2 c^4 - (W - V_0)^2)^{1/2} \tag{3.89}$$

This looks like it doesn't go to the correct limit as $V_0 \to 0$, but it does, because in that limit we must have a propagating solution so κ becomes purely imaginary. In (3.89) we are obliged to take the positive square root, in order that (3.88) becomes zero as $x \to \infty$. For $x > 0$, W takes on the same value as for $x < 0$, of course. However, for $x > 0$, it is made up of (something like) a rest mass energy, a potential energy and a kinetic energy. Now, let us stop to consider this result in conjunction with (3.83). Imagine what happens as we increase V_0 from 0. At first, it appears that everything behaves properly: as $W^2 \geq m^2 c^4$, κ is imaginary, and so we have propagating solutions. The first change comes when $V_0 \geq W - mc^2$. When the potential reaches this point, κ is zero, and as the potential rises, κ becomes real, so we have total reflection of the wave by the barrier, as one would expect. However, as V_0 approaches W, the difficulties start. The probability density approaches zero, and indeed becomes negative, even though $|\psi|^2$ does not. Also, we can see from (3.89) that κ takes on a maximum value of the Compton wavevector as V_0 approaches W. This is not the only peculiar thing that happens. One's quantum mechanical intuition (a notoriously unreliable thing) leads one to believe that, as V_0 increases, the penetration into the barrier will become less and less, although the wave will continue to be totally reflected. This is not the case according to the equations above. As V_0 increases above W, κ decreases, so the wave can penetrate further into the barrier. The final nail in the coffin of one's intuition is hammered home when V_0 has increased to so large a value that $V_0 > W + mc^2$. At that stage κ becomes imaginary, and in this limit of a barrier large compared with all other energies in the problem, including the energy of the incident particles, we have a propagating solution inside the barrier! Let us summarize our results so far.

- $V_0 < W - mc^2$: κ is imaginary leading to propagating solutions, i.e. the particle goes over the top of the barrier.
- $W - mc^2 < V_0 < W$: κ is real and increases in this range, so there is total reflection at the barrier. The probability density in this region of potential is positive but decreasing, i.e. the particle hits the barrier and is reflected from it.
- $W < V_0 < W + mc^2$: κ is still real but decreases in this range. The probability density in the barrier is negative, i.e. there is still reflection at the barrier, but something peculiar is happening inside the barrier.
- $V_0 > W + mc^2$: κ becomes imaginary again, so reflection at the barrier does not necessarily occur. There are wave-like solutions of the Klein–Gordon equation on both sides of the barrier.

To try to understand this apparently lunatic state of affairs, it helps to calculate the probability current density for $x > 0$ in the $V_0 \to \infty$ case. The current is given by the right hand side of equation (3.30). If we put the wavefunction (3.88) into this with $ik' = \kappa$, it is easy to see that

$$\mathbf{j} = -\frac{\hbar k'}{m}|\psi|^2 \tag{3.90}$$

For a very large potential and $x > 0$ the current is in the negative x-direction. It looks like a beam of particles coming in from $-\infty$ to zero. However, since $\rho < 0$, we interpret this as a beam of antiparticles being emitted from $x = 0$ and propagating towards $x = \infty$. This enables us to interpret our negative probability density to some extent. We can say that a negative probability for finding a particle is equivalent to a positive probability for finding the antiparticle. Of course, a negative probability still doesn't make sense, but at least we understand why the definition of probability is breaking down and we can give (3.83) a sensible meaning by multiplying the right hand side by e and calling it a charge density.

We are able to go further still. There have been no restrictions imposed on R or T. Let us force the wavefunctions and their first derivatives to be continuous at $x = 0$. As in the non-relativistic case this gives us

$$1 + R = T \tag{3.91a}$$

$$ik(1 - R) = -\kappa T \tag{3.91b}$$

These are easy enough to solve for R and T:

$$R = \frac{ik + \kappa}{ik - \kappa} \tag{3.92a}$$

$$T = \frac{2ik}{ik - \kappa} \tag{3.92b}$$

Now let $\kappa \to ik'$ with k' positive as we only have particles incident from the left, and look what happens:

$$R = \frac{k + k'}{k - k'} > 1 \qquad (3.93a)$$

$$T = \frac{2k}{k - k'} > 1 \qquad (3.93b)$$

From these equations it looks as if there is more wave reflected and more wave transmitted than is actually incident in the first place. The only way that all these apparent contradictions can be reconciled is if there are particle/antiparticle pairs created at the barrier. The created antiparticles find a barrier that is repulsive to particles but attractive to themselves. We can ask what minimum energy of the barrier is necessary to create a particle/antiparticle pair. Clearly, this will be when they both have zero kinetic energy. For a particle on the left of $x = 0$ in figure 3.3 this will be mc^2. For an antiparticle on the right of $x = 0$ it will be $mc^2 - V_0$, so the minimum possible size of barrier to create a particle/antiparticle pair is given by

$$V_{min} = 2mc^2 \qquad (3.94)$$

Actually this analysis leaves several questions unanswered. We are using a single-particle theory, and as soon as we can have creation of particles the theory becomes suspect. To treat the problem properly we should really include the electrostatic attraction between the particles and antiparticles. How does the probability of creation of a particle/antiparticle pair affect the potential seen by the incident particle? Is it actually necessary to have an incident particle for the creation to take place? These questions expose the inadequacy of the theory presented here and lead to quantum field theory, which is beyond the scope of this book.

3.7 The Radial Klein–Gordon Equation

As for the Schrödinger equation, if we want to discuss the quantum mechanics of atoms, it is convenient to express the Klein–Gordon equation in spherical polar coordinates (Baym 1967). We assume that the potential is dependent only on the radial distance away from some origin (Corinaldesi and Strocchi 1963)

$$V(r) = V(\mathbf{r}) \qquad (3.95)$$

and then the Klein–Gordon equation is

$$\left((W - V(r))^2 + \hbar^2 c^2 \nabla^2 - m^2 c^4 \right) \psi(\mathbf{r}) = 0 \qquad (3.96)$$

Now, in spherical polar coordinates

$$\nabla^2 = \frac{1}{r^2}\frac{\partial}{\partial r}\left(r^2\frac{\partial}{\partial r}\right) + \frac{1}{r^2\sin\theta}\frac{\partial}{\partial\theta}\left(\sin\theta\frac{\partial}{\partial\theta}\right) + \frac{1}{r^2\sin^2\theta}\frac{\partial^2}{\partial\phi^2} \tag{3.97}$$

As readable a derivation as you can get of this horrible operator is given by Eisberg and Resnick (1985). We adopt the same procedure as in the non-relativistic case to find the radial component of this equation: we separate the wavefunction into three functions, each dependent on just one of the coordinates

$$\psi(r,\theta,\phi) = R(r)\Theta(\theta)\Phi(\phi) \tag{3.98}$$

Note that ϕ here has nothing to do with the function ϕ defined in equation (3.44). That ϕ was a component of the two-component Klein–Gordon equation and does not appear in this section. The ϕ here is the usual angular coordinate in spherical polars. Substituting (3.97) and (3.98) into (3.96) gives a messy equation. After some juggling we are left with

$$\frac{1}{R(r)\Theta(\theta)}\left(\frac{(W-V)^2r^2\sin^2\theta}{\hbar^2c^2} + \sin^2\theta\frac{\partial}{\partial r}\left(r^2\frac{\partial}{\partial r}\right) + \right.$$
$$\left. \sin\theta\frac{\partial}{\partial\theta}\left(\sin\theta\frac{\partial}{\partial\theta}\right) - \frac{m^2c^2}{\hbar^2}r^2\sin^2\theta\right)R(r)\Theta(\theta) = -\frac{1}{\Phi(\phi)}\frac{\partial^2\Phi(\phi)}{\partial\phi^2} \tag{3.99}$$

Here we have an equation where ϕ does not appear on the left hand side and r and θ do not appear on the right hand side. The only way this equation can be valid for all values of r, θ and ϕ is if both sides of (3.99) are equal to a constant. We set this constant equal to m_l^2 and then

$$\frac{\partial^2\Phi(\phi)}{\partial\phi^2} = -m_l^2\Phi(\phi) \tag{3.100}$$

Substituting from equation (3.100) back into (3.99) followed by some trivial manipulations gives

$$\frac{(W-V(\mathbf{r}))^2r^2}{c^2} + \frac{\hbar^2}{R(r)}\frac{\partial}{\partial r}r^2\frac{\partial R(r)}{\partial r} - m^2c^2r^2$$
$$= \frac{m_l^2\hbar^2}{\sin^2\theta} - \frac{\hbar^2}{\Theta(\theta)}\frac{1}{\sin\theta}\frac{\partial}{\partial\theta}\sin\theta\frac{\partial\Theta(\theta)}{\partial\theta} \tag{3.101}$$

We can use the same trick again. The left hand side of (3.101) has no θ dependence and the right hand side has no r dependence. In parallel with the non-relativistic case we set both sides of this equation equal to a constant $l(l+1)\hbar^2$. There is no loss of generality in this choice, as we have not placed any restrictions on l at this stage. However, substituting this into (3.101) gives

$$\frac{m_l^2}{\sin^2\theta}\Theta(\theta) - \frac{1}{\sin\theta}\frac{\partial}{\partial\theta}\sin\theta\frac{\partial\Theta(\theta)}{\partial\theta} = l(l+1)\Theta(\theta) \qquad (3.102)$$

Equations (3.100) and (3.102) for the angular parts of the wavefunction are precisely the same as those obtained when the radial Schrödinger equation is derived. Therefore the angular part of the wavefunction for a Klein–Gordon particle in a central potential must be the same as that for a Schrödinger particle. That is, equations (3.100) and (3.102) are the defining equations for the spherical harmonics, and so the angular parts of the wavefunctions are the spherical harmonics. Simple properties of the spherical harmonics are discussed in appendix C.

This means that all the discussion about the angular dependence of the one-electron atom wavefunctions in non-relativistic quantum theory books carries over directly to the relativistic atom if the electron spin is ignored. Polar diagrams showing the directional dependence of the probability densities for the spinless electron atom can be drawn which are exactly the same as in non-relativistic quantum mechanics, so we do not need to reproduce them here. A good example of such diagrams is given by Eisberg and Resnick (1985). All the discussion of chemical bonding (Pettifor 1995) that follows from these diagrams also follows in the relativistic case.

Where relativity has its effect is in the radial part of the wavefunction. The radial Klein–Gordon equation is

$$\frac{(W - V(\mathbf{r}))^2 r^2}{c^2} + \frac{\hbar^2}{R(r)}\frac{\partial}{\partial r}r^2\frac{\partial R(r)}{\partial r} - m^2c^2r^2 = l(l+1)\hbar^2 \qquad (3.103)$$

and a little rearrangement gives

$$\frac{\partial^2 R(r)}{\partial r^2} + \frac{2}{r}\frac{\partial R(r)}{\partial r} + \left(\frac{(W - V(\mathbf{r}))^2}{\hbar^2 c^2} - \frac{m^2c^2}{\hbar^2} - \frac{l(l+1)}{r^2}\right)R(r) = 0 \quad (3.104)$$

Equation (3.104) is the relativistic generalization of the familiar radial Schrödinger equation. Together equations (3.100), (3.102) and (3.104) are all that is required to completely describe a Klein–Gordon particle in a spherical potential. It is instructive to examine the similarities and differences between equation (3.104) and the radial Schrödinger equation

$$\frac{\partial^2 R(r)}{\partial r^2} + \frac{2}{r}\frac{\partial R(r)}{\partial r} + \frac{2m}{\hbar^2}\left(E - V(r) - \frac{l(l+1)}{r^2}\right)R(r) = 0 \qquad (3.105)$$

Clearly the derivative terms and the term containing the orbital angular momentum quantum number are identical in both equations. In the Schrödinger equation it is the $l(l+1)$ term that determines the behaviour of the wavefunction near the origin (it is large as $r \to 0$). In the Klein–Gordon case for a Coulomb potential the $V(r)^2$ term will also be of order $1/r^2$ and hence will also affect the behaviour close to the origin. The total energy/potential term appearing in the radial Klein–Gordon equation is quadratic because of the existence of the antiparticles. We can still take the positive or negative square root to find the energy eigenvalues. So, from these two statements, it appears that the existence of antiparticles may affect the behaviour of the spin-zero electron wavefunction close to the origin. Finally, the Compton wavevector term in (3.104) is a correction for the rest mass energy.

3.8 The Spinless Electron Atom

In this section we work through an example application of the radial Klein–Gordon equation (3.104). It should come as no surprise that the example we choose is the Coulomb potential. There are good reasons for this. Firstly, it is the straight relativistic generalization of the Schrödinger case. The results of this section, when compared with the non-relativistic case, will show the effect on the energy levels of the fact that the orbiting particle actually obeys relativistic kinematics. Secondly, in a later chapter we examine the 'real' one-electron atom. The only difference between that case and the present one is that there the electron has spin $1/2$ whereas here it does not. Hence a comparison of that result with those in this section is a direct measure of the effect of spin on the energy levels. So, by going to the spinless electron atom as an intermediate step between the non-relativistic and the relativistic single-electron atom, we are able to separate the effects of spin from the effects of relativistic kinematics on the wavefunctions and energy levels. We will not go further than determining the energy levels and lowest eigenfunctions of the spinless electron in a central coulombic field as we discuss the calculation of observables for the hydrogen atom in chapter 8.

The model is a negatively charged spinless electron in orbit around an infinitely massive positively charged source of potential. Obviously the negatively charged electron will feel a potential

$$V(\mathbf{r}) = -\frac{Ze^2}{4\pi\epsilon_0 r} \tag{3.106}$$

Let us substitute this into (3.104) and multiply out the brackets.

$$\frac{\partial^2 R(r)}{\partial r^2} + \frac{2}{r}\frac{\partial R(r)}{\partial r}$$
$$+ \left(\frac{1}{\hbar^2}(\frac{W^2}{c^2} - m^2 c^2) - \frac{l(l+1) - Z^2\alpha^2}{r^2} + \frac{2WZ\alpha}{\hbar c r} \right) R(r) = 0$$
$$\tag{3.107}$$

where α is the fine structure constant. This looks very complicated. We can simplify it by collecting together constants:

$$\frac{\partial^2 R(r)}{\partial r^2} + \frac{2}{r}\frac{\partial R(r)}{\partial r} + \left(F + \frac{2H}{r} + \frac{G}{r^2}\right) R(r) = 0 \qquad (3.108)$$

where the replacements are the obvious ones. Next we change our space variable using

$$\rho = 2r\sqrt{-F} \qquad (3.109)$$

This may look unusual as it contains the square root of an apparently negative number. However, we are considering bound states, so W must be less than the rest mass energy, and therefore $W^2 - m^2c^4$ must also be less than 0. Hence $F < 0$ and we are really taking the square root of a positive number. Making this change of variable in (3.108) gives

$$\frac{\partial^2 R(\rho)}{\partial \rho^2} + \frac{2}{\rho}\frac{\partial R(\rho)}{\partial \rho} + \left(-\frac{1}{4} + \frac{H}{\rho\sqrt{-F}} + \frac{G}{\rho^2}\right) R(\rho) = 0 \qquad (3.110)$$

Now this equation is in a form where we can begin to solve it. We know the wavefunction has to tend to zero as $\rho \to \infty$, so let us try the solution

$$R(\rho) = e^{-\rho/2}u(\rho) \qquad (3.111)$$

Calculating the necessary derivatives of (3.111) and putting them into (3.110) leaves

$$\frac{\partial^2 u(\rho)}{\partial \rho^2} + \left(\frac{2}{\rho} - 1\right)\frac{\partial u(\rho)}{\partial \rho} + \left(\frac{H/\sqrt{-F} - 1}{\rho} + \frac{G}{\rho^2}\right) u(\rho) = 0 \qquad (3.112)$$

The next step is to expand $u(\rho)$ as a power series in ρ:

$$u(\rho) = \rho^s \sum_{\nu=0}^{\infty} a_\nu \rho^\nu \qquad (3.113)$$

Doing the necessary differentiations and substituting for $u(\rho)$ in (3.112) enables us to compare coefficients of particular powers of ρ and obtain a recursion relation for the coefficients a_ν.

$$a_\nu = \frac{(s + \nu - H/\sqrt{-F})a_{\nu-1}}{(s+\nu)^2 + s + \nu + G} \qquad (3.114)$$

If we set $\nu = 0$ in this equation and we know that $a_{\nu-1} = 0$ for $\nu = 0$ (by definition), then the only way to avoid having $a_\nu = 0$ for all ν (which would mean the particle didn't exist) is for the denominator of (3.114) to be zero as well. This is achieved if

$$s(s+1) = -G = l(l+1) - Z^2\alpha^2 \qquad (3.115)$$

Furthermore, we know that the wavefunction must be finite (or zero) at infinity, otherwise it would be unnormalizable. Hence there must be some value of v (which we call n') for which the bracketed part of the numerator in (3.114) is zero. This will ensure that $a_{n'}$ and all subsequent values of a_v are zero. The value of n' is defined by

$$s + n' + 1 - H/\sqrt{-F} = 0 \tag{3.116}$$

Hence we can substitute from (3.116) and (3.115) into (3.114) to put it in the more succinct form

$$a_v = \frac{v - n' - 1}{v(v + 2s + 1)} a_{v-1} \tag{3.117a}$$

A general a_v is given in terms of a_0 by

$$a_v = \frac{(-1)^v n'! (2s+1)!}{(n'-v)! v! (v+2s+1)!} a_0 \tag{3.117b}$$

Now a_0 is the only part of the wavefunction that remains unspecified. It can be chosen arbitrarily. Later on the wavefunction can be normalized and that will eliminate the arbitrariness. For now, we leave this parameter undefined. Now we can collect together our results from (3.111), (3.113), (3.116) and (3.117) to write our radial wavefunction as

$$R(\rho) = e^{-\rho/2} a_0 \rho^s \sum_{v=0}^{n'} \frac{(-1)^v n'! (2s+1)!}{(n'-v)! v! (v+2s+1)!} \rho^v \tag{3.118}$$

The value of s can be determined from the indicial equation (3.115). This is a quadratic for s which can be evaluated in the usual way to give

$$s = \pm((l + \tfrac{1}{2})^2 - Z^2\alpha^2)^{1/2} - \tfrac{1}{2} \tag{3.119}$$

There are two possible solutions here. We are only interested in the positive one, as the negative one will lead to the wavefunction in (3.118) diverging at the origin. If we take the non-relativistic limit of (3.119) we find $s = l$ in agreement with the Schrödinger case.

It doesn't take long to convince oneself that (3.118) can be written as

$$R(\rho) = e^{-\rho/2} a_0 \rho^s M(-n', 2s + 2, \rho) \tag{3.120}$$

where $M(a, b, x)$ is a confluent hypergeometric function whose properties are discussed in appendix B, and we have used equation (B.2). $M(a, b, x)$ is a finite polynomial only if a is a negative integer. Indeed, in this case, M is proportional to a Laguerre polynomial (Abramowitz and Stegun 1972). This is another good place to examine the non-relativistic limit. In this limit the fine structure constant is zero, so from (3.115) we can see that s must be an integer. From (3.116) we see that this means there will only be acceptable solutions when $-H^2/F$ is an integer. In that case the wavefunction becomes a Laguerre polynomial, which is the correct limit. That $-n'$ has to be a negative integer in the relativistic case forces the series $(B.2)$ to be finite. This is required for the wavefunction to be normalizable. This restriction also leads to the quantization condition for the energy. We can obtain the relativistic energy levels from (3.116). Neither s nor $-H^2/F$ has to be an integer in this case.

Inserting the definitions of F, G and H from (3.108) into (3.116), and carrying out some tedious algebra we obtain

$$W_{n'} = \frac{mc^2}{(1 + Z^2\alpha^2/(n' + ((l + 1/2)^2 - Z^2\alpha^2)^{1/2} + 1/2)^2)^{1/2}} \qquad (3.121)$$

This is not a very transparent form for the energy eigenvalues. However, by repeated use of the small x expansion

$$(1 + x)^p = 1 + px + \frac{p(p-1)}{2!}x^2 + \cdots \qquad (3.122)$$

we find

$$W - mc^2 = -\frac{Z^2 e^4 m}{(4\pi\epsilon_0)^2 2\hbar^2 n^2} + \frac{Z^4 e^8 m}{(4\pi\epsilon_0)^4 \hbar^4 c^2}\left(\frac{3}{8n^4} - \frac{1}{(2l+1)n^3}\right) \qquad (3.123)$$

where

$$n = n' + l + 1 \qquad (3.124)$$

is the relation between n' and the usual principal quantum number n. If you check equation (3.123), take care, it is very easy to lose terms you should retain (I speak from experience).

Equation (3.124) looks a bit odd. We don't expect n to depend on l as directly as this. Indeed our experience of the periodic table implies this would be an unreasonable (stupid) conclusion. Equation (3.124) is a bit misleading in this respect as n' is itself defined in terms of l. If we consider (3.116) and substitute for s from equation (3.115) and n' from equation (3.124) we obtain the correct relation between n and l. They only occur together in the relativistic corrections to the non-relativistic energy eigenvalues.

The first term in equation (3.123) is the familiar Rydberg form for the energy levels of hydrogen. The second term here is the correction to the Rydberg term to order $1/c^2$. Outside the brackets the prefactor of the second term is simply the square of the Rydberg energy divided by mc^2. In the non-relativistic formula the energy of a one-electron atom is independent of the l quantum number. Hence the energy of all l-shells for a given value of n are degenerate. We see that this is not the case for the relativistic theory. This degeneracy is lifted and the l-shells for a given n have different energies. However, the $2l + 1$ degeneracy associated with the m_l quantum number remains. The relativistic correction to the energy enters equation (3.123) at the $1/c^2$ level. Therefore we expect the separation between levels with the same n and different l to be very small compared with the separation between levels of different n.

Figure 3.4 is a schematic diagram of the energy levels in the spinless electron atom calculated relativistically. The energy levels from the non-relativistic theory are shown on the right of the figure. An atomic number of $Z = 64$ is chosen because we want Z to be large to maximize relativistic effects. However, if $Z\alpha > l + 1/2$, the denominator of (3.121) is imaginary. This occurs at $Z = 69$ for s-electrons so we cannot go above that value. The energy levels of the one-electron atom do not split as implied by figure 3.4 when observed experimentally. This is clear evidence that something is omitted from the description of the atom in this section. In chapter 8 we will see that the missing physics is the spin.

In figures $3.5a, b$ and c we show the radial part of the wavefunction calculated from equation (3.118). Also drawn on these diagrams are the equivalent non-relativistic radial wavefunctions. Note that the figures show $rR(r)$ rather than the wavefunction itself. There is good reason for this. We have actually glossed over a subtle point, as to mention it at the time would have distracted from the thread of the mathematics. Consider equation (3.119) for $l = 1, 2, 3, \cdots, \infty$: it is necessary to take the positive square root to get s positive. However, for $l = 0$, we find that s is negative whichever square root we take, which means that the wavefunction (3.120) is divergent at the origin. This does not matter provided that it does not diverge faster than $1/r$. What does matter is that the radial probability density does not diverge. To calculate any observable we have to do an integral in spherical polar coordinates involving something like the product of two wavefunctions. In this integral we always multiply by the infinitesimal volume $r^2 \sin\theta \, dr d\theta d\phi$. As long as the wavefunction diverges less rapidly than $1/r$ as $r \to 0$, there will be no divergence in observables. The $1s$ wavefunction tends to zero as $r \to \infty$, has a maximum at $r = s/\sqrt{-F}$, and decreases a little before rising to infinity as $r \to \infty$. Indeed any $l = 0$ wavefunction behaves similarly close to the origin. It is this divergence that is caused by the radial equation being quadratic in the potential, as mentioned when we compared equations (3.104) and (3.105). So the divergence is, indirectly, due to the existence of the negative energy states.

There is one other intriguing limit of the above theory which we have not discussed. Although no one-electron atom exists with $Z > 69$, there is no limit on Z in the above theory. However, consider again equation (3.119) for s. If $Z\alpha > l + 1/2$ we have to take the square root of a negative number and s becomes complex. The real part of ρ^s is then $\rho^{-1/2}$ which diverges as $\rho \to 0$. What happens in this case is that the

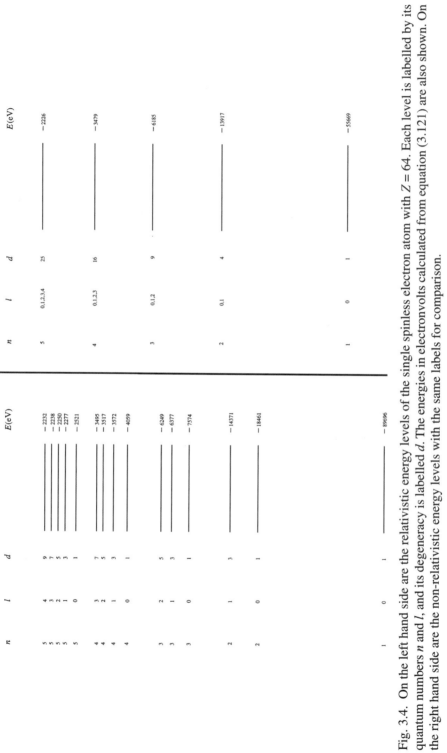

Fig. 3.4. On the left hand side are the relativistic energy levels of the single spinless electron atom with $Z = 64$. Each level is labelled by its quantum numbers n and l, and its degeneracy is labelled d. The energies in electronvolts calculated from equation (3.121) are also shown. On the right hand side are the non-relativistic energy levels with the same labels for comparison.

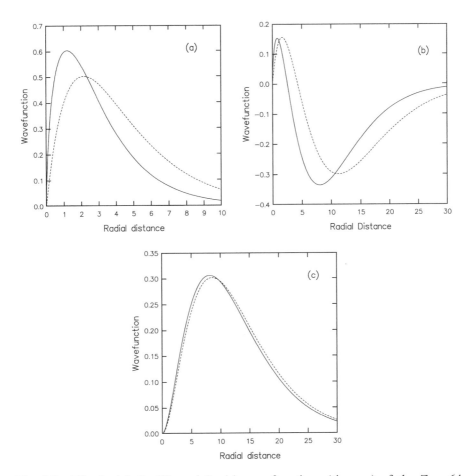

Fig. 3.5. The $1s$ (a), $2s$ (b), and $2p$ (c) wavefunctions (times r) of the $Z = 64$ spinless electron atom (full line), and their non-relativistic counterpart (dashed line). Distance is measured in units of the Compton wavelength and the wavefunction is calculated in relativistic units to emphasize the effects of relativity (see appendix D).

attractive Coulomb potential is so large that the kinetic energy of the orbiting electron (or the zero point energy in the $n = 1$ case) is insufficient to prevent it spiralling down into the nucleus. For $l = 0$ this means that the attractive potential energy must have fallen to less than half the rest mass energy at a distance of one Compton wavelength from the nucleus. To treat such atomic systems properly it is necessary to consider other factors as well, such as the finite extent of the nucleus.

Now that we have evaluated the eigenvalues and eigenfunctions for the spinless electron atom, it seems a shame to stop here and not calculate some observable quantities. We will do this for the *real* hydrogen atom, but we don't do it for this case for two reasons. Firstly, although we have derived an expression for the wavefunction in equation (3.120), we have not normalized it. The integration involved in the normalization is a very messy calculation. Secondly, we are going to calculate observables for the single-electron atom later on and it would be a duplication of effort for minimal extra insight. Therefore we will cease our description of a spinless particle in a fixed central Coulomb field here.

There is no such thing as a spinless electron, so the results described here are only of use for comparison with the non-relativistic case and with the relativistic one-electron atom including spin. However, with only the masses changed, the formalism described here could be applied to a an *atom* composed of a negative pion (which has spin zero) in a central Coulomb potential. One can conceive of creating such an atom (West 1958).

3.9 Problems

(1) Show that the Klein–Gordon equation is Lorentz invariant.
(2) In this chapter we have considered only interactions between particles that are electromagnetic in nature and distinguish between particles and antiparticles. Derive a Klein–Gordon equation for scalar interactions that treat particles and antiparticles alike.
(3) A spinless electron atom is placed in a constant magnetic field which is parallel to the x-axis. Derive an expression for the splitting of the energy levels due to the field.
(4) Equation (3.59) enables us to calculate the antiparticle wavefunction, given a particle wavefunction for spin-zero particles. Explain how electrically neutral particles fit into this picture. Is the theory capable of differentiating between particles such as the π^0 which are their own antiparticle and the K^0 which are neutral, but have another quantum number reversed in the antiparticle (in this case the strangeness)?
(5) Find the eigenvectors of the radial Klein–Gordon equation (equation (3.104)) for a potential defined by

$$V = \begin{cases} -V_0 & r \leq R \\ 0 & r > R \end{cases}$$

where V_0 is a positive number. Find an expression that determines the eigenvalues.

4

The Dirac Equation

In the previous chapter we discussed the direct relativistic generalization of the Schrödinger theory of quantum mechanics. It was shown that this leads to the Klein–Gordon equation from which some fundamental physics follows. Spin does not appear in either the Schrödinger or the Klein–Gordon theory. Indeed, the Klein–Gordon equation is only appropriate for particles with spin zero. Most of the particles encountered in everyday life (not that one actually encounters many fundamental particles in everyday life) are not spin zero. The most common ones, the neutron, proton and electron, all have spin $1/2$. In this chapter we find the equation to describe particles with spin $1/2$ and explore its properties. From the title of this chapter you will not be surprised to learn that it is known as the Dirac equation. The Dirac equation is more general than anything that has gone before, and therefore cannot really be derived. Much of the rest of this book involves looking for solutions of the Dirac equation under various circumstances.

Although the Dirac equation cannot really be derived from anything learnt up to the present level, some plausibility arguments for its existence can be given and a couple of these are presented in the first section. As in the Schrödinger theory, it is very useful to be able to write the Dirac equation in Hamiltonian form. This is also discussed in the first section. Much of the theory of spin-$1/2$ particles is written in terms of 4×4 matrices. In other textbooks there are several representations of the Dirac matrices. Throughout this book we use a common one, which is the most convenient for our purposes. The properties of these matrices and a brief mention of their relation to other representations is given in the second section. A relativistic quantum theory is not much use unless the fundamental equation looks the same in any Lorentz frame of reference. Therefore in the third section we show explicitly that the Dirac equation is Lorentz invariant and how it can be transformed from one frame to

99

another. Next we take the non-relativistic limit of the Dirac equation and show that this leads to the Pauli equation already mentioned in chapter 2. We can correct the Pauli equation to order $1/c^2$ and see that this yields some extra terms in the Pauli Hamiltonian. The non-relativistic limit is very instructive as it predicts the magnetic moment of the electron, and also we learn that this leads to the spin–orbit coupling so beloved by physics students everywhere. A less familiar form of the Dirac equation is then derived, which was first used in the theory of β-decay, but has not found much application in condensed matter physics. Here we derive it for later use in our discussion of a single electron in a constant magnetic field. After that we show how the probability density and current are defined in Dirac theory. This leads to a natural method of normalizing the Dirac four-component wavefunction. Next we consider the Gordon decomposition, which is a method of separating the probability density and current into a part similar to that for Klein–Gordon particles and a part arising from the internal state of the particle. Finally we consider a particle in an electromagnetic field and explain how the Dirac equation can be used to find the forces on such a particle. Hence we are able to display explicitly the connection between relativistic quantum theory and the classical theory of electromagnetism. The review article by Feshbach and Villars (1958) covers a lot of the material presented in this chapter.

4.1 The Origin of the Dirac Equation

In this section we present two *origins* of the Dirac equation, one based on symmetry arguments and the other on energetic considerations.

The Lorentz transformations unite time and space into a single four-dimensional quantity. The basic equation underlying relativistic quantum theory must reflect this unity, implying that there must be complete symmetry between the time and space parts of the equation. Clearly the Schrödinger equation does not satisfy this constraint because it is first order in the time derivatives and second order in the space derivatives. Let us assume that, like the Schrödinger equation, the Dirac equation will be linear in the time derivatives. Therefore it must also be linear in space derivatives. It is certainly desirable for the principle of superposition to hold in relativistic quantum theory, so we have to constrain the equation to be linear. We know from classical relativity that the energy of a free particle is given by

$$W^2 = p^2 c^2 + m^2 c^4 \tag{4.1}$$

If we make the operator substitutions in (4.1), $\mathbf{p} \rightarrow \hat{\mathbf{p}} = -i\hbar\nabla$ and $W \rightarrow \hat{W} = i\hbar\partial/\partial t$, and have them operate on a wavefunction, we find the

Klein–Gordon equation

$$\left(\nabla^2 - \frac{1}{c^2}\frac{\partial^2}{\partial t^2} - \frac{m^2 c^2}{\hbar^2}\right)\psi(\mathbf{r}, t) = 0 \tag{4.2}$$

If we insert a plane wave $\psi(\mathbf{r}, t) = \exp(i(\mathbf{p} \cdot \mathbf{r} - Wt)/\hbar)$, this equation reduces to (4.1). For a wave equation describing a spin $1/2$ particle we must also be able to substitute in a free-particle solution to obtain (4.1).

Above we have insisted on several necessary constraints on the form of the Dirac equation. Within these constraints let us allow the equation to be as general as possible. We allow the wavefunction to have N components where the value of N is as yet unspecified. Now we express the time derivative of one component of $\psi(\mathbf{r}, t)$ in the most general way conceivable (Dirac 1928a,b, Bjorken and Drell 1964)

$$\frac{1}{c}\frac{\partial \psi_i(\mathbf{r}, t)}{\partial t} = -\sum_{k=x,y,z}\sum_{n=1}^{N}\alpha_{i,n}^{k}\frac{\partial \psi_n}{\partial k} - \frac{imc}{\hbar}\sum_{n=1}^{N}\beta_{i,n}\psi_n(\mathbf{r}, t) \tag{4.3}$$

The left hand side of this equation is the time derivative of one element of $\psi(\mathbf{r}, t)$. The first term on the right hand side is a sum over all possible spatial derivatives of all components of the wavefunction and the second term on the right hand side is a linear combination of all the components of $\psi(\mathbf{r}, t)$. We have inserted the constants into equation (4.3) for the sole reason that they make the equation dimensionally correct. The factor i has been included in the final term with no loss of generality as the $\alpha_{i,n}^{k}$s and $\beta_{i,n}$s have not yet been defined. Equation (4.3) looks a little complicated, but we can make it look rather more simple by writing it in matrix form. If we let $\psi(\mathbf{r}, t)$ be a column vector with N elements then $\alpha_{i,n}^{k}$ and $\beta_{i,n}$ are, in general, elements of an $N \times N$ matrix. Furthermore, the sum over k in the first term on the right of (4.3) just amounts to a dot product. So the $\alpha_{i,n}^{k}$ can be written as a vector $\boldsymbol{\alpha}$ and (4.3) becomes

$$\frac{1}{c}\frac{\partial \psi(\mathbf{r}, t)}{\partial t} = -\tilde{\boldsymbol{\alpha}} \cdot \nabla \psi(\mathbf{r}, t) - \frac{imc}{\hbar}\tilde{\beta}\psi(\mathbf{r}, t) \tag{4.4}$$

The tilde over $\boldsymbol{\alpha}$ and β indicates that they are matrices. Next let us make the usual replacement $\hat{\mathbf{p}} \to \hbar\nabla/i$ and multiply the whole equation through by $c\hbar/i$. We end up with

$$i\hbar\frac{\partial}{\partial t}\psi(\mathbf{r}, t) = \hat{H}\psi(\mathbf{r}, t) = (c\tilde{\boldsymbol{\alpha}} \cdot \hat{\mathbf{p}} + \tilde{\beta}mc^2)\psi(\mathbf{r}, t) \tag{4.5}$$

This is one of the familiar forms of the Dirac equation. Of course, the $\tilde{\boldsymbol{\alpha}}$ and $\tilde{\beta}$ matrices are still undefined.

The Dirac matrices will be discussed in detail in the following section. Next we are going to arrive at the Dirac equation in a rather more straightforward manner which will also give us some clues about the properties of the Dirac matrices (Dirac 1958). Consider equation (4.1). To obtain a linear equation we write the square root of it in the following way (without worrying about the sign ambiguity when we take a square root):

$$W = \sqrt{p^2c^2 + m^2c^4} = \boldsymbol{\alpha} \cdot \mathbf{p}c + \beta mc^2 \tag{4.6}$$

Here, $\boldsymbol{\alpha}$ and β have to be such that when we square (4.6) we arrive back at (4.1). Clearly, β has to be a scalar and $\boldsymbol{\alpha}$ has to be a vector so that both terms on the right hand side of this equation are scalars. Let's square this explicitly.

$$W^2 = (\boldsymbol{\alpha} \cdot \mathbf{p}c + \beta mc^2)(\boldsymbol{\alpha} \cdot \mathbf{p}c + \beta mc^2)$$

$$= \alpha_x^2 p_x^2 c^2 + \alpha_y^2 p_y^2 c^2 + \alpha_z^2 p_z^2 c^2 + \boldsymbol{\alpha} \cdot \mathbf{p}\beta mc^3 + \beta\boldsymbol{\alpha} \cdot \mathbf{p}mc^3 + \beta^2 m^2 c^4 \tag{4.7}$$

$$+ c^2((\alpha_x\alpha_y + \alpha_y\alpha_x)p_xp_y + (\alpha_y\alpha_z + \alpha_z\alpha_y)p_yp_z + (\alpha_z\alpha_x + \alpha_x\alpha_z)p_zp_x)$$

For this to be consistent with (4.1) we require that

$$\alpha_x^2 = \alpha_y^2 = \alpha_z^2 = \beta^2 = I \tag{4.8a}$$

and

$$\alpha\beta + \beta\alpha = 0 \tag{4.8b}$$

$$\alpha_x\alpha_y + \alpha_y\alpha_x = \alpha_y\alpha_z + \alpha_z\alpha_y = \alpha_x\alpha_z + \alpha_z\alpha_x = 0 \tag{4.8c}$$

In words, this means that the squares of the $\boldsymbol{\alpha}$ and β operators have to be equal to the identity operator, and all components of $\boldsymbol{\alpha}$ and β have to anticommute with each other. Provided these conditions can be met, it is clear that (4.6) is a valid way of writing the energy. Furthermore, if we make the usual operator substitutions in (4.6), we obtain equation (4.5), which was derived from symmetry and Lorentz invariance considerations. No numbers exist that can satisfy (4.8a, b and c) simultaneously. However, it turns out that there are matrices that do so. Therefore we have again found that the Dirac equation is a matrix equation.

Finally, let us sum up what has been achieved in this section. Two plausibility arguments have been used to *derive* an equation of motion for a spin-1/2 particle, equation (4.5). Both these arguments led us to the same equation (I wouldn't have presented them if they had not) which is written in terms of some unknown quantities $\tilde{\alpha}$ and $\tilde{\beta}$. By considering the energy of a free particle we have found some constraints that must be obeyed by these quantities. These are given by equations (4.8).

Next we will separate the Dirac equation into time-dependent and time-independent parts, and hence derive the Hamiltonian form of the Dirac equation. In relativistic theory, time and space variables are linked in a way that they are not in non-relativistic theory. It is not so clear, intuitively, that separating time and space makes sense. In fact it does, because the Dirac equation, like the Schrödinger equation, contains only a first order time derivative. Let us take the equation in the form of (4.5)

$$i\hbar \frac{\partial}{\partial t} \psi(\mathbf{r}, t) = c(-i\hbar\tilde{\boldsymbol{\alpha}} \cdot \nabla + \tilde{\beta}mc)\psi(\mathbf{r}, t) \tag{4.9}$$

We have already seen that $\tilde{\boldsymbol{\alpha}}$ and $\tilde{\beta}$ are $N \times N$ matrices. Let us pre-empt ourselves just a little and state that they are, in fact, 4×4 matrices. In that case the wavefunction in (4.9) must, in general, be a 4-vector. Let us write it in the form

$$\psi(\mathbf{r}, t) = \psi(\mathbf{r})e^{-iWt/\hbar} = \begin{pmatrix} \psi_1(\mathbf{r}) \\ \psi_2(\mathbf{r}) \\ \psi_3(\mathbf{r}) \\ \psi_4(\mathbf{r}) \end{pmatrix} e^{-iWt/\hbar} \tag{4.10}$$

Substituting this into (4.9) and doing the time differentiation gives

$$\hat{H}\psi(\mathbf{r}) = (c\tilde{\boldsymbol{\alpha}} \cdot \hat{\mathbf{p}} + \tilde{\beta}mc^2)\psi(\mathbf{r}) = W\psi(\mathbf{r}) \tag{4.11}$$

Equation (4.11) defines the Hamiltonian for a free particle. W is the energy eigenvalue again. Instead of equation (4.10), we could have chosen the solution $\psi(\mathbf{r}, t) = \psi(\mathbf{r})e^{iWt/\hbar}$. In that case, we would have obtained an equation that looked like (4.11) except that on the right hand side W would be replaced by $-W$. Such a wavefunction would represent either a particle going backward in time or a particle in a negative energy state, like those we didn't think too much about when we took the square root in equation (4.6).

4.2 The Dirac Matrices

Equations (4.8) are constraints on the Dirac matrices. Our task now is to find the matrices that satisfy them. We require four matrices, all of which anticommute with one another. The matrices we have seen before that describe spin are the Pauli matrices of equation (2.29). It would be satisfying if the $\tilde{\boldsymbol{\alpha}}$ and $\tilde{\beta}$ could be written in terms of the Pauli matrices. Unfortunately the Pauli matrices are 2×2 and the maximum possible number of anticommuting 2×2 matrices is three. It is necessary to go to 4×4 matrices to find four that anticommute (Rose 1961, Feynman 1962, Baym 1967).

We will not derive a set of matrices satisfying the rules above, rather we will write them down and leave their verification as an exercise. There are actually several important sets of matrices in relativistic quantum theory, and we will write them down here and discuss their properties. It turns out that we can write the matrices in a representation in which the Pauli spin matrices play a role, and that is the representation we select. Firstly the identity matrix and $\tilde{\beta}$:

$$\tilde{I}_4 = \begin{pmatrix} 1 & 0 & 0 & 0 \\ 0 & 1 & 0 & 0 \\ 0 & 0 & 1 & 0 \\ 0 & 0 & 0 & 1 \end{pmatrix} = \begin{pmatrix} \tilde{I}_2 & 0 \\ 0 & \tilde{I}_2 \end{pmatrix} \qquad (4.12a)$$

$$\tilde{\beta} = \begin{pmatrix} 1 & 0 & 0 & 0 \\ 0 & 1 & 0 & 0 \\ 0 & 0 & -1 & 0 \\ 0 & 0 & 0 & -1 \end{pmatrix} = \begin{pmatrix} \tilde{I}_2 & 0 \\ 0 & -\tilde{I}_2 \end{pmatrix} \qquad (4.12b)$$

Henceforth we will label the identity matrix $\tilde{I}$ whether it is 2×2 or 4×4. Which it is should be obvious from the context. Next the α-matrices:

$$\tilde{\alpha}_x = \begin{pmatrix} 0 & 0 & 0 & 1 \\ 0 & 0 & 1 & 0 \\ 0 & 1 & 0 & 0 \\ 1 & 0 & 0 & 0 \end{pmatrix} = \begin{pmatrix} 0 & \tilde{\sigma}_x \\ \tilde{\sigma}_x & 0 \end{pmatrix} \qquad (4.13a)$$

$$\tilde{\alpha}_y = \begin{pmatrix} 0 & 0 & 0 & -i \\ 0 & 0 & i & 0 \\ 0 & -i & 0 & 0 \\ i & 0 & 0 & 0 \end{pmatrix} = \begin{pmatrix} 0 & \tilde{\sigma}_y \\ \tilde{\sigma}_y & 0 \end{pmatrix} \qquad (4.13b)$$

$$\tilde{\alpha}_z = \begin{pmatrix} 0 & 0 & 1 & 0 \\ 0 & 0 & 0 & -1 \\ 1 & 0 & 0 & 0 \\ 0 & -1 & 0 & 0 \end{pmatrix} = \begin{pmatrix} 0 & \tilde{\sigma}_z \\ \tilde{\sigma}_z & 0 \end{pmatrix} \qquad (4.13c)$$

Equations (4.12) and (4.13) are the matrices that appear in the Dirac equation (4.5) in what is known as the 'standard representation'. Check for yourself that conditions (4.8) hold and also show that

$$\tilde{\beta}^{-1}\tilde{\alpha}_i\tilde{\beta} = -\tilde{\alpha}_i \qquad (4.14)$$

The 4×4 Pauli spin matrices are also widely used. They are

$$\tilde{\sigma}_{4x} = \begin{pmatrix} 0 & 1 & 0 & 0 \\ 1 & 0 & 0 & 0 \\ 0 & 0 & 0 & 1 \\ 0 & 0 & 1 & 0 \end{pmatrix} = \begin{pmatrix} \tilde{\sigma}_x & 0 \\ 0 & \tilde{\sigma}_x \end{pmatrix} \qquad (4.15a)$$

$$\tilde{\sigma}_{4y} = \begin{pmatrix} 0 & -i & 0 & 0 \\ i & 0 & 0 & 0 \\ 0 & 0 & 0 & -i \\ 0 & 0 & i & 0 \end{pmatrix} = \begin{pmatrix} \tilde{\sigma}_y & 0 \\ 0 & \tilde{\sigma}_y \end{pmatrix} \qquad (4.15b)$$

$$\tilde{\sigma}_{4z} = \begin{pmatrix} 1 & 0 & 0 & 0 \\ 0 & -1 & 0 & 0 \\ 0 & 0 & 1 & 0 \\ 0 & 0 & 0 & -1 \end{pmatrix} = \begin{pmatrix} \tilde{\sigma}_z & 0 \\ 0 & \tilde{\sigma}_z \end{pmatrix} \tag{4.15c}$$

Again, from now on we will not differentiate in the notation between the 2×2 and the 4×4 Pauli matrices. The matrices (4.15) obey

$$\tilde{\sigma}_x^2 = \tilde{\sigma}_y^2 = \tilde{\sigma}_z^2 = \tilde{I} \tag{4.16}$$

Furthermore, they have angular momentum commutation rules

$$\tilde{\sigma}_i \tilde{\sigma}_j - \tilde{\sigma}_j \tilde{\sigma}_i = 2i\epsilon_{ijk}\tilde{\sigma}_k \tag{4.17}$$

where $\epsilon_{ijk} = 0$ unless $i \neq j \neq k$, $\epsilon_{ijk} = 1$ if $ijk = xyz$ or cyclic permutations thereof, and $\epsilon_{ijk} = -1$ otherwise. The anticommutators are

$$\{\tilde{\sigma}_i, \tilde{\sigma}_j\} = \tilde{\sigma}_i \tilde{\sigma}_j + \tilde{\sigma}_j \tilde{\sigma}_i = 2\tilde{I}\delta_{ij} \tag{4.18}$$

One further very useful property of both the α-matrices and the σ-matrices also arises from being able to write them in terms of the Pauli matrices. If **A** and **B** are arbitrary vectors, then

$$\tilde{\alpha} \cdot \mathbf{A} \, \tilde{\alpha} \cdot \mathbf{B} = \mathbf{A} \cdot \mathbf{B} + i\tilde{\sigma} \cdot \mathbf{A} \times \mathbf{B} \tag{4.19a}$$

$$\tilde{\sigma} \cdot \mathbf{A} \, \tilde{\sigma} \cdot \mathbf{B} = \mathbf{A} \cdot \mathbf{B} + i\tilde{\sigma} \cdot \mathbf{A} \times \mathbf{B} \tag{4.19b}$$

The α-matrices are called odd because they have non-zero elements only in the top right and the bottom left quadrants. The σ-matrices, on the other hand, are called even, because they have non-zero elements only in the top left and bottom right quadrants. It is not hard to convince oneself that $\tilde{\beta}$ commutes with all even matrices and anticommutes with all odd matrices. This is a property we will make full use of when it comes to separating the particle and antiparticle states in chapter 7.

There are commutation relations between the α- and the σ-matrices which will be useful later. We leave it to the problems for the reader to prove that

$$[\tilde{\sigma}_i, \tilde{\alpha}_i] = 0 \tag{4.20a}$$

$$\{\tilde{\sigma}_i, \tilde{\alpha}_j\} = 0 \qquad\qquad i \neq j \tag{4.20b}$$

$$\tilde{\alpha} \times \tilde{\alpha} = 2i\tilde{\sigma} \tag{4.21}$$

One further set of relativistic matrices is very often defined. They are the γ-matrices. There are several slightly differing definitions of the γ-matrices in the literature. We select a definition in agreement with that of Feynman (1962) and Bjorken and Drell (1964) for $\tilde{\gamma}_1$ to $\tilde{\gamma}_4$ and a slightly different definition for $\tilde{\gamma}_5$ (Rose 1961) because that simplifies our derivation of the radial Dirac equation later. The definitions are

$$\tilde{\gamma}_1 = \tilde{\beta}\tilde{\alpha}_x = \begin{pmatrix} 0 & 0 & 0 & 1 \\ 0 & 0 & 1 & 0 \\ 0 & -1 & 0 & 0 \\ -1 & 0 & 0 & 0 \end{pmatrix} = \begin{pmatrix} 0 & \tilde{\sigma}_x \\ -\tilde{\sigma}_x & 0 \end{pmatrix} \qquad (4.22a)$$

$$\tilde{\gamma}_2 = \tilde{\beta}\tilde{\alpha}_y = \begin{pmatrix} 0 & 0 & 0 & -i \\ 0 & 0 & i & 0 \\ 0 & i & 0 & 0 \\ -i & 0 & 0 & 0 \end{pmatrix} = \begin{pmatrix} 0 & \tilde{\sigma}_y \\ -\tilde{\sigma}_y & 0 \end{pmatrix} \qquad (4.22b)$$

$$\tilde{\gamma}_3 = \tilde{\beta}\tilde{\alpha}_x = \begin{pmatrix} 0 & 0 & 1 & 0 \\ 0 & 0 & 0 & -1 \\ -1 & 0 & 0 & 0 \\ 0 & 1 & 0 & 0 \end{pmatrix} = \begin{pmatrix} 0 & \tilde{\sigma}_z \\ -\tilde{\sigma}_z & 0 \end{pmatrix} \qquad (4.22c)$$

$$\tilde{\gamma}_4 = \tilde{\beta}\tilde{I} = \begin{pmatrix} 1 & 0 & 0 & 0 \\ 0 & 1 & 0 & 0 \\ 0 & 0 & -1 & 0 \\ 0 & 0 & 0 & -1 \end{pmatrix} = \begin{pmatrix} I & 0 \\ 0 & -I \end{pmatrix} \qquad (4.22d)$$

$$\tilde{\gamma}_5 = i\tilde{\gamma}_1\tilde{\gamma}_2\tilde{\gamma}_3\tilde{\gamma}_4 = \begin{pmatrix} 0 & 0 & -1 & 0 \\ 0 & 0 & 0 & -1 \\ -1 & 0 & 0 & 0 \\ 0 & -1 & 0 & 0 \end{pmatrix} = \begin{pmatrix} 0 & -\tilde{I} \\ -\tilde{I} & 0 \end{pmatrix} \qquad (4.22e)$$

Several other conventions may be found in the literature for $\tilde{\gamma}_5$, but the one above is most convenient for our purposes. One form of the Dirac equation was written down in equation (4.5). If we premultiply it by $i\tilde{\beta}/\hbar c$, it becomes

$$-\tilde{\beta}\frac{\partial}{\partial ct}\psi(\mathbf{r}, t) = \left(\tilde{\gamma} \cdot \hat{\nabla} + \frac{imc}{\hbar}\tilde{I} \right) \psi(\mathbf{r}, t) \qquad (4.23)$$

Now if we define the 4-vectors

$$\nabla^\mu = \frac{\partial}{\partial x^\mu} = \left(\frac{\partial}{\partial x}, \frac{\partial}{\partial y}, \frac{\partial}{\partial z}, \frac{1}{c}\frac{\partial}{\partial t} \right), \qquad \tilde{\gamma}^\mu = (\tilde{\gamma}_1, \tilde{\gamma}_2, \tilde{\gamma}_3, \tilde{\gamma}_4) \qquad (4.24)$$

we can write the Dirac equation in the rather succinct form

$$\left(\sum_{\mu=1}^{4} \tilde{\gamma}^\mu \nabla^\mu + iIk_C \right) \psi(\mathbf{r}, t) = 0 \qquad (4.25)$$

where k_C is the Compton wavevector.

The γ-matrices also obey some anticommutation rules. It is easy to verify that

$$\tilde{\gamma}_i\tilde{\gamma}_j + \tilde{\gamma}_j\tilde{\gamma}_i = 0 \qquad\qquad i \neq j \qquad (4.26)$$

All the anticommutators of the first four γ-matrices can be described by the matrix elements g^{ij}.

$$\tilde{\gamma}_i\tilde{\gamma}_j + \tilde{\gamma}_j\tilde{\gamma}_i = 2g^{ij}\tilde{I} \tag{4.27}$$

where g^{ij} is the fundamental tensor of equation (1.38). The matrix $\tilde{\gamma}_5$ anticommutes with all the other γ-matrices, i.e.

$$\{\tilde{\gamma}_5, \tilde{\gamma}_i\} = \tilde{\gamma}_5\tilde{\gamma}_i + \tilde{\gamma}_i\tilde{\gamma}_5 = 0 \tag{4.28a}$$

but commutes with all the α- and σ-matrices

$$[\tilde{\gamma}_5, \tilde{\alpha}_i] = [\tilde{\gamma}_5, \tilde{\sigma}_i] = 0 \tag{4.28b}$$

for all i. Triple products of the γ-matrices can also be taken and more relationships between them are found:

$$\begin{aligned} \tilde{\gamma}_5\tilde{\gamma}_1 &= i\tilde{\gamma}_2\tilde{\gamma}_3\tilde{\gamma}_4, & \tilde{\gamma}_5\tilde{\gamma}_2 &= -i\tilde{\gamma}_3\tilde{\gamma}_4\tilde{\gamma}_1 \\ \tilde{\gamma}_5\tilde{\gamma}_3 &= i\tilde{\gamma}_1\tilde{\gamma}_2\tilde{\gamma}_4, & \tilde{\gamma}_5\tilde{\gamma}_4 &= i\tilde{\gamma}_1\tilde{\gamma}_2\tilde{\gamma}_3 \end{aligned} \tag{4.29}$$

Sixteen matrices are required to form a complete set of 4×4 matrices. Such a complete set can be formed from the γ-matrices. These are given by equations (4.22), (4.12a) and (4.29), and the products $\tilde{\gamma}_1\tilde{\gamma}_2$, $\tilde{\gamma}_1\tilde{\gamma}_3$, $\tilde{\gamma}_2\tilde{\gamma}_3$, $\tilde{\gamma}_1\tilde{\gamma}_4$, $\tilde{\gamma}_2\tilde{\gamma}_4$ and $\tilde{\gamma}_3\tilde{\gamma}_4$. Any 4×4 matrix can be written as a linear combination of these sixteen matrices.

The matrices (4.12) and (4.13) were written down because they obey the conditions (4.8). However, they are not the only matrices that could have been deduced. Any matrix related to (4.12) and (4.13) by the transformations

$$\tilde{\alpha}' = \tilde{S}\tilde{\alpha}\tilde{S}^{-1}, \qquad \tilde{\beta}' = \tilde{S}\tilde{\beta}\tilde{S}^{-1} \tag{4.30}$$

also satisfies (4.8). This can be verified trivially by substitution. If we use such a representation in the Dirac equation we have

$$\psi = \tilde{S}^{-1}\psi' \tag{4.31}$$

and the Dirac equation becomes

$$i\hbar\frac{\partial\psi'}{\partial t} = \hat{H}'\psi' = i\hbar\tilde{S}\frac{\partial\psi}{\partial t} = \tilde{S}\hat{H}\tilde{S}^{-1}\psi' = \tilde{S}\hat{H}\psi \tag{4.32}$$

and cancelling the $\tilde{S}$ gives us back the Dirac equation in the original representation. We can ask what effect the different representations of our matrices and wavefunctions will have on the calculation of observables. Here we have to pre-empt ourselves and assume we already know how to calculate expectation values with 4-vector wavefunctions; there will be further discussion of this in section 4.6. In the new representation we assume an operator $\hat{O}$ can be written as a matrix and its expectation value is given by

$$\int \psi'^\dagger \hat{O}' \psi' d\mathbf{r} = \int \psi^\dagger \tilde{S}^\dagger \tilde{S} \hat{O} \tilde{S}^{-1} \tilde{S} \psi d\mathbf{r} = \int \psi^\dagger \tilde{S}^\dagger \tilde{S} \hat{O} \psi d\mathbf{r} = \int \psi^\dagger \hat{O} \psi d\mathbf{r} \quad (4.33)$$

In the last step here we have assumed that the matrix $\tilde{S}$ is unitary. We can turn this argument round and say that if observables are going to be independent of our matrix representation (as they must be) then the allowable representations are related to the standard representation by the transformations (4.30) and (4.31) provided $\tilde{S}$ is unitary.

Next we will do a particular example of the unitary transformation above. This is done by stating a transformation matrix and using that to transform the α-, β- and σ-matrices from one Lorentz frame to another (Baym 1967). As σ represents spin, an angular momentum, and the product of two vectors, we can assume it transforms as a second rank tensor, like the electromagnetic field. Therefore we will show how to transform one of the matrices and then complete the transformation of the Dirac matrices by analogy with the electromagnetic field transformation already described in chapter 1. The Lorentz transformation matrix for this case is

$$\tilde{T}_i = \exp(\tilde{\boldsymbol{\alpha}} \cdot \mathbf{u}/2) \quad (4.34a)$$

and so

$$\tilde{T}_i^{-1} = \exp(-\tilde{\boldsymbol{\alpha}} \cdot \mathbf{u}/2) \quad (4.34b)$$

Here we have the primed frame moving with velocity $\mathbf{v}$ relative to the unprimed frame. The vector $\mathbf{u}$ lies in the direction of this velocity and has magnitude is given by

$$u = \tanh^{-1} \frac{v}{c}, \qquad (v = |\mathbf{v}|) \quad (4.35a)$$

and hence

$$\cosh u = (1 - v^2/c^2)^{-1/2} \quad (4.35b)$$

Let us take the x-axis as the direction of the relative velocity, and firstly let us transform $\tilde{\sigma}_z$.

$$\tilde{\sigma}_z' = \tilde{T}_i \tilde{\sigma}_z \tilde{T}_i^{-1} = \exp(\tilde{\boldsymbol{\alpha}} \cdot \mathbf{u}/2)\tilde{\sigma}_z \exp(-\tilde{\boldsymbol{\alpha}} \cdot \mathbf{u}/2) = \exp(\tilde{\boldsymbol{\alpha}} \cdot \mathbf{u})\tilde{\sigma}_z$$
$$= (\cosh u + \tilde{\boldsymbol{\alpha}} \cdot \hat{\mathbf{u}} \sinh u)\tilde{\sigma}_z = \cosh u(1 + \tilde{\boldsymbol{\alpha}} \cdot \hat{\mathbf{u}} \tanh u)\tilde{\sigma}_z$$
$$= (1 - v^2/c^2)^{-1/2}(\tilde{\sigma}_z + \tilde{\boldsymbol{\alpha}} \cdot (\mathbf{v}/c)\tilde{\sigma}_z) \quad (4.36)$$

where we have used the commutation relations (4.20) and the series representation of the hyperbolic functions. A little further algebra gives

$$\tilde{\sigma}_z' = \frac{(\tilde{\sigma}_z - iv\tilde{\alpha}_y/c)}{(1 - v^2/c^2)^{1/2}} \quad (4.37)$$

This is indeed analogous to the transformation of the electromagnetic field given by equation (1.58), and verifies our choice (4.34). Similar analyses can be done for all components of $\tilde{\sigma}$ and $\tilde{\alpha}$ and the analogy with the electromagnetic field holds for all components. Hence the Dirac matrices are components of the second rank tensor

$$
\sigma^{\mu\nu} = \begin{pmatrix}
0 & -\tilde{\sigma}_z & \tilde{\sigma}_y & i\tilde{\alpha}_x \\
\tilde{\sigma}_z & 0 & -\tilde{\sigma}_x & i\tilde{\alpha}_y \\
-\tilde{\sigma}_y & \tilde{\sigma}_x & 0 & i\tilde{\alpha}_z \\
-i\tilde{\alpha}_x & -i\tilde{\alpha}_y & -i\tilde{\alpha}_z & 0
\end{pmatrix}
\tag{4.38}
$$

and so the Lorentz transformations of the α- and β-matrices can be written as

$$
(\sigma^{\mu\nu})' = \tilde{T}\sigma^{\mu\nu}\tilde{T}^{-1}
\tag{4.39}
$$

where $\tilde{T}$ is given by equation (4.34). We have not actually proved that the α- and σ-matrices transform like this, because the transformation (4.34) was written down without proof. However, we have shown that everything works out nice and consistently provided that (4.34) is a valid description of the Lorentz transformation properties of the Dirac matrices. Finally we need to see how to transform $\tilde{\beta}$. This is also straightforward because $\tilde{\beta}$ anticommutes with $\tilde{\alpha}$, and we find

$$
\tilde{\beta}' = \frac{\tilde{\beta} - \tilde{\beta}\mathbf{v}\cdot\tilde{\boldsymbol{\alpha}}/c}{(1 - v^2/c^2)^{1/2}}
\tag{4.40}
$$

From the above, the transformation properties of the γ-matrices can easily be deduced, and I leave this to you in the problems. Note that the original transformation defined by (4.34) has the same form as the rotation operator described in section 2.2.

4.3 Lorentz Invariance of the Dirac Equation

Above we have postulated, but not proven, the existence of the Dirac equation. The best we have been able to do is give some plausibility arguments for its existence. Clearly, under such circumstances, it is necessary to verify that the equation behaves correctly before using it in a problem solving role. One property the Dirac equation must have is that it takes the same form in all inertial frames (Dirac 1958, Baym 1967). Any equation for which this is not true would be of very limited applicability.

To prove that the Dirac equation is invariant under a Lorentz transformation we first have to recall some formulae from chapter 1. We have already defined the γ-matrix 4-vector (equation (4.24)) and the energy–momentum 4-vector (above equation (1.35)). Their scalar product (as defined by equation (1.37)) can easily be taken and must, obviously, give a scalar. Let us take the square of this scalar product and see what the result is. To do this we also rewrite the γ-matrices in terms of the α- and β-matrices.

$$(\tilde{\gamma}_\mu p^\mu)^2 = (\tilde{\beta} W/c - \tilde{\beta}\tilde{\alpha} \cdot \mathbf{p})^2 = (\tilde{\beta} W/c)^2 - \frac{W\tilde{\beta}}{c}(\tilde{\beta}\tilde{\alpha} + \tilde{\alpha}\tilde{\beta}) \cdot \mathbf{p} + (\tilde{\beta}\tilde{\alpha} \cdot \mathbf{p})^2$$
$$= W^2/c^2 - p^2 = m^2 c^2 \qquad (4.41)$$

where we have used equation (4.19) and the relations between the Dirac matrices (4.8). The quantity m in equation (4.41) is the rest mass of our particle. This is the same in any Lorentz frame. So, taking the square root of (4.41), we see that

$$\tilde{\gamma}_\mu p^\mu = (\tilde{\beta} W/c - \tilde{\beta}\tilde{\alpha} \cdot \mathbf{p}) = \pm mc \qquad (4.42)$$

Thus we have shown that this scalar product is Lorentz invariant. If we label quantities in one Lorentz frame with a prime and quantities in another without a prime we have

$$\tilde{\beta}' W'/c - \tilde{\beta}'\tilde{\alpha}' \cdot \mathbf{p}' = \tilde{\beta} W/c - \tilde{\beta}\tilde{\alpha} \cdot \mathbf{p} \qquad (4.43)$$

Now let us write the Dirac equation (4.9) in the primed reference frame:

$$i\hbar \frac{\partial}{\partial t'} \psi(\mathbf{r}', t') = c\left(\frac{\hbar}{i}\tilde{\alpha}' \cdot \nabla' + \tilde{\beta}' mc\right)\psi(\mathbf{r}', t') \qquad (4.44)$$

The next step is to make the usual operator substitutions in (4.43):

$$\tilde{\beta}' i\hbar \frac{\partial}{\partial t'} - \tilde{\beta}' \frac{\hbar c}{i}\tilde{\alpha}' \cdot \nabla' = \tilde{\beta} i\hbar \frac{\partial}{\partial t} - \tilde{\beta} \frac{c\hbar}{i}\tilde{\alpha} \cdot \nabla \qquad (4.45)$$

We can now multiply (4.44) by $\tilde{\beta}'$ and substitute for the operators from (4.45) into (4.44). Then multiplying through by $\tilde{\beta}$ gives a result that is identical to (4.9) if $(\mathbf{r}', t')$ represents the same space–time point as $(\mathbf{r}, t)$. The primed quantities $\mathbf{r}'$ and t' are related to $\mathbf{r}$ and t via a simple Lorentz transformation.

The argument above shows that the Dirac equation takes on the same form in all Lorentz frames. However, it would soon become clear that (4.44) is not the most convenient way of writing the Dirac equation in the new frame. This is because the α- and β-matrices take on different forms in different frames. It turns out to be more convenient if we keep the matrices the same in all frames. In that case, the wavefunctions have to differ between frames. As is shown in section 4.2, the Dirac matrices in one frame are related to those in any other frame via a transformation of the type (4.30). If we premultiply (4.44) by $\tilde{\beta}'$ and use these transformations we get

$$\tilde{T}_i \tilde{\beta} \tilde{T}_i^{-1} i\hbar \frac{\partial}{\partial t'} \psi(\mathbf{r}', t') = c\left(\frac{\hbar}{i} \tilde{T}_i \tilde{\beta} \tilde{T}_i^{-1} \tilde{T}_i \tilde{\alpha} \tilde{T}_i^{-1} \cdot \nabla' + mc\right) \psi(\mathbf{r}', t') \qquad (4.46)$$

Now if we premultiply this by $\tilde{\beta}\tilde{T}_i^{-1}$ we obtain

$$i\hbar\frac{\partial}{\partial t'}\psi'(\mathbf{r}',t') = \left(\frac{c\hbar}{i}\tilde{\alpha}\cdot\nabla' + \tilde{\beta}mc^2\right)\psi'(\mathbf{r}',t') \tag{4.47}$$

where

$$\psi'(\mathbf{r}',t') = \tilde{T}_i^{-1}\psi(\mathbf{r},t) \tag{4.48}$$

Thus equation (4.47) is a representation of the Dirac equation for which the matrices $\tilde{\alpha}$ and $\tilde{\beta}$ are the same in all frames. We have simply transformed to a representation where the difference in the matrices between frames is replaced by a difference in the wavefunctions between frames. The wavefunction in the primed frame is related to the wavefunction in the unprimed frame by equation (4.48). This shows that we can write the Dirac equation in a form such as (4.9) and it will be valid in any frame. The change in the wavefunction between one frame and another will be taken care of by changes in the potentials.

4.4 The Non-Relativistic Limit of the Dirac Equation

Another property we require of the Dirac equation is that it have a reasonable non-relativistic limit. It would be difficult to justify putting an equation at the heart of relativistic quantum theory if its non-relativistic limit did not make sense. In this section, then, we will give a detailed account of how to take this limit, and discuss the results obtained. Following Baym (1967), we start by writing down the Dirac equation in the presence of an electromagnetic field. We state without proof that this can be done in the same way as for the Schrödinger equation with the simple addition of the scalar potential and the replacement $\hat{\mathbf{p}} \to \hat{\mathbf{p}} - e\mathbf{A}$. In the penultimate section of this chapter we derive some results for particles in electromagnetic fields which confirm this procedure. Our Dirac equation becomes

$$i\hbar\frac{\partial\psi(\mathbf{r},t)}{\partial t} = \left(c\tilde{\alpha}\cdot\left(\frac{\hbar}{i}\nabla - e\mathbf{A}(\mathbf{r})\right) + \tilde{\beta}mc^2 + V(\mathbf{r})\right)\psi(\mathbf{r},t) \tag{4.49}$$

where $V(\mathbf{r})$ is related to the electric field $\mathbf{E}$ and the scalar potential Φ via $\mathbf{E} = -\nabla\Phi(\mathbf{r})$ and $V(\mathbf{r}) = e\Phi(\mathbf{r})$. We are going to proceed with the following strategy. Firstly, we look at the non-relativistic limit of (4.49). Secondly, relativity has a natural small parameter in it, $1/c$, and we consider corrections to the non-relativistic limit to order $1/c^2$. Equation (4.49) is rather abstract in itself and it is difficult to have any physical intuition about its solutions and properties. However, the corrections to the non-relativistic limit are amenable to direct physical interpretation. One feature of equation (4.49) worthy of note at this stage is that the external vector potential couples directly to $\tilde{\alpha}$ which is associated with internal degrees of freedom of the particle described by the 4-vector wavefunction.

To take the non-relativistic limit of (4.49) we write $\psi(\mathbf{r}, t)$ as

$$\psi(\mathbf{r}, t) = \begin{pmatrix} \phi(\mathbf{r}, t) \\ \chi(\mathbf{r}, t) \end{pmatrix} \tag{4.50}$$

where ϕ and χ are both two-component quantities. Substituting (4.50) into (4.49) and using the definition of the α-matrices (4.13) we find

$$i\hbar \frac{\partial \phi(\mathbf{r}, t)}{\partial t} = c \left(\frac{\hbar}{i} \nabla - e\mathbf{A}(\mathbf{r}) \right) \cdot \tilde{\sigma} \chi(\mathbf{r}, t) + \left(V(\mathbf{r}) + mc^2 \right) \phi(\mathbf{r}, t) \tag{4.51a}$$

$$i\hbar \frac{\partial \chi(\mathbf{r}, t)}{\partial t} = c \left(\frac{\hbar}{i} \nabla - e\mathbf{A}(\mathbf{r}) \right) \cdot \tilde{\sigma} \phi(\mathbf{r}, t) + \left(V(\mathbf{r}) - mc^2 \right) \chi(\mathbf{r}, t) \tag{4.51b}$$

We will require these equations shortly. Now, recall the solution of the free-particle Dirac equation (4.10). In the non-relativistic limit this becomes

$$\psi(\mathbf{r}, t) = \psi(\mathbf{r}) e^{-iWt/\hbar} \approx \psi(\mathbf{r}) e^{-imc^2 t/\hbar} \tag{4.52}$$

and this is a solution of

$$i\hbar \frac{\partial \psi(\mathbf{r}, t)}{\partial t} \approx mc^2 \psi(\mathbf{r}, t) \tag{4.53}$$

Therefore, the lower component of equation (4.50) obeys

$$i\hbar \frac{\partial \chi(\mathbf{r}, t)}{\partial t} \approx mc^2 \chi(\mathbf{r}, t) \tag{4.54}$$

Now we go back to (4.51b) and substitute for the time derivative from (4.54). As c is large we ignore terms of order c^0 and lower, and we obtain

$$\chi(\mathbf{r}, t) = \frac{1}{2mc} \left(\frac{\hbar}{i} \nabla - e\mathbf{A}(\mathbf{r}) \right) \cdot \tilde{\sigma} \phi(\mathbf{r}, t) \tag{4.55}$$

This equation tells us about the relative sizes of ϕ and χ. Evidently, the term in brackets is a measure of the momentum of the particle. Classically this would be given by $m\mathbf{v}$. This is divided by the $2mc$ outside the brackets. The m cancels and we are left with χ being of order v/c times ϕ, so χ is called the small component of the wavefunction and ϕ is called (you guessed it) the large component.

Now that we have an expression for χ in terms of ϕ we can substitute it into (4.51a) to remove χ from that equation and hence make it easier to solve. Clearly this is the right way round to proceed. From the above we see that $\chi \to 0$ in the extreme non-relativistic limit, so we would not obtain anything useful by deriving a non-relativistic expression for ϕ and substituting that into (4.51b). Eliminating χ from (4.51a) gives

$$i\hbar \frac{\partial \phi(\mathbf{r}, t)}{\partial t} = \frac{1}{2m} \left(\left(\frac{\hbar}{i} \nabla - e\mathbf{A}(\mathbf{r}) \right) \cdot \tilde{\sigma} \right)^2 \phi(\mathbf{r}, t) + (V(\mathbf{r}) + mc^2) \phi(\mathbf{r}, t) \tag{4.56}$$

This can be simplified using (4.19*b*)

$$\left(\left(\frac{\hbar}{i}\nabla - e\mathbf{A}\right)\cdot\tilde{\boldsymbol{\sigma}}\right)^2 \phi = \left(\frac{\hbar}{i}\nabla - e\mathbf{A}\right)^2 \phi - e\hbar\tilde{\boldsymbol{\sigma}}\cdot(\nabla\times\mathbf{A} + \mathbf{A}\times\nabla)\phi$$

$$= \left(\frac{\hbar}{i}\nabla - e\mathbf{A}\right)^2 \phi - e\hbar\tilde{\boldsymbol{\sigma}}\cdot\hat{\mathbf{B}}\phi \qquad (4.57)$$

Putting this back into (4.56)

$$i\hbar\frac{\partial\phi(\mathbf{r},t)}{\partial t} = \left(\frac{1}{2m}\left(\frac{\hbar}{i}\nabla - e\mathbf{A}(\mathbf{r})\right)^2 - \frac{e\hbar}{2m}\tilde{\boldsymbol{\sigma}}\cdot\hat{\mathbf{B}}(\mathbf{r}) + V(\mathbf{r}) + mc^2\right)\phi(\mathbf{r},t)$$

$$(4.58)$$

This is the Pauli equation which we have already encountered as a description of spin-1/2 particles in an electromagnetic field in chapter 2. It differs from the previous version of equation (2.55) in the mc^2 term. This only amounts to a redefinition of the zero of energy. Equation (4.58) is a very pleasing result. Not only does it show that the Dirac equation has the correct non-relativistic limit, but also the magnetic moment of the electron has fallen out of this formalism – we have not had to put it in, as is done *ad hoc* in the non-relativistic theory.

$$\mu = \frac{e\hbar}{2m} \qquad (4.59)$$

This was a major success for the Dirac theory of the electron. All the usual non-relativistic theory of atomic spectroscopy, the anomalous Zeeman effect and the interpretation of the Stern–Gerlach experiment follow from this result. Before the discovery of the Dirac equation such experiments were understood on the basis of a spin which could only be postulated rather arbitrarily.

It would be pleasing to be able to say that equation (4.58) also predicts the magnetic moment of the proton and neutron – after all, they are also spin-1/2 particles. However, this cannot be true for the neutron because it is uncharged, and it is also untrue for the proton. This failure is rather instructive. It is a first indication that protons and neutrons are composite particles, made up of three quarks (each of which is a spin-1/2 particle). Quarks should obey the Dirac equation, and their magnetic moment may be deduced from an equivalent expression to (4.58). In fact the gyromagnetic ratio g of a neutron turns out to be -3.8, and that of a proton is $+5.6$, compared with $g = 2$ for the electron obtained from equation (4.59). The proton and neutron feel the strong nuclear force which is mediated by gluons, and to describe their properties fully requires a more complete theory taking into account the gluon field.

We can continue the analysis above to find the lowest order relativistic corrections to the Pauli equation. Adding and subtracting $2mc^2\chi(\mathbf{r},t)$ to the right hand side of (4.51b) and a little rearranging gives

$$
\chi(\mathbf{r}, t) = \frac{1}{2mc} \left(\frac{\hbar}{i} \nabla - e\mathbf{A}(\mathbf{r}) \right) \cdot \tilde{\sigma} \phi(\mathbf{r}, t)
$$

$$
- \frac{1}{2mc^2} \left(i\hbar \frac{\partial}{\partial t} - V(\mathbf{r}) - mc^2 \right) \chi(\mathbf{r}, t) \tag{4.60}
$$

If we were to omit the final term here we would simply end up re-deriving equation (4.55). The first order correction to equation (4.55) comes from substituting (4.55) into equation (4.60) for χ on the right hand side. The nth order approximation would be to substitute equation (4.60) into itself $n - 1$ times and then substitute on the right hand side for χ from equation (4.55). Anyway, the first order approximation gives

$$
\chi(\mathbf{r}, t) = \frac{1}{2mc} \left(\frac{\hbar}{i} \nabla - e\mathbf{A}(\mathbf{r}) \right) \cdot \tilde{\sigma} \phi(\mathbf{r}, t)
$$

$$
- \frac{1}{4m^2 c^3} \left(i\hbar \frac{\partial}{\partial t} - V(\mathbf{r}) - mc^2 \right) \left(\frac{\hbar}{i} \nabla - e\mathbf{A}(\mathbf{r}) \right) \cdot \tilde{\sigma} \phi(\mathbf{r}, t) \tag{4.61}
$$

Following the same procedure as before we substitute this back into (4.51a) and obtain a rather long equation

$$
i\hbar \frac{\partial \phi(\mathbf{r}, t)}{\partial t} = \left(\frac{1}{2m} \left(\frac{\hbar}{i} \nabla - e\mathbf{A}(\mathbf{r}) \right)^2 - \frac{e\hbar}{2m} \tilde{\sigma} \cdot \hat{\mathbf{B}}(\mathbf{r}) + V(\mathbf{r}) + mc^2 \right) \phi(\mathbf{r}, t)
$$

$$
- \left(\frac{\hbar}{i} \nabla - e\mathbf{A}(\mathbf{r}) \right) \cdot \tilde{\sigma} \frac{\left(i\hbar \partial / \partial t - mc^2 - V(\mathbf{r}) \right)}{4m^2 c^2} \left(\frac{\hbar}{i} \nabla - e\mathbf{A}(\mathbf{r}) \right) \cdot \tilde{\sigma} \phi(\mathbf{r}, t) \tag{4.62}
$$

Comparison of this equation with equation (4.58) shows that the final term is the correction to the Pauli equation we have been looking for. It contains all the corrections to order v^2/c^2. To say the least, this term is not very easy to interpret. Luckily it can be rewritten in a more physically transparent form straightforwardly. To see this we have to evaluate the following commutator:

$$
\left[\left(i\hbar \frac{\partial}{\partial t} - mc^2 - V(\mathbf{r}) \right), \left(\frac{\hbar}{i} \nabla - e\mathbf{A}(\mathbf{r}) \right) \cdot \tilde{\sigma} \right] \phi(\mathbf{r}, t)
$$

$$
= \left(\left(i\hbar \frac{\partial}{\partial t} - mc^2 - V(\mathbf{r}) \right) \left(\frac{\hbar}{i} \nabla - e\mathbf{A}(\mathbf{r}) \right) \cdot \tilde{\sigma} \right) \phi(\mathbf{r}, t)
$$

$$
- \left(\left(\frac{\hbar}{i} \nabla - e\mathbf{A}(\mathbf{r}) \right) \cdot \tilde{\sigma} \left(i\hbar \frac{\partial}{\partial t} - mc^2 - V(\mathbf{r}) \right) \right) \phi(\mathbf{r}, t) \tag{4.63}
$$

The easiest part of this commutator to work out is the mc^2 because that commutes with everything so it can just be removed. The other terms are not quite so simple. The vector potential $\mathbf{A}(\mathbf{r})$ does not commute with the time derivative and the scalar potential does not commute with ∇. However, these are not too hard to evaluate. The product rule of differentiation and a little algebra yield

$$\left[\left(i\hbar \frac{\partial}{\partial t} - mc^2 - V(\mathbf{r}) \right), \left(\frac{\hbar}{i} \nabla - e\mathbf{A}(\mathbf{r}) \right) \cdot \tilde{\boldsymbol{\sigma}} \right] \phi(\mathbf{r}, t)$$

$$= i\hbar \left(-\nabla V(\mathbf{r}) - e\frac{\partial \mathbf{A}}{\partial t} \right) \cdot \tilde{\boldsymbol{\sigma}} \phi(\mathbf{r}, t) = i\hbar e \mathbf{E}(\mathbf{r}, t) \cdot \tilde{\boldsymbol{\sigma}} \phi(\mathbf{r}, t) \qquad (4.64)$$

where $\mathbf{E}$ is the electric field. Now we can eliminate the commutator between equations (4.64) and (4.63):

$$i\hbar e \mathbf{E}(\mathbf{r}, t) \cdot \tilde{\boldsymbol{\sigma}} \phi(\mathbf{r}, t) =$$

$$\left(\left(i\hbar \frac{\partial}{\partial t} - mc^2 - V(\mathbf{r}) \right) \left(\frac{\hbar}{i} \nabla - e\mathbf{A}(\mathbf{r}) \right) \cdot \tilde{\boldsymbol{\sigma}} \right) \phi(\mathbf{r}, t)$$

$$- \left(\left(\frac{\hbar}{i} \nabla - e\mathbf{A}(\mathbf{r}) \right) \cdot \tilde{\boldsymbol{\sigma}} \left(i\hbar \frac{\partial}{\partial t} - mc^2 - V(\mathbf{r}) \right) \right) \phi(\mathbf{r}, t) \qquad (4.65)$$

Next we go back to our relativistic correction to the Pauli equation (the last term in equation (4.62)) and substitute for the second two sets of brackets from equation (4.65):

$$\left(\frac{\hbar}{i} \nabla - e\mathbf{A}(\mathbf{r}) \right) \cdot \tilde{\boldsymbol{\sigma}} \left(i\hbar \frac{\partial}{\partial t} - mc^2 - V(\mathbf{r}) \right) \left(\frac{\hbar}{i} \nabla - e\mathbf{A}(\mathbf{r}) \right) \cdot \tilde{\boldsymbol{\sigma}} \phi(\mathbf{r}, t)$$

$$= \left(\left(\frac{\hbar}{i} \nabla - e\mathbf{A}(\mathbf{r}) \right) \cdot \tilde{\boldsymbol{\sigma}} \right)^2 \left(i\hbar \frac{\partial}{\partial t} - mc^2 - V(\mathbf{r}) \right) \phi(\mathbf{r}, t)$$

$$+ i\hbar e \left(\left(\frac{\hbar}{i} \nabla - e\mathbf{A}(\mathbf{r}) \right) \cdot \tilde{\boldsymbol{\sigma}} (\mathbf{E}(\mathbf{r}, t) \cdot \tilde{\boldsymbol{\sigma}}) \right) \phi(\mathbf{r}, t)$$

$$= \left(\frac{1}{2m} \left(\hat{\mathbf{p}} - e\mathbf{A}(\mathbf{r}) \right)^4 + e\hbar \tilde{\boldsymbol{\sigma}} \cdot (\mathbf{E}(\mathbf{r}, t) \times (\hat{\mathbf{p}} - e\mathbf{A}(\mathbf{r}))) \right) \phi(\mathbf{r}, t)$$

$$+ i\hbar e \left(\hat{\mathbf{p}} - e\mathbf{A}(\mathbf{r}) \right) \cdot \mathbf{E}(\mathbf{r}, t) \phi(\mathbf{r}, t) \qquad (4.66)$$

In the last step here we have used the Pauli equation to lowest order and have made liberal use of the identity (4.19). So our Pauli equation with corrections to order $1/c^2$ is

$$i\hbar \frac{\partial \phi(\mathbf{r}, t)}{\partial t} = \frac{1}{2m} \left(\frac{\hbar}{i} \nabla - e\mathbf{A}(\mathbf{r}) \right)^2 \phi(\mathbf{r}, t) - \frac{e\hbar}{2m} \tilde{\boldsymbol{\sigma}} \cdot \hat{\mathbf{B}}(\mathbf{r}) \phi(\mathbf{r}, t)$$

$$+ (V(\mathbf{r}) + mc^2) \phi(\mathbf{r}, t) - \frac{1}{8m^3c^2} (\mathbf{p} - e\mathbf{A}(\mathbf{r}))^4 \phi(\mathbf{r}, t)$$

$$- \frac{e\hbar}{4m^2c^2} \tilde{\boldsymbol{\sigma}} \cdot (\mathbf{E}(\mathbf{r}, t) \times (\mathbf{p} - e\mathbf{A}(\mathbf{r}))) \phi(\mathbf{r}, t)$$

$$- \frac{i\hbar e}{4m^2c^2} \mathbf{p} \cdot \mathbf{E}(\mathbf{r}, t) \phi(\mathbf{r}, t) \qquad (4.67)$$

One might think that this was the end of the story, we have found the Pauli equation and (4.67) is the Pauli equation with relativistic corrections to order $1/c^2$. Unfortunately this is not the end, and the reason is that the final term in (4.67) is not hermitian. This means it is not guaranteed to have real eigenvalues, which is disastrous for a Hamiltonian. The Dirac equation (4.49) is hermitian, so the non-hermiticity came about because of the approximations we have made. Let us assume that the solution to equation (4.49) was correctly normalized (we will discuss how to normalize a 4-vector wavefunction in a later section; here you should just accept that it can be done). So, using (4.50)

$$\int \psi^\dagger(\mathbf{r}, t)\psi(\mathbf{r}, t)d\mathbf{r} = \int (\phi^\dagger(\mathbf{r}, t)\phi(\mathbf{r}, t) + \chi^\dagger(\mathbf{r}, t)\chi(\mathbf{r}, t))d\mathbf{r} = 1 \qquad (4.68)$$

However, ϕ is defined in terms of χ in equation (4.54), so

$$\chi^\dagger(\mathbf{r}, t)\chi(\mathbf{r}, t) = \phi^\dagger(\mathbf{r}, t)\frac{1}{4m^2c^2}\left(\frac{\hbar}{i}\nabla - e\mathbf{A}(\mathbf{r})\right)^2 \phi(\mathbf{r}, t) \qquad (4.69)$$

but this expression for χ is correct only to order v^2/c^2, and so the normalization integral remains constant only to within terms of that order. Thus, in the non-relativistic limit, the integral we want to be constant is

$$\int \phi^\dagger(\mathbf{r}, t)\left(1 + \frac{1}{4m^2c^2}\left(\frac{\hbar}{i}\nabla - e\mathbf{A}(\mathbf{r})\right)^2\right) \phi(\mathbf{r}, t)d\mathbf{r} = 1 \qquad (4.70)$$

i.e. ignoring terms of order v^4/c^4

$$\int \left(\hat{M}\phi(\mathbf{r}, t)\right)^\dagger \left(\hat{M}\phi(\mathbf{r}, t)\right) d\mathbf{r} = 1 \qquad (4.71a)$$

where

$$\hat{M} = \left(1 + \frac{\left(\frac{\hbar}{i}\nabla - e\mathbf{A}(\mathbf{r})\right)^2}{8m^2c^2}\right) \qquad (4.71b)$$

All this tells us that the non-relativistic limit of the Dirac wavefunction needs to be renormalized. The correct non-relativistic wavefunction is related to the non-relativistic limit of the relativistic wavefunction by

$$\Xi(\mathbf{r}, t) = \left(1 + \frac{1}{8m^2c^2}\left(\frac{\hbar}{i}\nabla - e\mathbf{A}(\mathbf{r})\right)^2\right) \phi(\mathbf{r}, t) \qquad (4.72)$$

So we really require a wave equation for $\Xi(\mathbf{r}, t)$ rather than the one we have got for $\phi(\mathbf{r}, t)$. To obtain this we premultiply each term of (4.67) by $\hat{M}$ and take it through the equation. Luckily this commutes with most of the terms in (4.67). However, it does not commute with the time derivative or $V(\mathbf{r})$. Some algebra (quite a lot really) shows that if we work out the commutator and add to it the non-hermitian term from the Hamiltonian (4.67) (after taking out a common factor $1/8m^2c^2$) we obtain

$$\left(\left[\left(\frac{\hbar}{i}\nabla - e\mathbf{A}(\mathbf{r})\right)^2, \left(i\hbar\frac{\partial}{\partial t} - V(\mathbf{r})\right)\right] + 2i\hbar e\mathbf{p}\cdot\mathbf{E}(\mathbf{r},t)\right)\phi(\mathbf{r},t) \tag{4.73}$$
$$= -\hbar^2\nabla^2 V(\mathbf{r})\phi(\mathbf{r},t)$$

We can use this commutator to take $\hat{M}$ through equation (4.67). Then we can make the replacement (4.72). The result of all these operations is the non-relativistic limit of the Dirac equation in hermitian form, correct to order $1/c^2$. It is

$$i\hbar\frac{\partial\Xi(\mathbf{r},t)}{\partial t} = \hat{H}\Xi(\mathbf{r},t) \tag{4.74a}$$

with

$$\hat{H} = \frac{1}{2m}\left(\frac{\hbar}{i}\nabla - e\mathbf{A}(\mathbf{r})\right)^2 - \frac{e\hbar}{2m}\tilde{\sigma}\cdot\hat{\mathbf{B}}(\mathbf{r}) + (V(\mathbf{r}) + mc^2)$$
$$- \frac{1}{8m^3c^2}(\mathbf{p} - e\mathbf{A}(\mathbf{r}))^4 - \frac{e\hbar}{4m^2c^2}\tilde{\sigma}\cdot(\mathbf{E}(\mathbf{r},t)\times(\mathbf{p} - e\mathbf{A}(\mathbf{r})))$$
$$+ \frac{\hbar^2 e}{8m^2c^2}\nabla^2 V(\mathbf{r}) \tag{4.74b}$$

The interpretation of the first terms on the right hand side of this Hamiltonian is straightforward. The first term is the usual kinetic energy with the vector potential, the second is the interaction of our particle with the external field through its magnetic moment. Thirdly we have the potential and rest mass energies. These are followed by the new terms, all of order $1/c^2$, which correct the Pauli equation. We will discuss the physics underlying each of these in turn.

The first new term in the Hamiltonian is the one containing $\mathbf{p}^4$. This correction arises because within a relativistic theory the mass of the electron is a function of its velocity. In classical relativity the kinetic energy is given by equation (1.29) which is expanded to power of $1/c^2$ as shown in equations (1.30) and (1.31). Clearly there is a close similarity between the final term of (1.31b) and the mass–velocity correction in (4.67).

Let us group together the spin-dependent terms in (4.74b) to see the effect of the second correction to the Pauli Hamiltonian:

$$-\frac{e\hbar}{2m}\tilde{\sigma}\cdot\left(\hat{\mathbf{B}} + \frac{1}{2mc^2}(\mathbf{E}\times\hat{\mathbf{p}})\right) \approx -\frac{e\hbar}{2m}\tilde{\sigma}\cdot\left(\hat{\mathbf{B}} + \frac{\mathbf{E}\times\hat{\mathbf{p}}}{mc^2}\right) + \frac{e}{2mc^2}\hat{\mathbf{S}}\cdot(\mathbf{E}\times\hat{\mathbf{v}}) \tag{4.75}$$

The bracketed part of the first term on the right hand side here is the magnetic field felt by the electron in its rest frame, as can be seen immediately from equation (1.58*b*), and the whole term represents the interaction of the electron magnetic moment with that field. Comparison of the second term in (4.75) with (2.71) shows that it takes care of the contribution to the energy from the Thomas precession. Again we see that the relativistic quantum theory has explained on a deeper level a physical effect originally introduced on a semi-classical basis. So together the two spin-dependent terms in (4.74) lead to the usual spin–orbit coupling Hamiltonian, as shown in chapter 2. The fact that it falls out of relativistic quantum mechanics is an achievement that lends great credibility to the theory.

As is well known, spin–orbit coupling is responsible for much of the fine structure found in the spectra of atoms as well as a host of other physical phenomena. Among these are magnetocrystalline anisotropy and much of the dependence of spectroscopies on the polarization of the photons. The underlying reason that this term is so important a correction to non-relativistic quantum mechanics is that it breaks symmetries. We know that angular momentum has to be conserved. In non-relativistic quantum theory spin does not occur, so orbital angular momentum is conserved. Spin does not appear in the Schrödinger equation and so is always, by implication, zero. Hence it is trivially conserved. In relativistic quantum mechanics spin and orbital angular momentum are able to *talk* to one another and exchange angular momentum, and so only total angular momentum is conserved.

The final term in (4.74*b*) is known as the *Darwin* term. This is a contribution to the energy that has no non-relativistic analogue. It comes from the fact that the electron cannot be regarded as a point particle, but in relativistic theory is spread out over a volume of order the cube of the Compton wavelength $(\hbar/mc)^3$; in other words, the *Zitterbewegung* occurs in Dirac theory as well as in Klein–Gordon theory. This will be discussed further in chapter 7.

4.5 An Alternative Formulation of the Dirac Equation

In equations (4.5) and (4.25) we have already presented two alternative forms of the Dirac equation. Here we are going to derive a rather different version. This was first proposed by Feynman and Gell-mann (1958) as the fundamental form of the equation and applied by them to the theory of β-decay. In fact this formulation has not found much favour in condensed matter physics. Nonetheless, we present it here, and in chapter 9 we will apply it in an example.

Our starting point for this derivation is the observation by Feynman and Gell-mann that there is really no need for the Dirac equation to be a four-component equation. It is true that we need a two-component equation to describe the two possible spin states of the particle. However, with a two-component equation we can still allow the energy to be positive or negative, allowing four possible states in all. Let us write down the Dirac equation for a particle in an electromagnetic field.

$$i\hbar\frac{\partial\psi(\mathbf{r},t)}{\partial t} = \left(c\tilde{\boldsymbol{\alpha}}\cdot(\hat{\mathbf{p}}-e\mathbf{A}(\mathbf{r}))+\tilde{\beta}mc^2+V(\mathbf{r})\right)\psi(\mathbf{r},t) \tag{4.76}$$

Now we premultiply this equation by $\tilde{\beta}$ and rearrange it a bit, and we get

$$\left(c\tilde{\gamma}\cdot(\hat{\mathbf{p}}-e\mathbf{A}(\mathbf{r}))+mc^2+\tilde{\beta}e\Phi(\mathbf{r})+\frac{\hbar}{i}\tilde{\beta}\frac{\partial}{\partial t}\right)\psi(\mathbf{r},t)=0 \tag{4.77}$$

where $V(\mathbf{r}) = e\Phi(\mathbf{r})$. It turns out that this derivation is one in which it is much easier to work in terms of 4-vectors and we will do so. Using the definitions (4.24) and the energy–momentum and vector potential 4-vectors defined in chapter 1 we can rewrite the Dirac equation in 4-vector form as

$$\left(c\tilde{\gamma}^\mu(\hat{p}^\mu+eA_\mu)+mc^2\right)\psi(\mathbf{r},t)=0 \tag{4.78}$$

Next we make a definition

$$\psi(\mathbf{r},t) = \frac{1}{mc^2}\left(c\tilde{\gamma}^\mu(\hat{p}^\mu+eA_\mu)-mc^2\right)\chi(\mathbf{r},t)=0 \tag{4.79}$$

Equation (4.79) can be substituted into (4.78), and as mc^2 commutes with everything, the cross terms cancel when we multiply out the brackets so that we are left with

$$\left(c^2\tilde{\gamma}^\mu(\hat{p}^\mu+eA_\mu)\tilde{\gamma}^\nu(\hat{p}^\nu+eA_\nu)-m^2c^4\right)\chi(\mathbf{r},t)=0 \tag{4.80}$$

It is convenient to separate the contributions from $\mu = \nu$ from those with $\mu \neq \nu$ in this equation. Then with the use of (4.8) we have

$$(c^2(\gamma^\mu)^2(p^\mu+eA_\mu)^2+c^2\tilde{\gamma}_\mu\tilde{\gamma}_\nu(p^\mu+eA_\mu)(p^\nu+eA_\nu)-m^2c^4)\chi(\mathbf{r},t)=0 \tag{4.81}$$

The first and third terms here are what we require, the second needs some further manipulation. It is easy to see that (4.81) can be rewritten as

$$\left(c^2(\gamma^\mu)^2(p^\mu+eA_\mu)^2-m^2c^4+ec\tilde{\sigma}_{\mu\nu}F_{\mu\nu}\right)\chi(\mathbf{r},t)=0 \tag{4.82}$$

where

$$\tilde{\sigma}_{\mu\nu}=\frac{i\hbar}{2}(\tilde{\gamma}^\mu\tilde{\gamma}^\nu-\tilde{\gamma}^\nu\tilde{\gamma}^\mu),\qquad F_{\mu\nu}=c\left(\frac{\partial A_\nu}{\partial x^\mu}-\frac{\partial A_\mu}{\partial x^\nu}\right) \tag{4.83}$$

Now the important point about (4.82) is that $\tilde{\gamma}_5$ commutes through all terms in the brackets. So, if we have found the solution $\chi(\mathbf{r},t)$, then we must also have that $\tilde{\gamma}_5\chi(\mathbf{r}, t)$ is a solution. This can only be the case if the four-component wavefunction $\chi(\mathbf{r}, t)$ takes on one of the forms

$$\chi = \begin{pmatrix} a \\ b \\ a \\ b \end{pmatrix}, \qquad \chi = \begin{pmatrix} a \\ b \\ -a \\ -b \end{pmatrix} \tag{4.84}$$

Check for yourself that multiplying each of these forms for χ by $\tilde{\gamma}_5$ gives us back $\pm\chi$. In fact we can select only one of these solutions as that will make the other redundant. We choose the latter for which

$$\tilde{\gamma}_5\chi(\mathbf{r}, t) = \chi(\mathbf{r}, t) \tag{4.85}$$

Now in equation (4.79) we have written ψ in terms of χ. We can write χ in terms of ψ by premultiplying both sides of (4.79) by $\frac{1}{2}(\tilde{I} + \tilde{\gamma}_5)$. Using the commutators of $\tilde{\gamma}_5$ with the other γ-matrices we can easily see that

$$\chi(\mathbf{r}, t) = \frac{1}{2}(\tilde{I} + \tilde{\gamma}_5)\psi(\mathbf{r}, t) \tag{4.86}$$

Now if ψ looks like this

$$\psi = \begin{pmatrix} w \\ x \\ y \\ z \end{pmatrix} \tag{4.87}$$

equation (4.86) tells us that χ must take on the form

$$\chi = \frac{1}{2}\begin{pmatrix} 1 & 0 & -1 & 0 \\ 0 & 1 & 0 & -1 \\ -1 & 0 & 1 & 0 \\ 0 & -1 & 0 & 1 \end{pmatrix}\begin{pmatrix} w \\ x \\ y \\ z \end{pmatrix} = \begin{pmatrix} \frac{1}{2}(w - y) \\ \frac{1}{2}(x - z) \\ -\frac{1}{2}(w - y) \\ -\frac{1}{2}(x - z) \end{pmatrix} = \begin{pmatrix} \phi \\ -\phi \end{pmatrix} \tag{4.88}$$

where ϕ is a two-component wavefunction with the obvious definition. Clearly either the upper or the lower two components of χ are redundant. We could substitute (4.88) directly into (4.82) to get a two-component equation for χ, but it is easier to simplify (4.82) further first.

From the anticommutation relations of the γ-matrices we can write

$$\tilde{\sigma}^{\mu\nu} = i\hbar\tilde{\gamma}^\mu\tilde{\gamma}^\nu \tag{4.89}$$

Multiplying out these matrices and if necessary taking out a factor of $\tilde{\gamma}_5$ using (4.85) we can substitute into (4.82) to find

$$(c^2(\gamma^\mu)^2(p^\mu + eA_\mu)^2 - m^2c^4 + ec\hbar(\tilde{\sigma}_z F_{yx} + \tilde{\sigma}_x F_{zy} + \tilde{\sigma}_y F_{xz})$$
$$+ iec\hbar(\tilde{\sigma}_x F_{xt} + i\tilde{\sigma}_y F_{yt} + i\tilde{\sigma}_z F_{zt}))\chi(\mathbf{r}, t) = 0 \tag{4.90}$$

where in this equation the $\tilde{\sigma}$s have only one subscript as they are the 4×4 Pauli matrices. Using the definitions of $F_{\mu\nu}$ from chapter 1 we can write this as

$$\left(c^2(\gamma^\mu)^2(p^\mu + eA_\mu)^2 - m^2c^4 + ec\hbar\tilde{\sigma} \cdot (c\mathbf{B} + i\mathbf{E})\right)\chi(\mathbf{r}, t) = 0 \qquad (4.91)$$

Here the upper two components of the equation are completely decoupled from the lower two, and so we can use equation (4.22) to write it in two-component form as

$$\left(-c^2(\sigma^\mu)^2(p^\mu + eA_\mu)^2 - m^2c^4 + ec\hbar\tilde{\sigma} \cdot (c\mathbf{B} + i\mathbf{E})\right)\phi(\mathbf{r}, t) = 0 \qquad (4.92)$$

Finally let us put this back into the more familiar 3-vector form

$$\left(c^2(\hat{\mathbf{p}} - e\mathbf{A})^2 + m^2c^4 - ec\hbar\tilde{\sigma} \cdot (c\mathbf{B} + i\mathbf{E}) - \left(i\hbar\frac{\partial}{\partial t} - V(\mathbf{r})\right)^2\right)\phi(\mathbf{r}, t) = 0$$
$$(4.93)$$

This equation determines $\phi(\mathbf{r}, t)$ and hence from equation (4.88) it determines $\chi(\mathbf{r}, t)$. If we want to find the wavefunction in the original representation of the Dirac equation we can find it using equation (4.79). Note that if we had taken the solution $\tilde{\gamma}_5\chi = -\chi$ above, there would have been a slight variation in the signs in this section, but the physics contained in the resulting equation would have been identical. Finally we see that (4.93) is the Klein–Gordon equation, taking care of relativistic effects, plus the additional terms in the electric and magnetic field which couple to the spin. In other words, the effects on the wavefunction of relativity and spin are divided into separate terms in the Hamiltonian.

Equation (4.93) represents an alternative and rather unusual form of the Dirac equation. It is a two-component equation and has been derived in this form without any approximations. The cost of making it a two-component equation is that it is now second order in both time and space derivatives and so, in principle, is more difficult to solve than the original form of the Dirac equation. Equation (4.93) is also unusual in that the electric field appears explicitly in the equation rather than solely in the form of a potential. Furthermore, the coefficient of the electric field is defined as being complex. We will not discuss this form of the Dirac equation further here, but we will use it in chapter 9 to examine the behaviour of an electron in crossed electric and magnetic fields. The present form of the Dirac equation is well-suited to this problem as it leads easily to a relativistic description of Landau levels (McMurry 1993).

4.6 Probabilities and Currents

Now that we have explored the Dirac equation and seen that it has the correct limiting behaviour we can begin to have some faith in it. It is time to derive an expression for the probability density in terms of the wavefunction (Greiner 1990).

First in this section let us introduce some more notation. We will think of a 4-vector wavefunction $\psi(\mathbf{r})$. This is a column vector with four elements. The wavefunction $\psi^*(\mathbf{r})$ is also a column vector with four elements. Each element is the complex conjugate of the equivalent element in $\psi(\mathbf{r})$ and so $\psi^*(\mathbf{r})$ is the complex conjugate of $\psi(\mathbf{r})$. Next we define $\psi^T(\mathbf{r})$ which is the transpose of $\psi(\mathbf{r})$ so it is a row vector with four elements. Finally we define $\psi^\dagger(\mathbf{r})$ which is the transpose of $\psi^*(\mathbf{r})$.

Let us rewrite the Dirac equation in an electromagnetic field

$$i\hbar\frac{\partial}{\partial t}\psi(\mathbf{r},t) = (c\tilde{\boldsymbol{\alpha}}\cdot(\hat{\mathbf{p}} - e\mathbf{A}(\mathbf{r})) + V(\mathbf{r}) + \tilde{\beta}mc^2)\psi(\mathbf{r},t) \qquad (4.94)$$

Now we take the conjugate transpose of this equation and we have

$$-i\hbar\frac{\partial}{\partial t}\psi^\dagger(\mathbf{r},t) = (-c\hat{\mathbf{p}} - ec\mathbf{A}(\mathbf{r}))\cdot\psi^\dagger(\mathbf{r},t)\tilde{\boldsymbol{\alpha}} + \psi^\dagger(\mathbf{r},t)(mc^2\tilde{\beta} + V(\mathbf{r})) \quad (4.95)$$

Note that the 4×4 matrices are to the right of the wavefunctions which are row vectors in this equation. In deriving this we have used the facts that $\hat{\mathbf{p}}$ is a purely complex operator and the Dirac matrices satisfy

$$\tilde{\boldsymbol{\alpha}}^\dagger = \tilde{\boldsymbol{\alpha}}, \qquad \tilde{\beta}^\dagger = \tilde{\beta} \qquad (4.96)$$

The next step is to premultiply (4.94) by $\psi^\dagger(\mathbf{r},t)$ and to postmultiply (4.95) by $\psi(\mathbf{r},t)$. Then we subtract the latter equation from the former. The result of this can be seen immediately as

$$i\hbar\left(\psi^\dagger\frac{\partial\psi}{\partial t} + \frac{\partial\psi^\dagger}{\partial t}\psi\right) = \frac{\hbar c}{i}\left(\psi^\dagger\tilde{\boldsymbol{\alpha}}\cdot\nabla\psi + (\nabla\psi^\dagger)\cdot\tilde{\boldsymbol{\alpha}}\psi\right) \qquad (4.97)$$

Here we have dropped the $\mathbf{r}$ and t dependence of ψ for clarity. Now $\tilde{\boldsymbol{\alpha}}$ is just a matrix with no dependence on spatial variables, so both sides of (4.97) can be regarded as examples of the product rule of differentiation. Hence (4.97) can be written more succinctly as

$$\frac{\partial}{\partial t}(\psi^\dagger\psi) = -\nabla\cdot(\psi^\dagger c\tilde{\boldsymbol{\alpha}}\psi) \qquad (4.98)$$

This is just a continuity equation of the form

$$\frac{\partial\rho}{\partial t} = -\nabla\cdot\mathbf{j} \qquad (4.99)$$

if we identify the probability and current densities as

$$\rho = \psi^\dagger\psi, \qquad\qquad \mathbf{j} = c\psi^\dagger\tilde{\boldsymbol{\alpha}}\psi \qquad (4.100)$$

Now we have an expression for the probability density, and to make sense its integral over all space must be equal to one. This tells us how to normalize a four-component wavefunction. Using the notation of (4.10)

$$\int_{-\infty}^{\infty} \psi^\dagger \psi d\mathbf{r} = \int_{-\infty}^{\infty} (\psi_1^* \psi_1 + \psi_2^* \psi_2 + \psi_3^* \psi_3 + \psi_4^* \psi_4) d\mathbf{r} = 1 \qquad (4.101)$$

Note that the interpretation of the continuity equation (4.99) is different to the spin-zero case. Here we are able to define ρ as a probability density because $\psi^* \psi$ is, by definition, always positive, and therefore equation (4.99) says that the rate of change of probability density in a small volume of space is equal to the rate at which probability leaves that volume. In the Klein–Gordon case (equation (3.31)) ρ could only be interpreted as a charge density. With the definitions above, an expectation value can be written in analogy to the non-relativistic case as

$$\langle \hat{A} \rangle = \int \psi^\dagger \hat{A} \psi dV \qquad (4.102)$$

So, formally, Dirac quantum mechanics looks very similar to the non-relativistic theory.

4.7 Gordon Decomposition

We have derived expressions for the probability density and current for a spin-zero particle in equations (3.17), and for a spin-1/2 particle in equations (4.100). In section 3.2 we were able to interpret the time derivatives in terms of Lorentz–Fitzgerald contraction of a volume increasing the density of particles. However, the same is not apparent in (4.100). In the spin-1/2 case we have expressions for the probability and current densities that look remarkably like their non-relativistic counterparts. This is not very satisfactory because, at first sight, there is no reason why these quantities should differ from the spin-zero case. We would have expected the interpretation of the probability and current here to be the same as in Chapter 3. We will explain this now by describing, in detail, how to perform a Gordon decomposition, named after its discoverer, W. Gordon (1928) (yes, this is the Gordon who appears in the Klein–Gordon equation, but not in Clebsch–Gordan coefficients). Gordon decomposition is a method by which we can separate the probability and current densities into an outer part dependent on the usual parameters and a part dependent on the internal state of the particle. The outer part looks like equations (3.51) which we found for the Klein–Gordon equation. It differs only in the 4-vector nature of the wavefunction. Furthermore there is an interesting (really!) interpretation of the internal part (Baym 1967).

For a particle in an electromagnetic field, the Dirac equation and its adjoint are given by equations (4.94) and (4.95). We may premultiply (4.94), and postmultiply (4.95), by $\tilde{\beta}$. Then, recalling the property of the

β-matrix $\tilde{\beta}^\dagger = \tilde{\beta}^{-1} = \tilde{\beta}$, we rearrange the resulting equations to give

$$\psi(\mathbf{r}, t) = \frac{1}{mc^2} \left(\tilde{\beta} \left(i\hbar \frac{\partial}{\partial t} - V(\mathbf{r}) \right) \psi(\mathbf{r}, t) - c\tilde{\beta}\tilde{\alpha} \cdot (\hat{\mathbf{p}} - e\mathbf{A})\psi(\mathbf{r}, t) \right) \quad (4.103a)$$

$$\psi^\dagger(\mathbf{r}, t) = \frac{1}{mc^2} \left(\left(-i\hbar \frac{\partial}{\partial t} - V(\mathbf{r}) \right) \psi^\dagger(\mathbf{r}, t)\tilde{\beta} + c(-\hat{\mathbf{p}} - e\mathbf{A})\psi^\dagger(\mathbf{r}, t) \cdot \tilde{\beta}\tilde{\alpha} \right)$$
$$(4.103b)$$

where we have made use of the fact that the α-matrices are their own adjoint, and of the anticommutation relations of equation (4.8). Next, let us take (4.100) for the probability density and use (4.103) to decompose it as

$$\rho = \psi^\dagger \psi = \tfrac{1}{2}\psi^\dagger \psi + \tfrac{1}{2}\psi^\dagger \psi$$

$$= \frac{1}{2mc^2}\psi^\dagger \left(\tilde{\beta} \left(i\hbar \frac{\partial}{\partial t} - V(\mathbf{r}) \right) \psi - c\tilde{\beta}\tilde{\alpha} \cdot (\hat{\mathbf{p}} - e\mathbf{A})\psi \right)$$

$$+ \frac{1}{2mc^2} \left(\left(-i\hbar \frac{\partial}{\partial t} - V(\mathbf{r}) \right) \psi^\dagger \tilde{\beta} + c(-\hat{\mathbf{p}} - e\mathbf{A})\psi^\dagger \cdot \tilde{\beta}\tilde{\alpha} \right) \psi$$

$$= \frac{1}{2mc^2} \left(\overline{\psi}^\dagger \left(i\hbar \frac{\partial}{\partial t} - V(\mathbf{r}) \right) \psi + \left(-i\hbar \frac{\partial}{\partial t} - V(\mathbf{r}) \right) \overline{\psi}^\dagger \psi \right)$$

$$+ \frac{1}{2mc^2} \left(-\overline{\psi}^\dagger c\tilde{\alpha} \cdot (\hat{\mathbf{p}} - e\mathbf{A})\psi + c(-\hat{\mathbf{p}} - e\mathbf{A}) \cdot \overline{\psi}^\dagger \tilde{\alpha}\psi \right) = \rho_{\text{out}} + \rho_{\text{in}}$$
$$(4.104)$$

In order to fit this equation onto the page we have dropped the $\mathbf{r}$ and t dependence of ψ, and we have defined

$$\overline{\psi}(\mathbf{r}, t) = \tilde{\beta}\psi(\mathbf{r}, t), \qquad \overline{\psi}^\dagger(\mathbf{r}, t) = \psi^\dagger(\mathbf{r}, t)\tilde{\beta} \qquad (4.105)$$

The first term in (4.104) is in the form we want it. The second term can be simplified a bit. The term in the vector potential cancels and we are left with

$$\rho_{\text{in}} = \frac{1}{2mc} \left(-\overline{\psi}^\dagger(\mathbf{r}, t)\tilde{\alpha} \cdot \frac{\hbar}{i}\nabla\psi(\mathbf{r}, t) - \frac{\hbar}{i}\nabla \cdot \overline{\psi}^\dagger(\mathbf{r}, t)\tilde{\alpha}\psi(\mathbf{r}, t) \right)$$

$$= \frac{\hbar}{2mc}\nabla \cdot \left(\overline{\psi}^\dagger(\mathbf{r}, t)i\tilde{\alpha}\psi(\mathbf{r}, t) \right) \qquad (4.106)$$

So our final expression for the probability density is

$$\rho = \rho_{\text{out}} + \rho_{\text{in}} = \frac{1}{2mc^2}\overline{\psi}^\dagger(\mathbf{r}, t) \left(-i\hbar \frac{\partial}{\partial t} - V(\mathbf{r}) \right) \psi(\mathbf{r}, t)$$

$$+ \frac{1}{2mc^2} \left(-i\hbar \frac{\partial}{\partial t} - V(\mathbf{r}) \right) \overline{\psi}^\dagger(\mathbf{r}, t)\psi(\mathbf{r}, t) + \frac{\hbar}{2mc}\nabla \cdot \left(\overline{\psi}^\dagger(\mathbf{r}, t)i\tilde{\alpha}\psi(\mathbf{r}, t) \right)$$
$$(4.107)$$

A similar analysis can be gone through for the current. We will not reproduce the rather tedious algebra here, but will just quote the result:

$$\mathbf{j} = \mathbf{j}_{out} + \mathbf{j}_{in} = \frac{1}{2m}\left(\overline{\psi}^\dagger(\mathbf{r},t)(\hat{\mathbf{p}} - e\mathbf{A})\psi(\mathbf{r},t) - (\hat{\mathbf{p}} + e\mathbf{A})\overline{\psi}^\dagger(\mathbf{r},t)\psi(\mathbf{r},t)\right)$$
$$+ \frac{\hbar}{2m}\left(\nabla \times \overline{\psi}^\dagger(\mathbf{r},t)\tilde{\boldsymbol{\sigma}}\psi(\mathbf{r},t) - \frac{1}{c}\frac{\partial}{\partial t}\overline{\psi}^\dagger(\mathbf{r},t)i\tilde{\boldsymbol{\alpha}}\psi(\mathbf{r},t)\right)$$
$$(4.108)$$

Equations (4.107) and (4.108) are in an instructive form. The expressions for ρ_{out} and $\mathbf{j}_{out}$ are the same as the equivalent expressions found from the Klein–Gordon equation except that ψ^* has been replaced by $\overline{\psi}^\dagger$. Compare them with equations (3.52) and (3.17b). So we can interpret ρ_{out} and $\mathbf{j}_{out}$ as the probability and current densities due to the fact that the particle is described by relativistic quantum mechanics. This leaves ρ_{in} and $\mathbf{j}_{in}$ which must be associated with internal degrees of freedom of the particle. It is easy to see that these two can be written in the following form:

$$\rho_{in} = -\nabla \cdot \mathbf{P}, \qquad \mathbf{j}_{in} = \nabla \times \mathbf{M}_e + \frac{d\mathbf{P}}{dt} \qquad (4.109)$$

where

$$\mathbf{M}_e = \frac{\hbar}{2m}\overline{\psi}^\dagger(\mathbf{r},t)\tilde{\boldsymbol{\sigma}}\psi(\mathbf{r},t) \rightarrow \mathbf{M} = e\mathbf{M}_e = \mu_B\overline{\psi}^\dagger(\mathbf{r},t)\tilde{\boldsymbol{\sigma}}\psi(\mathbf{r},t) \qquad (4.110a)$$

$$\mathbf{P} = -\frac{\hbar}{2mc}\overline{\psi}^\dagger(\mathbf{r},t)i\tilde{\boldsymbol{\alpha}}\psi(\mathbf{r},t) \qquad (4.110b)$$

Now, these equations can be interpreted in a straightforward manner. Equation (4.110a) is obviously some sort of average value of the magnetic moment density at point $\mathbf{r}$. There is no time dependence in (4.110a), so this is the magnetic moment density for the particle in its rest frame. If we look at the same particle from a moving frame we will observe a combination of magnetic moment and electric moment, the latter described by $e\mathbf{P}$. Finally we note from (4.109) that ρ_{in} and $\mathbf{j}_{in}$ obey a continuity equation themselves and therefore so must ρ_{out} and $\mathbf{j}_{out}$. So the conclusion of all this is that the 4-vector nature of the Dirac equation and its solutions, which we have been obliged to introduce, is leading to some polarizability that is somehow intrinsic to the particle.

We will evaluate these quantities for a plane wave solution of the Dirac equation in the following chapter.

4.8 Forces and Fields

We will not dwell very long on the Dirac equation in electromagnetic fields here as we will be doing that throughout the book. However, it is worthwhile examining it briefly. Looking ahead we can show that some

familiar equations from classical electrodynamics arise naturally from the
Dirac Hamiltonian. The consistency this implies with earlier theories is a
convincing endorsement of the validity of the Dirac theory.

Consider the Dirac equation (4.94). This is an equation of the form

$$i\hbar\frac{\partial}{\partial t}\psi(\mathbf{r},t) = \hat{H}\psi(\mathbf{r},t) \tag{4.111}$$

We can work out the operator for the time rate of change of a quantity
from Heisenberg's equation of motion (equation (3.80)). Let us use this to
work out the time rate of change of some quantities when our Hamiltonian
is (4.94). We know that the usual momentum position commutator is

$$[\hat{p}_x, x] = \frac{\hbar}{i} \tag{4.112}$$

so we can evaluate the rate of change of position operator as

$$\frac{\widehat{d\mathbf{r}}}{dt} = \mathbf{v} = \frac{i}{\hbar}[\hat{H}, \hat{\mathbf{r}}] = c\tilde{\boldsymbol{\alpha}} \tag{4.113a}$$

We can also find the rate of change of momentum. Recalling that both
$V(\mathbf{r})$ and $\mathbf{A}(\mathbf{r})$ depend on spatial variables we can easily see that

$$\frac{d\mathbf{p}}{dt} = \frac{i}{\hbar}[\hat{H}, \hat{\mathbf{p}}] = -\nabla V(\mathbf{r}) + ec\nabla(\tilde{\boldsymbol{\alpha}} \cdot \mathbf{A}(\mathbf{r})) \tag{4.113b}$$

and finally

$$\frac{d\mathbf{A}}{dt} = \frac{\partial \mathbf{A}}{\partial t} + \frac{i}{\hbar}[\hat{H}, \hat{\mathbf{A}}] = \frac{\partial \mathbf{A}}{\partial t} + c(\tilde{\boldsymbol{\alpha}} \cdot \nabla)\mathbf{A} \tag{4.113c}$$

We can combine (4.113b) and (4.113c), and use Newton's second law to
evaluate an expression for the force felt by our particle:

$$\mathbf{F} = \frac{d}{dt}(\mathbf{p} - e\mathbf{A}) = -\nabla V(\mathbf{r}) + ec\nabla(\tilde{\boldsymbol{\alpha}} \cdot \mathbf{A}(\mathbf{r})) - e\frac{\partial \mathbf{A}}{\partial t} - ec(\tilde{\boldsymbol{\alpha}} \cdot \nabla)\mathbf{A}$$

$$= e\left(-\nabla\Phi - \frac{\partial \mathbf{A}}{\partial t}\right) + ec\left(\nabla(\tilde{\boldsymbol{\alpha}} \cdot \mathbf{A}) - \tilde{\boldsymbol{\alpha}} \cdot \nabla\mathbf{A}\right)$$

$$= e\mathbf{E} + e(c\tilde{\boldsymbol{\alpha}} \times (\nabla \times \mathbf{A})) = e\mathbf{E} + e\mathbf{v} \times \mathbf{B} \tag{4.114}$$

where we have used (4.113a) to substitute in $\mathbf{v}$, and equation (1.49) for $\mathbf{B}$,
and we have used the vector identity

$$\mathbf{A} \times (\mathbf{B} \times \mathbf{C}) = \mathbf{B}(\mathbf{A} \cdot \mathbf{C}) - \mathbf{C}(\mathbf{A} \cdot \mathbf{B}) \tag{4.115}$$

In (4.114) we have also used the definition below equation (4.49) for Φ
and we have used (1.50) for $\mathbf{E}$.

Obviously, the last line of equation (4.114) is the familiar expression
for the Lorentz force felt by a particle of charge e due to electric field
$\mathbf{E}$ and magnetic field $\mathbf{B}$. So we have shown that the response to an

electromagnetic field of a spin-1/2 particle is consistent with the classical theory – a pleasing result. A point to note about equation (4.114) is that a particle in an electric field experiences an increase in momentum. In relativistic theory, of course, this is not the same as an increase in velocity. A particle in an electric field for a *long* time will have a velocity that approaches the speed of light asymptotically. One might expect, therefore, that the expectation value of the velocity of the particle would be equal to c.

4.9 Gauge Invariance and the Dirac Equation

We have said that the Dirac equation in Hamiltonian form is given by

$$\left(c\tilde{\alpha} \cdot \left(\frac{\hbar}{i} \nabla - e\mathbf{A}(\mathbf{r}) \right) + \tilde{\beta} mc^2 + V(\mathbf{r}) \right) \psi(\mathbf{r}, t) = W\psi(\mathbf{r}, t) \tag{4.116}$$

However, in equation (1.51), we have defined gauge transformations on the potentials, which we have said do not change the fields and hence do not affect our calculation of any observable quantities. But it is the potentials, not the fields, that appear in the Dirac equation. A natural question to ask is what effect a gauge transformation will have on the wavefunctions and energy that appear in (4.116) (Corinaldesi and Strocchi 1963).

We will assume we have a steady state system with no time dependence. Then the scalar potential is unaffected by a gauge transformation of the type (1.51). However, the new vector potential is

$$\mathbf{A}'(\mathbf{r}) = \mathbf{A}(\mathbf{r}) - \nabla\theta(\mathbf{r}) \tag{4.117}$$

where θ is any continuously differentiable function. Mathematically the question we want to ask is how the solution to

$$\left(c\tilde{\alpha} \cdot \left(\frac{\hbar}{i} \nabla - e\mathbf{A}'(\mathbf{r}) \right) + \tilde{\beta} mc^2 + V(\mathbf{r}) \right) \psi'(\mathbf{r}, t) = W\psi'(\mathbf{r}, t) \tag{4.118}$$

is related to the solution of (4.116) if $\mathbf{A}'(\mathbf{r})$ and $\mathbf{A}(\mathbf{r})$ are related via equation (4.117). We have insisted that the energy eigenvalue W be the same for both equations. Let us define the answer to this, substitute it into (4.118) and show that it leads us back to (4.116). We choose

$$\psi'(\mathbf{r}, t) = e^{-ie\theta/\hbar} \psi(\mathbf{r}, t) \tag{4.119}$$

Now, operating on this with the momentum operator gives

$$\hat{\mathbf{p}}\psi'(\mathbf{r}, t) = e^{-ie\theta/\hbar} \hat{\mathbf{p}}\psi(\mathbf{r}, t) - e(\nabla\theta)e^{-ie\theta/\hbar} \psi(\mathbf{r}, t) \tag{4.120}$$

Hence

$$(\hat{\mathbf{p}} + e\nabla\theta)\psi'(\mathbf{r}, t) = e^{-ie\theta/\hbar} \hat{\mathbf{p}}\psi(\mathbf{r}, t) \tag{4.121}$$

Now, using (4.117) we can write (4.118) as

$$\left(c\tilde{\alpha}\cdot(\hat{\mathbf{p}}-e\mathbf{A}(\mathbf{r})+e\nabla\theta)+\tilde{\beta}mc^2+V(\mathbf{r})\right)\psi'(\mathbf{r},t)=W\psi'(\mathbf{r},t) \qquad (4.122)$$

Using (4.121) and (4.119) we can write this as

$$e^{-ie\theta/\hbar}\left(c\tilde{\alpha}\cdot\left(\frac{\hbar}{i}\nabla-e\mathbf{A}(\mathbf{r})\right)+\tilde{\beta}mc^2+V(\mathbf{r})\right)\psi(\mathbf{r},t)=We^{-ie\theta/\hbar}\psi(\mathbf{r},t)$$

$$(4.123)$$

where we have used the fact that $e^{-ie\theta/\hbar}$ commutes through all terms except the one containing the momentum operator directly. Now the exponentials on each side of this equation cancel and we regain (4.116).

This section proves that if we have a solution to the Dirac equation $\psi(\mathbf{r},t)$ with energy W, then, after an electromagnetic gauge transformation of the type (4.117), the solution of the new Dirac equation which has the same energy is related to the old solution by equation (4.119). The effect of the gauge transformation is to multiply the wavefunction by a phase factor. Therefore $\psi'^\dagger\psi'=\psi^\dagger\psi$ so the probability density, and hence observables, are not affected by the gauge transformation.

In this chapter the aim has been to familiarize readers with the Dirac equation and to present some arguments that lead us to have faith in it. The remainder of the book examines the Dirac equation and its solutions under a variety of circumstances.

4.10 Problems

(1) The neutrino is believed to be a massless particle. Show that a 2×2 equation is sufficient to describe such a particle and 'derive' two possible such equations in terms of the Pauli spin matrices. The neutrino spin is observed experimentally to be always directed antiparallel to its motion whereas the spin of the antineutrino is always directed parallel to its motion. Show that one of your equations (the Weyl equation) describes this and the other does not. What does this tell you?

(2) Verify that equations (4.12) and (4.13) satisfy equations (4.8). Also verify equations (4.14), (4.20) and (4.21).

(3) The Dirac matrices presented in this chapter are not unique. The Majorana representation of the Dirac matrices is one in which the α-matrices are all real and $\tilde{\beta}$ is purely imaginary. Write down this set of matrices and prove that they have all the necessary properties. Find the matrix $\tilde{S}$ which transforms the matrices from the standard representation to the Majorana representation using

$$\tilde{\alpha}_M=\tilde{S}\tilde{\alpha}\tilde{S}^{-1}, \qquad \tilde{\beta}_M=\tilde{S}\tilde{\beta}\tilde{S}^{-1}$$

(4) Derive the Lorentz transformations for the Dirac γ-matrices.

(5) Consider the matrices

$$\tilde{P}_\pm = \frac{1}{2}(\tilde{I} \pm \tilde{\beta})$$

Show that the determinant of $\tilde{P}_\pm$ is zero in any representation. Suggest an application for these matrices.

(6) The γ-matrices (including $\tilde{\gamma}_5$) all anticommute with one another. Are there any other matrices that anticommute with all of them?

5
Free Particles/Antiparticles

As in non-relativistic quantum theory, the simplest problem to solve in relativistic quantum theory is that of describing a free particle. Much can be learned from this case which will be of use in interpreting the topics covered in later chapters. Furthermore, some of the most profound features of relativistic quantum theory are well illustrated by the free particle, so it is a very instructive problem to consider in detail. Another advantage of the free-particle problem is that the mathematics involved in solving it is not nearly as involved as that necessary for solving problems involving particles under the influence of potentials.

Firstly we shall look briefly at the formulae for the current and probability density, then we shall go on to examine the solutions of the Dirac equation and investigate their behaviour. This leads us to a discussion of spin, the Pauli limit, and the relativistic spin operator. Next we consider the negative energy solutions and show how relativistic quantum theory predicts the existence of antiparticles. Some of the dilemmas this concept introduces and their resolution are discussed. At the end of the chapter we go back to the Klein paradox, and examine it for an incident spin-1/2 particle. We find that the Klein paradox exists for Dirac particles in exactly the same way as it existed for Klein–Gordon particles and has the same resolution and interpretation.

5.1 Wavefunctions, Densities and Currents

The Dirac equation for a particle in zero potential can be written

$$i\hbar\frac{\partial\psi}{\partial t} = \hat{H}\psi = W\psi \tag{5.1}$$

with

$$\hat{H} = c\tilde{\boldsymbol{\alpha}}\cdot\hat{\mathbf{p}} + \tilde{\beta}mc^2 \tag{5.2}$$

130

We know that the solution to (5.1) must be plane-wave-like, and that these solutions must also be eigenfunctions of the momentum operator

$$\hat{\mathbf{p}}\psi = \mathbf{p}\psi \tag{5.3}$$

Here $\mathbf{p}$ is just the momentum vector, a set of three numbers. We shall return to the momentum operator later. To describe a free particle, we look for a solution of (5.1) of the form

$$\psi = U(\mathbf{p})\exp(i(\mathbf{p}\cdot\mathbf{r} - Wt)/\hbar) \tag{5.4}$$

where $U(\mathbf{p})$ is a four-dimensional quantity. Before going on to look at the consequences of this solution, let us consider the probability density and current. Let us calculate the Gordon decomposed probability density (Gordon 1928) and current according to equations (4.107) and (4.108) for the wavefunction of equation (5.4). In fact, because of the exponential factors, $\mathbf{M}$ and $\mathbf{P}$ of equations (4.110) are constants, so we find

$$\rho_{in} = 0, \qquad\qquad \mathbf{j}_{in} = 0 \tag{5.5}$$

The outer solutions can be evaluated straightforwardly and we have

$$\rho_{out} = \frac{W}{mc^2} \tag{5.6a}$$

where we insist that W is the positive root of $W^2 = p^2c^2 + m^2c^4$ and

$$\mathbf{j}_{out} = \frac{\mathbf{p}c^2}{W}\rho_{out} = \mathbf{v}\rho_{out} \tag{5.6b}$$

Here we have solutions that are completely analogous to the Klein–Gordon case and $\mathbf{p}c^2/W$ is the well-known expression for velocity in relativistic mechanics of equation (1.26).

5.2 Free-Particle Solutions

We already know the solution to the relativistic free-particle problem. It is written down in equation (5.4). In this section we substitute this into the Dirac equation to find out what we can about $U(\mathbf{p})$, and what constraints must be placed upon U and $\mathbf{p}$ to ensure that the Dirac equation is satisfied. We are going to do this for a very specific case. A free particle must be moving with a constant velocity along a straight path. Here we will define our reference frame such that the particle is moving in the xy-plane. The z-axis is perpendicular to the direction of motion. There are two reasons for doing this – firstly, it simplifies the mathematics, and secondly, it is instructive later on to have the z-axis decoupled from the motion when we discuss the spin of the particle (Feynman 1962, or any other book containing a chapter on relativistic quantum theory).

We will separate the solution (5.4) into time-dependent and time-independent parts as described by equation (4.10). The time-dependent part is the usual complex exponential, but let us evaluate the details of the time-independent part.

$$\begin{pmatrix} \psi_1 \\ \psi_2 \\ \psi_3 \\ \psi_4 \end{pmatrix} = \begin{pmatrix} U_1 \\ U_2 \\ U_3 \\ U_4 \end{pmatrix} \exp(i\mathbf{p} \cdot \mathbf{r}/\hbar) \tag{5.7}$$

Let us substitute this into the Hamiltonian form of the Dirac equation.

$$(c\tilde{\boldsymbol{\alpha}} \cdot \hat{\mathbf{p}} + \tilde{\beta}mc^2 - W)\psi = 0 \tag{5.8}$$

Now

$$\hat{\mathbf{p}}\psi = \frac{\hbar}{i}\nabla\psi = \frac{\hbar}{i}\nabla U(\mathbf{p})\exp(i\mathbf{p} \cdot \mathbf{r}/\hbar) = \mathbf{p}U(\mathbf{p})\exp(i\mathbf{p} \cdot \mathbf{r}/\hbar) = \mathbf{p}\psi \tag{5.9}$$

so each time we apply this operator we bring down a factor of the appropriate component of $\mathbf{p}$. Now let us put this into the Dirac equation. Writing it out explicitly in matrix form, with $p_\pm = p_x \pm ip_y$,

$$\begin{pmatrix} mc^2 - W & 0 & 0 & cp_- \\ 0 & mc^2 - W & cp_+ & 0 \\ 0 & cp_- & -mc^2 - W & 0 \\ cp_+ & 0 & 0 & -mc^2 - W \end{pmatrix} \begin{pmatrix} U_1 \\ U_2 \\ U_3 \\ U_4 \end{pmatrix} e^{i\mathbf{p}\cdot\mathbf{r}/\hbar} = 0 \tag{5.10a}$$

or equivalently

$$(mc^2 - W)U_1 + c(p_x - ip_y)U_4 = 0 \tag{5.10b}$$

$$(mc^2 - W)U_2 + c(p_x + ip_y)U_3 = 0 \tag{5.10c}$$

$$c(p_x - ip_y)U_2 - (mc^2 + W)U_3 = 0 \tag{5.10d}$$

$$c(p_x + ip_y)U_1 - (mc^2 + W)U_4 = 0 \tag{5.10e}$$

We can combine (5.10b) and (5.10e) to obtain

$$\frac{U_1}{U_4} = \frac{c(p_x - ip_y)}{W - mc^2} = \frac{W + mc^2}{c(p_x + ip_y)} \tag{5.11}$$

Taking the middle and last part of this equation we can multiply through by $(W - mc^2)c(p_x + ip_y)$ to obtain

$$W^2 = p^2c^2 + m^2c^4 \tag{5.12}$$

This is the familiar expression for the energy of a free particle in classical relativity (equation (1.19)). It is not a surprise that we found this. It tells us that, provided the Dirac equation separates into time- and space-dependent parts as discussed earlier, we must choose $\mathbf{p}$ to satisfy (5.12) for the total energy. If we had used (5.10c and d) instead we would also have come up with this expression.

There are two linearly independent solutions of the Dirac equation with the same energy. This is obvious from the fact that in $(5.10b - e)$ we only have conditions on pairs of the Us; U_2 and U_3 are decoupled from U_1 and U_4. Note that this has happened for two reasons, firstly because of the assumed form of the wavefunction (5.4) and secondly because we have limited our particle to move in the xy-plane. Had we allowed the particle to move in the z-direction as well, the mathematics would have become a little more difficult. As it is we can write each of our Us with two components equal to zero. Let us label the two independent solutions with $\uparrow$ and $\downarrow$ superscripts:

$$\psi^\uparrow = U^\uparrow(\mathbf{p})\exp(i(\mathbf{p}\cdot\mathbf{r} - Wt)/\hbar) \propto \begin{pmatrix} W + mc^2 \\ 0 \\ 0 \\ c(p_x + ip_y) \end{pmatrix} \exp(i(\mathbf{p}\cdot\mathbf{r} - Wt)/\hbar)$$

$$(5.13a)$$

$$\psi^\downarrow = U^\downarrow(\mathbf{p})\exp(i(\mathbf{p}\cdot\mathbf{r} - Wt)/\hbar) \propto \begin{pmatrix} 0 \\ W + mc^2 \\ c(p_x - ip_y) \\ 0 \end{pmatrix} \exp(i(\mathbf{p}\cdot\mathbf{r} - Wt)/\hbar)$$

$$(5.13b)$$

Of course, these solutions are unnormalized at this stage.

Clearly these eigenfunctions of the Hamiltonian are also eigenfunctions of the momentum operator. As an example of this let us consider the x-component

$$\hat{p}_x\psi^\uparrow = \frac{\hbar}{i}\frac{\partial}{\partial x} \begin{pmatrix} W + mc^2 \\ 0 \\ 0 \\ c(p_x + ip_y) \end{pmatrix} \exp(i(\mathbf{p}\cdot\mathbf{r} - Wt)/\hbar) \qquad (5.14)$$

As x only appears in the exponential, doing the trivial differentiation brings down a factor $ip_x/\hbar$ and it is easy to see that we are left with

$$\hat{p}_x\psi^\uparrow = p_x\psi^\uparrow \qquad (5.15)$$

and the same result follows if we apply the momentum operator to $\psi^\downarrow$.

5.3 Free-Particle Spin

In the previous section we have found the two linearly independent solutions of the Dirac equation in free space. There must exist some operator that will distinguish between them (Feynman 1962). This operator must commute with the momentum operator. The astute reader will already have surmised what this operator represents. It is the spin orientation. Because our particle moves in the xy-plane, we can take the axis of quantization as the z-direction and the spin and motion are, to some extent, decoupled from one another. The total spin of our particle is $1/2$, but the z-component of spin can take on two possible values. It is straightforward to find the eigenvalues of the z-component of the spin operator

$$\hat{S}_z = \frac{\hbar}{2}\tilde{\sigma}_z = \frac{\hbar}{2}\begin{pmatrix} 1 & 0 & 0 & 0 \\ 0 & -1 & 0 & 0 \\ 0 & 0 & 1 & 0 \\ 0 & 0 & 0 & -1 \end{pmatrix} \tag{5.16}$$

Now, we are going to consider this operator acting on the wavefunctions (5.13) in the particle's rest frame, i.e. when $p_x = p_y = 0$ and as we are restricted to the xy-plane $\mathbf{p} = 0$. Furthermore, in the rest frame $W = \pm mc^2$. We must have $W = +mc^2$ in this case because otherwise the wavefunctions (5.13) would vanish in the particle's rest frame. So equations (5.13) become

$$\psi^\uparrow = \begin{pmatrix} 2mc^2 \\ 0 \\ 0 \\ 0 \end{pmatrix} e^{-iWt/\hbar} \tag{5.17a}$$

$$\psi^\downarrow = \begin{pmatrix} 0 \\ 2mc^2 \\ 0 \\ 0 \end{pmatrix} e^{-iWt/\hbar} \tag{5.17b}$$

These wavefunctions both represent particles (as we have taken $W = +mc^2$), but we can differentiate between them by operating with (5.16) on each. If we do that we obtain

$$\hat{S}_z\psi^\uparrow = \frac{\hbar}{2}\tilde{\sigma}_z\begin{pmatrix} 2mc^2 \\ 0 \\ 0 \\ 0 \end{pmatrix} e^{-iWt/\hbar} = \frac{\hbar}{2}\psi^\uparrow$$

$$\hat{S}_z\psi^\downarrow = \frac{\hbar}{2}\tilde{\sigma}_z\begin{pmatrix} 0 \\ 2mc^2 \\ 0 \\ 0 \end{pmatrix} e^{-iWt/\hbar} = -\frac{\hbar}{2}\psi^\downarrow \tag{5.18b}$$

So we can now fully interpret these free-particle wavefunctions. They are both positive energy states, but $\psi^\uparrow$ is a spin-up particle and $\psi^\downarrow$ is a spin-down particle. Although this operator has differentiated between the two eigenstates, note that these wavefunctions are only eigenfunctions of the z-component of spin in the rest frame. Even in the limit we have taken of a particle moving in the xy-plane and spin quantized along the z-axis, there is some profound relationship between the momentum and the spin. It turns out that if the particle is moving it is only possible to quantize the spin unambiguously, using the operator (5.16), in the direction of the motion. A more general form of the spin operator will be discussed later in the chapter.

The next step is to write equations (5.10) in 2×2 matrix form (Rose 1961). Recalling that $p_z = 0$ we can write

$$c\tilde{\sigma} \cdot \mathbf{p} = \begin{pmatrix} 0 & c(p_x - ip_y) \\ c(p_x + ip_y) & 0 \end{pmatrix} \tag{5.19}$$

and so equations (5.10) become

$$c\tilde{\sigma} \cdot \mathbf{p} \begin{pmatrix} U_1 \\ U_2 \end{pmatrix} = (W + mc^2) \begin{pmatrix} U_3 \\ U_4 \end{pmatrix} \tag{5.20a}$$

$$c\tilde{\sigma} \cdot \mathbf{p} \begin{pmatrix} U_3 \\ U_4 \end{pmatrix} = (W - mc^2) \begin{pmatrix} U_1 \\ U_2 \end{pmatrix} \tag{5.20b}$$

We can eliminate U_3 and U_4 in these equations:

$$c\tilde{\sigma} \cdot \mathbf{p}(W + mc^2)^{-1} c\tilde{\sigma} \cdot \mathbf{p} \begin{pmatrix} U_1 \\ U_2 \end{pmatrix} = (W - mc^2) \begin{pmatrix} U_1 \\ U_2 \end{pmatrix} \tag{5.21}$$

and this just leads again to the all too familiar expression (5.12). Consider the solutions (5.13). They are unnormalized, so we can multiply them by an arbitrary constant and not change their significance. We multiply each of them by $(W + mc^2)^{-1}$:

$$\psi^{\uparrow} = U^{\uparrow}(\mathbf{p}) \exp\left(i(\mathbf{p} \cdot \mathbf{r} - Wt)/\hbar\right) = \begin{pmatrix} 1 \\ 0 \\ 0 \\ \frac{c(p_x + ip_y)}{W + mc^2} \end{pmatrix} \exp(i(\mathbf{p} \cdot \mathbf{r} - Wt)/\hbar)$$

$$= \begin{pmatrix} \chi^{1/2} \\ \frac{c\tilde{\sigma} \cdot \mathbf{p}}{W + mc^2} \chi^{1/2} \end{pmatrix} \exp(i(\mathbf{p} \cdot \mathbf{r} - Wt)/\hbar) \tag{5.22a}$$

$$\psi^{\downarrow} = U^{\downarrow}(\mathbf{p}) \exp\left(i(\mathbf{p} \cdot \mathbf{r} - Wt)/\hbar\right) = \begin{pmatrix} 0 \\ 1 \\ \frac{c(p_x - ip_y)}{W + mc^2} \\ 0 \end{pmatrix} \exp(i(\mathbf{p} \cdot \mathbf{r} - Wt)/\hbar)$$

$$= \begin{pmatrix} \chi^{-1/2} \\ \frac{c\tilde{\sigma} \cdot \mathbf{p}}{W + mc^2} \chi^{-1/2} \end{pmatrix} \exp(i(\mathbf{p} \cdot \mathbf{r} - Wt)/\hbar) \tag{5.22b}$$

where the $\chi^{\pm 1/2}$ are two-component basis spinors first defined in equation (2.27). This has got us into a position where it is straightforward to normalize these eigenfunctions. The exponential can be normalized in the standard way. Then we require that $U^{\uparrow(\downarrow)}U^{\uparrow(\downarrow)} = 1$. Some simple multiplication of the coefficients leaves us with

$$\psi^\uparrow = \frac{1}{\sqrt{V}} \left(\frac{W + mc^2}{2W} \right)^{\frac{1}{2}} \left(\begin{array}{c} \chi^{1/2} \\ \frac{c\tilde{\sigma}\cdot\mathbf{p}}{W+mc^2}\chi^{1/2} \end{array} \right) \exp(i(\mathbf{p}\cdot\mathbf{r} - Wt)/\hbar) \qquad (5.23a)$$

$$\psi^\downarrow = \frac{1}{\sqrt{V}} \left(\frac{W + mc^2}{2W} \right)^{\frac{1}{2}} \left(\begin{array}{c} \chi^{-1/2} \\ \frac{c\tilde{\sigma}\cdot\mathbf{p}}{W+mc^2}\chi^{-1/2} \end{array} \right) \exp(i(\mathbf{p}\cdot\mathbf{r} - Wt)/\hbar) \qquad (5.23b)$$

These are our final results for the spin-up and spin-down wavefunctions. We have selected a particular normalization, but other normalizations are equally valid. Figure 5.1 shows the large and small components of the free-particle wavefunction for two momenta. As expected, wavelength decreases as energy increases. As you can see, even a wavefunction describing a free particle is a very complicated quantity when all components are drawn out explicitly. Note also that the minimum energy a free particle can have is $W = mc^2$. If the energy were below this the momentum would be complex and the exponents in equation (5.23) would become real. The wavefunction would then diverge at plus or minus infinity and hence would be unnormalizable (although we will qualify these statements later).

The probability current density is defined in equation (4.100). Let us find it for a unit volume for these wavefunctions. The exponentials cancel and we have

$$\mathbf{j} = U^\dagger c\tilde{\alpha} U$$

$$= \left(\frac{W + mc^2}{2W} \right) \left(\chi^{m\dagger}, c\frac{\tilde{\sigma}^\dagger \cdot \mathbf{p}}{W + mc^2}\chi^{m\dagger} \right) c \left(\begin{array}{cc} 0 & \tilde{\sigma} \\ \tilde{\sigma} & 0 \end{array} \right) \left(\begin{array}{c} \chi^m \\ c\frac{\tilde{\sigma}\cdot\mathbf{p}}{W+mc^2}\chi^m \end{array} \right)$$

$$= \frac{c^2}{2W}(\chi^{m\dagger}\tilde{\sigma}(\tilde{\sigma}\cdot\mathbf{p})\chi^m + \chi^{m\dagger}\tilde{\sigma}\cdot\mathbf{p}\tilde{\sigma}\chi^m) \qquad (5.24)$$

If we insert the matrices and evaluate this explicitly for, say, the x-component of the current we find

$$j_x = \frac{c^2 p_x}{W} \qquad (5.25)$$

and similarly for the y- and z-components. This is the quantum mechanical expression for the current density for a free particle. It is also the classical relativistic expression for velocity (equation (1.26)).

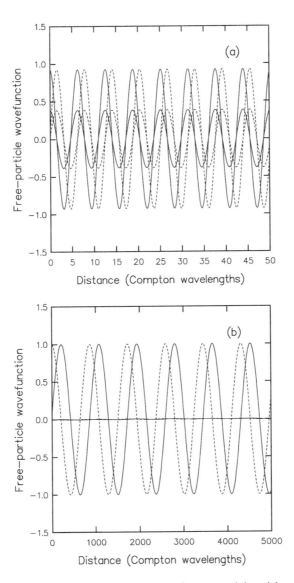

Fig. 5.1. (*a*) The free-particle wavefunction for a particle with momentum $\hbar k_{\mathrm{C}}$ where k_{C} is the Compton wavevector. The full lines are the real part of the large and small components of the wavefunction, the dotted lines are the imaginary part. (*b*) Same as (*a*) but the particle momentum is $\hbar/a_0$ where a_0 is the first Bohr radius. The structure of the small components is invisible on this scale. The figures are drawn using relativistic units.

Let us look at the non-relativistic limit of the wavefunctions (5.23). In this limit $W \to mc^2$ and $p_x, p_y \to 0$ so

$$U^\uparrow(\mathbf{p}) = \left(\frac{W + mc^2}{2W} \right)^{1/2} \begin{pmatrix} 1 \\ 0 \\ 0 \\ \frac{c(p_x + ip_y)}{W + mc^2} \end{pmatrix} \longrightarrow \begin{pmatrix} 1 \\ 0 \\ 0 \\ 0 \end{pmatrix} \qquad (5.26a)$$

$$U^\downarrow(\mathbf{p}) = \left(\frac{W + mc^2}{2W} \right)^{1/2} \begin{pmatrix} 0 \\ 1 \\ \frac{c(p_x - ip_y)}{W + mc^2} \\ 0 \end{pmatrix} \longrightarrow \begin{pmatrix} 0 \\ 1 \\ 0 \\ 0 \end{pmatrix} \qquad (5.26b)$$

This is the expected result. In the non-relativistic limit the bottom two components of the wavefunction disappear, and therefore it becomes possible to do quantum mechanics including spin with two components to the wavefunctions rather than four, i.e, we have recovered the Pauli formalism discussed in chapter 2. Because of the limit described by (5.26) the upper two elements of the wavefunction are known as the large component and the lower two are known as the small component. To calculate any measurable quantities we need to evaluate the expectation values of operators. An even operator will couple the two large components together and the two small components together, and the effect of the small components will be overwhelmed. However, an odd operator will couple the large and small components together, which will be a purely relativistic effect. So, if we want to *see* relativistic effects in nature, we should look at observables described by an odd operator. (See the discussion of magnetic dichroism in chapter 12 for example.)

As we have seen, the Pauli operator for the spin of an electron ($\frac{\hbar}{2}\tilde{\sigma}$) is insufficient to fully describe spin within Dirac theory. The Pauli operator is only appropriate when the momentum is zero or parallel to the particles spin. This can easily be seen if we write the eigenvalue equation explicitly

$$\tilde{\sigma}_z U(\mathbf{p}) = \begin{pmatrix} \tilde{\sigma}_z \chi^{\pm 1/2} \\ \frac{c\tilde{\sigma}_z \tilde{\sigma} \cdot \mathbf{p}}{W + mc^2} \chi^{\pm 1/2} \end{pmatrix} \qquad (5.27)$$

The upper component of this behaves appropriately, but $\tilde{\sigma}_z$ does not commute with $\tilde{\sigma} \cdot \mathbf{p}$ unless $\mathbf{p}$ is in the z-direction. So, a sensible question to ask is, 'What is the spin operator for a free particle moving in some general direction?' Much of the rest of this section is devoted to deriving the answer to this question. We will do this in two ways. Firstly, we will write down the spin eigenfunctions in the usual way and then rotate them using spin rotation matrices. Secondly, we will define a generalized spin operator, which is valid for any direction of the electron spin.

Rotations and Spinors

In this section we will show explicitly what happens to a spinor when we rotate the coordinate system and hence the axis of quantization. This is a fairly straightforward task, but we can use the results in chapter 6 when we discuss spin projection operators.

The strategy in this section will be as follows. Firstly, we will recall the Pauli spinors. Secondly, we will use the Pauli spin matrices to define a rotation operator for spinors which is closely akin to previously defined rotation operators (equation (2.22)). Thirdly, we will apply the rotation operator on a Pauli spinor explicitly to confirm that it describes the rotation we expect.

In chapter 2 we defined the Pauli spinors

$$\chi^{1/2} = \begin{pmatrix} 1 \\ 0 \end{pmatrix}, \qquad \chi^{-1/2} = \begin{pmatrix} 0 \\ 1 \end{pmatrix} \qquad (5.28)$$

which form a complete set in spin space (for $s = 1/2$), so any wavefunction for a spin-1/2 particle can be written as a linear combination of them.

$$\Phi = \sum_{m_s=\pm 1/2} \psi_{m_s} \chi^{m_s} \qquad (5.29)$$

where ψ_{m_s} contains all the space, time and energy dependence, etc., of the wavefunction.

In the above discussion we have implicitly assumed that the spin is quantized along the z-direction, so $\hat{S}_z$ is diagonal with eigenvalues $\pm\hbar/2$. Let us rotate the axes, so that $\hat{S}$ is diagonal in another direction. In section 2.2 we have seen that rotation operators can be written in terms of angular momentum operators. Let us write down the rotation operator in terms of spin angular momentum by analogy with equation (2.22)

$$\hat{R} = e^{-i\theta \cdot \hat{S}/\hbar} \qquad (5.30)$$

Now θ here is an angle of magnitude θ about an axis $\hat{n}$. We know that $\hat{S}$ can be written in terms of the Pauli matrices, so writing (5.30) as an infinite series (which is the only way it can be understood), we have

$$\hat{R} = 1 - \frac{i\theta \hat{n} \cdot \tilde{\sigma}}{2} + \frac{i^2\theta^2(\hat{n} \cdot \tilde{\sigma})^2}{2 \times 2!} - \frac{i^3\theta^3(\hat{n} \cdot \tilde{\sigma})^3}{2 \times 3!} + \cdots \qquad (5.31)$$

It follows immediately from equation (4.19) and the fact that $\hat{n}$ is a unit vector that $\hat{n} \cdot \tilde{\sigma}$ raised to any even power is the unit matrix. Therefore when it is raised to any odd power it must be either plus or minus the unit matrix. Inserting this into equation (5.31) tells us that $\hat{R}$ can be equal to $\exp(\pm i\theta/2)$. We can combine both signs of the exponent into one operator if we write it as a diagonal matrix:

$$\hat{\mathbf{R}} = \begin{pmatrix} e^{i\theta/2} & 0 \\ 0 & e^{-i\theta/2} \end{pmatrix} \tag{5.32}$$

So this matrix operating on a spinor rotates it through an angle θ around the direction defined by $\hat{\mathbf{n}}$.

Let us illustrate this explicitly with an arbitrary spinor χ of the form (5.29) which we rotate to make spinor χ' using (5.32)

$$\chi' = \begin{pmatrix} a' \\ b' \end{pmatrix} = \begin{pmatrix} e^{i\theta/2}a \\ e^{-i\theta/2}b \end{pmatrix} = \begin{pmatrix} e^{i\theta/2} & 0 \\ 0 & e^{-i\theta/2} \end{pmatrix} \begin{pmatrix} a \\ b \end{pmatrix} = \hat{\mathbf{R}}\chi \tag{5.33}$$

Now, as $\hat{\mathbf{S}}$ is diagonal along some arbitrary direction $\hat{\mathbf{q}}$, we can write

$$\frac{\hbar}{2}\tilde{\sigma} \cdot \hat{\mathbf{q}}\chi = q_s\chi \tag{5.34a}$$

where q_s are the eigenvalues. Writing equation (5.34a) in matrix form gives us

$$\frac{\hbar}{2}\begin{pmatrix} q_z & q_x - iq_y \\ q_x + iq_y & -q_z \end{pmatrix}\begin{pmatrix} a \\ b \end{pmatrix} = q_s\begin{pmatrix} a \\ b \end{pmatrix} \tag{5.34b}$$

Next, let us premultiply (5.34) by (5.32) and note that $\hat{\mathbf{R}}^\dagger\hat{\mathbf{R}} = \tilde{I}$. So

$$\frac{\hbar}{2}\begin{pmatrix} e^{i\theta/2} & 0 \\ 0 & e^{-i\theta/2} \end{pmatrix}\begin{pmatrix} q_z & q_x - iq_y \\ q_x + iq_y & -q_z \end{pmatrix}\begin{pmatrix} e^{-i\theta/2} & 0 \\ 0 & e^{i\theta/2} \end{pmatrix} \times$$
$$\begin{pmatrix} e^{i\theta/2} & 0 \\ 0 & e^{-i\theta/2} \end{pmatrix}\begin{pmatrix} a \\ b \end{pmatrix} = q_s\begin{pmatrix} e^{i\theta/2} & 0 \\ 0 & e^{-i\theta/2} \end{pmatrix}\begin{pmatrix} a \\ b \end{pmatrix} \tag{5.35}$$

Using the associative property of matrices we can multiply out the first three matrices on the left hand side of this equation and then we use the definition (5.33) to obtain

$$\frac{\hbar}{2}\begin{pmatrix} q_z & (q_x - iq_y)e^{i\theta} \\ (q_x + iq_y)e^{-i\theta} & -q_z \end{pmatrix}\chi' = q_s\chi' \tag{5.36}$$

Expanding the complex exponentials as trigonometric functions gives

$$\frac{\hbar}{2}\begin{pmatrix} q_z & q_-\cos\theta - iq_-\sin\theta \\ q_+\cos\theta - iq_+\sin\theta & -q_z \end{pmatrix}\chi' = q_s\chi' \tag{5.37a}$$

with

$$q_- = q_x - iq_y, \qquad q_+ = q_x + iq_y \tag{5.37b}$$

But we know the equations that represent a rotation about the z-axis from simple trigonometric arguments:

$$\begin{aligned} q_x' &= q_x\cos\theta + q_y\sin\theta \\ q_y' &= -q_x\sin\theta + q_y\cos\theta \\ q_z' &= q_z \end{aligned} \tag{5.38}$$

So we can substitute from (5.38) and (5.37b) into (5.37a) and we find

$$\frac{\hbar}{2}\begin{pmatrix} q'_z & q'_x - iq'_y \\ q'_x + iq'_y & -q'_z \end{pmatrix}\chi' = q_s\chi' \tag{5.39}$$

Equation (5.39) is the same as (5.34b), but written with respect to a rotated set of axes.

This formalism and example show that, given a Pauli wavefunction and an arbitrary direction $\hat{\mathbf{q}}$, we can form the operator $\hbar\tilde{\boldsymbol{\sigma}}\cdot\hat{\mathbf{q}}/2$ according to (5.34a) which gives us the component of the spin along the direction $\hat{\mathbf{q}}$. If we premultiply this equation by $\hat{\mathbf{R}}$ we find it corresponds to a rotation of both the spin direction and $\hat{\mathbf{q}}$ by an angle θ. However, as they have both been rotated by the same angle, the component of spin along $\hat{\mathbf{q}}$ should be unaffected by the rotation. This is verified by the fact that q_s is the same in equations (5.34b) and (5.39). None of this is very surprising.

Now let us pursue this line of thought further. We can find the eigenvalues q_s by explicit solution of equation (5.34b). This is a simple simultaneous equation and will have non-trivial solutions if

$$\begin{vmatrix} q_z - \frac{2}{\hbar}q_s & q_x - iq_y \\ q_x + iq_y & -q_z - \frac{2}{\hbar}q_s \end{vmatrix} = 0 \tag{5.40}$$

This can be solved: recalling that $\hat{\mathbf{q}}$ is a unit vector, we have

$$\frac{4}{\hbar^2}q_s^2 = q_z^2 + (q_x - iq_y)(q_x + iq_y) = q_z^2 + q_y^2 + q_x^2 = \hat{\mathbf{q}}^2 = 1 \tag{5.41}$$

Hence $q_s = \pm\hbar/2$. Now that we have the eigenvalues in (5.34b) we can find the corresponding eigenvectors. It is more convenient to work in spherical polar coordinates, so $q_x = \sin\theta\cos\phi$, $q_y = \sin\theta\sin\phi$ and $q_z = \cos\theta$, where θ and ϕ are now the usual angles in spherical polar coordinates. If the normalization condition for the spinors is chosen as

$$|a|^2 + |b|^2 = 1 \tag{5.42}$$

it is straightforward to show that satisfactory solutions are

$$\chi^+ = \begin{pmatrix} e^{-i\phi/2}\cos\theta/2 \\ e^{i\phi/2}\sin\theta/2 \end{pmatrix}, \qquad q_s = \hbar/2 \tag{5.43a}$$

$$\chi^- = \begin{pmatrix} -e^{-i\phi/2}\sin\theta/2 \\ e^{i\phi/2}\cos\theta/2 \end{pmatrix}, \qquad q_s = -\hbar/2 \tag{5.43b}$$

The relative phases here have been chosen so as to give the most symmetric form to (5.43), but other choices are equally valid. Equations (5.43) are the expressions we were looking for. They allow us to write the two-component wavefunctions relative to a set of axes other than the direction along which we quantize their spin. Furthermore, they are orthonormal, and in the limit that $\theta = \phi \to 0$ they become the usual spinors of equation (5.28).

A Generalized Spin Operator

To set up a generalized spin operator we are going to construct an arbitrary operator with specific properties, and then show that it is an operator which differentiates between the two spin states. It is central to this argument to recall that if an operator commutes with the Hamiltonian it represents a constant of the motion.

Consider any unit vector $\hat{\mathbf{r}}$. Let us evaluate the commutator

$$[\tilde{\boldsymbol{\sigma}} \cdot \hat{\mathbf{r}}, \hat{H}] = [\tilde{\boldsymbol{\sigma}} \cdot \hat{\mathbf{r}}, c\tilde{\boldsymbol{\alpha}} \cdot \hat{\mathbf{p}}] = 2ic\tilde{\boldsymbol{\alpha}} \cdot \hat{\mathbf{r}} \times \hat{\mathbf{p}} \qquad (5.44)$$

So this commutator is zero if $\hat{\mathbf{r}}$ and $\hat{\mathbf{p}}$ are parallel. Now let us evaluate the commutator

$$[\tilde{\beta}\tilde{\boldsymbol{\sigma}} \cdot \hat{\mathbf{r}}, \hat{H}] = [\tilde{\beta}\tilde{\boldsymbol{\sigma}} \cdot \hat{\mathbf{r}}, c\tilde{\boldsymbol{\alpha}} \cdot \mathbf{p}] = -2c\tilde{\beta}\gamma_5\hat{\mathbf{r}} \cdot \mathbf{p} \qquad (5.45)$$

This commutator is zero if $\hat{\mathbf{r}}$ and $\mathbf{p}$ are perpendicular. Equations (5.44) and (5.45) are not obvious, but can be verified easily (and tediously) by writing out the matrices explicitly. Now we are in a position to define a coordinate frame in terms of directions relative to the momentum. We can define appropriate operators along these directions to give us an operator that commutes with the Hamiltonian. Let us define the coordinate system with one axis defined by a unit vector in the direction of the momentum $\hat{\mathbf{p}}$ and the other two by unit vectors $\hat{\mathbf{r}}_1$ and $\hat{\mathbf{r}}_2$. These are guaranteed to be perpendicular to one another if we define them by the relations

$$\hat{\mathbf{r}}_1 \times \hat{\mathbf{r}}_2 = \hat{\mathbf{p}}, \qquad \hat{\mathbf{r}}_2 \times \hat{\mathbf{p}} = \hat{\mathbf{r}}_1, \qquad \hat{\mathbf{p}} \times \hat{\mathbf{r}}_1 = \hat{\mathbf{r}}_2 \qquad (5.46)$$

Such a coordinate system is illustrated in figure 5.2. Using it and the commutators (5.44) and (5.45) we can define a generalized spin operator

$$\hat{\mathbf{S}} = \frac{\hbar}{2}(\tilde{\boldsymbol{\sigma}} \cdot \hat{\mathbf{p}}\hat{\mathbf{p}} + \tilde{\beta}\tilde{\boldsymbol{\sigma}} \cdot \hat{\mathbf{r}}_1\hat{\mathbf{r}}_1 + \tilde{\beta}\tilde{\boldsymbol{\sigma}} \cdot \hat{\mathbf{r}}_2\hat{\mathbf{r}}_2) \qquad (5.47)$$

In conjunction with (5.44) and (5.45) this automatically gives us

$$[\hat{\mathbf{S}}, \hat{H}] = 0 \qquad (5.48)$$

As we have already seen, in the non-relativistic limit we need only consider the upper two components of the wavefunction. So in this limit, we can neglect the $\tilde{\beta}$s in (5.47) and then $\hat{\mathbf{S}}$ clearly reduces to the Pauli form. Now that we have set up this operator, we need to operate with it on the eigenfunctions (5.23) and see that it does indeed produce the right eigenvalues. Initially we have to do this component by component.

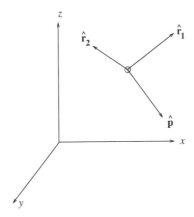

Fig. 5.2. The coordinate system used to define the generalized spin operator. The particle's motion is in the direction $\mathbf{p}$, and $\mathbf{r}_1$ and $\mathbf{r}_2$ are defined in equations (5.46).

$$\tilde{\boldsymbol{\sigma}} \cdot \hat{\mathbf{r}}_1 U^{\uparrow(\downarrow)} = \left(\frac{W + mc^2}{2W}\right)^{1/2} \begin{pmatrix} \tilde{\boldsymbol{\sigma}} \cdot \hat{\mathbf{r}}_1 \chi^{\pm 1/2} \\ \frac{c\tilde{\boldsymbol{\sigma}} \cdot \hat{\mathbf{r}}_1 \tilde{\boldsymbol{\sigma}} \cdot \hat{\mathbf{p}}}{W + mc^2} \chi^{\pm 1/2} \end{pmatrix}$$

$$= \left(\frac{W + mc^2}{2W}\right)^{1/2} \begin{pmatrix} \tilde{\boldsymbol{\sigma}} \cdot \hat{\mathbf{r}}_1 \chi^{\pm 1/2} \\ \frac{-icp\tilde{\boldsymbol{\sigma}} \cdot \hat{\mathbf{r}}_2}{W + mc^2} \chi^{\pm 1/2} \end{pmatrix} \tag{5.49a}$$

where we have made use of equations (4.19) and (5.46). Similarly

$$\tilde{\boldsymbol{\sigma}} \cdot \hat{\mathbf{r}}_2 U^{\uparrow(\downarrow)} = \left(\frac{W + mc^2}{2W}\right)^{1/2} \begin{pmatrix} \tilde{\boldsymbol{\sigma}} \cdot \hat{\mathbf{r}}_2 \chi^{\pm 1/2} \\ \frac{icp\tilde{\boldsymbol{\sigma}} \cdot \hat{\mathbf{r}}_1}{W + mc^2} \chi^{\pm 1/2} \end{pmatrix} \tag{5.49b}$$

$$\tilde{\boldsymbol{\sigma}} \cdot \hat{\mathbf{p}} U^{\uparrow(\downarrow)} = \left(\frac{W + mc^2}{2W}\right)^{1/2} \begin{pmatrix} \tilde{\boldsymbol{\sigma}} \cdot \hat{\mathbf{p}} \chi^{\pm 1/2} \\ \frac{cp\hat{p}}{W + mc^2} \chi^{\pm 1/2} \end{pmatrix} \tag{5.49c}$$

Combining (5.49a, b and c) according to the definition of $\hat{\mathbf{S}}$ in equation (5.47) and taking the z-component of the resulting expression gives

$$\hat{S}_z U^{\uparrow(\downarrow)} = \frac{\hbar}{2} \left(\frac{W + mc^2}{2W}\right)^{\frac{1}{2}} \begin{pmatrix} (\tilde{\boldsymbol{\sigma}} \cdot \hat{\mathbf{p}} \hat{p}_z + \tilde{\boldsymbol{\sigma}} \cdot \hat{\mathbf{r}}_1 \hat{r}_{1z} + \tilde{\boldsymbol{\sigma}} \cdot \hat{\mathbf{r}}_2 \hat{r}_{2z}) \chi^{\pm \frac{1}{2}} \\ \frac{cp}{W + mc^2} (\hat{p}_z + i\tilde{\boldsymbol{\sigma}} \cdot \hat{\mathbf{r}}_2 \hat{r}_{1z} - i\tilde{\boldsymbol{\sigma}} \cdot \hat{\mathbf{r}}_1 \hat{r}_{2z}) \chi^{\pm \frac{1}{2}} \end{pmatrix} \tag{5.50}$$

The upper component of (5.50) is easy to simplify. Each term is just the component of $\tilde{\boldsymbol{\sigma}}$ along one of the axes and then we take the z-component of that. So this just becomes $\tilde{\sigma}_z \chi^{\pm 1/2} = \pm \chi^{\pm 1/2}$. Next we consider the lower component of (5.50). This is rather more difficult to interpret. We need to look at the $\chi^{1/2}$ and $\chi^{-1/2}$ cases separately really, but we will only worry about $\chi^{1/2}$ and leave the other case to the problems (aren't you grateful?). For any vector **r** we have

$$\tilde{\sigma} \cdot \mathbf{r}\chi^{1/2} = \begin{pmatrix} r_z \\ r_x + ir_y \end{pmatrix} \tag{5.51}$$

Using this in the expression for the lower component of (5.50) gives

$$\hat{S}_z U^\uparrow =$$

$$\frac{\hbar}{2}\left(\frac{W+mc^2}{2W}\right)^{\frac{1}{2}}\left(\begin{array}{c} \chi^{1/2} \\ \frac{cp}{W+mc^2}\left(\begin{array}{c} \hat{p}_z \\ i((\hat{r}_{2x}+i\hat{r}_{2y})\hat{r}_{1z} - (\hat{r}_{1x}+i\hat{r}_{1y})\hat{r}_{2z}) \end{array}\right) \end{array}\right) \tag{5.52}$$

(because the z-components of the terms in $\mathbf{r}_1$ and $\mathbf{r}_2$ cancel). Now

$$i((\hat{r}_{2x}+i\hat{r}_{2y})\hat{r}_{1z} - (\hat{r}_{1x}+i\hat{r}_{1y})\hat{r}_{2z}) = i(\hat{r}_1 \times \hat{r}_2)_y + (\hat{r}_1 \times \hat{r}_2)_x$$
$$= \hat{p}_x + i\hat{p}_y \tag{5.53}$$

and substituting this back into (5.52) gives

$$\hat{S}_z U^\uparrow = \frac{\hbar}{2}\left(\frac{W+mc^2}{2W}\right)^{1/2}\left(\begin{array}{c} \chi^{1/2} \\ \frac{c\tilde{\sigma}\cdot\mathbf{p}}{W+mc^2}\chi^{1/2} \end{array}\right) \tag{5.54a}$$

That is,

$$\hat{S}_z U^\uparrow = \frac{\hbar}{2}U^\uparrow \tag{5.54b}$$

and if we do the same thing for $\chi^{-1/2}$ we find

$$\hat{S}_z U^\downarrow = -\frac{\hbar}{2}U^\downarrow \tag{5.54c}$$

We have obtained the results (5.54) regardless of the value of the momentum. This tells us that the z-component of $\hat{\mathbf{S}}$ takes on the same value whatever the momentum. The operator $\hat{\mathbf{S}}$ has been normalized so that the eigenvalues of $\hat{S}_z$ are $\pm\hbar/2$. The expectation value of $\hat{S}^2$ can be calculated for the states $U_{\uparrow(\downarrow)}$. We have to evaluate

$$\int \psi^\dagger \hat{S}^2 \psi d\mathbf{r} \tag{5.55}$$

The volume integral just gives us a volume which cancels with the $1/\sqrt{V}$s in (5.23). Hence we are left with

$$\langle\hat{S}^2\rangle_{\uparrow(\downarrow)} = \frac{\hbar^2}{4}(\langle U_{\uparrow(\downarrow)}|(\tilde{\sigma}\cdot\hat{\mathbf{p}})^2|U_{\uparrow(\downarrow)}\rangle + \langle U_{\uparrow(\downarrow)}|(\tilde{\beta}\tilde{\sigma}\cdot\hat{\mathbf{r}}_1)^2|U_{\uparrow(\downarrow)}\rangle$$
$$+ \langle U_{\uparrow(\downarrow)}|(\tilde{\beta}\tilde{\sigma}\cdot\hat{\mathbf{r}}_2)^2|U_{\uparrow(\downarrow)}\rangle) \tag{5.56}$$

The β-matrix commutes with the σ-matrices and we can use (4.19) to write the three matrix elements on the right hand side of (5.56) as the expectation values $\hat{\mathbf{p}}^2$, $\hat{\mathbf{r}}_1^2$ and $\hat{\mathbf{r}}_2^2$. The square of a unit vector is just unity, and the expectation value of unity is one. So

$$\langle \hat{\mathbf{S}}^2 \rangle_{\uparrow(\downarrow)} = \frac{\hbar^2}{4}(\langle \hat{\mathbf{p}}^2 \rangle + \langle \hat{\mathbf{r}}_1^2 \rangle + \langle \hat{\mathbf{r}}_2^2 \rangle) = \frac{3\hbar^2}{4} \tag{5.57}$$

This shows that the square of the spin operator we have defined in (5.47) has the expected length $3/4$ (in units of $\hbar^2$). Furthermore we have seen that this operator may be used to distinguish between the plane wave functions (5.23), pointing along the $+z$-direction for $\psi^\uparrow$ and along the $-z$-direction for $\psi^\downarrow$. This differentiation between the two cases is the origin of the phrases 'spin up' and 'spin down' frequently encountered in non-relativistic quantum mechanics and condensed matter physics, particularly in magnetism. As you have seen, the mathematical justification for that picture of the electron's spin is fairly complicated. However, it is necessary to understand spin at the level at which it is treated here to realize the limitations of that model.

The functions (5.23) are eigenvectors of the $\hat{\mathbf{S}}$ operator, and they have different eigenvalues. Obviously, this means that they must be orthogonal to one another; I leave it as an exercise for the reader to verify this. Finally, let me mention that the choice of the z-direction as the axis of quantization is arbitrary, but is motivated by convention and by the fact that the σ_z-matrix is the only one of the σ-matrices that is diagonal. Hence the mathematics is simplest in this case.

5.4 Negative Energy States, Antiparticles

In our solution of the free-particle Dirac equation we found the relative values of the coefficients U_1, U_2, U_3, and U_4 in equations (5.10). Following this we took the right hand expression in (5.11) to give us the ratio U_1/U_4. There was no reason to select this expression, we could just as easily have taken the central formula (Baym 1967). Had we done this we would have replaced equation (5.13) by the following

$$\psi^\downarrow \propto \begin{pmatrix} c(p_x - ip_y) \\ 0 \\ 0 \\ W - mc^2 \end{pmatrix} \exp(i(\mathbf{p} \cdot \mathbf{r} - Wt)/\hbar) \tag{5.58a}$$

$$\psi^\uparrow \propto \begin{pmatrix} 0 \\ c(p_x + ip_y) \\ W - mc^2 \\ 0 \end{pmatrix} \exp(i(\mathbf{p} \cdot \mathbf{r} - Wt)/\hbar) \tag{5.58b}$$

This pair of solutions is as valid as the pair given by equation (5.13). It is easy to see that one set of solutions may be obtained from the other simply by multiplying by a constant factor. Surprisingly, despite this, the two sets of solutions are definitely not equivalent. To show this explicitly we take the example of $U^\uparrow$ from (5.13a) and obtain the coefficient in (5.58a) simply by multiplying by $c(p_x - ip_y)$

$$\begin{pmatrix} W + mc^2 \\ 0 \\ 0 \\ c(p_x + ip_y) \end{pmatrix} \times c(p_x - ip_y) \longrightarrow (W + mc^2) \begin{pmatrix} c(p_x - ip_y) \\ 0 \\ 0 \\ W - mc^2 \end{pmatrix} \qquad (5.59)$$

where we have used (5.12). It appears that one solution is just a simple multiple of the other, but this is not the case. The easiest way to see this is to examine these solutions in the particle rest frame as we did for the particles in equation (5.17), i.e. when $p_x = p_y = p^2 = 0$. We find

$$V^\downarrow(\mathbf{p}) \propto \begin{pmatrix} c(p_x - ip_y) \\ 0 \\ 0 \\ W - mc^2 \end{pmatrix} \longrightarrow \begin{pmatrix} 0 \\ 0 \\ 0 \\ -2mc^2 \end{pmatrix} \qquad (5.60a)$$

$$V^\uparrow(\mathbf{p}) \propto \begin{pmatrix} 0 \\ c(p_x + ip_y) \\ W - mc^2 \\ 0 \end{pmatrix} \longrightarrow \begin{pmatrix} 0 \\ 0 \\ -2mc^2 \\ 0 \end{pmatrix} \qquad (5.60b)$$

Here we have had to take the negative square root of the expression for the energy to get non-trivial solutions. In equations (5.17) we took the positive square root. So the states described by having the large part of the wavefunction in the lower half of the four-component wavefunction have negative energies. The real origin of this property is equation (5.12) for the energy. It has both a positive and a negative square root. We could get one solution from the other in (5.59) because we used (5.12) which contains no information about sign of the energy. Now we have found all four independent solutions of the free-particle Dirac equation. There are positive and negative energy solutions and two possible spin directions.

Let us be absolutely clear about what we mean here:

$$\psi_p^\uparrow = \begin{pmatrix} 1 \\ 0 \\ 0 \\ 0 \end{pmatrix} e^{-iWt/\hbar} = \text{spin-up positive energy state in its rest frame.}$$

$$\psi_p^\downarrow = \begin{pmatrix} 0 \\ 1 \\ 0 \\ 0 \end{pmatrix} e^{-iWt/\hbar} = \text{spin-down positive energy state in its rest frame.}$$

$$\psi_n^\uparrow = \begin{pmatrix} 0 \\ 0 \\ 1 \\ 0 \end{pmatrix} e^{iWt/\hbar} = \text{spin-up negative energy state in its rest frame.}$$

$$\psi_n^\downarrow = \begin{pmatrix} 0 \\ 0 \\ 0 \\ 1 \end{pmatrix} e^{iWt/\hbar} = \text{spin-down negative energy state in its rest frame.}$$

The magnitudes here are arbitrary because they depend on the normalization. Note that in $\psi_p^\uparrow$ we could have the third component of the 4-vector non-zero and still have an eigenfunction of $\hat{S}_z$ defined by equation (5.16) with eigenvalue $\hbar/2$. Similarly we could have the first component of $\psi_n^\uparrow$ non-zero. Furthermore we could have the fourth component of $\psi_p^\downarrow$ and the second component of $\psi_n^\downarrow$ non-zero and still have an eigenfunction of $\hat{S}_z$ with eigenvalue $-\hbar/2$. By studying the derivation of our plane wave solutions in section 5.2 it can be seen that motion in the z-direction would correspond to the introduction of these components. However, motion in the x- and y-directions leads us to a situation in which our wavefunction is no longer an eigenfunction of $\hat{S}_z$. This shows unambiguously that, with this operator, we can really only define the spin parallel or antiparallel to the particle's momentum, and reveals the need for the generalized spin operator defined by equation (5.47). The normalized eigenfunctions for the negative energy states in a general frame are

$$\psi^{\uparrow-} = V^\uparrow(\mathbf{p})\exp(i(\mathbf{p}\cdot\mathbf{r}+Wt)/\hbar)$$

$$= \frac{1}{\sqrt{V}}\left(\frac{W+mc^2}{2W}\right)^{1/2}\begin{pmatrix} -\frac{c\bar{\sigma}\cdot\mathbf{p}}{W+mc^2}\chi^{1/2} \\ \chi^{1/2} \end{pmatrix}\exp(i(\mathbf{p}\cdot\mathbf{r}+Wt)/\hbar) \qquad (5.61a)$$

$$\psi^{\downarrow -} = V^{\downarrow}(\mathbf{p}) \exp(i(\mathbf{p} \cdot \mathbf{r} + Wt)/\hbar)$$

$$= \frac{1}{\sqrt{V}} \left(\frac{W + mc^2}{2W} \right)^{1/2} \left(\begin{array}{c} -\frac{c\bar{\sigma}\cdot\mathbf{p}}{W+mc^2}\chi^{-1/2} \\ \chi^{-1/2} \end{array} \right) \exp(i(\mathbf{p} \cdot \mathbf{r} + Wt)/\hbar)$$

$$(5.61b)$$

where W is a positive energy. Clearly, there will be some interpretational difficulties with the negative energy states. Here we mention a couple of the problems.

In the same way as was discussed for a Klein–Gordon particle, any negative energy state has lower energy than any positive energy state, so if a negative energy state is empty a positive energy particle should be able to decay down into it, emitting a photon. Indeed it should be able to continue doing this forever, with the only restrictions being the usual angular momentum selection rules. Obviously, any reasonable, consistent view of nature cannot allow the universe to disappear up its own posterior like this, and anyway the photons emitted as the particles disappear down to $-\infty$ are not observed.

In (5.25) we have written the current operator for a particle. In this equation we can take W as the negative square root and the velocity of the particle is then in the opposite direction to its momentum, clearly an idea that requires a lot of justification.

These difficulties were resolved by Dirac (1928b). One simply assumes that in a vacuum all the negative energy states are filled with electrons and all the positive energy states are empty, as illustrated schematically in figure 5.3. This is where we have to qualify our earlier statement that electrons must have $W > mc^2$. For $-mc^2 < W < mc^2$ we have a forbidden energy region, because the minimum energy any electron can have is its rest mass energy. Below this region down to $-\infty$ are all the filled negative energy states, and above the region up to $+\infty$ are states available for standard positive energy electrons to occupy. In this scheme the Pauli exclusion principle automatically explains why there are no transitions to negative energy states.

Now, suppose a negative energy electron is excited to a positive energy state. This will give a positive energy electron, but it will leave a hole in the 'sea' of negative energy states. The absence of a negative energy state will look the same as the existence of a positive energy state of the same mass. For electrons, the absence of the electronic charge will be equivalent to a charge of equal magnitude and opposite sign. This also means that the particle's velocity and momentum are parallel to one another. It is necessary to be careful in defining the spin of the hole. The absence of $\psi_n^{\uparrow}$ will leave us with a hole that has apparent spin down, and the absence of $\psi_n^{\downarrow}$ will leave us with a hole of spin up.

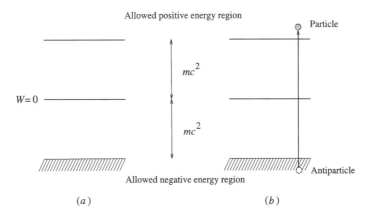

Fig. 5.3. (a) The energetically forbidden region of width $2mc^2$ separating the positive and negative energy states. This figure represents the vacuum with the shading indicating that the negative energy sea is filled. (b) Same as (a), but here a negative energy particle has been excited up to an empty state in the positive energy continuum. This creates a positive energy particle and the hole left behind is identified as the antiparticle.

If this picture is correct, relativistic quantum theory is predicting the existence of particles with very specific properties relative to the electron (or any other spin-1/2 particle). Such a particle was discovered, after the predictions of Dirac (1928b), by Anderson (1933) and dubbed the positron. Undoubtedly, this prediction was one of the greatest triumphs of relativistic quantum theory, but it raises as many questions as it answers. There is something unsatisfactory about this description of nature because the particle/antiparticle symmetry inherent in the Dirac equation is broken in an unexplained manner.

A more pressing problem this leaves us with is how to describe a vacuum properly. It appears to require very elaborate electrodynamics for a complete explanation, as any applied field should shift the negative energy states and hence polarize the vacuum. As there is an infinite number of these negative energy states the polarization of the vacuum should be infinite (Schwinger 1948, 1949, Tomonaga 1946).

As an example of this, consider a single electron in a vacuum. This electron has charge e and it will repel the nearby electrons that fill the negative energy states. So, in effect, it polarizes the vacuum around itself, making the surrounding area positive. If the charge density of the electron is $\rho_e(r)$ and the charge density of the polarized vacuum is $\rho_v(r)$ then any macroscopic measurement of the electronic charge is going to see the sum of the electronic charge and the positive vacuum polarization $\int(\rho_e(r) + \rho_v(r))d\mathbf{r} = e$. However if some test charge could get inside the radius of the vacuum polarization it would see the electronic charge as greater than e. Of course, the test charge would itself polarize the vacuum and so the problem is not as simple as stated here. Nonetheless, the effect of vacuum polarization can be observed. In hydrogen, the $l = 0$ wavefunctions get very close to the nucleus, and a small correction to their energy levels occurs, owing to the vacuum polarization.

So, although the discovery of antiparticles was one of the great successes of relativistic quantum mechanics, it was also the beginning of the end, and showed that the single-particle picture was insufficient even for the simplest problems. This and other considerations led to the development of quantum field theory, which is beyond the scope of this book.

This does not mean we can say no more about antiparticles, but it does mean we have to be careful not to overstep the bounds of validity of relativistic single-particle quantum mechanics which we are discussing in much of this book. Equations (5.61) give the wavefunction for the negative energy electron states. However, this is not the real particle and the correctly normalized wavefunctions for a positive energy positron are

$$\psi_C^{\uparrow(\downarrow)} = \pm \frac{1}{\sqrt{V}} \left(\frac{W + mc^2}{2W}\right)^{1/2} \left(\begin{array}{c} \chi^{\pm 1/2} \\ -\frac{c\tilde{\sigma}\cdot\mathbf{p}}{W+mc^2}\chi^{\pm 1/2} \end{array}\right) \exp(i(-\mathbf{p}\cdot\mathbf{r} - Wt)/\hbar)$$

$$(5.62a)$$

while the negative energy positron states are

$$\psi_C^{\uparrow(\downarrow)} = \pm \frac{1}{\sqrt{V}} \left(\frac{W + mc^2}{2W}\right)^{1/2} \left(\begin{array}{c} \frac{c\tilde{\sigma}\cdot\mathbf{p}}{W+mc^2}\chi^{\pm 1/2} \\ \chi^{\pm 1/2} \end{array}\right) \exp(i(-\mathbf{p}\cdot\mathbf{r} + Wt)/\hbar)$$

$$(5.62b)$$

where W is defined as being positive, i.e. it is the magnitude of the energy and $\mathbf{p}$ represents the true momentum of the particle, which we have chosen as opposite to that of the negative energy state of (5.61) for consistency within the hole theory. These wavefunctions are identical to the free-electron wavefunctions, as they should be. There is no difference between the Dirac equation for a free positron and the Dirac equation for a free electron. The charge has no effect on their properties as long as there are no fields present. The sign outside the wavefunction is an arbitrary phase factor which cannot be observed, of course, but which is left because it appears again when we discuss charge conjugation in chapter 6.

5.5 Classical Negative Energy Particles?

There is nothing quantum mechanical about equation (5.12) – it does not contain $\hbar$ – so one can ask whether negative energy particles can be created in non-quantum mechanics. Our experience is that antiparticle creation cannot occur in a situation that is governed by classical relativity

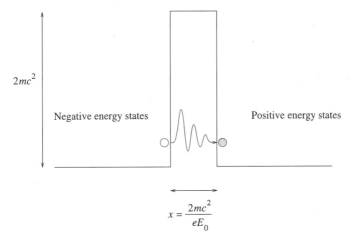

Fig. 5.4. The potential barrier of height $2mc^2$ and width $2mc^2/eE_0$ used in our discussion of particle/antiparticle pair production. A negative energy particle tunnels through the barrier to become a positive energy one and to leave behind an antiparticle.

and mechanics. Here we are going to make a handwaving estimate of the production rate of particle/antiparticle pairs assuming the process can be described as tunnelling through a potential barrier.* Such a view is suggested by our analysis of the Klein paradox earlier and by the picture we have of antiparticles being the hole in the negative energy *sea of electrons* that is left behind after a particle has been excited through the forbidden zone, as shown in figure 5.3.

Referring to figure 5.4, we require a potential barrier with height (in energy) sufficient to create a particle/antiparticle pair: it must be of height at least $2mc^2$. For the purposes of this argument let us assume that this potential barrier is due to a constant field given by E_0. This assumption will not materially affect our argument. We also have to separate the particle/antiparticle pair by a distance x so that they do not instantly annihilate. When the particle travels a distance x in the field E_0 the drop in potential energy (increase in kinetic energy) is given by eE_0x. Let us take the distance it is necessary to separate the particle/antiparticle pair as the distance required for the drop in potential energy to be equal to the rest mass energy. We could take another numerical factor times the rest mass energy, but that will not qualitatively affect our answer. So

$$eE_0x = 2mc^2 \tag{5.63}$$

---

* I would like to thank Professor J.M.F. Gunn for bringing this argument to my attention.

or

$$x = \frac{2mc^2}{eE_0} \tag{5.64}$$

The particle is excited from one side of the barrier (in the sea of negative energy states). It has to tunnel through the barrier of height $2mc^2$ and width $2mc^2/eE_0$. Once on the other side it is the particle and the hole it leaves behind is the antiparticle. The pair creation rate is equal to the tunnelling rate for the particle through the barrier.

This single-particle tunnelling rate is given in many quantum theory textbooks (Gasiorowicz 1974) as

$$T^2 \propto \exp\left(-2\sqrt{\frac{2m}{\hbar^2}(V-E)a}\right) \tag{5.65}$$

where the barrier has height V and width $2a$. Our only task is to calculate (5.65) for the particle. If we set $E = 0$ as we have no incident particle and substitute in our values of the barrier height and width we find

$$T^2 \propto \exp\left(-2\sqrt{\frac{2m}{\hbar^2}(2mc^2)}\frac{mc^2}{eE_0}\right) = \exp\left(-\frac{4mc^2}{e\lambda_C E_0}\right) \tag{5.66}$$

In (5.66), λ_C is the Compton wavelength. Now the point has been made, there will not be significant creation of particle/antiparticle pairs unless the potential changes by a significant fraction of the rest mass energy over a distance of order the Compton wavelength. This simple argument is in extremely good agreement with the much more sophisticated quantum field theory calculation (Itzykson and Zuber 1980) which gets the same functional form as equation (5.66), but replaces the factor 4 by π. Everything in (5.66) is a constant; let us expand λ_C and write it as

$$T^2 \propto \exp\left(-\frac{4m^2c^3/eE_0}{\hbar}\right) \tag{5.67}$$

Now if we take the classical limit $\hbar \to 0$, the rate of creation of particle/antiparticle pairs tends to zero. So, if the universe were governed by classical relativistic mechanics, antiparticles might have existed since the beginning, but they could not have been created since then.

5.6 The Klein Paradox Revisited

In chapter 3 we solved the Klein–Gordon equation for a particle incident upon a potential step. For large enough values of the potential step, we found propagating solutions which do not occur in non-relativistic theory. Here we re-examine the same problem with an incident spin-1/2 particle

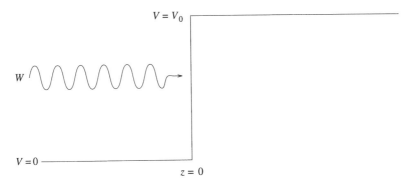

Fig. 5.5. The potential barrier problem that is solved to illustrate the Klein paradox for the Dirac equation. The particle wavefunction is found for all possible positive values of V_0 relative to W.

to see how the Klein paradox arises from the Dirac equation (Bjorken and Drell 1964, Greiner 1990). Our one-dimensional potential is

$$V(z) = 0 \qquad\qquad z < 0$$
$$V(z) = V_0 \qquad\qquad z > 0 \qquad\qquad (5.68)$$

This is illustrated in figure 5.5. We have chosen our one dimension as the z-axis this time as that means the incident particle can be an eigenfunction of the $\hat{S}_z$-operator. We will not solve this problem at $z = 0$, rather we will solve for $z < 0$ and $z > 0$ and match the solutions at $z = 0$. In fact, we have already done the hard work to solve the Dirac equation for this potential. On the left hand side of the barrier we have the free-particle solution

$$\psi(\mathbf{r}) = \begin{pmatrix} 1 \\ 0 \\ \frac{cp_{1z}}{W+mc^2} \\ 0 \end{pmatrix} e^{ip_{1z}z/\hbar} + R \begin{pmatrix} 1 \\ 0 \\ \frac{-cp_{1z}}{W+mc^2} \\ 0 \end{pmatrix} e^{-ip_{1z}z/\hbar} \qquad (5.69)$$

and for the transmitted wave the solution is just that of a free electron rigidly displaced by energy V_0, so to the right of the barrier we have

$$\psi(\mathbf{r}) = T \begin{pmatrix} 1 \\ 0 \\ \frac{cp_{2z}}{W-V_0+mc^2} \\ 0 \end{pmatrix} e^{ip_{2z}z/\hbar} \qquad (5.70)$$

with

$$p_{2z} = \frac{1}{c}((W - V_0)^2 - m^2c^4)^{1/2} \qquad (5.71)$$

Note that we have defined the incident particle as being spin up. There can be no spin flip at the barrier because then the second and fourth components of the wavefunction would be non-zero while the first and third were zero. There would then be no non-trivial way of matching the wavefunctions at $z = 0$. We have to conserve current and so we can match the solutions (5.69) and (5.70) at the boundary. The top row of the 4-vector equation gives

$$1 + R = T \tag{5.72a}$$

and the third row yields

$$1 - R = \frac{p_{2z}}{p_{1z}} \left(\frac{W + mc^2}{W - V_0 + mc^2} \right) T \tag{5.72b}$$

These are simultaneous equations for R and T and can be solved straightforwardly to give

$$T = \frac{2p_{1z}(W - V_0 + mc^2)}{p_{1z}(W - V_0 + mc^2) + p_{2z}(W + mc^2)} \tag{5.73a}$$

$$R = \frac{p_{1z}(W - V_0 + mc^2) - p_{2z}(W + mc^2)}{p_{1z}(W - V_0 + mc^2) + p_{2z}(W + mc^2)} \tag{5.73b}$$

Now, equation (5.71) leads us to exactly the same conclusions as we found in the Klein–Gordon case. There are several important regions of size of potential and the interpretation of them is given below equation (3.89). The transmission and reflection coefficients are given by equations (5.73). We see from (5.73b) that provided p_{1z} and p_{2z} are both positive (as they should be for particles incident from the left) we may for very large potential barriers find $R < -1$ and hence $|R|^2 > 1$, i.e. more wave reflected than incident again. The transmission coefficient T can also become greater than one and again this can only be understood on the basis of the creation of particle/antiparticle pairs at the boundary. What has happened here is that when the potential becomes $V_0 > 2mc^2$ the negative energy electrons are displaced in energy by a sufficient amount for them to be taken out of the vacuum and the holes left behind in the barrier are the antiparticles, i.e. the field is sufficiently strong to force the vacuum to decay.

5.7 Lorentz Transformation of the Free-Particle Wavefunction

In this section we take the wavefunction for a free particle moving in some general direction and perform a Lorentz transformation on it to move into the particle's rest frame (Baym 1967). This is simply an example showing how to do the Lorentz transformations described in the previous chapter. From (4.34) a Lorentz transformation acting on a wavefunction is

$$\psi(\mathbf{r}', t') = \hat{T}^{-1} \psi(\mathbf{r}, t) \tag{5.74}$$

where

$$\hat{T} = \exp(\tilde{\boldsymbol{\alpha}} \cdot \mathbf{u}/2) \tag{5.75}$$

with

$$u = \tanh^{-1} \frac{|\mathbf{v}|}{c} \tag{5.76a}$$

$$\cosh u = (1 - v^2/c^2)^{-1/2} = \gamma \tag{5.76b}$$

Here $\mathbf{v}$ is the velocity of the primed frame as measured by an observer in the unprimed frame, and u is the magnitude of the vector $\mathbf{u}$ in the direction of $\mathbf{v}$. Using (5.75) we can rewrite (5.74):

$$\psi(\mathbf{r}', t') = \exp(-\tilde{\boldsymbol{\alpha}} \cdot \mathbf{u}/2)\psi(\mathbf{r}, t) = (\cosh(u/2) - \tilde{\boldsymbol{\alpha}} \cdot \hat{\mathbf{u}} \sinh(u/2))\psi(\mathbf{r}, t) \tag{5.77}$$

Now, some elementary identities of the hyperbolic functions can be used to write

$$
\begin{aligned}
\psi(\mathbf{r}', t') &= \left(\left(\frac{\cosh u + 1}{2} \right)^{1/2} - \tilde{\boldsymbol{\alpha}} \cdot \hat{\mathbf{u}} \left(\frac{\cosh u - 1}{2} \right)^{1/2} \right) \psi(\mathbf{r}, t) \\
&= \left(\left(\frac{\gamma + 1}{2} \right)^{1/2} - \tilde{\boldsymbol{\alpha}} \cdot \hat{\mathbf{u}} \left(\frac{\gamma - 1}{2} \right)^{1/2} \right) \psi(\mathbf{r}, t)
\end{aligned}
\tag{5.78}
$$

The explicit form we are going to use for the wavefunction is that of equation (5.23), so

$$
\psi' = \frac{1}{\sqrt{2V}} \left((\gamma + 1)^{1/2} - \tilde{\boldsymbol{\alpha}} \cdot \hat{\mathbf{u}}(\gamma - 1)^{1/2} \right) \left(\frac{W + mc^2}{2W} \right)^{1/2}
$$

$$
\times \left(\frac{\chi^{m_s}}{\frac{c\tilde{\boldsymbol{\sigma}} \cdot \mathbf{p}}{W + mc^2} \chi^{m_s}} \right) e^{i(\mathbf{p} \cdot \mathbf{r} - Wt)/\hbar} \tag{5.79}
$$

where we have used the shorthand that ψ' is the wavefunction in the new frame. Now the term in the exponent is a Lorentz invariant, as can be seen by taking the 4-vector scalar product of the Lorentz energy–momentum transformations with the Lorentz space–time transformations.

Equation (5.79) can be simplified if we recall that this is a free particle and so its total energy is given in terms of γ by equation (1.21). Also we write the α-matrices in terms of the 2×2 Pauli spin matrices using (4.13). We also require equation (4.19) to show that $\tilde{\boldsymbol{\sigma}} \cdot \mathbf{p} \tilde{\boldsymbol{\sigma}} \cdot \hat{\mathbf{u}} = p$. This is because the velocity and the momentum vectors are (obviously) parallel and $\hat{\mathbf{u}}$ is a unit vector. After a little algebra we have

$$\psi' = \frac{1}{2\sqrt{V\gamma}} \left(\begin{pmatrix} (\gamma+1)\chi^{m_s} \\ \frac{c\tilde{\sigma}\cdot\mathbf{p}}{mc}\chi^{m_s} \end{pmatrix} - \frac{p}{mc} \begin{pmatrix} \frac{p}{mc(\gamma+1)}\chi^{m_s} \\ \tilde{\sigma}\cdot\hat{u}\chi^{m_s} \end{pmatrix} \right) \tag{5.80}$$

The lower terms in (5.80) cancel immediately, because p times the unit vector $\hat{u}$ is just the vector $\mathbf{p}$. It takes a little more work to simplify the upper term, but if we substitute for the momentum from $p^2c^2 = W^2 - m^2c^4$ and use equation (1.21) again we end up with

$$\psi' = \frac{1}{\sqrt{V\gamma}} \begin{pmatrix} \chi^{m_s} \\ 0 \end{pmatrix} e^{i(\mathbf{p}\cdot\mathbf{r}-Wt)/\hbar} \tag{5.81}$$

The wavefunction ψ' is still normalized. Its normalization simply represents the effect of Lorentz–Fitzgerald contraction along its direction of motion in the same way as we saw for the Klein–Gordon particle. Comparison of this equation with (5.23) shows that (5.81) just represents a free spin-up particle in its rest frame, i.e. $\mathbf{p} = 0$.

In this section we have shown how to perform a Lorentz transformation on a wavefunction. We have actually chosen a transformation to the particle rest frame which works out rather easily. As we have never specified the velocity of the particle, the question of how we chose the rest frame as the one to which we were transforming is rather subtle. The answer lies in the definition of the energy of the moving particle. We wrote $W = \gamma mc^2$ where m is actually the rest mass, the mass as viewed in the frame to which we are transforming. To transform to some arbitrary frame would require us to write the energy in the untransformed frame in terms of quantities measured in the new frame.

5.8 Problems

(1) Solve the Dirac equation for a particle that is free to move along the z-axis only. Find all four solutions and show that they are eigenfunctions of both the $\hat{S}_z$ operator and the $\hat{p}_z$ operator.

(2) For a positive energy, free-particle wavefunction show that the expectation values of $\tilde{\beta}$ and $\tilde{\alpha}$ are

$$\langle \tilde{\beta} \rangle = \gamma^{-1} \qquad\qquad \langle \tilde{\alpha} \rangle = \frac{\mathbf{v}}{c}$$

(3) Verify that the free-particle eigenfunctions of equation (5.23) are indeed orthogonal to one another.

(4) Starting from equation (5.50) verify equation (5.54c).

(5) Starting from the wavefunction for a stationary free particle, perform a Lorentz transformation of the type described in section (5.7) to find the wavefunction of the particle in the frame of reference of an observer moving with velocity $\mathbf{v}$ relative to the particle.

6

Symmetries and Operators

In this chapter we are going to discuss some rather esoteric topics. Despite this nature they are of fundamental significance in relativistic quantum theory and have profound consequences. Clearly we could discuss a lot of topics that routinely occur in non-relativistic quantum theory under such a chapter heading. We will not do this, but only consider topics of specific importance in relativistic quantum theory.

We will start this chapter by introducing a new type of operator known as a projection operator. This is an operator that acts on some wavefunction and projects out the part of the wavefunction corresponding to particular properties. In particular there are energy projection operators which can project out the positive or negative energy part of a wavefunction and spin projection operators which (surprise surprise!) project out the part of the wavefunction corresponding to a particular spin direction. Such operators form an essential part of the theory of high energy scattering. We will not be using them much in our discussion of scattering because our aim there is towards solid state applications. For further discussion of projection operators, the books by Rose (1961), Bjorken and Drell (1964) and Greiner (1990) are useful.

Secondly, we will look at some symmetries that occur in the Dirac equation. The symmetries we will consider are charge conjugation, time-reversal invariance, parity and combinations of them. Matrix operators representing these symmetries will be discussed and we will show how this leads to the Feynman view of the negative energy states as representing antiparticles moving backwards through time (Feynman 1949).

Thirdly, we will return to the topic of angular momentum. This follows on from our discussion in chapter 2. There we looked at angular momentum in the Pauli theory. Here we will generalize that work to a four-component representation of angular momentum in preparation for our discussion of the relativistic one-electron atom. Next, we will reproduce a simple argument relating the relativistic and non-relativistic kinetic energy operators for a particle in a magnetic field.

Finally a major section of this chapter is devoted to many-body theory. Clearly this is something we have to know about for a description of the electrons in condensed matter. In this chapter we look at formal aspects of the theory only; we will apply the theory in several later chapters. Here it will be necessary to revise non-relativistic theory, before we introduce the relativistic formulation. We will introduce second quantized operators and discuss their properties and applications. Then we will relate them to the Dirac field operators, and show how to generalize non-relativistic theory to describe relativistic particles. Finally we will see how this enables us to develop a more consistent interpretation of the negative energy states and antiparticles, which removes the requirement to introduce an infinite number of negative energy electrons to maintain the single-particle picture.

6.1 Non-Relativistic Spin Projection Operators

A spin projection operator is defined as an operator that acts as follows. If we have a general wavefunction containing both spin-up and spin-down character, a spin projection operator acting on that wavefunction will give us the part of the wavefunction corresponding to one spin direction and will filter away the part of the wavefunction corresponding to the other spin direction. The term spin projection operator should not be confused with the operator $\tilde{\sigma} \cdot \hat{\mathbf{q}}$. This latter operator, when used in an expectation value expression, just gives the average value of the spin (plus or minus a factor $\hbar/2$) along the direction $\hat{\mathbf{q}}$.

From the discussion in chapter 5 it is very easy to see what operator we should use to project out the spin character of a particle described by a spinor. Let us define the operators first, then show that they do project out one particular spin and then discuss their properties. The projection operators are

$$\hat{P}^{+} = \tfrac{1}{2}(1 + \tilde{\sigma} \cdot \hat{\mathbf{q}}) \tag{6.1a}$$

$$\hat{P}^{-} = \tfrac{1}{2}(1 - \tilde{\sigma} \cdot \hat{\mathbf{q}}) \tag{6.1b}$$

In chapter 5 we saw that the eigenvalues of $\tilde{\sigma} \cdot \hat{\mathbf{q}}$ are ± 1. If the eigenvalue of $\tilde{\sigma} \cdot \hat{\mathbf{q}}$ is $+1$ then $\hat{P}^{+}$ has eigenvalue $+1$ and $\hat{P}^{-}$ has eigenvalue 0, and *vice versa* if the eigenvalue of $\tilde{\sigma} \cdot \hat{\mathbf{q}}$ is -1. So we are claiming that, as $\tilde{\sigma} \cdot \hat{\mathbf{q}}$ is the operator whose eigenvalue is the value of the spin (in units of $\hbar/2$) along the direction $\hat{\mathbf{q}}$, $\hat{P}^{+}$ operating on an arbitrary spinor will project out the spin parallel to $\hat{\mathbf{q}}$ from that spinor and $\hat{P}^{-}$ will project out the spin antiparallel to $\hat{\mathbf{q}}$.

To show that this is so, let us take a spinor χ and operate on it with the projection operators (6.1). The result of this operation is an eigenfunction of $\tilde{\sigma} \cdot \hat{q}$, i.e, it is a state whose spin is purely along $\hat{q}$.

$$\tilde{\sigma} \cdot \hat{q} \hat{P}^+ \chi = \tilde{\sigma} \cdot \hat{q} \tfrac{1}{2}(1 + \tilde{\sigma} \cdot \hat{q})\chi = \tfrac{1}{2}(\tilde{\sigma} \cdot \hat{q} + (\tilde{\sigma} \cdot \hat{q})^2)\chi$$
$$= \tfrac{1}{2}(\tilde{\sigma} \cdot \hat{q} + 1)\chi = +1\hat{P}^+ \chi = +1\chi^\uparrow \tag{6.2a}$$
$$\tilde{\sigma} \cdot \hat{q} \hat{P}^- \chi = \tilde{\sigma} \cdot \hat{q} \tfrac{1}{2}(1 - \tilde{\sigma} \cdot \hat{q})\chi = \tfrac{1}{2}(\tilde{\sigma} \cdot \hat{q} - (\tilde{\sigma} \cdot \hat{q})^2)\chi$$
$$= \tfrac{1}{2}(\tilde{\sigma} \cdot \hat{q} - 1)\chi = -1\hat{P}^- \chi = -\chi^\downarrow \tag{6.2b}$$

Here the unit matrix is understood to multiply any terms where it is necessary and we have used equation (4.19) again.

We have defined these operators in (6.1) and shown that they are spin projection operators by proving that when operating on an arbitrary spinor they give an eigenfunction of $\tilde{\sigma} \cdot \hat{q}$. Now, let us look at some other simple properties of these operators. Firstly, it is trivial to see that

$$(\hat{P}^\pm)^2 = \hat{P}^\pm \tag{6.3}$$

The mathematical term for this property is idempotent. Furthermore, there is no overlap between the two operators as

$$\hat{P}^+ \hat{P}^- = \hat{P}^- \hat{P}^+ = 0 \tag{6.4}$$

Finally, as $\hat{P}^+$ projects out one spin direction and $\hat{P}^-$ projects out the other, adding up the results of both operations should give us the original wavefunction back:

$$(\hat{P}^+ + \hat{P}^-)\chi = (\tfrac{1}{2}(1 + \tilde{\sigma} \cdot \hat{n}) + \tfrac{1}{2}(1 - \tilde{\sigma} \cdot \hat{n}))\chi = \chi \tag{6.5}$$

as required. Finally we note from (6.2) and (6.4) that

$$\hat{P}^+ \chi^\downarrow = \hat{P}^- \chi^\uparrow = 0 \tag{6.6}$$

In many experiments it is necessary to detect the spin of a particle. Imagine we have a detector that will sense whether an incident particle is parallel or antiparallel to some direction $\hat{q}$. Let us write the wavefunction of the incident particle as

$$\Phi = \sum_{n=\uparrow,\downarrow} \psi_n(\mathbf{r})\chi^n \tag{6.7}$$

We can project out the component of Φ parallel (antiparallel) to $\hat{q}$ using $\hat{P}^+(\hat{P}^-)$. The fraction of the particles that the detector sees as being parallel (antiparallel) to $\hat{q}$ will be

$$F^{\pm q} = \frac{\int \Phi^\dagger(\mathbf{r})\hat{P}^\pm \Phi(\mathbf{r})d\mathbf{r}}{\int \Phi^\dagger(\mathbf{r})\Phi(\mathbf{r})d\mathbf{r}} \tag{6.8}$$

and, obviously, if we substitute in the operators we find

$$F^{+q} - F^{-q} = \frac{\int \psi_\uparrow^*(\mathbf{r})\psi_\uparrow(\mathbf{r})^3\mathbf{r} - \int \psi_\downarrow^*(\mathbf{r})\psi_\downarrow(\mathbf{r})d\mathbf{r}}{\int \psi_\uparrow^*(\mathbf{r})\psi_\uparrow(\mathbf{r})d\mathbf{r} + \int \psi_\downarrow^*(\mathbf{r})\psi_\downarrow(\mathbf{r})d\mathbf{r}} \tag{6.9}$$

This is a quantity often measured in experiments that involve either the electron or photon polarization.

What we have considered up to this point in this chapter is the non-relativistic spin projection operator. We now go on to look at how this formalism generalizes to the relativistic case.

6.2 Relativistic Energy and Spin Projection Operators

It turns out to be convenient to consider the energy projection operators rather than the spin projection operators first. These work in an analogous way to the spin projection operators described above. If we operate with the positive energy projection operator, the part of the wavefunction corresponding to negative energies is removed, and if we operate with the negative energy projection operator, the positive energy part is removed.

We will illustrate the energy projection operators using the free particle wavefunctions of chapter 5. For the positive energy states the wavefunction is of the form (equation (5.23))

$$\psi(\mathbf{r}, t) = U(\mathbf{p})e^{i(\mathbf{p}\cdot\mathbf{r} - Wt)/\hbar} \tag{6.10a}$$

and for the negative energy states (c.f. equation (5.61))

$$\psi(\mathbf{r}, t) = V(\mathbf{p})e^{i(\mathbf{p}\cdot\mathbf{r} + Wt)/\hbar} \tag{6.10b}$$

Now, let us define the operators (with the square root always positive)

$$\hat{\Gamma} = \frac{\tilde{\boldsymbol{\alpha}} \cdot \hat{\mathbf{p}} + \tilde{\beta}mc}{(\mathbf{p}^2 + m^2c^2)^{1/2}} \tag{6.11}$$

$$\hat{\Gamma}^{\pm} = \tfrac{1}{2}(\hat{I} \pm \hat{\Gamma}) \tag{6.12}$$

Note that the numerator of $\hat{\Gamma}$ is the total energy operator (divided by c), so it can be replaced by $i\hbar\partial/\partial ct$ (from the Dirac equation). Now we can operate on the states described by (6.10) with $\hat{\Gamma}$:

$$\hat{\Gamma}U(\mathbf{p})e^{i(\mathbf{p}\cdot\mathbf{r} - Wt)/\hbar} = \frac{1}{c(\mathbf{p}^2 + m^2c^2)^{1/2}}i\hbar\frac{\partial}{\partial t}U(\mathbf{p})e^{i(\mathbf{p}\cdot\mathbf{r} - Wt)/\hbar}$$

$$= \frac{1}{(c^2\mathbf{p}^2 + m^2c^4)^{1/2}}WU(\mathbf{p})e^{i(\mathbf{p}\cdot\mathbf{r} - Wt)/\hbar} = U(\mathbf{p})e^{i(\mathbf{p}\cdot\mathbf{r} - Wt)/\hbar} \tag{6.13a}$$

and similarly

$$\hat{\Gamma}V(\mathbf{p})e^{i(\mathbf{p}\cdot\mathbf{r} + Wt)/\hbar} = -V(\mathbf{p})e^{i(\mathbf{p}\cdot\mathbf{r} + Wt)/\hbar} \tag{6.13b}$$

and from this it is obvious that

$$\hat{\Gamma}^+ U(\mathbf{p})e^{i(\mathbf{p}\cdot\mathbf{r}-Wt)/\hbar} = U(\mathbf{p})e^{i(\mathbf{p}\cdot\mathbf{r}-Wt)/\hbar} \tag{6.14a}$$

$$\hat{\Gamma}^- U(\mathbf{p})e^{i(\mathbf{p}\cdot\mathbf{r}-Wt)/\hbar} = 0 \tag{6.14b}$$

$$\hat{\Gamma}^+ V(\mathbf{p})e^{i(\mathbf{p}\cdot\mathbf{r}+Wt)/\hbar} = 0 \tag{6.14c}$$

$$\hat{\Gamma}^- V(\mathbf{p})e^{i(\mathbf{p}\cdot\mathbf{r}+Wt)/\hbar} = V(\mathbf{p})e^{i(\mathbf{p}\cdot\mathbf{r}+Wt)/\hbar} \tag{6.14d}$$

So, if we have a wavefunction that is some linear combination of positive and negative energy states, operating on it with $\hat{\Gamma}^+$ will project out of the wavefunction only those states that have positive energy. Similarly $\hat{\Gamma}^-$ will project out only those states that have negative energy. These energy projection operators have the following properties:

$$(\hat{\Gamma}^\pm)^2 = \hat{\Gamma}^\pm, \qquad \hat{\Gamma}^+\hat{\Gamma}^- = \hat{\Gamma}^-\hat{\Gamma}^+ = 0, \qquad \hat{\Gamma}^2 = \hat{I} \tag{6.15}$$

We can conclude from this section that the energy projection operators are written in an analogous way to the non-relativistic spin projection operators and their properties are very similar.

As we saw in chapter 5 it is difficult to define the spin of an electron in a general state of motion. It was shown in section 5.3 that the spin of a moving electron can only be defined simply in the directions parallel and antiparallel to its motion. If we define the particle spin in these directions then the relativistic spin projection operator can be defined simply in analogy with the non-relativistic case as

$$P^\uparrow = \tfrac{1}{2}(1 - \tilde{\gamma}_5\tilde{\alpha}\cdot\hat{\mathbf{q}}) = \tfrac{1}{2}(1 + \tilde{\sigma}\cdot\hat{\mathbf{q}}) \tag{6.16a}$$

$$P^\downarrow = \tfrac{1}{2}(1 + \tilde{\gamma}_5\tilde{\alpha}\cdot\hat{\mathbf{q}}) = \tfrac{1}{2}(1 - \tilde{\sigma}\cdot\hat{\mathbf{q}}) \tag{6.16b}$$

where $\tilde{\sigma}$ are now the 4×4 Pauli spin matrices and $\hat{\mathbf{q}}$ is a unit vector in the direction of the particle motion. For the general spin operator of equation (5.47), the projection operator becomes rather more complicated (Rose 1961).

By combining (6.12) and (6.16) we have obtained a set of operators which will project out spin and energy components of the wavefunction. To state that explicitly: $P^\uparrow\Gamma^+$ operating on a wavefunction will give us the positive energy spin-up part of the wavefunction, $P^\downarrow\Gamma^+$ gives us the spin-down positive energy part, $P^\uparrow\Gamma^-$ gives the spin-up negative energy part and finally $P^\downarrow\Gamma^-$ gives the spin-down negative energy part. We have chosen the free particle/antiparticle wavefunctions here. These form a complete set, so any wavefunction can be written as a linear combination of them. In principle, therefore, the energy and spin projection operators can be used on any wavefunction.

6.3 Charge Conjugation

In chapter 5 we saw that the existence of negative energy states is a prediction of the Dirac equation and that there emerges from this the hole
theory where an antiparticle is interpreted as the absence of a negative
energy particle. Mathematically, we could just as easily have started with
the Dirac equation for the positron and constructed electrons as the absence of negative energy solutions to that equation. Clearly this symmetry
is present in the Dirac equation and in this section we put that symmetry
on a more formal mathematical basis.

Let us start by writing down the Dirac equation for an electron in an
electromagnetic field, i.e. equation (4.49)

$$i\hbar\frac{\partial\psi(\mathbf{r},t)}{\partial t} = \left(c\tilde{\alpha}\cdot\left(\frac{\hbar}{i}\nabla - e\mathbf{A}(\mathbf{r},t)\right) + \tilde{\beta}mc^2 + V(\mathbf{r}) \right)\psi(\mathbf{r},t) \qquad (6.17a)$$

Of course, this has negative energy solutions. Now, let us write down the
Dirac equation for a positron in the same electromagnetic field. Obviously
the sign of the charge of the particle differs so the sign of the term in
the vector potential is reversed. The mass is still positive so the rest mass
energy term in the Hamiltonian does not change. Finally, if the electron
finds a potential attractive the positron with opposite charge will find it
repulsive and *vice versa*, so the term in the scalar potential also reverses
sign. We are left with

$$i\hbar\frac{\partial\psi_C(\mathbf{r},t)}{\partial t} = \left(c\tilde{\alpha}\cdot\left(\frac{\hbar}{i}\nabla + e\mathbf{A}(\mathbf{r},t)\right) + \tilde{\beta}mc^2 - V(\mathbf{r}) \right)\psi_C(\mathbf{r},t) \qquad (6.17b)$$

where $\psi_C(\mathbf{r},t)$ is known as the charge conjugate state. This equation
describes the positron. On the hole theory we identify an antiparticle as
the absence of a negative energy state. Therefore there must be a one-
to-one correspondence between the negative energy solutions to (6.17a)
and the positive energy solutions to (6.17b). Our aim now is to find
an operator that will transform (6.17a) into (6.17b) and that will also
transform (6.17b) back into (6.17a). We have to change the sign of the
terms in the electromagnetic field and retain the sign in the terms for the
kinetic energy, rest mass energy and time dependence of the wavefunction.
The first step is to take the complex conjugate of (6.17a) and multiply by
-1. This gives

$$i\hbar\frac{\partial\psi^*(\mathbf{r},t)}{\partial t} = \left(c\tilde{\alpha}^*\cdot\left(\frac{\hbar}{i}\nabla + e\mathbf{A}(\mathbf{r},t)\right) - \tilde{\beta}^*mc^2 - V(\mathbf{r}) \right)\psi^*(\mathbf{r},t) \qquad (6.18)$$

The next step is to postulate the existence of a non-singular matrix which
we shall call $\tilde{C}$. We will premultiply (6.18) by this and insert $\tilde{C}^{-1}\tilde{C}$, the

unit matrix, into the middle. Recalling that $\tilde{\beta}$ is real gives

$$i\hbar\tilde{C}\frac{\partial\psi^*(\mathbf{r},t)}{\partial t} =$$
$$\tilde{C}\left(c\tilde{\alpha}^* \cdot \left(\frac{\hbar}{i}\nabla + e\mathbf{A}(\mathbf{r},t)\right) - \tilde{\beta}mc^2 - V(\mathbf{r})\right)\tilde{C}^{-1}\tilde{C}\psi^*(\mathbf{r},t) \qquad (6.19)$$

Now it is simply necessary to find the matrix $\tilde{C}$. In order to satisfy the condition that (6.19) be equivalent to (6.17b) we must require that $\tilde{C}$ satisfies the following:

$$\tilde{C}\tilde{\alpha}^*\tilde{C}^{-1} = \tilde{\alpha}, \qquad\qquad \tilde{C}\tilde{\beta}\tilde{C}^{-1} = -\tilde{\beta} \qquad (6.20)$$

A matrix that is consistent with (6.20) for all components of $\tilde{\alpha}$ and is non-singular (all its eigenvalues are non-zero and its inverse exists) is

$$\tilde{C} = \begin{pmatrix} 0 & 0 & 0 & 1 \\ 0 & 0 & -1 & 0 \\ 0 & -1 & 0 & 0 \\ 1 & 0 & 0 & 0 \end{pmatrix} = i\tilde{\beta}\tilde{\alpha}_y \qquad (6.21)$$

Note that we have just written this matrix down, we have not proved it uniquely satisfies all the necessary conditions. The matrix is its own inverse, which is a desirable property, because if we perform charge conjugation twice we expect to return to the original Hamiltonian and particle. If we perform the matrix multiplications in equation (6.19) explicitly we find

$$i\hbar\frac{\partial}{\partial t}\tilde{C}\psi^*(\mathbf{r},t) = \left(c\tilde{\alpha} \cdot \left(\frac{\hbar}{i}\nabla + e\mathbf{A}(\mathbf{r})\right) + \tilde{\beta}mc^2 - V(r)\right)\tilde{C}\psi^*(\mathbf{r},t) \qquad (6.22)$$

and making the identification

$$\psi_C(\mathbf{r},t) = \tilde{C}\psi^*(\mathbf{r},t) \qquad (6.23)$$

gives us back equation (6.17b). Thus, by devising the operator $\hat{\mathscr{C}}$=*take the complex conjugate and premultiply by* $i\tilde{\beta}\tilde{\alpha}_y$, we have discovered a mathematical way of transforming (6.17a) into (6.17b).

Let us check that the relation between the electron and positron wavefunctions implied by the above discussion is consistent with the free-particle wavefunctions we wrote down explicitly in chapter 5. Labelling the electron (ψ) and positron (ψ_C) wavefunctions ↑ (↓) for spin up (down) and +(−) for positive (negative) energy solutions, and lumping all the constants into one and calling it A we have, for a spin-up positron

$$\hat{\mathscr{C}}\psi_C^{\uparrow +} = \tilde{C}A^* \begin{pmatrix} 1 \\ 0 \\ \frac{cp_z}{W+mc^2} \\ \frac{cp_+}{W+mc^2} \end{pmatrix}^* e^{i\mathbf{p}\cdot\mathbf{r}/\hbar} = A \begin{pmatrix} \frac{cp_-}{W+mc^2} \\ -\frac{cp_z}{W+mc^2} \\ 0 \\ 1 \end{pmatrix} e^{i\mathbf{p}\cdot\mathbf{r}/\hbar} = \psi^{\downarrow -} \qquad (6.24a)$$

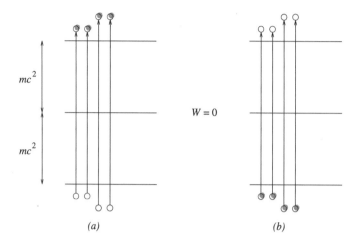

Fig. 6.1. (*a*) The (more or less) filled sea of negative energy electrons. A negative energy electron can be excited to positive energy states as shown. The hole left behind is the positron. (*b*) The charge conjugation operation shows that (mathematically) we could equally well consider the positron as the fundamental particle. A negative energy positron could be excited up to the positive energy states and the hole left behind would correspond to an electron.

and similarly for the spin-down positive energy electron

$$\hat{\mathscr{C}}\psi^{\downarrow+} = \tilde{C}A^* \begin{pmatrix} 0 \\ 1 \\ \frac{cp_-}{W+mc^2} \\ -\frac{cp_z}{W+mc^2} \end{pmatrix}^* e^{-i\mathbf{p}\cdot\mathbf{r}/\hbar} = -A \begin{pmatrix} \frac{cp_z}{W+mc^2} \\ \frac{cp_+}{W+mc^2} \\ 1 \\ 0 \end{pmatrix} e^{-i\mathbf{p}\cdot\mathbf{r}/\hbar} = -\psi_C^{\uparrow-}$$

$$(6.24b)$$

Here we have used the fact that A is real. This operator replaces a negative energy spin up (down) solution of (6.17*a*) with a positive energy spin-down (up) solution of (6.17*b*) and vice versa. So we can use this operator within the hole theory to give us the wavefunction for the antiparticle, if we know the wavefunction for the negative energy particle. From this analysis we see that the antiparticle has the same mass as, but opposite charge, momentum, spin and energy to, the particle, which is what we would expect within the hole theory.

Now that we have discovered a one-to-one mapping between the negative energy electrons and the positrons we can broaden our definition of the negative energy sea of electrons. We can see that an electron in a positive energy state is entirely equivalent to a positron in its equivalent negative energy state. There is a symmetrized sea of electrons and positrons which exist side by side. This is illustrated in figure 6.1. Mathematically there is no reason for assuming the electron is the fundamental particle. In the theory, both electron and positron have equal status. From the point of view of an observer on earth there appears to be an imbalance between the amount of matter and antimatter in the universe. However, it is very difficult to verify whether there is actually an imbalance if we consider the universe as a whole.

This symmetry of charge conjugation also tells us that the motion of an electron in a field $(\mathbf{A}(\mathbf{r}), V(\mathbf{r}))$ is identical to the motion of a positron in a field described by $(-\mathbf{A}(\mathbf{r}), -V(\mathbf{r}))$. Some readers may find this unsurprising. Indeed it is what would be expected on the basis of classical physics. The remarkable thing is that the positrons are actually predicted to have specific properties from charge conjugation and the hole theory. Clearly their discovery, with these properties, is one of the great successes of theoretical physics.

6.4 Time-Reversal Invariance

The physical meaning of time-reversal invariance is not quite so self-evident as that of charge conjugation. Perhaps the easiest way to think about it is to consider running a film of some simple physical process, involving electrons, forwards and backwards. It isn't strictly necessary to consider only electrons, but let's keep things as simple as we can. We can run the film forwards and we will see events occurring that are describable within the Dirac theory. Now, if we run the film backwards we will see all the events running backwards in time. Time-reversal invariance holds true if the dynamics underlying the time-reversed behaviour can also be described by the Dirac equation.

Time-reversal invariance is bound to be embodied in the Dirac equation if we can find some mathematical transformation that can be performed which leaves the Dirac equation invariant when we change $t \to -t$. The correct time-forward Dirac equation for a particle in an electromagnetic field is equation (6.17a). We can ask what this becomes if $t \to -t = t'$. Obviously the term that contains t will change sign and the t in the argument of the wavefunction will become t'. Under time reversal $\mathbf{r} \to \mathbf{r}'$ and $m \to m'$ so there is no change to the momentum and rest mass terms. The potential $V(\mathbf{r})$ will remain the same (like charges repel and unlike charges attract regardless of the direction of time). However, the vector potential $\mathbf{A}(\mathbf{r})$ reverses its sign if we reverse the direction of time. This is because the vector potential is generated by moving charges and the direction of motion of these charges will be reversed if time changes direction. So, the Dirac-like equation describing the time-reversed situation is

$$-i\hbar \frac{\partial \psi(\mathbf{r}, t')}{\partial t'} = \left(c\tilde{\boldsymbol{\alpha}} \cdot \left(\frac{\hbar}{i} \nabla + e\mathbf{A}(\mathbf{r}, t') \right) + \tilde{\beta} mc^2 + V(\mathbf{r}) \right) \psi(\mathbf{r}, t') \quad (6.25)$$

This is not the same as the Dirac equation (6.17a). So, now we are looking for an operation that will transform (6.25) into (6.17a). Let's not beat about the bush. To do this we take the complex conjugate of (6.25) and premultiply by

$$\tilde{T} = -\begin{pmatrix} 0 & -i & 0 & 0 \\ i & 0 & 0 & 0 \\ 0 & 0 & 0 & -i \\ 0 & 0 & i & 0 \end{pmatrix} = -i\tilde{\alpha}_x\tilde{\alpha}_z \tag{6.26}$$

This matrix is its own inverse, which is a useful property because we also require that it transform from the time-reversed state back to the time-forward state, as two time reversals must land you back in the time-forward state again. Let us perform this operation on equation (6.25):

$$i\hbar\frac{\partial}{\partial t'}\tilde{T}\psi^*(\mathbf{r},t') =$$
$$\tilde{T}\left(c\tilde{\alpha}^* \cdot \left(-\frac{\hbar}{i}\nabla + e\mathbf{A}(\mathbf{r},t')\right) + \tilde{\beta}^*mc^2 + V(\mathbf{r})\right)\tilde{T}^{-1}\tilde{T}\psi^*(\mathbf{r},t') \tag{6.27}$$

Now, it is easy to verify that

$$\tilde{T}\tilde{\alpha}^*\tilde{T}^{-1} = -\tilde{\alpha}, \qquad\qquad \tilde{T}\tilde{\beta}\tilde{T}^{-1} = \tilde{\beta} \tag{6.28}$$

and if we make the identification

$$\psi_T(\mathbf{r},t') = \tilde{T}\psi^*(\mathbf{r},t) \tag{6.29}$$

equation (6.27) takes the same form as equation (6.17a). So, our full time-reversal operator is: $\hat{\mathscr{T}}$=*take the complex conjugate and premultiply by the 4 × 4 matrix $\tilde{T}$ and let $t \to t' = -t$.* This operation leaves the Dirac equation invariant. Following the same route as we took in the case of the charge conjugation operator, let us act on a free-particle wavefunction with the time-reversal operator to see what we get. We will limit ourself to a particle moving in the z-direction only, for reasons which will become clear soon. The spin-up and spin-down free-particle wavefunctions for this case are given by equation (5.23) with $p_x = p_y = 0$, and again we will lump all the constants together into A. Now we operate on these wavefunctions with the time-reversal operator.

$$\hat{\mathscr{T}}\psi^\uparrow(p_z) = A\tilde{T}\begin{pmatrix} 1 \\ 0 \\ \frac{cp_z}{W+mc^2} \\ 0 \end{pmatrix}e^{-\frac{i}{\hbar}(\mathbf{p}\cdot\mathbf{r}-Wt')} = A\begin{pmatrix} 0 \\ i \\ 0 \\ \frac{icp_z}{W+mc^2} \end{pmatrix}e^{-\frac{i}{\hbar}(\mathbf{p}\cdot\mathbf{r}+Wt)}$$
$$= i\psi^\downarrow(-p_z) \tag{6.30a}$$

$$\hat{\mathscr{T}}\psi^{\downarrow}(p_z) = A\tilde{T}\begin{pmatrix} 0 \\ 1 \\ 0 \\ \frac{-cp_z}{W+mc^2} \end{pmatrix} e^{-\frac{i}{\hbar}(\mathbf{p}\cdot\mathbf{r}-Wt')} = A\begin{pmatrix} -i \\ 0 \\ \frac{icp_z}{W+mc^2} \\ 0 \end{pmatrix} e^{-\frac{i}{\hbar}(\mathbf{p}\cdot\mathbf{r}+Wt)}$$

$$= -i\psi^{\uparrow}(-p_z) \qquad (6.30b)$$

So a spin-up particle is mapped onto a spin-down particle with opposite momentum and *vice versa*. The i and $-i$ multiplying the final expressions in (6.30a) and (6.30b) are phase factors. When included in an expectation value expression they disappear. If such a phase factor does occur and the Hamiltonian is invariant under time reversal, then the energy eigenstates occur in time-reversed pairs. This is known as Kramers degeneracy. If we had included the x- and y-components of momentum in this example we would have got into difficulties with the definition of the spin direction of a moving electron again.

For electromagnetic interactions we have shown that if we film a physical process described by the Dirac equation, we can run the film backwards and we will still see a process that can be described by the Dirac equation. The particles in the time-forward case will be described by a wavefunction $\psi(\mathbf{r}, t)$ and those in the time-reversed case by $\psi_T(\mathbf{r}, t')$. So, time-reversal invariance is an intrinsic symmetry of the Dirac equation, and therefore Dirac particles must have this symmetry. This operation was first introduced by Wigner (1932).

6.5 Parity

The parity operator acting on a wavefunction or equation transforms the position coordinate to its negative, i.e. $\mathbf{r} \rightarrow -\mathbf{r} = \mathbf{r}'$. This operation is often described as taking a mirror image, although it is not really that. A plane mirror only inverts coordinates perpendicular to itself, whereas to get the parity operation we require a rotation about the normal to the mirror as well. To describe this symmetry we proceed in exactly the same way as we have for the charge conjugation and time-reversal symmetries. We will write down the Dirac equation after a parity operation and then find the operator that transforms the result back into the usual Dirac form.

After a parity operation the scalar potential remains the same, whereas the vector potential changes sign. This is a consequence of its vector nature. Of course, the momentum operator also changes sign and we have

$$i\hbar\frac{\partial\psi_P(\mathbf{r}', t)}{\partial t} = \left(c\tilde{\alpha}\cdot\left(-\frac{\hbar}{i}\nabla' + e\mathbf{A}(\mathbf{r}', t)\right) + \tilde{\beta}mc^2 + V(\mathbf{r}')\right)\psi_P(\mathbf{r}', t) \quad (6.31)$$

The signs of ∇ and $\mathbf{A}(\mathbf{r})$ have only changed if we measure them with respect to the untransformed set of space coordinates. If we measure them with respect to the parity-inverted coordinates we would expect to get back the original Dirac equation (6.17a). Now let us define the matrices

$$\tilde{P} = \begin{pmatrix} i & 0 & 0 & 0 \\ 0 & i & 0 & 0 \\ 0 & 0 & -i & 0 \\ 0 & 0 & 0 & -i \end{pmatrix} = i\tilde{\beta} \longrightarrow \tilde{P}^{-1} = \begin{pmatrix} -i & 0 & 0 & 0 \\ 0 & -i & 0 & 0 \\ 0 & 0 & i & 0 \\ 0 & 0 & 0 & i \end{pmatrix} = -i\tilde{\beta}$$

(6.32)

Now, an operator that transforms (6.31) into the form of (6.17a) is just *premultiply by* $\tilde{P}$:

$$i\hbar \frac{\partial}{\partial t} \tilde{P}\psi(\mathbf{r}', t) = \tilde{P} \left(c\tilde{\alpha} \cdot (i\hbar\nabla' + e\mathbf{A}(\mathbf{r}', t)) + \tilde{\beta}mc^2 + V(\mathbf{r}') \right) \tilde{P}^{-1}\tilde{P}\psi(\mathbf{r}, t)$$

(6.33)

and it is straightforward to see that

$$\tilde{P}\tilde{\alpha}\tilde{P}^{-1} = -\tilde{\alpha}, \qquad\qquad \tilde{P}\tilde{\beta}\tilde{P}^{-1} = \tilde{\beta} \qquad\qquad (6.34)$$

and making the identification

$$\psi_P(\mathbf{r}', t) = \tilde{P}\psi(\mathbf{r}, t) \qquad\qquad (6.35)$$

we have obtained (6.17a) as intended. So the full parity operator is $\hat{\mathscr{P}}=$ Let $\mathbf{r} \to -\mathbf{r} = \mathbf{r}'$ *and premultiply by* $\tilde{P}$. We have shown that the Dirac equation is invariant under the operation of inverting space. Any physical process described by the Dirac equation can be looked at in a mirror (a special mirror that inverts all three dimensions) and that process is also describable by the Dirac equation. The inverse is also true, we can use the parity operator to describe the unreflected events in the reflected coordinate system. Next we ask what effect the parity operator has on the free-particle wavefunction. There is really no need to write this out explicitly. It is clear that replacing $\mathbf{r}$ by $\mathbf{r}'$ in the free-particle wavefunction (5.23) and premultiplying by $\tilde{P}$ will give a phase factor multiplying a wavefunction for a free particle with the same spin and opposite momentum.

6.6 $\hat{\mathscr{C}}\hat{\mathscr{P}}\hat{\mathscr{T}}$

In the above three sections we have examined the charge conjugation, time-reversal and parity operators individually. In fact there is further insight to be gained into the symmetry of Dirac particles if we combine them, so in this section we consider them together.

Consider first the Dirac equation for an electron and act on it with the parity operator. This will give us back the Dirac equation with $\mathbf{r}$ replaced with $\mathbf{r}'$. Acting on this with the charge conjugation operator gives

$$i\hbar \frac{\partial \psi_{PC}(\mathbf{r}', t)}{\partial t} = \left(c\tilde{\alpha} \cdot \left(\frac{\hbar}{i}\nabla' + e\mathbf{A}(\mathbf{r}', t) \right) + \tilde{\beta}mc^2 - V(\mathbf{r}') \right) \psi_{PC}(\mathbf{r}', t) \quad (6.36)$$

Now let us write down the Dirac equation for the positron.

$$i\hbar\frac{\partial\psi_C(\mathbf{r},t)}{\partial t} = \left(c\tilde{\boldsymbol{\alpha}}\cdot\left(\frac{\hbar}{i}\nabla + e\mathbf{A}(\mathbf{r},t)\right) + \tilde{\beta}mc^2 - V(\mathbf{r})\right)\psi_C(\mathbf{r},t) \quad (6.37)$$

and we operate on this equation with the parity operator. This gives

$$i\hbar\frac{\partial\psi_{CP}(\mathbf{r}',t)}{\partial t} = \left(c\tilde{\boldsymbol{\alpha}}\cdot\left(\frac{\hbar}{i}\nabla' - e\mathbf{A}(\mathbf{r}',t)\right) + \tilde{\beta}mc^2 - V(\mathbf{r}')\right)\psi_{CP}(\mathbf{r}',t) \quad (6.38)$$

But (6.38) is identical to (6.36), so the space-inverted positron is the charge conjugate of the space-inverted electron, or, put mathematically, the charge conjugation operator and the parity operator commute.

Now let us look at what happens if we operate on the Dirac equation for the negative energy solutions with all three of these operators:

$$\hat{\mathscr{C}}\hat{\mathscr{P}}\hat{\mathscr{T}}\left(c\tilde{\boldsymbol{\alpha}}\cdot(\hat{\mathbf{p}} - e\mathbf{A}(\mathbf{r},t)) + \tilde{\beta}mc^2 + V(\mathbf{r},t)\right)(\hat{\mathscr{C}}\hat{\mathscr{P}}\hat{\mathscr{T}})^{-1}\hat{\mathscr{C}}\hat{\mathscr{P}}\hat{\mathscr{T}}\psi(\mathbf{r},t)$$
$$= -W\hat{\mathscr{C}}\hat{\mathscr{P}}\hat{\mathscr{T}}\psi(\mathbf{r},t) \quad (6.39)$$

where W is defined as being positive. First, consider the wavefunction.

$$\hat{\mathscr{P}}\hat{\mathscr{C}}\hat{\mathscr{T}}\psi(\mathbf{r},t) = \hat{\mathscr{P}}\hat{\mathscr{C}}(\tilde{T}\psi^*(\mathbf{r},t')) = \hat{\mathscr{P}}(\tilde{C}(i\tilde{\alpha}_x\tilde{\alpha}_z\psi^*(\mathbf{r},t'))^*)$$
$$= \hat{\mathscr{P}}(i\tilde{\beta}\tilde{\alpha}_y(-i\tilde{\alpha}_x\tilde{\alpha}_z\psi(\mathbf{r},t'))) = i\tilde{\beta}(i\tilde{\beta}\tilde{\alpha}_y(-i\tilde{\alpha}_x\tilde{\alpha}_z\psi(\mathbf{r}',t')))$$
$$= i\tilde{\alpha}_y\tilde{\alpha}_x\tilde{\alpha}_z\psi(\mathbf{r}',t') = -\tilde{\gamma}_5\psi(\mathbf{r}',t') = \psi_{\mathscr{C}\mathscr{P}\mathscr{T}}(\mathbf{r}',t') \quad (6.40)$$

With this substitution (6.39) becomes

$$\hat{\mathscr{C}}\hat{\mathscr{P}}\hat{\mathscr{T}}\left(c\tilde{\boldsymbol{\alpha}}\cdot(\hat{\mathbf{p}} - e\mathbf{A}(\mathbf{r},t))) + \tilde{\beta}mc^2 + V(\mathbf{r})\right)\hat{\mathscr{T}}^{-1}\hat{\mathscr{P}}^{-1}\hat{\mathscr{C}}^{-1}\psi_{\mathscr{C}\mathscr{P}\mathscr{T}}(\mathbf{r}',t')$$
$$= -W\psi_{\mathscr{C}\mathscr{P}\mathscr{T}}(\mathbf{r}',t') \quad (6.41)$$

Now, we already know that the time-reversal operator and the parity operator will give us the same Hamiltonian as appears in (6.41) with $\mathbf{r}$ replaced by $\mathbf{r}' = -\mathbf{r}$ and t replaced by $t' = -t$. We also know from earlier that the charge conjugation operator acting on this Hamiltonian will give us the Hamiltonian with the signs of the terms in the kinetic energy and rest mass reversed, but those in the potentials remaining the same. If we then multiply through by -1, (6.41) becomes

$$\left(c\tilde{\boldsymbol{\alpha}}\cdot\left(\frac{\hbar}{i}\nabla' + e\mathbf{A}(\mathbf{r}',t')\right) + \tilde{\beta}mc^2 - V(\mathbf{r}')\right)\psi_{\mathscr{C}\mathscr{P}\mathscr{T}}(\mathbf{r}',t') = W\psi_{\mathscr{C}\mathscr{P}\mathscr{T}}(\mathbf{r}',t')$$
$$(6.42)$$

What does this mathematics mean? You may well ask. We have started with the Dirac equation for a negative energy electron and by using these symmetries we have ended up with (6.42) which is the Dirac equation for a positron, but $\mathbf{r}'$ and t' are of opposite sign to $\mathbf{r}$ and t. So, using the $\hat{\mathscr{C}}\hat{\mathscr{P}}\hat{\mathscr{T}}$ we have transformed a negative energy electron into a positive energy

Table 6.1. Symmetry of various operators and physical quantities under charge conjugation, time reversal, parity and $\hat{\mathscr{C}}\hat{\mathscr{P}}\hat{\mathscr{T}}$.

Symmetry	$\hat{\mathscr{C}}$	$\hat{\mathscr{P}}$	$\hat{\mathscr{T}}$	$\hat{\mathscr{P}}\hat{\mathscr{C}}\hat{\mathscr{T}}$
Operator				
$\hat{\mathbf{r}}$	$+$	$-$	$+$	$-$
$\hat{\mathbf{p}}$	$-$	$-$	$-$	$-$
$\hat{\mathbf{r}} \times \hat{\mathbf{p}}$	$-$	$+$	$-$	$+$
ρ	$+$	$+$	$+$	$+$
$\hat{\mathbf{v}}$	$+$	$-$	$-$	$+$
e	$-$	$+$	$+$	$-$
$\mathbf{A}(\mathbf{r})$	$+$	$-$	$-$	$+$
$\Phi(\mathbf{r}) = V(\mathbf{r})/e$	$+$	$+$	$+$	$+$
$\tilde{\alpha}$	$+$	$-$	$-$	$+$
$\tilde{\beta}$	$-$	$+$	$+$	$-$

positron moving backwards in space-time. This view of the positron is fundamental, but is really only a model, and forms the basis of the positron theory formulated by Stückelberg (1941) and Feynman (1949). Let us check this by evaluating $\psi_{\mathscr{C}\mathscr{P}\mathscr{T}}(\mathbf{r}',t')$ using (6.40):

$$\psi_{\mathscr{C}\mathscr{P}\mathscr{T}}(\mathbf{r}',t') = -\tilde{\gamma}_5\psi(\mathbf{r}',t') = A \begin{pmatrix} 0 & \tilde{I} \\ \tilde{I} & 0 \end{pmatrix} \begin{pmatrix} \chi^{\pm 1/2} \\ \frac{c\tilde{\sigma}\cdot\mathbf{p}}{W+mc^2}\chi^{\pm 1/2} \end{pmatrix} e^{i(\mathbf{p}\cdot\mathbf{r}'-Wt')/\hbar}$$

$$= A \begin{pmatrix} \frac{c\tilde{\sigma}\cdot\mathbf{p}}{W+mc^2}\chi^{\pm 1/2} \\ \chi^{\pm 1/2} \end{pmatrix} e^{i(\mathbf{p}\cdot\mathbf{r}'-Wt')/\hbar} \tag{6.43}$$

This is the required result: comparison with (5.61) shows that we have indeed transformed a positron into a negative energy electron moving backwards in space and time.

One application of this view of antiparticles is the following. Equation (6.42) can be taken as an equation for a positron in a negative Coulomb potential. All the theory of the relativistic one-electron atom (which we examine in chapter 8) can be gone through for the antiparticle one-electron atom, i.e. a positron in the field of an antiproton. Equation (6.42) predicts that the spectrum associated with the antimatter one-electron atom will be identical to that for the usual one-electron atom. If it is not, the $\hat{\mathscr{C}}\hat{\mathscr{P}}\hat{\mathscr{T}}$ symmetry is broken, with fundamental consequences for our view of nature. The creation of such antimatter atoms and measurement of their spectra is a very sensitive test of this symmetry (Baur *et al.* 1996).

For the reader's convenience, in table 6.1 there is a summary showing how various operators and physical quantities behave under the charge conjugation, parity and time-reversal operations and $\mathscr{C}\mathscr{P}\mathscr{T}$. You should note that we have proved these symmetries only for particles experiencing electromagnetic interactions. Lee and Yang (1957) showed that these individual symmetries don't necessarily hold in interactions governed by the weak nuclear force. However, $\hat{\mathscr{C}}\hat{\mathscr{P}}\hat{\mathscr{T}}$ is still a valid symmetry in that case. It has been shown that if we only assume Lorentz invariance and the fact that spin-1/2 particles obey Fermi–Dirac statistics whereas spin-zero particles obey Bose–Einstein statistics, $\hat{\mathscr{C}}\hat{\mathscr{P}}\hat{\mathscr{T}}$ invariance is guaranteed (Ludders 1954).

6.7 Angular Momentum Again

In chapter 2 we discussed angular momentum within non-relativistic spin theory in some detail. Now that we have derived and examined the Dirac equation, it is time to think about angular momentum again, in particular its generalization to a 4×4 theory and, hence, its implications for Dirac particles. This, of course, will find much application when we come to discuss atoms. We are going to consider angular momentum in a slightly non-standard way. Usually authors discuss matrix angular momentum in terms of the well-known properties of the operators. Here we are going to write out the matrices explicitly. Although this is cumbersome, it is instructive to do it once.

The first step is to write down the full Dirac Hamiltonian in matrix form as we will want to refer to it several times later in this section:

$$\hat{H} = \begin{pmatrix} V + mc^2 & 0 & c(\hat{p}_z - eA_z) & c(\hat{p}_- - eA_-) \\ 0 & V + mc^2 & c(\hat{p}_+ - eA_+) & -c(\hat{p}_z - eA_z) \\ c(\hat{p}_z - eA_z) & c(\hat{p}_- - eA_-) & V - mc^2 & 0 \\ c(\hat{p}_+ - eA_+) & -c(\hat{p}_z - eA_z) & 0 & V - mc^2 \end{pmatrix}$$

(6.44)

where $\hat{p}_\pm = \hat{p}_x \pm i\hat{p}_y$ and $A_\pm = A_x \pm iA_y$ and where we have omitted the dependence of the potentials on the coordinate for brevity. Next we write the angular momentum operators in matrix form. The orbital angular momentum is simply the usual operator multiplied by the identity matrix:

$$\hat{L}_x = (\hat{y}\hat{p}_z - z\hat{p}_y)\tilde{I}_4, \qquad \hat{L}_y = (\hat{z}\hat{p}_x - x\hat{p}_z)\tilde{I}_4, \qquad \hat{L}_z = (\hat{x}\hat{p}_y - y\hat{p}_x)\tilde{I}_4$$

(6.45)

The spin operators have already been defined in 2×2 matrix form in equation (2.30). The 4×4 form of $\hat{S}_z$ was applied in chapter 5. In full the 4×4 spin operator takes on the unsurprising form

$$\hat{S}_x = \frac{\hbar}{2}\tilde{\sigma}_x, \qquad \hat{S}_y = \frac{\hbar}{2}\tilde{\sigma}_y, \qquad \hat{S}_z = \frac{\hbar}{2}\tilde{\sigma}_z$$

(6.46)

where the $\tilde{\sigma}_i$ are the 4×4 σ-matrices of equation (4.15). Now we can define the total angular momentum operator

$$\hat{\mathbf{J}} = (\hat{J}_x, \hat{J}_y, \hat{J}_z)$$

(6.47)

and, of course, this can be written in terms of the spin and orbital angular momentum. For illustrative purposes we will write the total angular momentum components out in full matrix form

$$\hat{J}_z =$$
$$\begin{pmatrix} \hat{x}\hat{p}_y - \hat{y}\hat{p}_x + \frac{\hbar}{2} & 0 & 0 & 0 \\ 0 & \hat{x}\hat{p}_y - \hat{y}\hat{p}_x - \frac{\hbar}{2} & 0 & 0 \\ 0 & 0 & \hat{x}\hat{p}_y - \hat{y}\hat{p}_x + \frac{\hbar}{2} & 0 \\ 0 & 0 & 0 & \hat{x}\hat{p}_y - \hat{y}\hat{p}_x - \frac{\hbar}{2} \end{pmatrix}$$

(6.48a)

$$\hat{J}_x = \begin{pmatrix} \hat{y}\hat{p}_z - \hat{z}\hat{p}_y & \hbar/2 & 0 & 0 \\ \hbar/2 & \hat{y}\hat{p}_z - \hat{z}\hat{p}_y & 0 & 0 \\ 0 & 0 & \hat{y}\hat{p}_z - \hat{z}\hat{p}_y & \hbar/2 \\ 0 & 0 & \hbar/2 & \hat{y}\hat{p}_z - \hat{z}\hat{p}_y \end{pmatrix}$$

(6.48b)

$$\hat{J}_y = \begin{pmatrix} \hat{z}\hat{p}_x - \hat{x}\hat{p}_z & -i\hbar/2 & 0 & 0 \\ i\hbar/2 & \hat{z}\hat{p}_x - \hat{x}\hat{p}_z & 0 & 0 \\ 0 & 0 & \hat{z}\hat{p}_x - \hat{x}\hat{p}_z & -i\hbar/2 \\ 0 & 0 & i\hbar/2 & \hat{z}\hat{p}_x - \hat{x}\hat{p}_z \end{pmatrix}$$

(6.48c)

Before examining the properties of these matrix operators in relation to the Dirac Hamiltonian we should first confirm that they are correct representations by checking that they obey the commutation relations of equation (2.4b). Here we will evaluate one such commutator explicitly and it will be apparent that the others will follow by cyclic permutation of the indices. From the definitions above it is straightforward to see that if we multiply them one way (being careful to keep the operators in the correct order at all times) we get

$$\hat{J}_x\hat{J}_y = \begin{pmatrix} \hat{L}_x\hat{L}_y + \frac{i\hbar^2}{4} & -\frac{i\hbar\hat{L}_+}{2} & 0 & 0 \\ \frac{i\hbar\hat{L}_-}{2} & \hat{L}_x\hat{L}_y - \frac{i\hbar^2}{4} & 0 & 0 \\ 0 & 0 & \hat{L}_x\hat{L}_y + \frac{i\hbar^2}{4} & -\frac{i\hbar\hat{L}_+}{2} \\ 0 & 0 & \frac{i\hbar\hat{L}_-}{2} & \hat{L}_x\hat{L}_y - \frac{i\hbar^2}{4} \end{pmatrix}$$

(6.49a)

and if we multiply them the other way we find

$$\hat{J}_y\hat{J}_x = \begin{pmatrix} \hat{L}_y\hat{L}_x - \frac{i\hbar^2}{4} & -\frac{i\hbar\hat{L}_+}{2} & 0 & 0 \\ \frac{i\hbar\hat{L}_-}{2} & \hat{L}_y\hat{L}_x + \frac{i\hbar^2}{4} & 0 & 0 \\ 0 & 0 & \hat{L}_y\hat{L}_x - \frac{i\hbar^2}{4} & -\frac{i\hbar\hat{L}_+}{2} \\ 0 & 0 & \frac{i\hbar\hat{L}_-}{2} & \hat{L}_y\hat{L}_x + \frac{i\hbar^2}{4} \end{pmatrix}$$

(6.49b)

Subtracting these two gives the commutator between $\hat{J}_x$ and $\hat{J}_y$:

$$[\hat{J}_x, \hat{J}_y] = \hat{J}_x\hat{J}_y - \hat{J}_y\hat{J}_x$$

$$= [\hat{L}_x, \hat{L}_y] \begin{pmatrix} 1 & 0 & 0 & 0 \\ 0 & 1 & 0 & 0 \\ 0 & 0 & 1 & 0 \\ 0 & 0 & 0 & 1 \end{pmatrix} + \frac{i\hbar^2}{2} \begin{pmatrix} 1 & 0 & 0 & 0 \\ 0 & -1 & 0 & 0 \\ 0 & 0 & 1 & 0 \\ 0 & 0 & 0 & -1 \end{pmatrix}$$

$$= i\hbar\hat{L}_z\tilde{I} + i\hbar\frac{\hbar}{2}\tilde{\sigma}_z = i\hbar(\hat{L}_z + \hat{S}_z) = i\hbar\hat{J}_z \tag{6.50}$$

An exactly analogous piece of mathematics will give the corresponding results for $[\hat{J}_y, \hat{J}_z]$ and $[\hat{J}_z, \hat{J}_x]$. So these matrix representations of the angular momentum operators clearly do obey the required commutators.

It is now just a matter of matrix multiplication and use of the position–momentum commutator to show that

$$[\hat{J}_i, \hat{H}] = 0 \tag{6.51}$$

where $i = x, y, z$. So total angular momentum commutes with the Dirac Hamiltonian and hence is a constant of the motion. When this commutator is evaluated it is easy to see the necessity of including the spin and orbital angular momenta to get something that commutes with $\hat{H}$. In fact we find

$$[\hat{L}_i, \hat{H}] = -[\hat{S}_i, \hat{H}] \neq 0 \tag{6.52}$$

Using these commutators and equation (2.4a) proves that total angular momentum is conserved in the fully relativistic quantum theory. However, spin angular momentum and orbital angular momentum are not conserved separately, as we found previously for Pauli theory with spin–orbit coupling.

The next step is to write the $\hat{\mathbf{J}}^2$ operator in matrix form:

$$\hat{\mathbf{J}}^2 = \hat{J}_x^2 + \hat{J}_y^2 + \hat{J}_z^2 = \left(\hat{L}_x + \frac{\hbar\tilde{\sigma}_x}{2}\right)^2 + \left(\hat{L}_y + \frac{\hbar\tilde{\sigma}_y}{2}\right)^2 + \left(\hat{L}_z + \frac{\hbar\tilde{\sigma}_z}{2}\right)^2$$

$$= \hat{\mathbf{L}}^2 + \frac{\hbar^2}{4}(\tilde{\sigma}_x^2 + \tilde{\sigma}_y^2 + \tilde{\sigma}_z^2) + \hbar(\hat{L}_x\tilde{\sigma}_x + \hat{L}_y\tilde{\sigma}_y + \hat{L}_z\tilde{\sigma}_z)$$

$$= \hat{\mathbf{L}}^2 + \hbar\hat{\mathbf{L}}\cdot\tilde{\boldsymbol{\sigma}} + \frac{3\hbar^2}{4} \tag{6.53}$$

where we have used the facts that $\tilde{\boldsymbol{\sigma}}$ and $\hat{\mathbf{L}}$ commute and the square of all the Pauli matrices is the identity matrix. Now, we know from chapter 2 that the eigenvalues of the total angular momentum operator are of the form $j(j+1)\hbar^2$. Next we define a new operator $\hat{K}$ such that

$$\hat{K}^2 = \hat{\mathbf{J}}^2 + \frac{\hbar^2}{4} \tag{6.54}$$

and the eigenvalues of this are

$$\hat{K}^2\psi = \left(\hat{\mathbf{J}}^2 + \frac{\hbar^2}{4}\right)\psi = (j(j+1)\hbar^2 + \frac{\hbar^2}{4} = \left(j+\frac{1}{2}\right)^2\hbar^2\psi \tag{6.55}$$

and so

$$\hat{K}\psi = \pm\left(j+\frac{1}{2}\right)\hbar\psi = -\kappa\hbar\psi \tag{6.56}$$

In chapter 2 we defined an operator $\hat{K}$, and the more awake reader may realise that (6.56) is simply the 4×4 equivalent of equation (2.87). From (6.53) and (6.54), and using the fact that $(\tilde{\sigma}\cdot\hat{\mathbf{L}})^2$ is not, as you might expect from (4.19), equal to $\hat{\mathbf{L}}^2$ because the components of $\hat{\mathbf{L}}$ do not commute, we find that

$$\hat{K}^2 = \hat{\mathbf{J}}^2 + \frac{\hbar^2}{4} = \hat{\mathbf{L}}^2 + \hbar\hat{\mathbf{L}}\cdot\tilde{\sigma} + \hbar^2 = \hat{\mathbf{L}}^2 - \hbar\hat{\mathbf{L}}\cdot\tilde{\sigma} + 2\hbar\hat{\mathbf{L}}\cdot\tilde{\sigma} + \hbar^2$$

$$= (\tilde{\sigma}\cdot\hat{\mathbf{L}})^2 + 2\hbar\hat{\mathbf{L}}\cdot\tilde{\sigma} + \hbar^2 = \tilde{\beta}^2(\tilde{\sigma}\cdot\hat{\mathbf{L}} + \hbar)^2 \tag{6.57}$$

Here the $\tilde{\beta}$ is not strictly necessary, but it is included by convention. Equation (6.57) is an alternative expression for $\hat{K}$ (see equation (2.86)). This operator is of fundamental importance in spherically symmetric problems, and naturally one thinks of atoms in this context. It can be seen immediately from (6.56) that the eigenvalue κ can take on all integral values except zero. Indeed κ as the eigenvalue of (6.56) has exactly the same relation to the other quantum numbers as those described in equations (2.126) and (2.127). The eigenvalue is chosen as negative for consistency with equation (2.87). For completeness we will write $\hat{K}$ as a matrix operator:

$$\hat{K} = \tilde{\beta}(\tilde{\sigma}\cdot\hat{\mathbf{L}} + \hbar) = \begin{pmatrix} \hat{L}_z + \hbar & \hat{L}_- & 0 & 0 \\ \hat{L}_+ & -\hat{L}_z + \hbar & 0 & 0 \\ 0 & 0 & -\hat{L}_z - \hbar & -\hat{L}_- \\ 0 & 0 & -\hat{L}_+ & \hat{L}_z - \hbar \end{pmatrix} \tag{6.58}$$

Now let us verify some commutators. Firstly a simple one: using (6.51) it is easy to see that

$$[\hat{\mathbf{J}}^2, \hat{H}] = \hat{\mathbf{J}}^2\hat{H} - \hat{H}\hat{\mathbf{J}}^2 = \hat{\mathbf{J}}[\hat{\mathbf{J}}, \hat{H}] - [\hat{H}, \hat{\mathbf{J}}]\hat{\mathbf{J}} = 0 \tag{6.59}$$

$$[\hat{K}^2, \hat{H}] = [\hat{\mathbf{J}}^2 + \hbar^2/4, \hat{H}] = 0 \tag{6.60}$$

It is not straightforward to show that

$$[\hat{K}, \hat{H}] = 0 \tag{6.61}$$

but this commutator is equal to zero, and if you don't believe it, feel free to work it out from equations (6.44) and (6.58). It is a long mathematical proof (I recommend you believe it). Finally, from the definition of $\hat{K}$ it is obvious that

$$[\hat{K}, \hat{\mathbf{J}}^2] = [\hat{K}, \hat{J}_z] = 0 \tag{6.62}$$

All these commutators prove two things: firstly, that $\hat{\mathbf{J}}^2$, $\hat{J}_z$ and $\hat{K}$ all represent quantities that are constants of the motion; secondly, eigenfunctions of the Hamiltonian are also eigenfunctions of these operators. When we come to discuss the relativistic hydrogen atom all these operators will be of central importance.

6.8 Non-Relativistic Limits Again

Let us write down the Dirac Hamiltonian for a free particle (yet again)

$$\hat{H} = c\tilde{\boldsymbol{\alpha}} \cdot \hat{\mathbf{p}} + \tilde{\beta}mc^2 \tag{6.63}$$

Now, suppose that the particle described by this Hamiltonian experiences a perturbation due to some electromagnetic effect. This may be an interaction with a photon or entering some region where it experiences a magnetic field. The effect on the eigenvalues and eigenvectors of $\hat{H}$ of such an occurrence may be calculated using perturbation theory. The perturbation Hamiltonian is found in the usual way by letting

$$\hat{\mathbf{p}} \to \hat{\mathbf{p}} - e\mathbf{A} \tag{6.64}$$

where $\mathbf{A}$ is the vector potential associated with the perturbation. Inserting this into (6.63) gives us the perturbation Hamiltonian

$$\hat{H}' = -ec\tilde{\boldsymbol{\alpha}} \cdot \mathbf{A} \tag{6.65}$$

This does not look like the more familiar non-relativistic expression. The Schrödinger Hamiltonian for a free particle is

$$\hat{H}_S = \frac{1}{2m}\hat{\mathbf{p}}^2 \tag{6.66}$$

and putting (6.64) in here, the Schrödinger perturbation Hamiltonian is

$$\hat{H}'_S = \frac{e}{2m}(\mathbf{A} \cdot \hat{\mathbf{p}} + \hat{\mathbf{p}} \cdot \mathbf{A}) \tag{6.67}$$

where we are assuming the vector potential is small so the term in $\mathbf{A}^2$ can be ignored. The question that arises is how (6.65) and (6.67) can possibly describe the same thing. In the Schrödinger case the Hamiltonian is quadratic in the momentum operator and when the perturbation is added, terms linear in the momentum operator are carried through into the perturbation. However, the relativistic Dirac Hamiltonian is linear in momentum, so no terms containing the momentum operator carry through into the perturbation.

Certainly, one can resolve this difference by going through the involved mathematics required to take the non-relativistic limit of the Dirac equation described in chapter 4, or by performing a Foldy–Wouthuysen transformation as described in chapter 7. However, there is a simple hand-waving kind of argument which can get us from (6.65) to (6.67) directly. We start by remembering that $\mathbf{A}$ is a position dependent vector field whereas $\tilde{\alpha}$ is just a 4×4 matrix containing simple numbers, so they obviously commute with one another. We also have to recall that $c\tilde{\alpha}$ is the velocity operator $\hat{\mathbf{v}}$, so

$$ec\tilde{\alpha} \cdot \mathbf{A} = \frac{e}{2}(c\tilde{\alpha} \cdot \mathbf{A} + \mathbf{A} \cdot c\tilde{\alpha}) = \frac{e}{2}(\hat{\mathbf{v}} \cdot \mathbf{A} + \mathbf{A} \cdot \hat{\mathbf{v}}) = \frac{e}{2m}(\hat{\mathbf{p}} \cdot \mathbf{A} + \mathbf{A} \cdot \hat{\mathbf{p}}) \quad (6.68)$$

As required, this is identical to the right hand side of (6.67). We have taken the non-relativistic limit in deriving (6.68) because we have used a non-relativistic formula for the momentum operator $\hat{\mathbf{p}} = m\hat{\mathbf{v}}$.

6.9 Second Quantization

In this section we will discuss the creation and annihilation operators for bosons and fermions, although it is the latter with which we will be principally concerned in subsequent chapters. Much, if not all, of the first part of this section could equally well be presented in a text on non-relativistic quantum theory. Our strategy will be as follows. We will introduce some general concepts in many-body theory and see how they lead us to postulate the two different kinds of particles, bosons and fermions. Then we will discuss bosons and fermions separately. After that we will do some more general work with second quantized operators. This will lead naturally to field operators. Among the many good books on this subject are those by Baym (1967) and Merzbacher (1970). Only in the final section will we see how this formalism is modified when applied to relativistic systems (see Sakurai (1967) for example).

Creation and annihilation operators act on a many-body wavefunction which is written in Fock space (where?). Such wavefunctions are called state vectors. Fock space is a mathematical space which contains all possible state vectors (so that's consistent then). These are a rather different way of writing wavefunctions to that which we have been using. State vectors are a way of writing general many-body wavefunctions in a manageable way using Dirac notation. They look like this:

$$|n_1, n_2, n_3, \cdots, n_i, \cdots\rangle \quad (6.69)$$

This is actually a wavefunction with some subtle assumptions underlying it. It postulates some operator, often, but not necessarily, a Hamiltonian $\hat{A}$ with eigenvalues A_i. Equation (6.69) states that there are n_1 particles with eigenvalue A_1, n_2 with eigenvalue A_2, and n_i particles with eigenvalue A_i.

Such a wavefunction assumes that the many-body system it describes can be written as some sort of sum of single-particle behaviour. i.e. it assumes that in the many-body system the particles retain a lot of their single-particle identity. This is by no means obvious in quantum mechanics. Equation (6.69) is a particularly easy way to write a many-body wavefunction because all the position and time dependence as well as constants and normalization are suppressed. Particular examples of Fock space wavefunctions are as follows.

The vacuum state is written as

$$|0, 0, \cdots, 0\rangle = |0\rangle \tag{6.70}$$

i.e. it contains no particles in any states. However, with our picture of the vacuum as a sea of filled negative energy states, some modification of this definition will be required later.

Single-particle states are

$$|0, 0, \cdots, 0, n_i = 1, 0, 0, \cdots, 0\rangle = |A_i\rangle \tag{6.71}$$

Here there is a single particle in quantum state i and none in any other state. We can see that the space of all single-particle wavefunctions is a very small subset of the whole Fock space.

Now we are going to define the second quantized operators a_i and $a_i^\dagger$. The first of these is known as an annihilation operator. It takes a many-body system with n_i particles in the ith state and relates it to the many-body system with $n_i - 1$ particles in the ith state. The operator $a_i^\dagger$ is known as a creation operator. It relates the many-body system with n_i particles in the ith state to the many body system with $n_i + 1$ particles in the ith state. In both cases the number of particles in all other states remains constant, i.e.

$$a_i|n_1, n_2, n_3, \cdots, n_i, \cdots\rangle = k|n_1, n_2, n_3, \cdots, n_i - 1, \cdots\rangle \tag{6.72}$$

$$a_i^\dagger|n_1, n_2, n_3, \cdots, n_i, \cdots\rangle = p|n_1, n_2, n_3, \cdots, n_i + 1, \cdots\rangle \tag{6.73}$$

where k and p are, as yet, unknown constants.

One very important feature of this approach to the quantum mechanics of many-body systems is how a unitary transformation from one basis to another is made. To see this explicitly let us introduce a second operator $\hat{B}$ with eigenvalues B_j and occupation numbers m_j. Then the transformation from the m to the n basis is written as

$$|A_i\rangle = \sum_j |B_j\rangle\langle B_j|A_i\rangle = \sum_j |B_j\rangle c_{ij} \tag{6.74}$$

where the transformation coefficients c_{ij} form a unitary matrix:

$$\sum_k c_{ik} c_{kj}^* = \delta_{ij} \tag{6.75}$$

Of course in writing (6.74) we have used the fact that the eigenfunctions of both $\hat{A}$ and $\hat{B}$ form a complete set. A transformation like (6.74) does not change the vacuum state. However, a state that is a one-particle state in the basis of eigenfunctions of $\hat{A}$ is, in general, not a one-particle state in the basis of eigenfunctions of $\hat{B}$. It is perfectly possible, though, to introduce single-particle creation and annihilation operators in the latter representation in exactly the same way as was done for the former:

$$b_i |m_1, m_2, m_3, \cdots, m_i, \cdots\rangle = k' |m_1, m_2, m_3, \cdots, m_i - 1, \cdots\rangle \tag{6.76}$$

$$b_i^\dagger |m_1, m_2, m_3, \cdots, m_i, \cdots\rangle = p' |m_1, m_2, m_3, \cdots, m_i + 1, \cdots\rangle \tag{6.77}$$

For single-particle states

$$a_i^\dagger |0\rangle = |0, \cdots, n_i = 1, 0, \cdots\rangle = \sum_j c_{ij} |0, \cdots, n_j = 1, 0, \cdots\rangle = \sum_j c_{ij} b_j^\dagger |0\rangle \tag{6.78}$$

and this is only satisfied if

$$a_i^\dagger = \sum_j c_{ij} b_j^\dagger \tag{6.79a}$$

and we can take the hermitian adjoint of this equation to give

$$a_i = \sum_j c_{ij}^* b_j \tag{6.79b}$$

Equations (6.79) have been shown to hold for single-particle states only, but we will assume that they hold over the whole of Fock space.

Now a creation operator only acts on one quantum state within the Fock space wavefunction. Therefore, if we were to act with two creation operators on different single-particle states we would expect the state we end up with to be independent of the order in which we apply the operators; the only thing that might change is the normalization, so for the particular states l and m we expect

$$a_l^\dagger a_m^\dagger |l, m\rangle = \lambda a_m^\dagger a_l^\dagger |l, m\rangle \tag{6.80}$$

Now we can use (6.79a) to transform this into the basis of eigenfunctions of $\hat{B}$. This gives

$$(a_l^\dagger a_m^\dagger - \lambda a_m^\dagger a_l^\dagger)|l, m\rangle = \sum_{p,q} c_{lp} c_{mq} (b_p^\dagger b_q^\dagger - \lambda b_q^\dagger b_p^\dagger)|p, q\rangle = 0 \tag{6.81}$$

As the transformation coefficients are arbitrary, except for the constraint of unitarity, this is only satisfied if

$$b_p^\dagger b_q^\dagger - \lambda b_q^\dagger b_p^\dagger = 0 \qquad (6.82a)$$

Of course, we could just as easily have switched round the l and m subscripts in (6.80) and then we would have arrived at

$$b_q^\dagger b_p^\dagger - \lambda b_p^\dagger b_q^\dagger = 0 \qquad (6.82b)$$

This can be substituted into (6.82a) to give

$$b_p^\dagger b_q^\dagger = \lambda^2 b_p^\dagger b_q^\dagger \qquad (6.83)$$

and so

$$\lambda = \pm 1 \qquad (6.84)$$

Substituting $\lambda = +1$ back into (6.80) gives us immediately

$$a_l^\dagger a_m^\dagger - a_m^\dagger a_l^\dagger = 0 \qquad (6.85a)$$

and the hermitian adjoint of this is

$$a_l a_m - a_m a_l = 0 \qquad (6.85b)$$

A similar argument can be made to give

$$a_l a_m^\dagger - a_m^\dagger a_l = \delta_{l,m} \qquad (6.85c)$$

If, instead, we allow the solution $\lambda = -1$ in (6.80) we have

$$a_l^\dagger a_m^\dagger + a_m^\dagger a_l^\dagger = 0 \qquad (6.86a)$$

and the hermitian adjoint of this is

$$a_l a_m + a_m a_l = 0 \qquad (6.86b)$$

and as previously

$$a_l a_m^\dagger + a_m^\dagger a_l = \delta_{l,m} \qquad (6.86c)$$

So the operators with $\lambda = +1$ commute and those with $\lambda = -1$ anticommute. This argument implies that there are two different types of particle. Those that obey (6.85) are known as Bose–Einstein particles or bosons for short, those that obey (6.86) are called Fermi–Dirac particles or fermions for short. For a system of bosons in equilibrium at temperature T with chemical potential μ the fraction of particles in the quantum state with energy W_i is

$$f(W_i - \mu) = \frac{1}{e^{(W_i - \mu)/k_B T} - 1} \qquad (6.87)$$

and for fermions the same quantity is

$$f(W_i - \mu) = \frac{1}{e^{(W_i - \mu)/k_B T} + 1} \tag{6.88}$$

With our understanding of the commutation relations for the creation and annihilation operators for bosons and fermions we can now assign consistent values to the constants k and p in (6.72) and (6.73).

Let us consider bosons first. Equations (6.72) and (6.73) can be written

$$a_i | n_1, n_2, n_3, \cdots, n_i, \cdots \rangle = \sqrt{n_i} | n_1, n_2, n_3, \cdots, n_i - 1, \cdots \rangle \tag{6.89}$$

$$a_i^\dagger | n_1, n_2, n_3, \cdots, n_i, \cdots \rangle = \sqrt{n_i + 1} | n_1, n_2, n_3, \cdots, n_i + 1, \cdots \rangle \tag{6.90}$$

and the commutation rules can immediately be seen. We haven't proved (6.85c) but with the definitions (6.89) and (6.90) we have

$$(a_i a_i^\dagger - a_i^\dagger a_i) | \cdots, n_i, \cdots \rangle$$
$$= a_i \sqrt{n_i + 1} | \cdots, n_i + 1, \cdots \rangle - a_i^\dagger \sqrt{n_i} | \cdots, n_i - 1, \cdots \rangle$$
$$= \sqrt{n_i + 1} \sqrt{n_i + 1} | \cdots, n, \cdots \rangle - \sqrt{n_i} \sqrt{n_i} | \cdots, n_i, \cdots \rangle$$
$$= | \cdots, n_i, \cdots \rangle \tag{6.91}$$

which is consistent with (6.85c). Now, a_i is the hermitian conjugate of $a_i^\dagger$ and so if $|\Psi\rangle$ is a general many-body wavefunction we can write

$$\langle \Psi | a_i a_i^\dagger | \Psi \rangle = \langle a_i^\dagger \Psi | a_i^\dagger \Psi \rangle = \langle a_i a_i^\dagger \Psi | \Psi \rangle \tag{6.92}$$

Next consider the operator $\hat{N} = a_i^\dagger a_i$ acting on an arbitrary state vector

$$\hat{N} | \cdots, n_i, \cdots \rangle = a_i^\dagger a_i | \cdots, n_i, \cdots \rangle = a_i^\dagger \sqrt{n_i} | \cdots, n_i - 1, \cdots \rangle$$
$$= \sqrt{n_i} a_i^\dagger | \cdots, n_i - 1, \cdots \rangle = n_i | \cdots, n_i, \cdots \rangle \tag{6.93}$$

So the operator $\hat{N}$ is a number operator, it tells us the number of particles in the ith quantum state. The existence of this operator was implied by our representation of the wavefunction in the form (6.69).

With the definitions we have so far we can see that any boson many-body state in Fock space can be constructed by repeated operation of the creation operators on the vacuum state:

$$|n_1, n_2, \cdots, n_i, \cdots \rangle = \frac{(a_1^\dagger)^{n_1}}{\sqrt{n_1!}} \frac{(a_2^\dagger)^{n_2}}{\sqrt{n_2!}} \cdots \frac{(a_i^\dagger)^{n_i}}{\sqrt{n_i!}} \cdots |0, 0, \cdots, 0, \cdots \rangle \tag{6.94}$$

Creation and annihilation operators that act on different states do not interfere with one another, as shown by the commutation relations (6.85). It doesn't matter which order we perform the operations in equation (6.94).

Now let us move on to consider the creation and annihilation operators for fermions. There are two important differences from the boson case.

Firstly, any state can only be occupied by one particle. Hence n_i in the state vector can only take on the values 0 and 1. Secondly, the operators anticommute, which leads to the antisymmetric nature of the many-fermion wavefunction. It would be useful to see explicitly how this antisymmetry comes about. By analogy with equation (6.94) we can build up a state vector for a many-fermion system by repeated operation of the fermion creation and annihilation operators

$$|n_1, n_2, \cdots, n_i, \cdots\rangle = (a_1^\dagger)^{n_1}(a_2^\dagger)^{n_2} \cdots (a_i^\dagger)^{n_i} \cdots |0, 0, \cdots, 0, \cdots\rangle \qquad (6.95)$$

with n_i always equal to zero or one. The factorials in (6.94) are not present here because $n_i!$ is always one. Using the anticommutation relations (6.85) it is immediately clear that if we interchange the order of any two of the creation operators in (6.95) the sign of the wavefunction on the left hand side is changed, so the antisymmetry is built in. Let us see how this works for simple cases.

Firstly consider the simplest case, a single particle state. The creation and annihilation operators acting on such a state are

$$a_1^\dagger|0\rangle = |1\rangle, \qquad a_1|1\rangle = |0\rangle, \qquad a_1^\dagger|1\rangle = a_1|0\rangle = 0 \qquad (6.96)$$

This is easy, it is just the same as (6.89) and (6.90) for bosons. As there is only one particle there is no need to worry about antisymmetry.

For two particles the situation is a bit more complicated. Inside the state vector we call the left hand state 1 and the right hand state 2. There are four possible states, and the creation and annihilation operators acting on them are

$$a_1^\dagger|0, 0\rangle = |1, 0\rangle, \quad a_1^\dagger|0, 1\rangle = |1, 1\rangle, \quad a_1^\dagger|1, 0\rangle = a_1^\dagger|1, 1\rangle = 0 \qquad (6.97a)$$
$$a_1|1, 1\rangle = |0, 1\rangle, \quad a_1|1, 0\rangle = |0, 0\rangle, \quad a_1|0, 1\rangle = a_1|0, 0\rangle = 0 \qquad (6.97b)$$
$$a_2^\dagger|0, 0\rangle = |0, 1\rangle, \quad a_2^\dagger|1, 0\rangle = -|1, 1\rangle, \quad a_2^\dagger|0, 1\rangle = a_2^\dagger|1, 1\rangle = 0 \qquad (6.97c)$$
$$a_2|1, 1\rangle = -|1, 0\rangle, \quad a_2|0, 1\rangle = |0, 0\rangle, \quad a_2|1, 0\rangle = a_2|0, 0\rangle = 0 \qquad (6.97d)$$

As you can see we have defined some negative signs here. This has been done to maintain the antisymmetry. The second equation in (6.97c) could be said to come from $a_2^\dagger|1, 0\rangle = a_2^\dagger a_1^\dagger|0, 0\rangle$ which is (6.95) for two particles with the operators reversed. That is why it has a negative sign. The negative signs also have to be consistent so the sign in the first equation of (6.97d) is there because it must undo the work of the equivalent creation operation in the second of (6.97c). Antisymmetry of the two-fermion state can be demonstrated explicitly by taking the state $|1, 1\rangle$ and removing the particle in state 1 using a_1. The particle in state 2 is removed from state 2 and put in state 1 using $a_1^\dagger a_2$. The first particle is then placed back in state 2 with $a_2^\dagger$, i.e.

$$a_2^\dagger a_1^\dagger a_2 a_1 |1, 1\rangle = a_2^\dagger a_1^\dagger a_2 |0, 1\rangle = a_2^\dagger a_1^\dagger |0, 0\rangle = a_2^\dagger |1, 0\rangle = -|1, 1\rangle \qquad (6.98)$$

Of course, this could also be done the other way round. We could remove the particle from state 2, move the particle from state 1 to state 2 and put back the remaining particle in state 1. This gives the same result as (6.98). You might think that the operation defined in (6.98) should yield the same result as

$$a_1^\dagger a_2^\dagger a_2 a_1 |1, 1\rangle = a_1^\dagger a_2^\dagger a_2 |0, 1\rangle = a_1^\dagger a_2^\dagger |0, 0\rangle = a_1^\dagger |0, 1\rangle = |1, 1\rangle \qquad (6.99)$$

but it doesn't. The reason for the lack of a sign change is that (6.99) represents the operation of removing a particle from state 1, removing a particle from state 2 and then putting the same particle back into state 2, and finally putting the original particle back into state 1 – that is, removing the electrons and putting them back into their original states. When using these operators for fermions it is very important to make sure that the order in which the operations are performed corresponds to the physical system you are trying to describe. A rule of thumb is that you continue to work with the same particle unless the operator cannot operate on it in its present state, then you go back to the most recently operated on particle that you can. Of course you are always subject to the restriction that no state can ever contain two or more particles.

Finally here, we will consider the three-state system without going into all the mathematical detail of the two-state system above. The wavefunction must change sign if we interchange any two particles regardless of whether the third state is filled or empty. So, in analogy with the case above, we can write the following:

$$
\begin{aligned}
a_1^\dagger |0, 1, 0\rangle &= -|1, 1, 0\rangle, & a_1|1, 1, 0\rangle &= -|0, 1, 0\rangle \\
a_2^\dagger |0, 0, 1\rangle &= -|0, 1, 1\rangle, & a_2|0, 1, 1\rangle &= -|0, 0, 1\rangle & (6.100) \\
a_3^\dagger |1, 0, 0\rangle &= -|1, 0, 1\rangle, & a_3|1, 0, 1\rangle &= -|1, 0, 0\rangle
\end{aligned}
$$

All other single creation or annihilation operators acting on the three-particle state either give zero or the appropriate state multiplied by $+1$. Again we will see what happens if we interchange two particles. If we interchange particles in states 1 and 2 the interchange operator is identical to the two-particle case.

$$a_2^\dagger a_1^\dagger a_2 a_1 |1, 1, 0\rangle = -a_2^\dagger a_1^\dagger a_2 |0, 1, 0\rangle = -a_2^\dagger a_1^\dagger |0, 0, 0\rangle$$
$$= -a_2^\dagger |1, 0, 0\rangle = -|1, 1, 0\rangle \qquad (6.101a)$$

Now, if the third state is filled we must still get the antisymmetry emerging. Let us see if that works:

$$a_2^\dagger a_1^\dagger a_2 a_1 |1, 1, 1\rangle = a_2^\dagger a_1^\dagger a_2 |0, 1, 1\rangle = -a_2^\dagger a_1^\dagger |0, 0, 1\rangle$$
$$= -a_2^\dagger |1, 0, 1\rangle = -|1, 1, 1\rangle \qquad (6.101b)$$

So it works again. It can be seen immediately that cyclic permutation of the subscripts 1, 2 and 3 will yield the same result. If you have time, try doing the same thing for other pairs of particles or with the operator with the subscripts reversed $a_2^\dagger a_1^\dagger a_2 a_1$. If you don't have time, do it anyway.

For four or more particles, working out which operators acting on which state vectors have negative prefactors becomes complicated. In condensed matter we are usually working with the order of 10^{23} particles, so working it out becomes impossible and is pointless. Useful quantities are usually written in terms of several creation and annihilation operators. For fermions the number operator for the number of particles in a single particle state i is given by

$$\hat{N}_i = a_i^\dagger a_i \qquad (6.102)$$

and the operator for the total number of fermions in the system is

$$\hat{N}_{\text{tot}} = \sum_i \hat{N}_i = \sum_i a_i^\dagger a_i \qquad (6.103)$$

Now we will go on to consider a few more of the general properties of second quantized operators. In particular we will see how to use them to calculate some useful quantities. The simplest case is an observable which can be regarded as the sum of single-particle observables. An example might be the total mass of the system. Anyway, operators representing such observables can be written using the number operator and single-particle eigenvalues as

$$\hat{A} = \sum_i A_i \hat{N}_i = \sum_i A_i a_i^\dagger a_i \qquad (6.104)$$

This can be transformed to the b representation above using equations (6.79). Let us do this explicitly:

$$\hat{A} = \sum_i A_i a_i^\dagger a_i = \sum_i A_i \sum_q c_{iq} b_q^\dagger \sum_r c_{ir}^* b_r = \sum_{q,r} b_q^\dagger b_r \sum_i c_{iq} A_i c_{ir}$$
$$= \sum_{q,r} b_q^\dagger b_r \sum_i \langle B_q | A_i \rangle A_i \langle A_i | B_r \rangle = \sum_{q,r} b_q^\dagger b_r \langle B_q | A | B_r \rangle \qquad (6.105)$$

with

$$A = \sum_i |A_i\rangle A_i \langle A_i| \qquad (6.106)$$

being just a definition which we use to keep the equations as short as possible. Operators like these are the simplest imaginable.

Let us go one step further and consider additive two-particle operators, the obvious example being any two particles' interaction such as their mutual potential energy. The total potential energy operator is the sum of the individual two-particle potential energy operators. If V_{ij} is the matrix element of the potential energy operator for two particles, one in state i and the other in state j, the operator representing the total potential energy is

$$\hat{V} = \frac{1}{2} \sum_{i,j \neq i} \hat{N}_i \hat{N}_j V_{ij} = \frac{1}{2} \sum_{i,j} (\hat{N}_i \hat{N}_j - \hat{N}_i \delta_{ij}) V_{ij} \qquad (6.107)$$

The quantity $P_{ij} = \hat{N}_i \hat{N}_j - \hat{N}_i \delta_{ij}$ is known as the pair correlation function. In a gas of fermions there is a tendency for the particles to avoid one another, because of the Pauli exclusion principle. The pair correlation function can be used to give a theoretical measure of this. Physically it tells us the probability of finding a particle in state j if we know there is an electron in state i. Using the Fermi–Dirac operator commutation relations we can see from (6.107) that

$$P_{ij} = a_i^\dagger a_j^\dagger a_j a_i \qquad (6.108)$$

So

$$\hat{V} = \frac{1}{2} \sum_{i,j} a_i^\dagger a_j^\dagger a_j a_i V_{ij} \qquad (6.109)$$

or, if we want to transform to the b representation

$$\hat{V} = \sum_{p,q,r,s} b_p^\dagger b_r^\dagger b_q b_s \langle p, r | V_{prqs} | q, s \rangle \qquad (6.110)$$

where

$$\langle p, r | V_{prqs} | q, s \rangle = \sum_{ij \neq i} \langle B_p | A_i \rangle \langle A_i | B_q \rangle V_{ij} \langle B_r | A_j \rangle \langle A_j | B_s \rangle \qquad (6.111)$$

For the potential energy operator we must have $V_{ij} = V_{ji}$ and so

$$\langle p, r | V | q, s \rangle = \langle r, p | V | s, q \rangle \qquad (6.112)$$

A matrix element such as this can be represented diagrammatically as shown in figure 6.2.

Now let us just briefly consider the type of problem for which a second quantized approach is appropriate. It will often be useful in a system where we have both localized and itinerant electrons. Imagine we have an atom whose outer electronic shell is partially filled and highly degenerate.

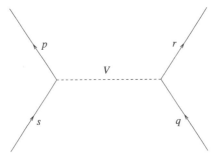

Fig. 6.2. Illustration of a two-particle matrix element.

The obvious example is a rare earth atom with its partially filled f-shell. A Hartree–Fock type of calculation can be carried out and the single-electron energy levels ϵ_i can be found. These are fully occupied up to some energy and above that they are unoccupied. The energy of the atom is described by the Hamiltonian

$$\hat{H}_0 = \sum_{i=1}^{\text{occ}} \epsilon_i a_i^\dagger a_i \tag{6.113}$$

Now suppose we immerse this atom in a system of free electrons. The free-electron energy bands will fill up to the Fermi energy, and free electrons may drop into the localized states. However, there is an energy barrier to this, which is the Coulomb repulsion of the electrons already localized on the atom. This can be represented by a term in the Hamiltonian

$$\hat{H}_C = \frac{1}{2} \sum_{i,j,i\neq j} U_{ij} a_i^\dagger a_i a_j^\dagger a_j = \frac{1}{2} \sum_{i,j,i\neq j} U_{ij} \hat{n}_i \hat{n}_j \tag{6.114}$$

Here U_{ij} represents the Coulomb repulsion between an electron in state i and one in state j. This is multiplied by the relevant number operators, and summed over all possible states. We can see that the Coulomb repulsion increases by some factor proportional to U_{ij} every time another electron is put into the localized state. Thus the energy necessary to add in another electron rises until we get to a position where the energy required would raise the localized level above the Fermi energy. At that point no further electrons can drop into the localized state. So, the occupation number of the localized state depends upon the Coulomb repulsion, as well as the single-particle properties that were calculated originally. The Hamiltonian $\hat{H} = \hat{H}_0 + \hat{H}_C$ is known as the Hubbard or Anderson–Hubbard Hamiltonian and finds application in many areas of condensed matter physics. It is the simplest imaginable way of including some electron–electron correlations into condensed matter theory.

6.10 Field Operators

In what we have covered so far on the topic of second quantization we have made an enormous assumption. We have assumed that the many-body system we have, and the operator used to define the state vector, have discrete eigenvalues. This is OK if we are working with particles in bound systems, but in many situations this is not the case and we need to consider the possibility that the single-particle basis we use is a continuous one. Probably the most familiar example of such a basis is the usual position coordinate $\mathbf{r}$. We will only be concerned with this case and will use it as our example throughout. The normalization condition that we use is

$$\langle \mathbf{r}' | \mathbf{r}'' \rangle = \delta(\mathbf{r}' - \mathbf{r}'') \tag{6.115}$$

The second quantization formalism can be transcribed to the continuous case quite straightforwardly, the principal difference being that the occupation number representation in Fock space is no longer much use.

Because we have a continuous variable, rather than a discrete one, the creation and annihilation operators are now represented as $\varphi_i^\dagger(\mathbf{r})$ and $\varphi_i(\mathbf{r})$ respectively. The operator $\varphi_i^\dagger(\mathbf{r})$ creates a particle in the quantum state labelled by i at position $\mathbf{r}$ and $\varphi_i(\mathbf{r})$ destroys a particle in state i at position $\mathbf{r}$. The φ_is are continuous operators with position and are called field operators. The commutation and anticommutation relations obeyed by these operators are the same as in the discrete case for both bosons and fermions. For brevity we will unite these into a single set of relations

$$\varphi_i(\mathbf{r}')\varphi_j(\mathbf{r}'') \pm \varphi_j(\mathbf{r}'')\varphi_i(\mathbf{r}') = 0 \tag{6.116a}$$

$$\varphi_i^\dagger(\mathbf{r}')\varphi_j^\dagger(\mathbf{r}'') \pm \varphi_j^\dagger(\mathbf{r}'')\varphi_i^\dagger(\mathbf{r}') = 0 \tag{6.116b}$$

$$\varphi_i(\mathbf{r}')\varphi_j^\dagger(\mathbf{r}'') \pm \varphi_j^\dagger(\mathbf{r}'')\varphi_i(\mathbf{r}') = \delta(\mathbf{r}' - \mathbf{r}'')\delta_{ij} \tag{6.116c}$$

where the upper signs refer to fermions and the lower ones refer to bosons.

Now suppose we want to define some many-body state. It may be written down in terms of the field operators as

$$|\mathbf{r}_1, i_1; \mathbf{r}_2, i_2; \cdots; \mathbf{r}_n, i_n\rangle = \frac{1}{\sqrt{n}} \varphi_{i_1}^\dagger(\mathbf{r}_1)\varphi_{i_2}^\dagger(\mathbf{r}_2) \cdots \varphi_{i_n}^\dagger(\mathbf{r}_n)|\Psi(0)\rangle \tag{6.117}$$

From the commutation relations (6.116) we can see immediately that for bosons this state is symmetric under the interchange of any two particles, whereas for fermions it is antisymmetric under such an interchange. Equation (6.117) describes a localized state with a particle in state i_1 at point $\mathbf{r}_1$, a particle in state i_2 at $\mathbf{r}_2$ etc. This state automatically has the right symmetry because of the anticommutation relations of the field

operators. A full *n*-particle state is given by a coherent superposition of
these localized states:

$$|\Psi\rangle = \int d\mathbf{r}_1 \cdots \int d\mathbf{r}_n \psi(\mathbf{r}_1, \cdots, \mathbf{r}_n)|\mathbf{r}_1, i_1; \mathbf{r}_2, i_2; \cdots; \mathbf{r}_n, i_n\rangle \qquad (6.118)$$

where $\psi(\mathbf{r}_1, \mathbf{r}_2, \cdots, \mathbf{r}_n)$ is the spatial part of the many-body wavefunction.
Equations (6.117) and (6.118) may be regarded in the following way. The
field operators create localized particles at the points $\mathbf{r}_1, \mathbf{r}_2, \cdots, \mathbf{r}_n$. These
are then multiplied by the relative phase $\psi(\mathbf{r}_1, \mathbf{r}_2, \cdots, \mathbf{r}_n)$. The integration
completes the many-particle state by defining it over all space. Here again,
the correct symmetry is guaranteed.

Now, let us go on to see how we use the field operators in the evaluation
of observables. The number operator is

$$\hat{n} = \sum_i \int \varphi_i^\dagger(\mathbf{r})\varphi_i(\mathbf{r})d\mathbf{r} \qquad (6.119)$$

For a continuous system we can define operators in the same way as for
the discrete case. An operator representing an observable that may be
regarded as a sum of single-particle terms can be written analogously to
(6.105) as

$$\mathscr{A} = \sum_{ij} \int \int \varphi_i^\dagger(\mathbf{r}')\langle \mathbf{r}', i|\hat{A}|\mathbf{r}'', j\rangle \varphi_j(\mathbf{r}'')d\mathbf{r}'d\mathbf{r}'' \qquad (6.120)$$

where the matrix element within the integral is a matrix element of the
single-particle operator $\hat{A}$ with single-particle wavefunctions.

For additive two-particle operators we can make a direct generalization
of equation (6.109) to give

$$\mathscr{V} = \sum_{ij}\sum_{kl} \int \int \int \int \varphi_i^\dagger(\mathbf{r}')\varphi_j^\dagger(\mathbf{r}'')\langle \mathbf{r}', i, \mathbf{r}'', j|\hat{V}|\mathbf{r}''', k, \mathbf{r}'''', l\rangle$$
$$\varphi_k(\mathbf{r}''')\varphi_l(\mathbf{r}'''')d\mathbf{r}'d\mathbf{r}''d\mathbf{r}'''d\mathbf{r}'''' \qquad (6.121)$$

Again the matrix element on the right hand side here is a matrix element
of a two-particle operator over two pairs of single-particle states.

This is all rather esoteric, so let us consider these operators in a bit
more detail by doing a specific example. We will work in a basis of
non-relativistic plane waves as they are by now very familiar and are
particularly simple to work with. The field operator for such a state
creates a plane wave at position $\mathbf{r}$. We take the spatial parts of our
wavefunctions in the form

$$\psi(\mathbf{r}) = \frac{1}{\sqrt{V}}e^{-i\mathbf{p}\cdot\mathbf{r}/\hbar} \qquad (6.122)$$

Now, a field operator creates a particle with spin s at point $\mathbf{r}$ which is in a superposition of these states, i.e.

$$\varphi_s^\dagger(\mathbf{r}) = \frac{1}{\sqrt{V}} \sum_{\mathbf{p}} e^{-i\mathbf{p}\cdot\mathbf{r}/\hbar} a_{\mathbf{p}s}^\dagger \tag{6.123a}$$

and the hermitian conjugate of this equation is

$$\varphi_s(\mathbf{r}) = \frac{1}{\sqrt{V}} \sum_{\mathbf{p}} e^{i\mathbf{p}\cdot\mathbf{r}/\hbar} a_{\mathbf{p}s} \tag{6.123b}$$

The single-particle part of these field operators should be regarded only as an amplitude and phase multiplying the appropriate creation or annihilation operator. We can invert (6.123a) by multiplying each side by $e^{i\mathbf{p}_1\cdot\mathbf{r}/\hbar}$ and integrating over all space. This is completely straightforward and gives

$$a_{\mathbf{p}_1 s}^\dagger = \frac{1}{\sqrt{V}} \int e^{i\mathbf{p}_1\cdot\mathbf{r}/\hbar} \varphi_s^\dagger(\mathbf{r}) d\mathbf{r} \tag{6.124a}$$

and similarly for (6.123b)

$$a_{\mathbf{p}_1 s} = \frac{1}{\sqrt{V}} \int e^{-i\mathbf{p}_1\cdot\mathbf{r}/\hbar} \varphi_s(\mathbf{r}) d\mathbf{r} \tag{6.124b}$$

Now the kinetic energy of a system of particles may be regarded as the sum of the kinetic energies of the single particles. So using equation (6.104) we may write the total kinetic energy operator as

$$\hat{T} = \sum_{\mathbf{p}s} \frac{p^2}{2m} a_{\mathbf{p}s}^\dagger a_{\mathbf{p}s} \tag{6.125}$$

Now we can substitute (6.124) directly into this expression to give

$$\hat{T} = \frac{1}{2m} \frac{1}{V} \sum_{\mathbf{p}s} \int \int \mathbf{p} e^{i\mathbf{p}\cdot\mathbf{r}/\hbar} \mathbf{p} e^{-i\mathbf{p}\cdot\mathbf{r}'/\hbar} \varphi_s^\dagger(\mathbf{r})\varphi_s(\mathbf{r}') d\mathbf{r}' d\mathbf{r}$$

$$= \frac{1}{2mV} \sum_{\mathbf{p}s} \int \int \left(\frac{\hbar}{i}\nabla e^{i\mathbf{p}\cdot\mathbf{r}/\hbar}\right)\left(\frac{-\hbar}{i}\nabla' e^{-i\mathbf{p}\cdot\mathbf{r}'/\hbar}\right) \varphi_s^\dagger(\mathbf{r})\varphi_s(\mathbf{r}') d\mathbf{r}' d\mathbf{r}$$

$$= -\frac{\hbar^2}{2m} \sum_s \int \nabla\varphi_s^\dagger(\mathbf{r})\nabla\varphi_s(\mathbf{r}) d\mathbf{r} \tag{6.126}$$

where the $\mathbf{r}$ and $\mathbf{r}'$ integrations have separated immediately and we have carried out an integration by parts. Similarly if we don't do the integral in equation (6.119) we have the number density operator at a point $\mathbf{r}$

$$\hat{n}(\mathbf{r}) = \sum_i \varphi_i^\dagger(\mathbf{r})\varphi_i(\mathbf{r}) \tag{6.127}$$

The remarkable thing about equations (6.126) and (6.127) is that they look identical to the single-particle expression for the expectation values of these operators. We can look at this the other way round, in many-body theory, single-particle wavefunctions have become operators which create and annihilate particles, while single-particle expectation values have generalized to become operators for the corresponding many-body observables. This is the historical reason for the name 'second quantization'. We will utilize this formalism extensively when we come to discuss many-electron systems in chapter 10.

6.11 Second Quantization in Relativistic Quantum Mechanics

Up to the present our discussion of second quantized operators could equally well have appeared in a text on non-relativistic quantum theory. Here we will generalize the previous discussion to a relativistic framework.

Using the arguments above, the total Hamiltonian for a system of free relativistic Dirac particles may be postulated to be

$$\hat{H} = \int \varphi^\dagger (c\tilde{\boldsymbol{\alpha}} \cdot \hat{\mathbf{p}} + \tilde{\beta}mc^2)\varphi d\mathbf{r} \tag{6.128}$$

Now, the operator in brackets here is a 4×4 operator, implying, in analogy with the single-particle picture, that the Dirac field operator is a four-component quantity. So we need to understand what is meant by this. The discussion at the end of the previous section, describing the field operators as creation or annihilation operators multiplied by a plane wave, gives us the clue as to how to proceed.

Consider the free-particle wavefunction of equations (5.23). It will be convenient to change the notation slightly and write U and V as a single symbol $U^{(j)}$. We will return to the original notation shortly. Now we can write the free-particle wavefunctions as

$$\psi^\uparrow = \frac{1}{\sqrt{V}} U^{(1)}(\mathbf{p}) e^{i\mathbf{p}\cdot\mathbf{r}/\hbar}, \qquad \psi^\downarrow = \frac{1}{\sqrt{V}} U^{(2)}(\mathbf{p}) e^{i\mathbf{p}\cdot\mathbf{r}/\hbar} \tag{6.129a}$$

and for the negative energy states (5.61) gives

$$\psi^{\uparrow -} = \frac{1}{\sqrt{V}} U^{(3)}(\mathbf{p}) e^{i\mathbf{p}\cdot\mathbf{r}/\hbar}, \qquad \psi^{\downarrow -} = \frac{1}{\sqrt{V}} U^{(4)}(\mathbf{p}) e^{i\mathbf{p}\cdot\mathbf{r}/\hbar} \tag{6.129b}$$

In the previous section we had to sum our field operators over spin states. Here we will sum over both spin and positive/negative energy states. Our field operators are then

$$\varphi(\mathbf{r}, t) = \frac{1}{\sqrt{V}} \sum_{\mathbf{p}} \sum_{j} a_{\mathbf{p}}^{(j)} U^{(j)}(\mathbf{p}) e^{i\mathbf{p}\cdot\mathbf{r}/\hbar} \tag{6.130a}$$

$$\varphi^{\dagger}(\mathbf{r}, t) = \frac{1}{\sqrt{V}} \sum_{\mathbf{p}} \sum_{j} a_{\mathbf{p}}^{(j)\dagger} U^{(j)\dagger}(\mathbf{p}) e^{-i\mathbf{p}\cdot\mathbf{r}/\hbar} \tag{6.130b}$$

The operators $a_{\mathbf{p}}^{(j)\dagger}$ and $a_{\mathbf{p}}^{(j)}$ create and annihilate a particle in the state with momentum $\mathbf{p}$ and index j. Note that in this relativistic theory we explicitly allow the creation and annihilation operators to have a time dependence, which we will evaluate below. Firstly we substitute (6.130) into (6.128) giving

$$\hat{H} = \frac{1}{V} \int \sum_{\mathbf{p}} \sum_{j} a_{\mathbf{p}}^{(j)\dagger} U^{(j)\dagger}(\mathbf{p}) e^{-i\mathbf{p}\cdot\mathbf{r}/\hbar} (c\tilde{\alpha} \cdot \hat{\mathbf{p}} + \tilde{\beta} mc^2) \times$$

$$\sum_{\mathbf{p}'} \sum_{j'} a_{\mathbf{p}'}^{(j')} U^{(j')}(\mathbf{p}') e^{i\mathbf{p}'\cdot\mathbf{r}/\hbar} d\mathbf{r}$$

$$= \sum_{\mathbf{p},\mathbf{p}'} \sum_{j,j'} \delta_{\mathbf{p}\mathbf{p}'} W' U^{(j)\dagger}(\mathbf{p}) U^{(j')}(\mathbf{p}') a_{\mathbf{p}}^{(j)\dagger} a_{\mathbf{p}'}^{(j')} \tag{6.131}$$

where we have operated on the right hand plane wave with the Hamiltonian to obtain its energy and the momentum delta function has arisen from the integration over $\mathbf{r}$. From chapter 5 we know that

$$U^{(j)\dagger}(\mathbf{p}) U^{(j')}(\mathbf{p}) = \delta_{jj'}, \qquad \sum_{j} U^{(j)}(\mathbf{p}) U^{(j)\dagger}(\mathbf{p}) = \tilde{I}_4 \tag{6.132}$$

The first of these enables us to simplify the Hamiltonian to a level where the terms have a very clear meaning:

$$\hat{H} = \sum_{\mathbf{p}} \left(\sum_{j=1,2} W a_{\mathbf{p}}^{(j)\dagger} a_{\mathbf{p}}^{(j)} - \sum_{j=3,4} W a_{\mathbf{p}}^{(j)\dagger} a_{\mathbf{p}}^{(j)} \right) \tag{6.133}$$

where we have used the fact that the energy of a plane wave is positive if $j = 1, 2$ and negative if $j = 3, 4$. Equation (6.133) makes good sense. It is the energy of the positive energy states multiplied by the number of those states minus the modulus of the energy of the negative energy states multiplied by the number of those, so it is clearly the total energy written as the sum of the energies of the individual particles. With the Hamiltonian in this form, the time dependence of the creation and annihilation operators can be derived straight away from Heisenberg's equation of motion.

$$
\frac{\partial a_{\mathbf{p}}^{(j)}(t)}{\partial t} = \frac{i}{\hbar}[\hat{H}, a_{\mathbf{p}}^{(j)}(t)] = -\frac{i}{\hbar} W a_{\mathbf{p}}^{(j)}(t) \qquad j = 1, 2
$$
$$
= +\frac{i}{\hbar} W a_{\mathbf{p}}^{(j)}(t) \qquad j = 3, 4
$$
(6.134a)

$$
\frac{\partial a_{\mathbf{p}}^{(j)\dagger}(t)}{\partial t} = \frac{i}{\hbar}[\hat{H}, a_{\mathbf{p}}^{(j)\dagger}(t)] = +\frac{i}{\hbar} W a_{\mathbf{p}}^{(j)\dagger}(t) \qquad j = 1, 2
$$
$$
= -\frac{i}{\hbar} W a_{\mathbf{p}}^{(j)\dagger}(t) \qquad j = 3, 4
$$
(6.134b)

from which we immediately deduce that

$$
\begin{array}{llr}
a_{\mathbf{p}}^{(j)}(t) = a_{\mathbf{p}}^{(j)}(0)e^{-iWt/\hbar} & j = 1, 2 & (6.135a) \\
a_{\mathbf{p}}^{(j)}(t) = a_{\mathbf{p}}^{(j)}(0)e^{+iWt/\hbar} & j = 3, 4 & (6.135b) \\
a_{\mathbf{p}}^{(j)\dagger}(t) = a_{\mathbf{p}}^{(j)\dagger}(0)e^{+iWt/\hbar} & j = 1, 2 & (6.135c) \\
a_{\mathbf{p}}^{(j)\dagger}(t) = a_{\mathbf{p}}^{(j)\dagger}(0)e^{-iWt/\hbar} & j = 3, 4 & (6.135d)
\end{array}
$$

It is common practice to move this time dependence from the $a, a^\dagger$s to the plane wave part of the field operators. The second quantized operators are then set to their value at $t = 0$. If we do this equations (6.130) can be rewritten as

$$
\varphi(\mathbf{r}, t) = \frac{1}{\sqrt{V}} \sum_{\mathbf{p}} \left(\sum_{j=1,2} a_{\mathbf{p}}^{(j)} U^{(j)}(\mathbf{p}) e^{i(\mathbf{p}\cdot\mathbf{r} - Wt)/\hbar} \right.
$$
$$
\left. + \sum_{j=3,4} a_{\mathbf{p}}^{(j)} U^{(j)}(\mathbf{p}) e^{i(\mathbf{p}\cdot\mathbf{r} + Wt)/\hbar} \right)
$$
(6.136a)

$$
\varphi^\dagger(\mathbf{r}, t) = \frac{1}{\sqrt{V}} \sum_{\mathbf{p}} \left(\sum_{j=1,2} a_{\mathbf{p}}^{(j)\dagger} U^{(j)\dagger}(\mathbf{p}) e^{-i(\mathbf{p}\cdot\mathbf{r} - Wt)/\hbar} \right.
$$
$$
\left. + \sum_{j=3,4} a_{\mathbf{p}}^{(j)\dagger} U^{(j)\dagger}(\mathbf{p}) e^{-i(\mathbf{p}\cdot\mathbf{r} + Wt)/\hbar} \right)
$$
(6.136b)

In equation (6.136) the creation and annihilation operators have no functional dependence, so the right hand sides of these equations are just superpositions of relativistic plane waves, i.e. they are solutions of the Dirac equation. However, it must be remembered that the creation and annihilation operators are present, and equations (6.136) are not wave-functions but are operators which act on state vectors in Fock space. The operator $a_{\mathbf{p}}^{(j)\dagger}$ acting on the vacuum state creates a single particle in Fock space, i.e. it gives us a particle, which is then operated on by all the rest of equation (6.136a) and summed over $\mathbf{p}$ and j to make that particle into a superposition of plane waves.

$$\varphi^\dagger(\mathbf{r},t)| \cdots, 0, \cdots \rangle = \frac{1}{\sqrt{V}} \sum_{\mathbf{p}} \sum_{j} U^{(j)}(\mathbf{p}) e^{i(\mathbf{p}\cdot\mathbf{r}-Wt)/\hbar} | \cdots, 1, \cdots \rangle \qquad (6.137)$$

It is easy to calculate the operators representing observables within this theory. We have already considered the Hamiltonian, and the momentum operator is:-

$$\hat{\mathbf{p}} = \frac{\hbar}{i} \int \varphi^\dagger(\mathbf{r},t) \nabla \varphi(\mathbf{r},t) d\mathbf{r}$$

$$= \frac{\hbar}{Vi} \int \sum_{\mathbf{p}} \sum_{j} a_{\mathbf{p}}^{(j)\dagger} U^{(j)\dagger}(\mathbf{p}) e^{\mp i\mathbf{p}\cdot\mathbf{r}/\hbar} \nabla \sum_{\mathbf{p}'} \sum_{j'} a_{\mathbf{p}'}^{(j')} U^{(j')}(\mathbf{p}') e^{\pm i\mathbf{p}'\cdot\mathbf{r}/\hbar} d\mathbf{r} \qquad (6.138)$$

$$= \sum_{\mathbf{p},\mathbf{p}'} \sum_{j,j'} a_{\mathbf{p}}^{(j)\dagger} U^{(j)\dagger}(\mathbf{p}) \mathbf{p}' a_{\mathbf{p}'}^{(j')} U^{j'}(\mathbf{p}') \delta_{\mathbf{p},\mathbf{p}'} = \sum_{\mathbf{p}} \sum_{j} \mathbf{p} a_{\mathbf{p}}^{(j)\dagger} a_{\mathbf{p}}^{(j)}$$

In later chapters we will require a definition of the electric current density operator. This is the probability current density operator at a point multiplied by the electronic charge. Following the reasoning of this section and using equation (4.100), the electric current density operator can be written down immediately as

$$\hat{\mathbf{J}}(\mathbf{r},t) = ec\varphi^\dagger(\mathbf{r},t)\tilde{\boldsymbol{\alpha}}\varphi(\mathbf{r},t) \qquad (6.139)$$

and other operators work out in much the same way, or if they are two-particle operators they follow the analogy with (6.110). Now, we will see how these operators work. We will explicitly work out the expectation value of the momentum for a two-particle state using equation (6.138). The expectation value of the momentum operator in (6.138) on a Fock space state containing two positive energy particles with momenta $\mathbf{p}_1$ and $\mathbf{p}_2$ is

$$\langle \cdots \mathbf{p}_1, j_1; \mathbf{p}_2, j_2; \cdots |\hat{\mathbf{p}}| \cdots; \mathbf{p}_1, j_1; \mathbf{p}_2, j_2; \cdots \rangle$$

$$= \langle \cdots; \mathbf{p}_1, j_1; \mathbf{p}_2, j_2; \cdots | \sum_{\mathbf{p}} \sum_{j} \mathbf{p} a_{\mathbf{p}}^{(j)\dagger} a_{\mathbf{p}'}^{(j')} | \cdots; \mathbf{p}_1, j_1; \mathbf{p}_2, j_2; \cdots \rangle \qquad (6.140)$$

$$= \langle \cdots; \mathbf{p}_1, j_1; \mathbf{p}_2, j_2; \cdots |\mathbf{p}_1 + \mathbf{p}_2| \cdots; \mathbf{p}_1, j_1; \mathbf{p}_2, j_2; \cdots \rangle$$

$$= (\mathbf{p}_1 + \mathbf{p}_2)\langle \cdots; \mathbf{p}_1, j_1; \mathbf{p}_2, j_2; \cdots | \cdots; \mathbf{p}_1, j_1; \mathbf{p}_2, j_2; \cdots \rangle = \mathbf{p}_1 + \mathbf{p}_2$$

Here the creation and annihilation operators form the number operator, so acting to the right it has an eigenvalue of $+1$ each time $\mathbf{p}$ and j correspond to occupied states, and it has eigenvalue zero otherwise. Therefore the summations go and we are left with just the eigenvalues $\mathbf{p}_1$ and $\mathbf{p}_2$ in the middle. These can be taken outside the matrix element, which then becomes equal to unity because of the assumed normalization.

In the above theory we can create a negative energy state, and the Hamiltonian, momentum operator, number operator etc. all work out satisfactorily. However, this formalism is not without its drawbacks. Firstly, we create negative energy electrons rather than the physical particle, which is the positron. Secondly, the vacuum state is not really a base state upon which all else is built. The vacuum state still contains an infinite number of filled negative energy states. There are, in fact, inconsistencies in the present formalism which we will mention at the end of our discussion of second quantization. To remedy these we will develop an alternative formalism in which the vacuum is really empty (if we cheat), all energies are positive, and we can create and destroy both electrons and positrons.

Recall that the creation of a positron of momentum **p** and spin up is the same thing, within the hole theory, as the removal (or annihilation) of a spin-down negative energy electron with momentum $-\mathbf{p}$. With this in mind we can define a new set of creation and annihilation operators in terms of the old ones.

$$a_{\mathbf{p}}^{(j)} = a_{\mathbf{p}}^{(j)}, \qquad a_{\mathbf{p}}^{(j)\dagger} = a_{\mathbf{p}}^{(j)\dagger} \qquad j = 1,2$$
$$b_{\mathbf{p}}^{(1)} = -a_{-\mathbf{p}}^{(4)\dagger}, \quad b_{\mathbf{p}}^{(2)} = a_{-\mathbf{p}}^{(3)\dagger}, \quad b_{\mathbf{p}}^{(1)\dagger} = -a_{-\mathbf{p}}^{(4)}, \quad b_{\mathbf{p}}^{(2)\dagger} = a_{-\mathbf{p}}^{(3)} \tag{6.141}$$

The a operators for $j = 1,2$ are the same as before, the b operators are related to the old operators in such a way that they represent the creation and annihilation of positrons rather than the annihilation and creation of negative energy states with opposite momenta and spin. This new set of second quantized operators obeys identical anticommutation relations to the old set:

$$\left\{ a_{\mathbf{p}}^{(j)}, b_{\mathbf{p'}}^{(j')} \right\} = \left\{ a_{\mathbf{p}}^{(j)}, b_{\mathbf{p'}}^{(j')\dagger} \right\} = \left\{ a_{\mathbf{p}}^{(j)\dagger}, b_{\mathbf{p'}}^{(j')} \right\} = \left\{ a_{\mathbf{p}}^{(j)\dagger}, b_{\mathbf{p'}}^{(j')\dagger} \right\} =$$
$$\left\{ a_{\mathbf{p}}^{(j)}, a_{\mathbf{p'}}^{(j')} \right\} = \left\{ a_{\mathbf{p}}^{(j)\dagger}, a_{\mathbf{p'}}^{(j')\dagger} \right\} = \left\{ b_{\mathbf{p}}^{(j)}, b_{\mathbf{p'}}^{(j')} \right\} = \left\{ b_{\mathbf{p}}^{(j)\dagger}, b_{\mathbf{p'}}^{(j')\dagger} \right\} = 0$$
$$\left\{ a_{\mathbf{p}}^{(j)}, a_{\mathbf{p'}}^{(j')\dagger} \right\} = \left\{ b_{\mathbf{p}}^{(j)}, b_{\mathbf{p'}}^{(j')\dagger} \right\} = \delta_{j,j'} \delta_{\mathbf{p},\mathbf{p'}} \tag{6.142}$$

If we make this definition of a new set of second quantized operators, we also have to redefine the 4-vector part of the wavefunctions $U^{(j)}$ consistently. We can determine their new form from equations (5.61) and (5.62)

$$U^{(j)}(\mathbf{p}) = U^{(j)}(\mathbf{p}) \qquad j = 1,2$$
$$V^{(1)}(\mathbf{p}) = -U^{(4)}(-\mathbf{p}) \qquad V^{(2)}(\mathbf{p}) = -U^{(3)}(-\mathbf{p}) \tag{6.143}$$

$$V^{(j)}(\mathbf{p})V^{(j')}(\mathbf{p}) = \delta_{j,j'} \qquad j = 1,2 \tag{6.144}$$

With these replacements, we can write the field operators of equation (6.136) as

$$\varphi(\mathbf{r},t) = \frac{1}{\sqrt{V}} \sum_{\mathbf{p}} \sum_{j=1,2} \left(a_{\mathbf{p}}^{(j)} U^{(j)}(\mathbf{p}) e^{i(\mathbf{p}\cdot\mathbf{r}-Wt)/\hbar} \right.$$

$$\left. + b_{\mathbf{p}}^{(j)\dagger} V^{(j)}(\mathbf{p}) e^{-i(\mathbf{p}\cdot\mathbf{r}-Wt)/\hbar} \right) \qquad (6.145a)$$

$$\varphi^{\dagger}(\mathbf{r},t) = \frac{1}{\sqrt{V}} \sum_{\mathbf{p}} \sum_{j=1,2} \left(a_{\mathbf{p}}^{(j)\dagger} U^{(j)\dagger}(\mathbf{p}) e^{-i(\mathbf{p}\cdot\mathbf{r}-Wt)/\hbar} \right.$$

$$\left. + b_{\mathbf{p}}^{(j)} V^{(j)\dagger}(\mathbf{p}) e^{i(\mathbf{p}\cdot\mathbf{r}-Wt)/\hbar} \right) \qquad (6.145b)$$

where we have used the fact that the momentum takes on both positive and negative values symmetrically, i.e. the sum over $\mathbf{p}$ runs over all possible directions. Now W is the energy of the positron or the electron and really is always positive. Note that while using this representation, if we write state vectors in the number representation, it is necessary to write them in the form $|a\cdots b\cdots\rangle$ where a and b are separate series of numbers. The numbers a represent the number of electrons in the given states and the numbers b represent the number of positrons in their states. We leave it as an exercise to prove that each element (α,β) of these Dirac field operators obeys the equal time anticommutation relations

$$\left\{ \varphi_{\alpha}(\mathbf{r},t)\varphi_{\beta}^{\dagger}(\mathbf{r}',t') + \varphi_{\beta}^{\dagger}(\mathbf{r}',t')\varphi_{\alpha}(\mathbf{r},t) \right\}_{t=t'} = \delta_{\alpha,\beta}\delta(\mathbf{r}-\mathbf{r}') \qquad (6.146)$$

Next let us write some operators in this representation. The Hamiltonian of equation (6.133) becomes

$$\hat{H} = \sum_{\mathbf{p}} \left(\sum_{j=1,2} W a_{\mathbf{p}}^{(j)\dagger} a_{\mathbf{p}}^{(j)} - \sum_{j=3,4} W a_{\mathbf{p}}^{(j)\dagger} a_{\mathbf{p}}^{(j)} \right)$$

$$= \sum_{\mathbf{p}} \left(\sum_{j=1,2} W a_{\mathbf{p}}^{(j)\dagger} a_{\mathbf{p}}^{(j)} - \sum_{j=3,4} W a_{-\mathbf{p}}^{(j)\dagger} a_{-\mathbf{p}}^{(j)} \right)$$

$$= \sum_{\mathbf{p}} \sum_{j=1,2} W \left(a_{\mathbf{p}}^{(j)\dagger} a_{\mathbf{p}}^{(j)} - b_{\mathbf{p}}^{(j)} b_{\mathbf{p}}^{(j)\dagger} \right)$$

$$= \sum_{\mathbf{p}} \sum_{j=1,2} W \left(a_{\mathbf{p}}^{(j)\dagger} a_{\mathbf{p}}^{(j)} + b_{\mathbf{p}}^{(j)\dagger} b_{\mathbf{p}}^{(j)} - 1 \right) \qquad (6.147)$$

The total charge operator is just the electronic charge times the number operator in the old representation. In the new representation it becomes

$$\hat{Q} = e \sum_{\mathbf{p}} \sum_{j=1,2} \left(a_{\mathbf{p}}^{(j)\dagger} a_{\mathbf{p}}^{(j)} - b_{\mathbf{p}}^{(j)\dagger} b_{\mathbf{p}}^{(j)} + 1 \right) \qquad (6.148)$$

The total momentum operator is

$$\hat{\mathbf{p}} = \sum_{\mathbf{p}} \sum_{j=1,2} \mathbf{p} \left(a_{\mathbf{p}}^{(j)\dagger} a_{\mathbf{p}}^{(j)} + b_{\mathbf{p}}^{(j)\dagger} b_{\mathbf{p}}^{(j)} \right) \tag{6.149}$$

Finally, let us look at the spin operator. We will think only about the z-component of spin. By analogy with earlier work we can expect the second quantized version of the spin operator to be

$$\hat{S}_z = \frac{\hbar}{2} \int \varphi^{\dagger}(\mathbf{r}, t) \tilde{\sigma}_z \varphi(\mathbf{r}, t) d\mathbf{r} \tag{6.150}$$

We know that the spin of the vacuum must be zero, so if we act on it with the spin operator we will get zero. This enables us to operate on the single-electron state in a subtle way. We write

$$\hat{S}_z |1, \cdots, 0\rangle = \hat{S}_z a_{\mathbf{p}}^{(j)\dagger} |0, \cdots, 0\rangle = \left[\hat{S}_z, a_{\mathbf{p}}^{(j)\dagger} \right] |0, \cdots, 0\rangle \tag{6.151}$$

So, to find the eigenvalues of $\hat{S}_z$ for a single-particle state, it is sufficient to find the action of the commutator of $\hat{S}_z$ and $a_{\mathbf{p}}^{(j)\dagger}$ on the vacuum state. The field operators (6.145) can be substituted into (6.150). Then the space integration removes one of the summations over $\mathbf{p}$. With the help of the anticommutation relations (6.142), the commutator $[\hat{S}_z, a_{\mathbf{p}}^{(j)\dagger}]$ can be reduced to

$$[\hat{S}_z, a_{\mathbf{p}}^{(j)\dagger}] = \frac{\hbar}{2} \sum_j U^{(j)\dagger}(\mathbf{p}) \tilde{\sigma}_z U^{(j)}(\mathbf{p}) \left(a_{\mathbf{p}}^{(j)\dagger} a_{\mathbf{p}}^{(j)} a_{\mathbf{p}}^{(j)\dagger} - a_{\mathbf{p}}^{(j)\dagger} a_{\mathbf{p}}^{(j)\dagger} a_{\mathbf{p}}^{(j)} \right) \tag{6.152}$$

The second term on the right hand side here gives zero when acting on the vacuum state. This can be seen immediately, we cannot destroy any electrons in the vacuum because there aren't any to destroy. The first term can act on the vacuum

$$a_{\mathbf{p}}^{(j)\dagger} a_{\mathbf{p}}^{(j)} a_{\mathbf{p}}^{(j)\dagger} |0, \cdots, 0\rangle = a_{\mathbf{p}}^{(j)\dagger} a_{\mathbf{p}}^{(j)} |1, \cdots, 0\rangle = 1 |1, \cdots, 0\rangle = |1, \cdots, 0\rangle \tag{6.153}$$

So

$$\hat{S}_z |1, \cdots, 0\rangle = \left[\hat{S}_z, a_{\mathbf{p}}^{(j)\dagger} \right] |0, \cdots, 0\rangle = \frac{\hbar}{2} \sum_j U^{(j)\dagger}(\mathbf{p}) \tilde{\sigma}_z U^{(j)}(\mathbf{p}) |1, \cdots, 0\rangle$$

$$\tag{6.154}$$

If we have only one particle there is no sum over j of course. Now let us look at the positron state. In exactly the same way we have

$$\hat{S}_z |0, \cdots, 1\rangle = \left[\hat{S}_z, b_{\mathbf{p}}^{(j)\dagger} \right] |0, \cdots, 0\rangle =$$

$$\frac{\hbar}{2} \sum_j V^{(j)\dagger}(\mathbf{p}) \tilde{\sigma}_z V^{(j)}(\mathbf{p}) \left(b_{\mathbf{p}}^{(j)} b_{\mathbf{p}}^{(j)\dagger} b_{\mathbf{p}}^{(j)\dagger} - b_{\mathbf{p}}^{(j)\dagger} b_{\mathbf{p}}^{(j)} b_{\mathbf{p}}^{(j)\dagger} \right) |0, \cdots, 0\rangle \tag{6.155}$$

The first term in the brackets here gives zero because it is trying to put two fermions in the same state. The second term gives

$$-b_{\mathbf{p}}^{(j)\dagger} b_{\mathbf{p}}^{(j)} b_{\mathbf{p}}^{(j)\dagger}|0,\cdots,0\rangle = -b_{\mathbf{p}}^{(j)\dagger} b_{\mathbf{p}}^{(j)}|1\rangle = -1|0,\cdots,1\rangle = -|0,\cdots,1\rangle \tag{6.156}$$

So

$$\hat{S}_z|0,\cdots,1\rangle = -\frac{\hbar}{2}\sum_j V^{(j)\dagger}(\mathbf{p})\tilde{\sigma} V^{(j)}|0,\cdots,1\rangle \tag{6.157}$$

It is now time to stop the rather relentless mathematics and start thinking about what this formulation of many-body quantum mechanics means.

Firstly consider the spin operators acting on the single-particle state in (6.154) and (6.157). Equation (6.154) tells us that for an electron with spin up we need the electron state to be an eigenvector of $\tilde{\sigma}_z$ with eigenvalue $+1$ and for an electron with spin down the electron state must be an eigenvector of $\tilde{\sigma}_z$ with eigenvector -1. On the other hand for a positron to have spin up we must have it in a state that is an eigenvector of $\tilde{\sigma}_z$ with eigenvalue -1 and for it to be spin down it must have eigenvalue $+1$. This is rather similar to the hole picture of antiparticles developed in chapter 5. There the absence of a negative energy state with spin s was interpreted as an antiparticle with spin $-s$. Hence, the present formalism is completely consistent with the earlier hole theory. This is exactly what we might have surmised from our definition of the positron creation and annihilation operators in (6.141).

You may well be asking yourselves why I bothered to introduce this positron operator. Up to that point we had described positrons as holes in the negative energy sea of electrons. OK, that is not a very elegant theory, but we haven't actually found anything wrong with it, have we? Well, there are several things wrong with that interpretation of the theory although we have not really pointed them out. The most damning evidence against this theory is as follows. The hole theory conserves electrons. If we create a positron we do it by exciting a negative energy electron up to a positive energy state. So a positron is an electron that is missing, and if we create a positron we necessarily create a positive energy electron as well. If we can find an example where a positron is created without an associated electron, the hole theory is sunk. In fact such an example is very easy to find. Positron beta-decay involves the following nuclear reaction

$$p \rightarrow n + \epsilon^+ + \nu$$

A positron is created, but where is the electron that used to occupy the now vacated negative energy state? It isn't there.

The second major objection to the hole theory is more philosophical. In earlier chapters we have tried to cling to a single-particle interpretation of the Dirac wavefunction (in fact we will continue to do this in some later chapters as well). However, we have also been forced to contradict this by introducing the sea of negative energy electrons – we have had to introduce an infinite number of electrons to maintain the single-electron picture. This is surely not a satisfactory state of affairs.

Even the positron picture we have introduced with the vacuum as a base state is not yet satisfactory, although it does have many desirable features. Consider the Hamiltonian (6.147) and the total charge operator (6.148). Clearly they both behave as they should under most circumstances. If we add either an electron or a positron to a system we increase the total energy by an amount W. If we add an electron the charge (with e negative) decreases by one unit of $|e|$, and if we add a positron it increases by one unit of $|e|$. However, look at what happens when we apply these operators to the vacuum state:

$$\hat{H}|0,\cdots,0\rangle = \sum_{\mathbf{p}} \sum_{j=1,2} W \left(a_{\mathbf{p}}^{(j)\dagger} a_{\mathbf{p}}^{(j)} + b_{\mathbf{p}}^{(j)\dagger} b_{\mathbf{p}}^{(j)} - 1 \right) |0,\cdots,0\rangle$$

$$= \sum_{\mathbf{p}} \sum_{j=1,2} W(0+0-1)|0,\cdots,0\rangle = -\infty|0,\cdots,0\rangle \quad (6.158)$$

$$\hat{Q} = e \sum_{\mathbf{p}} \sum_{j=1,2} (+1)|0,\cdots,0\rangle = +\infty|0,\cdots,0\rangle \quad (6.159)$$

The reason for the appearance of these infinities is that we have removed the negative energy sea without removing its infinite negative energy or charge. It is necessary to do this, of course, and it can be done easily by redefining our zeroes of energy and charge so that we can rewrite the Hamiltonian and total charge operators as

$$\hat{H} = \sum_{\mathbf{p}} \sum_{j=1,2} W \left(a_{\mathbf{p}}^{(j)\dagger} a_{\mathbf{p}}^{(j)} + b_{\mathbf{p}}^{(j)\dagger} b_{\mathbf{p}}^{(j)} \right) \quad (6.160)$$

$$\hat{Q} = e \sum_{\mathbf{p}} \sum_{j=1,2} \left(a_{\mathbf{p}}^{(j)\dagger} a_{\mathbf{p}}^{(j)} - b_{\mathbf{p}}^{(j)\dagger} b_{\mathbf{p}}^{(j)} \right) \quad (6.161)$$

The momentum and spin operators (6.149) and (6.150) are unaffected by this redefinition.

The positron operators have been introduced because the hole picture is not satisfactory for the reasons stated above. However, in condensed matter physics we are almost invariably interested in electromagnetic interactions where the number of electrons is conserved. Under these circumstances the hole picture of antiparticles can often be usefully applied.

6.12 Problems

(1) Show that the product of two commuting projection operators is another projection operator.

(2) Operate with the charge conjugation operator on the wavefunction of a free electron (including its time dependence) with negative energy in its rest frame, Interpret your solution.

(3) Derive the parity operator for the Pauli equation. Show that a term proportional to $\tilde{\sigma} \cdot \mathbf{E}$ in the Pauli equation would violate both parity and time-reversal invariance.

(4) Electrons are fermions and hence their creation and annihilation operators obey commutators (6.86). In the standard BCS theory (Bardeen *et al.* 1957) of superconductivity electrons pair up and the operators

$$c_{\mathbf{k}} = a_{-\mathbf{k}\downarrow} a_{\mathbf{k}\uparrow}, \qquad\qquad c_{\mathbf{k}}^{\dagger} = a_{\mathbf{k}\uparrow} a_{-\mathbf{k}\downarrow}$$

are important. Here $\mathbf{k}$ represents the wavevector of the electron and the arrow is its spin direction. Find the commutators $[c_{\mathbf{k}}, c_{\mathbf{k'}}]$, $\left[c_{\mathbf{k}}^{\dagger}, c_{\mathbf{k'}}^{\dagger}\right]$ and $\left[c_{\mathbf{k}}^{\dagger}, c_{\mathbf{k'}}\right]$. What does this tell you about the electron pairs?

(5) Prove the anticommutation relation (6.146).

7

Separating Particles from Antiparticles

As we have seen, in the non-relativistic limit, a Dirac particle (one with spin 1/2) is well described by the Pauli equation with a two-component wavefunction. The lower two components of the wavefunction become negligible in the non-relativistic limit if we are considering particles and the upper two if we are considering antiparticles. Clearly, this is as it should be. In the non-relativistic limit the particles and antiparticles separate. The question that then naturally arises is whether we can find a transformation in which the particles and antiparticles are separated for any value of the momentum. Foldy and Wouthuysen (1950) have shown that such a representation is possible. Furthermore, this representation is instructive and gives profound insight into the nature of Dirac particles. Therefore we shall discuss it in detail.

To understand how to find a representation in which we can eliminate two components of the wavefunction it is necessary to understand why the Dirac equation requires a four-component wavefunction in the first place. The reason is in the α-matrices. They are odd matrices. By this we mean that the top left and bottom right quadrants are zero and the top right and bottom left quadrants are non-zero. Therefore, the term $c\tilde{\alpha} \cdot \mathbf{p}$ in the Hamiltonian connects particle and antiparticle parts of the wavefunction. This means it is necessary to look for a representation in which the terms in the Hamiltonian that contain odd matrices are small. The Foldy–Wouthuysen (F–W) transformation is a very successful attempt to do this. Indeed, as we shall see later, it leads to the familiar Pauli equation and the relativistic corrections to it. All books on relativistic quantum theory contain a section on the F–W transformation, so any such books in the references may be consulted for another point of view on this subject.

In fact the F–W transformation for free particles looks rather different from that for particles in a field. For free particles it is an exact analytical transformation, but for particles in a field it is an expansion in powers of $1/mc^2$ which is only correct up to the order chosen. It turns out to be most convenient and least confusing to treat these two cases separately.

7.1 The Foldy–Wouthuysen Transformation for a Free Particle

Before going on to the F–W transformation, let us consider separating the upper and lower components of the wavefunction in a way that one might first think of, and show that it does not work. Firstly we write down the time-independent form of the free-particle Dirac equation

$$(c\tilde{\alpha} \cdot \hat{\mathbf{p}} + \tilde{\beta}mc^2)\psi(\mathbf{r}) = W\psi(\mathbf{r}) \tag{7.1}$$

Now we could try replacing $\psi(\mathbf{r})$ by $\frac{1}{2}(\tilde{I} + \tilde{\beta})\psi(\mathbf{r})$ where $\tilde{I}$ is the identity matrix. The lower two components of the wavefunction are forced to be zero. However, this cannot work because $\frac{1}{2}(\tilde{I} + \tilde{\beta})\psi(\mathbf{r})$ is not a solution of (7.1). To see this, premultiply each side of (7.1) by $\frac{1}{2}(\tilde{I} + \tilde{\beta})$. The right hand side works out all right, but one has to recall the commutation relations for the α-matrices with $\tilde{I}$ and $\tilde{\beta}$. The $\tilde{\alpha}$s commute with $\tilde{I}$ but anticommute with $\tilde{\beta}$. So when we commute through we find

$$\frac{1}{2}c\tilde{\alpha} \cdot \mathbf{p}(\tilde{I} - \tilde{\beta})\psi(\mathbf{r}) + \tilde{\beta}mc^2\frac{1}{2}(\tilde{I} + \tilde{\beta})\psi(\mathbf{r}) = W\frac{1}{2}(\tilde{I} + \tilde{\beta})\psi(\mathbf{r}) \tag{7.2}$$

Each term in (7.2) is a 4-vector. In the last two terms, only the upper two components of the 4-vector are non-zero because the bottom half of the 4×4 matrix $\frac{1}{2}(\tilde{I} + \tilde{\beta})$ is zero. In the first term, because of the change of sign in front of $\tilde{\beta}$, the lower two components are the ones that are non-zero. Thus the upper and lower components are not decoupled by this transformation. This little calculation illustrates two things. Firstly, it is necessary to be a good deal more subtle than this to decouple the upper and lower components of the wavefunction, and secondly, it is the symmetry of the α-matrices that leads to the relativistic wavefunctions being four-component, rather than two-component, objects.

Now let us look at the F–W transformation. The Dirac equation is

$$i\hbar\frac{\partial\psi(\mathbf{r})}{\partial t} = (c\tilde{\alpha} \cdot \hat{\mathbf{p}} + \tilde{\beta}mc^2)\psi(\mathbf{r}) = \hat{H}\psi(\mathbf{r}) \tag{7.3}$$

We want to make a transformation of the type

$$\psi'(\mathbf{r}) = S\psi(\mathbf{r}) = e^{iU}\psi(\mathbf{r}) \tag{7.4}$$

This transformed wavefunction fulfils the equation

$$\hat{H}'\psi'(\mathbf{r}) = i\hbar\frac{\partial\psi'(\mathbf{r})}{\partial t} \tag{7.5}$$

where

$$\hat{H}' = e^{iU}\hat{H}e^{-iU} - \hbar\frac{\partial U}{\partial t} \tag{7.6}$$

The truth of this is easily substantiated by substituting (7.6) and (7.4) into (7.5) and seeing that it reduces to (7.3). We are only going to be interested in the case when $\partial U/\partial t = 0$, so only the first term on the right hand side of (7.6) remains. Obviously, if we transform the Hamiltonian and wavefunction this way, we will also have to transform any other quantities we use, in particular, other operators. The next step is to choose a form of U such that $\exp(iU)$ separates the positive and negative energy solutions of (7.1). It was our heroes, Foldy and Wouthuysen, who achieved this. They suggested

$$U_{FW} = \frac{1}{i}\tilde{\beta}\frac{\tilde{\alpha}\cdot\hat{\mathbf{p}}}{|\hat{\mathbf{p}}|}\tan^{-1}\left(\frac{|\hat{\mathbf{p}}|c}{|\hat{H}| + mc^2}\right) \tag{7.7}$$

and

$$e^{\pm iU_{FW}} = \frac{|\hat{H}| + mc^2 \pm c\tilde{\beta}\tilde{\alpha}\cdot\hat{\mathbf{p}}}{[2|\hat{H}|(|\hat{H}| + mc^2)]^{1/2}} \tag{7.8}$$

To develop a transformation as complicated as this to give extra understanding of the nature of Dirac particles was an astonishing feat.

That (7.8) follows from (7.7) is not immediately obvious. Indeed the mathematics for getting from (7.7) to (7.8) is rather subtle. Therefore we reproduce it here in detail.

$$\begin{aligned}
e^{\pm iU_{FW}} &= \exp\left(\pm\frac{\tilde{\beta}\tilde{\alpha}\cdot\hat{\mathbf{p}}}{|\hat{\mathbf{p}}|}\tan^{-1}\left(\frac{|\hat{\mathbf{p}}|c}{|\hat{H}| + mc^2}\right)\right) \\
&= \cos\left(-i\tilde{\beta}\frac{\tilde{\alpha}\cdot\hat{\mathbf{p}}}{|\hat{\mathbf{p}}|}\tan^{-1}\left(\frac{|\hat{\mathbf{p}}|c}{|\hat{H}| + mc^2}\right)\right) \\
&\quad \pm i\sin\left(-i\tilde{\beta}\frac{\tilde{\alpha}\cdot\hat{\mathbf{p}}}{|\hat{\mathbf{p}}|}\tan^{-1}\left(\frac{|\hat{\mathbf{p}}|c}{|\hat{H}| + mc^2}\right)\right)
\end{aligned} \tag{7.9}$$

Next a subtlety occurs. If we expand $\cos\theta$ as a series we only obtain even powers of θ. It is easy to see that $(\tilde{\beta}\tilde{\alpha}\cdot\hat{\mathbf{p}}/|\hat{\mathbf{p}}|)^2 = -\tilde{I}$ from equation (4.19) and, since $-i^2 = -1$, when we expand the cosine term in (7.9) $(-i\tilde{\beta}\tilde{\alpha}\cdot\hat{\mathbf{p}}/|\hat{\mathbf{p}}|)^n$ will always give the unit matrix. Now if all even powers of $(-i\tilde{\beta}\tilde{\alpha}\cdot\hat{\mathbf{p}}/|\hat{\mathbf{p}}|)$ are one, it follows that all odd powers must be $(-i\tilde{\beta}\tilde{\alpha}\cdot\hat{\mathbf{p}}/|\hat{\mathbf{p}}|)$, so in the series expansion of the sine term in (7.9) this will come outside the summation and leave us with

$$e^{\pm iU_{FW}} = \cos\left[\tan^{-1}\left(\frac{|\hat{\mathbf{p}}|c}{|\hat{H}| + mc^2}\right)\right] + \tilde{\beta}\frac{\tilde{\alpha}\cdot\hat{\mathbf{p}}}{|\hat{\mathbf{p}}|}\sin\left[\tan^{-1}\left(\frac{|\hat{\mathbf{p}}|c}{|\hat{H}| + mc^2}\right)\right] \tag{7.10}$$

For simplicity let

$$X = \tan^{-1}\left(\frac{|\hat{\mathbf{p}}|c}{|\hat{H}| + mc^2}\right) \tag{7.11}$$

Therefore

$$\tan X = \frac{\sin X}{\cos X} = \frac{|\hat{\mathbf{p}}|c}{|\hat{H}| + mc^2} \tag{7.12}$$

But from $\sin^2 X + \cos^2 X = 1$, and (7.12), it can be seen that

$$\sin X = \frac{|\hat{\mathbf{p}}|c}{(\hat{\mathbf{p}}^2c^2 + (|\hat{H}| + mc^2)^2)^{\frac{1}{2}}}, \qquad \cos X = \frac{\hat{H} + mc^2}{(\hat{\mathbf{p}}^2c^2 + (|\hat{H}| + mc^2)^2)^{\frac{1}{2}}} \tag{7.13}$$

Substituting these back into (7.10), and remembering that

$$|\hat{H}| = \sqrt{\hat{\mathbf{p}}^2c^2 + m^2c^4} \tag{7.14}$$

soon leaves us with (7.8). It is important to remember that $|\hat{H}|$ is an operator. The next thing to do is prove that

$$e^{iU_{FW}}e^{-iU_{FW}} = 1 \tag{7.15}$$

This is fairly easily done:

$$
\begin{aligned}
e^{iU_{FW}}e^{-iU_{FW}} &= \frac{|\hat{H}| + mc^2 + c\tilde{\beta}\tilde{\alpha}\cdot\hat{\mathbf{p}}}{(2|\hat{H}|(|\hat{H}| + mc^2))^{1/2}} \frac{|\hat{H}| + mc^2 - c\tilde{\beta}\tilde{\alpha}\cdot\hat{\mathbf{p}}}{(2|\hat{H}|(|\hat{H}| + mc^2))^{1/2}} \\
&= \frac{(|\hat{H}| + mc^2)^2 - (c\tilde{\beta}\tilde{\alpha}\cdot\hat{\mathbf{p}})^2}{(2|\hat{H}|(|\hat{H}| + mc^2))} = \frac{\hat{\mathbf{p}}^2c^2 + 2m^2c^4 + 2|\hat{H}|mc^2 + c^2\hat{\mathbf{p}}^2}{(2|\hat{H}|(|\hat{H}| + mc^2))} \\
&= \frac{2|\hat{H}|^2 + 2|\hat{H}|mc^2}{(2|\hat{H}|(|\hat{H}| + mc^2))} = 1
\end{aligned}
\tag{7.16}
$$

where we have made copious use of (7.14). This is the required result. There is one further point to note about this transformation. Obviously a commutator of two operators $\hat{A}$ and $\hat{B}$ after they have been transformed to the F–W representation is

$$\left[e^{iU_{FW}}\hat{A}e^{-iU_{FW}}, e^{iU_{FW}}\hat{B}e^{-iU_{FW}}\right] = e^{iU_{FW}}[\hat{A}, \hat{B}]e^{-iU_{FW}} \tag{7.17}$$

i.e. the commutator of the transformed operators is equal to the transform of the commutators. This result will prove useful in the next section.

So, to summarize what we have done so far, we have postulated the transformation (7.8) with U_{FW} given by (7.7). We have shown that these definitions of U_{FW} are consistent with one another and we have verified (7.15). This proves that this transformation has the necessary mathematical properties and we are now ready to go ahead and use it.

7.2 Foldy–Wouthuysen Transformation of Operators

The next stage in our exploration of the Foldy–Wouthuysen formalism is to examine the way in which some of the important operators transform. Henceforth we shall label operators in the following way. Operators in the standard representation prior to a F–W transformation will just have a standard hat ($\hat{O}$). Operators in the F–W representation after an F–W transformation will be labelled with a hat and a subscript FW ($\hat{O}_{FW}$).

Let us start with the free-particle Hamiltonian. This is the key operator to consider. When transformed according to (7.6) it should display the separation of particles and antiparticles. Obviously it takes the form

$$\hat{H}_{FW} = \frac{|\hat{H}| + mc^2 + c\tilde{\beta}\tilde{\alpha}\cdot\mathbf{\hat{p}}}{(2|\hat{H}|(|\hat{H}| + mc^2))^{1/2}}(c\tilde{\alpha}\cdot\mathbf{\hat{p}} + \tilde{\beta}mc^2)\frac{|\hat{H}| + mc^2 - c\tilde{\beta}\tilde{\alpha}\cdot\mathbf{\hat{p}}}{(2|\hat{H}|(|\hat{H}| + mc^2))^{1/2}} \quad (7.18)$$

The most reliable way of finding a simple form of $\hat{H}_{FW}$ is to write everything out in explicit matrix form and multiply out the matrices that way. The relevant matrices are

$$|\hat{H}| + mc^2 \pm c\tilde{\beta}\tilde{\alpha}\cdot\mathbf{\hat{p}} = \begin{pmatrix} |\hat{H}| + mc^2 & 0 & \pm c\hat{p}_z & \pm c\hat{p}_- \\ 0 & |\hat{H}| + mc^2 & \pm c\hat{p}_+ & \mp c\hat{p}_z \\ \mp c\hat{p}_z & \mp c\hat{p}_- & |\hat{H}| + mc^2 & 0 \\ \mp c\hat{p}_+ & \pm c\hat{p}_z & 0 & |\hat{H}| + mc^2 \end{pmatrix} \quad (7.19a)$$

$$\hat{H} = (c\tilde{\alpha}\cdot\mathbf{\hat{p}} + \tilde{\beta}mc^2) = \begin{pmatrix} mc^2 & 0 & c\hat{p}_z & c\hat{p}_- \\ 0 & mc^2 & c\hat{p}_+ & -c\hat{p}_z \\ c\hat{p}_z & c\hat{p}_- & -mc^2 & 0 \\ c\hat{p}_+ & -c\hat{p}_z & 0 & -mc^2 \end{pmatrix} \quad (7.19b)$$

Here we have used the standard notation $\hat{p}_\pm = \hat{p}_x \pm i\hat{p}_y$. Now the transformation (7.18) is just a matter of multiplying out the matrices defined in (7.19) in the right order. It takes several pages of algebra, and it is very easy to make mistakes (believe me, I speak from experience). Therefore, I suggest you either do not do it, or, if possible, set it as an exercise for anyone who particularly irritates you. Anyway, one eventually finds

$$\hat{H}_{FW} = \tilde{\beta}|\hat{H}| = \begin{pmatrix} |\hat{H}| & 0 & 0 & 0 \\ 0 & |\hat{H}| & 0 & 0 \\ 0 & 0 & -|\hat{H}| & 0 \\ 0 & 0 & 0 & -|\hat{H}| \end{pmatrix} \quad (7.20)$$

This is an interesting result (honest): the Hamiltonian in this representation is diagonal. Equation (7.5) with $\hat{H}'$ given by (7.20) now has separable solutions. The upper (lower) two components represent the positive (negative) energy states. If we write

$$\psi'(\mathbf{r}) = \phi'(\mathbf{r}) + \chi'(\mathbf{r}) \quad (7.21a)$$

with

$$\phi'(\mathbf{r}) = \tfrac{1}{2}(\tilde{I} + \tilde{\beta})\psi'(\mathbf{r}), \qquad \chi'(\mathbf{r}) = \tfrac{1}{2}(\tilde{I} - \tilde{\beta})\psi'(\mathbf{r}) \qquad (7.21b)$$

then (7.5) reduces to two two-component equations:

$$i\hbar\frac{\partial\phi'(\mathbf{r})}{\partial t} = |\hat{H}|\phi'(\mathbf{r}), \qquad i\hbar\frac{\partial\chi'(\mathbf{r})}{\partial t} = -|\hat{H}|\chi'(\mathbf{r}) \qquad (7.22)$$

Rather than go on to discuss the implications of this result we are going to continue looking at the transformations and postpone the discussion until all the necessary mathematical results have been obtained.

Next we will consider the momentum operator $\hat{\mathbf{p}}$. This is a very easy operator to transform because it commutes with $|\hat{H}|$ and hence commutes with $\exp(\pm iU_{FW})$. This is trivial to see. In equation (7.8) we note that $\exp(\pm iU_{FW})$ contains constants and the momentum operator, in particular it does not contain the position operator, so $\hat{\mathbf{p}}$ is bound to commute with it. Therefore the F–W transform of the momentum operator is the momentum operator itself, i.e.

$$\hat{\mathbf{p}}_{FW} = \hat{\mathbf{p}} \qquad (7.23)$$

Now we will look at the F–W transformation of the position operator. This is a rather more difficult thing to get right, so we write out the derivation in a nauseating amount of detail. In this derivation we require the following commutator

$$[F(\hat{\mathbf{p}}), \hat{\mathbf{r}}] = \frac{\hbar}{i}\nabla_{\mathbf{p}}F(\hat{\mathbf{p}}) \qquad (7.24)$$

where $\nabla_{\mathbf{p}}$ is the momentum space gradient operator. Using (7.14) and (7.24) we have

$$[|\hat{H}|, \hat{\mathbf{r}}] = \frac{\hbar}{i}\frac{\hat{\mathbf{p}}c^2}{|\hat{H}|} \qquad (7.25)$$

The F–W transformation for the position operator is

$$\hat{\mathbf{r}}_{FW} = \frac{|\hat{H}| + mc^2 + c\tilde{\beta}\tilde{\alpha}\cdot\hat{\mathbf{p}}}{(2|\hat{H}|(|\hat{H}| + mc^2))^{1/2}}\hat{\mathbf{r}}\frac{|\hat{H}| + mc^2 - c\tilde{\beta}\tilde{\alpha}\cdot\hat{\mathbf{p}}}{(2|\hat{H}|(|\hat{H}| + mc^2))^{1/2}} \qquad (7.26)$$

Now, using (7.25) we have

$$\hat{\mathbf{r}}(|\hat{H}| + mc^2 - c\tilde{\beta}\tilde{\alpha}\cdot\hat{\mathbf{p}}) - (|\hat{H}| + mc^2 - c\tilde{\beta}\tilde{\alpha}\cdot\hat{\mathbf{p}})\hat{\mathbf{r}} = i\hbar\left(\frac{\hat{\mathbf{p}}c^2}{|\hat{H}|} - c\tilde{\beta}\tilde{\alpha}\right) \qquad (7.27)$$

We can substitute (7.27) into (7.26) to get the messy looking expression

$$\hat{\mathbf{r}}_{FW} = (|\hat{H}|^2 + |\hat{H}|mc^2)^{1/2} \ \hat{\mathbf{r}} \ \frac{1}{(|\hat{H}|^2 + |\hat{H}|mc^2)^{1/2}}$$

$$- \frac{1}{2(|\hat{H}|^2 + |\hat{H}|mc^2)}(|\hat{H}| + mc^2 + c\tilde{\beta}\tilde{\boldsymbol{\alpha}} \cdot \hat{\mathbf{p}})(i\hbar c\tilde{\beta}\tilde{\boldsymbol{\alpha}} - i\hbar c^2 \hat{\mathbf{p}}/|\hat{H}|)$$

(7.28)

The first term of this expression can be found using (7.24):

$$(|\hat{H}|^2 + |\hat{H}|mc^2)^{1/2} \ \hat{\mathbf{r}} \ \frac{1}{(|\hat{H}|^2 + |\hat{H}|mc^2)^{1/2}} = \hat{\mathbf{r}} + \frac{\hbar \hat{\mathbf{p}} c^2}{2i|\hat{H}|} \frac{2|\hat{H}| + mc^2}{|\hat{H}|^2 + |\hat{H}|mc^2}$$

(7.29)

Next we must multiply out the brackets in the second term of (7.28). This becomes quite involved. However, if we use the identity (4.19) we can simplify the resulting expression a little to

$$\hat{\mathbf{r}}_{FW} = \hat{\mathbf{r}} - \frac{i\hbar c\tilde{\beta}\tilde{\boldsymbol{\alpha}}}{2|\hat{H}|} + \frac{\hbar c^2(ic\tilde{\beta}(\tilde{\boldsymbol{\alpha}} \cdot \hat{\mathbf{p}})\hat{\mathbf{p}} - (\tilde{\boldsymbol{\sigma}} \times \hat{\mathbf{p}})|\hat{H}|)}{2|\hat{H}|^2(|\hat{H}| + mc^2)}$$

(7.30)

This is a mind-bogglingly complicated form for a position operator. Sadly, it is the price that we have to pay for diagonalizing the Hamiltonian.

There are several other important operators that can be transformed into the F–W representation. We will not go through the mathematics again, but leave that as an exercise for the reader. Here we merely quote the results. The velocity operator is given by

$$\hat{\mathbf{v}}_{FW} = e^{iU_{FW}}c\tilde{\boldsymbol{\alpha}}e^{-iU_{FW}} = c\tilde{\boldsymbol{\alpha}} + \frac{c^2\tilde{\beta}\hat{\mathbf{p}}}{|\hat{H}|} - \frac{c^3(\tilde{\boldsymbol{\alpha}} \cdot \hat{\mathbf{p}})\hat{\mathbf{p}}}{|\hat{H}|(|\hat{H}| + mc^2)}$$

(7.31)

The angular momentum operator is fairly simple:

$$\hat{\mathbf{L}}_{FW} = e^{iU_{FW}}(\hat{\mathbf{r}} \times \hat{\mathbf{p}})e^{-iU_{FW}} = \hat{\mathbf{r}}_{FW} \times \hat{\mathbf{p}}$$

(7.32)

and finally the spin operator becomes

$$\hat{\mathbf{S}}_{FW} = e^{iU_{FW}}\frac{\hbar\tilde{\boldsymbol{\alpha}} \times \tilde{\boldsymbol{\alpha}}}{4i}e^{-iU_{FW}} = \frac{\hbar}{2}\left(\tilde{\boldsymbol{\sigma}} + \frac{ic\tilde{\beta}\tilde{\boldsymbol{\alpha}} \times \hat{\mathbf{p}}}{|\hat{H}|} - \frac{c^2\hat{\mathbf{p}} \times (\tilde{\boldsymbol{\sigma}} \times \hat{\mathbf{p}})}{|\hat{H}|(|\hat{H}| + mc^2)}\right)$$

(7.33)

We are now in a position to discuss some of the physics of this transformation. A question that occurs is why we ever work in representations in which the particles and antiparticles are mixed when this representation exists in which they are not. The answer to this is twofold. Firstly, although the Hamiltonian and momentum operators are fairly simple in this representation, no other operators are, so calculating any observables becomes a daunting task. The second reason we don't work in this representation will be apparent when we transform the wavefunction later.

The fact that working in the F–W representation is mathematically complex does not mean we cannot learn from it. Indeed there is profound insight to be gained from a consideration of the form of the operators above. We develop this line of thought further in the next section.

7.3 Zitterbewegung

Let us consider the velocity operator. Up to now we have been working with an implicit contradiction which we have deliberately failed to point out. This concerns the non-relativistic limit of the velocity operator. In the non-relativistic Pauli theory the velocity operator is

$$\hat{\mathbf{v}} = \hat{\mathbf{p}}/m \qquad (7.34)$$

whereas in the Dirac theory it is given by

$$\hat{\mathbf{v}} = \frac{i}{\hbar}[\hat{H}, \hat{\mathbf{r}}] = c\tilde{\boldsymbol{\alpha}} \qquad (7.35)$$

In the former case the velocity can take on any value, but in the latter case the eigenvalues must be between $\pm c$, from the central postulates of relativity. Every non-zero element of the matrix operator (7.35) is of magnitude c. Despite this it is conceivable that (7.34) and (7.35) could represent the same quantity at low velocities. Consider how (7.35) works. The α-matrices are odd matrices. So, when placed in an expression for an expectation value, they multiply the large component of one wavefunction by the small component of the other. Now we have seen (below equation (4.55)) that the small component is of order $v/2c$ of the large component. If the wavefunctions are normalized then the large component of the wavefunction is of order 1, so in the expectation value expression (taking the x-component of velocity as our example) we can use the order of magnitude argument

$$\langle v_x \rangle = \int \psi^\dagger c\tilde{\alpha}_x \psi dx = \int \phi^* c\tilde{\sigma}_x \chi dx + \int \chi^* c\tilde{\sigma}_x \phi dx$$

$$\approx c \int \psi^* \frac{v}{2c} \psi dx + c \int \frac{v}{2c} \psi^* \psi dx = v \int \psi^* \psi dx \sim v \qquad (7.36)$$

where we have used the notation of chapter 4. As the velocity of the particle becomes larger, the small component of the wavefunction becomes larger and so the expectation value of velocity becomes larger. Furthermore, this argument shows that we are, at least approximately, limited to $v < c$ as we must be in a relativistic theory. So, although the operators (7.34) and (7.35) are very different, it is not completely unreasonable to identify them as representing the same quantity.

If we want to find the expectation value of the square of velocity we could just use $\langle \mathbf{v} \rangle^2$. However, we would also expect to be able to calculate it directly as $\langle \mathbf{v}^2 \rangle$ and just insert the square of (7.35) into an expectation value expression. But this does not work. The square of the α-matrices is the identity matrix. If that is inserted into the expectation value expression we get the normalization integral multiplied by c^2 regardless of our reference frame. This gives us that

$$\langle \mathbf{v}_x^2 \rangle^{1/2} = \pm c \tag{7.37}$$

and similarly for the y- and z-components. Clearly this is a ludicrous result. It gives us the unlikely sounding information that the electron always has at least the velocity of light (even in its rest frame). As if this wasn't awkward enough the x-, y-, and z-components of velocity commute with each other in non-relativistic theory, but don't in the relativistic case. So in Dirac theory they are not simultaneously measurable to arbitrary precision, but in the non-relativistic theory they are. What has gone wrong? It is impossible to believe that two such different operators representing the same physical quantity can both be correct. Indeed, such a contradiction is good reason to suspect that the whole theory is up the creek. However, those clever fellows Foldy and Wouthuysen come to our rescue again. The essential point is that these two apparently contradictory operators do not actually represent quite the same observable. This can be seen as follows.

We have shown that the F–W transform of the position operator is given by (7.30). We can ask what the operator in the original representation is that corresponds to $\hat{\mathbf{r}}_{FW} = \hat{\mathbf{r}}$ in the new representation. In other words, what do we get if we do a reverse F–W transformation of the position operator? The answer turns out to be similar to (7.30), but with some signs changed:

$$\hat{\mathbf{R}} = e^{-iU_{FW}} \hat{\mathbf{r}} e^{iU_{FW}} = \hat{\mathbf{r}} + \frac{i\hbar c \tilde{\beta} \tilde{\boldsymbol{\alpha}}}{2|\hat{H}|} - \frac{\hbar c^2 (ic\tilde{\beta}(\tilde{\boldsymbol{\alpha}} \cdot \mathbf{p})\mathbf{p} + (\tilde{\boldsymbol{\sigma}} \times \mathbf{p})|\hat{H}|)}{2|\hat{H}|^2(|\hat{H}| + mc^2)} \tag{7.38}$$

This is just as complicated as (7.30), but we can gain some insight from it by working out its time derivative. We do this using (7.24) and

$$\widehat{\frac{\partial A}{\partial t}} = \frac{i}{\hbar}[\hat{H}, \hat{A}] \tag{7.39}$$

The commutator will be evaluated in the F–W representation and then transformed to the standard representation as described by (7.17).

$$\frac{\partial}{\partial t}\hat{\mathbf{r}}_{FW} = \frac{i}{\hbar}[\hat{H}, \hat{\mathbf{r}}] = \frac{i}{\hbar}(\tilde{\beta}|\hat{H}|\hat{\mathbf{r}} - \hat{\mathbf{r}}\tilde{\beta}|\hat{H}|) = \frac{i}{\hbar}\frac{\hbar}{i}\frac{\partial}{\partial \mathbf{p}}\tilde{\beta}|\hat{H}| = \tilde{\beta}\frac{\mathbf{p}c^2}{|\hat{H}|} \tag{7.40}$$

as can easily be seen by actually operating with this commutator on a wavefunction. In (7.40) we have used the momentum space representation of the position operator $\hat{\mathbf{r}} = i\hbar\partial/\partial p$. Now let us transform (7.40) back from the Foldy–Wouthuysen representation to the standard representation.

$$\frac{\partial \hat{X}}{\partial t} = e^{-iU_{FW}} \frac{\hat{\mathbf{p}}c^2}{|\hat{H}|^2} \tilde{\beta} |\hat{H}| e^{iU_{FW}} = \frac{\hat{\mathbf{p}}c^2}{|\hat{H}|^2} e^{-iU_{FW}} \tilde{\beta} |\hat{H}| e^{iU_{FW}} = \frac{\hat{\mathbf{p}}c^2}{|\hat{H}|} \frac{\hat{H}}{|\hat{H}|} \quad (7.41)$$

Now the term $\hat{H}/|\hat{H}|$ here has eigenvalues $+1$ for particle wavefunctions and -1 for antiparticle wavefunctions. So, the time derivative of X is just $\hat{\mathbf{p}}c^2/|\hat{H}|$ for particles. In the non-relativistic limit it is easy to show that this reduces to $\hat{\mathbf{p}}/m$, as shown in equation (1.27).

Let us summarize what we have found out about these operators so far. We have seen that the standard velocity operator gives the electron speed as equal to the speed of light. However, the inverse F–W transform of the position operator given by (7.38) leads to a velocity operator that has a sensible non-relativistic limit. These two can be reconciled if we identify (7.41) with the average velocity of the particle. Then the motion of the electron can be divided into two parts as was done in chapter 3 for Klein–Gordon particles. Firstly there is the average velocity (7.41) and secondly there is very rapid oscillatory motion which ensures that if an instantaneous measurement of the velocity of the electron could be made it would give c. The rapid oscillatory motion is known as *Zitterbewegung* from the German for 'trembling motion'.

Note that we have introduced Foldy–Wouthuysen transformations because they separate the particle and antiparticle parts of the wavefunction. Also, doing these transformations leads from the velocity operator of (7.35) to that of (7.41), i.e. the Foldy–Wouthuysen transformation separates the average velocity from the *Zitterbewegung* for both particles and antiparticles. This implies that there is interference between the positive and negative energy states which leads to (7.35), and the Foldy–Wouthuysen transformation removes this. Such a conclusion is consistent with our interpretation of the spin-zero operators in chapter 3.

An analogous splitting of the position operator is also an unavoidable consequence of this transformation. The average position is given by $\hat{\mathbf{R}}$ but the exact position that would be measured in an instantaneous experiment is $\hat{\mathbf{r}}$ (both in the standard representation). This fits in with the *Zitterbewegung* – if there are average and absolute velocities there must also be average and absolute positions (within a given reference frame). This is the meaning of (7.38). The position operator is the usual position operator plus terms that correct this for the *Zitterbewegung*.

It is possible to go further in our examination of this peculiar motion *Zitterbewegung*. We know that the time derivative of any observable is given by (7.39). Let us evaluate the time derivative of the velocity operator, i.e. the acceleration. To keep the mathematics simple we will consider only the x-component of velocity, $\hat{v}_x = c\tilde{\alpha}_x$:

$$-i\hbar\frac{\partial}{\partial t}\hat{v}_x = [\hat{H}, c\tilde{\alpha}_x] \tag{7.42}$$

Now, $\tilde{\alpha}_x$ anticommutes with every term in the Hamiltonian except $c\tilde{\alpha}_x\hat{p}_x$. It is easy to show that

$$\tilde{\alpha}_x\hat{H} + \hat{H}\tilde{\alpha}_x = 2c\hat{p}_x \tag{7.43}$$

So

$$-i\hbar\frac{\partial}{\partial t}\hat{v}_x = c\hat{H}\tilde{\alpha}_x - c\tilde{\alpha}_x\hat{H} = 2c^2\hat{p}_x - 2c\tilde{\alpha}_x\hat{H} = 2c^2\hat{p}_x - 2\hat{v}_x\hat{H} \tag{7.44}$$

The Hamiltonian and the x-component of momentum are constants of the motion, so if we differentiate (7.44) with respect to time we find

$$i\hbar\frac{\partial^2\hat{v}_x}{\partial t^2} = 2\frac{\partial\hat{v}_x}{\partial t}\hat{H} \tag{7.45}$$

This is a differential equation for $\partial\hat{v}_x/\partial t$. It can be solved immediately:

$$\frac{\partial\hat{v}_x}{\partial t} = \hat{a}_x^0 e^{-2i\hat{H}t/\hbar} \tag{7.46}$$

where $\hat{a}_x^0$ is a constant determined by the boundary conditions. In general it is a matrix and must appear on the left hand side of the exponential because of the order in which the operators appear in (7.45). We now have an expression for $\partial\hat{v}_x/\partial t$ which we can substitute back into (7.44) to find

$$\hat{v}_x = \tfrac{1}{2}i\hbar\hat{a}_x^0 e^{-2i\hat{H}t/\hbar}\hat{H}^{-1} + c^2\hat{p}_x\hat{H}^{-1} \tag{7.47}$$

and hence, a simple integration gives

$$\hat{x} = -\frac{\hbar^2}{4}\hat{a}_x^0 e^{-2i\hat{H}t/\hbar}\hat{H}^{-2} + \hat{p}_x c^2\hat{H}^{-1}t \tag{7.48}$$

From (7.47) we again see that the x-component of velocity has two parts. The second term is the F–W term and is associated with the average motion of the particle, i.e. that given by classical relativistic formulae. The first term is the *Zitterbewegung* which oscillates extremely rapidly. In any macroscopic experimental time there will be a vast number of oscillations because of the $\hbar$ in the denominator of the exponent. Therefore only the average term can be observed. We can estimate the oscillation frequency from (7.47). In the electron rest frame the average velocity is zero, then $\hat{H} \rightarrow mc^2$ and the frequency will be $2mc^2/\hbar$. To see the *Zitterbewegung* we would have to perform a measurement of the velocity of the electron (involving two exceedingly accurate measurements of the electron position) over a timescale of order the inverse of this, i.e. in about 3×10^{-13} seconds! Clearly this is not a feasible experiment.

Another intriguing question to ask is what the amplitude of the oscillations is in the *Zitterbewegung*. We can find an order of magnitude estimate
of this from (7.48).

$$|\hat{x}_{\text{osc}}| = |\frac{\hbar^2}{4}\hat{a}_x^0 e^{-2i\hat{H}t/\hbar}\hat{H}^{-2}| = |\frac{i\hbar}{2}(\hat{p}_x c^2 \hat{H}^{-1} - \hat{v}_x)\hat{H}^{-1}| \qquad (7.49)$$

where we have substituted from (7.47) for the $a_x^0 \exp(-2i\hat{H}t/\hbar)$. Let us
consider the non-relativistic limit of this expression. The first term in the
brackets can be neglected compared with the second which has magni-
tude c. The $\hat{H}^{-1}$ outside the bracket is of order $(mc^2)^{-1}$. Putting these
together gives the amplitude of oscillation as

$$|\hat{x}_{\text{osc}}| \approx \frac{\hbar}{2mc} \qquad (7.50)$$

In this analysis we have worked in one dimension, but the mathematics
will follow through identically for the y- and z-components of momentum.
The result is that we find that the electron can no longer be considered as
a point particle, but is oscillating and hence is spread out over a volume
which is of the order of the cube of the Compton wavelength. As the
average velocity of the particle increases the amplitude of the oscillations
given by (7.49) decreases, becoming zero as the velocity approaches the
speed of light. Therefore, very high energy collision experiments which
imply the electron truly is a point particle are not necessarily inconsistent
with the view of the electron discussed here.

There is a rudimentary argument that ties *Zitterbewegung* in with another puzzling property of Dirac eigenfunctions. Consider the stationary
free particle/antiparticle wavefunctions of below equations (5.60). We will
take the first of these as our example. As a wavefunction to describe a
stationary spin-up particle it is fine. If we move to a frame of reference
where the particle is moving then one or both of the lower two elements
of the wavefunction become non-zero. It is almost as if the motion *mixes
in* some antiparticle character into the wavefunction. If we consider a
non-free particle, e.g. an atomic electron in the field of the nucleus, it
will always have a non-zero lower part of the wavefunction. Now it is
possible that an electron from the negative energy sea may be excited
in the field of the nucleus, or even in the field of the electron itself, i.e.
an electron/positron pair may be created. What can happen then is that
the original electron can decay into the hole in the negative energy sea,
leaving the excited electron to carry on the motion of the original one.
The negative energy part of the wavefunction is what allows this process
of exchange scattering to occur. Now let us appeal to the uncertainty
principle. The energy of the excited electron/positron pair is of order

$2mc^2$, so conservation of energy is violated by this amount. Such a violation is permitted in quantum mechanics, but only for a time interval allowed by the uncertainty principle $\Delta t \approx \hbar/2mc^2$. Next let us consider how far the excited electron can be from the original electron when the latter annihilates. Well, it cannot travel faster than the speed of light, so the maximum possible separation is $c\Delta t = \hbar/2mc$ which is of the order of the radius of the *Zitterbewegung* fluctuations. So this is further evidence that the trembling motion and finite extent of the electron are due to the interaction between the electron and the sea of negative energy particles.

So far, we have said little about angular momentum operators and their transforms (7.31) and (7.32). In the standard representation the orbital angular momentum is $(\hat{\mathbf{r}} \times \hat{\mathbf{p}})$ and the spin operator is $\hbar\hat{\boldsymbol{\sigma}}/2$. If we define these operators in the new F–W representation and transform them back to the standard representation as we have done for the position and velocity operators, we find they take the form

$$\hat{\mathbf{L}}_{\text{av}} = \hat{\mathbf{R}} \times \hat{\mathbf{p}} \tag{7.51}$$

$$\hat{\boldsymbol{\Sigma}} = \frac{\hbar}{2}\left(\tilde{\boldsymbol{\sigma}} - \frac{ic\tilde{\beta}(\tilde{\boldsymbol{\alpha}} \times \hat{\mathbf{p}})}{|\hat{H}|} - \frac{c^2\hat{\mathbf{p}} \times (\tilde{\boldsymbol{\sigma}} \times \hat{\mathbf{p}})}{|\hat{H}|(|\hat{H}| + mc^2)}\right) \tag{7.52}$$

These reverse-transformed operators can be regarded as the mean orbital and the mean spin angular momentum operators respectively. It is these operators that reduce to the conventional form in the Pauli limit.

There is good reason for mentioning the angular momentum operators in a section on *Zitterbewegung*. There is another conceptual problem with quantum theory which has been pushed under the carpet until now. That is the question of what is meant by spin. Spin is given this name because it obeys angular momentum commutation relations and because it is an intrinsic property of a particle. Beyond this, however, myriads of lecturers of quantum mechanics courses have told us not to think of spin as having a classical nature as in the spinning of a gyroscope. As the electron (and other particles) are usually treated as point particles this is sensible advice. What could be meant by the spin of a point? However, at the level of quantum theory we have now reached, we have seen that the electron does not exist at a point, even when 'stationary', but is spread out over a small volume. Clearly a particle occupying a small volume is capable of a quantum analogue of classical spin. Can the intrinsic spin be in any way said to be like our classical notions of spin? The answer to this question is 'almost', which is about as irritating as it could be. To see why this is so consider the following two arguments. The first is a hand-waving type of argument which can only be regarded as plausible because we know the result we are aiming towards. The second argument is somewhat more rigorous, but less conclusive.

Firstly, recall our discussion of velocity operators in the Klein–Gordon theory. We saw that the operator to find the speed of a particle is zero. In Dirac theory the same operator is the square of the velocity operator $c\tilde{\alpha}$, so the speed of a Dirac particle will come out as c. We have also found that the radius of the *Zitterbewegung* oscillations is of the same order in both the Klein–Gordon and Dirac cases and is given by equation (7.50). If we assume that the *Zitterbewegung* volume is spherical and that the point electron sits on the surface of this sphere executing circular motion we can find the angular momentum associated with the *Zitterbewegung*. For Klein–Gordon particles

$$|\mathbf{L}| = |m\mathbf{v} \times \mathbf{r}| = m|v||r| = m \times 0 \times r = 0 \qquad (7.53a)$$

For Dirac particles

$$|\mathbf{L}| = |m\mathbf{v} \times \mathbf{r}| = m|v||r| = m \times c \times \frac{\hbar}{2mc} = \frac{\hbar}{2} \qquad (7.53b)$$

So the angular momentum associated with *Zitterbewegung* for Klein–Gordon particles is zero, whereas that for Dirac particles is $\hbar/2$. Obviously there is a very strong temptation to associate this angular momentum with the spin. However, this discussion is based on rather unphysical arguments and should not be taken very seriously.

Next we present the second, rather more rigorous argument relating *Zitterbewegung* and spin. Consider the position operator (7.38) in the non-relativistic limit $|\mathbf{p}| \ll mc$. We divide this operator up into its odd and even parts, $\hat{\mathbf{R}}^-$ and $\hat{\mathbf{R}}^+$ respectively. The first term in the numerator can be ignored as it is of order $\mathbf{p}^2$ and we are left with

$$\hat{\mathbf{R}}^+ \approx \hat{\mathbf{r}} - \frac{\hbar(\tilde{\sigma} \times \hat{\mathbf{p}})}{4m^2c^2}, \qquad \hat{\mathbf{R}}^- \approx \frac{i\hbar\tilde{\beta}\tilde{\alpha}}{2mc} \qquad (7.54)$$

The average velocity operator is given by (7.41), though for our purposes we write it differently as the reverse transform of $c\tilde{\alpha}$

$$\hat{\mathbf{v}}_{av} = c\tilde{\alpha} - \frac{c^2\tilde{\beta}\mathbf{p}}{|\hat{H}|} + \frac{c^3(\tilde{\alpha} \cdot \mathbf{p})\mathbf{p}}{|\hat{H}|(|\hat{H}| + mc^2)} \qquad (7.55)$$

Dividing this into odd and even parts in the non-relativistic limit gives

$$\hat{\mathbf{v}}_+ \approx -\frac{\tilde{\beta}\mathbf{p}}{m}, \qquad \hat{\mathbf{v}}_- \approx c\tilde{\alpha} \qquad (7.56)$$

Note that the fact that the velocity operator has eigenvalues of $\pm c$ is found in the odd part of $\mathbf{v}$. Next we evaluate the even part of the angular momentum operator $\hat{\mathbf{J}} = m\hat{\mathbf{r}} \times \hat{\mathbf{v}}$. This is called $\hat{\mathbf{J}}$ rather than $\hat{\mathbf{L}}$ as in (7.32) because velocity and momentum are not equivalent quantities in relativistic mechanics. We cannot expect these two different angular momentum operators to be equivalent. The even part of $\hat{\mathbf{J}}$ is given by

$$\hat{\mathbf{J}} = -m(\hat{\mathbf{R}}^+ \times \hat{\mathbf{v}}_+ + \hat{\mathbf{R}}^- \times \hat{\mathbf{v}}_-) \tag{7.57}$$

where the minus sign is included by convention to give a positive angular momentum. Substituting from (7.54) and (7.56) into (7.57) gives

$$\hat{\mathbf{J}} = \tilde{\beta}\hat{\mathbf{r}} \times \hat{\mathbf{p}} - \frac{i\hbar\tilde{\beta}}{2}\tilde{\boldsymbol{\alpha}} \times \tilde{\boldsymbol{\alpha}} = \tilde{\beta}(\hat{\mathbf{L}} + 2\hat{\mathbf{S}}) \tag{7.58}$$

where we have used equation (4.21). Here $\hat{\mathbf{L}}$ and $\hat{\mathbf{S}}$ are the usual orbital and spin angular momentum operators. The product of the odd part of the position operator and the odd part of the velocity operator gives the spin multiplied by two. It is the odd part of the velocity operator that has eigenvalues $\pm c$ and forces us to postulate *Zitterbewegung*. It is also this part of the operator which leads, in the above argument, to the spin. So to some extent we can say that spin is the angular momentum associated with *Zitterbewegung*, i.e. we now view the electron as a point particle existing within a sphere with a volume of order the cube of the Compton wavelength. The point particle in this volume always has an instantaneous speed c and an acceleration given by equation (7.46).

Now consider the arguments of chapter 2 when we looked at the orbital moment of an atomic electron. If we think of our point electron as being in a circular orbit on the surface of its *Zitterbewegung* volume, we can find the magnetic moment associated with this motion (see figure 7.1). Noting the factor 2 in front of the spin in (7.58), we obtain

$$\mu_s = \frac{e}{m}\hat{\mathbf{S}} = \frac{2\mu_B}{\hbar}\hat{\mathbf{S}} \tag{7.59}$$

and we have predicted that the g-factor associated with spin is equal to 2 in accordance with simpler theories. Thus, the $\hat{\mathbf{J}}$ defined in (7.58) represents the angular momentum which is directly proportional to the total magnetic dipole moment. Unfortunately, this is not a very convincing argument because we have omitted the odd terms in our expression for $\hat{\mathbf{J}}$ quite arbitrarily. Nevertheless the above arguments do enable one to make the tentative hypothesis that the spin magnetic moment is the orbital magnetic moment of the *Zitterbewegung*, and the factor of 2 appears in the form of the gyromagnetic ratio, i.e. the point electron rushing round this small volume has an angular momentum associated with it that accounts for the macroscopically observed spin.

If we find the odd contributions to $\hat{\mathbf{J}}$ we get the angular momentum in the F–W representation and a correction to it which is proportional to the

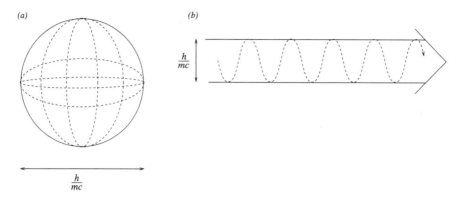

Fig. 7.1. (a) *Zitterbewegung* in the rest frame of the electron. The electron is a point particle which has oscillatory motion within a sphere of diameter the Compton wavelength. The dashed lines indicate possible orbits of the electron that will yield spin $\hbar/2$. (b) The *Zitterbewegung* in a different frame. The sphere of (a) is moving in the direction of the arrow. The electron executes oscillatory motion within the sphere to maintain a speed c at all times.

ratio of the kinetic energy to the rest mass energy. These contributions do not yield any further insight.

One final point to make in this section is on the mass of our particle. As it is moving at the speed of light one might expect it to have infinite mass. This is not true because, as we have seen, the motion is oscillatory and hence the particle does not remain in the same inertial frame.

7.4 Foldy–Wouthuysen Transformation of the Wavefunction

So far we have not examined the Foldy–Wouthuysen transform of the wavefunction. That omission is remedied in this section. We have to evaluate the wavefunction using equation (7.4).

$$\psi'(\mathbf{r}) = e^{iU_{FW}}\psi(\mathbf{r}) = \frac{|\hat{H}| + mc^2 + c\tilde{\beta}\tilde{\boldsymbol{\alpha}} \cdot \hat{\mathbf{p}}}{(2|\hat{H}|(|\hat{H}| + mc^2))^{1/2}}\psi(\mathbf{r}) \tag{7.60}$$

Now $\hat{H} = (c\tilde{\boldsymbol{\alpha}} \cdot \hat{\mathbf{p}} + \tilde{\beta}mc^2)$ and so $\tilde{\beta}\hat{H} = (c\tilde{\beta}\tilde{\boldsymbol{\alpha}} \cdot \hat{\mathbf{p}} + mc^2)$. So equation (7.60) becomes

$$\psi'(\mathbf{r}) = \frac{|\hat{H}| + \tilde{\beta}\hat{H}}{(2|\hat{H}|(|\hat{H}| + mc^2))^{1/2}}\psi(\mathbf{r}) \tag{7.61}$$

We can rewrite this as

$$\psi'(\mathbf{r}) = \left(\frac{|\hat{H}|}{2(|\hat{H}| + mc^2)}\right)^{1/2}(\tilde{I} \pm \tilde{\beta})\psi(\mathbf{r}) \tag{7.62}$$

The fraction on the right of (7.62) is a complicated operator, but it doesn't have a matrix structure. The $(\tilde{I} \pm \tilde{\beta})$ selects either the upper or the lower two elements of $\psi(\mathbf{r})$. Thus $\psi'(\mathbf{r})$ has either the upper or lower two components equal to zero. So we have separated it already. Interestingly, the consequences of the F–W transformation of the wavefunction do not end here. If we take a definite form for the wavefunction things soon get messy. To show this let us write our wavefunction as a general wavepacket

$$\psi(\mathbf{r}) = \int a(\mathbf{p}')e^{i\mathbf{p}'\cdot\mathbf{r}}d\mathbf{p}' \tag{7.63}$$

We can use the energy projection operators from chapter 6 to give

$$\frac{1}{2}\left(1 + \frac{c\tilde{\alpha}\cdot\hat{\mathbf{p}} + \tilde{\beta}mc^2}{|\hat{H}|}\right)\psi(\mathbf{r}) = \begin{array}{cc} \psi(\mathbf{r}) & \textit{positive energies} \\ 0 & \textit{negative energies} \end{array} \tag{7.64a}$$

$$\frac{1}{2}\left(1 - \frac{c\tilde{\alpha}\cdot\hat{\mathbf{p}} + \tilde{\beta}mc^2}{|\hat{H}|}\right)\psi(\mathbf{r}) = \begin{array}{cc} 0 & \textit{positive energies} \\ \psi(\mathbf{r}) & \textit{negative energies} \end{array} \tag{7.64b}$$

So if we define $\psi(\mathbf{r})$ as a sum of positive and negative energy states

$$\psi(\mathbf{r}) = \phi(\mathbf{r}) + \chi(\mathbf{r}) \tag{7.65}$$

we can write

$$\phi(\mathbf{r}) = \int \frac{1}{2}\left(1 + \frac{c\tilde{\alpha}\cdot\hat{\mathbf{p}}' + \tilde{\beta}mc^2}{|\hat{H}|}\right)a(\mathbf{p}')e^{i\mathbf{p}'\cdot\mathbf{r}}d\mathbf{p}' \tag{7.66a}$$

$$\chi(\mathbf{r}) = \int \frac{1}{2}\left(1 - \frac{c\tilde{\alpha}\cdot\hat{\mathbf{p}}' + \tilde{\beta}mc^2}{|\hat{H}|}\right)a(\mathbf{p}')e^{i\mathbf{p}'\cdot\mathbf{r}}d\mathbf{p}' \tag{7.66b}$$

From (7.65)

$$\psi'(\mathbf{r}) = \phi'(\mathbf{r}) + \chi'(\mathbf{r}) \tag{7.67}$$

Combining (7.62) with (7.66) and (7.67)

$$\phi'(\mathbf{r}) = \frac{1}{2}(\tilde{I} + \tilde{\beta})\int \left[\frac{|\hat{H}|}{2(|\hat{H}| + mc^2)}\right]^{\frac{1}{2}}\left[1 + \frac{c\tilde{\alpha}\cdot\hat{\mathbf{p}}' + \tilde{\beta}mc^2}{|\hat{H}|}\right]a(\mathbf{p}')e^{i\mathbf{p}'\cdot\mathbf{r}}d\mathbf{p}' \tag{7.68a}$$

$$\chi'(\mathbf{r}) = \frac{1}{2}(\tilde{I} - \tilde{\beta})\int \left[\frac{|\hat{H}|}{2(|\hat{H}| + mc^2)}\right]^{\frac{1}{2}}\left[1 - \frac{c\tilde{\alpha}\cdot\hat{\mathbf{p}}' + \tilde{\beta}mc^2}{|\hat{H}|}\right]a(\mathbf{p}')e^{i\mathbf{p}'\cdot\mathbf{r}}d\mathbf{p}' \tag{7.68b}$$

and so

$$\psi'(\mathbf{r}) = \int \left[\frac{|\hat{H}|}{2(|\hat{H}| + mc^2)}\right]^{\frac{1}{2}}\left[1 + \frac{\tilde{\beta}(c\tilde{\alpha}\cdot\hat{\mathbf{p}}' + \tilde{\beta}mc^2)}{|\hat{H}|}\right]a(\mathbf{p}')e^{i\mathbf{p}'\cdot\mathbf{r}}d\mathbf{p}' \tag{7.69}$$

Now $\psi(\mathbf{r})$ is given by (7.63) and from this it is easy to show that

$$a(\mathbf{p}') = \left(\frac{1}{2\pi}\right)^3 \int \psi(\mathbf{r}')e^{-i\mathbf{p}'\cdot\mathbf{r}'}d\mathbf{r}' \qquad (7.70)$$

Substituting from (7.70) into (7.69) gives

$$\psi'(\mathbf{r}) = \left(\frac{1}{2\pi}\right)^3 \times$$

$$\int \left[\frac{|\hat{H}|}{2(|\hat{H}| + mc^2)}\right]^{\frac{1}{2}} \left[1 + \frac{\tilde{\beta}(c\tilde{\alpha}\cdot\hat{\mathbf{p}}' + \tilde{\beta}mc^2)}{|\hat{H}|}\right] \int \psi(\mathbf{r}')e^{i\mathbf{p}'\cdot(\mathbf{r}-\mathbf{r}')}d\mathbf{r}'d\mathbf{p}' \qquad (7.71)$$

or, put more simply

$$\psi'(\mathbf{r}) = \frac{1}{8\pi^3} \int K(\mathbf{r}, \mathbf{r}')\psi(\mathbf{r}')d\mathbf{r}' \qquad (7.72)$$

where $K(\mathbf{r}, \mathbf{r}')$ is not a δ-function but is given by

$$K(\mathbf{r}, \mathbf{r}') = \int \left[\frac{|\hat{H}|}{2(|\hat{H}| + mc^2)}\right]^{\frac{1}{2}} \left[1 + \frac{\tilde{\beta}(c\tilde{\alpha}\cdot\hat{\mathbf{p}}' + \tilde{\beta}mc^2)}{|\hat{H}|}\right] e^{i\mathbf{p}'\cdot(\mathbf{r}-\mathbf{r}')}d\mathbf{p}' \quad (7.73)$$

So we see that the wavefunction transformation is an integral, rather than a point transformation. A solution at a single point in the standard representation depends upon the wavefunction in a volume of space in the F–W representation. This is not surprising; we have already deduced that the particle exists in a volume of order $(\hbar/mc)^3$. As we might guess $K(\mathbf{r}, \mathbf{r}')$ is also of this order. This non-locality is the second reason that we don't work in the F–W representation as a matter of course.

7.5 The F–W Transformation in an Electromagnetic Field

Much of what we have considered so far in this chapter has been applicable to free particles only. This is rather limiting as in general particles do experience forces and fields. Therefore, in this section, we consider a F–W transformation when the Hamiltonian contains field-dependent terms.

In this section we consider *weak* fields only. By *weak* we mean fields that contain energy small compared with $2mc^2$. Evidently, if the energy of the field is of this order we would have to worry about encountering the Klein paradox and any separation into positive and negative energy states would be of doubtful value. In the presence of electromagnetic interactions it is no longer possible to make a single unitary transform to a representation in which the Hamiltonian contains no odd operators. Instead we have to make a series of such transformations. It turns out that we can make an expansion in terms of $(1/mc^2)^n$, which can be terminated at any value of n. Because the rest mass energy is large compared with all other energies, such a series is rapidly convergent.

The general strategy for carrying out these transformations is as follows. We divide the Hamiltonian up into three terms

$$\hat{H} = \tilde{\beta}mc^2 + \Omega_e + \Omega_o \tag{7.74}$$

where the first term is the usual rest mass term and is large compared with the other two terms, Ω_e is an operator containing only even matrices and Ω_o contains only odd matrices. In the Dirac Hamiltonian Ω_o will contain the α-matrices (kinetic energy) term and Ω_e contains the scalar potential (which is multiplied by the identity matrix, an even matrix). The next step is to transform the Hamiltonian according to

$$\hat{H}_{FW} = e^{iU_{FW}} \hat{H} e^{-iU_{FW}} \tag{7.75}$$

where

$$U_{FW} = -\frac{i}{2mc^2}\tilde{\beta}\Omega_o \tag{7.76}$$

This is the general, as well as the specific, expression for U_{FW}. It is always $-i\tilde{\beta}/2mc^2$ times the sum of all the odd terms in the Hamiltonian. When the transformation (7.75) has been completed the new Hamiltonian contains even terms plus odd terms of order $1/mc^2$ or higher. These latter terms are used to give a new Ω_o in (7.76) and a second unitary transformation can be performed. The Hamiltonian that results from this contains odd terms of order $(1/mc^2)^2$ or higher. This process can continue to whatever power of $1/mc^2$ is required to give a satisfactory separation of the positive and negative energy states.

If we put (7.76) into (7.75) we have the exponential of a matrix, which is meaningless. Instead we have to expand the exponential as a series:

$$\hat{H}_{FW} = \hat{H} + i[U_{FW}, \hat{H}] + \frac{i^2}{2}[U_{FW}, [U_{FW}, \hat{H}]] + \cdots$$
$$+ \frac{i^n}{n!}[U_{FW}[U_{FW}[U_{FW} \cdots [U_{FW}, \hat{H}] \cdots]]] + \cdots \tag{7.77}$$

The validity of this equation is not immediately evident, although it is clear that the terms should get smaller because they contain ever ascending powers of U_{FW} each of which contains our small parameter $1/mc^2$. Equation (7.77) can easily be verified by noting that we can write the transformed Hamiltonian as a function of λ as follows. Later we will set $\lambda = 1$. Expanding $\hat{H}_{FW}(\lambda)$ in a Maclaurin series gives

$$\hat{H}_{FW}(\lambda) = e^{i\lambda U_{FW}}\hat{H}e^{-i\lambda U_{FW}} = \sum_{n=0}^{\infty}\frac{\lambda^n}{n!}\left(\frac{\partial^n \hat{H}_{FW}}{\partial \lambda^n}\right)_{\lambda=0} \tag{7.78}$$

Differentiating using the product rule gives

$$\frac{\partial \hat{H}_{FW}}{\partial \lambda} = iU_{FW}e^{i\lambda U_{FW}}\hat{H}e^{-i\lambda U_{FW}} - e^{i\lambda U_{FW}}\hat{H}iU_{FW}e^{-i\lambda U_{FW}}$$

$$= ie^{i\lambda U_{FW}}[U_{FW}, \hat{H}]e^{-i\lambda U_{FW}} \tag{7.79}$$

It is easy to see, once one has worked out (7.79), that

$$\frac{\partial^n \hat{H}_{FW}}{\partial \lambda^n} = i^n e^{i\lambda U_{FW}}[U_{FW}[U_{FW}[U_{FW}\cdots[U_{FW}, \hat{H}]\cdots]]]e^{-i\lambda U_{FW}} \tag{7.80}$$

If we set $\lambda = 1$ and insert (7.80) into (7.78), equation (7.77) is proven. Let us use (7.74) to evaluate the second term on the right of (7.77):

$$i[U_{FW}, \hat{H}] = i[U_{FW}, \tilde{\beta}mc^2] + i[U_{FW}, \Omega_e] + i[U_{FW}, \Omega_o] \tag{7.81}$$

We will calculate these in turn:

$$i[U_{FW}, \tilde{\beta}mc^2] = \tfrac{1}{2}[\tilde{\beta}\Omega_o, \tilde{\beta}] = -\Omega_o \tag{7.82}$$

since $\tilde{\beta}$ anticommutes with all odd matrices.

$$i[U_{FW}, \Omega_e] = \frac{\tilde{\beta}}{2mc^2}[\Omega_o, \Omega_e] \tag{7.83}$$

where we have used the fact that $\tilde{\beta}$ commutes with all even matrices. Equation (7.83) is a product of an odd and an even matrix, which is odd.

$$i[U_{FW}, \Omega_o] = \frac{1}{2mc^2}[\tilde{\beta}\Omega_o, \Omega_o] = \frac{1}{mc^2}\tilde{\beta}\Omega_o^2 \tag{7.84}$$

Here we have the product of two odd matrices, which is even. Let us add up all these terms according to (7.81)

$$i[U_{FW}, \hat{H}] = -\Omega_o + \frac{\tilde{\beta}}{2mc^2}[\Omega_o, \Omega_e] + \frac{1}{mc^2}\tilde{\beta}\Omega_o^2 \tag{7.85}$$

Now that we have found $[U_{FW}, \hat{H}]$ we can go on to evaluate the next terms in the series (7.77) until we have found all contributions of order less than $1/m^4c^8$. Let us just state the next few terms and leave their verification as an exercise.

$$-\frac{1}{2}[U_{\text{FW}},[U_{\text{FW}},\hat{H}]] = -\frac{\tilde{\beta}\Omega_\text{o}^2}{2mc^2} - \frac{1}{8m^2c^4}[\Omega_\text{o},[\Omega_\text{o},\Omega_\text{e}]] - \frac{\Omega_\text{o}^3}{2m^2c^4} \quad (7.86)$$

$$-\frac{i}{6}[U_{\text{FW}},[U_{\text{FW}},[U_{\text{FW}},\hat{H}]]] = \frac{\Omega_\text{o}^3}{6m^2c^4} - \frac{\tilde{\beta}}{6m^3c^6}\Omega_\text{o}^4$$
$$- \frac{\tilde{\beta}}{48m^3c^6}([\Omega_\text{o}^3,\Omega_\text{e}] + 3\Omega_\text{o}[\Omega_\text{e}\Omega_\text{o},\Omega_\text{o}])$$
$$\quad (7.87)$$

$$\frac{1}{24}[U_{\text{FW}},[U_{\text{FW}},[U_{\text{FW}},[U_{\text{FW}},\hat{H}]]]] = \frac{\tilde{\beta}\Omega_\text{o}^4}{24m^3c^6} \quad (7.88)$$

In equations (7.86–88) we have retained all terms to order $1/m^3c^6$. There are no further contributions from higher order, so it is time to collect them all together. Putting (7.74) and (7.85–88) into (7.77) we have

$$\hat{H}_{\text{FW}} = \tilde{\beta}mc^2 + \Omega_\text{e} + \frac{\tilde{\beta}\Omega_\text{o}^2}{2mc^2} - \frac{1}{8m^2c^4}[\Omega_\text{o},[\Omega_\text{o},\Omega_\text{e}]] - \frac{\tilde{\beta}}{8m^3c^6}\Omega_\text{o}^4$$
$$+ \frac{\tilde{\beta}}{2mc^2}[\Omega_\text{o},\Omega_\text{e}] - \frac{\Omega_\text{o}^3}{3m^2c^4} - \frac{\tilde{\beta}}{48m^3c^6}([\Omega_\text{o}^3,\Omega_\text{e}] + 3\Omega_\text{o}[\Omega_\text{e}\Omega_\text{o},\Omega_\text{o}])$$
$$\quad (7.89)$$

Here, only the final three terms are odd and they are of order $1/mc^2$ or higher. It is time to do a second F–W transformation. We proceed with an analogous method to above, this time with the Hamiltonian (7.89).

Using the same prescription as before, our transformation is given by

$$\hat{H}'_{\text{FW}} = e^{iU'_{\text{FW}}}\hat{H}_{\text{FW}}e^{-iU'_{\text{FW}}} \quad (7.90)$$

where

$$U'_{\text{FW}} = -\frac{i}{2mc^2}\tilde{\beta}\times$$
$$\left(\frac{\tilde{\beta}}{2mc^2}[\Omega_\text{o},\Omega_\text{e}] - \frac{\Omega_\text{o}^3}{3m^2c^4} - \frac{\tilde{\beta}}{48m^3c^6}([\Omega_\text{o}^3,\Omega_\text{e}] + 3\Omega_\text{o}[\Omega_\text{e}\Omega_\text{o},\Omega_\text{o}])\right)$$
$$\quad (7.91)$$

and we are going to evaluate the analogue of the expansion (7.77) with $\hat{H}_{\text{FW}}$ replaced by $\hat{H}'_{\text{FW}}$, $\hat{H}$ by $\hat{H}_{\text{FW}}$ and U_{FW} by U'_{FW}.

Now, again, we are only going to be interested in terms of order $1/m^3c^6$ and lower. When we calculate the commutator of (7.91) with the various terms in (7.89), it is clear that the first three terms and the first odd term will give contributions to $\hat{H}'_{FW}$ of the required order. Hence we need only consider those terms.

$$[iU'_{FW}, \tilde{\beta}mc^2] = -\frac{\tilde{\beta}}{2mc^2}[\Omega_o, \Omega_e] + \frac{\Omega_o^3}{3m^2c^4}$$

$$+ \frac{\tilde{\beta}}{48m^3c^6}([\Omega_o^3, \Omega_e] + 3\Omega_o[\Omega_e\Omega_o, \Omega_o]) \qquad (7.92)$$

This is very convenient, because when we add the Hamiltonian (7.89) and its commutators as prescribed in (7.77) we will see that (7.92) cancels the last three terms in (7.89). The second term in (7.89) contributes

$$[iU'_{FW}, \Omega_e] = \frac{1}{4m^2c^4}[[\Omega_o, \Omega_e], \Omega_e] - \frac{\tilde{\beta}}{6m^3c^6}[\Omega_o^3, \Omega_e] \qquad (7.93)$$

The third term gives

$$\left[iU'_{FW}, \frac{\tilde{\beta}\Omega_o^2}{2mc^2}\right] = -\frac{\tilde{\beta}}{8m^3c^6}([\Omega_o, \Omega_e]\Omega_o^2 + \Omega_o^2[\Omega_o, \Omega_e]) \qquad (7.94)$$

and the odd term

$$\left[iU'_{FW}, \frac{\tilde{\beta}}{2mc^2}[\Omega_o, \Omega_e]\right] = -\frac{\tilde{\beta}}{4m^3c^6}[\Omega_o, \Omega_e]^2 \qquad (7.95)$$

Collecting together (7.92 − 95)

$$[iU'_{FW}, \hat{H}'_{FW}] = -\frac{\tilde{\beta}}{2mc^2}[\Omega_o, \Omega_e] + \frac{\Omega_o^3}{3m^2c^4} + \frac{1}{4m^2c^4}[[\Omega_o, \Omega_e], \Omega_e]$$

$$+ \frac{\tilde{\beta}}{48m^3c^6}([\Omega_o^3, \Omega_e] + 3\Omega_o[\Omega_e\Omega_o, \Omega_o]) - \frac{\tilde{\beta}}{6m^3c^6}[\Omega_o^3, \Omega_e]$$

$$- \frac{\tilde{\beta}}{8m^3c^6}([\Omega_o, \Omega_e]\Omega_o^2 + \Omega_o^2[\Omega_o, \Omega_e]) - \frac{\tilde{\beta}}{4m^3c^6}[\Omega_o, \Omega_e]^2 \tag{7.96}$$

There is one contribution from the analogue of the third term on the right hand side of (7.77) which is of order $1/m^3c^6$

$$-\frac{1}{2}[U'_{FW}, [U'_{FW}, \tilde{\beta}mc^2]] = \frac{\tilde{\beta}}{8m^3c^6}[\Omega_o, \Omega_e]^2 \qquad (7.97)$$

Now we can insert (7.89), (7.96), and (7.97) into the analogue of (7.77) and we obtain

$$\hat{H}'_{FW} = \tilde{\beta}mc^2 + \Omega_e + \frac{\tilde{\beta}\Omega_o^2}{2mc^2} - \frac{1}{8m^2c^4}[\Omega_o, [\Omega_o, \Omega_e]] - \frac{\tilde{\beta}}{8m^3c^6}\Omega_o^4$$

$$- \frac{\tilde{\beta}}{8m^3c^6}[\Omega_o, \Omega_e]^2 + \frac{1}{4m^2c^4}[[\Omega_o, \Omega_e], \Omega_e] - \frac{\tilde{\beta}}{6m^3c^6}[\Omega_o^3, \Omega_e]$$

$$- \frac{\tilde{\beta}}{8m^3c^6}([\Omega_o, \Omega_e]\Omega_o^2 + \Omega_o^2[\Omega_o, \Omega_e]) \tag{7.98}$$

Equation (7.98) is the Hamiltonian after the second F–W transformation. In this Hamiltonian the final three terms are odd and they are of order $1/m^2c^4$ or higher. We require the Hamiltonian correct to order $1/m^3c^6$ and so we require another transformation:

$$\hat{H}''_{FW} = e^{iU''_{FW}}\hat{H}'e^{-iU''_{FW}} \tag{7.99}$$

and this time

$$U''_{FW} = -\frac{i}{8m^3c^6}\left(\tilde{\beta}[[\Omega_o,\Omega_e],\Omega_e]\right.$$
$$\left.-\frac{\tilde{\beta}}{48m^4c^8}(4[\Omega_o^3,\Omega_e] - 3([\Omega_o,\Omega_e]\Omega_o^2 + \Omega_o^2[\Omega_o,\Omega_e]))\right) \tag{7.100}$$

We only have to consider the commutator of this with the first two terms of $\hat{H}'_{FW}$, as all other terms will be at least of order $1/m^4c^8$. Following the same procedure as before we have

$$[iU''_{FW}, \hat{H}'_{FW}] = -\frac{1}{4m^2c^4}[[\Omega_o,\Omega_e],\Omega_e] + \frac{\tilde{\beta}}{6m^3c^6}[\Omega_o^3,\Omega_e]$$
$$+ \frac{\tilde{\beta}}{8m^3c^6}([\Omega_o,\Omega_e]\Omega_o^2 + \Omega_o^2[\Omega_o,\Omega_e]) + \frac{\tilde{\beta}}{8m^3c^6}[[[\Omega_o,\Omega_e],\Omega_e],\Omega_e] \tag{7.101}$$

When we add this to (7.98) several terms will cancel. Comparison of (7.101) and (7.100) shows that no higher order terms in the analogue of (7.77) will contribute, so our Hamiltonian after the third transformation is the sum of (7.98) and (7.101)

$$\hat{H}''_{FW} = \tilde{\beta}mc^2 + \Omega_e + \frac{\tilde{\beta}\Omega_o^2}{2mc^2} - \frac{1}{8m^2c^4}[\Omega_o, [\Omega_o, \Omega_e]]$$
$$- \frac{\tilde{\beta}}{8m^3c^6}\Omega_o^4 - \frac{\tilde{\beta}}{8m^3c^6}[\Omega_o,\Omega_e]^2 + \frac{\tilde{\beta}}{8m^3c^6}[[[\Omega_o,\Omega_e],\Omega_e]\Omega_e] \tag{7.102}$$

We have to perform one last transformation. This time

$$U'''_{FW} = -\frac{i}{2mc^2}\frac{1}{8m^3c^6}[[[\Omega_o,\Omega_e],\Omega_e]\Omega_e] \tag{7.103}$$

For this final F–W transformation we only have to find the commutator of U'''_{FW} with the first term of (7.102). This gives

$$[iU'''_{FW}, \hat{H}''_{FW}] = -\frac{\tilde{\beta}}{8m^3c^6}[[[\Omega_o,\Omega_e],\Omega_e]\Omega_e] \tag{7.104}$$

So, again using an analogue of (7.77), our final result for the F–W transformed Hamiltonian correct to order $1/m^3c^6$ is

$$\hat{H}'''_{FW} = \tilde{\beta}mc^2 + \Omega_e + \frac{\tilde{\beta}\Omega_o^2}{2mc^2} - \frac{1}{8m^2c^4}[\Omega_o, [\Omega_o, \Omega_e]] - \frac{\tilde{\beta}}{8m^3c^6}(\Omega_o^4 + [\Omega_o,\Omega_e]^2) \tag{7.105}$$

Following through the mathematics has shown us how this works. The Foldy–Wouthuysen transformation works in the way it does, without a proliferation of terms, because when you take the commutator of all the odd terms in the Hamiltonian with $\tilde{\beta}mc^2$, part of the answer is the negative of all the odd terms, because $\tilde{\beta}$ has the useful property that it anticommutes with all odd matrices and commutes with all even matrices. Hence the $(n+1)^{\text{th}}$ Foldy–Wouthuysen transformation produces terms that cancel the odd terms in the Hamiltonian generated by the nth Foldy–Wouthuysen transformation.

What we have seen so far in this section is a long and rather tedious derivation of (7.105), which doesn't look very enlightening, so why did we do it? The answer to this soon becomes clear when we apply this rather abstract formalism to the case of an electron in a time-independent electromagnetic field. In that case the Hamiltonian is

$$\hat{H} = c\tilde{\alpha} \cdot (\hat{\mathbf{p}} - e\mathbf{A}) + \tilde{\beta}mc^2 + V(\mathbf{r}) \tag{7.106}$$

Our odd and even matrices in this case are given by

$$\Omega_{\text{o}} = c\tilde{\alpha} \cdot (\hat{\mathbf{p}} - e\mathbf{A}), \qquad\qquad \Omega_{\text{e}} = V\tilde{I} \tag{7.107}$$

Henceforth we shall drop the identity matrix explicitly; where no matrix is written the identity matrix is assumed to multiply it. Now we want to substitute these into (7.105). Although we have retained terms to order $1/m^3c^6$ we are going to write the resulting Hamiltonian only to order $1/c^2$. The reason for this contorted way of getting a result to this order is that Ω_{o} for our case already contains c. So the Ω_{o}^4 term in (7.105) is really only correct to order $1/c^2$.

Evaluation of the first two terms in (7.105) for the Hamiltonian (7.106) is trivial. The third requires the use of the identity (4.19) and we obtain

$$\Omega_{\text{o}}^2 = c^2(\hat{\mathbf{p}} - e\mathbf{A})^2 - c^2e\hbar\tilde{\boldsymbol{\sigma}} \cdot \mathbf{B} \tag{7.108}$$

where $\mathbf{B} = \nabla \times \mathbf{A}$ is the applied magnetic field. Any readers feeling masochistic are welcome to evaluate the fourth term of equation (7.105) with the replacements equation (7.107). It involves a lot of algebra, so I suggest that you take my word for it that the result is given by

$$\frac{1}{8m^2c^4}[\Omega_{\text{o}},[\Omega_{\text{o}},\Omega_{\text{e}}]] = \frac{\hbar e}{4m^2c^2}\tilde{\boldsymbol{\sigma}} \cdot \mathbf{E} \times (\hat{\mathbf{p}} + e\mathbf{A}) - \frac{\hbar^2 e}{8m^2c^2}\nabla \cdot \mathbf{E} \tag{7.109}$$

where $\mathbf{E} = -\nabla V(\mathbf{r})$ is the applied electric field. The Ω_{o}^4 term in (7.105) can be evaluated by making use of our result for Ω_{o}^2 when we calculated the third term. From Maxwell's equations we know that terms involving $\mathbf{B}$ are of higher order in $1/c^2$ so they are ignored and we have

$$\frac{\tilde{\beta}}{8m^3c^6}\Omega_{\text{o}}^4 = \frac{\tilde{\beta}}{8m^3c^2}(\hat{\mathbf{p}} - e\mathbf{A})^4 \tag{7.110}$$

Finally the last term in (7.105) disappears with the substitutions (7.107) because Ω_e is just a multiple of the identity matrix. So, finally, (7.105) in its full glory becomes

$$\hat{H}_{FW}''' = \tilde{\beta}mc^2 + V(\mathbf{r}) + \frac{\tilde{\beta}}{2m}(\hat{\mathbf{p}} - e\mathbf{A})^2 - \frac{e\hbar}{2m}\tilde{\beta}\tilde{\sigma} \cdot \mathbf{B} + \frac{\hbar^2 e}{8m^2 c^2}\nabla \cdot \mathbf{E}$$

$$- \frac{\hbar e}{4m^2 c^2}\tilde{\sigma} \cdot \mathbf{E} \times (\hat{\mathbf{p}} - e\mathbf{A}) - \frac{\tilde{\beta}}{8m^3 c^2}(\hat{\mathbf{p}} - e\mathbf{A})^4 \qquad (7.111)$$

At last we have made the desired separation. One can see immediately that the only matrices that appear in this Hamiltonian are $\tilde{\sigma}$ and $\tilde{\beta}$ and we know that both of them are even, so equation (7.111) separates the four components of the Dirac wavefunction into two pairs. This is the result we were looking for (but what an effort to get an equation we have seen before!). It does not separate them into four equations, which would have been a bonus, because the σ-matrices are not diagonal. If we consider the upper two components of (7.111) only, we can replace the 4×4 σ-matrices with the 2×2 σ-matrices and we have a Hamiltonian for the positive energy states only:

$$\hat{H}_{p} = mc^2 + V(\mathbf{r}) + \frac{1}{2m}(\hat{\mathbf{p}} - e\mathbf{A})^2 - \frac{e\hbar}{2m}\tilde{\sigma} \cdot \mathbf{B} + \frac{\hbar^2 e}{8m^2 c^2}\nabla \cdot \mathbf{E}$$

$$- \frac{\hbar e}{4m^2 c^2}\tilde{\sigma} \cdot \mathbf{E} \times (\hat{\mathbf{p}} - e\mathbf{A}) - \frac{1}{8m^3 c^2}(\hat{\mathbf{p}} - e\mathbf{A})^4 \qquad (7.112a)$$

Of course, we can also write down the equation for the negative energy states only. It is:

$$\hat{H}_{a} = -mc^2 + V(\mathbf{r}) - \frac{1}{2m}(\hat{\mathbf{p}} - e\mathbf{A})^2 + \frac{e\hbar}{2m}\tilde{\sigma} \cdot \mathbf{B} + \frac{\hbar^2 e}{8m^2 c^2}\nabla \cdot \mathbf{E}$$

$$- \frac{\hbar e}{4m^2 c^2}\tilde{\sigma} \cdot \mathbf{E} \times (\hat{\mathbf{p}} + e\mathbf{A}) + \frac{1}{8m^3 c^2}(\hat{\mathbf{p}} - e\mathbf{A})^4 \qquad (7.112b)$$

Let us summarize what we have done in this section. A general relativistic Hamiltonian was divided into even and odd parts. Four F–W transformations were performed on this Hamiltonian. This was sufficient to ensure that odd terms in the Hamiltonian were of order $1/m^4 c^8$ or higher. The resulting Hamiltonian is given by (7.105). The next step was to use this formalism in an example. We took a particle in a steady electromagnetic field and substituted for the operators in (7.105). This leads us to a Hamiltonian (7.111) in which the positive and negative energy states are described separately. The equation for the positive energy particles is (7.112a), and that for the negative energy particles is (7.112b).

Compare equation (7.112a) with equation (4.74b). Clearly they are identical, and their interpretations are the same. The first three terms are obvious, they are the rest mass energy, the potential energy and the kinetic energy of the electron in an electromagnetic field respectively. The fourth term is the interaction of the magnetic field with the magnetic moment of the particle μ given by

$$\mu = -\frac{e\hbar}{2m} \tag{7.113}$$

We see that, in this approach too, the magnetic moment of the electron falls neatly out of the mathematics. This implies that the spin is in the Hamiltonian somewhere, and we have seen earlier that it can be loosely interpreted as the angular momentum associated with the *Zitterbewegung*. The fourth and sixth terms in (7.112a) may be combined as shown in equation (4.75) to give us the spin–orbit coupling and Thomas precession. The fifth and seventh terms are the Darwin term and the mass–velocity term, both of which have also been discussed earlier.

In chapter 2 we postulated the Pauli equation as an equation to describe non-relativistic particles with spin. In chapter 4 we saw that the non-relativistic limit of the Dirac Hamiltonian gives the Pauli equation, and this proved that the Pauli Hamiltonian is valid in the limit of low velocities. We were also able to correct the Pauli equation to order $1/c^2$. In the present section we have made no statement about the velocity of the particle. The only constraint required for this F–W expansion was that the rest mass energy be larger than any other energies in the Hamiltonian (not a very restrictive approximation). So here we have shown how to improve the accuracy of the Pauli Hamiltonian to any order.

The Foldy–Wouthuysen transformation was set up as a way of separating the Hamiltonian into positive and negative energy parts. It has led us to equations (7.112). The new feature here is that we have found a Pauli Hamiltonian for the negative energy states on an equal footing with that for the positive energy states. All the terms in (7.112a) have their analogue in (7.112b), the only difference being some sign changes. Note that (7.112b) is not the Hamiltonian for the antiparticle, it is the Hamiltonian for the negative energy state. From (7.112b) we see that a negative energy electron has (obviously) a negative rest mass energy, and experiences the same potential as a positive energy electron. It also contains the Darwin term with the same sign as the positive energy electron, implying that a negative energy particle undergoes *Zitterbewegung*. Equations (7.112) also show that the spin–orbit coupling in a negative energy atom would be the same as in a positive energy atom. The kinetic energy of the negative energy state is negative as expected, and finally the interaction of the negative energy particle spin with the applied field gives a contribution to the energy of opposite sign to that of the positive energy particle, so splitting of energy levels in the Zeeman effect would work the other way for negative energy atoms.

Finally, then, we have separated the Dirac Hamiltonian into two, one for the positive, and one for the negative, energy states. In doing this we have gained further insight into the nature and behaviour of Dirac particles. In particular we have seen more deeply into the *Zitterbewegung* and its relation to spin, and have made more rigorous our understanding of the negative energy states. This has culminated in equations (7.112).

There is a straightforward generalization of the above approach to the case of a time-dependent electromagnetic field (Bjorken and Drell 1964). Other transformations have been attempted (Chraplyvy 1953a,b, Barker and Glover 1955, Kursunoglu 1956) and the F–W scheme has been extended (Cini and Touschek 1958, Bose *et al.* 1959, Pac 1959a,b). However it has been shown that this F–W approach is essentially unique. Other approaches do not seem to teach us any new physics.

We can draw some general conclusions from this chapter (Costella and McKellar 1995). Note that in the standard Dirac representation we use minimal coupling $\hat{\mathbf{p}} \to \hat{\mathbf{p}} - e\mathbf{A}$ in the corresponding free particle equation. In any non-trivial transformation (one involving the momentum operator) we expect different results depending on whether we use minimal coupling and then transform, or transform and then use minimal coupling. There are an infinite number of representations of the Dirac equation, but two stand out above the others. Firstly there is the usual Dirac representation. This is important because the momentum operator appears in it linearly, and it is correct to use minimal coupling. In this representation the four components of the electron wavefunction are inextricably linked, and the electron is seen as a structureless charged point particle. In the second representation, known as the Newton–Wigner representation (1949), the positive and negative energy states are separate, and the electron wave equation looks like the classical equation. However, the equation of motion contains $\hat{\mathbf{p}}^2$, hence the coupling to the electromagnetic field is more complicated. Furthermore, in this second representation the electron is not a point particle, but acquires a radius, and an intrinsic magnetic moment. Each of these representations has unique advantages, and both stand out above all other representations.

Recently Thaller (1992) (see also Brouder *et al.* 1996) has thrown some doubt on the mathematical rigour of the Foldy–Wouthuysen transformation. Under some circumstances terms in the expansion of order $1/c^4$ and higher may be singular. However, what we have covered in this chapter reproduces the non-relativistic limit plus corrections, and seems to be valid provided the potentials do not become of order $2mc^2$.

7.6 Problems

(1) Evaluate equation (7.18) explicitly to show that $\hat{H}_{FW}$ is indeed a diagonal matrix.

(2) Prove for yourself that equations (7.31–33) are correct expressions for the velocity, angular momentum and spin operators in the Foldy–Wouthuysen representation.

(3) For a general wavepacket consisting of both positive and negative energy states show that the current density will exhibit *Zitterbewegung*.

(4) Verify the commutators (7.85–88).

8

One-Electron Atoms

In this chapter we are going to make a very detailed study of the one-electron atom using relativistic quantum mechanics. The one-electron atom is the simplest bound system that occurs in nature, and plays a central role in both classical and quantum theory. There are two reasons for this. Firstly, it is a model that can be solved exactly at many levels of theory – the classical Bohr atom, the non-relativistic quantum mechanical one-electron atom, and the relativistic one-electron atom with spin zero and spin 1/2. Secondly, most of our understanding of multielectron atoms, molecules and solids is based on this model.

Of course, a good description of the hydrogen atom is provided by non-relativistic Schrödinger theory. However, once the relativistic quantum theory had been discovered it was of great importance to validate the theory by describing hydrogen at least as well as Schrödinger theory. As we shall see, the Dirac theory was exceedingly successful in this, and furthermore, at several points in the theory the non-relativistic limit can be taken and the effect of relativity in one-electron atoms can be seen explicitly. This connection with the simpler theory will be made here wherever insight may be gained from doing so.

One of the key advances brought about by the advent of quantum theory was that it defined eigenfunctions as well as the eigenvalues predicted by the Bohr theory. If we have eigenfunctions we are able to do perturbation theory and examine such physical phenomena as the Zeeman and Stark effects. For reasons that will become apparent it is easier, in some respects, to do perturbation theory with the relativistic eigenfunctions than with the non-relativistic ones. Hence, towards the end of this chapter we will carry through a couple of examples which illustrate this. In particular we will consider the Zeeman effect and magnetic dichroism.

Antiparticles do not appear in this chapter. This is sensible because we do not expect negative energy states to play much role in determining

227

the properties of atoms. However, it is not quite so easy to justify this mathematically. The *Zitterbewegung* is still there, and the finite size of the electron due to this is completely ignored; the electron is assumed to feel the potential at a single point. Of course the solution presented in this chapter is entirely equivalent to that for a positron bound to an antiproton.

The one-electron atom is rather more complicated than other soluble models in quantum physics for two reasons: firstly it is three-dimensional and secondly it consists of two particles. The three-dimensional character of the atom means that it can have an angular momentum; this is true in a non-relativistic and a relativistic theory. Beyond this extra complication we also have the two-particle nature of the system, which leads to a problem in the relativistic theory that is not present in the non-relativistic theory. In non-relativistic theory we can model the atom as a nucleus of infinite mass being orbited by an electron, or we can consider the electron as having the reduced mass and orbiting about the centre of mass. In principle this is not possible within a relativistic theory because the reduced mass does not have a simple definition. There are two ways out of this in practice. Either we can ignore the nucleus and only regard it as a source of a static field, or we can introduce the reduced mass and not worry about the ambiguity because the nuclear motion can be taken as non-relativistic to a very good approximation.

With this latter approximation, the one-electron atom can be solved within relativistic quantum mechanics, and we present the solution in detail in this chapter. The derivation of the eigenvalues and eigenvectors of the one-electron atom is treated in much the same way in many texts. Here we go into a great deal of detail, and follow the work and notation of Rose (1961) fairly closely.

8.1 The Radial Dirac Equation

The Dirac equation for a spherical potential $V(\mathbf{r}) = V(r)$ is

$$(c\tilde{\boldsymbol{\alpha}} \cdot \hat{\mathbf{p}} + \tilde{\beta}mc^2 + V(r))\psi(\mathbf{r}) = W\psi(\mathbf{r}) \tag{8.1}$$

The origin is taken at the nucleus. For this problem it is most convenient to solve the Dirac equation in spherical polar coordinates. Obviously the $\tilde{\beta}mc^2$ term is independent of coordinate system and the potential is most easily written in radial form, so we only have to transform the kinetic energy term. To do this we use a vector identity

$$\nabla = \hat{\mathbf{r}}(\hat{\mathbf{r}} \cdot \nabla) - \hat{\mathbf{r}} \times (\hat{\mathbf{r}} \times \nabla) \tag{8.2}$$

where $\hat{\mathbf{r}}$ is a unit radial vector. This follows directly from equation (4.115), but proving it from first principles is good practice for you. We also need to recall that the orbital angular momentum operator is given by $\hat{\mathbf{L}} = \hat{\mathbf{r}} \times \hat{\mathbf{p}} = -i\hbar\mathbf{r} \times \nabla$. Here $\hat{\mathbf{r}}$ is the vector operator defining the position of the electron relative to the source of the field. For our case of a spherically symmetric potential, equation (8.2) simplifies because $\partial/\partial\theta$ and $\partial/\partial\phi$ are both equal to zero so $\mathbf{r} \cdot \nabla = r\partial/\partial r$. Thus, with some trivial manipulation, equation (8.2) can be written

$$\nabla = \hat{\mathbf{r}}\frac{\partial}{\partial r} - \frac{i}{\hbar}\frac{\hat{\mathbf{r}}}{|\mathbf{r}|} \times \hat{\mathbf{L}} \tag{8.3}$$

This is one way of writing the ∇ operator in spherical polar coordinates. Using it we write the relativistic kinetic energy operator in radial form

$$\tilde{\boldsymbol{\alpha}} \cdot \hat{\mathbf{p}} = -i\hbar\tilde{\boldsymbol{\alpha}} \cdot \nabla = -i\hbar\tilde{\boldsymbol{\alpha}} \cdot \hat{\mathbf{r}}\frac{\partial}{\partial r} - \frac{1}{|\mathbf{r}|}\tilde{\boldsymbol{\alpha}} \cdot \hat{\mathbf{r}} \times \hat{\mathbf{L}} \tag{8.4}$$

The operator $\hat{\mathbf{L}}$ is perpendicular to $\mathbf{r}$, so from (4.19) we can see that

$$\tilde{\boldsymbol{\alpha}} \cdot \hat{\mathbf{r}}\tilde{\boldsymbol{\alpha}} \cdot \hat{\mathbf{L}} = i\tilde{\boldsymbol{\sigma}} \cdot \hat{\mathbf{r}} \times \hat{\mathbf{L}} \tag{8.5}$$

We know that $\tilde{\gamma}_5$ commutes with the α- and the σ-matrices. Now, all these matrices are closely related to one another through $\tilde{\gamma}_5\tilde{\boldsymbol{\alpha}} = -\tilde{\boldsymbol{\sigma}}$ and $\tilde{\gamma}_5\tilde{\boldsymbol{\sigma}} = -\tilde{\boldsymbol{\alpha}}$. So we can postmultiply each side of (8.5) by $\tilde{\gamma}_5$ to find

$$i\tilde{\boldsymbol{\alpha}} \cdot \hat{\mathbf{r}}\tilde{\boldsymbol{\sigma}} \cdot \hat{\mathbf{L}} = -\tilde{\boldsymbol{\alpha}} \cdot \hat{\mathbf{r}} \times \hat{\mathbf{L}} \tag{8.6}$$

Now we can substitute (8.6) into the second term of (8.4)

$$\tilde{\boldsymbol{\alpha}} \cdot \hat{\mathbf{p}} = -i\hbar\tilde{\boldsymbol{\alpha}} \cdot \hat{\mathbf{r}}\frac{\partial}{\partial r} + \frac{i}{|\mathbf{r}|}\tilde{\boldsymbol{\alpha}} \cdot \hat{\mathbf{r}}\tilde{\boldsymbol{\sigma}} \cdot \hat{\mathbf{L}} \tag{8.7}$$

Next recall the K-operator defined in equation (6.58). This can be substituted into equation (8.7) to give

$$\tilde{\boldsymbol{\alpha}} \cdot \hat{\mathbf{p}} = -i\hbar\tilde{\alpha}_r\frac{\partial}{\partial r} + \frac{i}{|\mathbf{r}|}\tilde{\alpha}_r(\tilde{\beta}\hat{K} - \hbar) \tag{8.8}$$

with $\tilde{\alpha}_r = \tilde{\boldsymbol{\alpha}} \cdot \hat{\mathbf{r}}$. This is the form in which we want $\tilde{\boldsymbol{\alpha}} \cdot \mathbf{p}$ to substitute into the Dirac equation (8.1). Doing this and a little algebra leads to

$$\left(ic\tilde{\gamma}_5\tilde{\sigma}_r\left(\hbar\frac{\partial}{\partial r} + \frac{\hbar}{r} - \frac{\tilde{\beta}\hat{K}}{r}\right) + \tilde{\beta}mc^2 + V(r)\right)\psi(r) = W\psi(r) \tag{8.9}$$

This is the spherical polar form of the Dirac equation. We need to make use of equation (2.131a) and we are going to insist that the four-component eigenfunctions take on a form as similar as possible to that found when we discussed two-component Pauli spinors, i.e.

$$\psi_\kappa^{m_j}(\mathbf{r}) = \begin{pmatrix} g_\kappa(r)\chi_\kappa^{m_j}(\hat{\mathbf{r}}) \\ if_\kappa(r)\chi_{-\kappa}^{m_j}(\hat{\mathbf{r}}) \end{pmatrix} \tag{8.10}$$

where the $\chi_\kappa^{m_j}(\hat{\mathbf{r}})$ are the spin-angular functions of equation (2.131a). This is a reasonable form for the wavefunctions. We saw in chapter 3 that relativity without spin only changes the radial form of the wavefunctions. In chapter 2 when we examined the Pauli equation in a central field we found that spin without relativity introduced the Pauli spinors into the wavefunction. Here we have spin and relativity, and we presume that here as well relativity will not affect the angular part of the wavefunction while spin will introduce the Pauli spinors again. The reason for the negative κ and the i in the lower part of the wavefunction will become clear shortly.

This choice of eigenfunction means that the solutions are also eigenfunctions of $\hat{\mathbf{J}}^2$, $\hat{J}_z$ and $\hat{K}$. If we substitute (8.10) into (8.9) we obtain four equations, of course, but it turns out that there are really only two separate equations. Remembering that the eigenvalue of the $\hat{K}$ operator is $-\hbar\kappa$, these are

$$-c\hbar\left(\frac{\partial}{\partial r} + \frac{1}{r} + \frac{\kappa}{r}\right)g_\kappa(r)\chi_{-\kappa}^{m_j} + (W - V(r) + mc^2)f_\kappa(r)\chi_{-\kappa}^{m_j} = 0 \qquad (8.11a)$$

$$c\hbar\left(\frac{\partial}{\partial r} + \frac{1}{r} - \frac{\kappa}{r}\right)f_\kappa(r)\chi_\kappa^{m_j} + (W - V(r) - mc^2)g_\kappa(r)\chi_\kappa^{m_j} = 0 \qquad (8.11b)$$

Now the $\chi_\kappa^{m_j}$ cancel nicely and the role of the i in the solutions (8.10) is apparent. It was inserted to ensure that the radial part of the eigenfunction is manifestly real. We can write these equations in the form

$$\frac{\partial g_\kappa(r)}{\partial r} = -\frac{\kappa + 1}{r}g_\kappa(r) + \frac{1}{c\hbar}(W - V(r) + mc^2)f_\kappa(r) \qquad (8.12a)$$

$$\frac{\partial f_\kappa(r)}{\partial r} = \frac{\kappa - 1}{r}f_\kappa(r) - \frac{1}{c\hbar}(W - V(r) - mc^2)g_\kappa(r) \qquad (8.12b)$$

This is the usual form of the radial Dirac equation. It is sometimes useful to make the substitution $u_\kappa(r) = rg_\kappa(r)$ and $v_\kappa(r) = rf_\kappa(r)$. In that case equations (8.12) become

$$\frac{\partial u_\kappa(r)}{\partial r} = -\frac{\kappa}{r}u_\kappa(r) + \frac{1}{c\hbar}(W - V(r) + mc^2)v_\kappa(r) \qquad (8.13a)$$

$$\frac{\partial v_\kappa(r)}{\partial r} = \frac{\kappa}{r}v_\kappa(r) - \frac{1}{c\hbar}(W - V(r) - mc^2)u_\kappa(r) \qquad (8.13b)$$

Equations (8.13) are an equally valid form of the radial Dirac equation. Now we see that the form of the wavefunction chosen (8.10) is OK because it works, the equation for the radial part of the wavefunction only has dropped out of this analysis. Indeed we have shown that the form chosen for the angular part of the wavefunction is the correct one provided the radial part satisfies (8.12). Next we examine these radial functions.

8.2 Free Electron Solutions

Before going on to the Dirac hydrogen atom itself, it is instructive to examine the solutions of (8.13) when the potential is equal to zero. This is particularly so because the solutions we find will be useful when we discuss scattering theory and the eigenfunctions will resemble those found when we discussed relativistic plane waves in Cartesian coordinates.

Let us set the potential in (8.13) equal to zero and differentiate (8.13a) with respect to r:

$$\frac{\partial^2 u_\kappa(r)}{\partial r^2} = -\frac{\kappa}{r}\frac{\partial u_\kappa(r)}{\partial r} + \frac{\kappa}{r^2}u_\kappa(r) + \frac{1}{c\hbar}(W + mc^2)\frac{\partial v_\kappa(r)}{\partial r} \tag{8.14}$$

Now (8.13b) can be substituted into here for $\partial v_\kappa(r)/\partial r$ and we rearrange (8.13a) to find an expression for $v_\kappa(r)$

$$v_\kappa(r) = c\hbar(W + mc^2)^{-1}\left(\frac{\partial u_\kappa(r)}{\partial r} + \frac{\kappa}{r}u_\kappa(r)\right) \tag{8.15}$$

This is also substituted into (8.14), and some algebra eventually leaves us with a fairly simple looking equation:

$$\frac{\partial^2 u_\kappa(r)}{\partial r^2} = \left(\frac{\kappa(\kappa + 1)}{r^2} - \frac{p^2}{\hbar^2}\right)u_\kappa(r) \tag{8.16}$$

This is the radial free particle Dirac equation and can be solved rather easily (if you know where to look). The solution that is regular at the origin for $\kappa > 0$ is

$$u_\kappa(r) = Crj_l(kr) \tag{8.17a}$$

where C is an, as yet, unspecified normalization constant, $j_l(kr)$ is a spherical Bessel function of the first kind and $k = p/\hbar$ is the particle wavevector. For $\kappa < 0$ the solution is

$$u_\kappa(r) = Crn_l(kr) \tag{8.17b}$$

where $n_l(kr)$ is a spherical Neumann function. We can find $v(r)$ from these by substituting into (8.15) and using well-known recursion relations for the Bessel functions (Abramowitz and Stegun 1972). This is straightforward for the solution (8.17a). For (8.17b) we use

$$n_l(x) = (-1)^{l+1}j_{-l-1}(x) \tag{8.18}$$

However, these solutions are irregular at the origin and so are non-physical, so we will ignore them. Equations (8.17a) and (8.15) can be combined to give

$$\psi_\kappa^{m_j}(\mathbf{r}) = J_\kappa^{m_j}(\mathbf{r}) = \begin{pmatrix} j_l(kr)\chi_\kappa^{m_j}(\hat{\mathbf{r}}) \\ \frac{ikc\hbar S_\kappa}{W+mc^2}j_{\bar{l}}(kr)\chi_{-\kappa}^{m_j}(\hat{\mathbf{r}}) \end{pmatrix} \tag{8.19}$$

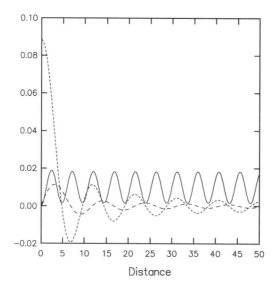

Fig. 8.1.   For an electron with energy $mc^2 + 10^5$ eV and $\kappa = -1$ the full line is the radial probability density discussed in the text, the short dashed line is the large component of the wavefunction and the long dashed line is the small component. All quantities are in relativistic units.

This is a relativistic plane wave in spherical polar coordinates. It is analogous to one term in the sum of the well-known expansion of exponentials in terms of spherical harmonics and plane waves (equation (C.21)).

Note that we have dropped the normalization in (8.19). Equation (8.19) is very similar to the form of the wavefunction found when we were discussing Dirac plane waves (equation (5.23)). This is hardly surprising as both are solutions of the Dirac equation with zero potential. The difference is in the replacement of the spinors (defined by m_s) in (5.23) by spin-angular functions (defined by κ and m_j) in (8.19). This indicates that the z-component of spin and total angular momentum play similar roles in the two cases. As we have seen (equation (2.147)) we can expand the spinors in terms of spin-angular functions and *vice versa*. In figures 8.1 and 8.2 we display the solutions of the radial Dirac equation for a free electron with energy 10^5 eV above its rest mass energy. Within each figure the normalizations are relatively correct, but the relative sizes of the peaks between the two diagrams are meaningless. As expected the radial probability density integrated over angles soon takes on a repeating oscillatory form, whereas the wavefunctions themselves slowly die away. It is necessary to select a large kinetic energy for the small component of the wavefunction to show up on this scale.

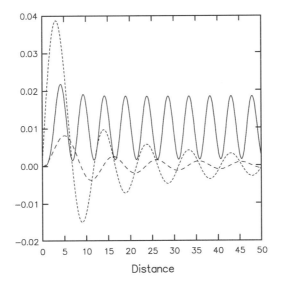

Fig. 8.2.   For an electron with energy $mc^2 + 10^5\,\text{eV}$ and $\kappa = -2$ the full line is the radial probability density discussed in the text, the short dashed line is the large component of the wavefunction and the long dashed line is the small component. All quantities are in relativistic units.

8.3 One-Electron Atoms, Eigenvectors and Eigenvalues

This is the central section of this chapter and is exceedingly mathematically involved. Although some readers may regard this as tedious, I make no apology for it. Part of the beauty of the subject of relativistic quantum theory is that such deep and complex mathematics leads to results that are fairly simple to understand physically and easily compared with experiment and the non-relativistic limit. We are going to consider the motion of an electron of mass m moving under the influence of the spherically symmetric Coulomb potential

$$V(r) = -\frac{Ze^2}{4\pi\epsilon_0 r} \tag{8.20}$$

To start this analysis let us write (8.13) with the form of the potential included explicitly. For ease of writing we put $\xi = Ze^2/(4\pi\epsilon_0\hbar c) = Z\alpha$:

$$\frac{\partial u_\kappa(r)}{\partial r} = -\frac{\kappa}{r}u_\kappa(r) + \left(W_\text{C} + \frac{\xi}{r} + k_\text{C}\right)v_\kappa(r) \tag{8.21a}$$

$$\frac{\partial v_\kappa(r)}{\partial r} = \frac{\kappa}{r}v_\kappa(r) - \left(W_\text{C} + \frac{\xi}{r} - k_\text{C}\right)u_\kappa(r) \tag{8.21b}$$

Here $k_C = mc/\hbar$ is the Compton wavevector, and $W_C = W/c\hbar$. Next we define a very unlikely looking transformation:

$$u_\kappa(r) = (k_C + W_C)^{1/2} e^{-\lambda r}(\phi_1 + \phi_2) \tag{8.22a}$$

$$v_\kappa(r) = (k_C - W_C)^{1/2} e^{-\lambda r}(\phi_1 - \phi_2) \tag{8.22b}$$

where $\lambda = (k_C^2 - W_C^2)^{1/2}$. Now we have to substitute (8.22) and the derivatives of (8.22) into (8.21). This gives two simultaneous differential equations for ϕ_1 and ϕ_2:

$$\frac{\partial \phi_1}{\partial r} + \frac{\partial \phi_2}{\partial r} = \left(\lambda - \frac{\kappa}{r}\right)(\phi_1 + \phi_2) + \left(W_C + \frac{\xi}{r} + k_C\right)\left(\frac{k_C - W_C}{k_C + W_C}\right)^{\frac{1}{2}}(\phi_1 - \phi_2) \tag{8.23a}$$

$$\frac{\partial \phi_1}{\partial r} - \frac{\partial \phi_2}{\partial r} = \left(\lambda + \frac{\kappa}{r}\right)(\phi_1 - \phi_2) - \left(W_C + \frac{\xi}{r} - k_C\right)\left(\frac{k_C + W_C}{k_C - W_C}\right)^{\frac{1}{2}}(\phi_1 + \phi_2) \tag{8.23b}$$

We rearrange (8.23b) to obtain an expression for $\partial \phi_2/\partial r$. This is then substituted into (8.23a) and we get a mind-bogglingly long equation. Luckily many things in that equation cancel. Actually it's not luck – we certainly wouldn't have come down this mathematical path if things didn't simplify fairly quickly. Anyway, we find

$$\lambda \frac{\partial \phi_1}{\partial r} - 2\lambda^2 \phi_1 + \frac{\kappa\lambda}{r}\phi_2 + \frac{\xi W_C}{r}\phi_1 + \frac{\xi k_C}{r}\phi_2 = 0 \tag{8.24a}$$

$$\lambda \frac{\partial \phi_2}{\partial r} + \frac{\kappa\lambda}{r}\phi_1 - \frac{\xi W_C}{r}\phi_2 - \frac{\xi k_C}{r}\phi_1 = 0 \tag{8.24b}$$

Let us change to a dimensionless variable $\rho = 2\lambda r$ and then we have

$$\frac{\partial \phi_1}{\partial \rho} = \left(1 - \frac{\xi W_C}{\lambda \rho}\right)\phi_1 - \left(\frac{\kappa}{\rho} + \frac{\xi k_C}{\lambda \rho}\right)\phi_2 \tag{8.25a}$$

$$\frac{\partial \phi_2}{\partial \rho} = \frac{\xi W_C}{\lambda \rho}\phi_2 + \left(-\frac{\kappa}{\rho} + \frac{\xi k_C}{\lambda \rho}\right)\phi_1 \tag{8.25b}$$

We have now got these equations into a form in which they can be solved, but this is still not simple. There is no nice well-defined function that is the solution of these equations. We have to solve them by substituting in a power series in ρ and then equating powers of ρ:

$$\phi_1(\rho) = \rho^s \sum_{m=0}^{\infty} \alpha_m \rho^m \tag{8.26a}$$

$$\phi_2(\rho) = \rho^s \sum_{m=0}^{\infty} \beta_m \rho^m \tag{8.26b}$$

There should be no confusion between the coefficients α_m and the α-matrices. The α-matrices do not appear in section 8.3 and the α coefficients appear nowhere else. Putting (8.26) into (8.25) is straightforward. Then if we compare coefficients of equal powers of ρ in each term we find the following recursion relations for the α and β coefficients.

$$\alpha_m(m+s) = \alpha_{m-1} - \frac{W_C\xi}{\lambda}\alpha_m - \left(\kappa + \frac{\xi k_C}{\lambda}\right)\beta_m \qquad (8.27a)$$

$$\beta_m(m+s) = \left(-\kappa + \frac{\xi k_C}{\lambda}\right)\alpha_m + \frac{W_C\xi}{\lambda}\beta_m \qquad (8.27b)$$

For $m > 0$ this determines all the α and β coefficients in terms of α_0 and β_0. However, as yet we do not know these, nor do we know the correct value of s. To find the latter we let $m = 0$ in equations (8.27). Then, by definition, from (8.26), $\alpha_{m-1} = 0$ and we have a pair of simultaneous equations for α_0 and β_0. These equations will have a non-trivial solution if the determinant of the coefficients is equal to zero.

$$\begin{vmatrix} s + W_C\xi/\lambda & \kappa + \xi k_C/\lambda \\ \kappa - \xi k_C/\lambda & s - W_C\xi/\lambda \end{vmatrix} = 0 \qquad (8.28)$$

i.e.

$$s^2 - \frac{W_C^2\xi^2}{\lambda^2} = \kappa^2 - \frac{\xi^2 k_C^2}{\lambda^2} \qquad (8.29)$$

But, by definition, $\lambda^2 = k_C^2 - W_C^2$ and so we are left with

$$s = +\sqrt{\kappa^2 - \xi^2} \qquad (8.30)$$

The usual argument for taking the positive square root is because the wavefunctions (8.26) must be regular at the origin. This is not actually the case. The wavefunctions may diverge as $r \to 0$ provided they do not do so faster than as $1/r$. As a factor r^2 is always included when we do an integral in spherical polar coordinates to calculate an observable, the observable will still take on a finite value. It is necessary to take the positive square root in (8.30) because that ensures the wavefunctions are always square integrable. The next stage in this analysis is to find α_m in terms of α_{m-1} and β_m in terms of β_{m-1}. This can be done using equations (8.27). Rearranging (8.27b) gives us

$$\frac{\beta_m}{\alpha_m} = \frac{\kappa - \xi k_C/\lambda}{W_C\xi/\lambda - m - s} = \frac{\kappa - \xi k_C/\lambda}{n' - m} \qquad (8.31)$$

where we define $n' = (W_C\xi/\lambda) - s$. The astute reader may guess the reason for making this definition. If not, it will soon become clear. We can use

(8.31) in (8.27*a*), and some simple algebra leads to

$$\alpha_m = -\frac{n' - m}{m(2s + m)}\alpha_{m-1} \tag{8.32}$$

We now have α_m in terms of α_{m-1}, but m just represents an arbitrary term in the expansions (8.26). Using (8.32) but reducing m by one means we can find α_{m-1} in terms of α_{m-2}. This process can be continued down to α_0 to give us

$$\alpha_m = (-1)^m \frac{(n' - 1)(n' - 2)\cdots(n' - m)}{m!(2s + 1)(2s + 2)\cdots(2s + m)}\alpha_0 \tag{8.33}$$

We can also use (8.31) to eliminate α_m and α_{m-1} in (8.27*a*). This gives us an expression for β_m in terms of β_{m-1}

$$\beta_m = -\frac{n' - m + 1}{m(2s + m)}\beta_{m-1} \tag{8.34}$$

and hence

$$\beta_m = (-1)^m \frac{n'(n' - 1)\cdots(n' - m + 1)}{m!(2s + 1)(2s + 2)\cdots(2s + m)}\beta_0 \tag{8.35}$$

Now the only quantities we don't know are α_0 and β_0. However, these are known in terms of each other from (8.31):

$$\beta_0 = \frac{\kappa - \xi k_C/\lambda}{n'}\alpha_0 \tag{8.36}$$

Now we can take an arbitrary value for, say, α_0 and then all α_m and β_m are known in terms of it. The arbitrariness can be removed later by our choice of normalization. Indeed the fact that the normalization is unspecified is the reason that the indeterminacy remains in the wavefunctions. To find the wavefunctions, we first substitute (8.33) and (8.35) into (8.26).

$$\phi_1(\rho) = \rho^s \alpha_0 \sum_{m=0}^{\infty} \frac{(1 - n')(2 - n')\cdots(m - n')}{m!(2s + 1)(2s + 2)\cdots(2s + m)}\rho^m$$

$$= \rho^s \alpha_0 \left(\frac{(1 - n')}{2s + 1}\rho + \frac{(1 - n')(2 - n')}{(2s + 1)(2s + 2)}\frac{\rho^2}{1!} + \cdots\right)$$

$$= \rho^s \alpha_0 M(1 - n', 2s + 1, \rho) \tag{8.37a}$$

$$\phi_2(\rho) = \rho^s \beta_0 \sum_{m=0}^{\infty}(-1)^m \frac{n'(n' - 1)\cdots(n' - m + 1)}{m!(2s + 1)(2s + 2)\cdots(2s + m)}\rho^m$$

$$= \rho^s \beta_0 \left(1 - \frac{n'}{2s + 1}\rho + \frac{n'(n' - 1)}{(2s + 1)(2s + 2)}\frac{\rho^2}{2!} + \cdots\right)$$

$$= \rho^s \beta_0 M(-n', 2s + 1, \rho) \tag{8.37b}$$

and using (8.36) we have

$$\phi_2(\rho) = \rho^s \frac{\kappa - \xi k_C/\lambda}{n'} \alpha_0 M(-n', 2s+1, \rho) \qquad (8.37c)$$

The solutions are written in terms of confluent hypergeometric functions. These functions, and their properties, are discussed in appendix *B*.

At this stage equations (8.37) can be compared with equation (3.120) for the spinless electron atom. Clearly the eigenfunctions for both cases have a lot in common. Equation (3.120) is not strictly comparable with (8.37) because the Klein–Gordon equation was solved in its one-component form whereas the analysis here is necessarily two-component.

Now we have essentially solved the one-electron atom problem and equations (8.37) are the hydrogenic wavefunctions. However, these are the general solutions, as we have not yet applied any boundary conditions. The boundary condition we use is the usual one for such problems: the wavefunction must remain finite as $r \to \infty$. To implement this we have to look at the asymptotic behaviour of the hypergeometric functions. This is given by equation (*B*.6). At large z this expression is dominated by the exponential and so diverges. Of course, we have made several transformations, we multiplied by r to get $u(r)$ and $v(r)$ from $g(r)$ and $f(r)$, then we made the transformations defined by (8.22) and finally we let $\rho = 2\lambda r$. One's first thought when faced with this divergence at infinity may be that we are so far from the actual wavefunction that a divergence in $\phi_1(\rho)$ and $\phi_2(\rho)$ does not necessarily imply an infinity in the wavefunction. However, in this problem, this would be an incorrect first thought. The way round this difficulty is to make judicious choices of the parameters n' and s in (8.37). As stated in appendix B, if $-n'$ is a negative integer and $2s + 1$ is not a negative integer it is easy to see from (*B*.2) that the confluent hypergeometric function becomes a simple polynomial of order n' in ρ. So

$$n' = 0, 1, 2, 3, \cdots \qquad (8.38)$$

We see here that n' can take on only integer values and it is closely related to the principal quantum number defined in non-relativistic quantum mechanics. There is a problem with the $n' = 0$ solution which requires separate comment from the $n' \neq 0$ case. This is because with $n' = 0$ the confluent hypergeometric function in (8.37a) diverges; the first parameter is not a negative integer. Therefore $\alpha_0 = 0$ and hence $\phi_1(\rho) = 0$. In this case α_0/n' in (8.37c) is just a normalization constant. In general for $n' = 0$, $\phi_2(\rho)$ is non-zero unless

$$\kappa = \frac{\xi k_C}{\lambda} \qquad (8.39)$$

As all quantities on the right hand side of equation (8.39) are positive by definition it can only be true for $\kappa \geq 1$.

We have now solved the Dirac equation for the radial parts of the hydrogenic wavefunctions. Let us combine the definitions of $u(r)$, $v(r)$ and ρ with equations (8.22) and (8.37) to give a final expression for $g(r)$ and $f(r)$. We end up with the horrendous looking formulae

$$f_\kappa(r) = 2\lambda(k_C - W_C)^{1/2} e^{-\lambda r} (2\lambda r)^{s-1} \alpha_0 \times$$
$$\left(M(1 - n', 2s + 1, 2\lambda r) - \frac{\kappa - \xi k_C/\lambda}{n'} M(-n', 2s + 1, 2\lambda r) \right) \qquad (8.40a)$$

$$g_\kappa(r) = 2\lambda(k_C + W_C)^{1/2} e^{-\lambda r} (2\lambda r)^{s-1} \alpha_0 \times$$
$$\left(M(1 - n', 2s + 1, 2\lambda r) + \frac{\kappa - \xi k_C/\lambda}{n'} M(-n', 2s + 1, 2\lambda r) \right) \qquad (8.40b)$$

Now we have the eigenvectors, the next task is to find the eigenvalues. To do this it is useful to define the principal quantum number

$$n = n' + |\kappa| \qquad (8.41)$$

(compare this with equation (3.124) for the spinless electron atom). When n' was defined earlier it was in the middle of finding the eigenvectors of equation (8.40), and, so as not to divert attention from the thrust of that argument, we omitted to point out that the definition of n' gives us an expression for the energy eigenvalues. We remedy that omission here. Recall that

$$n' = \frac{W_C \xi}{\lambda} - s = \frac{W_C \xi}{(k_C^2 - W_C^2)^{1/2}} - s \qquad (8.42)$$

Rearranging this and substituting for n' from (8.41) gives

$$W_C^2 = k_C^2 \left(1 + \frac{\xi^2}{(n - |\kappa| + s)^2} \right)^{-1} \qquad (8.43a)$$

and from the definitions of k_C and W_C we find

$$W = mc^2 \left(1 + \frac{\xi^2}{(n - |\kappa| + s)^2} \right)^{-1/2} \qquad (8.43b)$$

(compare this formula with equation (3.121) for the spinless electron atom). Equation (8.43) is the well-known formula for the energy levels of hydrogen-like atoms. Actually, describing (8.43) as well known is an overstatement. It is rather more true to say that to the very small fraction of the world's population that understands relativistic quantum theory this equation is familiar. Note that it depends on κ, a parameter associated with both spin and orbital angular momentum. When we examined the spinless electron atom we found that the l quantum number appeared in the formula for the energy eigenvalue. So the non-relativistic degeneracy of states with the same value of n but different l was broken. In this case we have $|\kappa|$ in the formula which is dependent on both the spin and orbital angular momenta, so the degeneracy in the single spinless electron atom is broken further.

The final step in this section is to make contact with the familiar non-relativistic eigenvalues of the one-electron atom. We do this in rather a lot of mathematical detail because the result will be the non-relativistic result plus relativistic corrections. These relativistic corrections are of interest in their own right, as we shall see. In this analysis we shall make liberal use of the expansion (Abramowitz and Stegun 1972):

$$(1+x)^p = 1 + px + \frac{p(p-1)}{2!}x^2 + \frac{p(p-1)(p-2)}{3!}x^3 + \cdots \qquad (8.44)$$

We have found s in equation (8.30). If we take a factor of κ outside the square root and expand (8.30) using (8.44) we obtain

$$s \approx \kappa \left(1 - \frac{\xi^2}{2\kappa^2} + \frac{3\xi^4}{8\kappa^4} \right) \qquad (8.45)$$

Inserting (8.45) into (8.43) and multiplying out the square of the denominator in the fraction in (8.43) gives

$$W \approx mc^2 \left(1 + \frac{\xi^2}{n^2} \left(1 - \frac{\xi^2}{n|\kappa|} + \frac{3\xi^4}{4n|\kappa|\kappa^2} + \frac{\xi^4}{4\kappa^2 n^2} \right)^{-1} \right)^{-1/2} \qquad (8.46)$$

if we retain terms of order ξ^4 and lower. Expanding the inner bracket of (8.46) using (8.44) leads to

$$W \approx mc^2 \left(1 + \frac{\xi^2}{n^2} \left(1 + \frac{\xi^2}{n|\kappa|} - \frac{3\xi^4}{4n|\kappa|\kappa^2} + \frac{3\xi^4}{4\kappa^2 n^2} \right) \right)^{-1/2} \qquad (8.47)$$

The reader will not be surprised to learn that we use the now all too familiar expansion (8.44) to simplify the term inside the outer brackets in (8.47), dropping terms of order ξ^6:

$$W \approx mc^2 \left(1 - \frac{\xi^2}{2n^2} + \xi^4 \left(\frac{3}{8n^4} - \frac{1}{2n^3|\kappa|} \right) \right) \qquad (8.48)$$

In non-relativistic quantum theory there is no rest mass energy, so the quantity we want to compare with the non-relativistic case is $W - mc^2$. Remembering the definition of ξ from above equation (8.21), equation

(8.48) can be written:

$$W - mc^2 = -\frac{Z^2 e^4 m}{(4\pi\epsilon_0)^2 \hbar^2 2n^2} + \frac{Z^4 e^8 m}{(4\pi\epsilon_0)^4 \hbar^4 c^2} \left(\frac{3}{8n^4} - \frac{1}{2n^3 |\kappa|} \right) \qquad (8.49)$$

(Compare this with equations (3.123) and (2.142).) Clearly the first term is identical with the non-relativistic energy levels for the one-electron atom. The lifting of degeneracies comes at the order E_{Ry}^2/mc^2, where E_{Ry} is the non-relativistic energy of the ground state of the one-electron atom. This analysis depends upon the expansion (8.44) which is only valid for small x. The fine structure constant is $\alpha \approx 1/137$ and for small Z this is clearly a good approximation. However, elements towards the higher end of the periodic table have $Z \sim 100$, so the assumption is not so good and accurate energies require us to retain higher order terms. Fortunately (?), nature does not provide us with atoms that have $Z > 104$, so the critical point of $Z > 137$ which would cause the series to diverge is not achieved. A schematic diagram of the energy levels of the one-electron atom with $Z = 92$ is shown in figure 8.3. This is compared with the non-relativistic energy levels and should also be compared with figure 3.4 for the $Z = 64$ one-electron atom without spin. Clearly the energy level splittings and degeneracies are strongly dependent upon which theory is applied.

A word should be said here about the negative energy states. The right hand side of equation (8.43) is a positive number, but the total energy is positive only because we have to add the rest mass energy to the negative potential energy. This is all fine. If, however, the potential well were so deep that the magnitude of the negative potential energy were greater than the rest mass energy of the electron, the analysis we have carried out would be in trouble because of interference from the negative energy states, and we would expect the trouble to manifest itself as the creation of particle/antiparticle pairs as in the Klein paradox.

Equation (8.49) is very similar to equation (3.123) for the spinless electron atom. The only difference between the two is that $l + 1/2$ in the denominator of the n^3 term is replaced by the combined spin-angular quantum number κ. This is the leading order effect of spin on the energy levels of hydrogen. When the spectrum of atomic hydrogen is observed experimentally, splittings consistent with equation (8.49), not with equation (3.123), are observed, hence the spectrum of hydrogen is strong evidence for the existence of electron spin.

We can compare the energy eigenvalues of the one-electron atom calculated in various ways. The correction term in $3/8n^4$ occurs in both the one-electron atom eigenvalues of equation (8.49) and in the spinless electron atom eigenvalues of equation (3.123). However, it does not occur when we treat spin–orbit coupling as a perturbation on the non-relativistic

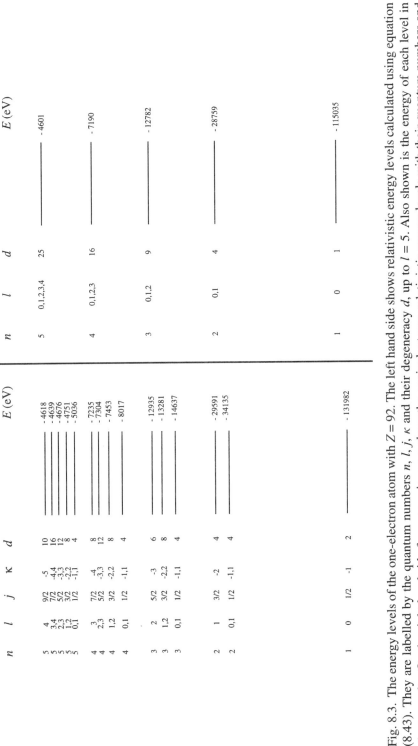

n	l	j	κ	d	E (eV)
5	4	9/2	-5	10	-4618
5	3,4	7/2	4,-4	16	-4639
5	2,3	5/2	-3,3	12	-4676
5	1,2	3/2	-2,2	8	-4751
5	0,1	1/2	-1,1	4	-5036
4	3	7/2	-4	8	-7235
4	2,3	5/2	-3,3	12	-7304
4	1,2	3/2	-2,2	8	-7453
4	0,1	1/2	-1,1	4	-8017
3	2	5/2	-3	6	-12935
3	1,2	3/2	-2,2	8	-13281
3	0,1	1/2	-1,1	4	-14637
2	1	3/2	-2	4	-29591
2	0,1	1/2	-1,1	4	-34135
1	0	1/2	-1	2	-131982

n	l	d	E (eV)
5	0,1,2,3,4	25	-4601
4	0,1,2,3	16	-7190
3	0,1,2	9	-12782
2	0,1	4	-28759
1	0	1	-115035

Fig. 8.3. The energy levels of the one-electron atom with $Z = 92$. The left hand side shows relativistic energy levels calculated using equation (8.43). They are labelled by the quantum numbers n, l, j, κ and their degeneracy d, up to $l = 5$. Also shown is the energy of each level in electronvolts. On the right hand side for comparison we show the equivalent non-relativistic energy levels with their quantum numbers and degeneracies.

one-electron atom in equation (2.142). Therefore this term can be ascribed to the mass–velocity and Darwin corrections to the Hamiltonian. The energy correction in $1/n^3$ occurs in all three expressions for the eigenvalues with different coefficients associated with angular momentum in each case, so it cannot be assigned to any single term in the Hamiltonian, but is clearly partially due to spin–orbit coupling.

Let us examine the physical consequences of equation (8.49). This equation shows that relativity will have its greatest effect on energy levels with low values of the principal quantum number. Indeed, by far the most significant effect is on the $1s$ level. Substituting the quantum numbers into (8.49) shows that the perturbation of the $1s$ level is over 10 times greater than the perturbation of the $2s$ level. The other thing to notice about (8.49) is that for a given orbital the relativistic shift is greater the larger Z becomes. Although in reality most atoms cannot be regarded as one-electron atoms, the implication of this result is that relativistic effects will play a greater role in atoms with a larger Z. This prediction is borne out by more sophisticated calculations and by experiment. An example is in the splitting of the $2p$ levels into $j = l + 1/2, (\kappa = -2)$ and $j = l - 1/2, (\kappa = 1)$ levels. The splitting is given by the last term in (8.49). For aluminium (a fairly light element, $Z = 13$), this gives rise to a doublet which is split by 1.3 eV. For iron ($Z = 26$) this splitting has risen to 21 eV (for real iron it is about 13 eV, but that (of course) is not a one-electron atom). By the time we reach a really heavy element, take uranium for example ($Z = 92$), it has risen to approximately 4500 eV. Here we have an example of the superiority of relativistic quantum theory over the non-relativistic theory. These splittings, and the trends in them, drop naturally out of the relativistic theory, but are only describable non-relativistically with the empirical introduction of spin–orbit coupling into the Hamiltonian.

The experimentally observed energy level scheme in hydrogen follows equation (8.43) closely, but not exactly. For example the $2s^{1/2}$ and $2p^{1/2}$ levels are not precisely degenerate. There is further hyperfine splitting due to two further effects which we do not treat here. The first is the splitting due to the interaction of the electron with the magnetic moment of the nucleus. The second is a purely quantum field theory effect which arises because the electromagnetic field can be treated as a superposition of quantum mechanical oscillators. The electron in the one-electron atom can interact with the zero point motion of this field and the interaction leads to a small change in the energy levels of the atom. This is known as the Lamb shift (Lamb and Retherford 1947, Welton 1948).

8.4 Behaviour of the Radial Functions

We have found the form of the radial wavefunctions given by (8.40). However, we have not stated anything about the normalization of the wavefunction. So far this has been determined by the value of α_0. We require that the probability of finding the electron somewhere in space is unity. The wavefunction describing our electron in a central field is given by (8.10). In this section we will drop the explicit angular dependence of the spin-angular functions. So

$$\int_0^\infty \psi^\dagger(\mathbf{r})\psi(\mathbf{r})d\mathbf{r} = \int \left(g_\kappa(r)\chi_\kappa^{m_j\dagger}, -if_\kappa(r)\chi_{-\kappa}^{m_j\dagger} \right) \left(\begin{matrix} g_\kappa(r)\chi_\kappa^{m_j} \\ if_\kappa(r)\chi_{-\kappa}^{m_j} \end{matrix} \right) d\mathbf{r} = 1 \quad (8.50)$$

Although the total normalization has to be unity, how it is divided between the radial parts and the spin-angular functions is arbitrary. Both $f_\kappa(r)$ and $g_\kappa(r)$ are always real and hence by far the most popular choice of normalization is

$$\int_0^\infty r^2(f_\kappa^2(r) + g_\kappa^2(r))dr = 1 \quad (8.51)$$

Once a choice of normalization such as this has been made, the arbitrariness of the choice of α_0 is eliminated.

In non-relativistic quantum mechanics it is well known that the radial parts of the single-electron atom wavefunctions are simply the Laguerre polynomials. Each radial eigenfunction is defined by the principal quantum number n and the orbital angular momentum quantum number l. The number of nodes – that is, the number of times the radial wavefunction passes through zero (excluding at the nucleus and at infinity) – is given in terms of the quantum numbers by $n - l - 1$. An obvious question to ask is whether there exists a similar prescription for the number of nodes of the wavefunctions in the relativistic quantum theory. In the relativistic theory there are two components of the radial wavefunction so we can certainly predict that any such rule is bound to be more complicated than the non-relativistic one. Indeed it is possible to go through a long and tedious mathematical analysis (Rose 1961) to discover several properties of the radial wavefunctions. However, the insight gained is hardly worth the mathematical complexity, so here we shall content ourselves with a simple summary of the principal general properties of the radial functions $g_\kappa(r)$ and $f_\kappa(r)$:

- For all values of κ: $f_\kappa(r)$ and $g_\kappa(r)$ have opposite signs as $r \to \infty$.
- For $\kappa < 0$: $f_\kappa(r)$ and $g_\kappa(r)$ have opposite signs as $r \to 0$.
- For $\kappa > 0$: $f_\kappa(r)$ and $g_\kappa(r)$ have the same sign as $r \to 0$.
- If $\kappa < 0$: the number of nodes in $f_\kappa(r)$ is equal to the number of nodes in $g_\kappa(r)$.

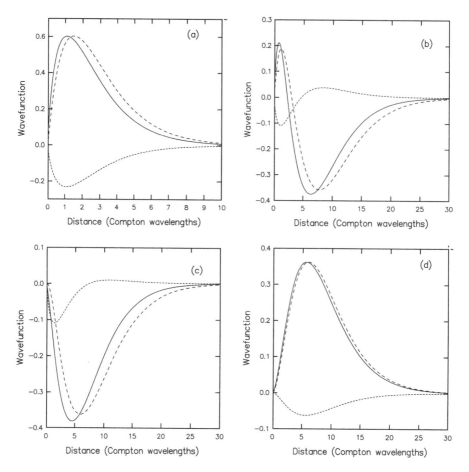

Fig. 8.4. (a) to (d) The relativistic $1s^{1/2}$, $2s^{1/2}$, $2p^{1/2}$, and $2p^{3/2}$ wavefunctions (times r) in the one-electron atom with $Z = 92$ as a function of distance from the nucleus. In each case the full line is the large component, the short dashed line is the small component and the long dashed line is the non-relativistic (a) $1s$, (b) $2s$, (c) $2p^{1/2}$ (d) $2p^{3/2}$ wavefunction. The wavefunctions for (a) to (d) are given by equations 8.53 a and b; c and d; e and f; and g and h respectively.

• If $\kappa > 0$: the number of nodes in $f_\kappa(r)$ exceeds the number of nodes in $g_\kappa(r)$ by one.

These rules can easily be confirmed by inspection of the eigenfunctions in figures 8.4a to d which show all the radial wavefunctions for the principal quantum numbers $n = 1, 2$ and spin-angular quantum numbers $\kappa = -1, 1$, and -2. These figures illustrate another noteworthy point in the comparison between the relativistic and non-relativistic quantum theory of atoms. In these figures, the peak of the large component of the relativistic wavefunction lies closer to the nucleus than that of the non-relativistic wavefunction. This fact implies that the screening of the nucleus by the electrons is more efficient in relativistic theory, a fact that is confirmed by more sophisticated calculations on multielectron atoms.

Equations (8.40) look exceedingly complicated, but they can be simplified to some extent. Incorporating a factor of $1/n'$ into α_0 enables us to rewrite (8.40) in the following form

$$f_\kappa(r) = 2\lambda(k_C - W_C)^{1/2}e^{-\lambda r}(2\lambda r)^{s-1}\alpha_0' \times$$
$$(n'M(1 - n', 2s + 1, 2\lambda r) - (\kappa - \xi k_C/\lambda)M(-n', 2s + 1, 2\lambda r)) \qquad (8.52a)$$

$$g_\kappa(r) = 2\lambda(k_C + W_C)^{1/2}e^{-\lambda r}(2\lambda r)^{s-1}\alpha_0' \times$$
$$(n'M(1 - n', 2s + 1, 2\lambda r) + (\kappa - \xi k_C/\lambda)M(-n', 2s + 1, 2\lambda r)) \qquad (8.52b)$$

These wavefunctions are not often written out explicitly in their full, normalized form, therefore we will write down explicit expressions for the lowest few eigenfunctions of the one-electron atom here For the 1s state

$$g(r) = \frac{(2\lambda)^{s+1/2}}{(2k_C\Gamma(2s + 1))^{1/2}}(k_C + W_C)^{1/2}r^{s-1}e^{-\lambda r}(b_0 + b_1 r) \qquad (8.53a)$$

$$f(r) = -\frac{(2\lambda)^{s+1/2}}{(2k_C\Gamma(2s + 1))^{1/2}}(k_C - W_C)^{1/2}r^{s-1}e^{-\lambda r}(a_0 + a_1 r) \qquad (8.53b)$$

For the 2s state

$$g(r) = \left(\frac{(2\lambda)^{2s+1}k_C(2s + 1)(k_C + W_C)}{2W_C(2W_C + k_C)\Gamma(2s + 1)}\right)^{\frac{1}{2}}r^{s-1}e^{-\lambda r}(b_0 + b_1 r) \qquad (8.53c)$$

$$f(r) = -\left(\frac{(2\lambda)^{2s+1}k_C(2s + 1)(k_C - W_C)}{2W_C(2W_C + k_C)\Gamma(2s + 1)}\right)^{\frac{1}{2}}r^{s-1}e^{-\lambda r}(a_0 + a_1 r) \qquad (8.53d)$$

For the $2p^{1/2}$ state

$$g(r) = \left(\frac{(2\lambda)^{2s+1}k_C(2s + 1)(k_C + W_C)}{2W_C(2W_C - k_C)\Gamma(2s + 1)}\right)^{\frac{1}{2}}r^{s-1}e^{-\lambda r}(b_0 + b_1 r) \qquad (8.53e)$$

$$f(r) = -\left(\frac{(2\lambda)^{2s+1}k_C(2s + 1)(k_C - W_C)}{2W_C(2W_C - k_C)\Gamma(2s + 1)}\right)^{\frac{1}{2}}r^{s-1}e^{-\lambda r}(a_0 + a_1 r) \qquad (8.53f)$$

Table 8.1. Parameters defining the eigenfunctions of the one-electron atom for the principal quantum numbers $n = 1$ and $n = 2$.

Orbital	$1s$	$2s$	$2p^{1/2}$	$2p^{3/2}$
n	1	2	2	2
n'	0	1	1	0
κ	-1	-1	1	-2
l	0	0	1	1
$\bar{l}$	1	1	0	2
s	$(1-\xi^2)^{1/2}$	$(1-\xi^2)^{1/2}$	$(1-\xi^2)^{1/2}$	$(4-\xi^2)^{1/2}$
W_C	$s\lambda/\xi$	$\xi k_C^2/2\lambda$	$\xi k_C^2/2\lambda$	$s\lambda/\xi$
a_0	1	$\frac{W_C}{k_C}+1$	$\frac{W_C}{k_C}$	1
a_1	0	$-\frac{\lambda}{2s+1}\left(1+\frac{2W_C}{k_C}\right)$	$\frac{\lambda}{2s+1}\left(1-\frac{2W_C}{k_C}\right)$	0
b_0	1	$\frac{W_C}{k_C}$	$\left(\frac{W_C}{k_C}-1\right)$	1
b_1	0	$-\frac{\lambda}{2s+1}\left(1+\frac{2W_C}{k_C}\right)$	$\frac{\lambda}{2s+1}\left(1-\frac{2W_C}{k_C}\right)$	0

and for the $2p^{3/2}$ state

$$g(r) = \frac{(2\lambda)^{s+1/2}}{(2k_C\Gamma(2s+1))^{1/2}}(k_C + W_C)^{1/2}r^{s-1}e^{-\lambda r}(b_0 + b_1 r) \qquad (8.53g)$$

$$f(r) = -\frac{(2\lambda)^{s+1/2}}{(2k_C\Gamma(2s+1))^{1/2}}(k_C - W_C)^{1/2}r^{s-1}e^{-\lambda r}(a_0 + a_1 r) \qquad (8.53h)$$

The values of the quantum numbers and the required parameters for each orbital are given in table 8.1. They can easily be evaluated using equation (8.30) and the definitions of λ and n'.

Equations (8.53) were obtained from (8.52) simply by substitution of the quantum numbers and normalization using the standard integral

$$\int_0^\infty x^{\nu-1}e^{-\mu x}dx = \frac{1}{\mu^\nu}\Gamma(\nu) \qquad (8.54)$$

(Gradshteyn and Ryzhik 1980) where $\Gamma(v)$ is the standard gamma function (Abramowitz and Stegun 1972).

All the $g(r)$ and $f(r)$ with $|\kappa| = 1$ have a divergence at the origin, but this is of no importance because the divergence is considerably weaker than $1/r$ so the wavefunction is still square integrable. The reason the divergence appears in the relativistic wavefunctions and not in the non-relativistic ones is as follows. Non-relativistically we have that $|V(r)| \ll mc^2$, and the small r behaviour is entirely determined by the $l(l+1)/r^2$ term in the Schrödinger Hamiltonian. This is not the case as $r \to 0$ in relativistic theory. If we rearrange (8.21a) to find an expression for $v_\kappa(r)$ and substitute that into (8.21b) we will obtain a second order equation for $u_\kappa(r)$ including a term ξ^2/r^2 which will also play a key role in determining the $r \to 0$ behaviour. This is the difference between the relativistic and non-relativistic theories that causes the weak divergence. The same thing occurred for the same reason in the theory of the spinless electron atom described in chapter 3.

In figures 8.4a to d we show the one-electron atom wavefunctions multiplied by r, for $Z = 92$, as a function of distance from the nucleus. The axes are in relativistic units which means the unit of distance is the Compton wavelength $\hbar/mc$. If we take the non-relativistic limit of these wavefunctions we find $rf(r) \to 0$ for all r and $rg(r)$ become the usual non-relativistic wavefunctions. The difference between the relativistic $rg(r)$ and the non-relativistic eigenfunctions shows up on these figures because we have plotted the wavefunctions in relativistic units. In any other familiar set of units this difference would be vanishingly small on figures this size.

Normalization yields a value for α_0'. For the 1s state this comes out as

$$\alpha_0' = -\left(\frac{\lambda}{4k_C\Gamma(2s+1)}\right)^{1/2} \tag{8.55}$$

Looking at (8.53) and the values of the parameters in table 8.1 enables us to say more about the radial functions. If $n' = 0$, as is true for the 1s and 2$p^{3/2}$ states, the ratio $f_\kappa(r)/g_\kappa(r)$ is a constant for all values of r, $f_\kappa(r)/g_\kappa(r) = -(k_C - W_C)^{1/2}/(k_C + W_C)^{1/2}$. It is clear that this will only happen for this special case of a particle in a Coulomb potential with $n' = 0$. Inserting the definitions of k_C and W_C into this expression for $f_\kappa(r)/g_\kappa(r)$ easily shows that this fraction is of order $1/c^2$. For this reason $f_\kappa(r)$ is usually known as the small component of the wavefunction and $g_\kappa(r)$ as the large component. Of course this is only a hand-waving argument; at values of r where $g_\kappa(r)$ is zero this fraction diverges. As an example, let us consider some simple properties of the 1s state. From the definition of λ and table 8.1 we have

$$\lambda^2 = k_C^2 - W_C^2 = k_C^2 - \frac{s^2\lambda^2}{\xi^2} = k_C^2(1-s^2) = k_C^2\xi^2 = \frac{m^2c^2}{\hbar^2}\frac{Z^2e^4}{(4\pi\epsilon_0)^2\hbar^2c^2} \quad (8.56)$$

Taking the positive square root of (8.56) to give a solution that makes sense in spherical polar coordinates, we obtain

$$\lambda = \frac{Ze^2m}{4\pi\epsilon_0\hbar^2} = \frac{Z}{a_0} \quad (8.57)$$

where a_0 is the Bohr radius. This λ is the coefficient of r in the negative exponential in the wavefunction. In the relativistic case this takes on exactly the same value of Z/a_0 as it does in the non-relativistic case, despite the profound differences between relativistic and non-relativistic theory. In particular the *Zitterbewegung* inherent in the theory has no effect on λ, nor does the fact that W_C contains the rest mass energy which does not appear in non-relativistic theory.

Next let us calculate the expectation value of the position for the 1s wavefunction. Integrating over the angles just gives a 4π which is cancelled by the $l = 0$ spherical harmonic in the spin-angular functions. Therefore we only consider the radial integrals here.

$$\langle \hat{r}_{-1} \rangle = \int_0^\infty r^3(f_{-1}^2(r) + g_{-1}^2(r))dr \quad (8.58)$$

Substituting equations (8.53a) into here, and using the standard integral (8.54) and the definition of α_0' (8.55), leads to

$$\langle \hat{r}_{-1} \rangle = \frac{2s+1}{2\lambda} \quad (8.59)$$

This result is easy to understand in terms of the non-relativistic limit. If $c \to \infty$, we set $\xi \to 0$, so $s \to |\kappa|$ and hence we have

$$\lim_{c\to\infty} \langle \hat{r}_{-1} \rangle = \frac{3a_0}{2} \quad (8.60)$$

This is exactly the result obtained in the non-relativistic case. However, as $s < |\kappa|$ the expectation value in the relativistic case is less than in the non-relativistic case. This reinforces the point made earlier that screening is more efficient in relativistic quantum theory.

8.5 The Zeeman Effect

The Zeeman effect is the observed splitting into several components of the spectral lines of an atom when it is placed in a steady magnetic field. For small fields the splitting is proportional to the applied field. The non-relativistic theory is divided into two, the normal Zeeman effect which can be described within a classical model of the atom, and the anomalous Zeeman effect which requires the existence of spin for a satisfactory theoretical description. As we have seen, spin is a relativistic quantity and so the anomalous Zeeman effect is, on a fundamental level, a relativistic effect, although it is possible to 'fix up' non-relativistic quantum theory to describe both effects without any reference to relativity. The fully relativistic theory of the Zeeman effect unites the normal and anomalous effects into a single theory. As in the non-relativistic case, we treat the Zeeman effect using first order perturbation theory (Rose 1961). Indeed, working through this treatment is a very good exercise, as much research in relativistic quantum theory is based on perturbation methods.

As stated in the introduction, the Zeeman effect is easier to treat relativistically than non-relativistically. This is because the effect of a magnetic field on the Hamiltonian is to make the replacement $\mathbf{p} \rightarrow \mathbf{p} - e\mathbf{A}$ where $\mathbf{A}$ is the vector potential associated with the applied field. In non-relativistic quantum mechanics the Hamiltonian depends upon momentum squared $(\mathbf{p}^2/2m)$ whereas in the relativistic theory the dependence is linear. As one does perturbation theory $\mathbf{A}$ is assumed to be small. In non-relativistic theory this smallness is invoked to ignore the term in $\mathbf{A}^2$ to make the problem tractable. In relativistic quantum theory this approximation becomes unnecessary.

We will examine the Zeeman effect for the case of the principal quantum number $n = 1, 2$ and the orbital angular momentum quantum number $l = 0, 1$ for the one-electron atom with $Z = 64$. If H_0 is the Hamiltonian for the electron in the atom with zero applied field and H' is the perturbation, then the full Hamiltonian may be written

$$\hat{H} = \hat{H}_0 + \hat{H}' \tag{8.61}$$

and for the special case of a constant applied field the vector potential can easily be written in terms of the magnetic field, giving

$$\hat{H}' = -ec\tilde{\boldsymbol{\alpha}} \cdot \mathbf{A} = \frac{ec}{2}\tilde{\boldsymbol{\alpha}} \cdot \mathbf{r} \times \mathbf{B} \tag{8.62}$$

Now, let us make life easy for ourselves and, in accordance with convention, define the z-axis as being the direction in which the uniform field is applied. In that case we can rearrange equation (8.62) using standard vector identities as

$$\hat{H}' = -\frac{ec}{2}B(\alpha_x y - \alpha_y x) \tag{8.63}$$

with $B = |\mathbf{B}| = B_z$. The next thing to do, which is left as an exercise, is to work out the commutators $[\hat{\mathbf{J}}^2, \hat{H}]$ and $[\hat{J}_z, \hat{H}]$. One finds that the full Hamiltonian does not commute with $\hat{\mathbf{J}}^2$, but does commute with $\hat{J}_z$. This tells us that the perturbation may mix states of different total angular momentum, but that matrix elements only exist between states with the same value of m_j. Matrix elements for the perturbation are

$$
\int_0^\infty \psi_\kappa^{m_j\dagger} \hat{H}' \psi_{\kappa'}^{m_j} d\mathbf{r} =
$$

$$
\int_0^\infty \begin{pmatrix} g_\kappa(r)\chi_\kappa^{m_j} \\ if_\kappa(r)\chi_{-\kappa}^{m_j} \end{pmatrix}^\dagger \frac{-eBc}{2}(\alpha_x y - \alpha_y x) \begin{pmatrix} g_\kappa(r)\chi_{\kappa'}^{m_j} \\ if_\kappa(r)\chi_{-\kappa'}^{m_j} \end{pmatrix} d\mathbf{r} = \tag{8.64}
$$

$$
-\frac{eBc}{2} \int_0^\infty \begin{pmatrix} g_\kappa(r)\chi_\kappa^{m_j} \\ if_\kappa(r)\chi_{-\kappa}^{m_j} \end{pmatrix}^\dagger \begin{pmatrix} 0 & (\tilde{\sigma} \times \mathbf{r})_z \\ (\tilde{\sigma} \times \mathbf{r})_z & 0 \end{pmatrix} \begin{pmatrix} g_\kappa(r)\chi_{\kappa'}^{m_j} \\ if_\kappa(r)\chi_{-\kappa'}^{m_j} \end{pmatrix} d\mathbf{r}
$$

where we have used equation (4.13) for the α-matrices. These matrix multiplications are straightforward, and lead immediately to a fairly simple expression for the matrix elements

$$
\int_0^\infty \psi_\kappa^{m_j\dagger} \hat{H}' \psi_{\kappa'}^{m_j} d\mathbf{r} = -\frac{ieBc}{2} R_{\kappa,\kappa'}^n H_{\kappa,\kappa'}^{m_j} \tag{8.65}
$$

where we have separated radial and angular integrals

$$
H_{\kappa,\kappa'}^{m_j} = \int \chi_\kappa^{m_j\dagger} (\tilde{\sigma} \times \mathbf{r})_z \chi_{-\kappa'}^{m_j} d\Omega \tag{8.66a}
$$

$$
R_{\kappa,\kappa'}^{(n)} = \int_0^\infty r^3 (g_\kappa f_{\kappa'} + f_\kappa g_{\kappa'}) dr \tag{8.66b}
$$

Here n represents the principal quantum number. In equation (8.66a) $\tilde{\sigma}$ are the 2×2 Pauli matrices, as we have omitted the zeros in the leading diagonal of (8.64), enabling us to reduce the dimensions of the equations from 4×4 to 2×2. The angular integrals are familiar from other areas of quantum mechanics and we evaluate them below.

With the earlier definition of χ^s it is easy to see that

$$
\tilde{\sigma}_x \chi^s = \chi^{-s}, \qquad \tilde{\sigma}_y \chi^s = i(-1)^{\frac{1}{2}-s} \chi^{-s} \tag{8.67}
$$

Next we recall the definitions of the spin-angular functions in equation (2.131a). Using them with (8.67) it is straightforward to see that (8.66a) becomes

$$
H_{\kappa,\kappa'}^{m_j} = i \left(\frac{8\pi}{3}\right)^{\frac{1}{2}} \times
$$

$$
\left(C(l\tfrac{1}{2}j; m_j - \tfrac{1}{2}, \tfrac{1}{2}) C(\bar{l}'\tfrac{1}{2}\bar{j}'; m_j + \tfrac{1}{2}, -\tfrac{1}{2}) \int d\Omega Y_l^{m_j-\frac{1}{2}*} Y_1^{-1} Y_{\bar{l}'}^{m_j+\frac{1}{2}} \right. \tag{8.68}
$$

$$
\left. + C(l\tfrac{1}{2}j; m_j + \tfrac{1}{2}, -\tfrac{1}{2}) C(\bar{l}'\tfrac{1}{2}\bar{j}'; m_j - \tfrac{1}{2}, \tfrac{1}{2}) \int d\Omega Y_l^{m_j+\frac{1}{2}*} Y_1^1 Y_{\bar{l}'}^{m_j-\frac{1}{2}} \right)
$$

The integrals over spherical harmonics in (8.68) are standard and often appear in quantum mechanics. They determine selection rules in many electronic transitions. They are given in appendix C and lead to

$$H^{m_j}_{\kappa,\kappa'} = i\left(2\frac{(2\bar{l}'+1)}{2l+1}\right)^{\frac{1}{2}} C(\bar{l}'1l;0,0)\times$$

$$\sum_s C(l\tfrac{1}{2}j;m_j-s,s)C(\bar{l}'\tfrac{1}{2}j';m_j+s,-s)C(\bar{l}'1l;m_j+s,-2s)$$

(8.69)

This expression determines which states need to be included in our perturbation theory. From it and the properties of the Clebsch–Gordan coefficients it is apparent that $H^{m_j}_{\kappa\kappa'}$ is zero unless

$$j'-j=0,\pm1, \qquad\qquad l-\bar{l}'=\pm1$$
$$l'+l \text{ is even}, \qquad\qquad l-l'=0,\pm2$$

Finding numerical values for $H^{m_j}_{\kappa\kappa'}$ is now just a matter of looking up the Clebsch–Gordan coefficients in tables 2.1 and 2.2. Doing this eventually leads to values for the elements diagonal in κ of

$$H^{m_j}_{\kappa,\kappa} = \frac{4i\kappa m_j}{4\kappa^2-1}$$

(8.70a)

and for the elements with different values of κ but the same value of l we have for positive κ

$$H^{m_j}_{-\kappa-1,\kappa} = -H^{m_j}_{\kappa,-\kappa-1} = i\frac{((l+\tfrac{1}{2})^2-m_j^2)^{1/2}}{2l+1}$$

(8.70b)

Some coupling is non-zero between orbitals with different values of l, for example states with the same value of m_j in the $s^{1/2}$ and $d^{3/2}$ orbital have non-zero matrix elements. However, these are usually well separated in energy and can be neglected.

The radial integrals in equation (8.66b) can also be done analytically. This is a long and tedious calculation which provides little extra insight. so it is not reproduced here. It is useful, however, to have the result of the radial integral for one case as it will be used later to illustrate the non-relativistic limit of the formalism discussed here. For simplicity we choose the diagonal $1s$ term which reduces to the simple form

$$R^{(1s)}_{-1,-1} = -\frac{1}{k_C}\left(s+\frac{1}{2}\right)$$

(8.71)

It is straightforward to put the quantum numbers into equations (8.70) and insert these into (8.65) to find the matrix elements of the radial functions.

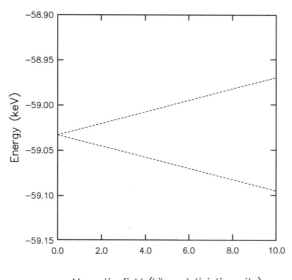

Magnetic field (kilo–relativistic units)

Fig. 8.5. The Zeeman effect in the 1s state of the one-electron atom with $Z = 64$. The energy is calculated from equations (8.43) and (8.72) and then the electron rest mass energy is subtracted.

In general we then have to set up and solve a secular determinant in the matrix elements. However, for s-states there is only one value of κ and the determinant is 1×1. So

$$E^{m_j} = W^n_{-1} - \frac{2eBc}{3} R^n_{-1,-1} m_j \qquad (8.72)$$

where the -1 subscript emphasizes that W^n also depends on the quantum number κ. The only difference between the splitting of the 1s and the 2s energy levels is in the radial integral. Clearly the splitting is linearly proportional to the applied field and the $m_j = 1/2$ and $m_j = -1/2$ levels are moved an equal energy in opposite directions away from the unperturbed level. Zeeman splitting of the 1s level of the one-electron atom with $Z = 64$ is shown in figure 8.5 as a function of applied field. On the figure the energy scale has had the electron rest energy subtracted from it. The Zeeman effect for the 2s level is included in figure 8.6.

For $2p^{3/2}$ states with $|m_j| = 3/2$ there is no coupling with the $j = 1/2$ state simply because $j = 1/2$ has no $|m_j| = 3/2$ state. Again the secular determinant is 1×1 and substituting in the quantum numbers gives

$$E^{\pm 3/2} = W^2_{-2} \mp \frac{2eBc}{5} R^{(2p)}_{-2,-2} \qquad (8.73)$$

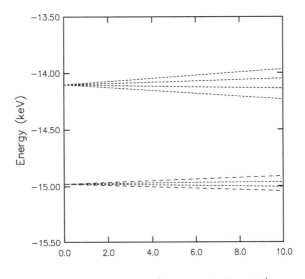

Magnetic field (kilo—relativistic units)

Fig. 8.6. The Zeeman effect in the $2s$ (long dashes) and $2p$ (short dashes) states of the one-electron atom with $Z = 64$. The energy is calculated from equations (8.43), (8.72), (8.73) and (8.76) and then the electron rest mass energy is subtracted. Note that the $2s$ and $2p^{1/2}$ levels are degenerate in zero field. This occurs because (8.43) depends on $|\kappa|$ rather than κ.

For the $m_j = \pm 1/2$ case the mathematics is slightly more difficult. In this case there is the possibility of coupling between the $2p^{1/2}$ and $2p^{3/2}$ levels. Therefore our secular determinant becomes 2×2:

$$\begin{vmatrix} W_1^2 + \langle \psi_1^{m_j} | \hat{H}' | \psi_1^{m_j} \rangle - E & \langle \psi_1^{m_j} | \hat{H}' | \psi_{-2}^{m_j} \rangle \\ \langle \psi_{-2}^{m_j} | \hat{H}' | \psi_1^{m_j} \rangle & W_{-2}^2 + \langle \psi_{-2}^{m_j} | \hat{H}' | \psi_{-2}^{m_j} \rangle - E \end{vmatrix} = 0 \qquad (8.74)$$

The matrix elements here can easily be evaluated using (8.65) as

$$\langle \psi_1^{m_j} | \hat{H}' | \psi_1^{m_j} \rangle = \frac{2eBc}{3} R_{11}^{(2p)} m_j \qquad (8.75a)$$

$$\langle \psi_{-2}^{m_j} | \hat{H}' | \psi_{-2}^{m_j} \rangle = -\frac{4eBc}{15} R_{-2-2}^{(2p)} m_j \qquad (8.75b)$$

$$\langle \psi_1^{m_j} | \hat{H}' | \psi_{-2}^{m_j} \rangle = \langle \psi_{-2}^{m_j} | \hat{H}' | \psi_1^{m_j} \rangle = \frac{eBc}{6} \left(\frac{9}{4} - m_j^2 \right)^{1/2} R_{1-2}^{(2p)} \qquad (8.75c)$$

The determinant (8.74) is a quadratic which can be solved in the usual way for W. We end up with a fairly dreadful looking expression.

$$E^{m_j} = \frac{1}{2}\left(W_1^2 + W_{-2}^2 + \frac{2}{3}eBcR_{1,1}m_j - \frac{4}{15}eBcR_{-2,-2}m_j \right.$$

$$\pm \left(\left(W_{-2}^2 - W_1^2 - \frac{2}{3}eBcR_{1,1}m_j - \frac{4}{15}eBcR_{-2,-2}m_j \right)^2 \right.$$

$$\left. \left. + \frac{e^2 B^2 c^2}{9}\left(\frac{9}{4} - m_j^2 \right) R_{1,-2}^2 \right)^{1/2} \right) \tag{8.76}$$

This is rather complicated: it has two solutions for each value of m_j, one corresponding to the level that has been split off from the $2p^{1/2}$ level and one corresponding to the level split off from the $2p^{3/2}$ level. It is easy to see in (8.76) that the correct limit is reached as $B \to 0$. Depending on which sign of the square root we take, we get either of the unperturbed $j = 1/2$ or $j = 3/2$ levels. Equation (8.76) will be valid as long as B is 'small', as we have used perturbation theory. Small here means small compared with the next higher energy scale in the problem. In this case that is the spin–orbit splitting of the 2p levels. Figure 8.6 shows the splitting of the 2s and 2p energy levels as a function of applied field.

Our final task is to check that in the non-relativistic limit the equations above reduce to the familiar ones for the anomalous Zeeman effect. This is rather difficult in general and so we content ourselves with examining the limit of the 1s level. In equation (8.71) we have evaluated the radial matrix element for the 1s electrons. Substituting this into (8.72) gives

$$E^{m_j} = W_{-1}^1 + \frac{2eBc}{3k_C}\left(s + \frac{1}{2} \right)m_j \tag{8.77}$$

Now, in the non-relativistic limit, $\xi \to 0$ and so $s \to |\kappa|$, so for this case $(s + 1/2) = 3/2$. Hence (8.77) becomes

$$E^{m_j} = W_{-1}^1 + \frac{\hbar eBcm_j}{mc} = W_{-1}^1 + g_s\hbar\omega_L m_j = W_{-1}^1 + g_s\mu_B B m_j \tag{8.78}$$

This is the usual expression for the anomalous Zeeman effect (see Eisberg and Resnick (1985), for example). The g-factor associated with spin has come out as $g_s = 2$, as we expect because there is no orbital angular momentum for the 1s state. The usual expression for the Larmor frequency ω_L given by equation (2.53) has also dropped neatly out of this formalism. Thus, for the 1s state at least, we can see that the familiar form of the anomalous Zeeman effect can be found easily in the non-relativistic limit.

8.6 Magnetic Dichroism

In this section we are going to discuss magnetic dichroism which is a purely relativistic effect. Essentially this is the dependence of the probability of some electronic transition on the state of polarization of the incident photon. The mathematics of this section is rather similar to that for the Zeeman effect, but it is somewhat more difficult. Nonetheless, this is a very important topic as dichroism is the basic physics underlying magneto-optical effects such as Faraday rotation and the Kerr effect, and dichroism in the single-electron atom is a model that can be solved analytically. Therefore we are going to go through it in detail. This section follows the work of Jenkins and Strange (1993).

For dichroism to occur in an atom it is necessary to break the spherical symmetry of the atom in some way. We have already done this. In the previous section on the Zeeman effect we selected a particular direction in space using a magnetic field. As we are already familiar with it let us break the spherical symmetry this way. We will illustrate dichroism in a one-electron atom in a magnetic field.

We will consider an electron being excited by an incident photon of the right energy from the $1s$ state up to the $2p^{3/2}$ and $2p^{1/2}$ states. The transition probability is given by the Fermi's golden rule

$$T_{fi} = \frac{2\pi}{\hbar}|\langle f|\hat{H}'|i\rangle|^2 n(E) \tag{8.79}$$

where $n(E)$ is the density of states available for the final state, H' represents the perturbation felt by the system due to the photon, and f and i represent the final and initial states respectively. Note that we have assumed this equation is valid in relativistic theory, we have not derived it. In fact it can be derived purely within relativistic quantum theory and this is shown explicitly in chapter 12. We need to evaluate the matrix element

$$\langle f|H'|i\rangle = \int \psi_\kappa^{m_j\dagger}(\mathbf{r})\hat{H}'\psi_{\kappa_i}^{m_j}(\mathbf{r})d\mathbf{r} \tag{8.80}$$

We saw in the previous section that a magnetic field couples states with the same value of m_j and l, but different values of κ. Therefore the perturbed wavefunction must be written

$$\psi_\kappa^{'m_j}(\mathbf{r}) = \sum_{\kappa'} a_{\kappa,\kappa'}^{m_j} \begin{pmatrix} g_{\kappa,\kappa'}(r)\chi_{\kappa'}^{m_j} \\ if_{\kappa,\kappa'}(r)\chi_{-\kappa'}^{m_j} \end{pmatrix}$$

$$= a_{\kappa,\kappa}^{m_j} \begin{pmatrix} g_{\kappa,\kappa}(r)\chi_\kappa^{m_j} \\ if_{\kappa,\kappa}(r)\chi_{-\kappa}^{m_j} \end{pmatrix} + a_{\kappa,-\kappa-1}^{m_j} \begin{pmatrix} g_{\kappa,-\kappa-1}(r)\chi_{-\kappa-1}^{m_j} \\ if_{\kappa,-\kappa-1}(r)\chi_{\kappa+1}^{m_j} \end{pmatrix} \tag{8.81}$$

Here we assume the unperturbed wavefunctions are normalized. The $a_{\kappa\kappa'}^{m_j}$ are determined from perturbation theory in a standard way with the assumption that the perturbed wavefunction is also normalized to unity. We omit the explicit derivation of this form for the perturbed wavefunction, as it distracts from the main thrust of this discussion. Each radial wavefunction has to be labelled by three quantum numbers now. Firstly there is m_j as before, secondly there is the left κ which labels the state the electron would be in in the absence of a magnetic field, thirdly there is the right κ which is the character of that particular component of the wavefunction. The left hand κ therefore has a dominant κ contribution and a secondary contribution from the $-\kappa - 1$ state.

The perturbative part of the Hamiltonian is given by

$$\hat{H}' = -ec\tilde{\alpha} \cdot \mathbf{A} \tag{8.82}$$

Here $\mathbf{A}$ is the vector potential associated with the incident photon. As is well known, electromagnetic radiation consists of oscillating electromagnetic fields which satisfy the wave equation. From Maxwell's equations it is easy to see that the vector potential associated with electromagnetic radiation also takes the form of a plane wave. Hence equation (8.82) takes the form

$$\hat{H}' = -ec\tilde{\alpha} \cdot \epsilon_\lambda e^{i\mathbf{q}\cdot\mathbf{r}} \tag{8.83}$$

where ϵ_λ is a vector describing the state of polarization of the photon. The next step is to make the *dipole approximation*. This involves assuming that the interaction between the electron and the photon is weak and so we can make the following simplification:

$$e^{i\mathbf{q}\cdot\mathbf{r}} = 1 + i\mathbf{q}\cdot\mathbf{r} - \frac{q^2 r^2}{2} + \cdots \approx 1 \tag{8.84}$$

Now we are going to calculate the matrix element (8.80) twice, once for incident right circularly polarized (RCP) and once for incident left circularly polarized (LCP) radiation. We can substitute (8.81), (8.83) and (8.84) into the matrix element (8.80) to get

$$\langle f|\hat{H}'|i\rangle = -iec\left(a_{\kappa,\kappa}^{m_j} \int r^2 g_{\kappa,\kappa}^{m_j}(r) f_{\kappa_i}^{m_{ji}}(r) H_{\kappa,-\kappa_i}^{m_j,m_{ji}} dr \right.$$
$$- a_{\kappa,\kappa}^{m_j} \int r^2 f_{\kappa,\kappa}^{m_j}(r) g_{\kappa_i}^{m_{ji}}(r) H_{-\kappa,\kappa_i}^{m_j,m_{ji}} dr$$
$$+ a_{\kappa,-\kappa-1}^{m_j} \int r^2 g_{\kappa,-\kappa-1}^{m_j}(r) f_{\kappa_i}^{m_{ji}}(r) H_{-\kappa-1,-\kappa_i}^{m_j,m_{ji}} dr$$
$$\left. - a_{\kappa,-\kappa-1}^{m_j} \int r^2 f_{\kappa,-\kappa-1}^{m_j}(r) g_{\kappa_i}^{m_{ji}}(r) H_{\kappa+1,\kappa_i}^{m_j,m_{ji}} dr \right) \tag{8.85}$$

where the limits on all the integrals are 0 and ∞. The quantum numbers with subscript i refer to the initial state, other quantum numbers refer to the final state. Note that, as in the Zeeman effect, only cross terms in f and g occur because the α-matrices are always off-diagonal. The $H_{\kappa_1,\kappa_2}^{m_1,m_2}$

contain all the angular integrals and their general form is

$$H_{\kappa\kappa'}^{m_j,m_j'} = \int \chi_\kappa^{m_j\dagger}(\tilde{\sigma} \cdot \epsilon_\lambda)\chi_{\kappa'}^{m_j'} d\Omega \tag{8.86}$$

where we have reduced the 4×4 nature of the equations to 2×2 in the same way as in the Zeeman effect. The reason for choosing the $1s$ state as the initial state is twofold. Firstly it is the natural one, being the ground state for hydrogen. Secondly, and more pragmatically, it has $\kappa = -1$ and there is no state with $-\kappa - 1$. Hence only the first term in the expansion (8.81) remains. Therefore we only have to label it with one value of κ. For the excited $2p$ levels there is no such simplification. Let us evaluate these angular parts of the matrix elements. We have to do this separately for right and left circularly polarized incident radiation. For LCP and RCP respectively we have

$$\epsilon_- = \frac{1}{\sqrt{2}}(1,-i,0), \qquad\qquad \epsilon_+ = \frac{1}{\sqrt{2}}(1,i,0) \tag{8.87}$$

where we have assumed the wavevector of the incident radiation points along the z-axis, parallel to the applied field. Equation (8.86) is analogous to (8.66a) in that they both determine selection rules, but the integral (8.86) is different (and somewhat easier to do). Therefore we can expect the selection rules to differ from those for the Zeeman effect. Substituting for the spin-angular functions gives

$$H_{\kappa\kappa'}^{m_j,m_j'} = \sqrt{2}C(l\tfrac{1}{2}j; m_j - \tfrac{1}{2}\tfrac{1}{2})C(l'\tfrac{1}{2}j'; m_j + \tfrac{1}{2} - \tfrac{1}{2})\int Y_l^{m_j - \frac{1}{2}} Y_{l'}^{m_j' + \frac{1}{2}} d\Omega \tag{8.88a}$$

for LCP and

$$H_{\kappa\kappa'}^{m_j,m_j'} = \sqrt{2}C(l\tfrac{1}{2}j; m_j + \tfrac{1}{2} - \tfrac{1}{2})C(l'\tfrac{1}{2}j'; m_j - \tfrac{1}{2}\tfrac{1}{2})\int Y_l^{m_j + \frac{1}{2}} Y_{l'}^{m_j' - \frac{1}{2}} d\Omega \tag{8.88b}$$

for RCP radiation. The angular integrals here are given in appendix C and we have

$$H_{\kappa\kappa'}^{m_j,m_j'} = \sqrt{2}C(l\tfrac{1}{2}j; m_j - \tfrac{1}{2}\tfrac{1}{2})C(l'\tfrac{1}{2}j'; m_j' + \tfrac{1}{2} - \tfrac{1}{2})\delta_{l,l'}\delta_{m_j - m_j',1} \tag{8.89a}$$

for LCP. For RCP we find

$$H_{\kappa\kappa'}^{m_j,m_j'} = \sqrt{2}C(l\tfrac{1}{2}j; m_j + \tfrac{1}{2} - \tfrac{1}{2})C(l'\tfrac{1}{2}j'; m_j' - \tfrac{1}{2} + \tfrac{1}{2})\delta_{l,l'}\delta_{m_j - m_j',-1} \tag{8.89b}$$

Here is the first sign of the magnetic dichroism. The two polarizations of the absorbed wave give a matrix element with different selection rules. We can substitute this into equation (8.85) to give an expression for the matrix element for the transition of interest. Let us develop some shorthand to simplify subsequent equations:

$$R1_{\kappa_1,\kappa_2,\kappa_i}^{m_j,m_{ji}} = a_{\kappa_1'\kappa_2}^{m_j} \int_0^\infty r^2 f_{\kappa_1,\kappa_2}^{m_j*}(r) g_{\kappa_i}^{m_{ji}} dr \tag{8.90a}$$

$$R2_{\kappa_1,\kappa_2,\kappa_i}^{m_j,m_{ji}} = a_{\kappa_1,\kappa_2}^{m_j} \int_0^\infty r^2 g_{\kappa_1,\kappa_2}^{m_j*}(r) f_{\kappa_i}^{m_{ji}} dr \tag{8.90b}$$

Substituting (8.89) and (8.90) into (8.85) gives

$$\langle f|\hat{H}'|i\rangle_{\mathrm{LCP}} = -\sqrt{2}iec\times$$
$$(C(l\tfrac{1}{2}j;m_j-\tfrac{1}{2}\tfrac{1}{2})C(\bar{l}_i\tfrac{1}{2}j_i;m_{ji}+\tfrac{1}{2},-\tfrac{1}{2})R2_{\kappa,\kappa,\kappa_i}^{m_jm_{ji}}\delta_{l,\bar{l}_i}\delta_{\Delta m_j,1}$$
$$- C(\bar{l}\tfrac{1}{2}jm_j-\tfrac{1}{2}\tfrac{1}{2})C(l_i\tfrac{1}{2}j_i;m_{ji}+\tfrac{1}{2},-\tfrac{1}{2})R1_{\kappa,\kappa,\kappa_i}^{m_jm_{ji}}\delta_{\bar{l},l_i}\delta_{\Delta m_j,1} \tag{8.91a}$$
$$+ C(l\tfrac{1}{2}j;m_j-\tfrac{1}{2}\tfrac{1}{2})C(\bar{l}_i\tfrac{1}{2}j_i;m_{ji}+\tfrac{1}{2},-\tfrac{1}{2})R2_{\kappa,-\kappa-1,\kappa_i}^{m_jm_{ji}}\delta_{l,\bar{l}_i}\delta_{\Delta m_j,1}$$
$$+ C(\bar{l}\tfrac{1}{2}j;m_j-\tfrac{1}{2}\tfrac{1}{2})C(l_i\tfrac{1}{2}j_i;m_{ji}+\tfrac{1}{2},-\tfrac{1}{2})R1_{\kappa,-\kappa-1,\kappa_i}^{m_jm_{ji}}\delta_{\bar{l},l_i}\delta_{\Delta m_j,1})$$

$$\langle f|\hat{H}'|i\rangle_{\mathrm{RCP}} = -\sqrt{2}iec\times$$
$$(C(l\tfrac{1}{2}j;m_j+\tfrac{1}{2}-\tfrac{1}{2})C(\bar{l}_i\tfrac{1}{2}j_i;m_{ji}-\tfrac{1}{2},\tfrac{1}{2})R2_{\kappa,\kappa,\kappa_i}^{m_jm_{ji}}\delta_{l,\bar{l}_i}\delta_{\Delta m_j,-1}$$
$$- C(\bar{l}\tfrac{1}{2}jm_j+\tfrac{1}{2}-\tfrac{1}{2})C(l_i\tfrac{1}{2}j_i;m_{ji}-\tfrac{1}{2},\tfrac{1}{2})R1_{\kappa,\kappa,\kappa_i}^{m_jm_{ji}}\delta_{\bar{l},l_i}\delta_{\Delta m_j,-1}$$
$$+ C(l\tfrac{1}{2}j;m_j+\tfrac{1}{2}-\tfrac{1}{2})C(\bar{l}_i\tfrac{1}{2}j_i;m_{ji}-\tfrac{1}{2},\tfrac{1}{2})R2_{\kappa,-\kappa-1,\kappa_i}^{m_jm_{ji}}\delta_{l,\bar{l}_i}\delta_{\Delta m_j,-1}$$
$$- C(\bar{l}\tfrac{1}{2}j;m_j+\tfrac{1}{2}-\tfrac{1}{2})C(l_i\tfrac{1}{2}j_i;m_{ji}-\tfrac{1}{2},\tfrac{1}{2})R1_{\kappa,-\kappa-1,\kappa_i}^{m_jm_{ji}}\delta_{\bar{l},l_i}\delta_{\Delta m_j,-1})$$
$$\tag{8.91b}$$

where we have used obvious notation and $\Delta m_j = m_j - m_{ji}$. The j in the first Clebsch–Gordan coefficient of each term refers to the second κ index of the radial integrals. Equations (8.91) are complicated enough. If we had used an initial state other than the s-state there would be eight rather than four terms in each of (8.91a and b). These equations look rather less forbidding if we insert the numbers explicitly for the Clebsch–Gordan coefficients. If we excite a 1s electron we have $l_i = 0, j_i = 1/2, \kappa_i = -1$ and $\bar{l}_i = 1$, and two possible values of $m_{ji} = \pm 1/2$. These two have to be treated separately. If we label the initial state by its κ_i and m_{ji} values and the final state by the κ and m_j values they would have in zero field, we have six non-zero matrix elements for 1s to 2p transitions:

$$\langle -2,\tfrac{1}{2}|\hat{H}'|-1,-\tfrac{1}{2}\rangle_{\mathrm{LCP}}$$
$$= iec\left(\frac{\sqrt{2}}{3}R2_{-2,1,-1}^{\frac{1}{2},-\frac{1}{2}} + \sqrt{2}R1_{-2,1,-1}^{\frac{1}{2},-\frac{1}{2}} - \frac{2}{3}R2_{-2,-2,-1}^{\frac{1}{2},-\frac{1}{2}}\right) \tag{8.92a}$$

$$\langle -2,\tfrac{3}{2}|\hat{H}'|-1,\tfrac{1}{2}\rangle_{\mathrm{LCP}} = -\frac{2iec}{\sqrt{3}}R2_{-2,-2,-1}^{\frac{3}{2},\frac{1}{2}} \tag{8.92b}$$

$\langle 1, \frac{1}{2} | \hat{H}' | -1, -\frac{1}{2} \rangle_{\text{LCP}}$

$$= iec \left(\frac{\sqrt{2}}{3} R2^{\frac{1}{2},-\frac{1}{2}}_{1,1,-1} + \sqrt{2} R1^{\frac{1}{2},-\frac{1}{2}}_{1,1,-1} - \frac{2}{3} R2^{\frac{1}{2},-\frac{1}{2}}_{1,-2,-1} \right) \tag{8.92c}$$

$\langle -2, -\frac{1}{2} | \hat{H}' | -1, \frac{1}{2} \rangle_{\text{RCP}}$

$$= iec \left(\frac{\sqrt{2}}{3} R2^{-\frac{1}{2},\frac{1}{2}}_{-2,1,-1} + \sqrt{2} R1^{-\frac{1}{2},\frac{1}{2}}_{-2,1,-1} + \frac{2}{3} R2^{-\frac{1}{2},\frac{1}{2}}_{-2,-2,-1} \right) \tag{8.92d}$$

$$\langle -2, -\frac{3}{2} | \hat{H}' | -1, -\frac{1}{2} \rangle_{\text{RCP}} = -\frac{2iec}{\sqrt{3}} R2^{-\frac{3}{2},-\frac{1}{2}}_{-2,-2,-1} \tag{8.92e}$$

$\langle 1, -\frac{1}{2} | \hat{H}' | -1, \frac{1}{2} \rangle_{\text{RCP}}$

$$= iec \left(\frac{\sqrt{2}}{3} R2^{-\frac{1}{2},\frac{1}{2}}_{1,1,-1} + \sqrt{2} R1^{-\frac{1}{2},\frac{1}{2}}_{1,1,-1} + \frac{2}{3} R2^{-\frac{1}{2},\frac{1}{2}}_{1,-2,-1} \right) \tag{8.92f}$$

In any real experiment a photon would be fired in at the appropriate energy, but we would be unable to tell whether it had excited the $m_{ji} = -1/2$ or the $m_{ji} = 1/2$ electron. Let us assume that an incident photon is equally likely to excite each of these electrons, and that there is no interference between these processes. Then we can just add the square of (8.92a) to that of (8.92b) and the square of (8.92d) to that of (8.92e) to get the transition rates to the $2p^{3/2}$ levels. If we assume the density of states in (8.79) is a δ-function for the hydrogen atom we have expressions for the $1s \to 2p^{3/2}$ transition rate for each polarization.

$$T^{\text{LCP}} = \frac{2\pi e^2 c^2}{\hbar} \delta(E_i - E_f + \hbar\omega) \times \left(\left(\frac{2}{\sqrt{3}} R2^{\frac{3}{2},\frac{1}{2}}_{-2,-2,-1} \right)^2 \right.$$

$$\left. + \left(\frac{\sqrt{2}}{3} R2^{\frac{1}{2},-\frac{1}{2}}_{-2,1,-1} + \sqrt{2} R1^{\frac{1}{2},-\frac{1}{2}}_{-2,1,-1} - \frac{2}{3} R2^{\frac{1}{2},-\frac{1}{2}}_{-2,-2,-1} \right)^2 \right) \tag{8.93a}$$

$$T^{\text{RCP}} = \frac{2\pi e^2 c^2}{\hbar} \delta(E_i - E_f + \hbar\omega) \times \left(\left(\frac{2}{\sqrt{3}} R2^{-\frac{3}{2},-\frac{1}{2}}_{-2,-2,-1} \right)^2 \right.$$

$$\left. + \left(\frac{\sqrt{2}}{3} R2^{-\frac{1}{2},\frac{1}{2}}_{-2,1,-1} + \sqrt{2} R1^{-\frac{1}{2},\frac{1}{2}}_{-2,1,-1} + \frac{2}{3} R2^{-\frac{1}{2},\frac{1}{2}}_{-2,-2,-1} \right)^2 \right) \tag{8.93b}$$

and similar expressions (with the first subscript on $R1$ and $R2$ changed from -2 to 1) for the $1s \to 2p^{1/2}$ transition. In general these two expressions are different, and the difference has two origins. Firstly the sign of $R2^{\frac{1}{2},-\frac{1}{2}}_{-2,-2,-1}$ is different for each case; secondly, in general, the radial matrix elements are different for each value of the m_j quantum numbers although

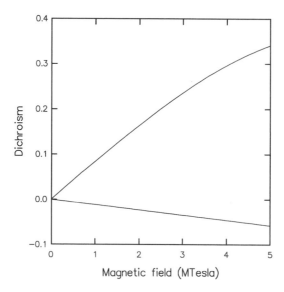

Fig. 8.7. Magnetic X-ray dichroism in the one-electron atom with $Z = 64$ as a function of applied magnetic field for excitation from the $1s$ to the $2p^{1/2}$ level (lower line) and $2p^{3/2}$ level (upper line).

even for heavy atoms this is likely to be a small effect. If there were no magnetic field present the first two terms on the right hand side of (8.93a and b) would disappear and the sign difference would disappear. Except at very large fields, where the perturbation theory breaks down anyway, (8.93a and b) will be dominated by the terms in $R_{-2,-2,-1}$. In general, finding the radial matrix elements is a numerical problem, although it can be done analytically for the one-electron atom. This is a time consuming and tedious task which we leave as an exercise for the reader. Now we are in a position to define the magnetic dichroism as

$$D = \frac{T_{fi}^{RCP} - T_{fi}^{LCP}}{T_{fi}^{RCP} + T_{fi}^{LCP}} \qquad (8.94)$$

It is meaningless to define dichroism of individual energy levels, so in (8.94) i means an average of all m_j values within the initial state multiplet and f means an average over all m_j in the final state multiplet. This assumes the applied field is small so the frequency profile of the incident beam is such that it cannot resolve the individual energy levels within a multiplet. The results of a calculation of the dichroism for a one-electron atom with atomic number $Z = 64$ and for excitations from the $1s$ level to the $2p^{1/2}$ and $2p^{3/2}$ levels are shown in figure 8.7.

Non-relativistically, the difference between the matrix elements $R_{1,1,-1}$ and $R_{-2,-2,-1}$ disappears. In fact the non-relativistic limit is difficult to examine here because as $c \to \infty$ the small component of the wavefunction disappears and all the radial integrals of (8.90) disappear as well. The reason for this is the off-diagonal nature of the α-matrices in the perturbation (8.82). Furthermore (8.82) is mathematically very different to the non-relativistic perturbation $\hat{H}' = -(e/2m)(\hat{\mathbf{p}} \cdot \hat{\mathbf{A}} + \hat{\mathbf{A}} \cdot \hat{\mathbf{p}}) + (e^2/2m)\hat{\mathbf{A}}^2$, although in chapter 6 we showed an argument relating the relativistic and non-relativistic forms of the perturbation (see equation (6.68)).

Although we have only looked at the single-electron atom, dichroism occurs in real permanent magnets, and measurement of the difference in absorption rate for LCP and RCP X-rays is an important probe of the magnetic properties of materials. The calculation performed above reproduces some of the dichroic behaviour found in real materials. In particular we see that dichroism is proportional to the applied field. In a real material the magnetic field is a way of representing the exchange interaction due to other electrons, and is much more complicated than the constant field used above. Nonetheless the proportionality still holds. In figure 8.10 the dichroism changes sign depending on whether we excite to the $2p^{1/2}$ or the $2p^{3/2}$ state. It is also observed in real materials that the states we excite to (or from) have a characteristic effect on the overall sign of the dichroism. Again in solids it is more complicated because we excite to an energy band rather than to a single orbital. In chapter 13 we show how to calculate dichroic effects in real materials.

8.7 Problems

(1) Evaluate the commutators $[\hat{\mathbf{J}}^2, \hat{H}]$ and $[\hat{J}_z, \hat{H}]$, where $\hat{H}$ is given by equation (8.61).
(2) The selection rules for right and left circularly polarized radiation are given by equations (8.89). Derive a selection rule for linearly polarized light. How could linear dichroism arise?
(3) If the total energy of an electron in a one-electron atom is greater than mc^2 we have a scattering state rather than a bound state. Solve the radial Dirac equation for such a state.
(4) Calculate the shift in the $2p$ energy levels of the one-electron atom, to first order in perturbation theory, if the atom is placed in a weak electric field (the Stark effect).
(5) Screening of the nucleus by electrons may be modelled by multiplying the bare Coulomb potential by $e^{-\mu r}$ where μ is a small number. Use first order perturbation theory to estimate the effect this might have on the energy levels of the one-electron atom.

(6) Find the solutions of the radial Dirac equation for a spherical flat potential well:

$$V(r) = -V_0 \qquad\qquad r \leq R$$
$$= 0 \qquad\qquad r > R$$

where V_0 is positive.

9

Potential Problems

I believe that physicists gain much of their physics *intuition* from solving simple model problems explicitly. Apart from the hydrogen atom, this is something that is not often done in relativistic quantum theory books (except the book by Greiner (1990)). In this chapter we set up and solve five simple models that have exact analytical solutions. Furthermore, they are all related to well-known non-relativistic counterparts, and most find application in many areas of physics, particularly solid state physics. These relationships will be pointed out and where appropriate we will also mention the applications and consequences of the models.

There are very few models in quantum mechanics that yield exact solutions, hence any that do are of fundamental interest, the hydrogen atom being, perhaps, the most famous example. The hydrogen atom with the potential $V(r) = -Ze^2/4\pi\epsilon_0 r$ is unique in that it can be solved analytically classically and in both non-relativistic and relativistic quantum theory, with and without spin. This has been discussed in detail in chapters 3 and 8.

The five examples we choose to consider here are the following. Firstly we will solve the Dirac equation for an electron in a one-dimensional well, a relativistic generalization of the non-relativistic particle in a box problem. Secondly we will look at the Dirac oscillator, which is not a relativistic version of the usual quantum mechanical harmonic oscillator, but is closely related to it. Thirdly we will consider the relativistic generalization of the Kronig–Penney model – a simple one-dimensional model of a solid which reproduces some of the properties of real solids, and which requires a relativistic generalization of Bloch's theorem. Fourthly we will look at an electron in constant perpendicular electric and magnetic fields and also take the relevant limits to observe the behaviour when the electric field is equal to zero, or the magnitudes of the fields are related by $|E| = c|B|$. This will enable us to find the relativistic generalization of some familiar non-

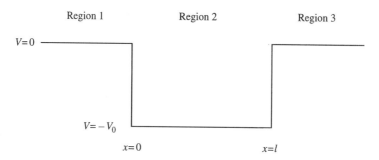

Fig. 9.1.   The potential well. Bound state solutions for this potential are found
in section 9.1. Here $V = 0$ corresponds to the rest mass energy of the particle
trapped in the box.

relativistic formulae. In particular we will derive relativistic expressions
for the properties of Landau levels. Finally we will analyse a non-linear
version of the Dirac equation which has soliton solutions.

  If you work through these five examples you will see that the basic
method of solving such problems is always the same. One writes down the
Dirac equation for the potential in question and tries to write down some
suitable boundary conditions. It is then just a question of manipulating
the equations into a form where they can be solved directly or where the
solutions correspond to some tabulated function. Indeed, this was also the
approach adopted when we solved the hydrogen atom.

9.1 A Particle in a One-Dimensional Well

In this section we will find the eigenvectors and eigenvalues for the particle
in a one-dimensional potential well. Our problem is slightly different
from a similar one which is tackled in great detail by Greiner (1990).
Surprisingly, I have not been able to find the particle in a box solved, within
relativistic quantum mechanics, anywhere else. This is a simple problem
in principle. We will define our potential box as shown in figure 9.1. Our
coordinate system is defined such that the origin falls at the left edge of
the box and we choose our one dimension to be in the x-direction. The
box has a length l, so the edges of the box occur at $x = 0$ and l. The
potential is zero, except for the region of the box, so we have three regions
of space to consider:

$$\textit{Region 1} \qquad V = 0 \qquad\qquad x < 0 \qquad\qquad (9.1a)$$

$$\textit{Region 2} \qquad V = -V_0 \qquad\quad 0 < x < l \qquad\quad (9.1b)$$

$$\textit{Region 3} \qquad V = 0 \qquad\qquad x > l \qquad\qquad (9.1c)$$

and V_0 is, of course, defined to be a positive real number. In regions 1 and 3 the wavefunctions are just solutions of the time-independent free-particle Dirac equation

$$(c\tilde{\alpha} \cdot \hat{\mathbf{p}} + \tilde{\beta}mc^2)\psi(x) = W\psi(x) \tag{9.2}$$

This has already been solved for particles free to move in the xy-plane in chapter 5, and the case discussed here of particles free to move only in the x-direction is just a restricted class of those solutions. In the box itself the Dirac equation takes the form

$$(c\tilde{\alpha} \cdot \hat{\mathbf{p}} - V_0 + \tilde{\beta}mc^2)\psi(x) = W\psi(x) \tag{9.3}$$

It doesn't take a genius to see that we can take V_0 over to the right hand side of this equation and then the solution to (9.3) is the same as the solution to (9.2) with W replaced by $W + V_0$. There can be no spin flip at the edges of the box, so we only need to work with one spin direction. Why this is so will soon become apparent. So, our general solutions are:

In region 1

$$\psi_1(x) = A_I \begin{pmatrix} 0 \\ 1 \\ 0 \\ \frac{cp_1}{W+mc^2} \end{pmatrix} e^{ip_1 x/\hbar} + A_{II} \begin{pmatrix} 0 \\ 1 \\ 0 \\ -\frac{cp_1}{W+mc^2} \end{pmatrix} e^{-ip_1 x/\hbar} \tag{9.4a}$$

in region 2

$$\psi_2(x) = B_I \begin{pmatrix} 0 \\ 1 \\ 0 \\ \frac{cp_2}{W+V_0+mc^2} \end{pmatrix} e^{ip_2 x/\hbar} + B_{II} \begin{pmatrix} 0 \\ 1 \\ 0 \\ \frac{-cp_2}{W+V_0+mc^2} \end{pmatrix} e^{-ip_2 x/\hbar} \tag{9.4b}$$

and in region 3

$$\psi_3(x) = C_I \begin{pmatrix} 0 \\ 1 \\ 0 \\ \frac{cp_3}{W+mc^2} \end{pmatrix} e^{ip_3 x/\hbar} + C_{II} \begin{pmatrix} 0 \\ 1 \\ 0 \\ \frac{-cp_3}{W+mc^2} \end{pmatrix} e^{-ip_3 x/\hbar} \tag{9.4c}$$

and the corresponding momenta are given by

$$p_1^2 c^2 = W^2 - m^2 c^4 \tag{9.5a}$$
$$p_2^2 c^2 = (W + V_0)^2 - m^2 c^4 \tag{9.5b}$$
$$p_3^2 c^2 = W^2 - m^2 c^4 = p_1^2 c^2 \tag{9.5c}$$

It is in (9.4) that we can see why there are going to be no spin flips at the boundaries of the potential. If they did occur the wavefunction (9.4b) would have the first and third component non-zero and the second and fourth zero. There would then be no values of the unknown coefficients (except zero) that would match such a wavefunction onto (9.4a) and (9.4c). Now the wavefunctions have to be continuous at the boundaries of the potential box so our boundary conditions are

$$\psi_1(0) = \psi_2(0) \tag{9.6a}$$

$$\psi_2(l) = \psi_3(l) \tag{9.6b}$$

So the problem is to find the set of constants A_I, A_{II}, B_I, B_{II}, C_I and C_{II} that satisfy these boundary conditions. The first step is to write out the boundary conditions in component form

$$A_I + A_{II} = B_I + B_{II} \tag{9.7a}$$

$$\frac{cp_1}{W + mc^2}(A_I - A_{II}) = \frac{cp_2}{W + V_0 + mc^2}(B_I - B_{II}) \tag{9.7b}$$

$$B_I e^{ip_2l/\hbar} + B_{II}e^{-ip_2l/\hbar} = C_I e^{ip_1l/\hbar} + C_{II}e^{-ip_1l/\hbar} \tag{9.7c}$$

$$\frac{cp_2}{W + V_0 + mc^2}\left(B_I e^{ip_2l/\hbar} - B_{II}e^{-ip_2l/\hbar}\right)$$
$$= \frac{cp_1}{W + mc^2}\left(C_I e^{ip_1l/\hbar} - C_{II}e^{-ip_1l/\hbar}\right) \tag{9.7d}$$

As a rule of thumb one can often reduce problems to their simplest and most fundamental form by writing them in terms of dimensionless quantities. To this end we define

$$\xi = \left(\frac{(W - mc^2)(W + V_0 + mc^2)}{(W + mc^2)(W + V_0 - mc^2)}\right)^{1/2} = \frac{p_1(W + V_0 + mc^2)}{p_2(W + mc^2)} \tag{9.8}$$

We can use ξ to write the boundary conditions in symmetric matrix form

$$\begin{pmatrix} A_I \\ A_{II} \end{pmatrix} = \frac{1}{2\xi}\begin{pmatrix} \xi + 1 & \xi - 1 \\ \xi - 1 & \xi + 1 \end{pmatrix}\begin{pmatrix} B_I \\ B_{II} \end{pmatrix} \tag{9.9a}$$

$$\begin{pmatrix} B_I \\ B_{II} \end{pmatrix} = \frac{1}{2}\begin{pmatrix} (1+\xi)e^{i(p_1-p_2)l/\hbar} & (1-\xi)e^{-i(p_1+p_2)l/\hbar} \\ (1-\xi)e^{i(p_1+p_2)l/\hbar} & (1+\xi)e^{-i(p_1-p_2)l/\hbar} \end{pmatrix}\begin{pmatrix} C_I \\ C_{II} \end{pmatrix} \tag{9.9b}$$

The next step is obvious, and you are sure to be delighted to see another rather long equation. We substitute (9.9b) into (9.9a) and obtain

$$\begin{pmatrix} A_I \\ A_{II} \end{pmatrix} = \frac{(1 - \xi^2)}{4\xi} \times$$
$$\begin{pmatrix} \frac{(1+\xi)}{(1-\xi)}e^{iql/\hbar} - \frac{(1-\xi)}{(1+\xi)}e^{ikl/\hbar} & e^{-ikl/\hbar} - e^{-iql/\hbar} \\ e^{ikl/\hbar} - e^{iql/\hbar} & \frac{(1+\xi)}{(1-\xi)}e^{-iql/\hbar} - \frac{(1-\xi)}{(1+\xi)}e^{-ikl/\hbar} \end{pmatrix}\begin{pmatrix} C_I \\ C_{II} \end{pmatrix} \tag{9.10}$$

where we have defined $q = p_1 - p_2$ and $k = p_1 + p_2$ for the sole reason of getting the matrix in (9.10) onto one line of the page. By following the above procedure we have reduced our problem from four equations with six unknowns to two equations with four unknowns. We can remove one unknown in one of two ways. Either we can say there is no wave incident from the right in which case $C_{II} = 0$, or we can use the normalization

$$\int \psi^\dagger \psi dx = 1 \tag{9.11}$$

which enables us to fix one of the unknowns if the others are found. Before proceeding it is useful to refer back to equation (9.5a). It is easy to see that, if $|W| > mc^2$, the momentum is real before and after the particle has interacted with the well. Wavefunctions that satisfy this criterion are called scattering states. If $|W| < mc^2$ the momentum outside the well must be a complex quantity, and these states are called bound states. It is these latter states that we shall examine here.

For p_1 imaginary, equations (9.4) are only satisfactory wavefunctions provided $A_I = C_{II} = 0$ (we are defining $p_0 = -ip_1$ with p_0 positive), otherwise the wavefunctions diverge as $x \to \pm\infty$ and hence are unnormalizable. This leads to a considerable simplification of equation (9.10). The top line of (9.10) becomes

$$\frac{1}{4\xi} \left((1 + \xi)^2 e^{-ip_2l/\hbar} - (1 - \xi)^2 e^{ip_2l/\hbar} \right) e^{ip_1l/\hbar} C_I = 0 \tag{9.12}$$

Now $C_I = 0$ is a solution to this. It is the solution when we have no particle present, so is fairly uninteresting. The non-trivial solution has

$$\left(\frac{1 + \xi}{1 - \xi} \right) e^{-ip_2l/\hbar} = \left(\frac{1 - \xi}{1 + \xi} \right) e^{ip_2l/\hbar} \tag{9.13}$$

From (9.8), if p_2 is real, ξ is imaginary, and the left side of (9.13) is the complex conjugate of the right side. If an expression is equal to its complex conjugate, it must be real. Hence

$$\mathrm{Im} \left(\frac{1 + \xi}{1 - \xi} \right) e^{-ip_2l/\hbar} = 0 \tag{9.14}$$

If we define $\xi = i\Xi$ with Ξ real we can rearrange (9.13) and substitute in the trigonometric form of the complex exponentials to find

$$\tan \frac{p_2l}{\hbar} = \frac{2\Xi}{1 - \Xi^2} \tag{9.15}$$

From the definitions of Ξ and p_0 we can use equation (9.8) to write

$$\tan \frac{p_2l}{\hbar} = \frac{2p_0p_2c^2(W + mc^2)(W + V_0 + mc^2)}{p_2^2c^2(W + mc^2)^2 - p_0^2c^2(W + V_0 + mc^2)} \tag{9.16}$$

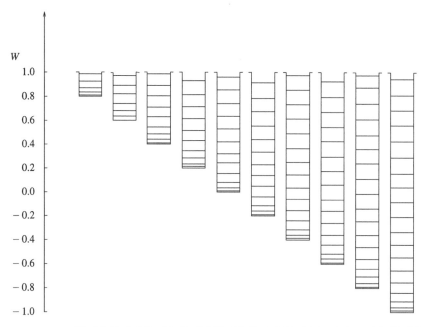

$V_0=0.2$ $V_0=0.4$ $V_0=0.6$ $V_0=0.8$ $V_0=1.0$ $V_0=1.2$ $V_0=1.4$ $V_0=1.6$ $V_0=1.8$ $V_0=2.0$

Fig. 9.2. Scale drawing of a series of potential wells of various depths (indicated by V_0) and their eigenvalue spectrum determined graphically from equation (9.19). The width of the well is 0.94 Bohr radii and all energies are written in relativistic units.

and using the following expressions for the momenta

$$p_2^2 c^2 = (W + V_0)^2 - m^2 c^4 = (W + V_0 + mc^2)(W + V_0 - mc^2) \quad (9.17a)$$

$$p_1^2 c^2 = -p_0^2 c^2 = W^2 - m^2 c^4 = (W + mc^2)(W - mc^2) \quad (9.17b)$$

we can rewrite (9.16) as

$$\tan \frac{p_2 l}{\hbar} = \frac{2 p_2 p_0 c^2}{(W + mc^2)(W + V_0 - mc^2) + (W - mc^2)(W + V_0 + mc^2)} \quad (9.18)$$

Multiplying out the brackets here shows that a lot of terms cancel. Using (9.5) again we can easily see that this reduces to

$$\frac{1}{p_2 c} \tan \frac{p_2 l}{\hbar} = -\frac{p_0 c}{p_0^2 c^2 - W V_0} \quad (9.19)$$

This equation defines the energy eigenvalues of the particle trapped in the potential well. It can only be solved graphically and we present a scale drawing of the wells and their eigenvalue spectra in figure 9.2. As one would expect the number of states entering the well increases as the well gets deeper. The level spacing is as one would expect on the basis of non-relativistic quantum theory. For the low energy levels the energy separation goes up as n^2 (where n just labels the state), and at higher energies the levels become more or less equally spaced.

9.2 The Dirac Oscillator

This is the second problem we are going to examine. Here we do not include a potential in the usual way, as we do for the non-relativistic quantum mechanical harmonic oscillator for example. Instead we start by postulating a rather odd form for a vector potential. The Dirac equation for a free particle is

$$(c\tilde{\boldsymbol{\alpha}} \cdot \mathbf{p} + \tilde{\beta}mc^2)\psi(\mathbf{r}) = W\psi(\mathbf{r}) \tag{9.20}$$

In (9.20) we can make the substitution first suggested by Ito *et al.* (1967)

$$\mathbf{p} \rightarrow \mathbf{p} - i\tilde{\beta}m\omega\mathbf{r} = \mathbf{p}_t \tag{9.21}$$

where m is the rest mass and ω is the classical frequency of the oscillator. In (9.21) the effective vector potential is purely complex and is a matrix. The motivation for introducing it is mathematical, in that it does lead to a soluble equation. Our reasons for saying that this vector potential describes an oscillator will become apparent shortly. Position is included here as a vector so we have a three-dimensional problem. We let

$$\psi(\mathbf{r}) = \begin{pmatrix} \psi_a(\mathbf{r}) \\ \psi_b(\mathbf{r}) \end{pmatrix} = \frac{1}{r} \begin{pmatrix} u(r)\chi_\kappa^{m_j}(\hat{\mathbf{r}}) \\ iv(r)\chi_{-\kappa}^{m_j}(\hat{\mathbf{r}}) \end{pmatrix} = \begin{pmatrix} g(r)\chi_\kappa^{m_j}(\hat{\mathbf{r}}) \\ if(r)\chi_{-\kappa}^{m_j}(\hat{\mathbf{r}}) \end{pmatrix} \tag{9.22}$$

in the same way as for the hydrogen atom. All the symbols take on the same meaning as in the previous chapter.

The Non-Relativistic Limit

Here we investigate the non-relativistic limit of the Dirac oscillator. This derivation follows closely the work of Moshinsky and Szczepaniak (1989). If we substitute (9.22) and (9.21) into (9.20) we can use equation (4.13) to get the following two simultaneous equations

$$W\psi_a(\mathbf{r}) = c\tilde{\boldsymbol{\sigma}} \cdot (\mathbf{p} + im\omega\mathbf{r})\psi_b(\mathbf{r}) + mc^2\psi_a(\mathbf{r}) \tag{9.23a}$$

$$W\psi_b(\mathbf{r}) = c\tilde{\boldsymbol{\sigma}} \cdot (\mathbf{p} - im\omega\mathbf{r})\psi_b(\mathbf{r}) - mc^2\psi_b(\mathbf{r}) \tag{9.23b}$$

In equation (9.22) $\psi_b(\mathbf{r})$ is the small component of the wavefunction which tends to zero in the non-relativistic limit. Therefore we rearrange (9.23b) to give an expression for $\psi_b(\mathbf{r})$ and substitute this into (9.23a). This is the standard way of taking the non-relativistic limit of the Dirac equation.

After some simple rearrangement and use of familiar properties of the spin matrices we find

$$(W^2 - m^2 c^4)\psi_a(\mathbf{r}) = c^2(\hat{\mathbf{p}}^2 + m^2\omega^2 r^2 - 3\hbar m\omega - 2m\omega\tilde{\sigma} \cdot \hat{\mathbf{L}})\psi_a(\mathbf{r}) \quad (9.24)$$

In the non-relativistic limit we can write $W \approx E + mc^2$. Therefore $W^2 - m^2c^4 \approx 2Emc^2$ if $E \ll mc^2$. Taking this limit and dividing through by mc^2 gives

$$\left(\frac{\hat{\mathbf{p}}^2}{2m} + \frac{1}{2}m\omega^2 r^2 - \frac{3}{2}\hbar\omega - \frac{2\omega}{\hbar}\hat{\mathbf{S}} \cdot \hat{\mathbf{L}}\right)\psi_a(\mathbf{r}) = E\psi_a(\mathbf{r}) \quad (9.25)$$

where $\hat{\mathbf{S}} = (\hbar/2)\tilde{\sigma}$. The first two terms on the left side of this equation are the terms that appear in the Hamiltonian of the non-relativistic harmonic oscillator, explaining why this 'potential' is called the Dirac oscillator. The third term is a constant which rigidly shifts all energy levels, but has no effect on the eigenfunctions. The final term on the right side is a spin–orbit coupling of strength $\omega/\hbar$. Such an expression for the strength is likely to be very large, having a factor of $\hbar$ in the denominator.

In summary, so far, we have seen that the non-relativistic limit of the Dirac equation when we make the substitution (9.21) is the usual harmonic oscillator with very strong spin–orbit coupling.

Of course the non-relativistic oscillator in spherical polars is a well-known problem in quantum mechanics (Dicke and Wittke 1974). The eigenfunctions satisfy

$$\left(\frac{\mathbf{p}^2}{2m} + \frac{1}{2}m\omega^2 r^2\right)\phi(\mathbf{r}) = \left(2n + l + \frac{3}{2}\right)\hbar\omega\phi(\mathbf{r}) = E\phi(\mathbf{r}) \quad (9.26)$$

with n equal to a non-negative integer. Obviously then

$$\left(\frac{\mathbf{p}^2}{2m} + \frac{1}{2}m\omega^2 r^2 - \frac{3}{2}\hbar\omega\right)\phi(\mathbf{r}) = (2n + l)\hbar\omega\phi(\mathbf{r}) \quad (9.27)$$

The radial part of the eigenfunctions of (9.26) and (9.27) is

$$R_{nl}(r) = A_{nl}\left(\left(\frac{m\omega}{\hbar}\right)^{\frac{1}{2}} r\right)^l e^{-m\omega r^2/2\hbar} M\left(\frac{-E}{2\hbar\omega} - \frac{l}{2} - \frac{3}{4}, l + \frac{3}{2}, \frac{m\omega r^2}{\hbar}\right)$$
$$(9.28)$$

where $M(a, b, x)$ is a confluent hypergeometric function. The quantization of the energy comes from demanding that the eigenfunctions vanish as $r \to \infty$ and hence the first expression in the argument of the confluent hypergeometric function must be equal to a negative integer or zero.

Clearly, then, a comparison of the non-relativistic limit of the Dirac oscillator with the standard three-dimensional harmonic oscillator will be instructive. If $l = 0$ the results must be identical, if $l \neq 0$ the discrepancy between the results will be an indication of the role of the spin–orbit term in equation (9.25).

Solution of the Dirac Oscillator

In this section we solve the Dirac equation with the substitutions (9.21) and (9.22). We outline the solution only and leave it as an exercise to fill in the details. Clear accounts of the solution of this problem are also given by Dominguez-Adame and González (1990) and by Moshinsky and Szczepaniak (1989). After making the substitutions we can follow the procedure for deriving the radial Dirac equation of section 8.1. The extra vector potential does not lead to any difficulties and we are left with

$$\frac{du(r)}{dr} = -\left(\frac{\kappa}{r} + \frac{m\omega r}{\hbar}\right)u(r) + \frac{1}{c\hbar}\left(W + mc^2\right)v(r) \qquad (9.29a)$$

$$\frac{dv(r)}{dr} = \left(\frac{\kappa}{r} + \frac{m\omega r}{\hbar}\right)v(r) - \frac{1}{c\hbar}\left(W - mc^2\right)u(r) \qquad (9.29b)$$

The most straightforward way to solve these equations is to rearrange (9.29a) to give an expression for $v(r)$ and substitute that into (9.29b) to give a single second order equation for $u(r)$:

$$\frac{d^2u(r)}{dr^2} - \left(\frac{l(l+1)}{r^2} + \frac{m^2\omega^2 r^2}{\hbar^2}\right)u(r) - \left(\frac{(2\kappa - 1)m\omega}{\hbar} - \frac{W^2 - m^2c^4}{\hbar^2 c^2}\right) = 0 \tag{9.30}$$

A similar equation exists for $v(r)$. Next we make a change of variable

$$y = \frac{m\omega r^2}{\hbar} \tag{9.31}$$

and then we try the solutions

$$u(y) = Ae^{-y/2}y^{(l+1)/2}\phi_1(y) \tag{9.32a}$$

$$v(y) = Be^{-y/2}y^{(l'+1)/2}\phi_2(y) \tag{9.32b}$$

where A and B are constants. Later a relation between A and B will be found, then the final indeterminacy can be removed by normalization. Substitution of these solutions into equation (9.30) yields

$$y\frac{d^2\phi_1(y)}{dy^2} + \left(l + \frac{3}{2} - y\right)\frac{d\phi_1(y)}{dy} - \frac{1}{2}\left(l + \frac{3}{2} - \mu\right)\phi_1(y) = 0 \tag{9.33}$$

and similarly for $\phi_2(y)$. Here

$$\mu = \frac{1}{2}\left(\frac{(W^2 - m^2c^4)}{mc^2\hbar\omega} - (2\kappa - 1)\right) \tag{9.34}$$

Equation (9.33) is the confluent hypergeometric equation which is discussed in appendix *B*. Hence

$$\phi_1(y) = M\left(\frac{1}{2}\left(l + \frac{3}{2} - \mu\right), l + \frac{3}{2}, y\right) \tag{9.35a}$$

$$\phi_2(y) = M\left(\frac{1}{2}\left(l' + \frac{3}{2} - \mu\right), l' + \frac{3}{2}, y\right) \tag{9.35b}$$

Next we substitute the solutions (9.32) with (9.35) back into (9.29) to find l' in terms of l and B in terms of A. For $j = l + 1/2$ we have

$$g(r) = Ae^{-m\omega r^2/2\hbar}\left(\left(\frac{m\omega}{\hbar}\right)^{1/2} r\right)^{j-1/2} M\left(-n-1, j+1, \frac{m\omega r^2}{\hbar}\right) \tag{9.36a}$$

$$f(r) = \frac{2c\hbar A r^{(j+\frac{1}{2})}}{W + mc^2}\frac{n+1}{j+1}\left(\frac{m\omega}{\hbar}\right)^{\frac{2j+3}{4}} e^{-m\omega r^2/2\hbar} M\left(-n, j+2, \frac{m\omega r^2}{\hbar}\right) \tag{9.36b}$$

and for $j = l - 1/2$

$$g(r) = Ae^{-m\omega r^2/2\hbar}\left(\left(\frac{m\omega}{\hbar}\right)^{1/2} r\right)^{j+1/2} M\left(-n, j+2, \frac{m\omega r^2}{\hbar}\right) \tag{9.37a}$$

$$f(r) = \frac{2c\hbar(j+1)A}{W + mc^2}\left(\frac{m\omega}{\hbar}\right)^{\frac{2j+1}{4}} e^{-m\omega r^2/2\hbar} r^{j-\frac{1}{2}} M\left(-n, j+1, \frac{m\omega r^2}{\hbar}\right) \tag{9.37b}$$

Here n is a non-negative integer, arising from the boundary condition that the wavefunction vanishes as $r \to \infty$. It also quantizes the energy eigenvalues.

Henceforth we will be primarily concerned with the $n = 0$ state. The eigenfunction for this state can be normalized according to equation (8.51). For $n = 0$ the normalization constant for $j = l + 1/2$ is

$$A = \left(\frac{2^{2j+1}m^3\omega^3}{\pi\hbar^3}\right)^{\frac{1}{4}} \times$$

$$\left(\frac{(2j+2)!!\hbar\omega mc^2}{(j+1)^2(W + mc^2)^2} + \frac{2j!!}{2} - \frac{(2j+2)!!}{2(j+1)} + \frac{(2j+4)!!}{8(j+1)^2}\right)^{-\frac{1}{2}} \tag{9.38}$$

and the constant for $j = l - 1/2$ is

$$A = \left(\frac{m^3\omega^3}{\pi\hbar^3}\right)^{\frac{1}{4}}\left(\frac{(2j)!!(j+1)^2\hbar\omega mc^2}{2^{(j-1/2)}(W + mc^2)^2} + \frac{(2j+2)!!}{2^{j+5/2}}\right)^{-\frac{1}{2}} \tag{9.39}$$

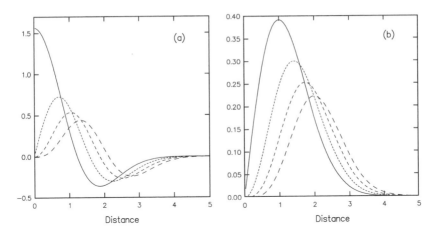

Fig. 9.3. The radial wavefunction for the Dirac oscillator with quantum numbers $n = 0$, $j = l + 1/2$ and $l = 0$ (full line), $l = 1$ (short dashed line), $l = 2$ (medium dashed line), and $l = 3$ (long dashed line). (*a*) $g(r)$, (*b*) $f(r)$. The frequency is 1 and all quantities are in relativistic units.

Figure 9.3*a* shows $g(r)$ for $n = 0$ for several different values of the l quantum number for $j = l + 1/2$. Figure 9.3*b* is the equivalent figure for $f(r)$. Note that $g(r)$ contains a node whereas $f(r)$ does not. Indeed the peak in $f(r)$ approximately corresponds with $g(r)$ passing through zero. Figure 9.4*a* shows $g(r)$ for several different values of the l quantum number for $j = l - 1/2$ and figure 9.4*b* is the equivalent diagram for $f(r)$. In this case neither $g(r)$ nor $f(r)$ contains a node. This is exactly the reverse of the behaviour of the radial component of the relativistic hydrogenic wavefunctions. In that case the numbers of nodes of $g(r)$ and $f(r)$ are equal for $j = l + 1/2$ and the number of nodes of $f(r)$ exceeds the number of nodes of $g(r)$ by one for $j = l - 1/2$. Figures 9.5*a* and 9.5*b* show the radial probability densities defined by

$$P(r) = 4\pi r^2 (f^2(r) + g^2(r)) \tag{9.40}$$

This shows the particle getting further from the origin as the angular momentum gets greater, which we might have expected. The $j = l + 1/2$ probability density has two peaks separated by a trough. At the trough $g(r)$ passes through zero, so the probability density here is entirely due to the small component of the wavefunction. Hence to move from one peak of the probability density to the other the particle goes 'via' the small component of the wavefunction. So, although $f(r)$ is small, it plays a key role in determining the behaviour of the Dirac oscillator. The $j = l - 1/2$ wavefunctions and probability densities are very similar to the equivalent

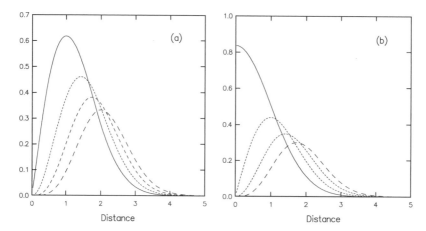

Fig. 9.4. The radial wavefunction for the Dirac oscillator with quantum numbers $n = 0$, $j = l - 1/2$ and $l = 1$ (full line), $l = 2$ (short dashed line), $l = 3$ (medium dashed line), and $l = 4$ (long dashed line), (a) $g(r)$, (b) $f(r)$. The frequency is 1 and all quantities are in relativistic units.

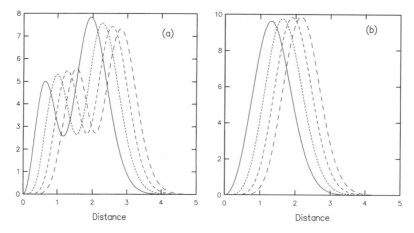

Fig. 9.5. The radial probability density $P(r)$ (equation (9.40)) for the Dirac oscillator with quantum numbers $n = 0$, and $l = 0$ (full line), $l = 1$ (short dashed line), $l = 2$ (medium dashed line), and $l = 3$ (long dashed line). (a) $j = l + 1/2$, (b) $j = l - 1/2$. The frequency is 1 and all quantities are in relativistic units.

quantities for the non-relativistic oscillator. This is not surprising because equation (9.37a) for the large component of the wavefunction has an identical functional form to (9.28) for the non-relativistic oscillator. The only difference between them is in the normalization, because we include the small component of the wavefunction in the normalization in the relativistic case.

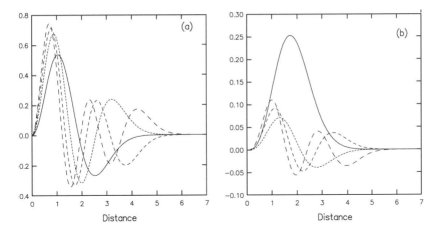

Fig. 9.6. The radial wavefunction with quantum numbers $j = l + 1/2$ with $l = 2$ and $n = 0$ (full line), $n = 1$ (short dashed line), $n = 2$ (medium dashed line), and $n = 3$ (long dashed line). (a) $g(r)$, (b) $f(r)$. The frequency is 1 and all quantities are in relativistic units.

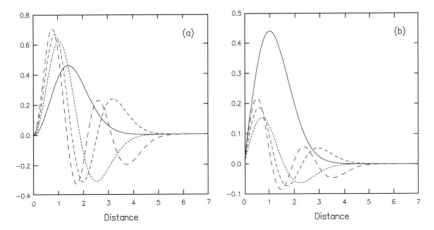

Fig. 9.7. The radial wavefunction with quantum numbers $j = l - 1/2$ with $l = 2$ and $n = 0$ (full line), $n = 1$ (short dashed line), $n = 2$ (medium dashed line), and $n = 3$ (long dashed line). (a) $g(r)$, (b) $f(r)$. The frequency is 1 and all quantities are in relativistic units.

In figures 9.6a and b we see how the $j = l + 1/2$ wavefunction changes with increasing n for constant angular momentum. Figures 9.7a and b are the equivalent diagrams for $j = l - 1/2$. As expected, as energy (n) increases the number of nodes in the wavefunction also increases. Comparison of figures 9.6b and 9.7b shows that the magnitude of the small component of the wavefunction is notably smaller for the $j = l + 1/2$ states than the $j = l - 1/2$ states. This is compensated by the $j = l + 1/2$ particle being further away from the origin, as is easily seen by comparison of the peak positions in the two figures. One can also see a remarkable similarity between $g(r)$ for quantum numbers n and $j = l + 1/2$, and $g(r)$ for quantum numbers $n+1$ and $j = l - 1/2$. On the scale of this figure they are close to identical, although they do differ slightly because differences in the small component of the wavefunction affect the large component through the normalization.

The energy eigenvalues of the Dirac oscillator come from insisting that the first parameter in the confluent hypergeometric functions of (9.35) is a negative integer. With μ given by equation (9.34), and the principal quantum number given by $N = 2n + l$ (see equation (9.27)) we have

$$W^2 = m^2c^4 + (2N - 2j + 1)\hbar\omega mc^2 \qquad j = l + 1/2 \qquad (9.41a)$$
$$W^2 = m^2c^4 + (2N + 2j + 3)\hbar\omega mc^2 \qquad j = l - 1/2 \qquad (9.41b)$$

There is a remarkable amount of degeneracy in these expressions. If q is an integer, all $j = l + 1/2$ states with $N \pm q, j \pm q$ have the same energy, i.e. the degeneracy for these states is infinite. For $j = l - 1/2$ all states with $N \pm q, j \mp q$ are degenerate. This degeneracy is small for low values of the quantum numbers, but increases with N. Clearly, for this case the degeneracy is not infinite because j must remain positive and $N \geq 0$. Another noteworthy point is that for $j = 1/2$ in (9.41a) there is no zero point energy. It is trivial to take the non-relativistic limit of equations (9.41) by letting $W \rightarrow E + mc^2$, then we have

$$E = (N - j + 1/2)\hbar\omega = 2n\hbar\omega \qquad\qquad j = l + 1/2 \qquad (9.42a)$$
$$E = (N + j + 3/2)\hbar\omega = (2n + 2l + 1)\hbar\omega \qquad j = l - 1/2 \qquad (9.42b)$$

Clearly, neither of these is quite the same as the non-relativistic oscillator of equation (9.27). However, comparison of (9.42a) with (9.27) is instructive. For $l = 0$ they yield the same eigenvalues, as they must. Furthermore, the lack of zero point energy for $j = l + 1/2$ can be understood because it has just been taken over to the left hand side of the Dirac equation. If $l \neq 0$ the average of (9.42a and b) is $E = (2n + l + 1/2)\hbar\omega$. This differs from the non-relativistic oscillator eigenvalue by $\hbar\omega$ which we attribute to the spin–orbit coupling. Finally note that we have not considered the negative energy states here, which have odd properties themselves (Benitez *et al.* 1990, Dominguez-Adame and Mèndez 1991).

Expectation Values and the Uncertainty Principle

To take our analysis of the Dirac oscillator further we will work out a few expectation values. Again this is something that is often done for the non-relativistic oscillator. To make the integrals as simple as possible we will work with the $n = 0$ wavefunction only and evaluate several expectation values for the Dirac oscillator for the case of $j = l + 1/2 = 1/2$ and $j = l - 1/2 = 1/2$. There is no problem in principle with the evaluation of the expectation values. The angular parts of the integrals are trivial and the radial parts simply require frequent use of the standard integral

$$\int_0^\infty x^{2n} e^{-ax^2} dx = \frac{1 \times 3 \times 5 \times \dots \times (2n-1)}{2^{n+1} a^n} \sqrt{\frac{\pi}{a}} \tag{9.43}$$

Applying this gives the following results. Feel free to check them. For $j = l + 1/2$ we have

$$\langle \mathbf{p} \rangle = 0, \qquad\qquad \langle \mathbf{r} \rangle = 0 \tag{9.44}$$

$$\langle r \rangle = \frac{3}{2} \left(\frac{\hbar}{\pi m \omega} \right)^{1/2} \left(1 + \frac{mc^2}{W} \right) + \frac{4}{3} \left(\frac{\hbar}{\pi m \omega} \right)^{1/2} \left(1 - \frac{mc^2}{W} \right) \tag{9.45}$$

$$\langle r^2 \rangle = \frac{7}{4} \left(\frac{\hbar}{m \omega} \right) \left(1 + \frac{mc^2}{W} \right) + \frac{5}{4} \left(\frac{\hbar}{m \omega} \right) \left(1 - \frac{mc^2}{W} \right) \tag{9.46}$$

$$\langle p^2 \rangle = \frac{7}{4} \hbar m \omega \left(1 + \frac{mc^2}{W} \right) + \frac{5}{4} \hbar m \omega \left(1 - \frac{mc^2}{W} \right) \tag{9.47}$$

and for $j = l - 1/2$

$$\langle \mathbf{p} \rangle = 0, \qquad\qquad \langle \mathbf{r} \rangle = 0 \tag{9.48}$$

$$\langle r \rangle = \frac{4}{3} \left(\frac{\hbar}{\pi m \omega} \right)^{1/2} \left(1 + \frac{mc^2}{W} \right) + \left(\frac{\hbar}{\pi m \omega} \right)^{1/2} \left(1 - \frac{mc^2}{W} \right) \tag{9.49}$$

$$\langle r^2 \rangle = \frac{5}{4} \left(\frac{\hbar}{m \omega} \right) \left(1 + \frac{mc^2}{W} \right) + \frac{3}{4} \left(\frac{\hbar}{m \omega} \right) \left(1 - \frac{mc^2}{W} \right) \tag{9.50}$$

$$\langle p^2 \rangle = \frac{5}{4} \hbar m \omega \left(1 + \frac{mc^2}{W} \right) + \frac{3}{4} \hbar m \omega \left(1 - \frac{mc^2}{W} \right) \tag{9.51}$$

If we make the usual identification of the velocity of the Dirac oscillator with the time derivative of the expectation value of position we can find $\langle v \rangle$ as the commutator of the Hamiltonian with position. This gives $\langle v \rangle = c$ as expected, indicating that the Dirac oscillator undergoes the *Zitterbewegung*. Above, equations (9.44) and (9.48) are the expectation values of vector quantities. We have taken the origin at the centre of the motion so it is not surprising these have turned out to be zero. Mathematically, these zeros arise from the orthonormality of the spherical harmonics. Equations (9.45) and (9.49) are expectation values of the radial distance from the centre, i.e. they are a measure of the amplitude of the oscillation, but they seem not to be as good a measure as $\langle r^2 \rangle$ which

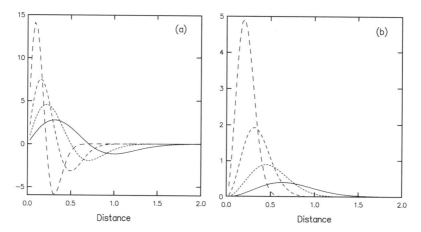

Fig. 9.8. The Dirac oscillator wavefunctions for a series of frequencies. The quantum numbers are $n = 0$, $j = l + 1/2$ and $l = 1$. The frequencies are $\omega = 5$ (full line), $\omega = 10$ (short dashed line), $\omega = 20$ (medium dashed line), $\omega = 50$ (long dashed line). (a) $g(r)$, (b) $f(r)$.

increases monotonically with l, as do the radial probability densities in figures 9.5. In the non-relativistic limit $(1 - mc^2/W) \to 0$ and $(1 + mc^2/W) \to 2$, hence the non-relativistic limit of these equations is simple to take. Remarkably, the non-relativistic limits of equations $(9.44 - 47)$ take on the same values as the $n = 1$, $l = 0$ expectation values of the non-relativistic harmonic oscillator (this is the second excited state – see equation (9.28)). Furthermore, the non-relativistic limits of equations $(9.48 - 51)$ take on the same values as the expectation values of the first excited state $n = 0$, $l = 1$ of the standard non-relativistic harmonic oscillator.

In figure 9.8 we show the Dirac oscillator wavefunctions for fixed quantum numbers and a series of frequencies. From equations (9.45) and (9.49) we see that we expect the particle to get closer to the origin at a rate proportional to $1/\sqrt{\omega}$. Clearly this is confirmed by the position of the peaks in the wavefunction as a function of ω in figure 9.8. In the high frequency limit the wavefunction and hence the probability density are well localized peaks. We can infer from this that the Dirac oscillator is trapped in a circular orbit around the origin.

One further limit of interest is the ultra-relativistic limit. In this case $(1 - mc^2/W) \to 1$ and $(1 + mc^2/W) \to 1$, and again the result of taking this limit can easily be seen from equations $(9.44 \to 51)$.

As is well known, the uncertainty in any observable $\hat{O}$ is defined by

$$\Delta \hat{O} = \left(\langle \hat{O}^2 \rangle - \langle \hat{O} \rangle^2 \right)^{1/2} \qquad (9.52)$$

With the expectation values calculated above we can evaluate the uncertainties in momentum and position, and hence work out the uncertainty principle (see appendix A) for the ground state of the Dirac oscillator. Inserting $\langle \mathbf{p} \rangle, \langle p^2 \rangle$ and $\langle \mathbf{r} \rangle, \langle r^2 \rangle$ into (9.52) gives long and not very illuminating expressions. However if we then look at the uncertainty in the non-relativistic limit we find

$$\Delta p \Delta r = \frac{7}{2}\hbar \geq \frac{1}{2}\hbar \qquad\qquad j = l + 1/2 \qquad\qquad (9.53)$$

$$\Delta p \Delta r = \frac{5}{2}\hbar \geq \frac{1}{2}\hbar \qquad\qquad j = l - 1/2 \qquad\qquad (9.54)$$

Clearly these versions of the uncertainty principle differ from the ground state of the non-relativistic oscillator solved directly. In that case

$$\Delta p \Delta r = \frac{3}{2}\hbar \qquad\qquad (9.55)$$

The reason for this difference is clear. Again it is due to the existence of the small component of the wavefunction. For the non-relativistic oscillator the radial part of the ground-state wavefunction is given by equation (9.28) with the first argument of the confluent hypergeometric function equal to zero. The large component of the Dirac oscillator wavefunction with $n = -1$ and $j = l + 1/2$ is well behaved, and corresponds to the ground state of the non-relativistic oscillator. However, this solution is inadmissible because the small component diverges with this boundary condition. Hence it is clear that the non-relativistic limit of the Dirac oscillator has a ground state and expectation values that look like the first excited state of the non-relativistic oscillator.

In the ultra-relativistic limit we can easily see that

$$\Delta p \Delta r = 3\hbar \qquad\qquad j = l + 1/2 \qquad\qquad (9.56)$$

$$\Delta p \Delta r = 2\hbar \qquad\qquad j = l - 1/2 \qquad\qquad (9.57)$$

In summary then we have examined some properties of the Dirac oscillator and the extent to which we can regard it as a relativistic generalization of the standard non-relativistic oscillator. We have seen that the $j = l - 1/2$ solutions are very similar to the non-relativistic oscillator. However, the ground state of the non-relativistic oscillator has no corresponding analogue in the Dirac oscillator, because of the different boundary conditions required for the Dirac oscillator and the existence of the small component of the wavefunction. Several expectation values have been found and using them in the uncertainty principle allows us to conclude that at small values of the quantum numbers the Dirac oscillator cannot be regarded as a direct relativistic generalization of the usual simple harmonic oscillator,

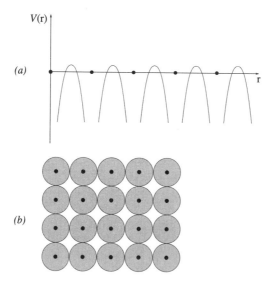

Fig. 9.9. (*a*) The potential on a one-dimensional lattice. The filled circles are the ions at their average positions. The curved lines are the periodic potential. The position of the horizontal axis relative to the potentials is of no significance. (*b*) The muffin tin model of the periodic potential on a two-dimensional lattice. The potential is spherically symmetric inside a sphere (the grey circles) surrounding each ion and flat outside the spheres (the unshaded areas). It is usual to choose the sphere radius as half the nearest neighbour distance.

but at larger values of the quantum numbers the small component of the wavefunction plays an increasingly insignificant role and the Dirac oscillator does behave in a very similar way to the non-relativistic oscillator. The Dirac oscillator has been discussed in detail by several authors (de Lange 1991, for example), and the energy spectrum can be described in terms of a hidden supersymmetry (Benitez *et al.* 1990). It has potential applications in models of quark confinement in particle physics (Su and Zhang 1984).

9.3 Bloch's Theorem

I haven't seen any book on relativistic quantum theory that even mentions Bloch's theorem. Conversely, I know of no elementary condensed matter text that fails to mention it, but it is always derived within Schrödinger's quantum mechanics (Ashcroft and Mermin 1976, Kittel 1986). Here, we show that it holds within relativistic quantum theory.

In figure 9.9*a* we show a schematic form of the potential on a one-dimensional lattice for which we may wish to solve the Dirac equation. Figure 9.9*b* shows one way of modelling the potential in a solid as a regular array of spherically symmetric potentials, surrounded by regions of flat potential, known as the muffin tin approximation. Clearly, any model of the potential must have the periodicity of the lattice. The potential energy of an electron at point $\mathbf{r}$ on the lattice is $V(\mathbf{r})$. By definition an infinite lattice potential energy obeys

$$V(\mathbf{r}) = V(\mathbf{r} + \mathbf{R}) \tag{9.58}$$

where $\mathbf{R}$ is a direct lattice vector. Because the potential is periodic with the lattice, it can be expanded as a Fourier series in the reciprocal lattice vectors, $\mathbf{G}$

$$V(\mathbf{r}) = \sum_{\mathbf{G}} V_{\mathbf{G}} e^{i\mathbf{G} \cdot \mathbf{r}} \tag{9.59}$$

The coefficients are given by

$$V_{\mathbf{G}} = \frac{1}{\Omega} \int_{\Omega} e^{-i\mathbf{G} \cdot \mathbf{r}} V(\mathbf{r}) d\Omega \tag{9.60}$$

where Ω is the volume of the unit cell. Now, the wavefunction $\psi(\mathbf{r})$ that we will use will be one to describe a single electron in a three-dimensional periodic potential. Let us write this wavefunction as a linear combination of relativistic plane waves

$$\psi(\mathbf{r}) = \sum_{\mathbf{q}} c_{\mathbf{q}} U(\mathbf{q}) e^{i\mathbf{q} \cdot \mathbf{r}} \tag{9.61}$$

Here the $U(\mathbf{q})$ are given by equation (5.26), where the particle momentum is $\mathbf{p} = \hbar \mathbf{q}$. With these definitions we can write the Dirac equation for our electron in a lattice as

$$\left(c\tilde{\boldsymbol{\alpha}} \cdot \hat{\mathbf{p}} + \tilde{\beta} mc^2 + \sum_{\mathbf{G}} V_{\mathbf{G}} e^{i\mathbf{G} \cdot \mathbf{r}} - W \right) \sum_{\mathbf{q}} c_{\mathbf{q}} U(\mathbf{q}) e^{i\mathbf{q} \cdot \mathbf{r}} = 0 \tag{9.62}$$

Let us consider the kinetic energy term first

$$c\tilde{\boldsymbol{\alpha}} \cdot \hat{\mathbf{p}} \psi(\mathbf{r}) = \frac{c\hbar}{i} \tilde{\boldsymbol{\alpha}} \cdot \nabla \sum_{\mathbf{q}} c_{\mathbf{q}} U(\mathbf{q}) e^{i\mathbf{q} \cdot \mathbf{r}} = \hbar c \sum_{\mathbf{q}} \tilde{\boldsymbol{\alpha}} \cdot \mathbf{q} c_{\mathbf{q}} U(\mathbf{q}) e^{i\mathbf{q} \cdot \mathbf{r}} \tag{9.63}$$

Next the potential energy term

$$\sum_{\mathbf{G}} V_{\mathbf{G}} e^{i\mathbf{G} \cdot \mathbf{r}} \sum_{\mathbf{q}} c_{\mathbf{q}} U(\mathbf{q}) e^{i\mathbf{q} \cdot \mathbf{r}} = \sum_{\mathbf{G}} \sum_{\mathbf{q}} V_{\mathbf{G}} c_{\mathbf{q}} U(\mathbf{q}) e^{i(\mathbf{G} + \mathbf{q}) \cdot \mathbf{r}} \tag{9.64}$$

Substituting (9.63) and (9.64) into equation (9.62) gives

$$\sum_{\mathbf{q}} \left(\left(c\hbar \tilde{\boldsymbol{\alpha}} \cdot \mathbf{q} + \tilde{\beta} mc^2 - W \right) c_{\mathbf{q}} U(\mathbf{q}) + \sum_{\mathbf{G}} V_{\mathbf{G}} c_{\mathbf{q} - \mathbf{G}} U(\mathbf{q} - \mathbf{G}) \right) e^{i\mathbf{q} \cdot \mathbf{r}} = 0 \tag{9.65}$$

where we have moved the origin of $\mathbf{q}$ by one reciprocal lattice vector in the final term. This has no effect as it is a sum over all reciprocal lattice vectors. Equation (9.65) must hold for every Fourier component, so

$$\left(c\hbar c_{\mathbf{q}} \tilde{\alpha} \cdot \mathbf{q} + \tilde{\beta} mc^2 - W \right) c_{\mathbf{q}} U(\mathbf{q}) + \sum_{\mathbf{G}} V_{\mathbf{G}} c_{\mathbf{q}-\mathbf{G}} U(\mathbf{q} - \mathbf{G}) = 0 \qquad (9.66)$$

This equation is just the Dirac equation for the given potential written in reciprocal space. It has changed from a differential equation in real space to a series of algebraic equations in reciprocal space. Equation (9.66) can be used to determine the coefficients $c_{\mathbf{q}}$. This equation tells us that the allowed values of $\mathbf{q}$ in (9.61) are restricted to a particular value and all those that differ from it by reciprocal lattice vectors. Equation (9.61) becomes

$$\psi_{\mathbf{q}}(\mathbf{r}) = \sum_{\mathbf{G}} c_{\mathbf{q}-\mathbf{G}} U(\mathbf{q} - \mathbf{G}) e^{i(\mathbf{q}-\mathbf{G})\cdot\mathbf{r}}$$

$$= \left(\sum_{\mathbf{G}} c_{\mathbf{q}-\mathbf{G}} U(\mathbf{q} - \mathbf{G}) e^{-i\mathbf{G}\cdot\mathbf{r}} \right) e^{i\mathbf{q}\cdot\mathbf{r}} = \phi_{\mathbf{q}}(\mathbf{r}) e^{i\mathbf{q}\cdot\mathbf{r}} \qquad (9.67)$$

The function $\phi_{\mathbf{q}}(\mathbf{r})$ is essentially a Fourier series over reciprocal lattice vectors, and is invariant under the transformation $\mathbf{r} \to \mathbf{r} + \mathbf{R}$. We can see this directly

$$\phi_{\mathbf{q}}(\mathbf{r} + \mathbf{R}) = \sum_{\mathbf{G}} c_{\mathbf{q}-\mathbf{G}} U(\mathbf{q} - \mathbf{G}) e^{-i\mathbf{G}\cdot(\mathbf{r}+\mathbf{R})}$$

$$= \sum_{\mathbf{G}} c_{\mathbf{q}-\mathbf{G}} U(\mathbf{q} - \mathbf{G}) e^{-i\mathbf{G}\cdot\mathbf{r}} e^{-i\mathbf{G}\cdot\mathbf{R}} = \sum_{\mathbf{G}} c_{\mathbf{q}-\mathbf{G}} U(\mathbf{q} - \mathbf{G}) e^{-i\mathbf{G}\cdot\mathbf{r}} = \phi_{\mathbf{q}}(\mathbf{r})$$

$$(9.68)$$

where the usual relation between direct and reciprocal lattice vectors holds:

$$e^{i\mathbf{G}\cdot\mathbf{R}} = 1 \qquad (9.69)$$

So (9.68) establishes the lattice periodicity of $\phi_{\mathbf{q}}(\mathbf{r})$ and hence (9.67) is a statement of Bloch's theorem.

We have derived Bloch's theorem using the symmetry of lattices and relativistic quantum theory, hence showing that Bloch's theorem is valid within a relativistic theory of condensed matter. In fact this is not surprising and the derivation above is very reminiscent of the non-relativistic derivation. The major difference is that we have had to carry $U(\mathbf{q})$ through the mathematics as it represents the four-component nature of the wavefunction. Bloch's theorem forms the basis of a great deal of our understanding of solid state physics, and in the following section we apply it in a simple one-dimensional model of a solid.

9.4 The Relativistic Kronig–Penney Model

In chapter 5 we examined the behaviour of a free electron. The free electron theory of metals explains, qualitatively, many properties of materials. However, such a model fails to shed any light on many other properties, for example, the difference between a conductor, a semiconductor and an insulator. As we know from elementary condensed matter physics the electrons in a solid are arranged in bands and these can be separated by energy gaps for which no wave-like solutions of the Schrödinger equation exist. Band theory has been remarkably successful in describing properties of materials as diverse as lattice constant, specific heat, magnetic properties, and superconducting transition temperatures, and in understanding the results of a vast array of spectroscopies. Here we introduce the relativistic theory of energy bands in the simplest way possible. We will examine the relativistic generalization of the Kronig–Penney model, i.e. a single electron in a one-dimensional periodic potential.

The non-relativistic Kronig–Penney model involves δ-function potentials. Such potentials in relativistic quantum theory lead to the Klein paradox. While this is interesting in itself (Fairbairn *et al.* 1973, McKellar and Stephenson 1987, Sutherland and Mattis 1981), it is not relevant to condensed matter physics, so we shall avoid this by looking only at potentials for which $V < 2mc^2$.

To solve the Kronig–Penney model as simply as possible we go through several stages. Firstly we need the one-dimensional Dirac equation. Next we solve the that equation for a potential step. Finally, we combine the solutions to the potential step as a method of solving the Dirac equation for a series of such steps. This is our Kronig–Penney model.

A One-Dimensional Time-Independent Dirac Equation

In this section we use the full time-independent Dirac equation (8.1) to deduce a one-dimensional version of the equation (we choose the x-direction) and then find the general solutions for a region of constant potential $V(x)$ (Glasser and Davison 1970). The x-component of the usual Dirac equation is

$$\left(\frac{c\hbar}{i}\tilde{\alpha}_x\frac{\partial}{\partial x} + \tilde{\beta}mc^2 + V(x)\right)\psi(x) = W\psi(x) \tag{9.70}$$

Now, if we write our wavefunction in the form of equation (5.7), we can decompose (9.70) into four simultaneous differential equations:

$$-i\hbar c \frac{\partial \psi_4}{\partial x} + mc^2 \psi_1 - (W - V(x))\psi_1 = 0 \qquad (9.71a)$$

$$-i\hbar c \frac{\partial \psi_3}{\partial x} + mc^2 \psi_2 - (W - V(x))\psi_2 = 0 \qquad (9.71b)$$

$$-i\hbar c \frac{\partial \psi_2}{\partial x} - mc^2 \psi_3 - (W - V(x))\psi_3 = 0 \qquad (9.71c)$$

$$-i\hbar c \frac{\partial \psi_1}{\partial x} - mc^2 \psi_4 - (W - V(x))\psi_4 = 0 \qquad (9.71d)$$

where we have used the definitions of the matrices (4.13). Now we let

$$\phi = \begin{pmatrix} \psi_1 \\ \psi_4 \end{pmatrix} \qquad \text{or} \qquad \phi = \begin{pmatrix} \psi_2 \\ \psi_3 \end{pmatrix} \qquad (9.72a)$$

and either of these can be written

$$\phi = \begin{pmatrix} \phi_1 \\ \phi_2 \end{pmatrix} \qquad (9.72b)$$

Then we can write

$$-i\hbar c \tilde{\sigma}_x \frac{\partial \phi}{\partial x} + mc^2 \tilde{\sigma}_z \phi = (W - V(x))\phi \qquad (9.73)$$

Equation (9.73) is a one-dimensional Dirac equation. Note that in one dimension we only require a two-component wavefunction because we only need two anticommuting matrices, rather than the four required in section 4.1. Equation (9.73) can be written in component form as

$$-i\hbar c \frac{\partial \phi_2}{\partial x} + mc^2 \phi_1 = (W - V(x))\phi_1 \qquad (9.74a)$$

$$-i\hbar c \frac{\partial \phi_1}{\partial x} - mc^2 \phi_2 = (W - V(x))\phi_2 \qquad (9.74b)$$

Now, we define a new energy $E = W - mc^2$ to eliminate the effect of the rest mass energy. As the rest mass energy is irrelevant in band theory we can adopt this approach. Then equations (9.74) become

$$-i\hbar c \frac{\partial \phi_2}{\partial x} = (E - V(x))\phi_1 \qquad (9.75a)$$

$$-i\hbar c \frac{\partial \phi_1}{\partial x} = (E - V(x) + 2mc^2)\phi_2 \qquad (9.75b)$$

In a region of constant potential, we can decouple these equations. Firstly we differentiate (9.75a) and substitute from (9.75b) for $\partial \phi_1 / \partial x$ to obtain

$$-\hbar^2 c^2 \frac{\partial^2 \phi_2}{\partial x^2} = (E - V)(E - V + 2mc^2)\phi_2 \qquad (9.76a)$$

Secondly we differentiate (9.75b) and substitute from (9.75a) for the first derivative of ϕ_2 to obtain

$$-\hbar^2 c^2 \frac{\partial^2 \phi_1}{\partial x^2} = (E - V)(E - V + 2mc^2)\phi_1 \qquad (9.76b)$$

These equations are now both of the same form:

$$\frac{\partial^2 \phi_{j(\rho)}}{\partial x^2} = -k_\rho^2 \phi_{j(\rho)} \qquad (9.77)$$

where

$$k_\rho^2 = \frac{(E - V)(E - V + 2mc^2)}{\hbar^2 c^2} \qquad (9.78)$$

Now, the subscripts are getting complicated, so let us state explicitly what they mean. In $\phi_{i(j)}$ (and $\alpha_{i(j)}$ and $\beta_{i(j)}$ below) the first subscript indicates that we are referring to the upper (1) or lower (2) component of the spinor as in equation (9.72b). The second subscript defines a particular region of space where the potential is flat and so the equations above are valid. This second subscript appears in brackets. Where a wavefunction-like quantity has only one subscript, which of the above two meanings it has will be indicated by the presence or absence of the brackets. Some quantities which do not depend on the wavefunction explicitly have only one subscript (e.g. k_ρ above and χ_ρ below). In these cases the subscript indicates the region of space where the potential is flat, and the subscript will not be bracketed. Equation (9.77) is a trivial differential equation whose solutions can be written down immediately:

$$\phi_{1(\rho)}(x) = \alpha_{1(\rho)} e^{ik_\rho x} + \beta_{1(\rho)} e^{-ik_\rho x} \qquad (9.79a)$$

$$\phi_{2(\rho)}(x) = \alpha_{2(\rho)} e^{ik_\rho x} + \beta_{2(\rho)} e^{-ik_\rho x} \qquad (9.79b)$$

Now we can substitute these solutions into (9.75) to find restrictions on the coefficients $\alpha_{j(\rho)}$ and $\beta_{j(\rho)}$. If we do this and compare coefficients of the exponentials we find that

$$\alpha_{2(\rho)} = \frac{(E - V)}{c\hbar k_\rho} \alpha_{1(\rho)} \qquad (9.80a)$$

$$-\beta_{2(\rho)} = \frac{(E - V)}{c\hbar k_\rho} \beta_{1(\rho)} \qquad (9.80b)$$

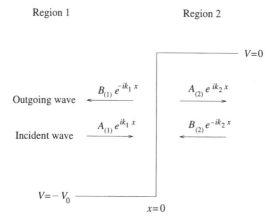

Fig. 9.10. A potential step with particles incident from both directions. The word wave is used in a rather general sense. Clearly, when the particle has energy below zero, it will be a decaying function inside the potential step.

and if we define

$$\chi_\rho = \frac{E - V}{c\hbar k_\rho} \qquad (9.81)$$

we can use (9.80) and (9.81) to rewrite (9.79) as

$$\begin{pmatrix} \phi_{1(\rho)} \\ \phi_{2(\rho)} \end{pmatrix} = \begin{pmatrix} 1 \\ \chi_\rho \end{pmatrix} \alpha_{1(\rho)} e^{ik_\rho x} + \begin{pmatrix} 1 \\ -\chi_\rho \end{pmatrix} \beta_{1(\rho)} e^{-ik_\rho x} \qquad (9.82)$$

i.e. we have eliminated one of the unknowns in equation (9.79).

A Potential Step

Here we are going to use the results of the previous section to find the behaviour of a free electron incident upon a step potential. This, of course, is a standard problem in non-relativistic quantum mechanics, and is illustrated in figure 9.10. We will solve the Dirac equation in both regions 1 and 2 where the potential is constant and then insist that the solutions match at the boundary at $x = 0$. We will use a scattering theory approach to solve this problem.

The general solutions to the Dirac equation in both regions of the potential in figure 9.10 are

$$\phi_{(1)}(x) = A_{(1)} e^{ik_1 x} + B_{(1)} e^{-ik_1 x} \qquad (9.83a)$$

$$\phi_{(2)}(x) = A_{(2)} e^{ik_2 x} + B_{(2)} e^{-ik_2 x} \qquad (9.83b)$$

as we have already seen in equations (9.79) and (9.82). Here the numerical subscripts in brackets represent the values of ρ, $\rho = 1$ to the left of the barrier and $\rho = 2$ to the right of the barrier. In region 1, $V = -V_0$, so from (9.78) we have

$$k_1^2 = \frac{(E + V_0 + 2mc^2)(E + V_0)}{c^2 \hbar^2} \tag{9.84a}$$

and in region 2, $V = 0$, so

$$k_2^2 = \frac{E(E + 2mc^2)}{c^2 \hbar^2} \tag{9.84b}$$

The $A_{(\rho)}$ and $B_{(\rho)}$ are two-component quantities which can be written in terms of previously defined coefficients as

$$A_{(\rho)} = \begin{pmatrix} \alpha_{1(\rho)} \\ \alpha_{2(\rho)} \end{pmatrix}, \qquad B_{(\rho)} = \begin{pmatrix} \beta_{1(\rho)} \\ \beta_{2(\rho)} \end{pmatrix} \tag{9.85}$$

The next step is to define a scattering matrix which simply relates the phase and amplitude of the outgoing wave to that of the incident wave

$$\begin{pmatrix} B_{(1)} \\ A_{(2)} \end{pmatrix} = \begin{pmatrix} S_{11} & S_{12} \\ S_{21} & S_{22} \end{pmatrix} \begin{pmatrix} A_{(1)} \\ B_{(2)} \end{pmatrix} \tag{9.86}$$

It turns out to be easier to solve this problem by writing (9.83) in terms of the matrix $\tilde{\mathbf{R}}$ which relates the amplitude and phase of the waves on one side of the barrier to those of the wave on the other side

$$\begin{pmatrix} A_{(2)} \\ B_{(2)} \end{pmatrix} = \begin{pmatrix} R_{11} & R_{12} \\ R_{21} & R_{22} \end{pmatrix} \begin{pmatrix} A_{(1)} \\ B_{(1)} \end{pmatrix} = \begin{pmatrix} S_{21} + \frac{S_{11}^2}{S_{12}} & -\frac{S_{11}}{S_{12}} \\ -\frac{S_{11}}{S_{12}} & S_{12}^{-1} \end{pmatrix} \begin{pmatrix} A_{(1)} \\ B_{(1)} \end{pmatrix} \tag{9.87}$$

where we have used equation (9.86), and we note that

$$\frac{S_{22}}{S_{12}} = -\frac{S_{11}}{S_{12}} \tag{9.88}$$

Equations (9.83) are the general solutions to the one-dimensional Dirac equation. Now it is time to invoke the boundary conditions to describe the step function of figure 9.10. We do this by insisting that the wavefunctions are continuous across the boundary at $x = 0$, i.e. we want $\phi_{(1)}(0) = \phi_{(2)}(0)$, and this implies that

$$A_{(1)} + B_{(1)} = A_{(2)} + B_{(2)} \tag{9.89}$$

From equation (9.82) this is equivalent to

$$\begin{pmatrix} \alpha_{1(1)} \\ \chi_1 \alpha_{1(1)} \end{pmatrix} + \begin{pmatrix} \beta_{1(1)} \\ -\chi_1 \beta_{1(1)} \end{pmatrix} = \begin{pmatrix} \alpha_{1(2)} \\ \chi_2 \alpha_{1(2)} \end{pmatrix} + \begin{pmatrix} \beta_{1(2)} \\ -\chi_2 \beta_{1(2)} \end{pmatrix} \tag{9.90}$$

Hence from the upper line of (9.90) we have

$$\alpha_{1(1)} + \beta_{1(1)} = \alpha_{1(2)} + \beta_{1(2)} \tag{9.91a}$$

and the lower line easily leads to

$$\beta_{1(2)} = \alpha_{1(2)} - \Gamma(\alpha_{1(1)} - \beta_{1(1)}) \qquad (9.91b)$$

where

$$\Gamma = \frac{\chi_1}{\chi_2} \qquad (9.92)$$

Next we substitute from (9.91b) into (9.91a) for $\beta_{1(2)}$ and we get

$$\beta_{1(1)} = \frac{2\alpha_{1(2)} - (1 + \Gamma)\alpha_{1(1)}}{1 - \Gamma} \qquad (9.93a)$$

and by eliminating $\beta_{1(1)}$ between (9.91a and b) we find

$$\beta_{1(2)} = \frac{2\Gamma\alpha_{1(1)} - \alpha_{1(2)}(\Gamma + 1)}{\Gamma - 1} \qquad (9.93b)$$

We can also write out equation (9.86) in component form

$$\beta_{1(1)} = S_{11}\alpha_{1(1)} + S_{12}\beta_{1(2)} \qquad (9.94a)$$

$$\beta_{2(1)} = S_{11}\alpha_{2(1)} + S_{12}\beta_{2(2)} \qquad (9.94b)$$

$$\alpha_{1(2)} = S_{21}\alpha_{1(1)} + S_{22}\beta_{1(2)} \qquad (9.94c)$$

$$\alpha_{2(2)} = S_{21}\alpha_{2(1)} + S_{22}\beta_{2(2)} \qquad (9.94d)$$

Next we rearrange (9.94c) to give

$$\beta_{1(2)} = \frac{\alpha_{1(2)} - S_{21}\alpha_{1(1)}}{S_{22}} \qquad (9.95a)$$

and we can eliminate $\beta_{1(2)}$ between (9.94a) and (9.95a). This yields

$$\beta_{1(1)} = \alpha_{1(1)}\left(S_{11} - \frac{S_{12}S_{21}}{S_{22}}\right) + \alpha_{1(2)}\frac{S_{12}}{S_{22}} \qquad (9.95b)$$

Now we can define the elements of the S-matrix by comparison of equations (9.93a) and (9.95b) and equations (9.93b) and (9.95a). It is simply necessary to equate the coefficients of $\alpha_{i(j)}$. This leads immediately to

$$S_{22} = \frac{1 - \Gamma}{1 + \Gamma} \qquad (9.96a)$$

$$S_{21} = \frac{2\Gamma}{1 + \Gamma} \qquad (9.96b)$$

$$S_{12} = \frac{2}{1 + \Gamma} \qquad (9.96c)$$

$$S_{11} = -\frac{1 - \Gamma}{1 + \Gamma} \qquad (9.96d)$$

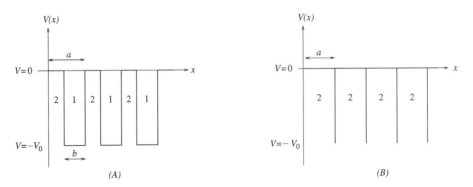

Fig. 9.11. (*A*) The Kronig–Penney potential showing region 1 where the potential is $-V_0$ and region 2 where the potential is zero. The unit cell size is *a* and the width of the potential well is *b*. (*B*) A schematic diagram of the Kronig–Penney potential used in the examples shown in figures 9.12 to 9.16. There is a very narrow potential well in each cell, otherwise the potential is zero.

Note that these definitions are in agreement with equation (9.88). Now, in equation (9.87) we have the *R*-matrix written in terms of the elements of the *S*-matrix, so now we are in a position to find the elements of the *R*-matrix in terms of Γ. Some simple algebra gives

$$\tilde{R} = \begin{pmatrix} R_{11} & R_{12} \\ R_{21} & R_{22} \end{pmatrix} = \frac{1}{2}\begin{pmatrix} 1+\Gamma & 1-\Gamma \\ 1-\Gamma & 1+\Gamma \end{pmatrix} \tag{9.97}$$

This is the expression we have been aiming towards in this section. It is a matrix defined by equation (9.87). Given the wavefunction on the left hand side of our potential step this matrix enables us to calculate the wavefunction on the right hand side of the step. It is this that we require to discuss the Kronig–Penney model in the following section. One final point to note about (9.87) is that we can premultiply both sides by $\tilde{R}^{-1}$ to give an expression for the wavefunction on the left side of the barrier given the wavefunction on the right side. The only condition required for this to be so is that $\tilde{R}^{-1}$ is well defined. It is easy to see that

$$\tilde{R}^{-1} = \frac{1}{2\Gamma}\begin{pmatrix} \Gamma+1 & \Gamma-1 \\ \Gamma-1 & \Gamma+1 \end{pmatrix} \tag{9.98}$$

A One-Dimensional Solid

In our model of a one-dimensional solid we are going to set up the crystal potential as a linear array of rectangular wells. This is shown schematically in figure 9.11*A*. From figures 9.10 and 9.11*A* we can see that at any point in the lattice there are, in general, components of the

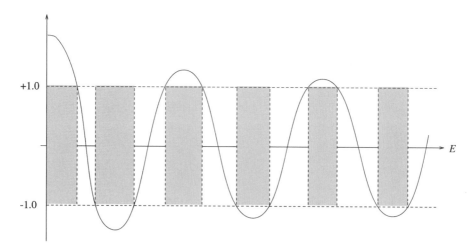

Fig. 9.12.   Figure showing the solution of equation (9.112). The dashed horizontal
lines at ± 1.0 represent the limits of the cosine on the left hand side of (9.112).
The full line is a schematic curve showing the behaviour of the right hand side
of equation (9.112) as a function of increasing energy, and is not bounded by
± 1.0. Therefore, where the full line takes on values outside this range there is no
wavelike solution of the Dirac equation. This is represented by the shaded areas.
At energies where there is no shading it is possible to satisfy equation (9.112) and
wavelike solutions of the Dirac equation exist.

wavefunction representing waves moving in both directions. Using (9.85)
in (9.79) we can write the wavefunction at point x in two-component form

$$\phi_\rho(x) = \begin{pmatrix} A_{(\rho)} e^{ik_\rho x} \\ B_\rho e^{-ik_\rho x} \end{pmatrix} \tag{9.99}$$

where we must keep in mind that each element of (9.99) is itself a two-
component quantity. Next we define a transfer matrix $\tilde{T}_d$ which will
translate the wavefunction a distance d through space provided it does
not have to pass through a change in potential.

$$\tilde{T}_d \phi_{(\rho)}(x) = \begin{pmatrix} e^{ik_\rho d} & 0 \\ 0 & e^{-ik_\rho d} \end{pmatrix} \begin{pmatrix} A_\rho e^{ik_\rho x} \\ B_\rho e^{-ik_\rho x} \end{pmatrix} = \begin{pmatrix} A_\rho e^{ik_\rho (x+d)} \\ B_\rho e^{-ik_\rho (x+d)} \end{pmatrix} \tag{9.100}$$

So now we have the R-matrices (equations (9.97) and (9.98)) which tell us
the wavefunction on one side of a discontinuity, given the wavefunction on
the other side, and the T-matrices which translate us through regions of
constant potential. With these matrices we can determine the wavefunction
at any point in the lattice of figure 9.11, given its form at any other point.
In particular, consider the wavefunction right at the right hand edge of
region 1. We can define a matrix $\tilde{G}_a$ which will translate it to the right

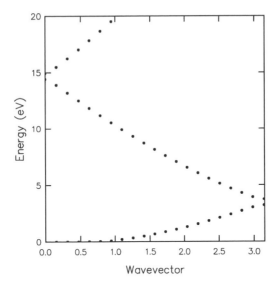

Fig. 9.13. Energy bands of the relativistic Kronig–Penney model for a one-dimensional lattice with $a = 6$ angstroms, $b = 6 \times 10^{-5}$ angstroms and $V = 0.05mc^2$. Note the existence of the energy gaps which correspond with energies where no wavelike solution of the Dirac equation exists.

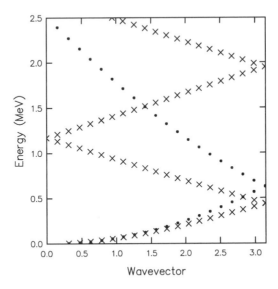

Fig. 9.14. Energy bands of the relativistic Kronig–Penney model for a one-dimensional lattice with $a = 2\lambda_C$, $b = 1.5 \times 10^{-2}\lambda_C$ and $V = mc^2$. The dots are the non-relativistic calculation and the crosses are the relativistic calculation. This clearly shows the relativistic reduction of the band width.

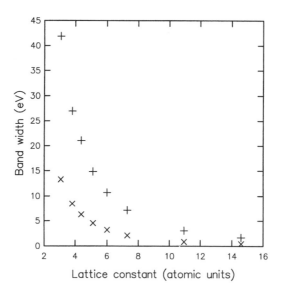

Fig. 9.15.   Energy band widths as a function of lattice constant for the lowest
($\times$) and second ($+$) energy bands in the relativistic Kronig–Penney model. To
produce this figure the lattice constant a was varied, b was always $10^{-5}a$ and the
potential was chosen such that $V_0 b = 4.1095 \times 10^{-4}$ relativistic units.

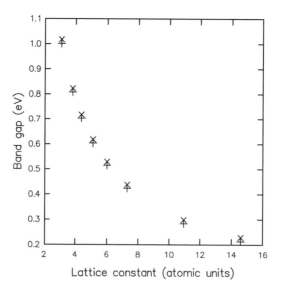

Fig. 9.16.   Energy gaps as a function of lattice constant for the lowest gap ($\times$)
and the second gap ($+$) in the relativistic Kronig–Penney model. To produce this
figure the lattice constant a was varied, b was always $10^{-5}a$ and the potential was
chosen such that $V_0 b = 4.1095 \times 10^{-4}$ relativistic units.

hand edge of region 1 in the neighbouring unit cell

$$\tilde{G}_a = \tilde{T}_b \tilde{R}^{-1} \tilde{T}_{a-b} \tilde{R} \tag{9.101}$$

Here $\tilde{R}$ gets the wavefunction over the barrier into region 2, $\tilde{T}_{a-b}$ translates the wavefunction through region 2 to the boundary with region 1, $\tilde{R}^{-1}$ gets the wavefunction from region 2, over the barrier to region 1 and $\tilde{T}_b$ translates the wave to the other side of region 1, and the wavefunction is back in a position equivalent to the one in which it started. If we label the first unit cell by the roman numeral I and the second by II we can write (Saxon and Hutner 1949)

$$\tilde{G}_a \phi_{(\rho),I} = \phi_{(\rho),II} \tag{9.102}$$

Our wavefunction exists in a lattice, so we can also impose the Bloch form upon it. If μ is our wavenumber we can write

$$e^{i\mu a} \phi_{(\rho),I} = \phi_{(\rho),II} \tag{9.103}$$

Let us write the 2×2 matrix $\tilde{G}_a$ as

$$\tilde{G}_a = \begin{pmatrix} G_{a11} & G_{a12} \\ G_{a21} & G_{a22} \end{pmatrix} \tag{9.104}$$

Next we eliminate $\phi_{(\rho),II}$ from (9.102) and (9.103), and choosing $x = 0$, write the resulting equation in component form using (9.99) and (9.104)

$$(G_{a11} - e^{i\mu a})A_\rho + G_{a12}B_\rho = 0 \tag{9.105a}$$

$$G_{a21}A_\rho + (G_{a22} - e^{i\mu a})B_\rho = 0 \tag{9.105b}$$

This has a non-trivial solution if the determinant of the coefficients of A_ρ and B_ρ is equal to zero, i.e. if

$$e^{2i\mu a} - (G_{a11} + G_{a22})e^{i\mu a} + G_{a11}G_{a22} - G_{a12}G_{a21} = 0 \tag{9.106}$$

It will be shown later that

$$\det \tilde{G}_a = 1 \tag{9.107}$$

Take my word for it, for now. Multiplying (9.106) through by $e^{-i\mu a}$ and using the definition of cosine in terms of complex exponentials leads to

$$\cos \mu a = \frac{1}{2}(G_{a11} + G_{a22}) = \frac{1}{2}\text{Trace } \tilde{G}_a \tag{9.108}$$

However, in equation (9.101) we have a definition of $\tilde{G}_a$ in terms of the input parameters into this problem. The R-matrices are given by equations (9.97) and (9.98). The transfer matrices for this case are

$$\tilde{T}_b = \begin{pmatrix} e^{ik_1 b} & 0 \\ 0 & e^{-ik_1 b} \end{pmatrix}, \qquad \tilde{T}_{a-b} = \begin{pmatrix} e^{ik_2(a-b)} & 0 \\ 0 & e^{-ik_2(a-b)} \end{pmatrix} \tag{9.109}$$

Multiplying all these matrices out according to (9.101) (Glasser and Davison 1970) gives

$$G_{a11} = \frac{1}{2\Gamma} e^{ik_1 b} (2\Gamma \cos k_2(a-b) + i(1+\Gamma^2) \sin k_2(a-b)) \qquad (9.110a)$$

$$G_{a12} = \frac{i}{2\Gamma} e^{ik_1 b} (1-\Gamma^2) \sin k_2(a-b) \qquad (9.110b)$$

$$G_{a21} = -\frac{i}{2\Gamma} e^{-ik_1 b} (1-\Gamma^2) \sin k_2(a-b) = G_{a12}^* \qquad (9.110c)$$

$$G_{a22} = \frac{1}{2\Gamma} e^{-ik_1 b} (2\Gamma \cos k_2(a-b) - i(1+\Gamma^2) \sin k_2(a-b)) = G_{a11}^* \qquad (9.110d)$$

It is fairly tedious to multiply out the brackets, but doing so, and using $\cos^2 \theta + \sin^2 \theta = 1$, yields

$$\det \tilde{G} = G_{a11} G_{a22} - G_{a12} G_{a21} = 1 \qquad (9.111)$$

Now, (9.110a and d) can be substituted into (9.108). Some simple manipulation leads to

$$\cos \mu a = \cos k_1 b \cos k_2 (a-b) - \frac{\Gamma^{-1} + \Gamma}{2} \sin k_1 b \sin k_2 (a-b) \qquad (9.112)$$

Equation (9.112) is the formula we have been deriving. It defines the allowed energy bands of our one-dimensional lattice, and agrees with an alternative derivation by Subramanian and Bhagwat (1971). The standard Kronig–Penney model, with δ-function potentials, comes from letting the square wells get progressively deeper and narrower. Equation (9.112) differs from one given by Fairbairn et al. (1973) because of the subtleties associated with the effect of δ-functions penetrating the negative energy states. Both Fairbairn's expression and equation (9.112) have the same non-relativistic limit. The left hand side of (9.112) is bounded by ± 1, but this is not the case for the right hand side. In regions where the magnitude of the right hand side is greater than 1 there are no wave-like solutions of the Dirac equation, as is shown graphically in figure 9.12.

Equation (9.112) is identical in form to its non-relativistic counterpart (Merzbacher 1970). The difference between the non-relativistic and relativistic Kronig–Penney models comes in the definitions of the parameters in (9.112). For example, the energy E is defined as $E = W - mc^2$ which is not the true non-relativistic energy unless $c \to \infty$. In figure 9.13 we show the lowest energy bands for a typical lattice constant and a series of almost δ-function wells. This figure shows that the allowed energy bands have exactly the same qualitative behaviour as is found on the basis of non-relativistic quantum theory. The bands are approximately parabolic, except at the Brillouin zone* boundaries where band gaps are opened up. Relativistic effects are too small to be observed on this figure. To illustrate the effect of relativity on the bands more dramatically we have plotted them for a one-dimensional crystal where the lattice constant is absurdly small, on the scale of real condensed matter, in figure 9.14. Note the energy scale is in MeV for this figure. It is clear from this picture that the band widths are dramatically reduced by relativity. The band gaps are also greatly reduced, although not in the linear way the band widths are. In the example chosen the lowest band gap (at π) reduces by a factor of about three, whereas the second gap (at 0) reduces by almost an order of magnitude. The relativistic reduction of the band width can also be observed in one-dimensional lattices of a size typical of crystalline materials in the following way. If one labels the bands by a band number, with the lowest band being number 1, then for high band numbers the relativistic bands appear rigidly below their non-relativistic counterparts owing to the cumulative effect of all the lower relativistic bands all being slightly narrower than the non-relativistic ones.

Figure 9.15 shows the widths of the lowest two energy bands as a function of lattice constant for a particular value of the potential. Clearly the band widths decrease as lattice constant increases, as one would expect. In figure 9.16 we show a similar diagram for the widths of the first two energy gaps. The band gaps also decrease with increasing lattice constant. However, the difference between the first two energy gaps is, to a first approximation, constant as a function of lattice constant.

We will stop our investigation of this model at this point. Extensions to diatomic (Sen Gupta 1974) and polyatomic (Dominguez-Adame 1989) crystals have been made. For further insight into this model the reader is referred to the original literature.

9.5 An Electron in Crossed Electric and Magnetic Fields

In chapter 4 we developed a two-component form of the Dirac equation (Feynman and Gell-Mann 1958) and in this example we use it to derive the eigenvectors and eigenvalues of an electron in a region of perpendicular constant electric and magnetic fields. Repeating (4.93), the equation is

$$\left(c^2(\hat{\mathbf{p}} - e\mathbf{A})^2 + m^2c^4 - ec\hbar\tilde{\boldsymbol{\sigma}} \cdot (c\mathbf{B} + i\mathbf{E})\right) \phi(\mathbf{r}, t) = (i\hbar\frac{\partial}{\partial t} - V(\mathbf{r}))^2\phi(\mathbf{r}, t)$$

$$(9.113)$$

---

* A volume in the reciprocal lattice may be defined by drawing lines connecting a reciprocal lattice point to all nearby reciprocal lattice points. The planes bisecting these lines enclose a volume in reciprocal space known as the Brillouin zone.

Following Lam (1970*a* and *b*) we start by setting up a frame of reference and we choose ours such that $\mathbf{B} = (0,0,B)$ and $\mathbf{E} = (0,E,0)$. With this choice of fields the potentials are

$$\mathbf{A} = \frac{1}{2}(-yB, xB, 0), \qquad \Phi = -yE \qquad (9.114a)$$

However, this turns out to be a slightly inconvenient form for the potentials and we can make a gauge transformation to a more useful form. Referring to equations (1.51) we can choose

$$\theta = \frac{1}{2}xyB \qquad (9.114b)$$

and our potentials in the new gauge become

$$\mathbf{A} = (-yB, 0, 0), \qquad \Phi = -yE \qquad (9.115)$$

We have chosen this gauge because y is the only space or time coordinate dependence in the potentials. Now we have to substitute (9.115) into (9.113) and we will assume a dependence on x, z, and t of the form

$$\phi(\mathbf{r}, t) = \phi(y)e^{i(p_x x + p_z z - Wt)/\hbar} = \begin{pmatrix} \phi_1(y) \\ \phi_2(y) \end{pmatrix} e^{i(p_x x + p_z z - Wt)/\hbar} \qquad (9.116)$$

where $\phi_{1(2)}(y)$ are single-component wavefunctions, p_x and p_z are the x- and z-components of momentum respectively and W is the total energy of the particle. Equation (9.113) becomes

$$\left(c^2(p_x + eyB)^2 + c^2\hat{p}_y^2 + c^2p_z^2 + m^2c^4 - \right.$$
$$\left. ec\hbar \left(\begin{pmatrix} cB & 0 \\ 0 & -cB \end{pmatrix} + \begin{pmatrix} 0 & E \\ -E & 0 \end{pmatrix} \right) \right) \phi(y) = (W + eyE)^2\phi(y) \qquad (9.117)$$

This is getting a bit complicated so we will simplify the notation by defining

$$\hat{Q} = c^2(p_x^2 + p_z^2 + \hat{p}_y^2) + m^2c^4 - 2ey(WE - c^2Bp_x) - W^2 + e^2y^2(c^2B^2 - E^2) \qquad (9.118)$$

Equation (9.117) can now be written in a much more succinct form

$$\left(\hat{Q} - ec^2\hbar B \right) \phi_1(y) - ec\hbar E\phi_2(y) = 0 \qquad (9.119a)$$

$$ec\hbar E\phi_1(y) + \left(\hat{Q} + ec^2\hbar B \right) \phi_2(y) = 0 \qquad (9.119b)$$

It is not straightforward to see how to proceed. Some very judicious changes of variable are required. Let

$$\eta = +(E^2 - c^2B^2)^{1/2} \qquad (9.120a)$$
$$\xi = (e\eta)^{1/2}(y - (c^2Bp_x - WE)/(e\eta^2)) \qquad (9.120b)$$

$$\tau = -\frac{i}{\hbar c}\xi^2 \tag{9.120c}$$

$$\zeta = \frac{1}{e\eta}\left(c^2(p_x^2 + p_z^2) + m^2c^4 - W^2 + \frac{(c^2Bp_x - WE)^2}{\eta^2}\right) \tag{9.120d}$$

In (9.120a) we are taking the positive square root. We have also assumed that both E and B are not equal to zero, although either one of them can be zero. Writing the operator $\hat{p}_y$ explicitly and making these substitutions in equation (9.119) leads to

$$\left(2ic\hbar\eta\left(\frac{\partial}{\partial\tau} + 2\tau\frac{\partial^2}{\partial\tau^2}\right) - i\hbar ce\eta\tau + e\eta\zeta - ec^2\hbar B\right)\phi_1(\tau) - ec\hbar E\phi_2(\tau) = 0 \tag{9.121a}$$

$$\left(2ic\hbar\eta\left(\frac{\partial}{\partial\tau} + 2\tau\frac{\partial^2}{\partial\tau^2}\right) - i\hbar ce\eta\tau + e\eta\zeta + ec^2\hbar B\right)\phi_2(\tau) + ec\hbar E\phi_1(\tau) = 0 \tag{9.121b}$$

Now let us try the solution

$$\phi_1(\tau) = e^{-\tau/2}u_1(\tau), \qquad \phi_2(\tau) = e^{-\tau/2}u_2(\tau) \tag{9.122}$$

Equations (9.121) then simplify further

$$\left(\tau\frac{\partial^2}{\partial\tau^2} + \left(\frac{1}{2} - \tau\right)\frac{\partial}{\partial\tau} - \frac{i\zeta}{4c\hbar} - \frac{1}{4} - \frac{cB}{4i\eta}\right)u_1(\tau) - \frac{E}{4i\eta}u_2(\tau) = 0 \tag{9.123a}$$

$$\frac{E}{4i\eta}u_1(\tau) + \left(\tau\frac{\partial^2}{\partial\tau^2} + \left(\frac{1}{2} - \tau\right)\frac{\partial}{\partial\tau} - \frac{i\zeta}{4c\hbar} - \frac{1}{4} + \frac{cB}{4i\eta}\right)u_2(\tau) = 0 \tag{9.123b}$$

These are still difficult to solve from first principles, but we will not take a rigorous mathematical stance on the solution. Rather we will write down a reasonable looking functional form for u_1 and see under what conditions it is a solution. We will insist that

$$u_1(\tau) = M\left(+\frac{i\zeta}{4c\hbar}, \frac{1}{2}, \tau\right) \tag{9.124}$$

where $M(a, b, z)$ is our old friend the confluent hypergeometric function. The reason for choosing this is apparent by comparing equation (9.123a) with the defining equation (B.1) of the confluent hypergeometric function in the appendix. Clearly there is a great deal of similarity between the two. If this is a correct form for $u_1(\tau)$ it must be true that

$$\tau\frac{\partial^2 u_1}{\partial\tau^2} + \left(\frac{1}{2} - \tau\right)\frac{\partial u_1}{\partial\tau} - \frac{i\zeta}{4c\hbar}u_1 = 0 \tag{9.125}$$

and for this to be consistent with (9.123a) we must have that

$$(i\eta + cB)u_1(\tau) + Eu_2(\tau) = 0 \tag{9.126a}$$

i.e.

$$\frac{u_1(\tau)}{u_2(\tau)} = -\frac{E}{i\eta + cB} \tag{9.126b}$$

So $u_1(\tau)$ and $u_2(\tau)$ differ from each other only by a multiplicative constant. To verify that these solutions really are valid, it is necessary to show that they are consistent with (9.123b) and we leave it as an exercise to show that

$$\begin{pmatrix} u_1(\tau) \\ u_2(\tau) \end{pmatrix} = A \begin{pmatrix} 1 \\ -\left(\frac{i\eta+cB}{E}\right) \end{pmatrix} M \left(+\frac{i\zeta}{4c\hbar}, \tfrac{1}{2}, \tau\right) = A \begin{pmatrix} M\left(\frac{i\zeta}{4c\hbar}, \tfrac{1}{2}, \tau\right) \\ M(\lambda_1, \tfrac{1}{2}, \tau) \end{pmatrix} \tag{9.127}$$

where A is the normalization constant and

$$\lambda_1 = \frac{i\zeta}{4c\hbar} + \frac{1}{4} - \frac{cB}{4i\eta} + \frac{E^2}{4i\eta(i\eta + cB)} \tag{9.128}$$

A second independent equation can be found if we insist that

$$u_1(\tau) = M \left(+\frac{i\zeta}{4c\hbar} + \tfrac{1}{2}, \tfrac{1}{2}, \tau\right) \tag{9.129}$$

and proceeding in the same way as above we find the solution

$$\begin{pmatrix} u_1(\tau) \\ u_2(\tau) \end{pmatrix} = C \begin{pmatrix} 1 \\ \left(\frac{i\eta-cB}{E}\right) \end{pmatrix} M \left(\frac{i\zeta}{4c\hbar} + \tfrac{1}{2}, \tfrac{1}{2}, \tau\right) = C \begin{pmatrix} M(\frac{i\zeta}{4c\hbar} + \tfrac{1}{2}, \tfrac{1}{2}, \tau) \\ M(\lambda_2, \tfrac{1}{2}, \tau) \end{pmatrix} \tag{9.130}$$

with C being the normalization constant and

$$\lambda_2 = \frac{i\zeta}{4c\hbar} + \frac{1}{4} - \frac{cB}{4i\eta} - \frac{E^2}{4i\eta(i\eta + cB)} \tag{9.131}$$

There are other solutions of this equation, but we will only require the two that we have found defined by equations (9.127) and (9.130). Up to this point we have simply been solving a differential equation, there has been no physics involved. Now that we have the wavefunctions in the form of a well-known function (which virtually nobody has ever heard of!), we can use it to calculate some useful quantities. We will make use of these mathematical results to derive the relativistic and quantum mechanical generalization of the familiar results for a particle in a magnetic field only and in crossed electric and magnetic fields.

An Electron in a Constant Magnetic Field

Let us start with the magnetic field case. If $E = 0$, equation (9.120a) tells us that η is purely imaginary. In that case ξ^2 is also purely imaginary and τ is real, from (9.120c). Referring to equation (9.125) and comparing it with equation (B.1) we see that it is a special case of the general form of Kummer's equation in that $b = 1/2$. However, it is a standard

mathematical identity that if $b = 1/2$ and $z = x^2$, the confluent hypergeometric function $M(a, b, z)$ reduces to a Hermite polynomial $H_n(x)$. Then
the wavefunctions looks very much like their non-relativistic counterparts
(Landau and Lifschitz 1977). If we want the wavefunction to tend to zero
as $z \to \infty$ (as we must for the wavefunction to be normalizable) we must
also have that $2n = -4a$ where n is a non-negative integer. Let us start by
simplifying equations (9.127) and (9.130) for the case when $E = 0$. Then
they become

$$\begin{pmatrix} u_1(\tau) \\ u_2(\tau) \end{pmatrix} = A \begin{pmatrix} 1 \\ 0 \end{pmatrix} M \left(\frac{i\zeta}{4c\hbar}, \tfrac{1}{2}, \tau \right) \tag{9.132a}$$

$$\begin{pmatrix} u_1(\tau) \\ u_2(\tau) \end{pmatrix} = C \begin{pmatrix} 0 \\ 1 \end{pmatrix} M \left(\frac{i\zeta}{4c\hbar} + \tfrac{1}{2}, \tfrac{1}{2}, \tau \right) \tag{9.132b}$$

This is suggestive, it implies that one of these solutions is spin up and the
other spin down. At this stage, though, this is not proven, as we are still
in the representation of the Dirac equation given in section 4.5.

The next step is to normalize the wavefunctions of equations (9.132). We
want to do this in the laboratory frame, and to have the wavefunctions
back in the usual representation. Therefore we have to refer back to
section 4.5. From equations (9.116), (9.122), (9.132) and (4.88) we can see
that

$$\chi^\uparrow(\mathbf{r}, t) = A e^{-\tau/2} \begin{pmatrix} 1 \\ 0 \\ -1 \\ 0 \end{pmatrix} M \left(\frac{i\zeta}{4c\hbar}, \tfrac{1}{2}, \tau \right) e^{i(p_x x + p_z z - Wt)/\hbar} \tag{9.133a}$$

$$\chi^\downarrow(\mathbf{r}, t) = C e^{-\tau/2} \begin{pmatrix} 0 \\ 1 \\ 0 \\ -1 \end{pmatrix} M \left(\frac{i\zeta}{4c\hbar} + \tfrac{1}{2}, \tfrac{1}{2}, \tau \right) e^{i(p_x x + p_z z - Wt)/\hbar} \tag{9.133b}$$

Now, as in the case of a plane wave, we have to normalize inside a cubic
box with sides of length L. Then if we insist that $\int \chi^\dagger \chi dV = 1$ we find

$$\chi_n^\uparrow(\mathbf{r}, t) = \frac{1}{L} \left(\frac{eB}{4\pi\hbar} \right)^{\frac{1}{4}} \begin{pmatrix} 1 \\ 0 \\ -1 \\ 0 \end{pmatrix} e^{-\frac{eB}{2\hbar}(y-K)^2} M(-n, \tfrac{1}{2}, \tau) e^{i(p_x x + p_z z - Wt)/\hbar}$$

$$\tag{9.134a}$$

$$\chi_n^\downarrow(\mathbf{r}, t) = \frac{1}{L} \left(\frac{eB}{4\pi\hbar} \right)^{\frac{1}{4}} \begin{pmatrix} 0 \\ 1 \\ 0 \\ -1 \end{pmatrix} e^{-\frac{eB}{2\hbar}(y-K)^2} M(-n, \tfrac{1}{2}, \tau) e^{i(p_x x + p_z z - Wt)/\hbar}$$

$$\tag{9.134b}$$

where

$$K = -\frac{p_x}{eB} \qquad (9.135)$$

In equations (9.134) we have normalized forms of the eigenfunctions defined in equation (4.88). If we insist that ψ of equation (4.87) is normalized to unity and the components of χ are defined in terms of the components of ψ as in (4.88) it is easy to show that

$$\psi_n^\uparrow(\mathbf{r}, t) = \frac{1}{L}\left(\frac{eB}{4\pi\hbar}\right)^{\frac{1}{4}} \begin{pmatrix} 1 \\ 0 \\ -1 \\ 0 \end{pmatrix} e^{-\frac{eB}{2\hbar}(y-K)^2} M(-n, \tfrac{1}{2}, \tau) e^{i(p_x x + p_z z - Wt)/\hbar}$$

$$(9.136a)$$

$$\psi_n^\uparrow(\mathbf{r}, t) = \frac{1}{L}\left(\frac{eB}{4\pi\hbar}\right)^{\frac{1}{4}} \begin{pmatrix} 0 \\ 1 \\ 0 \\ -1 \end{pmatrix} e^{-\frac{eB}{2\hbar}(y-K)^2} M(-n, \tfrac{1}{2}, \tau) e^{i(p_x x + p_z z - Wt)/\hbar}$$

$$(9.136b)$$

i.e. they are exactly the same as the χs. Finally we have the normalized form of the wavefunctions in the laboratory frame. Now we can operate on both of these with the $\hat{S}_z$ operator of equation (5.16) and it is trivial to see that (9.136a) has eigenvalue $\hbar/2$ and (9.136b) has eigenvalue $-\hbar/2$, as we surmised earlier. It is interesting to note that this is only true because we set $E = 0$. If the electric field had been non-zero, the zeros in the four-component part of the wavefunction would become non-zero and then these wavefunctions would no longer be eigenstates of $\hat{S}_z$.

Henceforth we will consider only the ground state of (9.136) for which $n = 0$. In that case the confluent hypergeometric function is 1 and so the eigenfunctions simplify considerably. We expect the electrons to be moving in a circular path. Let us find the radius of that path. This can be found simply by writing down the probability density and looking to see where it has a maximum. The result of this is independent of which of the two eigenfunctions we take.

$$\rho(\mathbf{r}, t) = \psi_n^\dagger(\mathbf{r}, t)\psi_n(\mathbf{r}, t) = \frac{1}{L^2}\left(\frac{eB}{\pi\hbar}\right)^{1/2} e^{-\frac{eB}{\hbar}(y^2 - 2Ky + K^2)} \qquad (9.137)$$

Differentiating this with respect to y brings down the derivative of the exponent. So, setting the derivative equal to zero, we have

$$\frac{\partial\rho(\mathbf{r}, t)}{\partial y} = -\frac{2eB}{\hbar}(y - K)\rho(\mathbf{r}, t) = 0 \qquad (9.138)$$

and this can only be generally true if

$$y = K = -\frac{p_x}{eB} \qquad (9.139)$$

This is exactly the classical result (except for the negative sign, but we are only interested in the magnitude of y). Equation (9.139) is true in Newtonian, and classical relativistic, theory, although, of course, the definition of momentum differs in these two cases. It tells us that an electron put into our field region at the origin with initial momentum p_x executes circular motion about point K with radius p_x/eB.

Now that we have the eigenfunctions (9.136), let's use them to calculate some observables for the $n = 0$ state. The actual calculation of the expectation values is fairly straightforward. We use (9.136) in an equation like (4.102). The limits on the integral are the size of the normalization box ($\pm L/2$ in all directions). In the y-direction we assume the box is so large that the limits can be replaced by $\pm\infty$. In these calculations the following standard integrals are required:

$$\int_{-\infty}^{\infty} e^{-ay^2+2by} dy = \sqrt{\frac{\pi}{a}} e^{b^2/4a} \tag{9.140a}$$

$$\int_{-\infty}^{\infty} ye^{-ay^2+2by} dy = \frac{b}{a}\sqrt{\frac{\pi}{a}} e^{b^2/a} \tag{9.140b}$$

$$\int_{-\infty}^{\infty} y^2 e^{-ay^2+2by} dy = \frac{1}{2a}\sqrt{\frac{\pi}{a}} \left(1+\frac{2b^2}{a}\right) e^{b^2/a} \tag{9.140c}$$

Firstly we find the expectation value of position. For $\langle \hat{x} \rangle$ and $\langle \hat{z} \rangle$ the integrations are trivial but they do not tell us anything. There is no x and z dependence in the integrand except the operator and the expectation values come out as the centre of the normalization box. However, for $\langle \hat{y} \rangle$ the calculation yields

$$\langle \hat{y} \rangle = \frac{p_x}{eB} \tag{9.141}$$

which is the non-relativistic expression for the coordinate of a particle with initial momentum p_x in a B-field, and is in full agreement with equation (9.139). The expectation values of the components of velocity are calculated using the operator given by (4.113a). This leads to

$$\langle \hat{v}_x \rangle = \langle \hat{v}_y \rangle = 0 \qquad\qquad \langle \hat{v}_z \rangle = c \tag{9.142}$$

This illustrates the *Zitterbewegung* again. It comes out as zero for the x- and y-components because of the circular motion. Next we will find the expectation values of the Cartesian components of momentum. As in equation (4.114) for the Lorentz force, we have to include the vector potential in our expectation value in the usual way. This is given by (9.115) and only has an x-component. Evaluation of the integrals gives

$$\langle \hat{p}_x - e\hat{A}_x \rangle = \langle \hat{p}_x + eB\hat{y} \rangle = 0 \qquad (9.143a)$$

$$\langle \hat{p}_y - e\hat{A}_y \rangle = \langle \hat{p}_y \rangle = 0 \qquad (9.143b)$$

$$\langle \hat{p}_z - e\hat{A}_z \rangle = \langle \hat{p}_z \rangle = p_{z_0} \qquad (9.143c)$$

So the average value of the z-component of momentum does not change from its initial value. The x- and y-components of both velocity and momentum are zero. This is consistent with the classical view of electron motion in a constant field. The electron moves in a circle in a plane perpendicular to the field direction. In circular motion the velocity at one point on the circle is the negative of the velocity on the diametrically opposed point, so the velocity and momentum should average to zero.

Finally let us find $\langle \hat{L}_z \rangle$. This is quite a long and tedious calculation, but it boils down to

$$\langle \hat{L}_z \rangle = \langle (\hat{\mathbf{r}} \times (\hat{\mathbf{p}} - e\hat{A}))_z \rangle = -\frac{\hbar}{2} \qquad (9.144)$$

This is quite a surprising result, but can be interpreted fairly readily if we divide the expectation value into two.

$$\langle (\hat{\mathbf{r}} \times \hat{\mathbf{p}})_z \rangle = \frac{p_x^2}{eB} \qquad (9.145a)$$

$$-e\langle (\hat{\mathbf{r}} \times \hat{A})_z \rangle = -\frac{p_x^2}{eB} - \frac{\hbar}{2} \qquad (9.145b)$$

Clearly equation (9.145a) represents the classical angular momentum of the particle. It is just $p_x \times \langle \hat{y} \rangle$. We have evaluated it for $n = 0$, but we would expect this for any value of n as we have circular motion. Now, the field term of (9.145b) exactly cancels (9.145a). This must occur for any value of n because angular momentum has to be quantized in units of $\hbar/2$. For the $n = 0$ eigenfunction, the Hermite polynomial is unity, so the wavefunction (9.136) has its peak at the peak of the exponential, i.e. at $\langle y \rangle = K$, the centre of the circular motion. This is to be expected for a particle with zero orbital angular momentum. It is analogous to what happens for the s-states in the non-relativistic hydrogen atom. For $n = 0$ the finite size of the orbital motion is due to the zero point energy and it is tempting to attribute the $\hbar/2$ to the free particle spin. But this is not correct, because the derivation of the eigenfunctions we used to calculate $\langle \hat{L}_z \rangle$ is invalid if $|E| = |B| = 0$. Equations (9.144) and (9.145) imply that the angular momentum for a particular value of n is constant regardless of the strength of the applied field. Hence, as $\langle \hat{y} \rangle$ decreases with increasing field, the true velocity of the particles must increase. For higher values of n the angular momentum will be greater, but the expectation value of position will still be given by (9.141), although in this case the expectation

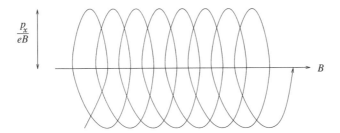

Fig. 9.17. Motion of electron in magnetic field.

value of position will not necessarily correspond with the peaks in the wavefunction. Finally we can conclude that the motion of the electron in the field is as shown in figure 9.17.

Next let us consider the energy eigenvalues of these two eigenfunctions. From the condition that $n = -2a$ above we have

$$i\zeta = -2c\hbar n \qquad (9.146a)$$

for the eigenvalue associated with (9.136a) and

$$i\zeta = -2c\hbar(n+1) \qquad (9.146b)$$

for the eigenvalue associated with (9.136b). The definition of ζ is given in equation (9.120d) and contains the energy eigenvalue W. As $E = 0$ this can be substituted into (9.146) to give expressions for the eigenvalues:

$$W_n^{\uparrow 2} = p_z^2 c^2 + m^2 c^4 + 2nc^2 \hbar e B \qquad (9.147a)$$
$$W_n^{\downarrow 2} = p_z^2 c^2 + m^2 c^4 + 2(n+1)c^2 \hbar e B \qquad (9.147b)$$

where the terms in p_x have cancelled. Let us take the non-relativistic limit of these two expressions. We do this in the usual way, by letting $W_n = E_n + mc^2$, where E_n is the non-relativistic energy, and we find

$$E_n^{\uparrow} = \frac{p_z^2}{2m} + n\frac{e\hbar}{m}B \qquad (9.148a)$$

$$E_n^{\downarrow} = \frac{p_z^2}{2m} + (n+1)\frac{e\hbar}{m}B \qquad (9.148b)$$

Equations (9.148) can be written in several different forms which highlight various aspects of these eigenvalues:

$$E_n^{\uparrow} = \frac{p_z^2}{2m} + 2n\mu_B B = \frac{p_z^2}{2m} + n\hbar\omega_c \qquad (9.149a)$$

$$E_n^{\downarrow} = \frac{p_z^2}{2m} + (2n+2)\mu_B B = \frac{p_z^2}{2m} + (n+1)\hbar\omega_c \qquad (9.149b)$$

The left hand expression in (9.149) shows that the difference in energy between a spin-up electron and a spin-down electron in a magnetic field (with the same value of n) is, in the non-relativistic limit,

$$\Delta E_n = 2\mu_B B \tag{9.150}$$

as we would expect. We have defined the cyclotron frequency

$$\omega_c = \frac{eB}{m} \tag{9.151}$$

The expressions on the right hand side of (9.149) are not familiar in themselves. However, if we take a particle for which $p_z = 0$ and average over the spin direction we get

$$E_n = (n + 1/2)\hbar\omega_c \tag{9.152}$$

This, of course, is the expression for the energy levels of a non-relativistic harmonic oscillator. More relevantly for us, it is also the non-relativistic expression for the energy eigenvalues associated with Landau levels, the quantized orbits of an electron in a magnetic field (Kittel 1963). In terms of the cyclotron frequency the energy eigenvalues (9.147) become

$$W_n^{\uparrow 2} = p_z^2 c^2 + m^2 c^4 + 2nmc^2\hbar\omega_c \tag{9.153a}$$
$$W_n^{\downarrow 2} = p_z^2 c^2 + m^2 c^4 + 2(n+1)mc^2\hbar\omega_c \tag{9.153b}$$

In fact our method of averaging over the spin to obtain (9.152) was too simplistic. Consider (9.149a) for the case $p_z = 0$ and $n = 0$. Obviously this leads to $E_n = 0$ although we know that a stationary spin-up particle in a magnetic field should have energy

$$E_g = -\mu_B B = -\frac{e\hbar}{2m} B = -\frac{1}{2}\hbar\omega_c \tag{9.154}$$

and we see that the orientational energy of the spin in the magnetic field exactly cancels the zero point energy associated with the orbital motion. For an electron with spin down the energy in the ground state should be $+\frac{1}{2}\hbar\omega_c$. So, the total energy in the ground state should be the zero point energy plus $\frac{1}{2}\hbar\omega_c$ and this is confirmed by (9.149b).

What is surprising about this analysis is that the energy due to the spin of the electron has precisely the same magnitude as the zero-point energy associated with the orbital motion. However, this is in accord with the earlier calculation of the expectation value of the z-component of angular momentum as $-\hbar/2$, the same magnitude as the spin. As we have seen earlier, it is the angular momentum that defines the value of n and hence the energy of the electron. Hence this cancellation is simply a consequence of the quantization of angular momentum, and the fact that both the

$-\hbar/2$ of equation (9.144) and the spin can, in some sense, be regarded as zero-point angular momenta.

We have considered a single electron in a magnetic field so far. In any material there are many electrons. If we apply a magnetic field to a metal, say, the electrons will fall into Landau levels and these turn out to be highly degenerate. This is because electrons with the same value of n, but different values of $\langle \hat{y} \rangle$, have the same energy. Furthermore, the x- and y-coordinates cannot both be well defined at the same time because of the uncertainty principle $\Delta x \Delta p_x \geq \hbar/2$, and $\langle \hat{y} \rangle$ is directly proportional to p_x via equation (9.139). So the degeneracy is equal to the number of possible values of $\langle \hat{y} \rangle$ which is equal to the number of possible values of $p_x = \hbar k_x$.

Let us consider the free electron theory of metals. Imagine the electrons are in a cubic sample with sides of length L. From elementary solid state physics we know that, in such a sample, the allowed values of k_x are separated by a length in reciprocal space of $2\pi/L$. So the separation in the allowed values of $\langle \hat{y} \rangle$ is

$$\Delta \langle \hat{y} \rangle = \frac{\hbar(k_x + 2\pi/L)}{eB} - \frac{\hbar k_x}{eB} = \frac{h}{eBL} \tag{9.155}$$

and, for each spin direction, the number of allowed values of $\langle \hat{y} \rangle$ in our lump of material is

$$N = \frac{L}{\Delta \langle \hat{y} \rangle} = \frac{eBL^2}{h} = \frac{e\Phi}{h} \tag{9.156}$$

where $\Phi = BL^2$ is the magnetic flux passing through the sample (McMurry 1993). Each Landau level contains N states. For $1\,\mathrm{cm}^3$ of material and a field of 1 tesla the degeneracy is $N \approx 10^{10}$. Note that this is a non-relativistic calculation of N, but, qualitatively, for an observer moving with respect to the cubic sample, $B \to \gamma B$ from (1.58b) and $L \to L/\gamma$ from (1.7) in the direction of motion only, so N will remain unchanged.

Our picture of the motion of free electrons in a magnetic field is now fairly complete. The expectation values calculated above indicate circular motion in a plane perpendicular to the field with a radius and angular frequency in full agreement with classical mechanics and with non-relativistic quantum mechanics. The energy levels are quantized in accordance with equations (9.153). We have also seen that the Landau levels are highly degenerate.

Let us do something that isn't often done in relativistic quantum theory books. We will calculate some real numbers. We will find, explicitly, the separation between energy eigenvalues for the Landau levels in both the relativistic theory from equations (9.153) and the non-relativistic theory from (9.152) to see if any effects of relativity are observable. Let us take a

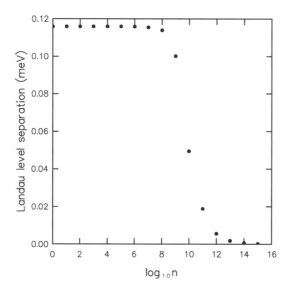

Fig. 9.18. Landau level separation as a function of $\log_{10} n$ for an electron in a magnetic field.

field of 1 tesla. In that case the non-relativistic separation between energy levels is

$$\Delta E = \hbar \omega_c = 1.855 \times 10^{-23} \, \text{J} = 1.159 \times 10^{-4} \, \text{eV} \qquad (9.157)$$

This is a constant separation independent of the size of the energy. In the relativistic case it is the square of the energy that is proportional to the quantum number. Therefore the eigenvalues are not equally spaced. Numerical difficulties make it difficult to calculate the level spacing for low values of n in (9.153) because the rest mass energy completely dominates the other terms. However, we find that

$$n \approx 10^6 \qquad \Delta W = 1.855 \times 10^{-23} \, \text{J} = 1.158 \times 10^{-4} \, \text{eV} \qquad (9.158a)$$

$$n \approx 10^8 \qquad \Delta W = 1.824 \times 10^{-23} \, \text{J} = 1.139 \times 10^{-4} \, \text{eV} \qquad (9.158b)$$

$$n \approx 10^9 \qquad \Delta W = 1.604 \times 10^{-23} \, \text{J} = 1.001 \times 10^{-4} \, \text{eV} \qquad (9.158c)$$

$$n \approx 10^{11} \qquad \Delta W = 3.012 \times 10^{-24} \, \text{J} = 1.880 \times 10^{-5} \, \text{eV} \qquad (9.158d)$$

Equations (9.158a − d) cover the important energy range, from $n \approx 10^6$ where the rest mass energy dominates equation (9.153) to $n \approx 10^{11}$ where the rest mass energy is dwarfed by energy associated with the circular motion. Figure 9.18 shows the spacing of the eigenvalues as a function of the logarithm of the principal quantum number n. At low values of the quantum number the spacing is constant at its non-relativistic value (the low values ($\log_{10} n < 5$) in the figure were found by extrapolation from the non-relativistic limit). When the energy due to the motion in the field is comparable with the rest energy of the electron the level spacing drops fairly dramatically (actually over about four orders of magnitude), and finally levels off, although it approaches zero logarithmically. Non-relativistically this graph would be flat at its low n value all the way across, so this is clearly an impressive example of the effect of relativity on the behaviour of electrons. To observe the decrease in Landau level spacing at low values of the principal quantum number would require a magnetic field of $10^9 - 10^{10}$ tesla!

Equations (9.153) show us that the lowest energy Landau level ($n = 0$) contains only electrons with their spin parallel to the field direction. After that the spin-up eigenvalue with quantum number n has the same energy as the spin-down eigenvalue with quantum number $n-1$. So for all except the lowest energy level we expect to find degenerate levels associated with spin-up and spin-down states. Now, from (9.151) and (9.153) we see that the energy levels are proportional to the applied field. Equation (9.156) tells us that the degeneracy is also directly proportional to the applied field. With this information we can predict what will happen as we increase the applied field. At some initial field all the Landau levels in a metal are filled up to, and including, the one with quantum number n_{max}. As the field is increased, the degeneracy of lower levels, both spin up and spin down, increases, so electrons with quantum number n_{max} are able to fall into lower Landau levels. This happens continuously until the n_{max} Landau level is empty. When that occurs the maximum energy of an electron drops suddenly by an amount given by an equation like (9.158). Then as the field increases further the energy of that level will increase until it too is empty and the energy jumps down again. This is shown schematically in figure 9.19. The drop into a lower energy state may occur with or without a spin flip. If the latter occurs, a mechanism for getting rid of the necessary angular momentum must be available, such as exchanging it with an electron flipping the other way. The highest electron energy varies periodically with $1/B$. So in a metal with free or nearly free electrons in a magnetic field the band structure is broken up in the direction of the field as the electrons fall into Landau levels. As a function of increasing field the mean energy of the highest electrons oscillates about the zero field Fermi energy.

Many properties of metals depend upon the behaviour of the nearly free electrons close to the Fermi energy. Nearly free electrons can be treated as free electrons with an effective mass m^* (Kittel 1986). So all the equations of the free-electron theory of metals may be applied (as a

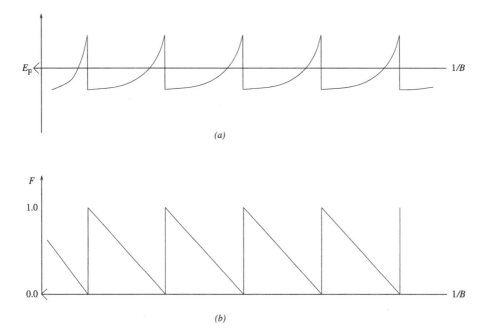

(a)

(b)

Fig. 9.19.   (a) Schematic figure showing the energy of the highest filled Landau level as a function of $1/B$. The horizontal line represents the Fermi energy in zero field. The energy and degeneracy gradually increase as a function of field until the top level is empty, then there is a sudden drop in the energy of the highest level. (b) The fraction of filled states in the highest occupied Landau level also as a function of $1/B$ and drawn to correspond with (a). Note that B increases to the right in both these figures.

first approximation) with m replaced by m^*. Any property of metals that depends upon the behaviour of the conduction electrons may be expected to display some oscillatory behaviour in a magnetic field. One of the most famous examples of this is the oscillation in the magnetization of a metal as a function of $1/B$ which is known as the de Haas–van Alphen effect. As stated earlier the energy spacing of the Landau levels is of the same order as the thermal energy, so to observe the de Haas–van Alphen effect requires low temperatures and high fields so that $\hbar\omega_c \gg k_B T$.

Another well-known illustration of the properties of Landau levels is provided by the quantum Hall effect. In elementary solid state physics the Hall effect is the potential difference that appears across two faces of a conductor when a current flows across it because of a transverse magnetic field (Kittel 1986). The Hall resistance is given by

$$\rho = R_H B = \frac{1}{n_e e} B \qquad (9.159)$$

where n_e is the density of carriers of charge e and R_H is known as the Hall coefficient. One expects the density of carriers to be constant, so the Hall resistance should rise linearly with applied field. However, in very thin (a few atomic layers) films of semiconductors, which are effectively two-dimensional, the Hall resistance does not rise linearly, rather a series of plateaus is observed separated by very sharp rises, so that the graph of ρ against field looks more like a staircase. The plateaus occur at values of ρ given by

$$\rho = \frac{h}{le^2} \tag{9.160}$$

where l is an integer. Substituting this into (9.159) gives

$$n_e = \frac{leB}{h} \tag{9.161}$$

If our two-dimensional sample is square with sides of length L the total number of charge carriers must be

$$N = L^2 n_e = \frac{leBL^2}{h} \tag{9.162}$$

Compare this with equation (9.156). Equation (9.162) is just the right number of electrons to fill up to, and including, the lth Landau level. So each plateau in the Hall resistance corresponds to a completely filled Landau level.

In conclusion to this section, then, any property of a metal depending on the energy of the electrons close to the Fermi level will show oscillatory behaviour as a function of applied field. At low fields non-relativistic theory is entirely adequate to describe the physics involved, but in extremely large fields, when $n\hbar\omega_c \approx mc^2$, we expect the behaviour of the electrons to diverge from the non-relativistic predictions. In particular although the degeneracy is unaffected by relativity the Landau levels get closer together and hence the plateaus in the quantum Hall resistance will be closer together than the non-relativistic theory predicts, and their separation will eventually go, logarithmically, to zero. Also, the frequency of the de Haas–van Alphen oscillations will become more rapid.

An Electron in a Field for which $|E| = c|B|$

Now we are going to consider a further limit of the equations in the first part of section 9.5. This is another exactly soluble problem. Here we will content ourselves with finding the eigenfunctions only, in the case when

$$|E| = c|B| \tag{9.163}$$

With this constraint $\hat{Q}$ of equation (9.118) reduces to

$$\hat{Q} = c^2(p_x^2 + \hat{p}_y^2 + p_z^2) - 2eycB(W - cp_x) + m^2c^4 - W^2 \qquad (9.164)$$

and using (9.163) we find that equations (9.119) become

$$\left(\hat{Q} - ec^2\hbar B\right)\phi_1(y) - ec^2\hbar B\phi_2(y) = 0 \qquad (9.165a)$$

$$ec^2\hbar B\phi_1(y) + \left(\hat{Q} + ec^2\hbar B\right)\phi_2(y) = 0 \qquad (9.165b)$$

If we add these two equations we get

$$\hat{Q}\phi_1(y) + \hat{Q}\phi_2(y) = 0 \qquad (9.166)$$

This can only be true if $\phi_1(y) = -\phi_2(y)$, and substituting this into (9.165) gives

$$\hat{Q}\phi_1(y) = 0 \qquad (9.167)$$

Remembering that $\hat{p}_y$ in (9.164) is an operator whereas p_x and p_z are eigenvalues we can write

$$\frac{\partial^2}{\partial y^2}\phi_1(y) - (a + by)\phi_1(y) = 0 \qquad (9.168)$$

where

$$a = (p_x^2 + p_z^2 + m^2c^2 - W^2/c^2)/\hbar^2, \qquad b = \frac{2eB}{c\hbar^2}(cp_x - W) \quad (9.169)$$

The next step is to make the change of variable

$$\xi = \left(\frac{a}{b} + y\right)b^{1/3} \qquad (9.170)$$

and then equation (9.168) becomes

$$\frac{\partial^2}{\partial \xi^2}\phi_1(\xi) - \xi\phi_1(\xi) = 0 \qquad (9.171)$$

This equation is the the one we were looking for: it is the defining equation for the Airy function Ai(ξ) (Abramowitz and Stegun 1972). Substituting this solution back into (9.116) yields

$$\phi(\mathbf{r}, t) = \begin{pmatrix} \mathrm{Ai}((\frac{a}{b} + y)b^{1/3}) \\ -\mathrm{Ai}((\frac{a}{b} + y)b^{1/3}) \end{pmatrix} e^{i(p_x x + p_z z - Wt)/\hbar} \qquad (9.172)$$

and hence, from equation (4.88)

$$\chi(\mathbf{r}, t) = \begin{pmatrix} \mathrm{Ai}((\frac{a}{b} + y)b^{1/3}) \\ -\mathrm{Ai}((\frac{a}{b} + y)b^{1/3}) \\ -\mathrm{Ai}((\frac{a}{b} + y)b^{1/3}) \\ \mathrm{Ai}((\frac{a}{b} + y)b^{1/3}) \end{pmatrix} e^{i(p_x x + p_z z - Wt)/\hbar} \qquad (9.173)$$

In this case the electron is not bound and the energy levels are not quantized (Lam 1970a). The Airy function has fascinating properties. Its asymptotic forms are

$$\lim_{\xi \to \infty} \text{Ai}(\xi) \approx \frac{1}{2} \xi^{-1/4} e^{-(2/3)\xi^{3/2}} \tag{9.174}$$

and

$$\lim_{\xi \to -\infty} \text{Ai}(\xi) = |\xi|^{-1/4} \sin \left(\frac{2}{3} |\xi|^{3/2} + \pi/4 \right) \tag{9.175}$$

so as $\xi \to \infty$ the Airy function diminishes exponentially, and as $\xi \to -\infty$ it is oscillatory. If b is negative this indicates that as $y \to -\infty$ the wavefunction decreases exponentially. This is what we would expect; the electron should be accelerated in the direction of the field, and the probability of finding it behind its classically expected position should decrease exponentially. As y becomes large and positive the probability density is oscillatory and the motion of the electron maps out a cycloid. Qualitatively, such behaviour can be more easily understood on the basis of the Lorentz transformations (1.58). Initially the electron is accelerated by the electric field. However, the faster the electron moves, the greater the magnetic field it feels in the direction perpendicular to its motion. When this field becomes great enough the electron will have a tendency to become trapped in Landau levels. The average velocity parallel to the electric field then decreases and hence the magnetic field felt by the electron decreases, and it can be accelerated by the electric field again. Obviously, the energy of the electron cannot go on increasing forever; as an accelerating charge it can emit its energy as bremsstrahlung.

9.6 Non-Linear Dirac Equations, the Dirac Soliton

So far we have solved the Dirac equation in cases where the potentials are simple functions of position. The Dirac equation has been linear and relatively easy to solve. Here we are going to solve a non-linear Dirac equation. Our reasons for doing this are twofold. Firstly, the model chosen is interesting for good physical reasons in its own right, and secondly, in the following chapter we will be solving rather complicated non-linear versions of the Dirac equation. The model we are going to look at here is one of the simplest non-linear Dirac equations (Nogami and Toyama 1992), so this will serve as some sort of introduction to the following chapter.

Let us write the Dirac equation with a scalar potential (yet again):

$$i\hbar \frac{\partial}{\partial t} \psi(\mathbf{r}, t) = (c\tilde{\boldsymbol{\alpha}} \cdot \hat{\mathbf{p}} + \tilde{\beta} mc^2 + V(\mathbf{r})) \psi(\mathbf{r}, t) \tag{9.176}$$

To make life as easy as possible we will work in one dimension. As we have seen earlier, in that case we only require two anticommuting matrices so the Dirac equation reduces to two-component form. For this example we choose the y-direction as our coordinate as that simplifies the mathematics by making the Dirac equation real. We will introduce our non-linearity by making the potential $V(y)$ a function of our wavefunction

$$V(y) = -g(\psi^\dagger(y, t)\tilde{\beta}\psi(y, t))\tilde{\beta} \tag{9.177}$$

and the Dirac equation can now be written in matrix form as

$$
i\hbar \frac{\partial}{\partial t}\psi(y, t)
$$
$$
= \left(c\hbar \begin{pmatrix} 0 & -1 \\ 1 & 0 \end{pmatrix} \frac{\partial}{\partial y} + \begin{pmatrix} 1 & 0 \\ 0 & -1 \end{pmatrix} mc^2 - g(\psi^\dagger(y, t)\tilde{\beta}\psi(y, t))\tilde{\beta} \right) \psi(y, t) \tag{9.178}
$$

where g is just a coupling constant (with units of energy times distance) which defines the strength of the potential. This is a scalar non-linearity which describes a soliton.* We will examine the Dirac soliton in its rest frame, in which case the wavefunction can be written

$$\psi(y, t) = \phi(y)e^{-iWt/\hbar} \tag{9.179}$$

This means that $V(y)$ is independent of time and the Dirac equation can be separated in the usual way.

$$
\left(c\hbar \begin{pmatrix} 0 & -1 \\ 1 & 0 \end{pmatrix} \frac{\partial}{\partial y} + \begin{pmatrix} 1 & 0 \\ 0 & -1 \end{pmatrix} mc^2 - g(\phi^\dagger(y)\tilde{\beta}\phi(y))\tilde{\beta} \right) \phi(y) = W\phi(y) \tag{9.180}
$$

Equation (9.180) is horrendous to solve. Let's take the coward's way out and guess a form for $\phi(y)$ and then substitute it in to see under what circumstances it actually is a solution of (9.180):

$$\phi(y) = \frac{(kW)^{1/2}}{mc^2 + W\cosh 2ky} \begin{pmatrix} (mc^2 + W)^{1/2}\cosh ky \\ -(mc^2 - W)^{1/2}\sinh ky \end{pmatrix} \tag{9.181}$$

I didn't actually guess this, I read it in a paper (Nogami *et al.* 1995). The first thing to do is evaluate $V(y)$, and this comes out fairly simply as

$$V(y) = -\frac{gkW}{(mc^2 + W\cosh 2ky)} \begin{pmatrix} 1 & 0 \\ 0 & -1 \end{pmatrix} \tag{9.182}$$

Next, we substitute (9.181) and (9.182) into (9.180). This becomes a bit messy, but there is no problem in doing it. The derivative of $\phi(y)$ is straightforward to evaluate using the quotient rule. The equations get too

---

* A soliton is a single solitary wave which travels through a medium without dissipation (Drazin and Johnson 1989).

long for a while to write out in this book. However, after substitution of (9.181) and (9.182) into the Dirac equation we let $ky = x$, multiply through by $(mc^2 + W \cosh 2x)^2$, and then write out the upper and lower components of the Dirac equation separately:

$$0 = \hbar ck(mc^2 - W)^{\frac{1}{2}}(mc^2 \cosh x + W \cosh 2x \cosh x - 2W \sinh x \sinh 2x)$$

$$+ W(mc^2 + W)^{\frac{1}{2}} \left[\frac{m^2 c^4}{W} + mc^2 \cosh 2x - gk - mc^2 - W \cosh 2x \right] \cosh x$$

$$(9.183a)$$

$$0 = \hbar ck(mc^2 + W)^{\frac{1}{2}}(mc^2 \sinh x + W \cosh 2x \sinh x - 2W \cosh x \sinh 2x)$$

$$+ W(mc^2 - W)^{\frac{1}{2}} \left[\frac{m^2 c^4}{W} + mc^2 \cosh 2x - gk + mc^2 + W \cosh 2x \right] \sinh x$$

$$(9.183b)$$

These can be reduced a lot further if we multiply equation (9.183a) by $(mc^2+W)^{1/2} \sinh ky$ and (9.183b) by $(mc^2-W)^{1/2} \cosh ky$ and then subtract the latter from the former. We then obtain the relatively simple equation:

$$2\hbar ck W(m^2 c^4 - W^2)^{1/2} \sinh 2ky - 2gk W^2 \cosh ky \sinh ky = 0 \quad (9.184)$$

Finally, use of simple hyperbolic function identities and some further cancellation leads to

$$W = mc^2 \left(1 + \frac{g^2}{4\hbar^2 c^2} \right)^{-1/2} \quad (9.185)$$

So equation (9.181) is a solution of the non-linear Dirac equation (9.180) provided the energy eigenvalue is written in terms of the coupling constant g according to equation (9.185). But the solution is not yet complete because there has been no definition of k. We can rearrange (9.185) and define the wavevector k by

$$\frac{g^2 W^2}{4\hbar^2 c^2} = m^2 c^4 - W^2 = p^2 c^2 = \hbar^2 k^2 c^2 \quad (9.186)$$

and so

$$k = \frac{g W}{2\hbar^2 c^2} \quad (9.187)$$

Next we need to consider the normalization of (9.179). We want to show that it satisfies

$$\int_{-\infty}^{\infty} \psi^\dagger(y, t)\psi(y, t)dy = \int_{-\infty}^{\infty} \phi^\dagger(y)\phi(y)dy = 1 \quad (9.188)$$

This is actually very difficult to do from first principles. However, if we note that $\cosh x$ is an even function of x and make use of a very obscure standard integral:

$$\int_0^\infty \frac{A \cosh x + 1}{(A + \cosh x)^2} dx = 1 \tag{9.189}$$

we can verify equation (9.188) with the wavefunction (9.179) and (9.181) in a few lines.

Finally let us calculate the expectation value of the energy in this state.

$$\langle \hat{H} \rangle = \int_{-\infty}^\infty \phi^\dagger(y) \hat{H} \phi(y) dy$$

$$= \int_{-\infty}^\infty \left(\phi^\dagger(y) \left(-ic\hbar\tilde{\sigma}_y \frac{\partial}{\partial y} + \tilde{\sigma}_z mc^2 \right) \phi(y) - \frac{g}{2} (\phi^\dagger(y) \tilde{\beta} \phi(y))^2 \right) dy \tag{9.190}$$

When we put the wavefunctions into this integral there is an amazing amount of cancellation and it finally reduces to

$$\langle \hat{H} \rangle = 2mc^2 kW \int_0^\infty \frac{dy}{mc^2 + W \cosh 2ky} \tag{9.191}$$

This is a standard integral which can be looked up, and we find

$$\langle \hat{H} \rangle = \frac{2mc^3\hbar}{g} \ln \left(\frac{mc^2 + W + \hbar ck}{mc^2 + W - \hbar ck} \right) \tag{9.192}$$

Now, we assumed our soliton was at rest at the beginning of this analysis, so this energy is something like the rest mass energy of the soliton. Rearranging the constants a bit means we can write (9.192) as

$$\langle Mc^2 \rangle = \frac{2m^2 c^4 \lambda_C}{g} \ln \left(\frac{mc^2 + W + \hbar ck}{mc^2 + W - \hbar ck} \right) \tag{9.193}$$

You might well be wondering what all this means. Well, the probability density associated with our soliton solution (equation (9.181)) is shown in figure 9.20. Although we were initially considering a particle of mass m, letting the potential depend upon the wavefunction explicitly as we have done here has led to the physical state under consideration being some sort of excitation of the basic single-particle wavefunction. Allowing the potential to depend upon the wavefunction has given the solution the freedom to 'excite' itself to a state in which it has effective mass M.

It is a very good exercise to perform a Lorentz transformation of the type described in section 5.7 on the wavefunction above and repeat the above calculation to find the energy of a moving Dirac soliton. If one does this one has to do a very large amount of tricky mathematics, but one eventually finds that

$$\langle \hat{H} \rangle = \gamma Mc^2, \qquad \langle \hat{p}_y \rangle = \frac{v \langle \hat{H} \rangle}{c^2} \tag{9.194}$$

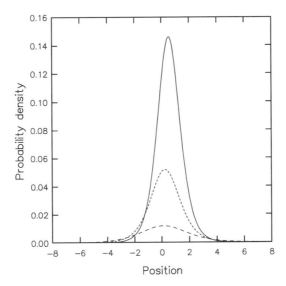

Fig. 9.20. The probability density for the Dirac soliton for $g = 5$ (full line), $g = 2$ (short dashed line), and $g = 1$ (long dashed line). Note the slight offset of the maximum from $y = 0$ (the probability density is not an even function) and the slight lack of symmetry about the peak. All quantities are in relativistic units.

where γ is the usual relativistic factor of equation (1.5). Equations (9.194) are just the expressions we would expect for a free particle of mass M. So, we have the amazing result that a particle of mass m under the potential (9.183) behaves like a free particle of mass M. Although there is a potential barrier (well) here, the particle passes right through it as though it wasn't there, i.e. we have a reflectionless quantum mechanical potential (Stahlhofen 1994, Nogami and Toyama 1992)!

This example may have been defined mathematically only, but it serves its purpose well, illustrating vividly that physical situations described by non-linear Dirac equations can, in principle, show an enormous range of different physical behaviours.

9.7 Problems

(1) Suppose that in the particle in a box problem of section 9.1 the particle was incident from the left with energy above the level of the box. Then we have

$$\frac{C_I}{A_I} = \sqrt{T}e^{-i\delta}$$

Find expressions for the transmission coefficient T and the phase shift δ in terms of the parameters defined in section 9.1.

(2) Re-analyse the particle in a one-dimensional potential well if the depth of the potential well exceeds $2mc^2$.

(3) Calculate the expectation value $\langle \hat{\mathbf{p}} \cdot \hat{\mathbf{r}} \rangle$ for the ground state of the Dirac oscillator and show that it is independent of whether $j = l + 1/2$ or $j = l - 1/2$ for $j = 1/2$.

(4) Verify equations (9.127) to (9.131).

(5) Solve the Dirac equation for an electron in an electric field E. How would you normalize the eigenfunctions?

(6) Perform a Lorentz transformation of the type described in section 5.7 on the wavefunction (9.181) to find the soliton wavefunction in a frame moving with velocity $\mathbf{v}$ relative to the rest frame of the soliton. Verify that (9.194) is the correct expression for the energy of the soliton in this frame.

(7) Find the eigenvectors and eigenvalues of a particle obeying the Dirac equation with a linear scalar potential depending on the z-coordinate only $V(r) = Az$ where A is a constant.

10

More Than One Electron

With the exception of the latter half of chapter 6 we have, up to this point, been discussing single-particle quantum mechanics. This is easy (although you may not think so). One can certainly gain a lot of insight and understanding from a single-particle theory. However, in real life there are very few (no) situations in physics in which a single-particle theory is able to paint the whole picture. Any real physical process involves the interaction of many particles. In fact even that is a vast simplification. Really any physical process involves the interaction of all particles. Even the gravitational attraction due to an electron at the other end of the universe is felt by an electron on earth.

To describe all particles in a calculation is, of course, absurd. You would have to include the particles of the paper you are writing the calculation down on and the particles in your brain thinking about the many-body problem. However, many-body theory on a more limited scale is feasible. In this chapter we discuss two ways of going beyond the one-electron approximation. Actually, we are not going very far beyond the one-electron approximation and you will see what is meant by that soon.

This chapter is essentially divided into two halves. The first half deals with the relativistic generalization of the Hartree–Fock method. After a short description of the interaction of two electrons beyond the simple Coulomb potential, we discuss two-electron and many-electron wavefunctions. Then we go on to use the variational principle to derive relativistic equations to enable us to calculate the electronic structure of atoms. The implementation of this theory is briefly described. In the second half of this chapter we present a discussion of a many-body theory, known as density functional theory, which has found much favour in condensed matter physics. We will discuss the simplest non-relativistic theory first, and then generalize that to magnetic systems. Then we develop the theory further to describe both non-magnetic and magnetic systems within a relativistic framework. This approach of writing down the non-relativistic theory as an introduction to the relativistic one has not been the general philosophy of this book, but some aspects of the theory are not clear in a relativistic context, and it would not be possible to discuss them without reference to the non-relativistic theory. In a final section we show some examples of relativistic density functional calculations.

Before we get started on the physics we need to say a word about notation. Some operators and variables in this chapter will be labelled with a subscript number. For example a wavefunction $\psi(r_2)$ means the wavefunction is a function of position and the position variable is r_2. We call r_2 a space. An operator

$$\nabla_2 = \left(\frac{d}{dx_2}, \frac{d}{dy_2}, \frac{d}{dz_2} \right) \tag{10.1}$$

can be defined and this only acts within the space $r_2 = (x_2, y_2, z_2)$, not in any other space such as $r_1 = (x_1, y_1, z_1)$. Let us introduce the spatial part of an n-particle wavefunction $\Phi(r_1, r_2, \cdots, r_n, t)$. For consistency with the principles of quantum mechanics we insist that this wavefunction means that $|\Phi(r_1, r_2, \cdots, r_n, t)|^2 dr_1 dr_2 \cdots dr_n$ is the probability of finding particle 1 in the infinitesimal region dr_1 around point r_1, particle 2 in the infinitesimal region dr_2 around r_2, etc. at time t. We will assume a fixed number of particles, so integrating the probability density over all the variables must give unity. Also, in this chapter many complicated integrals occur. To simplify equations the limits on integrals have been omitted. Any integral over vector r has limits $-\infty$ to $+\infty$, and any integral over scalar r has limits 0 to $+\infty$, unless otherwise stated.

10.1 The Breit Interaction

In electrostatics the interaction of two static point charges separated by distance r_{12} is described exactly by Coulomb's law

$$V(r) = \frac{e_1 e_2}{4\pi\epsilon_0 r_{12}} \tag{10.2}$$

I am sure you are relieved to see an uncharacteristically simple equation. Things will soon get more complicated again! In any experiment the charges are moving and equation (10.2) is only an approximation. Wherever there are moving charges there are currents which lead to a magnetic interaction between the charges, and retardation effects as well (Moller 1931, 1932). The Breit interaction is an approximate description of these effects (Breit 1929, 1930, 1932). The reader should note that much more elegant and rigorous derivations of this interaction exist than the physically transparent one presented here. Pyykkö (1978) has discussed the effect these interactions have in calculations of atomic and molecular electronic structure.

In chapter 12 we will derive the 'golden rule' within relativistic quantum theory, but here we will take it for granted. The golden rule states that the rate at which a perturbed system in state $\langle\psi_i|$ will make the transition to state $\langle\psi_f|$ is

$$P_{fi} = \frac{2\pi}{\hbar}|\langle\psi_f|\hat{H}'|\psi_i\rangle|^2\delta(E_f - E_i) = \frac{2\pi}{\hbar}|S_{fi}|^2\delta(E_f - E_i) \qquad (10.3)$$

Now we will evaluate $\hat{H}'$ for two (otherwise) free electrons scattering from one another. We know that $\hat{H}'$ contains the Coulomb energy. The current–current interaction is somewhat more difficult to evaluate, particularly as it depends on the relative directions of the currents, but under certain circumstances it can be written (Duffin 1973)

$$V_c(r_{12}) = -\frac{1}{c^2}\frac{\mathbf{j}_1\cdot\mathbf{j}_2}{4\pi\epsilon_0 r_{12}} \qquad (10.4)$$

This is just the magnetic interaction of two particles moving parallel to one another. Combining (10.4) with the electrostatic potential (10.2), and using (4.113a) for the current operator, we have

$$V(r_{12}) = \frac{e^2}{4\pi\epsilon_0 r_{12}}(1 - \tilde{\boldsymbol{\alpha}}_1\cdot\tilde{\boldsymbol{\alpha}}_2) \qquad (10.5)$$

Including time in the free-particle wavefunctions in the usual way does not take account of the fact that the electromagnetic interaction is mediated by photons and hence travels at the speed of light. So we should include in the time dependence a term taking account of the time light takes to travel from one electron to the other, i.e. $t' = r_{12}/c$ (this effect is known as retardation). This is shown in figure 10.1. The energy carried by the photon is $W_3 - W_1$ and so the time factor we need to include in the matrix element S_{fi} in equation (10.3) above is $\exp(i(W_3 - W_1)r_{12}/\hbar c)$. We will use wavefunctions of the form (5.23) for the jth electron with momentum $\mathbf{p}_j$, spin s_j and energy W_j. A general two-particle matrix element looks like this

$$S_{fi} = \int\int \psi^\dagger(\mathbf{p}_4, s_4, \mathbf{r}_2)\psi^\dagger(\mathbf{p}_3, s_3, \mathbf{r}_1)\frac{e^2}{4\pi\epsilon_0 r_{12}}(1 - \tilde{\boldsymbol{\alpha}}_1\cdot\tilde{\boldsymbol{\alpha}}_2)$$
$$\times e^{i(W_3 - W_1)r_{12}/\hbar c}\psi(\mathbf{p}_2, s_2, \mathbf{r}_2)\psi(\mathbf{p}_1, s_1, \mathbf{r}_1)d\mathbf{r}_1 d\mathbf{r}_2 \qquad (10.6)$$

The next step is to expand the retardation exponential as a series

$$e^{i(W_3 - W_1)r_{12}/\hbar c} = 1 + \frac{i(W_3 - W_1)r_{12}}{\hbar c} - \frac{(W_3 - W_1)^2 r_{12}^2}{2\hbar^2 c^2} + \cdots \qquad (10.7)$$

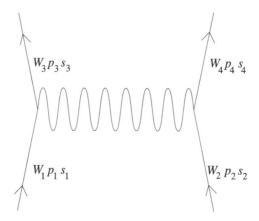

Fig. 10.1. Illustration of the Breit interaction. Two electrons with energies W_1 and W_2 interact (repel) one another by exchange of a photon. Thus the interaction travels at the speed of light and is not instantaneous, as assumed in classical electromagnetism. Later on the electrons have energies W_3 and W_4.

Now, we will replace the exponential in (10.6) with each term in (10.7) in turn, and see what we get for the perturbative Hamiltonian in (10.3). Substituting in 1 is easy, we get

$$\hat{H}'_1 = \frac{e^2}{4\pi\epsilon_0 r_{12}} (1 - \tilde{\alpha}_1 \cdot \tilde{\alpha}_2) \tag{10.8}$$

This term includes the Coulomb repulsion and the current–current interaction, and the correction to the electrostatic term here is sometimes referred to as the magnetic part of the Breit interaction.

If we replace the exponential in (10.6) by the second term in (10.7), r_{12} cancels and the integral $\mathbf{r}_1$ gives $\mathbf{p}_3 = \mathbf{p}_1$ because of the orthonormality of the wavefunctions. Therefore $W_3 = W_1$ and this term vanishes. The first correction to the Coulomb interaction from retardation comes when we substitute the third term in equation (10.7) for the exponential in (10.6):

$$S^3_{fi} = -\int\int \psi^\dagger(\mathbf{p}_4, s_4, \mathbf{r}_2)\psi^\dagger(\mathbf{p}_3, s_3, \mathbf{r}_1)\frac{e^2}{4\pi\epsilon_0}(1 - \tilde{\alpha}_1 \cdot \tilde{\alpha}_2)$$

$$\times \frac{(W_3 - W_1)^2 r_{12}}{2\hbar^2 c^2}\psi(\mathbf{p}_2, s_2, \mathbf{r}_2)\psi(\mathbf{p}_1, s_1, \mathbf{r}_1)d\mathbf{r}_1 d\mathbf{r}_2 \tag{10.9}$$

The term in $\tilde{\alpha}_1 \cdot \tilde{\alpha}_2$ links the large and small parts of the wavefunction and so will give a small contribution compared with the 1 which links the two large components. We are multiplying this by a term which is of order $1/c^2$, so we can justify omitting the current term at this level of approximation. The interaction depicted in figure 10.1 is symmetric in the two electrons, so it makes sense to write (10.9) such that it is symmetric in the energies. We do this using the conservation of energy

$$W_3 - W_1 = W_2 - W_4 \tag{10.10}$$

Hence

$$(W_3 - W_1)^2 = (W_3 - W_1)(W_2 - W_4) = W_3 W_2 - W_1 W_2 - W_3 W_4 + W_1 W_4 \tag{10.11}$$

The reason for writing the energy term in this way is that we can use the Dirac equation to replace the energies

$$(c\tilde{\alpha}_2 \cdot \hat{\mathbf{p}}_2 + \tilde{\beta}_2 mc^2)\psi(\mathbf{p}_4, s_4, \mathbf{r}_2) = W_4 \psi(\mathbf{p}_4, s_4, \mathbf{r}_2) \tag{10.12}$$

and similarly for other energies. Writing the first two products in (10.11) in this way (it is crucial to keep the operators in the right order here):-

$$r_{12}(W_3 W_2 - W_1 W_2) = (c\tilde{\alpha}_1 \cdot \hat{\mathbf{p}}_1 + \tilde{\beta}_1 mc^2)^\dagger r_{12}(c\tilde{\alpha}_2 \cdot \hat{\mathbf{p}}_2 + \tilde{\beta}_2 mc^2)$$
$$- r_{12}(c\tilde{\alpha}_1 \cdot \hat{\mathbf{p}}_1 + \tilde{\beta}_1 mc^2)(c\tilde{\alpha}_2 \cdot \hat{\mathbf{p}}_2 + \tilde{\beta}_2 mc^2) \tag{10.13}$$
$$= \hbar^2 c^2 (r_{12}\tilde{\alpha}_1 \cdot \nabla_1 \tilde{\alpha}_2 \cdot \nabla_2 - \tilde{\alpha}_1 \cdot \nabla_1 r_{12} \tilde{\alpha}_2 \cdot \nabla_2) + mc^3 \{\tilde{\alpha}_1 \cdot \hat{\mathbf{p}}_1\} r_{12} \tilde{\beta}_2$$

In general, operators act on everything to the right of themselves including the wavefunctions. However, in the last line here we have put $\tilde{\alpha}_1 \cdot \hat{\mathbf{p}}_1$ in curly brackets to indicate that it only acts on the r_{12} immediately to its right. Also note that we have examples here of what was stated earlier in this chapter, that the matrices and ∇-operator have subscripts relating them to individual electrons. Therefore we can commute any operators that have different subscripts. We can do a similar thing to (10.13) for the other two energy products and combine that with (10.13) to give

$$r_{12}(W_3 W_2 - W_1 W_2 - W_3 W_4 + W_1 W_4) = \hbar^2 c^2 (r_{12}\tilde{\alpha}_1 \cdot \nabla_1 \tilde{\alpha}_2 \cdot \nabla_2$$
$$- \tilde{\alpha}_1 \cdot \nabla_1 r_{12} \tilde{\alpha}_2 \cdot \nabla_2 + \tilde{\alpha}_1 \cdot \nabla_1 \tilde{\alpha}_2 \cdot \nabla_2 r_{12} - \tilde{\alpha}_2 \cdot \nabla_2 r_{12} \tilde{\alpha}_1 \cdot \nabla_1) \tag{10.14}$$

This can be inserted into equation (10.9) and then the product rule can be used to simplify the resulting expression as much as possible, giving

$$S_{fi}^3 = -\frac{1}{2} \int \int \psi^\dagger(\mathbf{p}_4, s_4, \mathbf{r}_2) \psi^\dagger(\mathbf{p}_3, s_3, \mathbf{r}_1) \frac{e^2}{4\pi\epsilon_0} \{\tilde{\alpha}_1 \cdot \nabla_1\} \{\tilde{\alpha}_2 \cdot \nabla_2\} r_{12}$$
$$\psi(\mathbf{p}_2, s_2, \mathbf{r}_2)\psi(\mathbf{p}_1, s_1, \mathbf{r}_1) d\mathbf{r}_1 d\mathbf{r}_2 \tag{10.15}$$

The gradient operators in (10.15) act only on r_{12}, so the contribution from the third term in the expansion (10.7) to the Hamiltonian is

$$\hat{H}_3' = -\frac{1}{2}\frac{e^2}{4\pi\epsilon_0} \{\tilde{\alpha}_1 \cdot \nabla_1\} \{\tilde{\alpha}_2 \cdot \nabla_2\} r_{12} \tag{10.16}$$

So, to second order, the Hamiltonian in (10.3) is

$$\hat{H}' = \hat{H}_1' + \hat{H}_3' = \frac{e^2}{4\pi\epsilon_0} \left(\frac{1 - \tilde{\alpha}_1 \cdot \tilde{\alpha}_2}{r_{12}} - \frac{1}{2} \{\tilde{\alpha}_1 \cdot \nabla_1\} \{\tilde{\alpha}_2 \cdot \nabla_2\} r_{12} \right) \quad (10.17)$$

The differentiations involved in this equation are messy, but present no problem in principle, and (10.17) can be written in the alternative form

$$\hat{H}' = \frac{e^2}{4\pi\epsilon_0} \left(\frac{1}{r_{12}} - \frac{1}{2} \left(\frac{\tilde{\alpha}_1 \cdot \tilde{\alpha}_2}{r_{12}} + \frac{\tilde{\alpha}_1 \cdot \mathbf{r}_{12} \tilde{\alpha}_2 \cdot \mathbf{r}_{12}}{r_{12}^3} \right) \right) \quad (10.18)$$

This is the usual form found in the literature (Pyykkö 1978) for the Breit interaction potential for two electrons. The first term is the Coulomb repulsion and the next terms are the corrections for retardation and the current interactions. There are three things to note about equation (10.18). Firstly, it was derived without considering the exchange interaction, which must also be included for a full description of two interacting electrons. Secondly, this equation has been derived perturbatively. It is incorrect to use it beyond first order in perturbation theory. Thirdly, equation (10.18) does not include Planck's constant, so it has a classical analogue. If we replace $c\tilde{\alpha}$ in (10.18) with $\mathbf{v}$ we obtain (Jackson 1962)

$$\hat{H}' = \frac{e^2}{4\pi\epsilon_0} \left(\frac{1}{r_{12}} - \frac{1}{2c^2} \left(\frac{\mathbf{v}_1 \cdot \mathbf{v}_2}{r_{12}} + \frac{(\mathbf{v}_1 \cdot \mathbf{r}_{12})(\mathbf{v}_2 \cdot \mathbf{r}_{12})}{r_{12}^3} \right) \right) \quad (10.19)$$

This is the Darwin interaction of classical electromagnetism.

10.2 Two Electrons

Most of the problems we have dealt with in previous chapters have been single-particle problems, i.e. one particle in some external potential. In the previous section we made a gigantic assumption, and deliberately did not point it out, we wrote the wavefunction representing a two-particle state as a product of two single-particle wavefunctions. It is by no means clear how to do this if our wavefunctions have four components. Although we will work in general terms in this section, we are really looking towards a description of two electrons in the field of a nucleus. The most famous example is helium, but others exist, such as H^-, Li^+ and Be^{2+}. Here we are going to be considering the spatial part of the wavefunction defined as ψ in equation (6.118). There we worked with plane waves, whereas here we are thinking about atomic electrons, so we use a different symbol Φ to represent this quantity. Some of this section is very similar to the non-relativistic theory, and some of it is very different (mainly because of the four-component nature of the wavefunction). We will emphasize these differences wherever it is appropriate. Let us write down the full

Hamiltonian for two relativistic electrons:

$$\hat{H} = \frac{\hbar c}{i}\tilde{\alpha}_1 \cdot \nabla_1 + \frac{\hbar c}{i}\tilde{\alpha}_2 \cdot \nabla_2 + \tilde{\beta}_1 m_1 c^2 + \tilde{\beta}_2 m_2 c^2 + V(\mathbf{r}_1) + V(\mathbf{r}_2)$$

$$+ \frac{e^2}{4\pi\epsilon_0}\left(\frac{1}{r_{12}} - \frac{1}{2}\left(\frac{\tilde{\alpha}_1 \cdot \tilde{\alpha}_2}{r_{12}} + \frac{\tilde{\alpha}_1 \cdot \mathbf{r}_{12}\tilde{\alpha}_2 \cdot \mathbf{r}_{12}}{r_{12}^3}\right)\right) \qquad (10.20)$$

Here the first two terms are (approximately) the kinetic energies of the individual electrons. The next two are the rest mass energies, the fifth and sixth terms represent the interaction of the electrons with an external potential and the final term represents the interaction of the electrons derived above, the Coulomb and Breit interaction. Without the Breit interaction this Hamiltonian is nowhere near being Lorentz invariant. The Breit interaction introduces approximate Lorentz invariance to first order in quantum electrodynamical effects (Rose 1961). For a completely invariant equation one has to go to the Bethe–Salpeter equation (Salpeter and Bethe 1951, Bethe and Salpeter 1957) which includes all orders of quantum electrodynamical interaction between the two electrons.

Next we introduce the exchange operator $\hat{P}_{12}$ which should be familiar from non-relativistic theory. When it acts on a two-particle state, $\hat{P}_{12}$ interchanges the space and spin coordinates of particle 1 and particle 2

$$\hat{P}_{12}\Phi(\mathbf{r}_1, \mathbf{r}_2) = \Phi(\mathbf{r}_2, \mathbf{r}_1) \qquad (10.21)$$

Consider the exchange operator acting on the Hamiltonian (10.20)

$$\hat{P}_{12}\hat{H}\Phi(\mathbf{r}_1, \mathbf{r}_2) = \hat{P}_{12}\left(\frac{\hbar c}{i}\tilde{\alpha}_1 \cdot \nabla_1 + \frac{\hbar c}{i}\tilde{\alpha}_2 \cdot \nabla_2 + \tilde{\beta}_1 m_1 c^2 + \tilde{\beta}_2 m_2 c^2\right.$$

$$\left. + V(\mathbf{r}_1) + V(\mathbf{r}_2) + U(|\mathbf{r}_1 - \mathbf{r}_2|)\right)\Phi(\mathbf{r}_1, \mathbf{r}_2)$$

$$= \left(\frac{\hbar c}{i}\tilde{\alpha}_2 \cdot \nabla_2 + \frac{\hbar c}{i}\tilde{\alpha}_1 \cdot \nabla_1 + \tilde{\beta}_2 m_2 c^2 + \tilde{\beta}_1 m_1 c^2\right.$$

$$\left. + V(\mathbf{r}_2) + V(\mathbf{r}_1) + U(|\mathbf{r}_2 - \mathbf{r}_1|)\right)\Phi(\mathbf{r}_2 \cdot \mathbf{r}_1)$$

$$= \hat{H}\hat{P}_{12}\Phi(\mathbf{r}_1, \mathbf{r}_2) \qquad (10.22)$$

where we have lumped the Coulomb and Breit terms together in $U(|\mathbf{r}_1 - \mathbf{r}_2|)$. The commutation relation described by (10.22) works only because the interaction term depends on the magnitude of $\mathbf{r}_1 - \mathbf{r}_2$ and not its direction or absolute value. This commutator tells us that the quantity described by the exchange operator is conserved.

Note that equation (10.21) is not an eigenvalue equation, it does not have the same function on the right and left hand sides. However, it does have eigenvalues and we can find out what they are by applying the exchange operator twice:

$$\hat{P}_{12}^2\Phi(\mathbf{r}_1,\mathbf{r}_2) = \hat{P}_{12}\Phi(\mathbf{r}_2,\mathbf{r}_1) = \Phi(\mathbf{r}_1,\mathbf{r}_2) \qquad (10.23)$$

Now the left and right hand sides of (10.23) do form an eigenvalue equation and we can see that the eigenvalue of $\hat{P}_{12}^2$ is 1. Therefore the eigenvalues of $\hat{P}_{12}$ are ± 1.

To go further with this line of thought we have to make a postulate that the two-particle wavefunction can be written as some combination of single-particle wavefunctions. The simplest combination we can think of is a direct product

$$\Phi_1(\mathbf{r}_1,\mathbf{r}_2) = \psi_a(\mathbf{r}_1)\psi_b(\mathbf{r}_2) \qquad (10.24a)$$

or equally validly

$$\Phi_2(\mathbf{r}_1,\mathbf{r}_2) = \psi_a(\mathbf{r}_2)\psi_b(\mathbf{r}_1) \qquad (10.24b)$$

where the subscripts a and b label the quantum state in which the particle resides, i.e. they represent the quantum numbers. We have still not addressed the question of what the product of two four-component wavefunctions actually means. We can put it off no longer. The product $\Phi_j(\mathbf{r}_1,\mathbf{r}_2)$ is a sixteen-component spinor wavefunction. It is a direct product of two four-component single-particle wavefunctions. That is, every element of the first four-component spinor is multiplied by every element of the second to give a sixteen-component quantity. That is the simplest way that a two-particle wavefunction can be constructed that allows a decomposition into single-particle states like (10.24). The direct product is not commutative, i.e. $\psi_a(\mathbf{r}_2)\psi_b(\mathbf{r}_1) \neq \psi_b(\mathbf{r}_1)\psi_a(\mathbf{r}_2)$. It takes too much space to write out the full two-particle wavefunction. However, if we take single-particle wavefunctions of the form (5.7), a single element of the two-particle wavefunction may be written in terms of them as (Breit 1929)

$$\Phi_{jk}(\mathbf{r}_1,\mathbf{r}_2) = \psi_j(\mathbf{r}_1)\psi_k(\mathbf{r}_2) = U_j U_k e^{i(\mathbf{p}_1\cdot\mathbf{r}_1 + \mathbf{p}_2\cdot\mathbf{r}_2)/\hbar} \qquad (10.25)$$

where j and k run from 1 to 4. As we will see in a moment, the order in which the sixteen elements appear in the spinor wavefunction doesn't matter. The question that immediately comes to mind now is 'What has become of the matrices in the Hamiltonian (10.20) if the two-particle wavefunction has sixteen, rather than four, components?' As you might guess, the answer to this is that they have become 16×16 matrices which have matrix elements defined as follows

$$(\tilde{\alpha}_1)_{mnpq} = (\tilde{\alpha})_{mn}\delta_{pq}, \qquad (\tilde{\beta}_1)_{mnpq} = (\tilde{\beta})_{mn}\delta_{pq}$$
$$(\tilde{\alpha}_2)_{mnpq} = \delta_{mn}(\tilde{\alpha})_{pq}, \qquad (\tilde{\beta}_2)_{mnpq} = \delta_{mn}(\tilde{\beta})_{pq} \qquad (10.26)$$

where the subscripts m, n define the spinor component of the first particle and p, q define that of the second particle. To make this absolutely clear let us consider a specific example $(\tilde{\alpha}_1)_{mnpq}$ acting on a two-particle wavefunction. If $p = q$ we are selecting a single element of the four-component spinor for the second electron. In the two-particle wavefunction that element appears in four elements, once multiplying each of the elements of the four-component spinor of the first electron. Those four elements of the two-particle wavefunction can be acted on by $\tilde{\alpha}_1$, which, according to the definition (10.26), amounts to multiplying them by $\tilde{\alpha}$. Now p and q can each take on the same four values, so there are four possibilities for them to be equal. That makes $\tilde{\alpha}_1$ a sixteen-component matrix and defines it consistently whatever order we choose to write the sixteen components of equation (10.25). A similar discussion describes how to operate with the other 16×16 matrices.

In fact it is usually unnecessary to work with the sixteen-component wavefunctions and 16×16 matrices directly. If we make a decomposition like (10.24), then $\tilde{\alpha}_1$ and $\tilde{\beta}_1$ act only on the spinor space of particle 1, and $\tilde{\alpha}_2$ and $\tilde{\beta}_2$ act only on the spinor space of particle 2. Therefore they can be reduced back to their 4×4 form, provided we retain the subscript indicating which space they operate within.

We would expect all the physics that we can calculate to be the same, independent of which electron goes in which state. This is only true, of course, if the electrons are truly indistinguishable. There is plenty of evidence to suggest that electrons are indistinguishable. If they are not they must have some property that makes them distinct and one would expect experiments on distinguishable electrons to produce different results. This does not occur. Now, although equations (10.24) are a simple product of single-particle wavefunctions, they are not acceptable two-particle wavefunctions because they are not eigenfunctions of the exchange operator. However, we can construct eigenfunctions of $\hat{P}_{12}$ from (10.24):

$$\Phi_s(\mathbf{r}_1, \mathbf{r}_2) = \frac{1}{\sqrt{2}}(\psi_a(\mathbf{r}_1)\psi_b(\mathbf{r}_2) + \psi_a(\mathbf{r}_2)\psi_b(\mathbf{r}_1)) \qquad (10.27a)$$

$$\Phi_a(\mathbf{r}_1, \mathbf{r}_2) = \frac{1}{\sqrt{2}}(\psi_a(\mathbf{r}_1)\psi_b(\mathbf{r}_2) - \psi_a(\mathbf{r}_2)\psi_b(\mathbf{r}_1)) \qquad (10.27b)$$

Here the $1/\sqrt{2}$ is included to preserve the normalization and the subscripts s and a stand for symmetric and antisymmetric respectively. From these definitions it is easy to see that

$$\hat{P}_{12}\Phi_s(\mathbf{r}_1, \mathbf{r}_2) = (+1)\Phi_s(\mathbf{r}_1, \mathbf{r}_2) \qquad (10.28a)$$

$$\hat{P}_{12}\Phi_a(\mathbf{r}_1, \mathbf{r}_2) = (-1)\Phi_a(\mathbf{r}_1, \mathbf{r}_2) \qquad (10.28b)$$

As discussed in chapter 6 the separation of many-particle wavefunctions into those that are symmetric and those that are antisymmetric defines whether the particles are bosons or fermions. The fact that the exchange operator commutes with the Hamiltonian reflects the fact that this symmetry will be conserved in any interaction (provided the number of particles is conserved). Clearly the antisymmetric wavefunction (10.27*b*) disappears when two electrons in the same state get close together, leading to the Pauli principle, whereas the symmetric wavefunction has a maximum when this occurs. Thus, in relativistic quantum theory the exchange force is postulated, following the same reasoning as in non-relativistic quantum theory.

For what follows we are going to be concerned with electrons, which are fermions, so we are considering an antisymmetric wavefunction. Therefore we drop the subscript a, and antisymmetry is to be assumed. Next we will discuss in detail how to calculate expectation values for two-particle relativistic wavefunctions like those of (10.27*b*). This will enable us to discuss the expectation values of many-particle wavefunctions as a generalization of the two-particle case later on.

There are three types of operator that commonly occur in condensed matter physics: constant operators, single-particle operators and two-particle operators. Higher order operators can be written down, but one very rarely has recourse to them and so we will not consider them here.

We will start with the simplest case of a constant operator. An obvious case is the unit operator which we use implicitly when normalizing a wavefunction. Consider (10.27*b*) and its overlap integral with

$$\Phi'(\mathbf{r}_1, \mathbf{r}_2) = \frac{1}{\sqrt{2}} (\psi'_c(\mathbf{r}_1)\psi'_d(\mathbf{r}_2) - \psi'_c(\mathbf{r}_2)\psi'_d(\mathbf{r}_1)) \tag{10.29}$$

The overlap integral is evaluated as follows

$$
\begin{aligned}
\langle \Phi | \Phi' \rangle &= \tfrac{1}{2} \int \int (\psi_a^\dagger(\mathbf{r}_1)\psi_b^\dagger(\mathbf{r}_2) - \psi_a^\dagger(\mathbf{r}_2)\psi_b^\dagger(\mathbf{r}_1)) \\
&\quad \times (\psi'_c(\mathbf{r}_1)\psi'_d(\mathbf{r}_2) - \psi'_c(\mathbf{r}_2)\psi'_d(\mathbf{r}_1))d\mathbf{r}_1 d\mathbf{r}_2 \\
&= \tfrac{1}{2} \int \psi_a^\dagger(\mathbf{r}_1)\psi'_c(\mathbf{r}_1)d\mathbf{r}_1 \int \psi_b^\dagger(\mathbf{r}_2)\psi'_d(\mathbf{r}_2)d\mathbf{r}_2 \\
&\quad - \tfrac{1}{2} \int \psi_a^\dagger(\mathbf{r}_1)\psi'_d(\mathbf{r}_1)d\mathbf{r}_1 \int \psi_b^\dagger(\mathbf{r}_2)\psi'_c(\mathbf{r}_2)d\mathbf{r}_2 \\
&\quad - \tfrac{1}{2} \int \psi_a^\dagger(\mathbf{r}_2)\psi'_d(\mathbf{r}_2)d\mathbf{r}_2 \int \psi_b^\dagger(\mathbf{r}_1)\psi'_c(\mathbf{r}_1)d\mathbf{r}_1 \\
&\quad + \tfrac{1}{2} \int \psi_a^\dagger(\mathbf{r}_2)\psi'_c(\mathbf{r}_2)d\mathbf{r}_2 \int \psi_b^\dagger(\mathbf{r}_1)\psi'_d(\mathbf{r}_1)d\mathbf{r}_1
\end{aligned} \tag{10.30}
$$

All of these integrals can be calculated straightforwardly in principle. Equation (10.30) is a long equation, but it simplifies considerably if the primed and unprimed single-particle functions are the same and if we assume they form an orthonormal set of functions, i.e.-

$$\int \psi_i^\dagger(\mathbf{r})\psi_j(\mathbf{r})d\mathbf{r} = \delta_{ij} \tag{10.31}$$

Equation (10.30) will then be zero unless $a = c$ and $b = d$, and then we have

$$\langle \Phi | \Phi \rangle = 1 \tag{10.32}$$

The second type of operator is the single-particle operator such as the $c\tilde{\boldsymbol{\alpha}} \cdot \hat{\mathbf{p}}$ term in the Dirac equation. To find the matrix element of such an operator we follow the same procedure as for the constant operator

$$
\begin{aligned}
\langle \Phi | \hat{f}(\mathbf{r}_1) | \Phi' \rangle &= \int \int \Phi^\dagger(\mathbf{r}_1, \mathbf{r}_2)\hat{f}(\mathbf{r}_1)\Phi'(\mathbf{r}_1, \mathbf{r}_2)d\mathbf{r}_1 d\mathbf{r}_2 \\
&= \tfrac{1}{2}\int \int (\psi_a^\dagger(\mathbf{r}_1)\psi_b^\dagger(\mathbf{r}_2) - \psi_a^\dagger(\mathbf{r}_2)\psi_b^\dagger(\mathbf{r}_1)) \\
&\quad \times \hat{f}(\mathbf{r}_1)(\psi_c'(\mathbf{r}_1)\psi_d'(\mathbf{r}_2) - \psi_c'(\mathbf{r}_2)\psi_d'(\mathbf{r}_1))d\mathbf{r}_1 d\mathbf{r}_2 \\
&= \tfrac{1}{2}\int \psi_a^\dagger(\mathbf{r}_1)\hat{f}(\mathbf{r}_1)\psi_c'(\mathbf{r}_1)d\mathbf{r}_1 \int \psi_b^\dagger(\mathbf{r}_2)\psi_d'(\mathbf{r}_2)d\mathbf{r}_2 \\
&\quad - \tfrac{1}{2}\int \psi_a^\dagger(\mathbf{r}_1)\hat{f}(\mathbf{r}_1)\psi_d'(\mathbf{r}_1)d\mathbf{r}_1 \int \psi_b^\dagger(\mathbf{r}_2)\psi_c'(\mathbf{r}_2)d\mathbf{r}_2 \\
&\quad - \tfrac{1}{2}\int \psi_a^\dagger(\mathbf{r}_2)\psi_d'(\mathbf{r}_2)d\mathbf{r}_2 \int \psi_b^\dagger(\mathbf{r}_1)\hat{f}(\mathbf{r}_1)\psi_c'(\mathbf{r}_1)d\mathbf{r}_1 \\
&\quad + \tfrac{1}{2}\int \psi_a^\dagger(\mathbf{r}_2)\psi_c'(\mathbf{r}_2)d\mathbf{r}_2 \int \psi_b^\dagger(\mathbf{r}_1)\hat{f}(\mathbf{r}_1)\psi_d'(\mathbf{r}_1)d\mathbf{r}_1 \tag{10.33}
\end{aligned}
$$

This is lengthy, but straightforward, to evaluate once the single-particle wavefunctions are known. Equation (10.33) is a general formula which simplifies if we are looking for an expectation value. Then the primed and unprimed states are the same, $a = c$ and $b = d$, and the expectation value of $\hat{f}(\mathbf{r})$ for the two-particle state is

$$\langle \Phi | \hat{f}(\mathbf{r}_1) | \Phi \rangle = \tfrac{1}{2}\int \psi_a^\dagger(\mathbf{r}_1)\hat{f}(\mathbf{r}_1)\psi_a(\mathbf{r}_1)d\mathbf{r}_1 + \tfrac{1}{2}\int \psi_b^\dagger(\mathbf{r}_1)\hat{f}(\mathbf{r}_1)\psi_b(\mathbf{r}_1)d\mathbf{r}_1 \tag{10.34a}$$

Similarly, if we had had a single-particle operator in the space of $\mathbf{r}_2$ we would have found the analogous expression

$$\langle \Phi | \hat{f}(\mathbf{r}_2) | \Phi \rangle = \tfrac{1}{2}\int \psi_a^\dagger(\mathbf{r}_2)\hat{f}(\mathbf{r}_2)\psi_a(\mathbf{r}_2)d\mathbf{r}_2 + \tfrac{1}{2}\int \psi_b^\dagger(\mathbf{r}_2)\hat{f}(\mathbf{r}_2)\psi_b(\mathbf{r}_2)d\mathbf{r}_2 \tag{10.34b}$$

As the volumes spanned by $\mathbf{r}_1$ in (10.34a) and $\mathbf{r}_2$ in (10.34b) are the same, the corresponding integrals will be equal.

Finally we will consider two-particle operators, the most obvious example being the Coulomb repulsion operator between two electrons. We follow exactly the same procedure as for the simpler types of operator

$$\langle \Phi | \hat{f}(\mathbf{r}_1, \mathbf{r}_2) | \Phi' \rangle = \int \int \Phi^\dagger(\mathbf{r}_1, \mathbf{r}_2) \hat{f}(\mathbf{r}_1, \mathbf{r}_2) \Phi'(\mathbf{r}_1, \mathbf{r}_2) d\mathbf{r}_1 d\mathbf{r}_2$$

$$= \tfrac{1}{2} \int \int (\psi_a^\dagger(\mathbf{r}_1)\psi_b^\dagger(\mathbf{r}_2) - \psi_a^\dagger(\mathbf{r}_2)\psi_b^\dagger(\mathbf{r}_1))$$

$$\times \hat{f}(\mathbf{r}_1, \mathbf{r}_2)(\psi_c'(\mathbf{r}_1)\psi_d'(\mathbf{r}_2) - \psi_c'(\mathbf{r}_2)\psi_d'(\mathbf{r}_1))d\mathbf{r}_1 d\mathbf{r}_2$$

$$= \tfrac{1}{2} \int \psi_a^\dagger(\mathbf{r}_1)(\int \psi_b^\dagger(\mathbf{r}_2)\hat{f}(\mathbf{r}_1, \mathbf{r}_2)\psi_d'(\mathbf{r}_2)d\mathbf{r}_2)\psi_c'(\mathbf{r}_1)d\mathbf{r}_1$$

$$- \tfrac{1}{2} \int \psi_a^\dagger(\mathbf{r}_2)(\int \psi_b^\dagger(\mathbf{r}_1)\hat{f}(\mathbf{r}_1, \mathbf{r}_2)\psi_c'(\mathbf{r}_1)d\mathbf{r}_1)\psi_d'(\mathbf{r}_2)d\mathbf{r}_2$$

$$- \tfrac{1}{2} \int \psi_a^\dagger(\mathbf{r}_1)(\int \psi_b^\dagger(\mathbf{r}_2)\hat{f}(\mathbf{r}_1, \mathbf{r}_2)\psi_c'(\mathbf{r}_2)d\mathbf{r}_2)\psi_d'(\mathbf{r}_1)d\mathbf{r}_1$$

$$+ \tfrac{1}{2} \int \psi_a^\dagger(\mathbf{r}_2)(\int \psi_b^\dagger(\mathbf{r}_1)\hat{f}(\mathbf{r}_1, \mathbf{r}_2)\psi_d'(\mathbf{r}_1)d\mathbf{r}_1)\psi_c'(\mathbf{r}_2)d\mathbf{r}_2$$

$$(10.35)$$

We can simplify this expression (thank goodness) by noting that the $\hat{f}(\mathbf{r}_1, \mathbf{r}_2)$ must be symmetric in $\mathbf{r}_1$ and $\mathbf{r}_2$, so interchanging them has no effect on it. Recall that $\mathbf{r}_1$ and $\mathbf{r}_2$ cover the same region of space so the integrals in the first and fourth terms in (10.35) will be identical and so will the integrals in the second and third terms. Therefore we can reduce the four terms to two. If we let the primed and unprimed functions be the same, and let $a = c$ and $b = d$ again, this leads to

$$\langle \Phi | \hat{f}(\mathbf{r}_1, \mathbf{r}_2) | \Phi' \rangle = \int \int \psi_a^\dagger(\mathbf{r}_1)\psi_b^\dagger(\mathbf{r}_2)\hat{f}(\mathbf{r}_1, \mathbf{r}_2)\psi_a(\mathbf{r}_1)\psi_b(\mathbf{r}_2)d\mathbf{r}_1 d\mathbf{r}_2$$

$$- \int \int \psi_a^\dagger(\mathbf{r}_1)\psi_b^\dagger(\mathbf{r}_2)\hat{f}(\mathbf{r}_1, \mathbf{r}_2)\psi_a(\mathbf{r}_2)\psi_b(\mathbf{r}_1)d\mathbf{r}_1 d\mathbf{r}_2$$

$$= J - K \qquad (10.36)$$

The first of these integrals is known as a direct integral and the second as an exchange integral. The latter arises because of the antisymmetric nature of the wavefunction which came about because we demanded that the wavefunction be an eigenstate of the exchange operator.

It is the formalism described in this section that we would use to calculate the energy levels of the helium atom, for example. Ignoring the Breit interaction, the Hamiltonian for the two electrons in the potential of a helium nucleus is

$$\hat{H} = c\tilde{\alpha} \cdot \hat{\mathbf{p}}_1 + \tilde{\beta}m_1c^2 + c\tilde{\alpha} \cdot \hat{\mathbf{p}}_2 + \tilde{\beta}m_2c^2 - \frac{Ze^2}{4\pi\epsilon_0 r_1} - \frac{Ze^2}{4\pi\epsilon_0 r_2} + \frac{e^2}{4\pi\epsilon_0 |\mathbf{r}_1 - \mathbf{r}_2|}$$

$$(10.37)$$

We could start with the hydrogenic wavefunctions (e.g. equation (8.53)) as our single-particle states, as is often done in non-relativistic theory, and evaluate expectation values of the single-particle operators $c\tilde{\alpha} \cdot \hat{\mathbf{p}}_i$, $\tilde{\beta}m_ic^2$ and $Ze^2/4\pi\epsilon_0 r_i$ using equation (10.34). The final term in equation (10.37) is the Coulomb repulsion between the two electrons which could be evaluated using equation (10.36). Although the uncertainty introduced by using such a wavefunction would dwarf the relativistic effects we are trying to describe, this illustrates, in principle, how such a relativistic calculation of the energy levels of a two-electron atom could go. Anyway, it would always be possible to improve the calculation by using the variational principle on the parameters in the wavefunctions, as is done non-relativistically (Dicke and Wittke 1974), to yield accurate energy eigenvalues for helium.

10.3 Many-Electron Wavefunctions

In this section we generalize the discussion of the previous section to systems containing many electrons. The wavefunction of equation (10.27b) can be written as a 2×2 determinant. We have seen that a two-particle wavefunction is a 4^2-component quantity, and this is the start of a trend. An N-particle wavefunction has 4^N components and the corresponding Dirac matrices are $4^N \times 4^N$, so things soon become unmanageably large. However, as before, we do not need to work with the full wavefunction directly, but can decompose it and write it in terms of its constituent single-particle wavefunctions. The simplest form for an N-body wavefunction that has all the right symmetry properties (i.e. obeys the Pauli exclusion principle) is an $N \times N$ determinantal wavefunction

$$\Phi(\mathbf{r}_1, \mathbf{r}_2, \cdots, \mathbf{r}_N) = \frac{1}{\sqrt{N!}} \begin{vmatrix} \psi_a(\mathbf{r}_1) & \psi_a(\mathbf{r}_2) & \cdots & \psi_a(\mathbf{r}_N) \\ \psi_b(\mathbf{r}_1) & \psi_b(\mathbf{r}_2) & \cdots & \psi_b(\mathbf{r}_N) \\ \cdots & \cdots & \cdots & \cdots \\ \psi_N(\mathbf{r}_1) & \psi_N(\mathbf{r}_2) & \cdots & \psi_N(\mathbf{r}_N) \end{vmatrix} \qquad (10.38)$$

This form for the wavefunction is only exact in the limit when the electrons are non-interacting. However, it turns out to be a good approximation for the wavefunction of closed shell atoms, and linear combinations of such determinants can be used to give an accurate account of elements that have open outer electron shells. Each term in the expansion of equation (10.38) is a 4^N-component quantity. Care must be taken when expanding (10.38) in terms of single-particle states because changing the order of the single-particle wavefunctions when taking their direct product changes the position of the elements of the many-particle wavefunction. Hence products must always be taken consistently. With a determinantal wavefunction like that in equation (10.38) it is then sufficient to keep the quantum state label a, b, etc. in the same order (in our case alphabetical order). As in the two-electron case, we want to know how to take matrix elements of such wavefunctions. Here we will assume from the beginning that the single-electron wavefunctions form a complete set of orthonormal functions. We want to evaluate matrix elements of the type

$$
\langle\Phi|\hat{f}|\Phi\rangle = \frac{1}{N!} \int \int \cdots \int
\begin{vmatrix}
\psi_a^\dagger(\mathbf{r}_1) & \psi_a^\dagger(\mathbf{r}_2) & \cdots & \psi_a^\dagger(\mathbf{r}_N) \\
\psi_b^\dagger(\mathbf{r}_1) & \psi_b^\dagger(\mathbf{r}_2) & \cdots & \psi_b^\dagger(\mathbf{r}_N) \\
\cdots & \cdots & \cdots & \cdots \\
\psi_N^\dagger(\mathbf{r}_1) & \psi_N^\dagger(\mathbf{r}_2) & \cdots & \psi_N^\dagger(\mathbf{r}_N)
\end{vmatrix} \times
$$

$$
\hat{f}
\begin{vmatrix}
\psi_a(\mathbf{r}_1) & \psi_a(\mathbf{r}_2) & \cdots & \psi_a(\mathbf{r}_N) \\
\psi_b(\mathbf{r}_1) & \psi_b(\mathbf{r}_2) & \cdots & \psi_b(\mathbf{r}_N) \\
\cdots & \cdots & \cdots & \cdots \\
\psi_N(\mathbf{r}_1) & \psi_N(\mathbf{r}_2) & \cdots & \psi_N(\mathbf{r}_N)
\end{vmatrix} d\mathbf{r}_1 d\mathbf{r}_2 \cdots d\mathbf{r}_n
\tag{10.39}
$$

Equation (10.39) is immensely complicated to evaluate for a system with more than a few electrons. However, it can be simplified by making use of an argument due to Löwdin (1955) which is described in detail by Slater (1960, 1963). If we expand the first determinant there will be one term that is the direct product of the elements in the leading diagonal. The contribution of that to the matrix element of (10.39) will be

$$
\langle\Phi|\hat{f}|\Phi\rangle_1 = \frac{1}{N!} \int \int \cdots \int \psi_a^\dagger(\mathbf{r}_1)\psi_b^\dagger(\mathbf{r}_2)\cdots\psi_N^\dagger(\mathbf{r}_N)
$$

$$
\hat{f}
\begin{vmatrix}
\psi_a(\mathbf{r}_1) & \psi_a(\mathbf{r}_2) & \cdots & \psi_a(\mathbf{r}_N) \\
\psi_b(\mathbf{r}_1) & \psi_b(\mathbf{r}_2) & \cdots & \psi_b(\mathbf{r}_N) \\
\cdots & \cdots & \cdots & \cdots \\
\psi_N(\mathbf{r}_1) & \psi_N(\mathbf{r}_2) & \cdots & \psi_N(\mathbf{r}_N)
\end{vmatrix} d\mathbf{r}_1 d\mathbf{r}_2 \cdots d\mathbf{r}_N
\tag{10.40}
$$

All of the $N!$ terms we obtain in the determinant expansion will give the same result. Let us prove this by taking a different term in the expansion of the determinant and showing that its contribution to the matrix element is equal to that of (10.40). We choose a term that illustrates our point simply. It is the same as (10.40) except that the arguments $\mathbf{r}_1$ and $\mathbf{r}_2$ have been swapped round in the first two single-particle wavefunctions, and the sign of this term will be negative.

$$
\langle\Phi|\hat{f}|\Phi\rangle_2 = \frac{1}{N!} \int \int \cdots \int -\psi_a^\dagger(\mathbf{r}_2)\psi_b^\dagger(\mathbf{r}_1)\cdots\psi_N^\dagger(\mathbf{r}_N)
$$

$$
\hat{f}
\begin{vmatrix}
\psi_a(\mathbf{r}_1) & \psi_a(\mathbf{r}_2) & \cdots & \psi_a(\mathbf{r}_N) \\
\psi_b(\mathbf{r}_1) & \psi_b(\mathbf{r}_2) & \cdots & \psi_b(\mathbf{r}_N) \\
\cdots & \cdots & \cdots & \cdots \\
\psi_N(\mathbf{r}_1) & \psi_N(\mathbf{r}_2) & \cdots & \psi_N(\mathbf{r}_N)
\end{vmatrix} d\mathbf{r}_2 d\mathbf{r}_1 \cdots d\mathbf{r}_N
\tag{10.41}
$$

The proof rests on two facts. Firstly, interchanging rows and columns in a determinant can only change the sign and not the magnitude of the determinant. Secondly, the $\mathbf{r}_j$ in the matrix elements are dummy variables, which are integrated over with the same integration limits, so we can change their labels with impunity. Let us rewrite (10.41) with the names of the labels $\mathbf{r}_1$ and $\mathbf{r}_2$ interchanged.

$$\langle\Phi|\hat{f}|\Phi\rangle_2 = \frac{1}{N!} \int \int \cdots \int -\psi_a^\dagger(\mathbf{r}_1)\psi_b^\dagger(\mathbf{r}_2)\cdots\psi_N^\dagger(\mathbf{r}_N)$$

$$\hat{f} \begin{vmatrix} \psi_a(\mathbf{r}_2) & \psi_a(\mathbf{r}_1) & \cdots & \psi_a(\mathbf{r}_N) \\ \psi_b(\mathbf{r}_2) & \psi_b(\mathbf{r}_1) & \cdots & \psi_b(\mathbf{r}_N) \\ \cdots & \cdots & \cdots & \cdots \\ \psi_N(\mathbf{r}_2) & \psi_N(\mathbf{r}_1) & \cdots & \psi_N(\mathbf{r}_N) \end{vmatrix} d\mathbf{r}_1 d\mathbf{r}_2 \cdots d\mathbf{r}_N \qquad (10.42)$$

Note that $\hat{f}$ must be symmetric in the coordinates, so letting $\mathbf{r}_{1(2)} \to \mathbf{r}_{2(1)}$ will have no effect on it. Now, if we switch the first and second columns in the determinant in (10.42) we get the determinant of equation (10.40) with an extra minus sign which cancels the minus sign at the front of the expression. Then we have exactly equation (10.40) again. The same will happen with any other term in the first determinant expansion, there will always be two positive or two negative signs which will cancel. As there are $N!$ terms in the expansion of the first determinant the result for the matrix element (10.39) is just $N!$ times equation (10.40), i.e.

$$\langle\Phi|\hat{f}|\Phi\rangle = \int \int \cdots \int \psi_a^\dagger(\mathbf{r}_1)\psi_b^\dagger(\mathbf{r}_2)\cdots\psi_N^\dagger(\mathbf{r}_N)\times$$

$$\hat{f} \begin{vmatrix} \psi_a(\mathbf{r}_1) & \psi_a(\mathbf{r}_2) & \cdots & \psi_a(\mathbf{r}_N) \\ \psi_b(\mathbf{r}_1) & \psi_b(\mathbf{r}_2) & \cdots & \psi_b(\mathbf{r}_N) \\ \cdots & \cdots & \cdots & \cdots \\ \psi_N(\mathbf{r}_1) & \psi_N(\mathbf{r}_2) & \cdots & \psi_N(\mathbf{r}_N) \end{vmatrix} d\mathbf{r}_1 d\mathbf{r}_2 \cdots d\mathbf{r}_N \qquad (10.43)$$

The simplification (10.43) enables us to describe matrix elements for determinantal wavefunctions for the different types of operators discussed in the two-electron case relatively simply.

First consider the case where $\hat{f}$ is a constant. We can take the operator outside the integral. The leading diagonal term in the expansion of the determinant in (10.43) gives a product of single-electron integrals such as

$$D = \int \psi_j^\dagger(\mathbf{r}_i)\psi_j(\mathbf{r}_i)d\mathbf{r}_i \qquad (10.44)$$

i.e. every space will contain a wavefunction premultiplied by its hermitian conjugate. By construction all these integrals are unity so the term in the leading diagonal of the determinant in (10.43) yields $1^N = 1$. All other terms in the expansion of the determinant yield zero when multiplied by the initial product in (10.43) and integrated, because there will always be at least one space that has a single-particle wavefunction premultiplied by the hermitian conjugate of a different single-particle wavefunction. By orthogonality that integral will be zero and so the whole term will be zero. So, for an operator which is a constant, the integral gives unity and the matrix element will take the numerical value of the constant operator

$$\langle \Phi | \hat{f} | \Phi \rangle = f \qquad (10.45)$$

Clearly this is analogous to the two-particle case (equation (10.32)).

Next consider the case when $\hat{f}$ in equation (10.43) is a single-particle operator, $\hat{f}(\mathbf{r}_1)$ say. We can see how this will work by inspection of equation (10.43). The product multiplied by the leading diagonal of the determinant will give us $N-1$ normalization integrals like equation (10.44), multiplying a single-particle expectation value of the type seen in equation (10.34). The normalization integrals will all be unity. So we are left with the single-particle expectation value. When the product multiplies any other term in the expansion of the determinant there will always be integrals that vanish because of orthonormality, so the expectation value of a single-particle operator with a determinantal wavefunction is

$$\langle \Phi | \hat{f}(\mathbf{r}_1) | \Phi \rangle = \int \psi_i^\dagger(\mathbf{r}_1) \hat{f}(\mathbf{r}_1) \psi_i(\mathbf{r}_1) d\mathbf{r}_1 \qquad (10.46)$$

Now, it is rare that we need to calculate the properties of an individual particle among many, even if that has any meaning in quantum mechanics. Equation (10.46) looks a bit odd because of this, it picks out the (arbitrary) i^{th} state on the right hand side. What we really need is a sum over occupied states of terms like (10.46). So, a more appropriate quantity is

$$\langle \Phi | \sum_{i=1}^{N} \hat{f}(\mathbf{r}) | \Phi \rangle = \sum_{i=1}^{N} \int \psi_i^\dagger(\mathbf{r}) \hat{f}(\mathbf{r}) \psi_i(\mathbf{r}) d\mathbf{r} \qquad (10.47)$$

To make a direct comparison of this with the two-electron case we have to add equations (10.34a and b). The first integrals in these two expressions are identical and so are the second integrals because $\mathbf{r}_1$ and $\mathbf{r}_2$ span the same region of space. When this addition is done it is immediately apparent that (10.47) is a direct generalization of (10.34).

Next we need an expression for the expectation value of a two-electron operator in equation (10.43). This is quite difficult to see by inspection. The operator is now $\hat{f}(\mathbf{r}_1, \mathbf{r}_2)$. If we take the leading diagonal of the determinant, our matrix element will be of the form

$$J = \int \int \psi_a^\dagger(\mathbf{r}_1) \psi_b^\dagger(\mathbf{r}_2) \hat{f}(\mathbf{r}_1, \mathbf{r}_2) \psi_a(\mathbf{r}_1) \psi_b(\mathbf{r}_2) d\mathbf{r}_1 d\mathbf{r}_2 \qquad (10.48)$$

There is one further term in the determinant that will contribute to the expectation value. That is the term which is the same as the leading diagonal except that $\psi_a(\mathbf{r}_1)\psi_b(\mathbf{r}_2)$ is replaced by $\psi_a(\mathbf{r}_2)\psi_b(\mathbf{r}_1)$. That term will give us the exchange integral

$$K = -\int \int \psi_a^\dagger(\mathbf{r}_1) \psi_b^\dagger(\mathbf{r}_2) \hat{f}(\mathbf{r}_1, \mathbf{r}_2) \psi_a(\mathbf{r}_2) \psi_b(\mathbf{r}_1) d\mathbf{r}_1 d\mathbf{r}_2 \qquad (10.49)$$

The best way to get a feel for the correctness of this is to work it out explicitly for a 3×3 determinant. One can then see that this argument is correct for 3×3 determinants and how it goes for larger ones. Again, this expression looks odd because it pulls two arbitrary functions ψ_a and ψ_b out of the determinant. What we normally want is the sum of this quantity over all pairs of particles, i.e.

$$\langle \Phi | \sum_{i>j} \hat{f}(\mathbf{r}_1, \mathbf{r}_2) | \Phi \rangle =$$

$$\sum_{i>j} \int \int \psi_i^\dagger(\mathbf{r}_1) \psi_j^\dagger(\mathbf{r}_2) \hat{f}(\mathbf{r}_1, \mathbf{r}_2) \psi_i(\mathbf{r}_1) \psi_j(\mathbf{r}_2) d\mathbf{r}_1 d\mathbf{r}_2$$

$$- \sum_{i>j} \int \int \psi_i^\dagger(\mathbf{r}_1) \psi_j^\dagger(\mathbf{r}_2) \hat{f}(\mathbf{r}_1, \mathbf{r}_2) \psi_j(\mathbf{r}_1) \psi_i(\mathbf{r}_2) d\mathbf{r}_1 d\mathbf{r}_2 \quad (10.50)$$

It can be seen immediately that this is a generalization of equation (10.36) to the many-electron case.

10.4 The Many-Electron Hamiltonian

A one-electron Hamiltonian contains three terms, the kinetic energy, the rest mass energy and the potential energy felt by the electron due to an external potential. A two-electron Hamiltonian is written down in equation (10.37) and contains seven terms (and even that is approximate). For a many-electron atom, uranium with 92 electrons for instance, the number of terms in the Hamiltonian is absolutely gigantic. It goes without saying that any attempt to solve such a Hamiltonian exactly would be doomed to failure. In this section we will discuss the many-body Hamiltonian and the approximations necessary to make its solution tractable.

A convenient way to write down the Hamiltonian for an N-electron atom is as follows (Bethe and Salpeter 1957, Layzer and Bachall 1962):

$$\hat{H} = \hat{H}_1 + \hat{H}_2 + \hat{H}_3 + \cdots + \hat{H}_N \quad (10.51)$$

where $\hat{H}_1$ contains the terms in the Hamiltonian that are single-particle-like, $\hat{H}_2$ are two-particle interactions, etc. Writing the Hamiltonian in this form is rather natural from both the mathematical and physical point of view. In general the first term is likely to contain terms that make the largest contribution to the energy, the second term the next largest etc. In virtually all applications we do not go beyond the second term in the series, and we will ignore all higher order terms from now on.

For a system of N Dirac particles the first term in equation (10.51) is given by

$$\hat{H}_1 = \sum_{i=1}^{N} \hat{H}_D(i) \quad (10.52)$$

where $\hat{H}_D(i)$ is the Hamiltonian for a single electron

$$\hat{H}_D(i) = c\tilde{\alpha}_i \cdot \hat{\mathbf{p}}_i + \tilde{\beta}_i m_i c^2 + v^{\text{ext}}(\mathbf{r}_i) \tag{10.53}$$

The first two terms here are the kinetic energy and rest mass energy of the electron. The final term represents the effect of an external potential. For the case of an atom $v^{\text{ext}}(\mathbf{r}_i)$ would be the potential at point $\mathbf{r}_i$ due to the presence of the nucleus. Usually this takes on the Coulomb form shown in (10.37). However, that is not always adequate for accurate calculations for heavy atoms because of the finite size of the nucleus. In some cases the nucleus may be large enough for inner s- and p-electron shells to penetrate it significantly, affecting the potential those electrons feel. Smith and Johnson (1967) have shown that if the nucleus is regarded as a uniformly charged sphere of radius R we may write

$$V(\mathbf{r}) = -\frac{Ze^2}{4\pi\epsilon_0 R}\frac{1}{2}\left(3 - \frac{r^2}{R^2}\right) \qquad r \leq R$$

$$= -\frac{Ze^2}{4\pi\epsilon_0 r} \qquad r > R \tag{10.54}$$

and for a nucleus containing A nucleons a good estimate of the nuclear radius is given by $R = 1 \cdot 48 \times 10^{-15} A^{1/3}$ metres.

The second term in (10.51) is dominated by the electron's Coulomb repulsion, and also includes the Breit correction to that. The Breit interaction is of order α^2 of the Coulomb interaction. It is derived perturbatively, and one is on safe mathematical ground in solving the Hamiltonian without it, and including it at the end of the calculation as a perturbation.

The approximate Hamiltonian for a many-electron atom that we will be considering in subsequent sections is

$$\hat{H} = \sum_{i=1}^{N}\left(c\tilde{\alpha}_i \cdot \hat{\mathbf{p}}_i + \tilde{\beta}_i m_i c^2 - \frac{Ze^2}{4\pi\epsilon_0 r_i}\right) + \sum_{i>j}^{N}\frac{e^2}{4\pi\epsilon_0|\mathbf{r}_i - \mathbf{r}_j|} \tag{10.55}$$

where the first summation is over all electrons and the second is over all pairs of electrons. Now we have all the ingredients necessary to derive the equations for calculating the electronic structure of atoms.

10.5 Dirac–Hartree–Fock Integrals

The strategy of the Dirac–Hartree–Fock method for calculating the electronic structure of atoms is to set up an expression for the expectation value of the Hamiltonian, and then to minimize it with respect to variations in the wavefunctions. This is true in the relativistic and non-relativistic theory. However, in relativistic theory there are some complications beyond those found in the non-relativistic theory. The problem that concerns us here is the fact that we have a four-component wavefunction and we want to minimize the total energy with respect to variations in all four components independently, while preserving the normalization. Non-relativistic Hartree–Fock theory is discussed in great detail by Slater (1960, 1963), and Swirles (1935, 1936) developed the relativistic version of the method. However, our treatment follows the papers of Grant (1961, 1965, 1970), but uses a somewhat simpler mathematical approach.

To find the total energy of an N-electron atom we have to find the expectation value of the Hamiltonian above for determinantal wavefunctions. We have seen how to do this in section 10.3, so we can just write down

$$\langle \Phi | \hat{H} | \Phi \rangle = \sum_{i=1}^{\text{occ}} \int \psi_i(\mathbf{r}) \left(c\tilde{\alpha}_i \cdot \hat{\mathbf{p}}_i + \tilde{\beta}_i m_i c^2 - \frac{Ze^2}{4\pi\epsilon_0 r_i} \right) \psi_i(\mathbf{r}_i) d\mathbf{r}_i$$

$$+ \frac{e^2}{4\pi\epsilon_0} \sum_{i>j} \int \int \psi_i^\dagger(\mathbf{r}_1) \psi_j^\dagger(\mathbf{r}_2) \frac{1}{|\mathbf{r}_1 - \mathbf{r}_2|} \psi_i(\mathbf{r}_1) \psi_j(\mathbf{r}_2) d\mathbf{r}_1 d\mathbf{r}_2$$

$$- \frac{e^2}{4\pi\epsilon_0} \sum_{i>j} \int \int \psi_i^\dagger(\mathbf{r}_1) \psi_j^\dagger(\mathbf{r}_2) \frac{1}{|\mathbf{r}_1 - \mathbf{r}_2|} \psi_i(\mathbf{r}_2) \psi_j(\mathbf{r}_1) d\mathbf{r}_1 d\mathbf{r}_2$$

$$(10.56a)$$

where the first term is summed over all occupied states and the other terms are summed over pairs of electrons. It is common to write (10.56a) in the shorthand form

$$E_\mathrm{T} = \langle \Phi | \hat{H} | \Phi \rangle = \sum_i E_i + \sum_{i>j} (J - K) \qquad (10.56b)$$

where J is the direct Coulomb integral and K is the exchange integral. Now the single-particle wavefunctions take on the usual form for solutions of the Dirac equation in a central field (see equation (8.10))

$$\psi_\kappa^{m_j}(\mathbf{r}) = \begin{pmatrix} g_\kappa(r) \chi_\kappa^{m_j}(\hat{\mathbf{r}}) \\ if_\kappa(r) \chi_{-\kappa}^{m_j}(\hat{\mathbf{r}}) \end{pmatrix} = \frac{1}{r} \begin{pmatrix} u_\kappa(r) \chi_\kappa^{m_j}(\hat{\mathbf{r}}) \\ iv_\kappa(r) \chi_{-\kappa}^{m_j}(\hat{\mathbf{r}}) \end{pmatrix} \qquad (10.57)$$

In this form we note that the energy depends only on the radial part of the wavefunction, not on the spin-angular functions. Therefore it will be convenient if we can integrate the spin-angular functions away and leave ourselves with integrals involving $u(r)$ and $v(r)$ only. That is what we will do. Firstly we will consider single-particle and two-particle integrals in turn. The latter subsection will be divided into two, one dealing with the direct Coulomb integral and one with the exchange integral.

In (10.57) the wavefunction is labelled with the quantum numbers κ and m_j. In an atom with many electrons we also have to consider the principal quantum number n explicitly. Henceforth in our discussion of

Hartree–Fock theory, where appropriate, we will replace the subscript κ with the letters a, b, c or d which will represent both n and κ.

Single-Particle Integrals

To find an expression for the total energy of an atom, we need the expectation value of the kinetic energy, rest mass energy and external potential energy in equation (10.56). The kinetic energy is by far the least easy of these, so we will deal with that first.

We want to evaluate $\langle \psi | c\tilde{\alpha} \cdot \hat{\mathbf{p}} | \psi \rangle$. Luckily we have already done most of the work involved in writing this matrix element in terms of $u_a(r)$ and $v_a(r)$ in chapter 8. From equation (8.9) we see that we can write this matrix element as

$$\langle \psi_a^{m_j} | c\tilde{\alpha} \cdot \hat{\mathbf{p}} | \psi_a^{m_j} \rangle = \int \psi_a^{m_j\dagger}(\mathbf{r}) i c \tilde{\gamma}_5 \sigma_r \left(\hbar \frac{\partial}{\partial r} + \frac{\hbar}{r} - \tilde{\beta} \frac{\hat{K}}{r} \right) \psi_a^{m_j}(\mathbf{r}) d\mathbf{r} \quad (10.58)$$

Evaluating (10.58) is now just a question of substituting in the matrices, and the wavefunctions in the form (10.57). This is a little tedious, but presents no difficulty in practice. We can integrate over angles and use the orthonormality of the spin-angular functions to give

$$\langle \psi_a^{m_j} | c\tilde{\alpha} \cdot \hat{\mathbf{p}} | \psi_a^{m_j} \rangle = c\hbar \left(\int v_a(r) \left(\frac{\partial u_a(r)}{\partial r} + \frac{\kappa}{r} u_a(r) \right) dr \right.$$
$$\left. - \int u_a(r) \left(\frac{\partial v_a(r)}{\partial r} - \frac{\kappa}{r} u_a(r) \right) dr \right) \quad (10.59)$$

Next we would like the expectation value of the rest mass energy. The matrix $\tilde{\beta}$ is diagonal so this is easy. It should be clear by inspection that

$$\langle \psi_a^{m_j} | \tilde{\beta} mc^2 | \psi_a^{m_j} \rangle = mc^2 \int \psi_a^{m_j\dagger}(\mathbf{r}) \tilde{\beta} \psi_a^{m_j}(\mathbf{r}) d\mathbf{r}$$

$$= mc^2 \int (u_a^2(r) - v_a^2(r)) dr \quad (10.60)$$

and finally, the nuclear potential term. We use the coulombic form here. This integral is modified if the potential (10.54) is appropriate.

$$\left\langle \psi_a^{m_j} \left| \frac{Ze^2}{4\pi\epsilon_0 r} \right| \psi_a^{m_j} \right\rangle = \frac{Ze^2}{4\pi\epsilon_0} \int \psi_a^{m_j\dagger}(\mathbf{r}) \frac{1}{r} \psi_a^{m_j}(\mathbf{r}) d\mathbf{r}$$

$$= \frac{Ze^2}{4\pi\epsilon_0} \int \frac{1}{r} (u_a^2(r) + v_a^2(r)) dr \quad (10.61)$$

Equations (10.59), (10.60) and (10.61) give the total single-particle energy for one electron in an atom. Note that none of these equations contain m_j. Therefore we may write the total single-particle energy for all the electrons in an atom as a sum over all occupied shells a of equations (10.59), (10.60) and (10.61) with each term in the sum multiplied by n_a, the number of occupied m_j states for that value of a, i.e.

$$\langle \Phi | \hat{H}_1 | \Phi \rangle = \sum_{a,m_j} \int \psi_a^{m_j\dagger}(\mathbf{r}) \hat{H}_D(\kappa, m_j) \psi_a^{m_j}(\mathbf{r}) d\mathbf{r}$$

$$= \sum_a n_a \left(mc^2 \int (u_a^2(r) - v_a^2(r)) dr - \frac{Ze^2}{4\pi\epsilon_0} \int \frac{1}{r}(u_a^2(r) + v_a^2(r)) dr \right.$$

$$\left. +c\hbar \int \left(v_a(r) \left(\frac{\partial u_a(r)}{\partial r} + \frac{\kappa_a}{r} u_a(r) \right) - u_a(r) \left(\frac{\partial v_a(r)}{\partial r} - \frac{\kappa_a}{r} u_a(r) \right) \right) dr \right)$$

(10.62)

This equation expresses the single-particle part of the total energy in terms of the radial part of the single-particle wavefunctions only. It has the drawback that it contains the rest mass energy of the electrons, which is gigantic compared with atomic energies. An approximate way of removing rest mass energies is to subtract $\sum_a n_a mc^2 \int (u_a^2(r) + v_a^2(r)) dr$ from (10.62). This leaves an expression for the energy which, when evaluated, can be compared directly with a non-relativistic calculation.

Two-Particle Integrals

As we have seen, in Hartree–Fock theory there are two types of two-particle integral, the direct Coulomb, and the exchange, term. Here we will develop a general two-particle integral and then specialize to evaluate each type separately. A general two-particle Coulomb integral is

$$\langle a, b | \frac{e^2}{4\pi\epsilon_0 |\mathbf{r}_{12}|} | c, d \rangle = \int \int \psi_a^\dagger(\mathbf{r}_1) \psi_b^\dagger(\mathbf{r}_2) \frac{e^2}{4\pi\epsilon_0 |\mathbf{r}_{12}|} \psi_c(\mathbf{r}_1) \psi_d(\mathbf{r}_2) d\mathbf{r}_1 d\mathbf{r}_2$$

(10.63)

This is a nasty integral because the $1/r_{12} = 1/|\mathbf{r}_1 - \mathbf{r}_2|$ becomes infinite as $\mathbf{r}_2 \to \mathbf{r}_1$. However, there is a famous expansion, which no-one's ever heard of, that we can use to get round this problem:

$$\frac{1}{|\mathbf{r}_1 - \mathbf{r}_2|} = \sum_{l=0}^\infty \sum_{m=-l}^{+l} \left(\frac{4\pi}{2l+1} \right) \frac{r_<^l}{r_>^{l+1}} Y_l^{m*}(\hat{\mathbf{r}}_<) Y_l^m(\hat{\mathbf{r}}_>)$$

(10.64)

where $\mathbf{r}_<$ is the smaller of $\mathbf{r}_1$ and $\mathbf{r}_2$, and $\mathbf{r}_>$ is the larger. (Note the fact that the left hand side of this expression is unaffected if we interchange $\mathbf{r}_1$ and $\mathbf{r}_2$, so the right hand side must also be unaffected.) Now our task is to substitute (10.64), and the wavefunctions (10.57), into (10.63). After multiplying out the 4-component wavefunctions we find

$$\langle a,b|\frac{e^2}{4\pi\epsilon_0|\mathbf{r}_1-\mathbf{r}_2|}|c,d\rangle =$$

$$\frac{e^2}{4\pi\epsilon_0}\int d\mathbf{r}_1\int d\mathbf{r}_2\sum_{l=0}^{\infty}\sum_{m=-l}^{+l}\left(\frac{4\pi}{2l+1}\right)\frac{r_<^l}{r_>^{l+1}}Y_l^{m*}(\hat{\mathbf{r}}_<)Y_l^m(\hat{\mathbf{r}}_>)\times$$

$$\left(u_a(r_1)u_c(r_1)u_b(r_2)u_d(r_2)\chi_{\kappa_a}^{m_{ja}\dagger}(\hat{\mathbf{r}}_1)\chi_{\kappa_c}^{m_{jc}}(\hat{\mathbf{r}}_1)\chi_{\kappa_b}^{m_{jb}\dagger}(\hat{\mathbf{r}}_2)\chi_{\kappa_d}^{m_{jd}}(\hat{\mathbf{r}}_2)\right.$$

$$+u_a(r_1)u_c(r_1)v_b(r_2)v_d(r_2)\chi_{\kappa_a}^{m_{ja}\dagger}(\hat{\mathbf{r}}_1)\chi_{\kappa_c}^{m_{jc}}(\hat{\mathbf{r}}_1)\chi_{-\kappa_b}^{m_{jb}\dagger}(\hat{\mathbf{r}}_2)\chi_{-\kappa_d}^{m_{jd}}(\hat{\mathbf{r}}_2)$$

$$+v_a(r_1)v_c(r_1)u_b(r_2)u_d(r_2)\chi_{-\kappa_a}^{m_{ja}\dagger}(\hat{\mathbf{r}}_1)\chi_{-\kappa_c}^{m_{jc}}(\hat{\mathbf{r}}_1)\chi_{\kappa_b}^{m_{jb}\dagger}(\hat{\mathbf{r}}_2)\chi_{\kappa_d}^{m_{jd}}(\hat{\mathbf{r}}_2)$$

$$\left.+v_a(r_1)v_c(r_1)v_b(r_2)v_d(r_2)\chi_{-\kappa_a}^{m_{ja}\dagger}(\hat{\mathbf{r}}_1)\chi_{-\kappa_c}^{m_{jc}}(\hat{\mathbf{r}}_1)\chi_{-\kappa_b}^{m_{jb}\dagger}(\hat{\mathbf{r}}_2)\chi_{-\kappa_d}^{m_{jd}}(\hat{\mathbf{r}}_2)\right) \quad (10.65)$$

This is a dreadful looking expression, and it gets worse! Next we expand the spin-angular functions of equation (2.131a) and substitute them into (10.65). The resulting equation can be shortened with a judicious choice of notation. Let us rename the wavefunctions as follows. Let $u(r) = R_{(+1)}$ and $v(r) = R_{(-1)}$. The reason for this will become clear shortly. We will define the new quantities $\beta_a = \pm 1$ and $\beta_b = \pm 1$. We recall the definition of S_κ in equation (2.129), and define

$$\lambda_a = j_a + \tfrac{1}{2}S_{\kappa_a}\beta_a, \qquad \lambda_b = j_b + \tfrac{1}{2}S_{\kappa_b}\beta_b$$
$$\lambda_c = j_c + \tfrac{1}{2}S_{\kappa_c}\beta_a, \qquad \lambda_d = j_d + \tfrac{1}{2}S_{\kappa_d}\beta_b \qquad (10.66)$$

where j_i is the j quantum number associated with κ_i. These are just definitions (Grant 1970), but we can use them to expand (10.65), and keep it relatively short (even so, it's still pretty long):

$$\langle a,b|\frac{e^2}{4\pi\epsilon_0|\mathbf{r}_1-\mathbf{r}_2|}|c,d\rangle = \frac{e^2}{4\pi\epsilon_0}\int d\mathbf{r}_1\int d\mathbf{r}_2\sum_{l=0}^{\infty}\sum_{m=-l}^{+l}\left(\frac{4\pi}{2l+1}\right)\frac{r_<^l}{r_>^{l+1}}$$

$$Y_l^{m*}(\hat{\mathbf{r}}_<)Y_l^m(\hat{\mathbf{r}}_>)\sum_{\beta_a=\pm 1}\sum_{\beta_b=\pm 1}R_{a(\beta_a)}(r_1)R_{c(\beta_a)}(r_1)R_{b(\beta_b)}(r_2)R_{d(\beta_b)}(r_2)$$

$$\left(C(\lambda_a\tfrac{1}{2}j_a;m_{ja}-\tfrac{1}{2},\tfrac{1}{2})C(\lambda_c\tfrac{1}{2}j_c;m_{jc}-\tfrac{1}{2},\tfrac{1}{2})Y_{\lambda_a}^{m_{ja}-\frac{1}{2}*}(\hat{\mathbf{r}}_1)Y_{\lambda_c}^{m_{jc}-\frac{1}{2}}(\hat{\mathbf{r}}_1)+\right.$$

$$\left.C(\lambda_a\tfrac{1}{2}j_a;m_{ja}+\tfrac{1}{2},-\tfrac{1}{2})C(\lambda_c\tfrac{1}{2}j_c;m_{jc}+\tfrac{1}{2},\tfrac{1}{2})Y_{\lambda_a}^{m_{ja}+\frac{1}{2}*}(\hat{\mathbf{r}}_1)Y_{\lambda_c}^{m_{jc}+\frac{1}{2}}(\hat{\mathbf{r}}_1)\right).$$

$$\left(C(\lambda_b\tfrac{1}{2}j_b;m_{jb}-\tfrac{1}{2},\tfrac{1}{2})C(\lambda_d\tfrac{1}{2}j_d;m_{jd}-\tfrac{1}{2},\tfrac{1}{2})Y_{\lambda_b}^{m_{jb}-\frac{1}{2}*}(\hat{\mathbf{r}}_2)Y_{\lambda_d}^{m_{jd}-\frac{1}{2}}(\hat{\mathbf{r}}_2)+\right.$$

$$\left.C(\lambda_b\tfrac{1}{2}j_b;m_{jb}+\tfrac{1}{2},-\tfrac{1}{2})C(\lambda_d\tfrac{1}{2}j_d;m_{jd}+\tfrac{1}{2},-\tfrac{1}{2})Y_{\lambda_c}^{m_{jc}+\frac{1}{2}*}(\hat{\mathbf{r}}_2)Y_{\lambda_d}^{m_{jd}+\frac{1}{2}}(\hat{\mathbf{r}}_2)\right)$$

$$(10.67)$$

We can reduce this considerably further by summing over $s_a = \pm\frac{1}{2}$ and $s_b = \pm\frac{1}{2}$. If we multiply out the brackets in (10.67) and separate the radial and angular integrals it becomes

$$\langle a, b | \frac{e^2}{4\pi\epsilon_0 |\mathbf{r}_1 - \mathbf{r}_2|} | c, d \rangle = \sum_{\beta_a = \pm 1} \sum_{\beta_b = \pm 1} \sum_{s_a = \pm\frac{1}{2}} \sum_{s_b = \pm\frac{1}{2}}$$

$$\sum_{l=0}^{\infty} \frac{4\pi}{2l+1} \sum_{m=-l}^{+l} C(\lambda_a \tfrac{1}{2} j_a; m_{j_a} - s_a, s_a) C(\lambda_b \tfrac{1}{2} j_b; m_{j_b} - s_b, s_b)$$

$$C(\lambda_c \tfrac{1}{2} j_c; m_{j_c} - s_a, s_a) \int Y_{\lambda_a}^{m_{j_a} - s_a *}(\hat{\mathbf{r}}_1) Y_{\lambda_c}^{m_{j_c} - s_a}(\hat{\mathbf{r}}_1) Y_l^{m *}(\hat{\mathbf{r}}_1) d\hat{\mathbf{r}}_1$$

$$C(\lambda_d \tfrac{1}{2} j_d; m_{j_d} - s_b, s_b) \int Y_l^m(\hat{\mathbf{r}}_2) Y_{\lambda_b}^{m_{j_b} - s_b *}(\hat{\mathbf{r}}_2) Y_{\lambda_d}^{m_{j_d} - s_b}(\hat{\mathbf{r}}_2) d\hat{\mathbf{r}}_2$$

$$\frac{e^2}{4\pi\epsilon_0} \int \int R_{a(\beta_a)}(r_1) R_{c(\beta_a)}(r_1) R_{b(\beta_b)}(r_2) R_{d(\beta_b)}(r_2) \frac{r_<^l}{r_>^{l+1}} dr_1 dr_2 \qquad (10.68)$$

There is a subtlety here. We have assigned $r_< = r_1$. This identification can be made for the following reason. The expansion (10.64) must be symmetric in $\mathbf{r}_1$ and $\mathbf{r}_2$, because it doesn't make any difference to the left hand side if we swap them round. It is only the summation over l and m that is symmetric and so has no angular dependence. Each individual term in the summation can (and does) have an angular dependence. The integral of three spherical harmonics in (10.68) will give different answers depending on whether we assign $r_<$ and $r_>$ to r_1 or r_2. However, when we go on to do the summations over l and m, the sum will be independent of which way round we made the assignment.

Now we have got about as far as we can go for a general two-particle matrix element. Next we will calculate the direct integral and the exchange integral separately.

The Direct Coulomb Integral

For the direct Coulomb integral, equation (10.68) simplifies considerably because $c = a$ and $d = b$. We have

$$J = \langle a, b | \frac{e^2}{4\pi\epsilon_0 |\mathbf{r}_1 - \mathbf{r}_2|} | a, b \rangle = \frac{e^2}{4\pi\epsilon_0} \sum_{\beta_a = \pm 1} \sum_{\beta_b = \pm 1} \sum_{s_a = \pm\frac{1}{2}} \sum_{s_b = \pm\frac{1}{2}}$$

$$\sum_{l=0}^{\infty} \frac{4\pi}{2l+1} \sum_{m=-l}^{+l} \int \int R_{a(\beta_a)}^2(r_1) R_{b(\beta_b)}^2(r_2) \frac{r_<^l}{r_>^{l+1}} dr_1 dr_2 \qquad (10.69)$$

$$C^2(\lambda_a \tfrac{1}{2} j_a; m_{j_a} - s_a, s_a) \int Y_{\lambda_a}^{m_{j_a} - s_a *}(\hat{\mathbf{r}}_1) Y_{\lambda_a}^{m_{j_a} - s_a}(\hat{\mathbf{r}}_1) Y_l^{m *}(\hat{\mathbf{r}}_1) d\hat{\mathbf{r}}_1$$

$$C^2(\lambda_b \tfrac{1}{2} j_b; m_{j_b} - s_b, s_b) \int Y_l^m(\hat{\mathbf{r}}_2) Y_{\lambda_b}^{m_{j_b} - s_b *}(\hat{\mathbf{r}}_2) Y_{\lambda_b}^{m_{j_b} - s_b}(\hat{\mathbf{r}}_2) d\hat{\mathbf{r}}_2$$

To simplify this further we need to do the angular integrals. They are products of three spherical harmonics and can be done using equations (C.16) and (C.19) from appendix C. For our particular case we have the added simplification that $l'' = l$ and $m'' = m$ in (C.19). So

$$
\int Y_{\lambda_a}^{m_{ja}-S_a *}(\hat{\mathbf{r}}_1) Y_l^{m*}(\hat{\mathbf{r}}_<) Y_{\lambda_a}^{m_{ja}-S_a}(\hat{\mathbf{r}}_1) d\hat{\mathbf{r}}_1 =
$$

$$
\left(\frac{2l+1}{4\pi}\right)^{\frac{1}{2}} C(l_1 l l_1; m_1 m) C(l_1 l l_1; 00) \qquad (10.70)
$$

Now, by the triangle condition (2.108a) for Clebsch–Gordan coefficients, we have that the only value of l for which the right hand side of (10.70) is non-zero is $l = 0$. This means that m is also equal to zero, and hence we can eliminate the sums over l and m in (10.69). The numerical value of the integral in (10.70) is $1/\sqrt{4\pi}$. In (10.69) there is actually a product of two of these integrals, so all the angular integrals in (10.69) reduce to $1/4\pi$ and (10.69) becomes

$$
J = \frac{e^2}{4\pi\epsilon_0} \sum_{\beta_a=\pm1} \sum_{\beta_b=\pm1} \sum_{s_a=\pm\frac{1}{2}} \sum_{s_b=\pm\frac{1}{2}} C^2(\lambda_a \tfrac{1}{2} j_a; m_{j_a} - s_a, s_a)
$$

$$
\times C^2(\lambda_b \tfrac{1}{2} j_b; m_{j_b} - s_b, s_b) \int\int R_{a(\beta_a)}^2(r_1) R_{b(\beta_b)}^2(r_2) \frac{1}{r_>} dr_1 dr_2 \qquad (10.71)
$$

From table (2.2) we can see that

$$
C^2(l\tfrac{1}{2}j; m_j - \tfrac{1}{2}, \tfrac{1}{2}) + C^2(l\tfrac{1}{2}j; m_j + \tfrac{1}{2}, -\tfrac{1}{2}) = 1 \qquad (10.72)
$$

We can re-expand the summations over s_a and s_b in equation (10.71) and use (10.72) to see that

$$
J = \frac{e^2}{4\pi\epsilon_0} \sum_{\beta_a=\pm1} \sum_{\beta_b=\pm1} \int\int R_{a(\beta_a)}^2(r_1) R_{b(\beta_b)}^2(r_2) \frac{1}{r_>} dr_1 dr_2 \qquad (10.73)
$$

Finally we can also get rid of the summations over β_a and β_b. This enables us to put $u(r)$ and $v(r)$ back into the equations giving

$$
J = \frac{e^2}{4\pi\epsilon_0} \int\int (u_a^2(r_1) + v_a^2(r_1))(u_b^2(r_2) + v_b^2(r_2)) \frac{1}{r_>} dr_1 dr_2 = F_0(a,b) \quad (10.74a)
$$

with

$$
F_l(a,b) = \frac{e^2}{4\pi\epsilon_0} \int\int (u_a^2(r_1) + v_a^2(r_1))(u_b^2(r_2) + v_b^2(r_2)) \frac{r_<^l}{r_>^{l+1}} dr_1 dr_2 \quad (10.74b)
$$

This is in full agreement with the derivation of Grant (1961, 1970). As in the non-relativistic case, the radial integral does not depend on the m_j quantum number. Therefore we can find the Coulomb energy for all N

electrons in the atom as a sum over shells (labelled by the value of their quantum numbers a) weighted by the number of electrons in each shell.

$$J_T = \sum_a \left(\frac{1}{2} n_a(n_a - 1)F_0(a, a) + \frac{1}{2}\sum_{b \neq a} n_a n_b F_0(a, b) \right) \tag{10.75}$$

The first term in equation (10.75) is the interaction of all electrons in a shell with all the other electrons in that shell (because $F_0(a, a)$ describes the interaction of one pair of electrons and $\frac{1}{2}n_a(n_a - 1)$ is the number of pairs in a shell containing n_a electrons). The second term is the interaction of all electrons in the a-shell with all electrons in the b-shell. The half outside the summation avoids counting every pair of electrons twice.

Equation (10.75) is what we were looking for in this section. We have written the direct Coulomb integral in the expectation value of the many-body Hamiltonian in terms of the radial part of the single-particle wave-functions only. Now we go on to do the same for the exchange integral.

The Exchange Integral

To evaluate the exchange integral we need to take the special case of equation (10.68) when $c = b$ and $d = a$. It becomes

$$K = \langle a, b | \frac{e^2}{4\pi\epsilon_0 |\mathbf{r}_1 - \mathbf{r}_2|} | b, a \rangle = \sum_{\beta_a = \pm 1} \sum_{\beta_b = \pm 1} \sum_{s_a = \pm \frac{1}{2}} \sum_{s_b = \pm \frac{1}{2}}$$

$$\sum_{l=0}^{\infty} \frac{4\pi}{2l + 1} \sum_{m=-l}^{+l} C(\lambda_b \tfrac{1}{2} j_b; m_{j_b} - s_a, s_a) C(\lambda_a \tfrac{1}{2} j_a; m_{j_a} - s_b, s_b)$$

$$C(\lambda_a \tfrac{1}{2} j_a; m_{j_a} - s_a, s_a) \int Y_{\lambda_a}^{m_{j_a} - s_a *}(\hat{\mathbf{r}}_1) Y_{\lambda_b}^{m_{j_b} - s_a}(\hat{\mathbf{r}}_1) Y_l^{m*}(\hat{\mathbf{r}}_1) d\hat{\mathbf{r}}_1$$

$$C(\lambda_b \tfrac{1}{2} j_b; m_{j_b} - s_b, s_b) \int Y_l^m(\hat{\mathbf{r}}_2) Y_{\lambda_b}^{m_{j_b} - s_b *}(\hat{\mathbf{r}}_2) Y_{\lambda_a}^{m_{j_a} - s_b}(\hat{\mathbf{r}}_2) d\hat{\mathbf{r}}_2$$

$$\frac{e^2}{4\pi\epsilon_0} \int \int R_{a(\beta_a)}(r_1) R_{b(\beta_a)}(r_1) R_{b(\beta_b)}(r_2) R_{a(\beta_b)}(r_2) \frac{r_<^l}{r_>^{l+1}} dr_1 dr_2 \tag{10.76}$$

The first thing to do here is the angular integrals. Again they can be evaluated immediately using equations (C.16) and (C.19) from appendix C. Now, as can be seen by inspection of (C.19), the integral of three spherical harmonics in (10.76) will be zero unless

$$m = m_{j_a} - m_{j_b} \tag{10.77}$$

So we can remove the sum over m in (10.76) and replace m wherever it occurs using (10.77). If we also substitute for the angular integrals in

(10.76) from appendix C we have, after a little rearrangement,

$$K = \frac{e^2}{4\pi\epsilon_0}(-1)^{m_{ja}-m_{jb}} \sum_{\beta_a=\pm 1}\sum_{\beta_b=\pm 1}\sum_{l=0}^{\infty} C(\lambda_b l \lambda_a;0,0)C(\lambda_a l \lambda_b;0,0)$$

$$\int\int R_{a(\beta_a)}(r_1)R_{b(\beta_a)}(r_1)R_{b(\beta_b)}(r_2)R_{a(\beta_b)}(r_2)\frac{r_<^l}{r_>^{l+1}}dr_1 dr_2$$

$$\times I_{\lambda_a,\lambda_b,l}^{m_{ja},m_{jb}} I_{\lambda_b,\lambda_a,l}^{m_{jb},m_{ja}} \tag{10.78}$$

where we have defined

$$I_{\lambda_2,\lambda_1,l}^{m_{j_2},m_{j_1}} = \sum_{s=\pm\frac{1}{2}} C(\lambda_2\tfrac{1}{2}j_2;m_{j_2}-s,s)$$

$$\times C(\lambda_1\tfrac{1}{2}j_1;m_{j_1}-s,s)C(\lambda_1 l \lambda_2;m_{j_1}-s,m_{j_2}-m_{j_1}) \tag{10.79}$$

because the two summations over s_a and s_b have the same form. It is these we simplify next. We work with the general sum (10.79) and will substitute the result back into the specific examples in equation (10.78) later.

Using Clebsch–Gordan symmetry relations we can write $I_{\lambda_2,\lambda_1,l}^{m_{j_2},m_{j_1}}$ as

$$I_{\lambda_2,\lambda_1,l}^{m_{j_2},m_{j_1}} = (-1)^{\lambda_1+\lambda_2-l-j_2-m_{j_2}}\left(\frac{2j_1+1}{2l+1}\right)^{1/2} \times$$

$$\sum_{s=\pm\frac{1}{2}} C(j_1\tfrac{1}{2}\lambda_1;-m_{j_1},s)C(\tfrac{1}{2}\lambda_2 j_2;s,m_{j_2}-s)C(\lambda_1\lambda_2 l;s-m_{j_1},m_{j_2}-s)$$

$$\tag{10.80}$$

Equation (10.80) is in a form where we can use the recoupling formula, discussed in chapter 2, to remove the summation over s. Using the symmetry relations for the Racah coefficients (equations (2.125)) we find

$$I_{\lambda_2,\lambda_1,l}^{m_{j_2},m_{j_1}} = (-1)^{\lambda_2-m_{j_2}-1/2} \times \left(\frac{(2j_1+1)(2j_2+1)}{2l+1}\right)^{1/2} \times$$

$$(2\lambda_1+1)^{1/2}W(\lambda_1 j_1 \lambda_2 j_2;\tfrac{1}{2}l)C(j_1 j_2 l;-m_{j_1},m_{j_2}) \tag{10.81}$$

Now, expressions for I_{s_a} and I_{s_b} can be put into (10.78) and we get

$$K = \sum_{l=0}^{\infty}(-1)^{m_{ja}-m_{jb}} \sum_{\beta_a=\pm 1}\sum_{\beta_b=\pm 1} d^l(j_a,m_{ja},j_b,m_{jb})d^l(j_b,m_{jb},j_a,m_{ja})$$

$$\tag{10.82}$$

$$\times \frac{e^2}{4\pi\epsilon_0}\int\int R_{a(\beta_a)}(r_1)R_{b(\beta_a)}(r_1)R_{b(\beta_b)}(r_2)R_{a(\beta_b)}(r_2)\frac{r_<^l}{r_>^{l+1}}dr_1 dr_2$$

where we have defined the coefficients

$$d^l(j_1, m_{j_1}, j_2, m_{j_2}) = (-1)^{\lambda_1 - m_{j_1} - 1/2} \left(\frac{(2j_1 + 1)(2j_2 + 1)}{2l + 1} \right)^{1/2} \times$$

$$(2\lambda_1 + 1)^{1/2} C(j_1 j_2 l; -m_{j_1}, m_{j_2}) W(\lambda_1 j_1 \lambda_2 j_2; \tfrac{1}{2} l) C(\lambda_1 l \lambda_2; 0, 0)$$

(10.83)

Let us just consider the Racah coefficient and the final Clebsch–Gordan coefficient in (10.83). We use the recoupling formula (2.123*b*) again and the Clebsch–Gordan symmetry relations (again, it's not boring, honest!) to write them as

$$W(\lambda_1 j_1 \lambda_2 j_2; \tfrac{1}{2} l) C(\lambda_1 l \lambda_2; 0, 0)$$

$$= (-1)^{\lambda_1} \left(\frac{(2\lambda_2 + 1)}{(2l + 1)(2j_1 + 1)(2j_2 + 1)} \right)^{1/2} C(j_1 j_2 l; -\tfrac{1}{2}, \tfrac{1}{2}) \times$$

$$(C(\lambda_1 \tfrac{1}{2} j_1; 0, -\tfrac{1}{2}) C(\lambda_2 \tfrac{1}{2} j_2; 0, -\tfrac{1}{2}) + C(\lambda_1 \tfrac{1}{2} j_1; 0, +\tfrac{1}{2}) C(\lambda_2 \tfrac{1}{2} j_2; 0, +\tfrac{1}{2}))$$

(10.84)

We know the algebraic expressions for the Clebsch–Gordan coefficients from table 2.2. Writing them in terms of *j* rather than *l* and substituting them into equation (10.84) enables us to write it more simply as

$$W(\lambda_1 j_1 \lambda_2 j_2; \tfrac{1}{2} l) C(\lambda_1 l \lambda_2; 0, 0) = (-1)^{\lambda_1} \left(\frac{(2\lambda_2 + 1)}{(2l + 1)(2j_1 + 1)(2j_2 + 1)} \right)^{\frac{1}{2}}$$

$$\times \left(\frac{(2j_1 + 1)(2j_2 + 1)}{(2\lambda_1 + 1)(2\lambda_2 + 1)} \right)^{\frac{1}{2}} C(j_1 j_2 l; -\tfrac{1}{2}, \tfrac{1}{2})$$

(10.85)

Putting this back into (10.83) and doing some cancelling gives

$$d^l(j_1, m_{j_1}, j_2, m_{j_2}) = (-1)^{m_{j_1} + 1/2} \times$$

$$\frac{((2j_1 + 1)(2j_2 + 1))^{1/2}}{(2l + 1)} C(j_1 j_2 l; -\tfrac{1}{2}, \tfrac{1}{2}) C(j_1 j_2 l; -m_{j_1}, m_{j_2})$$

(10.86)

For cases of interest to us m_{j_1} is a half-integer, so $(-1)^{m_{j_1} + 1/2}$ must be ± 1 and $d^l(j_1, m_{j_1}, j_2 m_{j_2})$ is real. With this expression for the coefficients it is easy to use Clebsch–Gordan symmetry relations (again) to show

$$d^l(j_2, m_{j_2}, j_1, m_{j_1}) = (-1)^{m_{j_2} - m_{j_1}} d^l(j_1, m_{j_1}, j_2, m_{j_2})$$

(10.87)

These coefficients have now become independent of λ_1 and λ_2, so they can go outside the summations over β_a and β_b when we substitute for

$d^l(j_1, m_{j_1}, j_2, m_{j_2})$ in (10.82). Doing this gives

$$K = \frac{e^2}{4\pi\epsilon_0} \sum_{l=0}^{\infty} (d^l(j_a, m_{j_a}, j_b, m_{j_b}))^2 \sum_{\beta_a=\pm 1} \sum_{\beta_b=\pm 1}$$

$$\int\int R_{a(\beta_a)}(r_1)R_{b(\beta_a)}(r_1)R_{b(\beta_b)}(r_2)R_{a(\beta_b)}(r_2)\frac{r_<^l}{r_>^{l+1}} dr_1 dr_2 \qquad (10.88)$$

We can get rid of the summations over β_a and β_b by returning to the symbols $u(r)$ and $v(r)$ to represent the radial components of the wavefunctions. The equation can be shortened still further by defining

$$b^l(j_a, m_{j_a}, j_b, m_{j_b}) = (d^l(j_a, m_{j_a}, j_b, m_{j_b}))^2 \qquad (10.89)$$

So our final expression for the exchange integral for two electrons is

$$K = \sum_{l=0}^{\infty} b^l(j_a, m_{j_a}, j_b, m_{j_b})G_l(a, b) \qquad (10.90a)$$

where $G_l(a, b)$ is the radial integral

$$G_l(a, b) = \frac{e^2}{4\pi\epsilon_0} \int\int (u_a(r_1)u_b(r_1) + v_a(r_1)v_b(r_1)) \times$$

$$(u_a(r_2)u_b(r_2) + v_a(r_2)v_b(r_2))\frac{r_<^l}{r_>^{l+1}} dr_1 dr_2 \qquad (10.90b)$$

We simplify $b^l(j_a, m_{j_a}, j_b, m_{j_b})$ by writing it explicitly using (10.86) and the symmetry of the Clebsch–Gordan coefficients (yet again). As we have seen, $m_{j_a} + 1/2$ is an integer, so $(-1)^{2m_{j_a}+1} = 1$ and we find

$$b^l(j_a, m_{j_a}, j_b, m_{j_b}) = \left(\frac{2j_b+1}{2l+1}\right) C^2(j_a j_b l; -\tfrac{1}{2}, \tfrac{1}{2})C^2(j_b l j_a; m_{j_b}, m_{j_a} - m_{j_b})$$

$$(10.91)$$

Note that m_{j_b} only appears in the last Clebsch–Gordan coefficient here. Let us evaluate the exchange integral for the case of an electron with quantum numbers j_a, m_{j_a} interacting with a complete shell of electrons with quantum number j_b and all possible values of m_{j_b}. The radial integral in (10.90) has no m_{j_b} dependence, so, it is sufficient to evaluate

$$\sum_{m_{j_b}=-j_b}^{j_b} b^l(j_a, m_{j_a}, j_b, m_{j_b}) =$$

$$\left(\frac{2j_b+1}{2l+1}\right) C^2(j_a j_b l; -\tfrac{1}{2}, \tfrac{1}{2}) \sum_{m_{j_b}=-j_b}^{j_b} C^2(j_b l j_a; m_{j_b}, m_{j_a} - m_{j_b}) \qquad (10.92)$$

But from equation (2.115) we find

$$\sum_{m_{j_b}=-j_b}^{j_b} b^l(j_a, m_{j_a}, j_b, m_{j_b}) = \tfrac{1}{2}(2j_b + 1)\Gamma^l_{j_a,j_b} \tag{10.93a}$$

$$\Gamma^l_{j_a,j_b} = \frac{2}{2l+1} C^2(j_a j_b l; -\tfrac{1}{2}, \tfrac{1}{2}) \tag{10.93b}$$

We are discussing a full shell of electrons so $(2j_b + 1)$ is the number of electrons in the b-shell. With these definitions the exchange energy between an electron in the κ_a-shell and all the electrons in the κ_b-shell is

$$K = \sum_{l=0}^{\infty} \tfrac{1}{2}(2j_b + 1)\Gamma^l_{j_a,j_b} G_l(a, b) \tag{10.94}$$

Finally, we write the total exchange energy of the atom as a sum over electron shells. Let us consider a closed shell atom. The exchange energy between electrons in the same shell can be estimated from (10.94)

$$K = \sum_{l=0}^{\infty} \tfrac{1}{2}(2j_a + 1)\Gamma^l_{j_a,j_a} G_l(a, a) \tag{10.95}$$

This is multiplied by $\tfrac{1}{2}n_a$ to give the total exchange energy in one shell and avoid double counting. A single determinant wavefunction is only appropriate for an atom with full shells or full shells plus one electron. so if we multiply (10.95) by $(n_a - 1)/(2j)$ (one for a full shell, zero for a shell with one electron) we can sum (10.95) over occupied shells. The exchange energy between one electron and electrons in a different shell is given by (10.94). To get the exchange energy of the atom we sum over shells, weighted by the number of electrons in each shell, and multiply by $\tfrac{1}{2}$ to avoid double counting. The total exchange energy for an atom is

$$K_{\mathrm{T}} = \sum_{a} \left(\tfrac{1}{2}n_a \frac{n_a - 1}{2j_a} \sum_{l=0}^{\infty} \tfrac{1}{2}(2j_a + 1)\Gamma^l_{j_a,j_a} G_l(a, a) \right.$$
$$\left. + \tfrac{1}{2}\sum_{b \neq a} n_a \sum_{l=0}^{\infty} \tfrac{1}{2}(2j_b + 1)\Gamma^l_{j_a,j_b} G_l(a, b) \right) \tag{10.96}$$

This is our final expression for the exchange energy. It is written simply in terms of single-particle wavefunctions, quantum numbers and occupation numbers.

10.6 The Dirac–Hartree–Fock Equations

In the previous sections we have derived expressions for the components of the energy of an atom in terms of the single-electron radial wavefunctions.

Let us bring all these together to find an expression for the total energy of an atom (Swirles 1935, 1936). For the Hamiltonian of equation (10.55) and from equations (10.56), (10.62), (10.75), and (10.96), this is given by

$$
\begin{aligned}
E_T = \\
\sum_a n_a \Bigg(c\hbar \Bigg(\int v_a(r) \left(\frac{\partial u_a(r)}{\partial r} + \frac{\kappa_a}{r} u_a(r) \right) dr \\
- \int u_a(r) \left(\frac{\partial v_a(r)}{\partial r} - \frac{\kappa_a}{r} u_a(r) \right) dr \Bigg) \\
- \frac{Ze^2}{4\pi\epsilon_0} \int \frac{1}{r} (u_a^2(r) + v_a^2(r)) dr \\
+ mc^2 \int (u_a^2(r) - v_a^2(r)) dr \Bigg) \\
- \sum_a \Bigg(\frac{1}{2} n_a \frac{n_a - 1}{2 j_a} \sum_{l=0}^{\infty} \frac{1}{2}(2j_a + 1) \Gamma^l_{j_a, j_a} F_l(a, a) \\
- \sum_{b \neq a} n_a \sum_{l=0}^{\infty} \frac{1}{4}(2j_b + 1) \Gamma^l_{j_a, j_b} G_l(a, b) \Bigg) \\
+ \sum_a \Bigg(\frac{1}{2} n_a(n_a - 1) F_0(a, a) + \frac{1}{2} \sum_{b \neq a} n_a n_b F_0(a, b) \Bigg) \qquad (10.97)
\end{aligned}
$$

where we have used the fact that $F_l(a, a) = G_l(a, a)$, and the wavefunctions are constrained to be normalized such that

$$
I_{a,b} = \int (u_a^*(r) u_b(r) + v_a^*(r) v_b(r)) dr = \delta_{a,b} \qquad (10.98)
$$

The next step is to derive the Dirac–Hartree–Fock equations by minimizing the expectation value (10.97) with respect to variations in $u_a(r)$ and $v_a(r)$, subject to maintaining the normalization. Equation (10.97) has been set up for many-electron atoms. However, it is instructive to minimize it for a one-electron atom, and we will do that first.

The One-Electron Atom

In the one-electron limit there is no Coulomb repulsion or exchange energy between electrons (obviously). The factors of $n_a - 1$ take care of that within the occupied shell and the occupancy of all other shells is zero. As there is only one occupied shell the summations over shells also disappear and we only have to worry about the angular momentum quantum numbers.

Equation (10.97) can then be rearranged slightly to

$$E_T^1 = \int v_\kappa(r) \left(\left(\frac{\partial u_\kappa(r)}{\partial r} + \frac{\kappa}{r} u_\kappa(r) \right) c\hbar - v_\kappa(r)(mc^2 - V(r)) \right) dr$$
$$- \int u_\kappa(r) \left(\left(\frac{\partial v_\kappa(r)}{\partial r} - \frac{\kappa}{r} v_\kappa(r) \right) c\hbar - u_\kappa(r)(mc^2 + V(r)) \right) dr \qquad (10.99)$$

Now, let us work out the change in this energy, ΔE_T^1, if we vary $u(r)$ while everything else remains constant.

$$\Delta E_T^1 = \int v_\kappa(r) \left(\left(\frac{\partial \Delta u_\kappa(r)}{\partial r} + \frac{\kappa}{r} \Delta u_\kappa(r) \right) c\hbar \right) dr$$
$$- \int \Delta u_\kappa(r) \left(\left(\frac{\partial v_\kappa(r)}{\partial r} - \frac{\kappa}{r} v_\kappa(r) \right) c\hbar - 2u_\kappa(r)(mc^2 + V(r)) \right) dr \qquad (10.100)$$

and the variation in the normalization is

$$\Delta I = 2 \int (\Delta u_\kappa(r) u_\kappa(r) + \Delta v_\kappa(r) v_\kappa(r)) dr \qquad (10.101)$$

The right way to minimize a quantity subject to a constraint is to use the Lagrange multipliers method, i.e. we demand that

$$\Delta E_T^1 - \epsilon \Delta I = 0 \qquad (10.102)$$

It is straightforward to set this up. The only term in (10.100) that gives trouble is the derivative of $\Delta u_\kappa(r)$ and this can be solved by integrating by parts

$$\int v_\kappa(r) \frac{\partial \Delta u_\kappa(r)}{\partial r} dr = [v_\kappa(r) \Delta u_\kappa(r)]_0^\infty - \int \Delta u_\kappa(r) \frac{\partial v_\kappa(r)}{\partial r} dr \qquad (10.103)$$

The first term on the right hand side is zero provided we demand that $\Delta u_\kappa(r)$ is zero at $r = 0$ and $r = \infty$. Putting (10.103) into (10.100) and then putting (10.100) and (10.101) into (10.102) gives

$$\int \Delta u_\kappa(r) \left(2c\hbar \left(-\frac{\partial v_\kappa(r)}{\partial r} + \frac{\kappa}{r} v_\kappa(r) \right) + 2(mc^2 + V(r) - \epsilon) u_\kappa(r) \right) dr = 0 \qquad (10.104)$$

This equation must be true for an arbitrary variation $\Delta u_\kappa(r)$. That is only the case if the quantity in square brackets is equal to zero, i.e. if

$$\frac{\partial v_\kappa(r)}{\partial r} = \frac{\kappa}{r} v_\kappa(r) - \frac{1}{c\hbar} (\epsilon - V(r) - mc^2) u_\kappa(r) \qquad (10.105a)$$

A similar calculation, where we vary $v_\kappa(r)$ in (10.99), yields

$$\frac{\partial u_\kappa(r)}{\partial r} = -\frac{\kappa}{r} u_\kappa(r) + \frac{1}{c\hbar} (\epsilon - V(r) + mc^2) v_\kappa(r) \qquad (10.105b)$$

But these are just the single-particle radial Dirac equations first introduced in equation (8.13). This is rather reassuring, as it means that the Dirac–Hartree–Fock method becomes exact in the one-electron limit.

The Many-Electron Atom

Once we get away from the one-electron atom, the problem becomes to minimize the full expression for the energy (10.97) subject to the normalization (10.98). If we vary one of the radial wavefunctions the first terms in (10.98) will behave in the same way as in the one-electron case. Now, in addition to these we will have to deal with the direct and exchange Coulomb terms. To find the variation in these it is necessary to find the variation in the integrals $F_l(a, b)$ and $G_l(a, b)$. These are easy to work out from their definitions (10.74b) and (10.90b). For small variations $\Delta u_a(r)$ in $u_a(r)$ and $\Delta v_a(r)$ in $v_a(r)$ we find

$$\Delta F_l(a, b) = \frac{2e^2}{4\pi\epsilon_0} \int \int (u_a(r_1)\Delta u_a(r_1) + v_a(r_1)\Delta v_a(r_1))$$
$$\times (u_b^2(r_2) + v_b^2(r_2))\frac{r_<^l}{r_>^{l+1}} dr_1 dr_2 \qquad (10.106)$$

$$\Delta F_l(a, a) = \frac{4e^2}{4\pi\epsilon_0} \int \int (u_a(r_1)\Delta u_a(r_1) + v_a(r_1)\Delta v_a(r_1))$$
$$\times (u_a^2(r_2) + v_a^2(r_2))\frac{r_<^l}{r_>^{l+1}} dr_1 dr_2 \qquad (10.107)$$

$$\Delta G_l(a, b) = \frac{2e^2}{4\pi\epsilon_0} \int \int (u_b(r_1)\Delta u_a(r_1) + v_b(r_1)\Delta v_a(r_1))$$
$$\qquad\qquad\qquad (10.108)$$
$$\times (u_a(r_2)u_b(r_2) + v_a(r_2)v_b(r_2))\frac{r_<^l}{r_>^{l+1}} dr_1 dr_2$$

and we define the following integral to simplify later expressions

$$Y_l(a, b, r) = \frac{e^2}{4\pi\epsilon_0} r \int (u_a(r_2)u_b(r_2) + v_a(r_2)v_b(r_2))\frac{r_<^l}{r_>^{l+1}} dr_2 \qquad (10.109)$$

Now, following Slater (1960), we set up the equation

$$\Delta E_T - \sum_a \epsilon_{a,a}\Delta I_{a,a} - \sum_{a,b}{}' (\epsilon_{a,b}\Delta I_{a,b} + \epsilon_{b,a}\Delta I_{b,a}) = 0 \qquad (10.110)$$

It is somewhat more complicated this time as there are a lot more undetermined multipliers. We choose $\epsilon_{a,b} = \epsilon_{b,a}^*$ so that the last two terms in equation (10.110) are complex conjugates, although we have chosen the wavefunctions real, so that the Lagrange multipliers will also be real.

Equation (10.110) shows that a change in the wavefunction that preserves orthogonality and normalization will leave $\Delta E_T = 0$, which is what we require. It is straightforward, in principle, to vary the wavefunctions in (10.97) and use (10.106–109) in (10.110) to obtain the minimum condition. This is three or four pages of algebra, so we will show one or two of the intermediate steps. The single-particle part of (10.97) works out as shown in the previous section. We will introduce the symbol $\sum'$ which means when we sum over pairs, every pair is only summed once, not twice as the summations in (10.111) below would imply if the prime were absent from the summation sign. If we vary the wavefunctions then, the terms in the change in energy that contain $\Delta u_a(r)$ are

$$\Delta E_T = \sum_a n_a \left(\int \Delta u_a(r_1) \left(2c\hbar \left(-\frac{\partial v_a(r_1)}{\partial r_1} + \frac{\kappa_a}{r_1} v_a(r_1) \right) \right. \right.$$

$$+ 2(mc^2 + V(r_1)) u_a(r_1) + \sum_b{}' 2n_b \frac{1}{r_1} Y_0(b,b,r_1) u_a(r_1)$$

$$- \frac{n_a - 1}{2j_a} \sum_{l=0}^{\infty} (2j_a + 1) \Gamma_{j_a,j_a}^l \frac{1}{r_1} Y_l(a,a,r_1) u_a(r_1)$$

$$\left. \left. - \sum_{b \neq a}{}' (2j_b + 1) \sum_{l=0}^{\infty} \Gamma_{j_a,j_b}^l \frac{1}{r_1} Y_l(a,b,r_1) u_b(r_1) \right) dr_1 \right) \qquad (10.111)$$

Here the first $\sum'$ does include $b = a$ and for that case $n_b = n_a - 1$. We have brought together in this term the two terms in F_0 from equation (10.97). A similar expression to (10.111) for the coefficients of $\Delta v_a(r)$ can also be written down. Equation (10.111) can be inserted into (10.110) along with the normalization condition of equation (10.98). Then we have

$$\sum_a n_a \left(\int \Delta u_a(r_1) \left[2c\hbar \left(-\frac{\partial v_a(r_1)}{\partial r_1} + \frac{\kappa_a}{r_1} v_a(r_1) \right) \right. \right.$$

$$+ 2(mc^2 + V(r_1) - \epsilon_{a,a}) u_a(r_1) + \sum_b{}' 2n_b \frac{1}{r_1} Y_0(b,b,r_1) u_a(r_1)$$

$$- \frac{n_a - 1}{2j_a} \sum_{l=0}^{\infty} (2j_a + 1) \Gamma_{j_a,j_a}^l \frac{1}{r_1} Y_l(a,a,r_1) u_a(r_1)$$

$$- \sum_{b \neq a}{}' (2j_b + 1) \sum_{l=0}^{\infty} \Gamma_{j_a,j_b}^l \frac{1}{r_1} Y_l(a,b,r_1) u_b(r_1)$$

$$\left. \left. - \sum_{b \neq a}{}' \epsilon_{a,b} u_b(r_1) \right] dr_1 \right) = 0 \qquad (10.112)$$

We can simplify these expressions further. The potential $V(r_1)$ is the usual nuclear Coulomb potential felt by the electron. It multiplies $u_a(r_1)$. Two other terms in this equation also multiply $u_a(r_1)$. They can all be combined together into an effective potential

$$U_a(r) = \frac{Ze^2}{4\pi\epsilon_0 r_1} - {\sum_b}' n_b \frac{1}{r_1} Y_0(b, b, r_1)$$

$$+ \frac{1}{2} \frac{n_a - 1}{2 j_a} \sum_{l=0}^{\infty} (2 j_a + 1) \Gamma^l_{j_a, j_a} \frac{1}{r_1} Y_l(a, a, r_1) \qquad (10.113)$$

The terms in this potential can easily be understood. It is a potential felt by an electron in shell a. The first term is the nuclear potential, the second is the direct Coulomb interaction due to all other electrons, and the third is an effective exchange potential due to all other electrons in the a-shell. Equation (10.113) can be inserted directly into (10.112), then we note that the variation $\Delta u_a(r)$ (the subscript 1 on r is no longer needed so we remove it) is arbitrary, so this equation can only be satisfied if the expression in the inner square brackets is equal to zero. Dividing it through by two gives

$$c\hbar \left(\frac{\partial v_a(r)}{\partial r} - \frac{\kappa_a}{r} v_a(r) \right) + (\epsilon_{a,a} + U_a(r) - mc^2) u_a(r) +$$

$$\frac{1}{2} {\sum_{b \neq a}}' \sum_{l=0}^{\infty} (2 j_b + 1) \Gamma^l_{j_a, j_b} \frac{1}{r} Y_l(a, b, r) u_b(r) + {\sum_{b \neq a}}' \epsilon_{a,b} u_b(r) \delta_{\kappa_a, \kappa_b} = 0$$

$$(10.114a)$$

If we had looked at $\Delta v_a(r)$ instead we would have found

$$c\hbar \left(\frac{\partial u_a(r)}{\partial r} + \frac{\kappa_a}{r} u_a(r) \right) - (\epsilon_{a,a} + U_a(r) + mc^2) v_a(r) -$$

$$\frac{1}{2} {\sum_{b \neq a}}' \sum_{l=0}^{\infty} (2 j_b + 1) \Gamma^l_{j_a, j_b} \frac{1}{r} Y_l(a, b, r) v_b(r) - {\sum_{b \neq a}}' \epsilon_{a,b} v_b(r) \delta_{\kappa_a, \kappa_b} = 0$$

$$(10.114b)$$

Equations (10.114) are what we have been aiming at in this chapter. They are the relativistic Hartree–Fock equations for the electronic structure of many-electron atoms. There is a pair of Hartree–Fock equations for each electron in the atom, as $U_a(r)$ is different for each electron, or at least for each electron shell. The effective potential $U_a(r)$ here has a different sign to the potential in (10.105) because it has been defined that way. The term in $Y_l(a, b, r)$ derives from the exchange energy between the electron being described and the electrons in all other shells. The δ-function in the final term is non-zero if the angular momentum quantum numbers of the two states are equal, but the principal quantum numbers differ.

In the present formulation we have removed any m_j dependence and so the number of pairs of equations has been reduced to the number of occupied shells. This is a very useful economy when it comes to the implementation of the Hartree–Fock method. In a magnetic field, the energy levels would split and the present formalism would be inappropriate.

The nature of the Lagrange multipliers requires some explanation. As we will see shortly $\epsilon_{a,a}$ are easy to interpret. A physical description of the off-diagonal multipliers is more difficult. Nonetheless, in what follows we make an attempt and present expressions for them.

10.7 Koopmans' Theorem

Our final task in the formal theory of the Dirac–Hartree–Fock method is to identify the physical meaning of the Lagrange multipliers. It is not difficult to do this, particularly as we have the one-electron atom for guidance. In that case we found that the Lagrange multiplier corresponded to the energy eigenvalue of the electronic state, or, more physically, to the energy needed to remove the electron from the nucleus to infinity.

It turns out to be convenient to consider the diagonal and off-diagonal Lagrange multipliers separately. We consider first the diagonal ones. What we need is an expression for $\epsilon_{a,a}$ in terms of simple quantities. Such an expression can be derived directly from the relativistic Hartree–Fock equations. We multiply (10.114a) by $u_a(r)$ and (10.114b) by $-v_a(r)$. Next we add the resulting equations and integrate the sum of them over r. This is straightforward, but (unfortunately!) all the intervening equations are too long to write down, so we will just present the end result

$$
\epsilon_{a,a} = c\hbar \int \left(v_a(r) \left[\frac{\partial u_a(r)}{\partial r} + \frac{\kappa_a}{r} u_a(r) \right] - u_a(r) \left[\frac{\partial v_a(r)}{\partial r} - \frac{\kappa_a}{r} v_a(r) \right] \right) dr
$$

$$
- \int \frac{Ze^2}{4\pi\epsilon_0 r} (u_a^2(r) + v_a(r)^2) dr + mc^2 \int (u_a^2(r) - v_a(r)^2) dr
$$

$$
- \frac{1}{2} \left(\frac{n_a - 1}{2j_a} \right) (2j_a + 1) \sum_{l=0}^{\infty} \Gamma_{j_a,j_a}^l F_l(a, a) + \sum_{b\neq a}' n_b F_0(a, b)
$$

$$
- \frac{1}{2} \sum_{b\neq a}' \sum_{l=0}^{\infty} (2j_b + 1) \Gamma_{j_a,j_b}^l G_l(a, b) \tag{10.115}
$$

The first four terms here are the one-particle energy of an electron in state a. From (10.97) we see that the remaining terms are the reduction in the interaction energy if we remove an electron in state a. Therefore $\epsilon_{a,a}$ is the energy needed to remove one electron in state a from the atom to infinity, provided the other electron states are unaffected by the removal. This is Koopmans' theorem (Koopmans 1933). This shows that the diagonal Lagrange multipliers do have physical meaning. Differences between the Lagrange multipliers and experimental ionization energies are a measure of how much electronic rearrangement goes on as a result of the ionization.

Finally we must consider the off-diagonal Lagrange multipliers. To find an expression for these we multiply (10.114a) by $u_c(r)$ and (10.114b) by $-v_c(r)$, where $c \neq a$. Then we add the two equations and integrate over all space. All the intervening steps involve gigantic equations. However, the procedure is straightforward and we end up with

$$\epsilon_{a,c} =$$

$$mc^2 \int [u_a(r)u_c(r) - v_a(r)v_c(r)]dr - \int \frac{Ze^2}{4\pi\epsilon_0 r}[u_a(r)u_c(r) + v_a(r)v_c(r)]dr$$

$$+ c\hbar \int \left(v_c(r)\left(\frac{\partial u_a(r)}{\partial r} + \frac{\kappa_a}{r}u_a(r) \right) - u_c(r)\left(\frac{\partial v_a(r)}{\partial r} - \frac{\kappa_a}{r}v_a(r) \right) \right) dr$$

$$+ \sum_{b \neq a,c}' n_b R^0(babc) + (n_a - 1)R^0(a,a,a,c) + n_c \frac{2j_c}{2j_c + 1}R^0(c,a,c,c)$$

$$- \frac{1}{2}\left(\frac{n_a - 1}{2j_a} \right)(2j_a + 1)\sum_{l=0}^{\infty} \Gamma^l_{j_a,j_a} R^l(a,a,a,c)$$

$$- \frac{1}{2}\sum_{l=0}^{\infty}\left(\sum_{b \neq a,c}'(2j_b + 1)\Gamma^l_{j_a,j_b} R^l(a,b,b,c) - (2j_c + 1)\Gamma^l_{j_a,j_c} R^l(a,c,c,c) \right)$$

$$\tag{10.116a}$$

where we have defined

$$R^l(a,b,c,d) = \frac{e^2}{4\pi\epsilon_0} \int \int (u_a(r_1)u_c(r_1) + v_a(r_1)v_c(r_1)) \times$$

$$\tag{10.116b}$$

$$(u_b(r_2)u_d(r_2) + v_b(r_2)v_d(r_2)) \frac{r_<^l}{r_>^{l+1}} dr_1 dr_2$$

As we insisted that $\epsilon_{a,b} = \epsilon^*_{a,b}$, the Lagrange multipliers form a hermitian matrix, which can be diagonalized with a unitary transformation. An alternative, and computationally simpler, expression for the off-diagonal Lagrange multipliers is provided by Grant (1970), although it requires a longer derivation, so we do not reproduce it here. Equation (10.116) is difficult to interpret. If we make a local approximation for the exchange energy, such as that suggested by Slater which is based upon the exchange energy of a free electron gas, the wavefunctions are automatically orthogonal and the off-diagonal Lagrange multipliers are not needed.

10.8 Implementation of the Dirac–Hartree–Fock Method

Now the procedure for implementing the Dirac–Hartree–Fock method is clear. One starts with an initial set of occupation numbers n_a and wavefunctions $u_a^{(1)}(r)$ and $v_a^{(1)}(r)$. These could be hydrogenic, for example. The

Table 10.1. Coefficients $\Gamma^l_{j_a, j_b}$ for various values of the quantum numbers. These values are from the paper by Grant (1961). Where no value is given the value of l is disallowed by the selection rules.

j_a	j_b	l						
		0	1	2	3	4	5	6
1/2	1/2	1	1/3	—	—	—	—	—
3/2	1/2	—	1/3	1/5	—	—	—	—
5/2	1/2	—	—	1/5	1/7	—	—	—
7/2	1/2	—	—	—	1/7	1/9	—	—
3/2	3/2	1/2	1/30	1/10	9/70	—	—	—
5/2	3/2	—	1/5	1/35	2/35	2/21	—	—
7/2	3/2	—	—	9/70	1/42	5/126	5/66	—
5/2	5/2	1/3	1/105	8/105	8/315	2/63	50/693	—
7/2	5/2	—	1/7	1/105	1/21	5/231	5/231	25/429
7/2	7/2	1/4	4/315	5/84	36/693	9/308	75/4004	25/1716

Lagrange multipliers can be estimated, or determined from (10.115) and (10.116). The potential $U(r)$ is calculated from (10.113), and the coefficients of $u_b(r)$ and $v_b(r)$ in (10.114) can all be found from the initial guess at the wavefunctions. There are a pair of equations like (10.114) for every occupied shell within the atom. These can all be solved simultaneously to give our second guess at the wavefunctions $u_a^{(2)}(r)$ and $v_a^{(2)}(r)$. This process can then be repeated *ad nauseam* until the wavefunctions from the $(n+1)^{\text{th}}$ iteration are the same as those from the nth (usually to within a previously defined tolerance). When this has been achieved we say we have a self-consistent solution of the Dirac–Hartree–Fock equations. Such a calculation is computationally intensive, but, for atoms at least, is fairly routine with present-day computers. In equations (10.114) the $\Gamma^l_{j_a, j_b}$ can be calculated from (10.93b), but, for completeness, we reproduce them for low values of the quantum numbers in table 10.1.

In a book such as this it is impossible to give a comprehensive survey of the literature on relativistic Hartree–Fock calculations and their implementation (Desclaux *et al.* 1971). For further details a good place to start is the book edited by Wilson *et al.* (1991). Here we simply point out some important features of the implementation of the relativistic Hartree–Fock method, and show a few illustrative results.

In what follows we note some points that facilitate the numerical solution of the Dirac–Hartree–Fock equations.

It is not very efficient to have the potential and wavefunctions defined on a grid of equally spaced points. This is because the inner electrons are very closely bound to the nucleus, whereas the outer electrons are relatively broadly spaced. A grid of equally spaced points would put very few points on the tightly bound wavefunctions or an unnecessarily large number in the region where the outer electron wavefunctions are large. To take care of this it is convenient to define a logarithmic grid

$$t = \log r \qquad\qquad (10.117)$$

This puts more points close to the origin and fewer points at greater distance from the origin. It is then a trivial matter to change the variables in equations (10.113) and (10.114).

The differential equations (10.114) have to be integrated from the origin out to some radius (or back). The radius is chosen so that all wavefunctions have decayed to a very small fraction of their peak value (effectively at a radius of infinity). The radius necessary depends on the accuracy required. To initiate the integration a series solution close to the origin can be made. Near the nucleus the wavefunctions will be similar to the hydrogenic ones, and a series solution similar to that found in chapter 8 for the one-electron atom may be used. The arbitrary constant that will remain after this procedure may be removed by normalization.

It is necessary to have an initial estimate of the diagonal Lagrange multipliers. There are some technical ways of estimating these and homing in on the self-consistent value (Grant 1970). Here we just point out that if $\epsilon_{a,a}$ is chosen too high (less negative than the self-consistent value) there will be too many nodes in the wavefunction and if it is too low there will be too few nodes.

Once we have gone through one iteration of the Dirac–Hartree–Fock method we have a new set of wavefunctions. However, direct use of them in the next iteration is often unstable. It is often necessary to take a mixture of the new wavefunctions and the old

$$u^{(n+1)}(r) = cu^{(n+1)'}(r) + (1-c)u^n(r) \qquad 0 < c \le 1 \qquad (10.118)$$

and similarly for $v(r)$. Here $u^{(n+1)'}(r)$ are the wavefunctions that were produced in the nth iteration, $u^n(r)$ are the wavefunctions input into the nth iteration and $u^{(n+1)}(r)$ are the wavefunctions to initiate the $(n+1)^{\text{th}}$ iteration. The choice of the mixing parameter c depends on the potential, the orbital, and even how close to self-consistency the wavefunctions are. Choosing it judiciously is a mixture of guesswork and experience, and is discussed by Mayers and O'Brien (1968).

The degree of accuracy required in the solution of the Dirac–Hartree–Fock equations depends on the application for which one is solving them.

Table 10.2. Calculations of the Breit interaction energy (in Rydbergs) for the ground state of He, illustrating the similarity between the perturbative and variational approaches (from Grant 1986).

	MCDF	Basis sets
Coulomb energy	−5.72362672	−5.72362665
Breit correction × 10^{-5}		
perturbative	1.27552498	1.27552608
variational		1.275492
Total energy		
perturbative		−5.723499094152
variational		−5.723499097572

For very small shifts in energy levels we should also have included the Breit interaction in our development of equations (10.114) (Cooper 1965). However, for many applications it is sufficient to treat it as a perturbation after equations (10.114) have been solved self-consistently. Grant (1986) has shown that the difference between using a perturbative approach at the end of the calculation, and treating it in the variational part of the calculation is tiny and of no consequence for any normal application in condensed matter physics. To illustrate this, table 10.2 shows that for helium the Breit correction to the total energy is of order 10^{-5}–10^{-6} of the Coulomb energy, and uncertainties in it due to the use of a perturbative approach are in the fifth or sixth significant figure. There are two types of calculations in this table, a direct solution using multiconfigurational Dirac–Fock codes (MCDF) and a method where the wavefunctions are expanded in terms of some convenient basis functions. The contribution to the total energy from the Breit interaction is shown to be very small for some other examples of closed shell elements in the calculations of Kim (1967) reproduced in table 10.3. As can be seen from this table, the Breit correction increases roughly in proportion to the total energy itself.

If a less precise solution is sufficient the off-diagonal Lagrange multipliers can be set equal to zero, as they turn out to be small compared with the diagonal contributions. The equations can be solved without them. If necessary, they can be included in one final iteration.

In equations (10.114) the most difficult term to deal with in practice is the exchange term. The coefficient of a term in r_1 involves integrals over r_2. A term like this, where an electron wavefunction depends on the potential at positions where the electron isn't, is called non-local. However, we can make local approximations to the exchange energy and

Table 10.3. Calculations of the total energy for helium, beryllium, and neon. Calculations are amde using the relativistic Hartree–Fock method (Kim 1967), and the correction to it due to the Breit interaction is shown. All energies are in Rydbergs.

Atom	Variational energy	Breit correction	Total energy
He	−5.723676	0.00016	−5.723516
Be	−29.15180	0.00231	−29.14949
Ne	−257.3838	0.02380	−257.3600

some of these are discussed in sections 10.13 and 10.17. A particularly popular one is to evaluate the integrals in (10.114) explicitly, making the gross approximation that the electrons are plane waves. Such an approach (known as Slater's approximation) gives surprisingly good results, as is shown later in table 10.5.

In the previous sections we have derived the bare Dirac–Hartree–Fock equations. However, it is not computationally efficient to solve them directly. A standard technique to make the solution more efficient is to expand the wavefunctions $u_a(r)$ and $v_a(r)$ in terms of some basis functions. This was first proposed in molecular orbital theory by Roothan (1951) and Hall (1951), and applied to the relativistic theory by Kim (1967) and Kagawa (1975, 1980). The idea is to write the wavefunctions as

$$u_{n\kappa}(r) = \sum_i \xi_{n\kappa i} y_{\kappa i}(r) \tag{10.119a}$$

$$v_{n\kappa}(r) = \sum_j \eta_{n\kappa j} y_{\kappa j}(r) \tag{10.119b}$$

where $\xi_{n\kappa i}$ and $\eta_{n\kappa j}$ are coefficients to be determined and $y_{\kappa k}(r)$ are some standard basis functions, often chosen to be Slater-type functions

$$y_{\kappa k} = a_k r^n e^{-\zeta_k r} \tag{10.120a}$$

or Gaussians

$$y_{\kappa k} = b_k r^n e^{-\zeta_k r^2} \tag{10.120b}$$

where the integer $n \geq |\kappa|$, and a_k and b_k can be used for normalization. The appropriate value of ζ_k has to be determined and has been the subject of much debate (Wilson 1983, quoted by Grant 1986). A complicated prescription for them is given by Grant (1986). One can certainly imagine

Table 10.4. Dirac–Hartree–Fock calculations for neon (Grant 1986). The table shows the convergence of the energy eigenvalues and total energy as a function of basis set size. All energies are in Rydbergs.

	N	Total energy	ϵ_{1s}	ϵ_{2s}	$\epsilon_{2p_{3/2}}$	$\epsilon_{2p_{3/2}}$
Basis	4	−215.192	−48.10	−3.238	−1.948	−1.942
sets	5	−243.882	−59.974	−3.799	−1.756	−1.746
	6	−255.810	−64.960	−3.877	−1.708	−1.700
	7	−253.3536	−65.6114	−3.871672	−1.705674	−1.696568
	8	−257.383360	−65.634142	−3.871628	−1.705674	−1.696550
	9	−257.383908	−65.634994	−3.871692	−1.705658	−1.696532
MCDF		−257.383938	−65.634940	−3.871692	−1.705658	−1.696533

that for the Slater orbitals at least an appropriate choice might be

$$\zeta = \frac{Z}{na_0} \tag{10.121}$$

for an element with atomic number Z, where n is the principal quantum number and a_0 is the Bohr radius. The advantage of this basis set approach is that derivatives and integrals can be found analytically and what remains is a set of equations for the coefficients $\xi_{n\kappa i}$ and $\eta_{n\kappa j}$. Solving these is a much faster and computationally efficient task than solving the original equations. Of course, one still has to ask questions about the size of the basis set, but calculations can be converged to great accuracy with a rather modest number of basis functions. Table 10.4 illustrates this for the ground state of neon. Here we see that only nine basis set functions are sufficient to get agreement to at least six significant figures between basis set methods and direct solution of the Dirac–Hartree–Fock equations using finite difference methods

We have derived the relativistic Hartree–Fock equations (10.114) without giving much thought to the negative energy states. In a way this is sensible as we don't expect the negative energy states to play much role in determining the electronic properties of atoms, molecules and solids. On the other hand, the theory is incomplete if they are excluded. When the Dirac–Hartree–Fock equations were originally solved, finite difference methods were used and no problems with the negative energy states arose. However, when basis set methods were applied, catastrophic results followed. This was because the Dirac–Hartree–Fock equations were derived using the variational principle to minimize the total energy. As negative energy states have a lower total energy than positive energy states the minimization procedure often mixed negative energy states in with the positive energy ones, leading to disaster (Kutzelnigg 1984). It has been shown by Grant (1986) that correct implementation of the boundary conditions eliminates this problem. In particular the expectation value of the potential energy due to the nucleus must remain finite despite its singularity at the origin. Then there is a sufficient separation of the positive and negative energy states for this problem not to occur. Grant goes on to show that the earlier methods of solution of the relativistic Hartree–Fock equations, which essentially ignored negative energy states, are consistent with the interpretation of positrons as being holes in the negative energy sea of electrons.

Above, we have included only the simplest interactions in the derivation, but nowadays it is possible to perform much more sophisticated calculations, even to the extent of calculating the electroweak coupling constant (Hartley and Sandars 1991).

In table 10.5 we show some results for the energy eigenvalues of mercury. Mercury is chosen simply because fairly comprehensive data for it are easily available. The first three columns in the table are the quantum numbers labelling the electron state. Column four, labelled NRHFS, contains the energy eigenvalues, calculated self-consistently using the non-relativistic Hartree–Fock method with Slater exchange. In column five are the results of the Dirac–Hartree–Fock method with Slater's non-relativistic approximation for the exchange energy. Column six displays the results of the Dirac–Hartree–Fock calculations where the exchange energy is calculated without approximation as discussed above. In the final column are the experimental values of the ionization energies for the appropriate shells. Several points can be noted from these results. Obviously, comparison of the columns shows that relativistic calculations are markedly superior to non-relativistic ones. In general the eigenvalues calculated non-relativistically are of the order of 10% too low. It is also clear that relativistic effects in these heavy atoms are much greater than any uncertainty introduced by the use of an approximate expression for the exchange energy. The middle two columns show the effect of making such an approximation, and clearly it is small. The correspondence between the experimental ionization energies and the eigenvalues is remarkable. This confirms Koopmans' theorem. The eigenvalues were introduced as purely mathematical parameters in the minimization procedure. It truly is miraculous that such quantities should correspond so closely to the results of real experiments.

In table 10.6 we show the total energy for terbium in two possible electronic configurations decomposed into its component parts. This displays the relative magnitude of the different contributions to the energy

Table 10.5. Magnitude of the one-electron energy eigenvalues for the mercury atom (Sources: Herman and Skillman 1963, Liberman *et al.* 1965, Coulthard 1967, Desclaux 1973 (quoted by Desclaux and Kim 1975), Bearden and Burr 1967). The eigenvalues are all negative, of course. All energies are in Rydbergs. For an explanation of the columns, see the text.

n	l	j	NRHFS	DFHS	DHF	EXP
1	0	1/2	5535.8	6130.2	6148.5	6107.9
2	0	1/2	932.0	1090.3	1100.5	1090.6
2	1	1/2	896.9	1047.8	1053.7	1044.8
2	1	3/2	896.9	903.0	910.3	903.0
3	0	1/2	220.5	260.1	266.2	261.7
3	1	1/2	204.1	240.7	245.3	241.0
3	1	3/2	204.1	208.5	213.1	209.3
3	2	3/2	173.3	176.0	178.9	175.3
3	2	5/2	173.3	169.1	172.0	168.7
4	0	1/2	47.9	57.9	61.3	58.8
4	1	1/2	40.8	49.5	52.3	49.8
4	1	3/2	40.8	41.7	44.4	42.0
4	2	3/2	27.4	27.8	29.6	27.8
4	2	5/2	27.4	26.3	28.1	26.4
4	3	5/2	9.39	8.3	8.9	7.51
4	3	7/2	9.39	8.0	8.6	7.23
5	0	1/2	7.37	9.25	10.21	8.84
5	1	1/2	5.01	6.42	7.08	5.91
5	1	3/2	5.01	5.01	5.68	4.23
5	2	3/2	1.27	1.17	1.30	1.23
5	2	5/2	1.27	1.02	1.15	1.09
6	0	1/2	0.57	0.70	0.66	0.77

in equation (10.97). This table also shows the degree of numerical accuracy required to differentiate reliably between the configurations, as we add several large positive and large negative energies and these partially cancel each other.

Once a self-consistent solution has been obtained, the energy eigenvalues are known. Ionization energies may be obtained from Koopmans' theorem or by comparison with a second Hartree–Fock calculation for the ionized atom. We can also use equations (10.97) and (10.115) to calculate the relative sizes and importance of various terms in the total energy and potential. All sorts of quantum electrodynamical effects can perturb

Table 10.6. Decomposition of the contributions to the total energy for two
possible electronic configurations of terbium. Energies are in Rydbergs.

Contribution	$4f^8 5d^1 6s^2$	$4f^9 6s^2$
Sum of energy eigenvalues	-14646.3432	-14612.2832
Kinetic energy	25878.7286	25878.6312
Electron − nucleus potential energy	-58048.0798	-58082.2415
Electron − electron potential energy	9251.7914	9286.8762
Exchange energy	-735.4298	-736.6880
Total energy	-23652.9898	-23653.5952

Table 10.7. The ionization energy for a $1s$ electron
in mercury. The relative importance of the various
contributions are shown. All energies are in
Rydbergs.

Contribution	Energy
Electrostatic	-6141.65
Magnetic	24.09
Self-energy	14.65
Vacuum polarization	-3.29
Retardation	-1.77
Correlation	-0.08
Total	-6108.05

the energy levels, apart from those already discussed – we might want
to consider the effects of vacuum polarization, electron self-energy and
correlation for example. Table 10.7 shows the contributions to the energy
for the $1s$ electron in mercury. I have calculated the direct and exchange
energy and the other contributions are quoted from the work of Desiderio
and Johnson (1971). This table gives an indication of the magnitude of
the various contributions.

In figure 10.2 we show the $1s$ wavefunction for platinum calculated self-
consistently with the relativistic Hartree–Fock theory and with the non-
relativistic Hartree–Fock theory. Note that the $1s$ wavefunction calculated
relativistically peaks closer to the nucleus than the the one calculated
in non-relativistic theory. This is generally true, screening of the nuclear
potential is more efficient in relativistic theory. We have already seen this
in the one-electron atom.

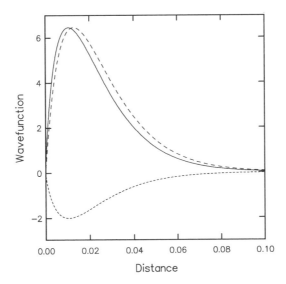

Fig. 10.2. The 1s wavefunction in platinum calculated non-relativistically (long dashes), and the large (full line) and small (short dashes) components calculated relativistically. All quantities are calculated in atomic units. Note that the relativistic wavefunction peaks 18% closer to the nucleus in the relativistic calculation than in the non-relativistic calculation. All other orbitals must remain orthogonal to the 1s state, so they shrink as well leading to a more efficient screening of the nucleus. The nucleus is then better shielded from the outer electron shells which, in turn, means that these shells expand.

One amusing consequence of this better screening is the following. As is well known, much of the world's economy is strongly dependent upon the price of gold. The reasons that gold holds its unique position are historical, but are, at least partially, due to its attractive colour. If the world were governed by non-relativistic quantum mechanics, as opposed to relativistic quantum mechanics, gold would actually have the silver appearance common to many metallic elements, owing to the less efficient screening of the nucleus by the core electrons, and therefore it would be less likely to hold such a central economic position. This would undoubtedly lead to the world's economy being radically different. For this reason, if for no other, relativistic quantum mechanics has had a profound effect on the decisions of governments and hence on the everyday life of ordinary people. The effect of relativity is easily seen in gold's ionization energy of 9.22 eV, compared with 7.574 eV for silver and 7.724 eV for copper which are in the same column of the periodic table.

Of course, it is not only energies and wavefunctions that can be determined. Once one has these quantities many others become straightforward

Table 10.8. Average radii (in atomic units) of the electron shells in mercury calculated using non-relativistic (NRHF) and relativistic (DHF) Hartree–Fock theory. Sources: Desclaux 1973, Froese-Fischer 1973, Desclaux and Kim 1975. The inner shells undergo the relativistic contraction, and outer shells undergo an expansion due to the greater shielding of the nucleus by the inner shells.

n	l	j	NRHF	DHF
1	0	1/2	0.01892	0.01659
2	0	1/2	0.07962	0.06922
2	1	1/2	0.06734	0.05700
2	1	3/2	0.06734	0.06563
3	0	1/2	0.2004	0.1798
3	1	1/2	0.1911	0.1704
3	1	3/2	0.1911	0.1861
3	2	3/2	0.1670	0.1622
3	2	5/2	0.1670	0.1670
4	0	1/2	0.4393	0.3990
4	1	1/2	0.4433	0.4016
4	1	3/2	0.4433	0.4340
4	2	3/2	0.4503	0.4416
4	2	5/2	0.4503	0.4525
4	3	5/2	0.4692	0.4767
4	3	7/2	0.4692	0.4832
5	0	1/2	1.010	0.9152
5	1	1/2	1.095	0.9871
5	1	3/2	1.095	1.079
5	2	3/2	1.433	1.431
5	2	5/2	1.433	1.499
6	0	1/2	3.328	2.843

to determine, such as electron orbital radii (Grant 1970) and spectroscopic cross sections (Holm 1988). Orbital radii for mercury are shown in table 10.8. For the inner electrons the velocity is large and hence the inertial mass increases. As mass appears in the denominator of the expression for the Bohr radius, the distance of the shell from the nucleus decreases. This is particularly clear in the 1*s* electron. Other *s* shells have to maintain their orthogonality to the 1*s* state through the oscillations in the radial part of the wavefunction and they also contract. The same occurs for the *p*-electrons. However, for the 3*d* electrons the relativistic contraction is negligible, and the dominant effect in the outer *d*-electrons is the fact that more efficiently screening by inner electrons allows the outer ones to get further away from the nucleus.

Although we have presented the theory for atoms and ions, the Dirac–Hartree–Fock method is not restricted to these systems. There are many research groups round the world performing Hartree–Fock calculations on molecules (Desclaux and Pyykkö 1974, Pyykkö 1978, Malli and Oreg 1975, Wilson *et al.* 1991 for example). In particular bond lengths and angles may be calculated with great accuracy.

One of the great things about theoretical physics is that it is not constrained by reality in the same way as experimental physics. In an entertaining series of calculations, Mann and Waber (1970) have performed relativistic Hartree–Fock calculations for superheavy elements with atomic numbers between 118 and 131, and hence have been able to comment on the probable behaviour of the 5g electron shell.

A final interesting possibility may be considered. The binding energy of a molecule can be found from the difference in total energy between self-consistent relativistic Hartree–Fock calculations on the molecule and corresponding calculations on the constituent atoms individually. Some molecules are predicted not to form if the calculations are performed non-relativistically, but are stable if the calculations are done relativistically. An example of such a relativistically-permitted molecule is $Tl_2Pt(CN)_4$ (Nagle *et al.* 1988). The existence of such ions and molecules is surveyed by Pyykkö (1991).

The Dirac–Hartree equations, where the wavefunction is a product, rather than a determinant, of single particle wavefunctions, are equations (10.114) with the exchange terms removed. Solutions of the relativistic Hartree method may be used to provide good initial wavefunctions for the relativistic Hartree–Fock method.

The one drawback of the Hartree–Fock method is that it does not include correlation of the electrons, although exchange may be treated exactly. Indeed, the correlation energy may be defined as the difference between an *exact* Hartree–Fock calculation and the exact result.

10.9 Introduction to Density Functional Theory

Before proceeding to discuss density functional theory specifically it is necessary to discuss some general aspects of the theory and to define a couple of important quantities. To start with let us just briefly define a functional. The total energy functional, for example, is written $E[n(\mathbf{r})]$. The square brackets are the standard way of showing that a function depends upon a function rather than on a variable. That is, $E[n(\mathbf{r})]$ depends on $n(\mathbf{r})$ in a particular way, regardless of how $n(\mathbf{r})$ varies with respect to $\mathbf{r}$. Functional derivatives and integrals can then be defined in analogy with the simpler case of a quantity being a function of a variable. For further details of the mathematics of functionals the reader is referred to a standard book on the topic (Sewell 1987 or Wan 1993 for example).

Now we introduce the quantity commonly called r_s. This is a measure of electron density in units of length. In an electron gas of number density n it defines the average spacing between electrons

$$\frac{4\pi}{3} nr_s^3 = 1 \qquad (10.122)$$

The sphere of radius r_s contains, on average, one electron. The radius r_s is usually written in dimensionless form as r_s/a_0 where a_0 is the Bohr radius. Henceforth we define it in this way. Note that the low density limit corresponds to very high values of r_s and *vice versa*. For very low densities the electron gas forms an ordered structure known as the Wigner crystal. This is very counter-intuitive behaviour, which occurs for $r_s > 100$; it finds no application in condensed matter physics.

Another salient parameter in relativistic density functional theory is

$$\beta = \frac{\hbar k_F}{mc} \qquad (10.123)$$

This is, essentially, the actual momentum of an electron divided by mc, which is the classical momentum of a particle travelling at the speed of light. The parameter β is some sort of measure of the importance of relativity in a calculation and it emerges that β is related to the electron density by

$$\beta = \frac{1}{71.4r_s} \qquad (10.124)$$

A density at which β approaches unity is one where relativistic effects are important. Such a density would correspond to $r_s = 0.014$, and to give you an idea of the order of such effects this corresponds to $6.15 \times 10^{35} \, \text{m}^{-3}$, which is fairly large. Ramana and Rajagopal (1983) state that the charge density at the nucleus in metallic indium ($Z = 49$) gives $\beta \approx 1$, so one would certainly expect relativity to play a key role there.

Density functional theory is a general theoretical framework which enables us to calculate the ground state energy E_g of any condensed matter system consisting of electrons and some external potential. In condensed matter the external potential is usually that due to the nuclei in the system. By applying density functional theory with various boundary conditions, we can find energy differences which can be directly related to physical observables. This point will be amplified in the following chapter. Central to the theory are two theorems. The first of these relates the total energy of the system directly to the electron density (relativistically the 4-current density). Defining observables in terms of the electron density rather than wavefunctions means that some phase information about the electrons may be lost, but this is irrelevant for most condensed matter problems. The second theorem is that the ground state energy can be found by minimizing the energy with respect to variations in the charge and current densities. These theorems lead to a rather remarkable simplification of the many-electron problem. They enable us to map the interacting electron problem onto a well-defined equivalent system of non-interacting electrons. We will see how this works presently.

It is not immediately obvious that the total energy is dependent only upon the electron density. One could certainly imagine E_g also depending on other properties of the electron gas, the single particle density matrix, for example. However, we will shortly reproduce an argument that implies that it really does depend only on the density.

In our discussion of density functional theory, we adopt an approach of giving a detailed discussion of the non-relativistic theory, and then repeating that for the relativistic theory. The reason for this is that there will emerge an important quantity known as the exchange–correlation energy which has a neat physical interpretation in non-relativistic theory, but is very difficult to describe physically in relativistic theory. To give any insight into this quantity, an understanding of the non-relativistic theory is necessary.

Density functional theory is a mathematical framework to describe a system of N interacting electrons in an external potential. In the following sections we put this statement on a more mathematical footing.

10.10 Non-Relativistic Density Functional Theory

Density functional theory is a many-body theory and the correct way to set it up is to invoke the methods described in chapter 6. This was first done in a series of papers by Hohenberg, Kohn and Sham (1964, 1965, 1966), and detailed reviews have been written by Gunnarsson (1979), Kohn and Vashishta (1982), and Von-Barth (1983). The Hamiltonian for a system of N interacting electrons is

$$\hat{H} = \hat{T} + \hat{U} + \hat{V} = \hat{H}_0 + \hat{V} \qquad (10.125)$$

where $\hat{T}$ represents the kinetic energy of the electrons, $\hat{U}$ represents their Coulomb repulsion and $\hat{V}$ is their interaction with the external field. It is convenient to define $\hat{H}_0$ which is the Hamiltonian of the electron system in the absence of an external potential. Written in the Dirac notation the ground state energy of the electron system is

$$E_g = \langle \Psi | \hat{H}_0 + \hat{V} | \Psi \rangle \qquad (10.126)$$

and the electron density is

$$n(\mathbf{r}) = \langle \Psi | \hat{n}(\mathbf{r}) | \Psi \rangle \tag{10.127}$$

Here Ψ is the wavefunction representing the whole system, and may be a determinant, a sum of determinants, or some other representation of the many-body wavefunction. In this theory its exact form remains undefined. Therefore we give it a different symbol to that used for the many-body wavefunction in Hartree–Fock theory. The operator $\hat{n}(\mathbf{r})$ represents the electron density at point $\mathbf{r}$. Mathematically this is given in terms of the field operators by equation (6.127)

$$\hat{n}(\mathbf{r}) = \sum_s \varphi_s^\dagger(\mathbf{r}) \varphi_s(\mathbf{r}) d\mathbf{r} \tag{10.128}$$

where we have explicitly summed over the particle spins s. We can also write down the expression for the energy operators in (10.125) in terms of the field operators as shown in equations (6.120), (6.121), and (6.126).

$$\hat{U} = \frac{1}{2} \frac{e^2}{4\pi\epsilon_0} \sum_{s,s'} \int \int \varphi_s^\dagger(\mathbf{r}) \varphi_{s'}^\dagger(\mathbf{r}') \frac{1}{|\mathbf{r} - \mathbf{r}'|} \varphi_{s'}(\mathbf{r}') \varphi_s(\mathbf{r}) d\mathbf{r} d\mathbf{r}' \tag{10.129}$$

$$\hat{V} = \sum_s \int v^{\text{ext}}(\mathbf{r}) \varphi_s^\dagger(\mathbf{r}) \varphi_s(\mathbf{r}) d\mathbf{r} \tag{10.130}$$

$$\hat{T} = -\frac{\hbar^2}{2m} \sum_s \int \nabla \varphi_s^\dagger(\mathbf{r}) \nabla \varphi_s(\mathbf{r}) d\mathbf{r} \tag{10.131}$$

These equations are self-explanatory. Equation (10.129) represents the Coulomb repulsion between the electrons. The factor $1/2$ outside the summation occurs because the sum counts each pair of electrons twice. Equation (10.130) represents the effect of the external potential v^{ext} and (10.131) is the kinetic energy operator.

Equations (10.129), (10.130) and (10.131) are the operators representing the various components of the energy. As discussed in chapter 6 they look like single-particle expectation values, but are, in fact, operators. The integrands act to give the contribution to the total energy due to a single point (or in the case of the Coulomb repulsion, a pair of points). These are then integrated over space to give the energy of the whole system. Now, the next stage is to use these as operators in a Schrödinger equation for the total wavefunction of the system, Ψ.

If it is assumed that the ground state is non-degenerate, then both Ψ and $n(\mathbf{r})$ are uniquely determined by the external potential. It is simply necessary to solve a Schrödinger equation to find them. We are now going to prove the remarkable result that the reverse is also true. If we know $n(\mathbf{r})$, then $v^{\text{ext}}(\mathbf{r})$ and Ψ are uniquely determined. In other words we will prove that an $n(\mathbf{r})$ uniquely determines the Hamiltonian for the electron gas.

This argument proceeds by assuming the statement above is untrue and deriving contradictory results. Let us assume we have an external potential $v^{\text{ext}}(\mathbf{r})$ and a ground state Ψ which, in turn, lead to a ground state energy E and density $n(\mathbf{r})$. Then let us assume we have a different potential $v^{\text{ext}'}(\mathbf{r})$ with ground state Ψ' and energy E', which leads to the same charge density $n(\mathbf{r})$. So, as the ground state energy is the minimum energy the system can have, the following inequality must hold true

$$E = \langle\Psi|\hat{H}|\Psi\rangle \quad < \quad \langle\Psi'|\hat{H}|\Psi'\rangle = \langle\Psi'|\hat{H}' - \hat{V}' + \hat{V}|\Psi'\rangle$$

$$= \langle\Psi'|\hat{H}'|\Psi'\rangle - \langle\Psi'|\sum_s \int (v^{\text{ext}'}(\mathbf{r}) - v^{\text{ext}}(\mathbf{r}))\varphi_s^\dagger(\mathbf{r})\varphi_s(\mathbf{r})d\mathbf{r}|\Psi'\rangle$$

$$= E' - \int (v^{\text{ext}'}(\mathbf{r}) - v^{\text{ext}}(\mathbf{r}))\langle\Psi'|\sum_s \varphi_s^\dagger(\mathbf{r})\varphi_s(\mathbf{r})|\Psi'\rangle d\mathbf{r}$$

$$= E' - \int (v^{\text{ext}'}(\mathbf{r}) - v^{\text{ext}}(\mathbf{r}))\langle\Psi'|\hat{n}(\mathbf{r})|\Psi'\rangle d\mathbf{r} \qquad (10.132)$$

where we have simply inserted equation (10.130) and used the definition of the number density operator. The matrix element in the last line here is the actual number density of electrons and, by construction, this is $n(\mathbf{r})$ regardless of whether we are in the state Ψ or Ψ'.

So, because the two densities are the same we have

$$E < E' - \int (v^{\text{ext}'}(\mathbf{r}) - v^{\text{ext}}(\mathbf{r}))n(\mathbf{r})d\mathbf{r} \qquad (10.133)$$

Similarly

$$E' = \langle\Psi'|\hat{H}'|\Psi'\rangle \quad < \quad \langle\Psi|\hat{H}'|\Psi\rangle = \langle\Psi|\hat{H} - \hat{V} + \hat{V}'|\Psi\rangle \qquad (10.134)$$

where we have just interchanged the primed and unprimed quantities. This gives us

$$E' < E - \int (v^{\text{ext}}(\mathbf{r}) - v^{\text{ext}'}(\mathbf{r}))n(\mathbf{r})d\mathbf{r} \qquad (10.135)$$

If we now add (10.133) and (10.135) we have

$$E + E' < E' + E \qquad (10.136)$$

and this is clearly an absurd result. It means that something we have assumed in the derivation of it is incorrect. Besides the usual general postulates of quantum theory, the only assumption in this derivation was that two different potentials could lead to the same charge density. By obtaining (10.136) we have shown that this is not so. We can therefore state that it is impossible for two different external potentials to give rise to the same ground state density distribution $n(\mathbf{r})$. That is, the ground state density uniquely determines the external potential.

The above discussion is rather mathematical. A slightly more physically intuitive argument can be given as follows. It is based on the idea that the total energy is a convex function of the external potential.* Suppose we have our two potentials $v_0^{\text{ext}}(\mathbf{r})$ and $v_1^{\text{ext}}(\mathbf{r})$, and the difference between them is in some sense small. They have associated ground state energies $E_g(0)$ and $E_g(1)$ respectively. We can define a whole family of potentials between the two as follows

$$v_\lambda^{\text{ext}}(\mathbf{r}) = v_0^{\text{ext}}(\mathbf{r}) + \lambda \Delta v(\mathbf{r}) \tag{10.137}$$

where λ varies between 0 and 1, and Δv is the difference between v_0^{ext} and v_1^{ext}. This changes the potential energy operator for the complete electron system defined in (10.130) by $\lambda \Delta V$. Now using standard non-degenerate second order perturbation theory we can write the ground state energy for any of this family of potentials as

$$E_g(\lambda) = E_g(0) + \lambda \langle \Psi_0 | \Delta V | \Psi_0 \rangle + \lambda^2 \sum_{n \neq 0} \frac{|\langle \Psi_0 | \Delta V | \Psi_n \rangle|^2}{E_g(0) - E_n} \tag{10.138}$$

where we have used the state Ψ_0 associated with v_0^{ext} as our reference state. Consider the second order term in (10.138): the numerator is a positive number by definition and the denominator is always negative because the ground state energy $E_g(0)$ is the lowest allowable. Hence, the coefficient of λ^2 in (10.138) is negative. If we set

$$A_0 = -\sum_{n \neq 0} \frac{|\langle \Psi_0 | \Delta V | \Psi_n \rangle|^2}{E_g(0) - E_n} \tag{10.139}$$

so that A_0 is a positive number, we can differentiate (10.138) to give

$$\frac{\partial E_g(\lambda)}{\partial \lambda} = \langle \Psi_0 | \Delta V | \Psi_0 \rangle - 2\lambda A_0 \tag{10.140}$$

and as specific cases of this we have for $\lambda = 0$

$$\left[\frac{\partial E_g(\lambda)}{\partial \lambda} \right]_{\lambda=0} = \langle \Psi_0 | \Delta V | \Psi_0 \rangle \tag{10.141}$$

and for $\lambda = 1$

$$\left[\frac{\partial E_g(\lambda)}{\partial \lambda} \right]_{\lambda=1} = \langle \Psi_0 | \Delta V | \Psi_0 \rangle - 2A_0 \quad < \quad \left[\frac{\partial E_g(\lambda)}{\partial \lambda} \right]_{\lambda=0} \tag{10.142}$$

---

* I would like to thank M.J. Gillan for bringing this argument to my attention.

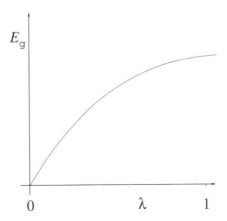

Fig. 10.3. Convexity of the ground state energy as a function of λ which leads to equation (10.147).

We could, of course, have used Ψ_1 as our reference state. In that case our ground state energy would be

$$E_{\mathrm{g}}(\lambda) = E_{\mathrm{g}}(1) + (\lambda - 1)\langle\Psi_1|\Delta V|\Psi_1\rangle + (\lambda - 1)^2 \sum_{n\neq 0} \frac{|\langle\Psi_1|\Delta V|\Psi_n\rangle|^2}{E_{\mathrm{g}}(1) - E_n} \quad (10.143)$$

and differentiating this with respect to λ gives

$$\frac{\partial E_{\mathrm{g}}(\lambda)}{\partial \lambda} = \langle\Psi_1|\Delta V|\Psi_1\rangle - 2(\lambda - 1)A_1 \quad (10.144)$$

where A_1 has the obvious definition in analogy with equation (10.139). If we set $\lambda = 1$ in here, the second term is zero and we can substitute (10.141) and (10.144) into the inequality (10.142) to give

$$\langle\Psi_1|\Delta V|\Psi_1\rangle \quad < \quad \langle\Psi_0|\Delta V|\Psi_0\rangle \quad (10.145)$$

Now, this tells us that the total energy of our family of potentials is a convex function of λ – the slope at $\lambda = 0$ is greater than the slope at $\lambda = 1$. This is sketched in figure 10.3. Recall from (10.132) that the expectation value of ΔV can also be written in terms of the electron densities so that an equivalent way of writing (10.145) is

$$\int \Delta v(\mathbf{r})n_1(\mathbf{r})d\mathbf{r} \quad < \quad \int \Delta v(\mathbf{r})n_0(\mathbf{r})d\mathbf{r} \quad (10.146)$$

or, if we define $\Delta n(\mathbf{r}) = n_1(\mathbf{r}) - n_0(\mathbf{r})$

$$\int \Delta v(\mathbf{r})\Delta n(\mathbf{r})d\mathbf{r} < 0 \quad (10.147)$$

This is an important inequality. An approximate interpretation of it is that increasing the potential results in a decrease in electron density. The important conclusion to be derived from (10.147) is that $\Delta n(\mathbf{r})$ cannot be zero, i.e. if we have two states associated with different potentials the difference between their charge densities cannot be equal to zero. The only proviso to this argument is that we have used non-degenerate perturbation theory, so we assume that the two states are not degenerate.

10.11 The Variational Principle and the Kohn–Sham Equation

We already know that $n(\mathbf{r})$ is uniquely determined by $v^{\mathrm{ext}}(\mathbf{r})$ and in the section above we have seen that $v^{\mathrm{ext}}(\mathbf{r})$ is uniquely determined by $n(\mathbf{r})$. This implies that the many-body wavefunction and hence the total energy and other observables can be regarded as being uniquely determined by $n(\mathbf{r})$. This is the first theorem of density functional theory.

(1) The total ground state energy of a system of interacting electrons is a unique functional of the electron density.

In particular, from this theorem we can write the total energy as

$$E_g[n(\mathbf{r})] = \int v^{\mathrm{ext}}(\mathbf{r})n(\mathbf{r})d\mathbf{r} + T[n(\mathbf{r})] + G[n(\mathbf{r})] \tag{10.148}$$

where $T[n(\mathbf{r})]$ is the ground state expectation value of kinetic energy of a non-interacting system of electrons with the same density as the interacting system. The value of $T[n(\mathbf{r})]$ is not necessarily the same as the kinetic energy of the real interacting system. The function $G[n(\mathbf{r})]$ is the expectation value of the internal potential energy of the system. To find the ground state density we have to invoke the second theorem:

(2) The ground state energy associated with a given external potential is found by minimizing the total energy functional with respect to changes in the electron density while the number of particles is held fixed. The density that yields the minimum total energy is the ground state density.

Next we need to look at the actual mathematics of performing this minimization of the total energy. As well as having the total energy expression (10.148) we are subject to the constraint that the total number of electrons in our system does not change:

$$\int n(\mathbf{r})d\mathbf{r} = N \tag{10.149}$$

where N is the total number of electrons in the system and

$$n(\mathbf{r}) = \langle \Psi | \hat{n}(\mathbf{r}) | \Psi \rangle = \sum_{i=1}^{N} \psi_i^*(\mathbf{r})\psi_i(\mathbf{r}) \tag{10.150}$$

As in the Hartree–Fock method, the way to deal with this minimization problem is to use the Lagrange multiplier method. We will do this, but we will not have to plough through all the necessary mathematics because of our understanding of the physics we are trying to describe. We will minimize the total energy with respect to variations in the electron density. This is one of the essential differences between density functional theory and Hartree–Fock theory where we minimize with respect to variations in the wavefunctions.

Realizing that the change in a function (we take the internal potential energy as our example) δG due to changes in the electron density is given in terms of a functional derivative as

$$\delta G = \int \frac{\delta G[n(\mathbf{r})]}{\delta n(\mathbf{r})} \delta n(\mathbf{r}) d\mathbf{r} \tag{10.151}$$

we can follow this minimization procedure. So, taking infinitesimal variations of (10.148) according to the definition (10.151) and setting the resulting expression equal to zero, we end up with

$$\int \delta n(\mathbf{r}) \left[v^{\text{ext}}(\mathbf{r}) + \frac{\delta T[n(\mathbf{r})]}{\delta n(\mathbf{r})} + \frac{\delta G[n(\mathbf{r})]}{\delta n(\mathbf{r})} - \epsilon \right] d\mathbf{r} = 0 \tag{10.152}$$

where ϵ is the undetermined Lagrange multiplier. Now we want this to be valid for arbitrary variations in density $\delta n(\mathbf{r})$, and that can only be true if the part of the integrand in square brackets is equal to zero, i.e. if

$$v^{\text{ext}}(\mathbf{r}) + \frac{\delta T[n(\mathbf{r})]}{\delta n(\mathbf{r})} + \frac{\delta G[n(\mathbf{r})]}{\delta n(\mathbf{r})} = \epsilon \tag{10.153}$$

An equation like (10.153), which expresses the condition that a quantity does not vary when small variations are made in the function upon which it depends, is known as an Euler equation. Now, you may think that this Euler equation does not get us much further forward, especially as ϵ which appears in it is undetermined and apparently arbitrary. That is not the case, and we can use our prior knowledge of quantum mechanics to understand why. Suppose that the electrons in our system had been moving independently of one another (i.e. they had only their kinetic energy and their interaction with the external potential). In that case $G[n(\mathbf{r})]$ would be zero and (10.153) would become

$$v^{\text{ext}}(\mathbf{r}) + \frac{\delta T[n(\mathbf{r})]}{\delta n(\mathbf{r})} = \epsilon \tag{10.154}$$

But we know how to treat systems of N non-interacting electrons quantum mechanically. The many-body Schrödinger equation can be reduced by separation of the variables to N single-particle Schrödinger equations. and the many-body wavefunction is then simply a product of the single-particle wavefunctions. Of course, this approach ignores the antisymmetric nature of the many-body wavefunction, but, as we will show in section 10.13, this can be taken care of at a later stage in the calculation. So we can write down the Schrödinger equation for each electron individually

$$\left(-\frac{\hbar^2}{2m}\nabla^2 + v^{\text{ext}}(\mathbf{r})\right)\psi_i = \epsilon_i\psi_i \qquad (10.155)$$

and we can get an exact solution for the single-particle wavefunctions. Our electrons actually obey equation (10.153) which includes their mutual interactions, but, by defining the effective potential

$$v^{\text{eff}}(\mathbf{r}) = v^{\text{ext}}(\mathbf{r}) + \frac{\delta G[n(\mathbf{r})]}{\delta n(\mathbf{r})} \qquad (10.156)$$

we can make equation (10.153) formally identical to (10.154) and so the interacting electrons obey a Schrödinger equation which is formally identical to (10.155)

$$\left(-\frac{\hbar^2}{2m}\nabla^2 + v^{\text{eff}}(\mathbf{r})\right)\psi_i = \epsilon_i\psi_i \qquad (10.157)$$

Hence, we have reduced our complicated many-body problem to a set of single-electron-like equations. The Schrödinger equation containing $v^{\text{eff}}(\mathbf{r})$ which gives us the density distribution of the interacting electron system is generally known as a Kohn–Sham equation, and ψ_i are the Kohn–Sham orbitals. The quantities ψ_i and ϵ_i are not the wavefunction and energy of real electrons, they are simply auxiliary quantities used to calculate the electron density and total energy.

We still need an expression for the effective potential that we can actually use in calculations and we will build that up now. One major contribution to $G[n(\mathbf{r})]$ will certainly be the Coulomb repulsion between the electrons. So we can write

$$G[n(\mathbf{r})] = \frac{1}{2}\frac{e^2}{4\pi\epsilon_0}\int\int\frac{n(\mathbf{r})n(\mathbf{r}')}{|\mathbf{r}-\mathbf{r}'|}d\mathbf{r}d\mathbf{r}' + E_{\text{xc}}[n(\mathbf{r})] \qquad (10.158)$$

Here $E_{\text{xc}}[n(\mathbf{r})]$ is called the exchange–correlation energy functional, and it is this quantity that contains our ignorance. It contains everything not included in the kinetic energy and Coulomb energy terms. We have to make an approximation for it, which we will discuss in more detail later on. Firstly we substitute (10.158) into (10.156) so that we have

$$v^{\text{eff}}(\mathbf{r}) = v^{\text{ext}}(\mathbf{r}) + \frac{e^2}{4\pi\epsilon_0}\int\frac{n(\mathbf{r}')}{|\mathbf{r}-\mathbf{r}'|}d\mathbf{r}' + \frac{\delta E_{\text{xc}}[n(\mathbf{r})]}{\delta n(\mathbf{r})} \qquad (10.159)$$

Now let us assume for the moment that we have an adequate approximation for the exchange–correlation energy functional defined in (10.158). We then have a set of self-consistent equations to solve. The starting point is to make an initial (educated) guess at the effective potential of (10.159). Then equations (10.157) can be solved using standard single-particle theory for all the electrons (we will be discussing a method of doing this on a lattice in the next chapter). The resulting ψ_i eigenfunctions can be used in (10.150) to calculate the number density and that density can be used in (10.159) to recalculate the effective potential. One can then solve (10.157) again and continue going round this loop of equations until the potential at one iteration is the same as the potential at the previous iteration. We then have a self-consistent solution. In practice such a process is rarely stable, and it is often necessary to iterate more slowly towards convergence by adding just a small percentage of the potential generated on iteration i to that used in iteration $i-1$ to get a suitable input potential to iteration $i+1$.

We can use the definition (10.158) to find a convenient form for the total energy $E_g[n(\mathbf{r})]$ of equation (10.148).

$$E_g[n(\mathbf{r})] = \int v^{\text{ext}}(\mathbf{r})n(\mathbf{r})d\mathbf{r} + T[n(\mathbf{r})]$$
$$+ \frac{1}{2}\frac{e^2}{4\pi\epsilon_0}\int\int\frac{n(\mathbf{r})n(\mathbf{r}')}{|\mathbf{r}-\mathbf{r}'|}d\mathbf{r}d\mathbf{r}' + E_{\text{xc}}[n(\mathbf{r})] \qquad (10.160)$$

This is still not an easy expression to evaluate even if we know the exchange–correlation energy. It is not easy to see how the kinetic energy functional may be calculated accurately, but fortunately we defined it as the kinetic energy of some equivalent system of non-interacting electrons. Therefore, the total energy associated with our set of one-electron equations (10.157) can be written as

$$\sum_{i-1}^{N}\epsilon_i = T[n(\mathbf{r})] + \int v^{\text{eff}}(\mathbf{r})n(\mathbf{r})d\mathbf{r} = T[n(\mathbf{r})] + \int v^{\text{ext}}(\mathbf{r})n(\mathbf{r})d\mathbf{r}$$
$$+ \frac{e^2}{4\pi\epsilon_0}\int\int\frac{n(\mathbf{r})n(\mathbf{r}')}{|\mathbf{r}-\mathbf{r}'|}d\mathbf{r}'d\mathbf{r} + \int\frac{\delta E_{\text{xc}}[n(\mathbf{r})]}{\delta n(\mathbf{r})}n(\mathbf{r})d\mathbf{r} \qquad (10.161)$$

We can use this expression to substitute into (10.160) for $T[n(\mathbf{r})]$. This gives us

$$E_g[n(\mathbf{r})] = \sum_{i=1}^{N}\epsilon_i - \frac{1}{2}\frac{e^2}{4\pi\epsilon_0}\int\int\frac{n(\mathbf{r})n(\mathbf{r}')}{|\mathbf{r}-\mathbf{r}'|}d\mathbf{r}'d\mathbf{r}$$
$$- \int\frac{\delta E_{\text{xc}}[n(\mathbf{r})]}{\delta n(\mathbf{r})}n(\mathbf{r})d\mathbf{r} + E_{\text{xc}}[n(\mathbf{r})] \qquad (10.162)$$

Now, an interesting point to note is that, unlike in Hartree–Fock theory, the statistics obeyed by the particles we are considering have not been mentioned up to now. In condensed matter we are, of course, interested primarily in electrons, but the formalism so far is equally applicable to bosons and to fermions. It is in the exchange–correlation energy that the effect of the statistics will be approximately taken into account. A further noteworthy point is that when $n(\mathbf{r})$ is known the theory can be made local, unlike Hartree–Fock theory.

10.12 Density Functional Theory and Magnetism

Many materials are magnetic, and so far in our discussion of density functional theory we have not included any possibility of the symmetry breaking necessary for magnetism to occur. If magnetism is weak we can include it in the density functional Hamiltonian directly as an effective magnetic field. This is a 'spin-only' theory which is reasonably easy to implement and has found much favour in condensed matter physics (Von-Barth and Hedin 1972). We describe this theory briefly in the following section. If the magnetic field is sufficiently strong that the possibility of inducing orbital currents cannot be ignored, we must invoke a more sophisticated version of density functional theory and we will go on to describe that in more detail. The latter theory is more important from our point of view because it is the non-relativistic limit of the full relativistic density functional theory we describe in section 10.14.

Density Functional Theory in a Weak Magnetic Field

In a 'spin-only' non-relativistic magnetic system we have two separate charge densities, $n_\uparrow(\mathbf{r})$ representing the spin-up electron density, and $n_\downarrow(\mathbf{r})$ representing the spin-down density. It is certainly desirable to extend the density functional theory to include magnetism in this way. This is done by extending the Hamiltonian (10.125) to include a term that couples to an external field:

$$\hat{H} = \hat{T} + \hat{U} + \hat{V} \tag{10.163a}$$

where $\hat{V}$ is now more complicated

$$\hat{V} = \int \left(v^{\text{ext}}(\mathbf{r})\langle\Psi|\hat{n}(\mathbf{r})|\Psi\rangle - \mu_B \mathbf{B}^{\text{ext}}(\mathbf{r}) \cdot \langle\Psi|\hat{\mathbf{m}}(\mathbf{r})|\Psi\rangle \right) d\mathbf{r} \tag{10.163b}$$

The difference between this and equation (10.130) is that $\hat{V}$ now contains a term coupling the spin density $\mathbf{m}(\mathbf{r})$ (but not the orbital motion) to an external magnetic field. The ground state energy now depends upon two independent variables, the density given by (10.150) which is the sum of the spin-up and spin-down contributions, and the spin density given by

$$m(\mathbf{r}) = n_\uparrow(\mathbf{r}) - n_\downarrow(\mathbf{r}) = \sum_i^{\text{occ}} \left(|\psi_{i\uparrow}|^2 - |\psi_{i\downarrow}|^2 \right) \tag{10.164}$$

where we have implicitly chosen the z-direction as the direction of the field and as the axis of quantization. Then equation (10.148) becomes

$$E[n(\mathbf{r}), m(\mathbf{r})] = T[n(\mathbf{r}), m(\mathbf{r})] + G[n(\mathbf{r}), m(\mathbf{r})]$$
$$+ \int \left(v^{\text{ext}}(\mathbf{r})n(\mathbf{r}) - \mu_B B^{\text{ext}}(\mathbf{r})m(\mathbf{r}) \right) d\mathbf{r} \qquad (10.165)$$

In actual fact the Kohn–Sham arguments of previous sections go through in the magnetic case in much the same way as in the non-magnetic case. Obviously it is necessary to minimize the total energy with respect to variations in both the total and spin densities subject to the constraint that the total number of particles remains constant. This procedure is discussed in detail by Von-Barth and Hedin (1972), Gunnarsson (1979) and Kohn and Vashishta (1982). The single-particle-like equation we derive from spin-density functional theory is

$$\left(-\frac{\hbar^2}{2m}\nabla^2 + v^{\text{eff}}(\mathbf{r}) - \mu_B \tilde{\sigma}_z B^{\text{eff}}(\mathbf{r}) \right) \psi_i = \epsilon_i \psi_i \qquad (10.166a)$$

where $\tilde{\sigma}_z$ is the 2×2 Pauli matrix, the ψ_i are two-component columns containing $\psi_{i\uparrow}$ and $\psi_{i\downarrow}$, and

$$v^{\text{eff}}(\mathbf{r}) = v^{\text{ext}}(\mathbf{r}) + \frac{e^2}{4\pi\epsilon_0} \int \frac{n(\mathbf{r}')}{|\mathbf{r} - \mathbf{r}'|} d\mathbf{r}' + \frac{\delta E_{\text{xc}}[n(\mathbf{r}), m(\mathbf{r})]}{\delta n(\mathbf{r})} \qquad (10.166b)$$

and

$$B_{\text{eff}}(\mathbf{r}) = B^{\text{ext}}(\mathbf{r}) - \frac{1}{\mu_B} \frac{\delta E_{\text{xc}}[n(\mathbf{r}), m(\mathbf{r})]}{\delta m(\mathbf{r})} \qquad (10.166c)$$

In (10.166a) the 2×2 identity matrix is understood to multiply the first two terms in the Hamiltonian. Equations (10.150), (10.164) and (10.166) form a self-consistent set of equations for a magnetic system, (see Weinberger 1990 for example). We have written them in terms of magnetization and charge densities. There is a completely equivalent formulation in terms of spin-up and spin-down densities, rather than total density and spin density. Because of the derivative with respect to the spin density in (10.166c) it is possible for magnetic solutions to these equations to occur even when the external magnetic field is set equal to zero. When such solutions occur with lower total energy than non-magnetic solutions the theory is predicting spontaneous magnetic ordering.

Practical schemes for implementing this theory require knowledge of the exchange–correlation energy in terms of the number and magnetization densities. In practice this involves us in making a good approximation. How we do this will be discussed in section 10.13.

Density Functional Theory in a Strong Magnetic Field

Here we present a different version of non-relativistic density functional theory where the external magnetic field is strong and orbital currents cannot be ignored. This theory is the true non-relativistic limit of the full relativistic density functional theory to be presented later. Furthermore, the presence of the magnetic field introduces an electron current, and a current also arises in the relativistic theory. The formulation in this section was originally derived by Vignale and Rasolt (1987, 1988).

In a magnetic field, the Hamiltonian for a system of N electrons is

$$\hat{H} = \hat{T} + \hat{U} + \hat{V} + \hat{W} \tag{10.167}$$

where $\hat{T}$, $\hat{U}$ and $\hat{V}$ are given by equations (10.129) to (10.131) and the magnetic field introduces the extra term

$$\hat{W} = e \sum_s \int \hat{\mathbf{j}}_{ps}(\mathbf{r}) \cdot \mathbf{A}_s(\mathbf{r}) d\mathbf{r} + \frac{e^2}{2m} \sum_s \int n_s(\mathbf{r}) \mathbf{A}_s^2(\mathbf{r}) d\mathbf{r} \tag{10.168}$$

The vector potential felt by the electrons is written as being dependent on spin. In fact there is only one vector potential and $\mathbf{A}_\downarrow(\mathbf{r}) = \mathbf{A}_\uparrow(\mathbf{r}) = \mathbf{A}(\mathbf{r})$. This artificial separation is necessary for the mathematics to follow, but is removed at the end of the calculation. The number density operator is given by equation (6.127) and the paramagnetic current density operator for electrons of spin s is

$$\hat{\mathbf{j}}_{ps} = \frac{\hbar}{2im} \left(\varphi_s^*(\mathbf{r}) \nabla \varphi_s(\mathbf{r}) - \varphi_s(\mathbf{r}) \nabla \varphi_s^*(\mathbf{r}) \right) \tag{10.169}$$

The full current density operator is

$$\hat{\mathbf{j}} = \hat{\mathbf{j}}_{ps} + \frac{e}{m} \hat{n}_s(\mathbf{r}) \mathbf{A}_s(\mathbf{r}) \tag{10.170}$$

The full current density operator and the number density operator are related by a continuity equation of the type (4.99). Of course, we also have the constraint that the total number of particles must remain constant.

In this version of density functional theory the first Hohenberg–Kohn theorem on page 367 generalizes to state that the potentials $v_s^{\text{ext}}(\mathbf{r})$ and $\mathbf{A}_s(\mathbf{r})$ are uniquely determined by the number and current density distributions $n_s(\mathbf{r})$ and $\mathbf{j}_{ps}(\mathbf{r})$. The proof proceeds in the way we have already seen. We assume that there are two different external potentials $v_s^{\text{ext}}(\mathbf{r})$, $\mathbf{A}_s(\mathbf{r})$ and $v_s^{\text{ext}'}(\mathbf{r})$, $\mathbf{A}_s'(\mathbf{r})$ which yield the same ground state densities $n_s(\mathbf{r})$ and $\mathbf{j}_{ps}(\mathbf{r})$. The corresponding Hamiltonians are $\hat{H}$ and $\hat{H}'$ and the associated total energies and ground state wavefunctions are E, E' and Ψ, Ψ'. With the Hamiltonian above, and the fact that the total energy is a minimum for the true ground state density, it follows that

$$E = \langle \Psi | \hat{H} | \Psi \rangle < \langle \Psi' | \hat{H} | \Psi' \rangle$$

$$= E' + \sum_s \int n_s(\mathbf{r})(v_s^{\text{ext}}(\mathbf{r}) - v_s^{\text{ext}'}(\mathbf{r}))d\mathbf{r}$$

$$+ e \sum_s \int \mathbf{j}_{ps}(\mathbf{r}) \cdot (\mathbf{A}_s(\mathbf{r}) - \mathbf{A}'_s(\mathbf{r}))d\mathbf{r}$$

$$+ \frac{e^2}{2m} \sum_s \int n_s(\mathbf{r})(\mathbf{A}_s^2(\mathbf{r}) - \mathbf{A}_s'^2(\mathbf{r}))d\mathbf{r} \qquad (10.171)$$

A second inequality is found by interchanging the primed and unprimed quantities, and adding the two inequalities gives us (10.136) again, which is a contradiction. So our assumption that two different potentials may lead to the same number and current densities must be false.

The next step is to write down the ground state energy functional.

$$E[n'_s(\mathbf{r}), \mathbf{j}'_{ps}(\mathbf{r})] = T_s[n'_s(\mathbf{r}), \mathbf{j}'_{ps}(\mathbf{r})] + \frac{e^2}{4\pi\epsilon_0} \frac{1}{2} \int \int \frac{n'(\mathbf{r})n'(\mathbf{r}')}{|\mathbf{r} - \mathbf{r}'|} d\mathbf{r} d\mathbf{r}'$$

$$+ E_{\text{xc}}[n'_s(\mathbf{r}), \mathbf{j}'_{ps}(\mathbf{r})] + \sum_s \int n'_s(\mathbf{r})v_s^{\text{ext}}(\mathbf{r})d\mathbf{r}$$

$$+ e \sum_s \int \mathbf{j}'_{ps}(\mathbf{r}) \cdot \mathbf{A}_s(\mathbf{r})d\mathbf{r} + \frac{e^2}{2m} \sum_s \int n'_s(\mathbf{r})\mathbf{A}_s^2(\mathbf{r})d\mathbf{r}$$

$$(10.172)$$

It is this quantity that we want to minimize with respect to variations in $n'_s(\mathbf{r})$ and $\mathbf{j}'_{ps}(\mathbf{r})$. We will approach the minimization from a slightly different direction to the non-magnetic case. Recall that $T_s[n'_s(\mathbf{r}), \mathbf{j}'_{ps}(\mathbf{r})]$ is the kinetic energy functional for the non-interacting electron system with the same number and current density as the full interacting system. By definition this means it can be written

$$T_s[n'_s(\mathbf{r}), \mathbf{j}'_{ps}(\mathbf{r})] = -\frac{\hbar^2}{2m} \left\langle \psi_0[n'_s(\mathbf{r}), \mathbf{j}'_{ps}(\mathbf{r})] \left| \sum_i^N \nabla_i^2 \right| \psi_0[n'_s(\mathbf{r}), \mathbf{j}'_{ps}(\mathbf{r})] \right\rangle$$

$$(10.173)$$

where $\psi_0[n'_s(\mathbf{r}), \mathbf{j}'_{ps}(\mathbf{r})]$ is the full ground state wavefunction of N non-interacting electrons, with the given densities. Because the particles are non-interacting, this wavefunction can be written exactly as a Slater determinant of the N one-electron wavefunctions $\psi_{is}[n'_s(\mathbf{r}), \mathbf{j}'_{ps}(\mathbf{r})] = \psi'_{is}(\mathbf{r})$ which are the N lowest energy solutions of the single-particle Schrödinger equation, and where s represents the electron spin and i just labels the electron and can be taken to represent its quantum numbers (other than spin).

$$\left(\frac{1}{2m}\left(\frac{\hbar}{i}\nabla - eA'_s(\mathbf{r}) \right)^2 + v_s^{eff'}(\mathbf{r}) \right) \psi'_{is}(\mathbf{r}) = \epsilon'_{is}\psi'_{is}(\mathbf{r}) \tag{10.174}$$

where $A'_s(\mathbf{r}) = A'_s[n'_s(\mathbf{r}), \mathbf{j}'_{ps}(\mathbf{r})]$ and $v_s^{eff'}(\mathbf{r}) = v_s^{eff'}[n'_s(\mathbf{r}), \mathbf{j}'_{ps}(\mathbf{r})]$ are effective single-particle vector and scalar potentials which we have yet to determine. The number density is given in terms of these functions by equation (10.150), and the current density is

$$\mathbf{j}'_{ps}(\mathbf{r}) = \frac{\hbar}{2im}\sum_{i=1}^{N_s}\left(\psi'^*_{is}(\mathbf{r})\nabla\psi'_{is}(\mathbf{r}) - \psi'_{is}(\mathbf{r})\nabla\psi'^*_{is}(\mathbf{r}) \right) \tag{10.175}$$

Now, from elementary quantum mechanics, the total energy of this system of non-interacting particles is just the sum of the energy eigenvalues. Their kinetic energy, then, is just the total energy minus the potential energy in the field, so

$$T_s[n'_s(\mathbf{r}), \mathbf{j}'_{ps}(\mathbf{r})] = \sum_s\sum_{i=1}^N \epsilon'_i - \sum_s\int n'_s(\mathbf{r})v_s^{eff'}(\mathbf{r})d\mathbf{r}$$

$$- e\sum_s\int \mathbf{j}_{ps}(\mathbf{r})\cdot A'_s(\mathbf{r})d\mathbf{r} - \frac{e^2}{2m}\sum_s\int n'_s(\mathbf{r})A_s'^2(\mathbf{r})d\mathbf{r} \tag{10.176}$$

We can put this into equation (10.172) to give us an expression for the total energy which we can minimize.

$$E[n'_s(\mathbf{r}), \mathbf{j}'_{ps}(\mathbf{r})] = \sum_s\sum_{i=1}^N \epsilon'_{is} - \sum_s\int n'_s(\mathbf{r})(v_s^{ext}(\mathbf{r}) - v_s^{eff'}(\mathbf{r}))d\mathbf{r}$$

$$+ \frac{e^2}{4\pi\epsilon_0}\frac{1}{2}\int\int \frac{n'(\mathbf{r})n'(\mathbf{r}')}{|\mathbf{r} - \mathbf{r}'|}d\mathbf{r}d\mathbf{r}' + \frac{e^2}{2m}\sum_s\int n'_s(\mathbf{r})(A_s^2(\mathbf{r}) - A'_s(\mathbf{r}))d\mathbf{r}$$

$$+ e\sum_s\int \mathbf{j}'_{ps}(\mathbf{r})(A_s(\mathbf{r}) - A'_s(\mathbf{r}))d\mathbf{r} + E_{xc}[n'_s(\mathbf{r}), \mathbf{j}'_{ps}(\mathbf{r})] \tag{10.177}$$

Now we have an expression for the total energy which we can minimize with respect to variations in $n'_s(\mathbf{r})$ and $\mathbf{j}'_{ps}(\mathbf{r})$ independently. This is a bit more complicated than the non-relativistic case because the quantities ϵ'_{is}, $A'_s(\mathbf{r})$ and $v_s^{eff'}(\mathbf{r})$ are functionals of these densities. However, we are lucky because, from (10.174), the derivatives of ϵ'_{is} exactly cancel the derivatives of $A'_s(\mathbf{r})$ and $v_s^{eff'}(\mathbf{r})$. So the variational principle leads simply to the following expressions. Variations with respect to $\mathbf{j}'_{ps}(\mathbf{r})$ give

$$\mathbf{A}'_s(\mathbf{r}) = \mathbf{A}_s(\mathbf{r}) + \frac{1}{e}\frac{\delta E_{\mathrm{xc}}[n'_s(\mathbf{r}),\mathbf{j}'_{ps}(\mathbf{r})]}{\delta \mathbf{j}'_{ps}(\mathbf{r})}$$

$$= \mathbf{A}(\mathbf{r}) + \mathbf{A}_{\mathrm{sxc}}(\mathbf{r}) \tag{10.178}$$

and varying the number density $n'_s(\mathbf{r})$ gives

$$v_s^{\mathrm{eff}\,'}(\mathbf{r}) = v^{\mathrm{ext}}(\mathbf{r}) + V_{\mathrm{sH}}(\mathbf{r}) + V_{\mathrm{sxc}}(\mathbf{r}) + \frac{e^2}{2m}(\mathbf{A}_s^2(\mathbf{r}) - \mathbf{A}_s'^{\,2}(\mathbf{r}))$$

$$= v^{\mathrm{ext}}(\mathbf{r}) + V_{\mathrm{sH}}(\mathbf{r}) + V_{\mathrm{sxc}}(\mathbf{r}) + \frac{e^2}{2m}\left(\mathbf{A}^2(\mathbf{r}) - (\mathbf{A}(\mathbf{r}) + \mathbf{A}_{\mathrm{sxc}}(\mathbf{r}))^2\right)$$

(10.179)

where in the last lines of (10.178) and (10.179) we have removed the artificial spin dependence of the vector potential, and the spin-dependent Hartree and exchange–correlation potentials are

$$V_{\mathrm{sH}}(\mathbf{r}) = \frac{e^2}{4\pi\epsilon_0}\frac{1}{2}\int\frac{n'_s(\mathbf{r})}{|\mathbf{r}-\mathbf{r}'|}d\mathbf{r}, \quad V_{\mathrm{sxc}}(\mathbf{r}) = \frac{\delta E_{\mathrm{xc}}[n'_s(\mathbf{r}),\mathbf{j}'_{ps}(\mathbf{r})]}{\delta n'_s(\mathbf{r})} \tag{10.180}$$

These are the potentials we put into the Schrödinger equation (10.174). They have been determined by insisting that the total energy takes on an extremal (hopefully minimum) value. Therefore, solving the equations (10.174), (10.150), (10.175), (10.178) and (10.179) self-consistently is a method of determining the total energy and density of a system of electrons in a strong magnetic field described by $\mathbf{B}(\mathbf{r}) = \nabla \times \mathbf{A}(\mathbf{r})$. A point worthy of note is that when we put equations (10.178) and (10.179) into (10.174) the quadratic terms in $\mathbf{A}(\mathbf{r}) + \mathbf{A}_{\mathrm{sxc}}(\mathbf{r})$ cancel, so the Schrödinger-like equation is linear in the effective vector potential, although the term $e^2\mathbf{A}^2(\mathbf{r})/2m$ does remain in the effective scalar potential. The total ground state energy is found by inserting (10.178) and (10.179) into (10.177) and using the self-consistently determined densities

$$E_{\mathrm{g}} = \sum_s\sum_{i=1}^{N}\epsilon'_{is} - \frac{e^2}{4\pi\epsilon_0}\frac{1}{2}\int\int\frac{n(\mathbf{r})n(\mathbf{r}')}{|\mathbf{r}-\mathbf{r}'|}d\mathbf{r}d\mathbf{r}' - \sum_s\int n_s(\mathbf{r})V_{\mathrm{sxc}}(\mathbf{r})d\mathbf{r}$$

$$+ e\sum_s\int \mathbf{j}_{ps}(\mathbf{r})\cdot\mathbf{A}_{\mathrm{sxc}}(\mathbf{r})d\mathbf{r} + E_{\mathrm{xc}}[n'_s(\mathbf{r}),\mathbf{j}'_{ps}(\mathbf{r})] \tag{10.181}$$

It can be shown that this theory is gauge invariant (an exercise for the reader). Solving this density functional theory on a lattice using Bloch's theorem presents considerable difficulties because the field will not have the periodicity of the lattice, although methods of solution have been suggested by Vignale and Rasolt (1988).

10.13 The Exchange–Correlation Energy

So far we have assumed that we can find some expression in terms of the density that will be a sufficient approximation for the exchange–correlation energy. In this section we will examine the physical meaning of this energy and then, later on, we will mention some of the approximations used to determine it. The interpretation of the exchange–correlation energy presented here is non-relativistic. There is no such explanation in the relativistic case. This is one reason why it is necessary to understand the non-relativistic theory prior to learning about relativistic density functional theory.

A prerequisite for this section is to recall the definition of the pair correlation function from equation (6.107). Here we write it in the form $n(\mathbf{r}_1, \mathbf{r}_2)$ in which it is frequently known as the joint probability density. This is the probability of finding an electron at $\mathbf{r}_1$ and simultaneously finding an electron at $\mathbf{r}_2$. Note that in a relativistic theory this quantity cannot be defined so easily as simultaneity depends upon your reference frame, and this is one reason why this discussion cannot be made in a relativistic context. If the electrons were free particles $n(\mathbf{r}_1, \mathbf{r}_2)$ would not depend upon the separation of $\mathbf{r}_1$ and $\mathbf{r}_2$. However, as we have already stated, there is no such thing as a free particle (so a fair proportion of this book is about things that don't exist), and this quantity will always depend upon the separation of the particles. For electrons, as $\mathbf{r}_1 \to \mathbf{r}_2$ the joint density must tend to zero because of the antisymmetry of the wavefunction. As $\mathbf{r}_1 - \mathbf{r}_2 \to \infty$ the two electrons can hardly interact at all, so $n(\mathbf{r}_1, \mathbf{r}_2) \to n(\mathbf{r}_1)n(\mathbf{r}_2)$. A proper evaluation of the joint density involves working out

$$n(\mathbf{r}_1, \mathbf{r}_2) = \langle \Psi | \hat{n}(\mathbf{r}_1)\hat{n}(\mathbf{r}_2) | \Psi \rangle \qquad (10.182)$$

where the density operator is defined in (10.128).

A popular way to think about the exchange–correlation energy intuitively is in terms of the exchange–correlation hole. This is a region of depleted electron density surrounding each electron owing to the Coulomb repulsion and the Pauli exclusion principle. The way to gain an understanding of the exchange–correlation hole is via a mathematical trick similar to the one we used when discussing convexity.

We will think of a typical density functional problem with a large number of interacting electrons where the energy operators are defined by equations (10.129) to (10.131). Consider in particular equation (10.129) for the Coulomb repulsion between the electrons. Let us multiply it by a factor λ which can take values between 0 and 1, so $\lambda = 1$ corresponds to the real system.

$$\hat{U}_\lambda = \frac{1}{2}\frac{\lambda e^2}{4\pi\epsilon_0}\sum_{s,s'}\int\int\varphi_s^\dagger(\mathbf{r})\varphi_{s'}^\dagger(\mathbf{r}')\frac{1}{|\mathbf{r}-\mathbf{r}'|}\varphi_{s'}(\mathbf{r}')\varphi_s(\mathbf{r})d\mathbf{r}d\mathbf{r}' \qquad (10.183)$$

Now, let us postulate (and it is a postulate) that we can construct a family of external potentials $v_\lambda^{\text{ext}}(\mathbf{r})$ such that the density $n(\mathbf{r})$ of the system is the same for any value of λ.

$$\hat{V}_\lambda = \sum_{i=1}^N v_\lambda^{\text{ext}}(\mathbf{r}_i) \qquad (10.184)$$

The ground state energy associated with any value of λ is

$$E_\lambda = \langle\Psi_\lambda|\hat{T} + \hat{U}_\lambda + \hat{V}_\lambda|\Psi_\lambda\rangle \qquad (10.185)$$

Now we differentiate this with respect to λ

$$\frac{\partial E_\lambda}{\partial\lambda} = \left\langle\frac{\partial\Psi_\lambda}{\partial\lambda}\left|\hat{T} + \hat{U}_\lambda + \hat{V}_\lambda\right|\Psi_\lambda\right\rangle$$
$$+ \langle\Psi_\lambda|\frac{\partial\hat{U}_\lambda}{\partial\lambda} + \frac{\partial\hat{V}_\lambda}{\partial\lambda}|\Psi_\lambda\rangle + \left\langle\Psi_\lambda\left|\hat{T} + \hat{U}_\lambda + \hat{V}_\lambda\right|\frac{\partial\Psi_\lambda}{\partial\lambda}\right\rangle \qquad (10.186)$$

Since the Hamiltonian is hermitian and Ψ_λ is an eigenfunction of $\hat{T} + \hat{U}_\lambda + \hat{V}_\lambda$ with eigenvalue E_λ, we can combine the first and third terms to write (10.186) as

$$\frac{\partial E_\lambda}{\partial\lambda} = \langle\Psi_\lambda|\frac{\partial\hat{U}_\lambda}{\partial\lambda} + \frac{\partial\hat{V}_\lambda}{\partial\lambda}|\Psi_\lambda\rangle + E_\lambda\frac{\partial}{\partial\lambda}\langle\Psi_\lambda|\Psi_\lambda\rangle \qquad (10.187)$$

Now, the second term here is just the derivative of the normalization integral, which obviously doesn't depend on λ, so it is zero. We can simplify (10.187) further by noting that

$$\frac{\partial\hat{U}_\lambda}{\partial\lambda} = \hat{U} = \frac{1}{\lambda}\hat{U}_\lambda \qquad (10.188)$$

Then

$$\frac{\partial E_\lambda}{\partial\lambda} = \lambda^{-1}\langle\Psi_\lambda|\hat{U}_\lambda|\Psi_\lambda\rangle + \int n(\mathbf{r})\frac{\partial v_\lambda^{\text{ext}}(\mathbf{r})}{\partial\lambda}d\mathbf{r} \qquad (10.189)$$

Note that it is not necessary to differentiate the density in (10.189) because it has been defined as being independent of λ in the construction of this argument. We are now in a position to integrate both sides of this equation with respect to λ

$$E_{\lambda=1} - E_{\lambda=0} = \int_0^1 \lambda^{-1} \langle \Psi_\lambda | \hat{U}_\lambda | \Psi_\lambda \rangle d\lambda$$

$$+ \int n(\mathbf{r}) v_{\lambda=1}^{\text{ext}}(\mathbf{r}) d\mathbf{r} - \int n(\mathbf{r}) v_{\lambda=0}^{\text{ext}}(\mathbf{r}) d\mathbf{r} \qquad (10.190)$$

However, $\lambda = 0$ implies from (10.183) that there is no interaction between the electrons at all. For this case the total energy (10.185) can be written

$$E_{\lambda=0} = \int n(\mathbf{r}) v_{\lambda=0}^{\text{ext}}(\mathbf{r}) d\mathbf{r} + T[n(\mathbf{r})] \qquad (10.191)$$

Now we can add (10.191) to (10.190) to get

$$E_{\lambda=1} = \int_0^1 \lambda^{-1} \langle \Psi_\lambda | \hat{U}_\lambda | \Psi_\lambda \rangle d\lambda + \int n(\mathbf{r}) v_{\lambda=1}^{\text{ext}}(\mathbf{r}) d\mathbf{r} + T[n(\mathbf{r})] \qquad (10.192)$$

We can also write down equation (10.160) for the case $\lambda = 1$:

$$E_{\lambda=1} = \int v_{\lambda=1}^{\text{ext}}(\mathbf{r}) n(\mathbf{r}) d\mathbf{r} + T[n(\mathbf{r})]$$

$$+ \frac{1}{2} \frac{e^2}{4\pi\epsilon_0} \int \int \frac{n(\mathbf{r})n(\mathbf{r}')}{|\mathbf{r} - \mathbf{r}'|} d\mathbf{r} d\mathbf{r}' + E_{\text{xc}}[n(\mathbf{r})] \qquad (10.193)$$

Note that the kinetic energy terms in (10.192) and (10.193) are the same because they are functionals of $n(\mathbf{r})$ only, which is the same regardless of the value of λ. Next we subtract (10.192) from (10.193) to get

$$E_{\text{xc}}[n(\mathbf{r})] = \int_0^1 \lambda^{-1} \langle \Psi_\lambda | \hat{U}_\lambda | \Psi_\lambda \rangle d\lambda - \frac{1}{2} \frac{e^2}{4\pi\epsilon_0} \int \int \frac{n(\mathbf{r})n(\mathbf{r}')}{|\mathbf{r} - \mathbf{r}'|} d\mathbf{r} d\mathbf{r}' \qquad (10.194)$$

The second term in (10.194) may be rewritten as

$$\frac{1}{2} \frac{e^2}{4\pi\epsilon_0} \int \int \frac{n(\mathbf{r})n(\mathbf{r}')}{|\mathbf{r} - \mathbf{r}'|} d\mathbf{r} d\mathbf{r}' = \frac{1}{2} \frac{e^2}{4\pi\epsilon_0} \int_0^1 \frac{d\lambda}{\lambda} \int \int \lambda \frac{n(\mathbf{r})n(\mathbf{r}')}{|\mathbf{r} - \mathbf{r}'|} d\mathbf{r} d\mathbf{r}' \qquad (10.195)$$

and so we can combine the terms in (10.194) as

$$E_{\text{xc}}[n(\mathbf{r})] = \int_0^1 \lambda^{-1} \left(\langle \Psi_\lambda | \hat{U}_\lambda | \Psi_\lambda \rangle - \frac{1}{2} \frac{e^2}{4\pi\epsilon_0} \int \int \lambda \frac{n(\mathbf{r})n(\mathbf{r}')}{|\mathbf{r} - \mathbf{r}'|} d\mathbf{r} d\mathbf{r}' \right) d\lambda$$

$$(10.196)$$

Now we can define $\langle \Psi_\lambda | \hat{U}_\lambda | \Psi_\lambda \rangle$ exactly, in terms of the joint probability density discussed above. Putting the operator (10.182) into this matrix element explicitly gives

$$\langle \Psi_\lambda | \hat{U}_\lambda | \Psi_\lambda \rangle = \frac{1}{2} \frac{e^2}{4\pi\epsilon_0} \int \int \lambda \frac{n(\mathbf{r}, \mathbf{r}')}{|\mathbf{r} - \mathbf{r}'|} d\mathbf{r} d\mathbf{r}' \qquad (10.197)$$

from which we finally obtain

$$E_{xc}[n(\mathbf{r})] = \int_0^1 \frac{1}{2} \frac{e^2}{4\pi\epsilon_0} \int \int \frac{d\mathbf{r}d\mathbf{r}'}{|\mathbf{r} - \mathbf{r}'|}(n(\mathbf{r}, \mathbf{r}') - n(\mathbf{r})n(\mathbf{r}'))d\lambda \qquad (10.198)$$

Now, this equation has a rather clear physical meaning. It indicates that the exchange–correlation energy is the difference between the true interaction energy of the electrons and the Hartree electrostatic energy of the electronic charge densities, i.e. it represents deviations from the average probability of finding an electron at $\mathbf{r}$ and $\mathbf{r}'$ simultaneously. Let's take a factor $n(\mathbf{r})$ outside the part of the integrand of (10.198) in brackets. This is not the most difficult piece of mathematics in this book. We get

$$n(\mathbf{r})n_{xc}(\mathbf{r}, \mathbf{r}') = n(\mathbf{r}) \left(\frac{n(\mathbf{r}, \mathbf{r}')}{n(\mathbf{r})} - n(\mathbf{r}') \right) \qquad (10.199)$$

The term in brackets in (10.199) called $n_{xc}(\mathbf{r}, \mathbf{r}')$ is the depletion in density at $\mathbf{r}'$ due to the presence of an electron at $\mathbf{r}$, i.e. a 'hole' in the density. This enables us to make an entertaining and interesting interpretation of the exchange–correlation energy of (10.198) as representing the Coulomb attraction between the electron at $\mathbf{r}$ and its own exchange–correlation hole, i.e.

$$E_{xc}[n(\mathbf{r})] = \int_0^1 \frac{1}{2} \frac{e^2}{4\pi\epsilon_0} \int \int \frac{d\mathbf{r}d\mathbf{r}'}{|\mathbf{r} - \mathbf{r}'|}n(\mathbf{r})n_{xc}(\mathbf{r}, \mathbf{r}')d\lambda \qquad (10.200)$$

The quantity $n_{xc}(\mathbf{r}, \mathbf{r}')$ represents a depletion of the charge density. Thus the exchange–correlation energy is negative as we expect.

This clear and intuitive interpretation of the exchange–correlation energy forms the basis of much that has been written about density functional theory (Harris and Jones 1974). It has, for example, been found that $E_{xc}[n(\mathbf{r})]$ is independent of the shape of the hole and that the displaced charge is exactly the negative of one electronic charge (Von Barth 1983). We will turn now to present some explicit expressions for the exchange–correlation energy that can be used in real calculations.

Although the exchange–correlation energy is not known accurately in general, it can be calculated very accurately for one special case, when the probability density $n(\mathbf{r})$ is constant, i.e. when the external potential $v(\mathbf{r})$ is independent of position.

To obtain a useful expression for the exchange–correlation energy it is invariably necessary to make the Local Density Approximation (LDA). In this approximation it is assumed that the exchange–correlation energy per electron $\epsilon_{xc}[n(\mathbf{r})]$ in a non-uniform system of electrons is related to the density $n(\mathbf{r})$ at point $\mathbf{r}$ in the same way as it is in the uniform density system. The exchange–correlation energy per unit volume is then

$n(\mathbf{r})\epsilon_{xc}[n(\mathbf{r})]$ and we can find the total exchange–correlation energy of the system by integrating over volume

$$E_{xc}[n(\mathbf{r})] = \int n(\mathbf{r})\epsilon_{xc}[n(\mathbf{r})]d\mathbf{r} \qquad (10.201)$$

The physical meaning of this equation is clear. In performing an integral we divide the volume up into infinitesimal volumes and sum the contributions to the integral from all of them. Here we assume that in each infinitesimal volume the density is approximately constant and hence the exchange–correlation energy will approximately take on the value for the homogeneous electron gas in that infinitesimal volume.

We can then define the exchange–correlation potential, using the expression for it in equation (10.159), as

$$V_{xc}(\mathbf{r}) = \frac{\delta E_{xc}[n(\mathbf{r})]}{\delta n(\mathbf{r})} = \frac{\delta}{\delta n(\mathbf{r})} \int n(\mathbf{r}')\epsilon_{xc}[n(\mathbf{r}')]d\mathbf{r}' \qquad (10.202a)$$

It is also common to define

$$\mu_{xc}[n(\mathbf{r})] = \frac{\delta}{\delta n(\mathbf{r})} n(\mathbf{r})\epsilon_{xc}[n(\mathbf{r})] = \epsilon_{xc}[n(\mathbf{r})] + n(\mathbf{r})\frac{\delta\epsilon[n(\mathbf{r})]}{\delta n(\mathbf{r})} \qquad (10.202b)$$

which is the exchange–correlation contribution to the chemical potential of a uniform system.

So, having made this approximation, we only require $\epsilon_{xc}[n(\mathbf{r})]$. To find this the common methods are to use some modification of the Hartree–Fock approximation for the exchange, to rely on Thomas–Fermi theory (Gunnarsson 1979), or to perform quantum Monte-Carlo calculations on the homogeneous electron gas. Some common approximations are given below.

In this section (only) we follow convention and write formulae in atomic units. For the relationships between unit systems see appendix D.

The simplest approximation for $\epsilon_{xc}[n(\mathbf{r})]$ is to take the exact expression for the exchange energy from Hartree–Fock theory and to evaluate it using non-relativistic free particle wave wavefunctions. This is known as the Slater approximation (Slater 1960). It has been often used in Hartree–Fock theory. To account for correlation a simple approach is to multiply this exchange energy by a constant α which can vary between $2/3$ and 1, i.e

$$\epsilon_{xc}[n(\mathbf{r})] = -6\alpha\left(\frac{3n(\mathbf{r})}{8\pi}\right)^{1/3} \qquad (10.203)$$

This is a simple approximation which gives remarkably accurate results, as was shown in table 10.5 for example.

When we include correlation explicitly no simple analytic expression for $\epsilon_{xc}[n(\mathbf{r})]$ can be obtained. There are a couple of quantities that appear frequently in exchange–correlation approximations, these are

$$G(x) = (1 + x^3)\ln\left(1 + \frac{1}{x}\right) + \frac{x}{2} - x^2 - \frac{1}{3}, \qquad \eta = \left(\frac{4}{9\pi}\right)^{1/3} \qquad (10.204)$$

One common approach, adopted by Hedin and Lundqvist (1971), is to separate the exchange and correlation contributions

$$\epsilon_{xc}[n(\mathbf{r})] = \epsilon_x[n(\mathbf{r})] + \epsilon_c[n(\mathbf{r})] \qquad (10.205a)$$

Hedin and Lundqvist took $\alpha = 1$ in (10.203) for the exchange contribution and

$$\epsilon_c[n(\mathbf{r})] = -CG(x) \qquad (10.205b)$$

where

$$x = \frac{r_s}{A}, \qquad\qquad C = \left(\frac{2B}{\pi\eta A}\right)^{1/3}$$
$$A = 21, \qquad\qquad B = 0.7734 \qquad\qquad\qquad (10.205c)$$

and r_s is defined by equation (10.122).

In a magnetic system we have to consider the exchange–correlation energy of the spin-polarized electron gas. In this case a popular approximation is due to Von Barth and Hedin (1972) who split the exchange and correlation terms up according to (10.205a) and set

$$\epsilon_x[n(\mathbf{r})] = -6\left(\frac{3}{4\pi}\right)^{1/3} \frac{n_\uparrow^{4/3}(\mathbf{r}) + n_\downarrow^{4/3}(\mathbf{r})}{n(\mathbf{r})} \qquad (10.206)$$

For a paramagnetic system this becomes equation (10.203). We also define x as the fraction of spin-up electrons, so $x = 1/2$ is the paramagnetic state and

$$\gamma = \frac{4a}{3(1-a)}, \qquad a = 2^{-1/3}, \qquad f(x) = (1-a)^{-1}(x^{4/3} + (1-x)^{4/3} - a) \qquad (10.207a)$$

Now if we write superscript F meaning the magnetic state and superscript P meaning the paramagnetic state we define

$$\mu_x^P = \gamma(\epsilon_x^F[n(\mathbf{r})] - \epsilon_x^P[n(\mathbf{r})]), \qquad v_c = \gamma(\epsilon_c^F[n(\mathbf{r})] - \epsilon_c^P[n(\mathbf{r})]) \qquad (10.207b)$$

It turns out that it is more convenient to parameterize the exchange and correlation energies in terms of r_s than in terms of $n(\mathbf{r})$, as

$$\epsilon_x[r_s] = \epsilon_x^P + \gamma^{-1}\mu_x^P f(x) \qquad (10.207c)$$
$$\epsilon_c[r_s] = \epsilon_c^P + \gamma^{-1}v_c f(x) \qquad (10.207d)$$

with

$$\epsilon_x^P = -\frac{0.9163}{r_s}, \qquad\qquad \mu_x^P = -\frac{4}{3}\frac{0.9163}{r_s} \qquad (10.207e)$$

$$\epsilon_c^P = -c^P G\left(\frac{r_s}{r^P}\right), \qquad\qquad \epsilon_c^F = -c^F G\left(\frac{r_s}{r^F}\right) \qquad (10.207f)$$

and

$$c^P = 0.0504, \qquad c^F = 0.0254, \qquad r^P = 30, \qquad r^F = 75 \qquad (10.207g)$$

For further discussion of exchange–correlation approximations see the review articles on density functional theory in the references.

10.14 Relativistic Density Functional Theory (RDFT)

The non-relativistic density functional theory discussed above is not sufficient to describe materials containing heavy elements. Exactly where in the periodic table relativistic effects become important is rather subjective and depends upon the particular application of the calculations. There certainly comes a point where one requires a relativistic generalization of the theory discussed above for a numerically accurate description of condensed matter. Relativistic density functional theory was first introduced by Rajagopal and Callaway (1973) who were mainly interested in the effects of spin on the density functional formalism. The relativistic nature of the theory, and particularly of the exchange–correlation energy, was emphasized later by Rajagopal (1978), Ramana and Rajagopal (1979) and Macdonald and Vosko (1979). Good summaries of the theory are given by Freeman *et al.* (1985) and by Weinberger (1990).

In the above paragraph the previous sections are described as non-relativistic, but this is barely true. Much of what has been done so far has not depended upon the explicit form of the Hamiltonian and will go through in exactly the same way in a relativistic theory. With a suitable redefinition of $\hat{U}$, $\hat{V}$ and $\hat{T}$, and subtraction of the rest mass energy, the above theory follows through with very little modification in a fully relativistic regime. The interpretation of the exchange–correlation energy becomes rather more complicated as it now includes the Breit interaction and several other subtle relativistic effects. However, one ends up with an effective single-particle Dirac equation with an effective potential given by the usual terms.

In our discussion of relativistic density functional theory we will be making direct application of the theory of section 6.11 of the Dirac, rather than the Schrödinger, field operators.

RDFT with an External Scalar Potential

In this section we will develop a limited relativistic density functional theory which represents many of the relativistic effects found in condensed matter without bringing the full complexity of quantum electrodynamics to bear on the problem. This theory was first discussed by MacDonald and Vosko (1979). Here we will only outline this approach as a gentle introduction to the full-blown theory.

In this theory then, the many-body Hamiltonian can be written in an exactly analogous way to equation (10.125)

$$\hat{H} = \hat{T} + \hat{U} + \hat{V} = \hat{H}_0 + \hat{V} \tag{10.208}$$

Here we will define the operators in accordance with equations (10.129) to (10.131)

$$\hat{U} = \frac{1}{2}\frac{e^2}{4\pi\epsilon_0} \int \int \varphi^\dagger(\mathbf{r})\varphi^\dagger(\mathbf{r}')\frac{1}{|\mathbf{r}-\mathbf{r}'|}\varphi(\mathbf{r}')\varphi(\mathbf{r})d\mathbf{r}d\mathbf{r}' \tag{10.209}$$

$$\hat{V} = \int v^{\text{ext}}(\mathbf{r})\varphi^\dagger(\mathbf{r})\varphi(\mathbf{r})d\mathbf{r} \tag{10.210}$$

$$\hat{T} = \int \varphi^\dagger(\mathbf{r})(c\tilde{\boldsymbol{\alpha}}\cdot\hat{\mathbf{p}} + \tilde{\beta}mc^2)\varphi(\mathbf{r})d\mathbf{r} \tag{10.211}$$

These are completely analogous to equations (10.129) to (10.131). The principle difference from the non-relativistic case is the inclusion of the rest mass energy in the definition of $\hat{T}$. The $\varphi^\dagger(\mathbf{r})$ and $\varphi(\mathbf{r})$ are the Dirac field operators given by equation (6.136). There is no summation over spin in the relativistic case, as that is included in the four-component nature of the field operators and in the summations in their definitions. The number density operator at point $\mathbf{r}$ for the system is

$$\hat{n}(\mathbf{r}) = \varphi^\dagger(\mathbf{r})\varphi(\mathbf{r}) \tag{10.212}$$

The electron density at this point is then

$$n(\mathbf{r}) = \langle\Psi|\hat{n}(\mathbf{r})|\Psi\rangle \tag{10.213}$$

where Ψ is the many-body state vector. With these definitions the argument from the paragraph above equation (10.132) to that below (10.136) follows through identically in this relativistic case. So again we have the result that it is impossible for two different external potentials to give rise to the same ground state density distribution $n(\mathbf{r})$.

The theorems upon which this version of density functional theory is based are identical to those of the non-relativistic theory. We assume that we can write the relativistic total energy functional as

$$W[n(\mathbf{r})] = \int v^{\text{ext}}(\mathbf{r})n(\mathbf{r})d\mathbf{r} + T[n(\mathbf{r})] + G[n(\mathbf{r})] \tag{10.214}$$

and the ground state total energy $W_g[n(\mathbf{r})]$ is found by minimizing equation (10.214) with respect to variations in $n(\mathbf{r})$ subject to the constraint

$$\int n(\mathbf{r})d\mathbf{r} = N \qquad (10.215)$$

where N is the total number of electrons in the system. This is a more serious constraint in relativistic theory than in non-relativistic theory because it precludes the formation of electron/positron pairs, which can occur at high energies in relativistic quantum theory. In reality, though, such processes are exceedingly unlikely in condensed matter applications.

We can now follow through an argument identical to that of equations (10.152) to (10.157) to obtain a set of effective single-particle Dirac equations to describe the many-electron problem. These are

$$(c\tilde{\boldsymbol{\alpha}} \cdot \hat{\mathbf{p}} + \tilde{\beta}mc^2 + v^{\text{eff}}(\mathbf{r}))\psi_i(\mathbf{r}) = w_i\psi_i \qquad (10.216)$$

with

$$n(\mathbf{r}) = \sum_{i=1}^{N} \psi_i^{\dagger}(\mathbf{r})\psi_i(\mathbf{r}) \qquad (10.217)$$

and

$$v^{\text{eff}}(\mathbf{r}) = v^{\text{ext}}(\mathbf{r}) + \frac{e^2}{4\pi\epsilon_0} \int \frac{n(\mathbf{r}')}{|\mathbf{r}-\mathbf{r}'|}d\mathbf{r}' + \frac{\delta E_{\text{xc}}[n(\mathbf{r})]}{\delta n(\mathbf{r})} \qquad (10.218)$$

and the total ground state energy is

$$W_g[n(\mathbf{r})] = \sum_{i=1}^{N} w_i - \frac{1}{2}\frac{e^2}{4\pi\epsilon_0}\int \frac{n(\mathbf{r})n(\mathbf{r}')}{|\mathbf{r}-\mathbf{r}'|} + E_{\text{xc}}[n(\mathbf{r})] - \int n(\mathbf{r})\frac{\delta E_{\text{xc}}}{\delta n(\mathbf{r})}d\mathbf{r} \quad (10.219)$$

Now $\psi(\mathbf{r})$ and $\psi^{\dagger}(\mathbf{r})$ are effective single-particle four-component wavefunctions. Note that we have no positron states here so the summation in the first term on the right hand side of this equation is over electron states with positive energy only. Apart from the obvious replacement of an effective single-particle Schrödinger equation with an effective single-particle Dirac equation, the main difference between this formalism and the non-relativistic case is in the exchange–correlation energy. In principle, in relativistic theory it contains not only the Coulomb repulsion and the Pauli principle effects, but also magnetic interactions between the electrons and retardation of the Coulomb interaction to all orders, so going well beyond the second order Breit interaction. Hence, it is much more difficult to give a simple interpretation of the exchange–correlation energy in the way that was done non-relativistically in section 10.13. An illustration of this theory is given in section 10.18.

RDFT with an External Vector Potential

For a full description of the properties of a system of interacting electrons and photons one should really work within the theory of quantum electrodynamics. Here we will do this, although the theory looks very similar to the relativistic quantum theory because we will not quantize the electromagnetic field. This section follows the work of Rajagopal and Callaway (1973), which was later amplified by Ramana and Rajagopal (1983) and has been reviewed by Rajagopal (1980). Our Dirac field operators are again given by equation (6.136). The many-body system containing electrons and photons obeys the Dirac equation

$$\left(i\hbar \frac{\partial}{\partial t} - \hat{H} \right) |\Psi\rangle = 0 \tag{10.220}$$

where the Hamiltonian for the system is

$$\hat{H} = \hat{H}_0 + \hat{H}_{\text{em}} + \hat{U} + \hat{V} \tag{10.221}$$

with

$$\hat{H}_0 = \int \varphi^\dagger(\mathbf{r})(c\tilde{\boldsymbol{\alpha}} \cdot \hat{\mathbf{p}} + \tilde{\beta}mc^2)\varphi(\mathbf{r})d\mathbf{r} \tag{10.222}$$

$$\hat{H}_{\text{em}} = \frac{1}{4}F_{\mu\nu}F^{\mu\nu} - \int \hat{\mathbf{J}}(\mathbf{r}) \cdot \mathbf{A}(\mathbf{r})d\mathbf{r} \tag{10.223}$$

$$\hat{U} = \frac{1}{2} \int \int \varphi^\dagger(\mathbf{r})\varphi^\dagger(\mathbf{r}')\hat{H}_{\text{B}}\varphi(\mathbf{r}')\varphi(\mathbf{r})d\mathbf{r}d\mathbf{r}' \tag{10.224}$$

$$\hat{V} = -ec \int \varphi^\dagger(\mathbf{r})\tilde{\boldsymbol{\alpha}} \cdot \mathbf{A}^{\text{ext}}(\mathbf{r})\varphi(\mathbf{r})d\mathbf{r} + \int \varphi^\dagger(\mathbf{r})v(\mathbf{r})\varphi(\mathbf{r})d\mathbf{r} \tag{10.225a}$$

Equation (10.222) represents the kinetic and rest mass energies of our system, the single-particle-like part of the Hamiltonian. In (10.223) $\hat{H}_{\text{em}}$ is the total energy density of a system of photons plus the interaction of the photon vector potential $\mathbf{A}(\mathbf{r})$ with the electron current density $\hat{\mathbf{J}}(\mathbf{r})$ and $F_{\mu\nu}$ is the antisymmetric field tensor of equation (1.55). The first term in $\hat{H}_{\text{em}}$ is essentially the Poynting vector written in tensor form. In (10.224) $\hat{U}$ represents the full interaction between the electrons. In fact this term contains the electron–electron interaction to all orders, not just the Coulomb repulsion and Breit interaction (although the symbol $\hat{H}_{\text{B}}$ was chosen to indicate the Breit interaction). Finally $\hat{V}$ is the interaction of the electrons with an unquantized external 4-potential. The first term in (10.225a) is what we would expect when we include a vector potential in the Dirac equation. However, recalling the definition of the current density operator (6.139), we can see that the integrand can also be written as a product of the current density operator and the vector potential.

Now this expression for the interaction with the external potential is somewhat cumbersome and can be written much more succinctly in terms of 4-vectors. Recalling the definition of the current and potential 4-vectors from section 1.4, and the definition of the scalar product of two 4-vectors from equation (1.37), we can combine the two terms in (10.225a) into one as

$$\hat{V} = e \int \hat{J}^\mu(\mathbf{r}) A_\mu^{\text{ext}}(\mathbf{r}) d\mathbf{r} \tag{10.225b}$$

For our purposes $A_\mu^{\text{ext}}(\mathbf{r})$ contains the static Coulomb field due to the nuclei. Now the current density of the ground state is

$$J^\mu = \langle \Psi | \hat{J}^\mu | \Psi \rangle \tag{10.226}$$

However, the four components of J^μ are not all independent because they have to obey the conservation of probability density:

$$\nabla \cdot \mathbf{J} = -\frac{\partial \rho}{\partial t} \tag{10.227a}$$

or, in 4-vector form

$$\frac{\partial J^\mu}{\partial x^\mu} = 0 \tag{10.227b}$$

It is now possible to repeat the argument of section 10.10 to show that the total energy is a unique functional of $J^\mu(\mathbf{r})$. We will do this explicitly. Suppose we have a non-degenerate ground state Ψ in an external 4-potential A_μ^{ext} that leads to a ground state 4-current density J^μ and ground state energy W_g. Then let us assume that a different 4-potential $A_\mu^{'\text{ext}}$ exists that leads to the same current density J^μ with ground state Ψ'. Because the ground state energy is the minimum that the system can have, the following inequality holds

$$W_g = \langle \Psi | \hat{H} | \Psi \rangle \quad < \quad \langle \Psi' | \hat{H} | \Psi' \rangle = \langle \Psi' | \hat{H}' - \hat{V}' + \hat{V} | \Psi' \rangle \tag{10.228}$$

So, if the two current densities are the same, we have

$$W_g < W_g' - e \int J^\mu(\mathbf{r})(A_\mu^{'\text{ext}}(\mathbf{r}) - A_\mu^{\text{ext}}(\mathbf{r})) d\mathbf{r} \tag{10.229}$$

Similarly

$$W_g' = \langle \Psi' | \hat{H}' | \Psi' \rangle \quad < \quad \langle \Psi | \hat{H}' | \Psi \rangle = \langle \Psi | \hat{H} - \hat{V} + \hat{V}' | \Psi \rangle \tag{10.230}$$

where we have just interchanged the primed and unprimed quantities. This gives us

$$W_g' < W_g - e \int J^\mu(\mathbf{r})(A_\mu^{\text{ext}}(\mathbf{r}) - A_\mu^{'\text{ext}}(\mathbf{r})) d\mathbf{r} \tag{10.231}$$

If we now add (10.229) and (10.231) we have

$$W_g + W'_g < W_g + W'_g \tag{10.232}$$

This, obviously, is a contradiction and implies that our assumption that two different external 4-potentials A_μ^{ext} could lead to the same 4-current density is incorrect. That is, it implies that the external 4-potential is a unique functional of the 4-current density. Some authors (Ramana and Rajagopal (1983), for instance) go through the argument above having subtracted the rest mass energy of the electrons at the beginning. As the rest mass energy is a constant, the argument above is unaffected by such a modification.

There is one further point to make about this derivation. We have seen in chapter 1 that the potentials only determine the fields to within a gauge transformation. For 4-potentials related by a gauge transformation we must obtain $W_g = W'_g$. Therefore, we must consider all potentials connected via a gauge transformation as being equivalent.

The Dirac–Kohn–Sham Equation

In the previous section we have seen that the 4-current density $J^\mu(\mathbf{r})$ uniquely determines the 4-potential. If the 4-potential is uniquely determined, the Hamiltonian (10.221) is completely defined. Hence, the total energy of the system is uniquely determined by $J^\mu(\mathbf{r})$. This leads to the first theorem of relativistic density functional theory

(1) The total energy per unit volume of a system described by the Hamiltonian (10.221) can be written as a functional of the expectation value of the 4-current density.

So we can write

$$W[J^\mu(\mathbf{r})] = T[J^\mu(\mathbf{r})] + G[J^\mu(\mathbf{r})] - e \int J^\mu(\mathbf{r}) A_\mu^{\text{ext}}(\mathbf{r}) d\mathbf{r} \tag{10.233}$$

Here, the first term $T[J^\mu(\mathbf{r})]$ represents the relativistic kinetic energy, plus the rest mass energy of a system of non-interacting electrons with the same 4-current density as the real system, plus the energy density of the photons and the interaction between them. The second term, $G[J^\mu(\mathbf{r})]$, is the internal potential energy of the real system,

$$T[J^\mu(\mathbf{r})] = \langle \Psi | \hat{H}_0 + \hat{H}_{\text{em}} | \Psi \rangle, \qquad G[J^\mu(\mathbf{r})] = \langle \Psi | \hat{U} | \Psi \rangle \tag{10.234}$$

The constraint that the number of particles N remains constant can be written in terms of J^μ as

$$\frac{1}{c} \int J^4(\mathbf{r}) d\mathbf{r} = N \tag{10.235}$$

Again we will use the Lagrange multiplier method. We will look for the condition that variations in J^μ result in no variation in the total energy, subject to the condition (10.235). If w is the Lagrange multiplier we have

$$\delta W[J^\mu(\mathbf{r})] = \int \delta J^\mu(\mathbf{r}) \left[eA_\mu^{\text{ext}}(\mathbf{r}) + \frac{\delta T[J^\mu(\mathbf{r})]}{\delta J^\mu(\mathbf{r})} + \frac{\delta G[J^\mu(\mathbf{r})]}{\delta J^\mu(\mathbf{r})} - \frac{w}{c} \right] d\mathbf{r} = 0$$

(10.236)

Now, following the non-relativistic argument, $\delta J^\mu(\mathbf{r})$ is an arbitrary small change in $J^\mu(\mathbf{r})$, so the only way in which the integral in (10.236) can be zero is if the part of the integrand in square brackets is equal to zero:

$$eA_\mu^{\text{ext}}(\mathbf{r}) + \frac{\delta T[J^\mu(\mathbf{r})]}{\delta J^\mu(\mathbf{r})} + \frac{\delta G[J^\mu(\mathbf{r})]}{\delta J^\mu(\mathbf{r})} = \frac{w}{c}$$

(10.237)

This is our Euler–Lagrange equation for the system of interacting relativistic electrons and photons. It is here that we invoke a second theorem

(2) The non-degenerate ground state of an inhomogeneous interacting system of N relativistic electrons can be described by a set of N single-particle effective Dirac equations with a suitably defined single-particle-like scalar and vector potential.

Now suppose that the electrons described by (10.237) have no interaction with one another. Then (10.237) would become

$$eA_\mu^{\text{ext}}(\mathbf{r}) + \frac{\delta T[J^\mu(\mathbf{r})]}{\delta J^\mu(\mathbf{r})} - \frac{w}{c} = 0$$

(10.238)

We know how to deal with such a system of N relativistic non-interacting particles. The wavefunction can be written as a direct product of N four-component single-particle wavefunctions and we have a $4^N \times 4^N$ dimensional Dirac equation. In this equation we can separate the variables to give N four-component single-particle Dirac equations. Now, as in the non-relativistic argument, equation (10.237) will become completely analogous to (10.238) if we define an effective external 4-potential

$$A_\mu^{\text{eff}}[J^\mu(\mathbf{r})] = A_\mu^{\text{ext}}(\mathbf{r}) + \frac{1}{e} \frac{\delta G[J^\mu(\mathbf{r})]}{\delta J^\mu(\mathbf{r})}$$

(10.239)

i.e. we are transferring the interaction between the electrons to an effective external 4-potential. Now we require a reasonable expression for $G[J^\mu(\mathbf{r})]$. As before we can factor out the Coulomb electrostatic interaction between the electrons and an equivalent term in the current (as suggested by equation (10.4)):

$$G[J^\mu(\mathbf{r})] = \frac{1}{2} \frac{e^2}{4\pi\epsilon_0} \int \int \frac{n(\mathbf{r})n(\mathbf{r}')}{|\mathbf{r} - \mathbf{r}'|} d\mathbf{r} d\mathbf{r}'$$

$$- \frac{1}{2} \frac{e^2}{4\pi\epsilon_0 c^2} \int \int \frac{\mathbf{J}(\mathbf{r}) \cdot \mathbf{J}(\mathbf{r}')}{|\mathbf{r} - \mathbf{r}'|} d\mathbf{r} d\mathbf{r}' + E_{\text{xc}}[J^\mu(\mathbf{r})]$$

(10.240)

and the relativistic exchange–correlation functional $E_{xc}[J^\mu(\mathbf{r})]$ is defined by this equation. We can substitute (10.240) directly into (10.239), but we get more physically intuitive results by dividing (10.239) into two parts, one that depends on the usual density and one that depends on the current density.

$$v^{\text{eff}}(\mathbf{r}) = v^{\text{ext}}(\mathbf{r}) + \frac{\delta G[J^\mu(\mathbf{r})]}{\delta n(\mathbf{r})} \tag{10.241a}$$

$$\mathbf{A}^{\text{eff}}(\mathbf{r}) = \mathbf{A}^{\text{ext}}(\mathbf{r}) + \frac{1}{e}\frac{\delta G[J^\mu(\mathbf{r})]}{\delta \mathbf{J}(\mathbf{r})} \tag{10.241b}$$

In (10.241a) we can separate out the instantaneous Coulomb interaction between the electrons and it then looks very much like the non-relativistic effective potential

$$v^{\text{eff}}(\mathbf{r}) = v^{\text{ext}}(\mathbf{r}) + \frac{e^2}{4\pi\epsilon_0}\int\frac{n(\mathbf{r}')}{|\mathbf{r} - \mathbf{r}'|}d\mathbf{r}' + \frac{\delta E_{xc}[n(\mathbf{r}), \mathbf{J}(\mathbf{r})]}{\delta n(\mathbf{r})} \tag{10.242}$$

The terms in the current density combine to give us the effective vector potential

$$\mathbf{A}^{\text{eff}}(\mathbf{r}) = \mathbf{A}^{\text{ext}}(\mathbf{r}) + \frac{e}{4\pi\epsilon_0 c^2}\int\frac{\mathbf{J}(\mathbf{r}')}{|\mathbf{r} - \mathbf{r}'|}d\mathbf{r}' + \frac{1}{e}\frac{\delta E_{xc}[n(\mathbf{r}), \mathbf{J}(\mathbf{r})]}{\delta \mathbf{J}(\mathbf{r})} \tag{10.243}$$

The terms associated with $\hat{H}_{\text{em}}$ can be subtracted out as being irrelevant in condensed matter, and the effective single-particle Dirac equation we obtain is

$$(c\tilde{\boldsymbol{\alpha}} \cdot (\hat{\mathbf{p}} - e\mathbf{A}^{\text{eff}}(\mathbf{r})) + \tilde{\beta}mc^2 + v^{\text{eff}}(\mathbf{r}))\psi_i(\mathbf{r}) = w_i\psi_i(\mathbf{r}) \tag{10.244}$$

with

$$n(\mathbf{r}) = \sum_{i=1}^{N} \psi_i^\dagger(\mathbf{r})\psi_i(\mathbf{r}) \tag{10.245}$$

$$\mathbf{J}(\mathbf{r}) = c\sum_{i=1}^{N} \psi_i^\dagger(\mathbf{r})\tilde{\boldsymbol{\alpha}}\psi_i(\mathbf{r}) \tag{10.246}$$

Equation (10.244) is known as the Dirac–Kohn–Sham equation. It looks like the usual single-particle Dirac equation, but w_i and $\psi_i(\mathbf{r})$ are not single-particle energies and wavefunctions, they are simply quantities which we can use to build up the 4-current density and the total energy. Equations (10.242) to (10.246) have to be solved self-consistently. As stated earlier, the exchange–correlation energy functional $E_{xc}[n(\mathbf{r}), \mathbf{J}(\mathbf{r})]$ is a much more complex object than its non-relativistic counterpart. In principle this complicated quantity can be evaluated from relativistic quantum Monte Carlo calculations. We briefly mention some of the common approximations for the relativistic exchange–correlation energy in section 10.17.

Our final task in this section is to write down an expression for the total energy in a relativistic framework. This is given by (10.233) initially. We can substitute (10.240) into this to give

$$W[J^\mu(\mathbf{r})] = T[J^\mu(\mathbf{r})] + \frac{1}{2}\frac{e^2}{4\pi\epsilon_0}\int\int\frac{n(\mathbf{r})n(\mathbf{r}')}{|\mathbf{r}-\mathbf{r}'|}d\mathbf{r}d\mathbf{r}' + E_{xc}[J^\mu(\mathbf{r})]$$

$$- e\int J^\mu(\mathbf{r})\cdot A_\mu^{ext}(\mathbf{r})d\mathbf{r} - \frac{1}{2}\frac{e^2}{4\pi\epsilon_0 c^2}\int\int\frac{\mathbf{J}(\mathbf{r})\cdot\mathbf{J}(\mathbf{r}')}{|\mathbf{r}-\mathbf{r}'|}d\mathbf{r}d\mathbf{r}'$$

(10.247)

The kinetic energy functional is hard to calculate, but we can use a similar trick to the non-relativistic case to eliminate it from the total energy expression. The total energy associated with our set of one-electron equations (10.244) is

$$\sum_{i=1}^{N}w_i = T[J^\mu(\mathbf{r})] + \int v^{eff}(\mathbf{r})n(\mathbf{r})d\mathbf{r} - e\int\mathbf{A}^{eff}(\mathbf{r})\cdot\mathbf{J}(\mathbf{r})d\mathbf{r}$$

(10.248)

This expression can be used to eliminate $T[J^\mu(\mathbf{r})]$ from (10.247). That will leave us with an expression for $W[J^\mu(\mathbf{r})]$ in terms of quantities we can calculate. The resulting expression for the total energy is

$$W[n(\mathbf{r}),\mathbf{J}(\mathbf{r})] = \sum_{i=1}^{N}w_i - \frac{1}{2}\frac{e^2}{4\pi\epsilon_0}\int\int\frac{n(\mathbf{r})n(\mathbf{r}')}{|\mathbf{r}-\mathbf{r}'|}d\mathbf{r}d\mathbf{r}' + E_{xc}[n(\mathbf{r}),\mathbf{J}(\mathbf{r})]$$

$$- \int\frac{\delta E_{xc}[n(\mathbf{r}),\mathbf{J}(\mathbf{r})]}{\delta n(\mathbf{r})}n(\mathbf{r})d\mathbf{r} + \frac{1}{e}\int\frac{\delta E_{xc}[n(\mathbf{r}),\mathbf{J}(\mathbf{r})]}{\delta\mathbf{J}(\mathbf{r})}\mathbf{J}(\mathbf{r})d\mathbf{r}$$

$$+ \frac{1}{2c^2}\frac{e^2}{4\pi\epsilon_0}\int\int\frac{\mathbf{J}(\mathbf{r})\cdot\mathbf{J}(\mathbf{r}')}{|\mathbf{r}-\mathbf{r}'|}d\mathbf{r}d\mathbf{r}'$$

(10.249)

So, to find the ground state energy of some relativistic electron system we have to solve equations (10.242) to (10.246) self-consistently and use the resulting charge and current densities in equation (10.249). The only problem that still remains (in principle – there are plenty of problems in practice I can assure you) is the evaluation of a suitable expression for the exchange–correlation energy.

Equation (10.249) does not contain the vector potential explicitly, and from equation (10.246) it is clear that $\mathbf{J}(\mathbf{r})$ has no explicit dependence on the vector potential. Therefore it is immediately apparent that the total energy functional is a gauge-invariant quantity.

This theory provides a first principles framework for calculating the properties of condensed matter where internal magnetic effects are important. It is the relativistic generalization of the strong field theory discussed on page 373. The external magnetic field breaks the initial magnetic degeneracies, but beyond that it is not required for the description of spontaneously magnetic materials. In a ferromagnet, for example, the internal field is described by the last two terms in equation (10.243), although these are not easy to interpret in terms of quantities familiar from the theory of solid state magnetism. Finally we point out that relativistic density functional theory is a misnomer, it should really be called 4-current density functional theory.

10.15 An Approximate Relativistic Density Functional Theory

An alternative method of taking relativistic effects into account in density functional theory has been developed by Rajagopal and Callaway (1973) and emphasised by MacDonald and Vosko (1979), and this alleviates some of the practical problems involved in implementing the above theory. In this approach we are able to derive equations that look more like the familiar non-relativistic density functional expressions for magnetic systems.

We start again with equation (10.233) for the total energy, but instead of proceeding directly to the Euler–Lagrange equation as before, we make a Gordon decomposition of the current (but not the electron density) in the final term. The Gordon decomposition of the current is given by equation (4.108), and so

$$
\begin{aligned}
\int J^\mu(\mathbf{r}) A_\mu^{\text{ext}}(\mathbf{r}) d\mathbf{r} = &-\langle \Psi | \int \varphi^\dagger(\mathbf{r}) \varphi(\mathbf{r}) v^{\text{ext}}(\mathbf{r}) d\mathbf{r} | \Psi \rangle \\
&+ \langle \Psi | \frac{e}{2m} \int \left(\tilde{\beta} \varphi^\dagger(\mathbf{r})(\hat{\mathbf{p}} - e\mathbf{A}(\mathbf{r})) \varphi(\mathbf{r}) \right) \cdot \mathbf{A}^{\text{ext}}(\mathbf{r}) d\mathbf{r} | \Psi \rangle \\
&- \langle \Psi | \frac{e}{2m} \int \left((\hat{\mathbf{p}} + e\mathbf{A}(\mathbf{r})) \varphi^\dagger(\mathbf{r}) \tilde{\beta} \varphi(\mathbf{r}) \right) \cdot \mathbf{A}^{\text{ext}}(\mathbf{r}) d\mathbf{r} | \Psi \rangle \\
&+ \langle \Psi | \frac{e\hbar}{2m} \int \left(\nabla \times (\tilde{\beta} \varphi^\dagger(\mathbf{r}) \tilde{\sigma} \varphi(\mathbf{r})) \right) \cdot \mathbf{A}^{\text{ext}}(\mathbf{r}) d\mathbf{r} | \Psi \rangle \\
&+ \langle \Psi | \frac{e\hbar}{2m} \int \left(\frac{1}{c} \frac{\partial}{\partial t} \tilde{\beta} \varphi^\dagger(\mathbf{r}) i\tilde{\alpha} \varphi(\mathbf{r}) \right) \cdot \mathbf{A}^{\text{ext}}(\mathbf{r}) d\mathbf{r} | \Psi \rangle
\end{aligned}
$$

$$(10.250)$$

On the right hand side of (10.250) all the terms have a clear meaning. The first term is the conventional interaction between the electron density and the external scalar potential. The next two terms express the contribution of the conventional and displacement currents, the fourth term represents the spin density and the final term denotes the electric moment. So, let us make some definitions

$$n(\mathbf{r}) = \langle \Psi | \varphi^\dagger(\mathbf{r}) \varphi(\mathbf{r}) | \Psi \rangle \tag{10.251a}$$

$$m(\mathbf{r}) = \frac{e\hbar}{2m} \langle \Psi | \varphi^\dagger(\mathbf{r}) \tilde{\beta} \tilde{\sigma} \varphi(\mathbf{r}) | \Psi \rangle \tag{10.251b}$$

$$\mathscr{J}(\mathbf{r}) = \langle \Psi | \frac{e}{2m} \left[\tilde{\beta} \varphi^\dagger(\mathbf{r}) (\hat{\mathbf{p}} - e\mathbf{A}(\mathbf{r})) \varphi(\mathbf{r}) \right] | \Psi \rangle$$
$$- \langle \Psi | \frac{e}{2m} \left[(\hat{\mathbf{p}} - e\mathbf{A}(\mathbf{r})) \tilde{\beta} \varphi^\dagger(\mathbf{r}) \varphi(\mathbf{r}) \right] | \Psi \rangle \tag{10.251c}$$

$$\mathbf{g}(\mathbf{r}) = \frac{ie\hbar}{2m} \langle \Psi | \tilde{\beta} \varphi^\dagger(\mathbf{r}) \tilde{\alpha} \varphi(\mathbf{r}) | \Psi \rangle \tag{10.251d}$$

With the help of some vector identities in the spin density term we can write (10.233) as

$$W[J_\mu(\mathbf{r})] = T[J_\mu(\mathbf{r})] + G[J_\mu(\mathbf{r})]$$
$$+ \int \left(n(\mathbf{r}) v(\mathbf{r}) - \mathbf{m}(\mathbf{r}) \cdot \mathbf{B}(\mathbf{r}) + \mathscr{J}(\mathbf{r}) \cdot \mathbf{A}^{\mathrm{ext}}(\mathbf{r}) - \frac{1}{c} \frac{\partial \mathbf{g}(\mathbf{r})}{\partial t} \cdot \mathbf{A}^{\mathrm{ext}}(\mathbf{r}) \right) d\mathbf{r} \tag{10.252}$$

At this stage we are going to make an approximation to make this version of density functional theory more tractable. If we ignore the terms in the external vector potential, which means physically that we neglect diamagnetic effects, the last two terms in the integral in (10.252) go and we are left with

$$W[J_\mu(\mathbf{r})] = T[J_\mu(\mathbf{r})] + G[J_\mu(\mathbf{r})] + \int (n(\mathbf{r}) v(\mathbf{r}) - \mathbf{m}(\mathbf{r}) \cdot \mathbf{B}(\mathbf{r})) \, d\mathbf{r} \tag{10.253}$$

Diamagnetic effects in metals are usually very small and so this approximation is often a good one. Equation (10.253) is formally identical to equation (10.165), the starting point for non-relativistic 'spin-only' density functional theory. The electron density, current density and spin density are independent variables (except to the extent that equation (10.227) has to be obeyed), and so we can follow through the Kohn–Sham argument again. The resulting single-particle-like equations are

$$(c\tilde{\alpha} \cdot \hat{\mathbf{p}} + \tilde{\beta} mc^2 + v^{\mathrm{eff}}(\mathbf{r}) - \mathbf{m}(\mathbf{r}) \cdot \mathbf{B}^{\mathrm{eff}}(\mathbf{r})) \psi_i(\mathbf{r}) = w_i \psi_i \tag{10.254}$$

with

$$n(\mathbf{r}) = \sum_{i=1}^{N} \psi_i^\dagger(\mathbf{r}) \psi_i(\mathbf{r} \tag{10.255}$$

$$\mathbf{m}(\mathbf{r}) = \mu_B \sum_{i=1}^{N} \psi_i^\dagger(\mathbf{r}) \tilde{\beta} \tilde{\sigma} \psi_i(\mathbf{r}) \tag{10.256}$$

and

$$v^{\text{eff}}(\mathbf{r}) = v(\mathbf{r}) + \frac{e^2}{4\pi\epsilon_0} \int \frac{n(\mathbf{r}')}{|\mathbf{r} - \mathbf{r}'|} d\mathbf{r}' + \frac{\delta E_{\text{xc}}[n(\mathbf{r}), \mathbf{m}(\mathbf{r})]}{\delta n(\mathbf{r})} \qquad (10.257)$$

$$\mathbf{B}^{\text{eff}}(\mathbf{r}) = \mathbf{B}^{\text{ext}}(\mathbf{r}) + \frac{\delta E_{\text{xc}}[n(\mathbf{r}), \mathbf{m}(\mathbf{r})]}{\delta \mathbf{m}(\mathbf{r})} \qquad (10.258)$$

This theory provides a basis upon which to calculate the magnetic properties of condensed matter. It is more approximate than the full current theory, but is easier to implement and has the advantage that it is written in terms of quantities familiar from the theory of magnetism. There is indeed a striking similarity between the theory of this section and the non-relativistic version in section 10.12. It should be remembered that we have coupled the field to the spin of the electrons, but not to their orbital motion; only the dipole moment part of the magnetic coupling is retained. Under these circumstances it is conventional to preface any observables calculated within this theory with the adjective 'spin-only'. Eschrig *et al.* (1985) have pointed out that this is not a valid approximation for rare earth and actinide materials and for materials in strong magnetic fields.

10.16 Further Development of RDFT

We have now taken the formal development of the relativistic density functional theory as far as we are going to. However, there have been extensions of relativistic density functional theory in several directions and in this short section we will make a brief mention of them. For a fuller discussion the reader is referred to the original literature.

All the density functional theory discussed in this book is to describe atoms, molecules and solids at zero temperature, but a finite temperature version of the theory does exist and is discussed by Kohn and Vashishta (1982) for the non-relativistic case and by Rajagopal (1980) for the relativistic regime. This theory is based on the work of Mermin (1965). It is very similar in its formal derivation to the zero-temperature version. Instead of the total energy, a grand potential is defined. At a temperature T this grand potential is completely determined by a density distribution which in turn is defined by $v^{\text{ext}}(\mathbf{r}) - \mu$ where $v^{\text{ext}}(\mathbf{r})$ is the external potential and μ is the chemical potential. One then assumes that the grand potential is a minimum when $n(\mathbf{r})$ takes on the correct value. A good summary of the non-relativistic theory is provided by Weinberger (1990).

There also exists a time-dependent version of density functional theory which is rather more complicated and is discussed in detail by Rajagopal (1994) and references therein.

Another version of density functional theory was developed by Oliveira, *et al.* (1988) to describe superconducting materials. This theory is not developed entirely from first principles because a form for the electron pairing has to be assumed, and so it can say nothing about the electron pairing mechanism. Nevertheless, when a particular form for pairing has been assumed, the theory can be implemented to calculate total energies and other ground state properties.

The strong magnetic field theory of Vignale, Rasolt and others discussed earlier has proven rather profitable in explaining several phenomena in high magnetic fields. For example, the properties of electron–hole droplets (Vignale *et al.* 1992), the Wigner crystal (Vignale 1993) and the quantum hall effect (Geller and Vignale 1994) have been studied with some success. Scaling and virial theorems have been derived by Erhard and Gross (1996) within this theory.

Jansen (1988) has discussed magneto-crystalline anisotropy in terms of relativistic density functional theory. It is well known that magneto-crystalline anisotropy arises because the spin–orbit interaction couples the electron spin direction to the crystal lattice. However, Jansen has shown that the Breit interaction also introduces some spin–orbit-like terms which will contribute to the anisotropy. There exists a Hartree approximation to the Breit interaction which can be calculated within relativistic density functional theory (see Györffy *et al.* (1991) for example). It is

$$E_B^H = \frac{e^2}{8\pi\epsilon_0} \int \int d\mathbf{r} d\mathbf{r}' \left(\frac{\mathbf{m}(\mathbf{r}) \cdot \mathbf{m}(\mathbf{r}')}{|\mathbf{r} - \mathbf{r}'|^3} - 3\frac{((\mathbf{r} - \mathbf{r}' \cdot \mathbf{m}(\mathbf{r}))((\mathbf{r} - \mathbf{r}' \cdot \mathbf{m}(\mathbf{r}'))}{|\mathbf{r} - \mathbf{r}'|^5} \right)$$
$$- \frac{4\pi}{3} \frac{e^2}{4\pi\epsilon_0} \int d\mathbf{r} \mathbf{m}(\mathbf{r}) \cdot \mathbf{m}(\mathbf{r}) \qquad (10.259)$$

Standard relativistic density functional theory calculations lump the Breit interaction into the exchange–correlation term, which may be a too gross approximation for calculating very small quantities such as magneto-crystalline anisotropy energies. The effect of the Breit interaction is also discussed in detail by Ramana *et al.* (1982).

We now go on to a brief discussion of the exchange–correlation potential in relativistic density functional theory.

10.17 Relativistic Exchange–Correlation Functionals

All the relativistic density functional theory we have done up to now will not be much use if we don't have a suitable expression for the exchange–correlation energy. As a first approximation one can always use the non-relativistic approximations detailed in section 10.13. However, this is difficult to justify rigorously. Relativistic quantum Monte Carlo calculations have been performed (Wilson and Györffy 1995, Kenny *et al.* 1996) to make an estimate of the exchange–correlation energy of a gas of constant density $\epsilon_{xc}[J_\mu(\mathbf{r})]$. We can then attempt to make a local approximation. For the case of a system of relativistic electrons in a scalar potential Macdonald *et al.* (1982) define the exchange–correlation potential as

$$\mu_{xc}[n(\mathbf{r})] = -\frac{2}{\pi \alpha r_s} \left(\zeta^{RX}(r_s) + 0.0545 r_s \ln(1 + 11.4/r_s) \right) \tag{10.260a}$$

where the potential is in Rydbergs. The parameter $\zeta^{RX}(r_s)$ is 1 in the non-relativistic case which will be indicated by the symbol NX henceforth. In the relativistic theory (RX) it is

$$\zeta^{RX}(r_s) = -\frac{1}{2} + \frac{3 \ln(\beta + (1 + \beta^2)^{1/2})}{2\beta(1 + \beta^2)^{1/2}} \tag{10.260b}$$

where r_s and β are given by equations (10.122) and (10.123) respectively, and α is the fine structure constant. Macdonald *et al.* also performed calculations for which the correlation part of equation (10.260a) was set equal to zero (XO) and we will present some of their results in section 10.18.

For the case of an external magnetic field, an expression given by Macdonald (1983) and Xing Xu *et al.* (1984) (see Cortona 1985) is as follows. We let

$$\xi = \frac{|m(\mathbf{r})|}{n(\mathbf{r})} \tag{10.261a}$$

and then we define a parameter z with

$$x = (1 + z)^{1/3}, \qquad y = (1 - z)^{1/3} \tag{10.261b}$$

$$\xi = \frac{1}{2\beta^3} \left(\frac{1}{3}\beta^3 x^3 + \beta x(1 + \beta^2 x^2)^{1/2} - \sinh^{-1}(\beta x) \right)$$
$$- \frac{1}{2\beta^3} \left(\frac{1}{3}\beta^3 y^3 + \beta y(1 + \beta^2 y^2)^{1/2} - \sinh^{-1}(\beta y) \right) \tag{10.261c}$$

The total exchange energy is written as the sum of a Coulombic and a transverse part

$$\epsilon_x[n(\mathbf{r})] = \epsilon_x^C[n(\mathbf{r})] + \epsilon_x^{tr}[n(\mathbf{r})] \tag{10.261d}$$

and

$$\epsilon_x^C[n(\mathbf{r})] = -\frac{3}{4\pi}n(\mathbf{r})E_F \frac{\alpha}{\beta} \left(x^4 + y^4 - \frac{1}{9}\beta^2(x^6 + y^6) \right) \tag{10.261e}$$

$$\epsilon_x^{tr}[n(\mathbf{r})] = -\frac{3}{4\pi}n(\mathbf{r})E_F \frac{\alpha}{\beta}\frac{10}{9}\beta^2 \left(1 + \frac{7}{60}(x^3 - y^3)^2 \right) \tag{10.261f}$$

where E_F is the Fermi energy. This approximation is also discussed by Weinberger *et al.* (1990). Expressions for the correlation energy in a

relativistic formalism are very complex. This is because no local approximation for it exists. Hence, implementation of it is much more difficult
than in the non-relativistic regime. We will not reproduce the complicated
equations here, but instead refer the reader to the discussion by Ramana
and Rajagopal (1981, 1983).

10.18 Implementation of RDFT

Although density functional theory is a very powerful technique, it is not
without its problems and limitations. Englisch and Englisch (1983) and
Levy (1982) have pointed out that it has been tacitly assumed that a density that is interacting v-representable (i.e. it is the density of the ground
state of an interacting system of electrons in the external potential v) is also
non-interacting v-representable (i.e. it is the density of the ground state of
a non-interacting system of electrons in some local external potential). If
the functions chosen to represent the density do not possess these properties there is no guarantee that the density functional method will work.
Although this is a formal limitation in the derivation of the equations of
density functional theory, in practice it seems to present no difficulty.

   For a system containing N electrons, the highest ϵ_i is very close to
the first ionization energy of the system. The reason for this is that the
Coulomb repulsion between all the electrons contains $\frac{1}{2}N(N-1)$ terms.
By removing one electron we only change $(N-1)$ of those terms, so the
error incurred by treating the highest ϵ_i as the ionization energy goes to
zero proportionally to $(1/N)$ as $N \to \infty$.

   The multipliers ϵ_i rigorously have no meaning. However, one can con-
strue them as one-electron-like energies in the interpretation of spectroscopies and the de Haas–van Alphen effect, and get excellent agreement
between experiment and theory. It is very easy not to realize how miraculous this is!

   Density functional theory with the local density approximation gives
good agreement with experiment for a vast range of material properties
such as crystal structure, zero-temperature magnetic properties, specific
heats, and all sorts of defect and surface energies. But the theory cannot
do everything. Perhaps the most famous limitation is that it cannot get
semiconductor band gaps right. Naively one would expect this to be given
by the difference in energy between the highest occupied state and the
lowest unoccupied state. It is easy to see that this is not the case by considering the Euler equation (10.237). That equation is valid for the ground
state energy. If we promote the highest electron up to the next value of w
the right hand side of that equation must change by a constant. Therefore
the left hand side must also change by a constant. As the external potential is not affected by the electronic configuration, it is the functional

derivatives of T and G that must change. By identifying the band gap as the difference between the highest occupied and lowest unoccupied bands we ignore this change which is an important contribution to the band gap.

Another problem which density functional theory in the above form cannot treat properly is that of superconductivity. Although a density functional theory for superconductors has been derived (Oliveira *et al.* 1988), the electron pairing mechanism is not calculated within the theory. In BCS theory the pairing interaction is not electronic in origin, so a theory of the electron gas is not the right place to start describing it. However, several electronic mechanisms of pairing have been suggested. They all depend directly on a very special interaction between two electrons. It is highly unlikely that such interactions could be described properly by some average density, even if the pairing mechanism was electronic in origin.

There are other physical situations where the local density approximation (LDA) may fall down, or at least lead to problems. Van der Waals forces have some non-local character, and will not be fully describable within the LDA. Also, just above a metal surface is a region of very low charge density. This can causes problems for the LDA because there is just not enough density for the depletion necessary to form the exchange–correlation hole.

Relativistic density functional theory has not been explored very extensively, mainly because there is no simple local approximation for the correlation energy. However, we can show a few examples of its implementation. In table 10.9 we display the binding energies of atomic mercury. The first column gives the density functional eigenvalues. The second column gives the binding energies for the electrons. In the third column are the results of Dirac–Hartree–Fock calculations for the energy eigenvalues and the final column gives the experimental ionization energies. This table illustrates several things. Firstly it shows that the density functional eigenvalues, although rigorously meaningless, are remarkably similar to the single-electron eigenvalues of Hartree–Fock theory and to the ionization energies. This is a fairly general feature of the DFT eigenvalues. The electron binding energies are calculated by finding the total energy of the neutral atom in one calculation, and the total energy of the ion with the pertinent electron missing, and then taking the difference of the two. These were calculated using the Hedin–Lundqvist approximation for the exchange–correlation energy. If we compare this column with those obtained by Hartree–Fock theory and by experiment, we see that density functional theory uniformly does better than Hartree–Fock theory in reproducing ionization energies. It is easy to understand why this is so. The Hartree–Fock theory treats exchange exactly and ignores correlation. Density functional theory treats both exchange and correlation

Table 10.9. Magnitude of the binding energies for the electrons in the mercury atom. The first column (DFTE) is the density functional Lagrange multipliers, the second (BEVH) is the binding energies calculated using density functional theory and the Hedin–Lundqvist approximation for exchange–correlation (see text). The third column (DHF) (Desclaux 1973, quoted by Desclaux and Kim 1975) is the corresponding eigenvalues calculated using Dirac–Hartree–Fock theory, and the final column is the experimental ionization energies (Bearden and Burr 1967). All energies are in Rydbergs.

n	l	j	DFTE	BEVH	DHF	Exp
1	0	1/2	6099.1	6144.5	6148.5	6107.9
2	0	1/2	1080.2	1090.3	1100.5	1090.6
2	1	1/2	1036.5	1047.4	1053.7	1044.8
2	1	3/2	893.5	902.5	910.3	903.0
3	0	1/2	256.3	258.7	266.2	261.7
3	1	1/2	236.6	239.2	245.3	241.0
3	1	3/2	205.0	207.2	213.1	209.3
3	2	3/2	172.3	175.1	178.9	175.3
3	2	5/2	165.5	168.2	172.0	168.7
4	0	1/2	56.3	57.4	61.3	58.8
4	1	1/2	47.9	48.9	52.3	49.8
4	1	3/2	40.2	41.2	44.4	42.0
4	2	3/2	26.4	27.4	29.6	27.8
4	2	5/2	25.0	26.0	28.1	26.4
4	3	5/2	7.26	8.17	8.9	7.51
4	3	7/2	6.95	7.85	8.6	7.23
5	0	1/2	8.73	9.27	10.21	8.84
5	1	1/2	5.94	6.46	7.08	5.91
5	1	3/2	4.57	5.07	5.68	4.23
5	2	3/2	0.86	1.28	1.30	1.23
5	2	5/2	0.72	1.13	1.15	1.09
6	0	1/2	0.56	0.84	0.66	0.77

approximately, and it is quite easy for the uncertainty introduced by the approximation for exchange to be outweighed by an approximate value of the correlation.

In table 10.10 we show the various contributions to the total energy of ^{92}U and several extreme lithium-like ions, as calculated by Das *et al.* (1980). In particular we see that the transverse and coulombic contri-

Table 10.10. Various contributions to the total energy of ^{92}U and several Li-like atoms. E_x^{tr} and E_x^C are the transverse and Coulomb part of the total exchange energy E_x. E_H is the Hartree energy, E_V is the potential energy due to the electron-nucleus interaction. E_K is the kinetic energy and E_T is the total energy. (Source: Das *et al.* (1980)). All energies are in Rydbergs.

	^{92}U	W^{71+}	Mo^{39+}	Fe^{23+}	C^{3+}
E_x^{tr}	113.19	29.580	5.302	1.248	0.012
E_x^C	−884.892	−114.484	−56.708	−34.146	−7.144
E_x	−771.704	−76.904	−51.406	−32.898	−7.132
E_H	20903.256	278.704	143.288	85.842	17.884
E_V	−144509.176	−29067.354	−8228.246	−3035.390	−148.758
E_K	68395.354	23487.333	4159.706	1500.958	68.538
E_T	−55982.27	−13207.332	−3976.658	−1481.488	−69.468

butions to the exchange–correlation energy differ in sign. However, the effects of relativity are also clear in the lithium sequence of ions. We see this in the fact that the transverse part of the exchange–correlation energy becomes an increasingly significant fraction of the coulombic part as the mass increases. The trends in the other contributions to the total energy are easy to understand. The nuclear potential energy becomes more negative as the ionic charge becomes greater because the electrostatic attraction between the nucleus and the electrons is greater and $|\mathbf{r} - \mathbf{r}'|$ in the denominator of the Coulomb potential becomes smaller. The Hartree energy becomes bigger as a consequence, because if the electrons are pulled closer to the nucleus, they are also pulled closer to each other. Finally, the electron kinetic energy must increase, because their velocity must increase to keep them from falling into the nucleus. In this table it is also evident that the numerical value of the total energy depends upon several very large negative and positive contributions which partially cancel each other. Obviously for such a calculation to be reliable very robust and accurate numerical and computational methods are required.

MacDonald *et al.* have performed an exhaustive investigation into the electronic properties of palladium and platinum (1981) and the noble metals (1982) using relativistic density functional theory. The Fermi energy is lowered by relativity in these elements, and this has knock-on effects on the density of states at the Fermi level and on the Fermi surface. We illustrate this by reproducing their results for the Fermi surface areas of platinum in table 10.11. This is a more interesting case than palladium because the effect of relativity is significant, and for some cross sections of

Table 10.11. Areas of cross sections of the Fermi surface of Pt calculated using relativistic density functional theory. The first column just labels the section of Fermi surface. In the following columns are the areas calculated using a non-relativistic exchange approximation (XO), including exchange and correlation (NX), and including a relativistic correction to the exchange approximation (RX) as well (see equation (10.260); Macdonald *et al.* (1981)). These can be compared with the areas measured experimentally which are given in the final column. All areas are quoted in atomic units.

Section	$A(\text{XO})$	$A(\text{NX})$	$A(\text{RX})$	$A(\text{Exp})$
$< 100 >$	0.765	0.743	0.761	0.770
$< 100 >_\epsilon$	1.931	1.883	1.922	1.890
$< 100 >_\alpha$	0.069	0.078	0.069	0.074
$< 110 >$	0.851	0.824	0.846	0.857
$< 111 >$	0.684	0.665	0.678	0.678

Table 10.12. Relativistic effects on the Fermi energy and magnetic moments of iron, cobalt and nickel calculated using the potentials of Moruzzi *et al.* (1978) and calculated by Ebert *et al.* (1988a). Energies are in Rydbergs and magnetic moments are in units of the Bohr magneton.

	Fe			Co			Ni		
	R	NR	Exp	R	NR	Exp	R	NR	Exp
E_F	0.743	0.756		0.708	0.727		0.654	0.683	
μ_s	2.08	2.15	2.13	1.51	1.56	1.52	0.60	0.59	0.57
μ_l	0.056		0.080	0.069		0.142	0.046		0.050
μ_T	2.14	2.15	2.21	1.58	1.56	1.66	0.65	0.59	0.62

the Fermi surface the relativistic correction to the exchange energy has an effect of roughly the same size as the correlation correction but of opposite sign. Therefore they largely cancel each other and give results very close to the exchange-only potential. The effect of using a relativistic exchange approximation can be seen by comparing the columns A(NX) and A(RX), and we see that the difference can be as great as 10%. Furthermore, in all cases the Fermi surface areas given by the calculation including correlation and the relativistic exchange energy agree with experiment as well as, or better than, when correlation and the non-relativistic exchange approximation are used.

In table 10.12 we show some results for the magnetic transition metals iron, cobalt, and nickel. Column R shows the relativistic calculation, NR is the non-relativistic calculation and Exp are the experimental results, where appropriate. The non-relativistic calculation is fully self-consistent within non-relativistic density functional theory. The relativistic calculation uses the same non-relativistic potential function in a single iteration of relativistic density functional theory. It is not fully self-consistent, but for these light elements the uncertainty introduced by this is very small. The first row shows the lowering of the Fermi energy caused by relativity. The next rows show the spin μ_s, orbital μ_l and total μ_T magnetic moments of the element. Obviously the orbital contribution in the non-relativistic theory is identically equal to zero. The separation of the experimental orbital and spin magnetic moments is subject to some uncertainty. Although the total moment calculated relativistically is marginally better than that calculated non-relativistically, there is still some difference between theory and experiment. The reasons for this could be the exchange–correlation approximation, or the shape approximations made for the potential, or possibly the lack of self-consistency within relativistic density functional theory.

Cortona *et al.* (1985) have calculated the properties of the trivalent rare earth ions using non-relativistic density functional theory, and both non-spin-polarized and spin-polarized relativistic density functional theory. The results are shown in table 10.13. All the f-level eigenvalues are shown although many of them are not occupied, of course.

Several things can be observed from this table. The first line is a fully non-relativistic calculation of the energy eigenvalue. These are really only reproduced for later comparison. The second two lines show the relativistic non-spin-polarized results. There are two sets of energy levels here for the sixfold degenerate $j = 5/2$ and the eightfold degenerate $j = 7/2$ f-levels. The spin–orbit separation of these levels increases monotonically as we proceed along the periodic table from 0.0248 Rydbergs to 0.0560 Rydbergs. This is what one might naively expect from equation (2.64) for the one-electron atom. In that case the coefficient of $\hat{\mathbf{S}} \cdot \hat{\mathbf{L}}$ depends upon the spatial derivative of $V(\mathbf{r})$ which contains the atomic number Z as a constant. This example shows that qualitatively, at least, such behaviour continues through to the many-electron atom. It is, of course, also well established experimentally using spectroscopic methods.

In part (*c*) of table 10.13 there is magnetic Zeeman, as well as spin–orbit, splitting of the energy levels and so all degeneracies are removed. Note that here the energy levels have been split into two sets of seven levels rather than the spin–orbit splitting into sets of six and eight levels. This is because the magnetism dominates the spin–orbit interaction and splits them into seven spin-up and seven spin-down levels. Cortona *et al.* also analysed the

Table 10.13. (*a*) 4*f* eigenvalues from non-spin-polarized non-relativistic calculations. (*b*) 4*f* eigenvalues from non-spin-polarized relativistic calculations. (*c*) 4*f* eigenvalues from spin-polarized relativistic calculations. All energies are in Rydbergs.

	$_{58}\text{Ce}^{3+}$	$_{59}\text{Pr}^{3+}$	$_{60}\text{Nd}^{3+}$	$_{61}\text{Pm}^{3+}$	$_{62}\text{Sm}^{3+}$	$_{63}\text{Eu}^{3+}$	$_{64}\text{Gd}^{3+}$
				(*a*)			
$l = 3$	2.2514	2.3412	2.4212	2.4932	2.5584	2.6176	2.6714
				(*b*)			
$J = 5/2$	2.0248	2.1086	2.1812	2.2450	2.3012	2.3508	2.3992
$J = 7/2$	2.0000	2.0790	2.1472	2.2062	2.2570	2.3010	2.3432
				(*c*)			
$J_z = 5/2$	2.0478	2.1554	2.2514	2.3378	2.4162	2.4882	2.5544
$J_z = 3/2$	2.0450	2.1518	2.2470	2.3328	2.4104	2.4816	2.5470
$J_z = 1/2$	2.0420	2.1480	2.2424	2.3274	2.4044	2.4746	2.5392
$J_z = -1/2$	2.0388	2.1440	2.2376	2.3218	2.3980	2.4676	2.5312
$J_z = -3/2$	2.0350	2.1396	2.2326	2.3160	2.3914	2.4600	2.5228
$J_z = -5/2$	2.0306	2.1346	2.2272	2.3100	2.3846	2.4522	2.5140
$J_z = -7/2$	2.0250	2.1292	2.2212	2.3034	2.3772	2.4440	2.5070
$J_z = -5/2$	1.9908	2.0426	2.0798	2.1058	2.1226	2.1316	2.1336
$J_z = -3/2$	1.9864	2.0378	2.0744	2.0998	2.1160	2.1240	2.1252
$J_z = -1/2$	1.9828	2.0334	2.0694	2.0942	2.1096	2.1168	2.1172
$J_z = 1/2$	1.9796	2.0294	2.0648	2.0888	2.1034	2.1100	2.1096
$J_z = 3/2$	1.9766	2.0256	2.0604	2.0838	2.0976	2.1034	2.1022
$J_z = 5/2$	1.9738	2.0220	2.0560	2.0788	2.0920	2.0970	2.0952
$J_z = 7/2$	1.9714	2.0186	2.0520	2.0740	2.0866	2.0910	2.0882

value of the spin contribution to the magnetic moment of the ions. This is defined as the sum of the mean values of the Bohr magneton times $\tilde{\sigma}_z$ for the occupied states. They obtained values of $0.97\mu_B$ for $_{58}\text{Ce}^{3+}$ and $6.96\mu_B$ for $_{64}\text{Gd}^{3+}$. These results are in excellent agreement with Hund's first rule which tells us to fill up orbitals with aligned spins first. So the magnetic moment per ion is essentially determined by the number of 4*f* electrons in this theory, as we know it should be from experiment.

We will cease our discussion of the results of relativistic density functional calculations with these few examples. In the following chapter we describe scattering theory and show how to use it to implement density functional theory in solids, so further results of relativistic density functional theory will be shown in that context.

11

Scattering Theory

The scattering of fast particles is an important tool in many fields of physics. In particular, virtually all that is known about elementary particles is a result of the interpretation of scattering experiments. In condensed matter physics as well, the bulk of our understanding of materials on a microscopic level comes from the scattering of neutrons, photons and electrons. Neutrons are used to determine crystal structures and to probe the dynamical properties of solids; photons are used in a plethora of spectroscopies to elucidate the details of the electronic and magnetic structure. Some of these will be discussed in chapter 12. Electrons can be used to determine the behaviour of surface plasmons and also to look at electronic transitions.

For these and many other reasons, an understanding of the quantum theory of scattering is of key importance for a theoretical physicist. Therefore in this chapter we develop relativistic scattering theory from scratch. This chapter doesn't assume any knowledge of non-relativistic scattering theory although such knowledge will aid your understanding.

In this chapter it has been necessary to include some mathematical preliminaries. Therefore the first three sections are an introduction to Green's functions and their uses. Later in the chapter we have some follow-up sections on free-particle and scattering Green's functions. Following these we derive expressions for the scattering particle wavefunction. As anyone who knows will tell you, there are only four interesting numbers in scattering theory, these are 0, 1, 2 and ∞. Consequently we discuss zero-site scattering (a free particle), single-site scattering and infinite-site scattering. From the latter a method of solving the density functional equations on a lattice can be derived, and this is described in detail. It is not much use compulsively calculating Green's functions and not applying them to something. So after the scattering theory we show how to use the Green's function to calculate quantities that are important in condensed matter physics such as densities of states and magnetic moments. At the end of the chapter we cover a relativistic indirect exchange interaction which is a source of magnetic anisotropy (and which is the two-site scattering).

11.1 Green's Functions

Much of scattering theory is written in terms of Green's functions, a mathematical device which seems to have been designed to terrorize generations of physics research students. In this section we introduce Green's functions in as simple a way as possible, doing the minimum mathematics necessary to enable the reader to understand the theory that follows. For a full discussion of Green's functions in condensed matter physics see Rickayzen (1980) and Inkson (1983). Their use in scattering theory is discussed by Weinberger (1990). Consider a Hamiltonian equation

$$\hat{H}_0(\mathbf{r})\phi(\mathbf{r}) = E\phi(\mathbf{r}) \tag{11.1}$$

Here $\hat{H}_0(\mathbf{r})$ is a general hermitian operator. The Green's function for this equation is defined by

$$(E - \hat{H}_0(\mathbf{r}))G_0(\mathbf{r}, \mathbf{r}', E) = \delta(\mathbf{r} - \mathbf{r}') \tag{11.2}$$

Equation (11.2) does not uniquely define a Green's function. For any particular application we have to insist that it be subject to the same boundary conditions as the solution $\phi(\mathbf{r})$. As an example consider the case when $\hat{H}_0(\mathbf{r}) = -\nabla^2$ and $E = k^2$, then (11.1) becomes

$$(\nabla^2 + k^2)\phi(\mathbf{r}) = 0 \tag{11.3}$$

The reason for using this example is that it will lead to a result that we will use in our derivation of the relativistic free-particle Green's function later. The Green's function for equation (11.3) is defined by

$$(k^2 + \nabla^2)G_0(\mathbf{r}, \mathbf{r}', \mathbf{k}) = \delta(\mathbf{r} - \mathbf{r}') \tag{11.4}$$

If we consider a spherically symmetric solution the angular parts of the ∇ operator give zero and the solution to (11.4) is

$$G_0(\mathbf{r}, \mathbf{r}', \mathbf{k}) = \frac{e^{ikR}}{4\pi R} \qquad R = |\mathbf{r} - \mathbf{r}'| \tag{11.5}$$

Proving this involves a contour integral, which is performed in detail by Merzbacher (1970). Equation (11.2) defines the Green's function in terms of the Hamiltonian and eigenvalue for the particular problem described by equation (11.1). You might expect (and you would be right) that the Green's function can be written in terms of the eigenvalues and eigenvectors of the Hamiltonian. Labelling the individual eigenvectors by $\phi_n(\mathbf{r})$ and the corresponding eigenvalues by E_n we have

$$(\hat{H}_0(\mathbf{r}) - E_n)\phi_n(\mathbf{r}) = 0 \qquad (11.6)$$

Assuming we normalize the wavefunctions in the usual way, the $\phi_n(\mathbf{r})$ (n indicates positive and negative energies) form a complete set of orthonormal functions, i.e.

$$\sum_n \phi_n^\dagger(\mathbf{r})\phi_n(\mathbf{r}') = \delta(\mathbf{r} - \mathbf{r}') \qquad (11.7a)$$

Equation (11.7) implies that we can write any function as

$$f(\mathbf{r}) = \sum_n a_n \phi_n(\mathbf{r}) \qquad (11.7b)$$

The Green's function $G_0(\mathbf{r}, \mathbf{r}', E)$ has two position variables, so quite generally, we may write

$$G_0(\mathbf{r}, \mathbf{r}', E) = \sum_{n,n'} a_{nn'} \phi_n(\mathbf{r})\phi_{n'}^\dagger(\mathbf{r}') \qquad (11.8)$$

Substituting this into equation (11.2) gives

$$(E - \hat{H}_0(\mathbf{r}))G_0(\mathbf{r}, \mathbf{r}', E) = \sum_{n,n'} a_{nn'}(E - \hat{H}_0(\mathbf{r}))\phi_n(\mathbf{r})\phi_{n'}^\dagger(\mathbf{r}')$$

$$= \sum_{n,n'} a_{nn'}(E - E_n)\phi_n(\mathbf{r})\phi_{n'}^\dagger(\mathbf{r}') = \delta(\mathbf{r} - \mathbf{r}') \qquad (11.9)$$

Comparing this with equation (11.7) we can say that

$$a_{nn'} = \frac{1}{E - E_n}\delta_{nn'} \qquad (11.10)$$

is a possible solution. Substituting this back into (11.8) gives

$$G_0(\mathbf{r}, \mathbf{r}', E) = \sum_n \frac{\phi_n(\mathbf{r})\phi_n^\dagger(\mathbf{r}')}{E - E_n} \qquad (11.11)$$

The derivation of this expression has not depended upon the explicit form of the eigenfunctions. Therefore the Green's function for any Hamiltonian may be written in similar form to (11.11).

So far all we have done is define a mathematical entity which we have called the Green's function. Now let us consider its applications. Its principal use in scattering theory is as a method of solving inhomogeneous differential equations. It is easy to see how this works. Suppose we have a Hamiltonian

$$\hat{H}(\mathbf{r}) = \hat{H}_0(\mathbf{r}) + V(\mathbf{r}) \qquad (11.12)$$

and we want to solve the equation

$$(E - \hat{H}_0(\mathbf{r}))\psi(\mathbf{r}) = V(\mathbf{r})\psi(\mathbf{r}) \tag{11.13}$$

where E is an eigenvalue of the Hamiltonian $\hat{H}_0$. The solution to equation (11.13) can be written in the following form

$$\psi(\mathbf{r}) = \phi(\mathbf{r}) + \int G_0(\mathbf{r}, \mathbf{r}', E)V(\mathbf{r}')\psi(\mathbf{r}')d\mathbf{r}' \tag{11.14}$$

The easiest way to see this is to operate on both sides of (11.14) with $E - \hat{H}_0(\mathbf{r})$. Doing this

$$(E - \hat{H}_0(\mathbf{r}))\psi(\mathbf{r}) = (E - \hat{H}_0(\mathbf{r}))\phi(\mathbf{r})$$
$$+ \int (E - \hat{H}_0(\mathbf{r}))G_0(\mathbf{r}, \mathbf{r}', E)V(\mathbf{r}')\psi(\mathbf{r}')d\mathbf{r}'$$
$$= 0 + \int \delta(\mathbf{r} - \mathbf{r}')V(\mathbf{r}')\psi(\mathbf{r}')d\mathbf{r}' = V(\mathbf{r})\psi(\mathbf{r}) \tag{11.15}$$

where we have used (11.1). This verifies equation (11.14). Equation (11.14) is known as the Dyson equation for the eigenfunction. Essentially, it is the Dirac (or Schrödinger) equation written in integral form. It is possible to derive a Dyson equation for the Green's function as well. The derivation is fairly similar to the derivation of (11.14), but is slightly more complicated. Let's do it anyway. Suppose we have two equations for the Green's functions of $\hat{H}_0$ and $\hat{H}$. Then clearly

$$(E - \hat{H}_0(\mathbf{r}))G_0(\mathbf{r}, \mathbf{r}', E) = \delta(\mathbf{r} - \mathbf{r}')$$
$$(E - \hat{H}_0(\mathbf{r}) - V(\mathbf{r}))G(\mathbf{r}, \mathbf{r}', E) = \delta(\mathbf{r} - \mathbf{r}') \tag{11.16}$$

where the energy parameter E of the first equation is chosen to correspond with that of the second equation. The solution to the second of these equations is given in terms of the solution to the first by

$$G(\mathbf{r}, \mathbf{r}', E) = G_0(\mathbf{r}, \mathbf{r}', E) + \int G_0(\mathbf{r}, \mathbf{r}'', E)V(\mathbf{r}'')G(\mathbf{r}'', \mathbf{r}', E)d\mathbf{r}'' \tag{11.17}$$

To prove this we operate on both sides of (11.17) with $(E - \hat{H}_0(\mathbf{r}))$. The left hand side becomes

$$(E - \hat{H}_0(\mathbf{r}))G(\mathbf{r}, \mathbf{r}', E) = (E - \hat{H}_0(\mathbf{r}) - V(\mathbf{r}) + V(\mathbf{r}))G(\mathbf{r}, \mathbf{r}', E)$$
$$= \delta(\mathbf{r} - \mathbf{r}') + V(\mathbf{r})G(\mathbf{r}, \mathbf{r}', E) \tag{11.18a}$$

and the right hand side is

$$(E - \hat{H}_0(\mathbf{r}))\left(G_0(\mathbf{r}, \mathbf{r}', E) + \int G_0(\mathbf{r}, \mathbf{r}'', E)V(\mathbf{r}'')G(\mathbf{r}'', \mathbf{r}', E)d\mathbf{r}'' \right)$$
$$= \delta(\mathbf{r} - \mathbf{r}') + \int \delta(\mathbf{r} - \mathbf{r}'')V(\mathbf{r}'')G(\mathbf{r}'', \mathbf{r}', E)d\mathbf{r}''$$
$$= \delta(\mathbf{r} - \mathbf{r}') + V(\mathbf{r})G(\mathbf{r}, \mathbf{r}', E) \tag{11.18b}$$

Clearly (11.18*a*) and (11.18*b*) are equal. As E can take on any value this proves that (11.17) is indeed the solution of the second of equations (11.16). As you can see (11.17) looks very similar to (11.14), and indeed it is. This is the Dyson equation for the Green's function.

The Dyson equation (11.17) defines the Green's function in terms of itself. So we can substitute for $G(\mathbf{r}, \mathbf{r}'', E)$ on the right hand side of (11.17) from the left hand side of (11.17). This can be done *ad nauseam*.

$$G(\mathbf{r}, \mathbf{r}', E) = G_0(\mathbf{r}, \mathbf{r}', E) + \int G_0(\mathbf{r}, \mathbf{r}'', E) V(\mathbf{r}'') G(\mathbf{r}'', \mathbf{r}', E) d\mathbf{r}''$$

$$= G_0(\mathbf{r}, \mathbf{r}', E) + \int G_0(\mathbf{r}, \mathbf{r}'', E) V(\mathbf{r}'') G_0(\mathbf{r}'', \mathbf{r}', E) d\mathbf{r}'' + \tag{11.19}$$

$$\int \int G_0(\mathbf{r}, \mathbf{r}''', E) V(\mathbf{r}''') G_0(\mathbf{r}''', \mathbf{r}'', E) V(\mathbf{r}'') G(\mathbf{r}'', \mathbf{r}', E) d\mathbf{r}''' d\mathbf{r}''$$

and we can substitute for $G(\mathbf{r}'', \mathbf{r}', E)$ etc. as many times as we want. An analogous expression can be written down for the wavefunction from repeated use of (11.14). This series defines the Born approximation. If we stop the series at the first term we are making the 1st Born approximation, at the second term the 2nd Born approximation, etc. The expansion (11.19) is also known as a Brillouin–Wigner perturbation series.

11.2 Time-Dependent Green's Functions

In many cases in scattering theory we do not have the time independence which has been assumed in the theory up to now. Often we have to solve a time-dependent equation of the form

$$i\hbar \frac{\partial}{\partial t} \Psi(\mathbf{r}, t) = \hat{H}(\mathbf{r}) \Psi(\mathbf{r}, t) \tag{11.20a}$$

This is just equation (4.49). The solution to it is

$$\Psi(\mathbf{r}, t) = \psi(\mathbf{r}) e^{-iEt/\hbar} \tag{11.20b}$$

In analogy with the time-independent case we define the time-dependent Green's function using

$$\left(i\hbar \frac{\partial}{\partial t} - \hat{H}(\mathbf{r}) \right) G(\mathbf{r}, \mathbf{r}', t - t') = \hbar \delta(\mathbf{r} - \mathbf{r}') \delta(t - t') \tag{11.21}$$

Here $\hbar$ appears on the right side of the equation to get the dimensions right. The Green's function defined here is the Fourier transform of the energy-dependent Green's function

$$G(\mathbf{r}, \mathbf{r}', t - t') = \frac{1}{2\pi} \int G(\mathbf{r}, \mathbf{r}', E) e^{-iE(t-t')/\hbar} dE \tag{11.22}$$

It is easy to show by substitution that (11.22) is a solution to (11.21) provided $G(\mathbf{r}, \mathbf{r}', E)$ obeys an equation like (11.2). We can rewrite (11.22) using the eigenfunction expansion (11.11)

$$G(\mathbf{r}, \mathbf{r}', t - t') = \frac{1}{2\pi} \int \sum_n \frac{\psi_n(\mathbf{r})\psi_n^\dagger(\mathbf{r}')}{E - E_n} e^{-iE(t-t')/\hbar} dE \qquad (11.23)$$

The reason for writing this equation in this unwieldy form is that it highlights the fact that the integral is undefined at real energies because of the singularities at E_n and has to be evaluated as a contour integral. The fact that the Green's function has these singularities at the eigenvalues of the governing equation is what makes it so useful. We are free to choose our contour and this leads us to define two new Green's functions

$$G^R(\mathbf{r}, \mathbf{r}', E) = \lim_{\epsilon \to 0} \sum_n \frac{\psi_n(\mathbf{r})\psi_n^\dagger(\mathbf{r}')}{E - E_n + i\epsilon} \qquad (11.24)$$

which is known as the retarded Green's function and can be Fourier transformed to give

$$\begin{aligned} G^R(\mathbf{r}, \mathbf{r}', t - t') &= \sum_n \psi_n(\mathbf{r})\psi_n^\dagger(\mathbf{r}')e^{-iE_n(t-t')/\hbar} \qquad (t > t') \\ &= 0 \qquad\qquad\qquad\qquad\qquad\qquad (t < t') \end{aligned} \qquad (11.25)$$

and

$$G^A(\mathbf{r}, \mathbf{r}', E) = \lim_{\epsilon \to 0} \sum_n \frac{\psi_n(\mathbf{r})\psi_n^\dagger(\mathbf{r}')}{E - E_n - i\epsilon}, \qquad (11.26)$$

which is the advanced Green's function, with Fourier transform

$$\begin{aligned} G^A(\mathbf{r}, \mathbf{r}', t - t') &= i \sum_n \psi_n(\mathbf{r})\psi_n^\dagger(\mathbf{r}')e^{iE_n(t-t')/\hbar} \qquad (t < t') \\ &= 0 \qquad\qquad\qquad\qquad\qquad\qquad\quad (t > t') \end{aligned} \qquad (11.27)$$

These Green's functions are used to fit time-like boundary conditions to a problem. They will be of use when we discuss multiple scattering theory to determine the order in which scattering events occur.

11.3 The T-Operator

The T-operator is a useful quantity to define in scattering theory. In this short section we define it and show how it is related to what has gone before. The importance of the T-operator will become apparent later.

The T-matrix is defined in many ways. Two equivalent definitions are

$$V(\mathbf{r})G(\mathbf{r}, \mathbf{r}_1, E) = \int T(\mathbf{r}, \mathbf{r}_2, E)G_0(\mathbf{r}_2, \mathbf{r}_1, E)d\mathbf{r}_2 \qquad (11.28a)$$

$$G(\mathbf{r}, \mathbf{r}_1, E)V(\mathbf{r}_1) = \int G_0(\mathbf{r}, \mathbf{r}_2, E)T(\mathbf{r}_2, \mathbf{r}_1, E)d\mathbf{r}_2 \qquad (11.28b)$$

Substituting directly into (11.17) gives

$$G(\mathbf{r}, \mathbf{r}', E) = G_0(\mathbf{r}, \mathbf{r}', E) +$$
$$\int\int G_0(\mathbf{r}, \mathbf{r}'', E) T(\mathbf{r}'', \mathbf{r}_2, E) G_0(\mathbf{r}_2, \mathbf{r}', E) d\mathbf{r}'' d\mathbf{r}_2 \qquad (11.29)$$

This shows that the T-operator describes all possible scattering in the system as it relates the free-particle Green's function to the full scattering Green's function. Next we assert that the T-matrix is given by

$$T(\mathbf{r}, \mathbf{r}', E) = V(\mathbf{r})\delta(\mathbf{r} - \mathbf{r}') + \int V(\mathbf{r}) G_0(\mathbf{r}, \mathbf{r}_1, E) T(\mathbf{r}_1, \mathbf{r}', E) d\mathbf{r}_1 \qquad (11.30)$$

and it is left as an exercise for the reader to substitute (11.30) into (11.29) to obtain (11.19) and thus verify equation (11.30). Equation (11.30) is similar to (11.17) and is a Dyson equation for the T-operator.

11.4 The Relativistic Free-Particle Green's Function

The description of Green's functions above is all very well, but it is not much use unless we can calculate the Green's function for the initial unperturbed Hamiltonian $\hat{H}_0$. Therefore, to this end, we are going to calculate the Green's function for a relativistic free particle. Obviously, from what has gone before, this is a solution of

$$(c\tilde{\alpha} \cdot \hat{\mathbf{p}} + \tilde{\beta}mc^2 - W) G_0(\mathbf{r}, \mathbf{r}') = -\delta(\mathbf{r} - \mathbf{r}')\tilde{I} \qquad (11.31)$$

and, defining $R = |\mathbf{R}|$ and $\mathbf{R} = \mathbf{r} - \mathbf{r}'$, the solution is

$$G_0(\mathbf{r}, \mathbf{r}', W) = -\frac{1}{\hbar^2 c^2}(c\tilde{\alpha} \cdot \hat{\mathbf{p}} + \tilde{\beta}mc^2 + W)\frac{e^{ipR/\hbar}}{4\pi R} \qquad (11.32)$$

which is easy to verify. From equation (5.12), and the properties of the α- and β-matrices, we find the non-relativistic limit of equation (11.31):

$$(\nabla^2 + k^2)\frac{e^{ikR}}{4\pi R} = \delta(\mathbf{r} - \mathbf{r}') \qquad (11.33)$$

where $k = p/\hbar$. This is the same as equations (11.4–5).

Next we are going to put the Green's function (11.32) in a partial wave representation. This is easy for the diagonal parts of the Green's function, but not so for the off-diagonal parts. Therefore the strategy adopted is as follows. Two different 'derivations' of the diagonal terms for $|\mathbf{r}| > |\mathbf{r}'|$ are presented and they agree with one another (they'd better). The off-diagonal components are then found by analogy and a plausibility argument. This derivation is also presented by Rose (1961).

Equation (11.31) tells us that $G_0(\mathbf{r}, \mathbf{r}', W)$ is a 4×4 object. Let us write it in the form

$$G_0(\mathbf{r}, \mathbf{r}', W) = \begin{pmatrix} G_{011}(\mathbf{r}, \mathbf{r}', W) & G_{012}(\mathbf{r}, \mathbf{r}', W) \\ G_{021}(\mathbf{r}, \mathbf{r}', W) & G_{022}(\mathbf{r}, \mathbf{r}', W) \end{pmatrix} \qquad (11.34)$$

where each $G_{0nm}(\mathbf{r}, \mathbf{r}', W)$ is now 2×2. Let us also introduce another of those well-known expansions that virtually nobody has ever heard of

$$\frac{e^{ikR}}{4\pi R} = ik \sum_{lm} h_l(kr) j_l(kr') Y_l^m(\hat{\mathbf{r}}) Y_l^{m*}(\hat{\mathbf{r}}') \quad (11.35)$$

which is valid for $r > r'$. We just substitute this into (11.32). Then we note that in (11.32) the $W + \tilde{\beta} mc^2$ is diagonal and will only occur in G_{011} and G_{022} whereas $c\tilde{\alpha} \cdot \hat{\mathbf{p}}$ is only off-diagonal and will only occur in G_{012} and G_{021}. So for the diagonal terms we have

$$G_{011}(\mathbf{r}, \mathbf{r}', W) = -\frac{ik(W + mc^2)}{\hbar^2 c^2} \sum_{lm} h_l(kr) j_l(kr') Y_l^m(\hat{\mathbf{r}}) Y_l^{m*}(\hat{\mathbf{r}}') I_2 \quad (11.36a)$$

$$G_{022}(\mathbf{r}, \mathbf{r}', W) = -\frac{ik(W - mc^2)}{\hbar^2 c^2} \sum_{lm} h_l(kr) j_l(kr') Y_l^m(\hat{\mathbf{r}}) Y_l^{m*}(\hat{\mathbf{r}}') I_2 \quad (11.36b)$$

These can be transformed to the (κ, m_j) representation using equation (2.131b) to give

$$G_{011}(\mathbf{r}, \mathbf{r}', W) = -\frac{ik(W + mc^2)}{\hbar^2 c^2} \sum_{\kappa m_j} h_l(kr) j_l(kr') \chi_\kappa^{m_j}(\hat{\mathbf{r}}) \chi_\kappa^{m_j\dagger}(\hat{\mathbf{r}}') \quad (11.37a)$$

$$G_{022}(\mathbf{r}, \mathbf{r}', W) = -\frac{ik(W - mc^2)}{\hbar^2 c^2} \sum_{\kappa m_j} h_l(kr) j_l(kr') \chi_\kappa^{m_j}(\hat{\mathbf{r}}) \chi_\kappa^{m_j\dagger}(\hat{\mathbf{r}}') \quad (11.37b)$$

where $\chi_\kappa^{m_j}(\hat{\mathbf{r}})$ are the usual spin-angular functions. Clearly

$$G_{022}(\mathbf{r}, \mathbf{r}', W) = \frac{W - mc^2}{W + mc^2} G_{rm011}(\mathbf{r}, \mathbf{r}', W) \quad (11.38)$$

This completes the first part of our strategy for finding the partial wave representation of the Green's function. Next we consider $r > r'$ again and find the Green's function rather differently. Obviously, in this case equation (11.31) reduces to the free-particle Dirac equation. In chapter 8 the free-particle Dirac equation was solved in a partial wave basis. For completeness the solutions are repeated here

$$J_\kappa^{m_j}(\mathbf{r}) = \left(\frac{(W + mc^2)}{\hbar^2 c^2}\right)^{1/2} \begin{pmatrix} j_l(kr) \chi_\kappa^{m_j}(\hat{\mathbf{r}}) \\ \frac{ikc\hbar S_\kappa}{W + mc^2} j_{\bar{l}}(kr) \chi_{-\kappa}^{m_j}(\hat{\mathbf{r}}) \end{pmatrix} \quad (11.39a)$$

$$H_\kappa^{m_j}(\mathbf{r}) = \left(\frac{(W + mc^2)}{\hbar^2 c^2}\right)^{1/2} \begin{pmatrix} h_l(kr) \chi_\kappa^{m_j}(\hat{\mathbf{r}}) \\ \frac{ikc\hbar S_\kappa}{W + mc^2} h_{\bar{l}}(kr) \chi_{-\kappa}^{m_j}(\hat{\mathbf{r}}) \end{pmatrix} \quad (11.39b)$$

where $h_l(x)$ is a spherical Hankel function of the first kind. These are 4-vectors, of course, and the normalization is arbitrary. Let us form a product of them and sum the result over κ and m_j. Again we get a 4×4 entity and we label the quadrants of it in the same way as (11.34):

$$\left[H_\kappa^{m_j}(\mathbf{r}) J_\kappa^{m_j \dagger}(\mathbf{r'}) \right]_{11} = \frac{(W + mc^2)}{\hbar^2 c^2} \sum_{\kappa m_j} h_l(kr) j_l(kr') \chi_\kappa^{m_j}(\hat{\mathbf{r}}) \chi_\kappa^{m_j \dagger}(\hat{\mathbf{r'}}) \quad (11.40a)$$

$$\begin{aligned}
\left[H_\kappa^{m_j}(\mathbf{r}) J_\kappa^{m_j \dagger}(\mathbf{r'}) \right]_{22} &= \frac{(W - mc^2)}{\hbar^2 c^2} \sum_{\kappa m_j} h_{\bar{l}}(kr) j_{\bar{l}}(kr') \chi_{-\kappa}^{m_j}(\hat{\mathbf{r}}) \chi_{-\kappa}^{m_j \dagger}(\hat{\mathbf{r'}}) \\
&= \frac{(W - mc^2)}{\hbar^2 c^2} \sum_{\kappa m_j} h_l(kr) j_l(kr') \chi_\kappa^{m_j}(\hat{\mathbf{r}}) \chi_\kappa^{m_j \dagger}(\hat{\mathbf{r'}})
\end{aligned} \quad (11.40b)$$

We have been able to make the last step because the summation is over all κ so changing our term from $-\kappa$ to κ has no effect on the sum.

Comparison of (11.40a and b) shows that these quantities obey an equation analogous to (11.38). Furthermore, comparison of (11.40) and (11.37) shows that for the diagonal components

$$G_{0nn}(\mathbf{r}, \mathbf{r'}, W) = -ik \sum_{\kappa m_j} \left[H_\kappa^{m_j}(\mathbf{r}) J_\kappa^{m_j \dagger}(\mathbf{r'}) \right]_{nn} \quad (11.41)$$

This works out similarly for the off-diagonal elements of the Green's function, then straightforward algebra leads to

$$\begin{aligned}
G_{012}(\mathbf{r}, \mathbf{r'}, W) &= -G_{021}(\mathbf{r}, \mathbf{r'}, W) \\
&= \frac{k^2}{\hbar c} \sum_{\kappa m_j} S_\kappa h_{\bar{l}}(kr) j_l(kr') \chi_{-\kappa}^{m_j}(\hat{\mathbf{r}}) \chi_\kappa^{m_j \dagger}(\hat{\mathbf{r'}})
\end{aligned} \quad (11.42)$$

Combining the diagonal and off-diagonal parts of the Green's function (11.41) and (11.42) we get simply

$$\begin{aligned}
G_0(\mathbf{r}, \mathbf{r'}, W) &= -ik \sum_{\kappa m_j} H_\kappa^{m_j}(\mathbf{r}) J_\kappa^{m_j \dagger}(\mathbf{r'}) && (r > r') \\
&= -ik \sum_{\kappa m_j} H_\kappa^{m_j}(\mathbf{r'}) J_\kappa^{m_j \dagger}(\mathbf{r}) && (r < r')
\end{aligned} \quad (11.43)$$

This completes our 'derivation'. We have written down the solution for $r < r'$ simply by interchanging them in (11.41) and (11.42). Note that the radial parts of the wavefunctions have played no role in the derivation of the Green's function, so to derive the Green's function for any spherical potential we need only replace the spherical Bessel and Hankel functions by the relevant radial functions.

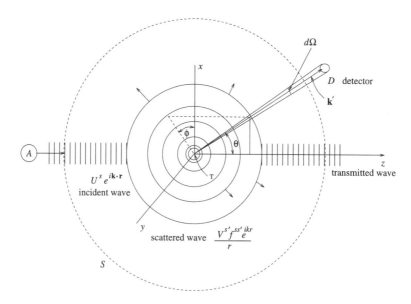

Fig. 11.1. A scattering experiment. A plane wave generated far away at an accelerator A is incident upon target T from the left. There is a probability of transmission through the target, or of scattering to a spherical wave state through angle $\Omega = (\theta, \phi)$. If the particle is scattered within a solid angle $d\Omega$ of Ω it is counted in the detector D. The scattered wave has reached its asymptotic value at the sphere S.

11.5 The Scattered Particle Wavefunction

In scattering theory we are describing an experiment which is shown pictorially in figure 11.1. A free particle is incident upon a target. There is a probability of scattering, or of being transmitted through the target. The scattered particle is usually described as a spherical wave. In this section we develop the mathematical formalism to write down the wavefunction of the scattered particle. As we are interested in condensed matter, we are considering scattering of electrons by a potential.

We start by expanding (11.14) in a Born series

$$\psi(\mathbf{r}) = \phi(\mathbf{r}) + \int G_0(\mathbf{r}, \mathbf{r}', E) V(\mathbf{r}') \psi(\mathbf{r}') d\mathbf{r}'$$

$$= \phi(\mathbf{r}) +$$

$$\int \left(G_0(\mathbf{r}, \mathbf{r}', E) + \int G_0(\mathbf{r}, \mathbf{r}'', E) V(\mathbf{r}'') G_0(\mathbf{r}'', \mathbf{r}', E) d\mathbf{r}'' + \cdots \right) V(\mathbf{r}') \phi(\mathbf{r}')$$

$$= \phi(\mathbf{r}) + \int G(\mathbf{r}, \mathbf{r}', E) V(\mathbf{r}') \phi(\mathbf{r}') d\mathbf{r}' \qquad (11.44)$$

Then we use the definition (11.28) to give

$$\psi(\mathbf{r}) = \phi(\mathbf{r}) + \int\int G_0(\mathbf{r}, \mathbf{r}', W)t(\mathbf{r}', \mathbf{r}'', W)\phi(\mathbf{r}'')d\mathbf{r}'d\mathbf{r}'' \tag{11.45}$$

Equation (11.45) can be understood in scattering theory terms (Arola 1991). The wavefunction $\phi(\mathbf{r})$ represents the incident particle and the second term is the perturbation of the wavefunction due to the potential. Let us try to reduce (11.45) to a more physically transparent form. Firstly we know that we want the unperturbed state to be a free particle, hence $\phi(\mathbf{r})$ is given by equation (5.23). Lumping all the four-component character into U^s where $1 \leq s \leq 4$ labels the spin and sign of the energy (as does j in equation (6.132)), we can write (11.45) as

$$\psi(\mathbf{r}) = U^s e^{i\mathbf{k}\cdot\mathbf{r}} + \int\int G_0(\mathbf{r}, \mathbf{r}', W)t(\mathbf{r}', \mathbf{r}'', W)U^s e^{i\mathbf{k}\cdot\mathbf{r}''}d\mathbf{r}'d\mathbf{r}'' \tag{11.46}$$

In relativistic, as opposed to non-relativistic, quantum theory the particle can be scattered with a flip in its spin. This possibility is included in (11.46) even though s' does not appear explicitly on the right hand side. The Green's function is given by (11.32). Let us look at its asymptotic behaviour when $r \gg r'$. In that case $|\mathbf{r} - \mathbf{r}'| \approx r - \mathbf{r}\cdot\mathbf{r}'/r$ and so

$$\psi(\mathbf{r}) = U^s e^{i\mathbf{k}\cdot\mathbf{r}} - \frac{1}{\hbar^2 c^2}(c\tilde{\boldsymbol{\alpha}}\cdot\mathbf{p} + \tilde{\beta}mc^2 + W)\frac{e^{ikr}}{4\pi r} \times$$
$$\sum_{s'} U^{s'}\int\int U^{s'\dagger}e^{-ik(\mathbf{r}\cdot\mathbf{r}')/|r|}t(\mathbf{r}', \mathbf{r}'', W)U^s e^{i\mathbf{k}\cdot\mathbf{r}''}d\mathbf{r}'d\mathbf{r}'' \tag{11.47}$$

where we have used equation (6.132) to insert the identity matrix. Next, inspection of the exponent in the integral shows that $k\mathbf{r}/|r|$ is a vector in the direction $\mathbf{r}$ of length k. Let us call this $\mathbf{k}'$.

$$\psi(\mathbf{r}) = U^s e^{i\mathbf{k}\cdot\mathbf{r}} + \sum_{s'} f^{s's}(\mathbf{k}', \mathbf{k})\frac{e^{ikr}}{r} \tag{11.48}$$

where

$$f^{s's}(\mathbf{k}', \mathbf{k}) = -\frac{1}{4\pi}\frac{1}{2mc^2}(c\tilde{\boldsymbol{\alpha}}\cdot\mathbf{p} + \tilde{\beta}mc^2 + W)U^{s'}(\mathbf{k}')t^{s's}(\mathbf{k}', \mathbf{k}) \tag{11.49a}$$

is a four-component quantity and

$$t^{s's}(\mathbf{k}', \mathbf{k}) = \frac{2m}{\hbar^2}\int\int U^{s'\dagger}(\mathbf{k}')e^{-i\mathbf{k}'\cdot\mathbf{r}'}t(\mathbf{r}', \mathbf{r}'', W)U^s(\mathbf{k})e^{i\mathbf{k}\cdot\mathbf{r}''}d\mathbf{r}'d\mathbf{r}''$$
$$= (4\pi)^2 \sum_{lm}\sum_{l'm'} i^{l-l'}Y_{l'}^{m'}(\mathbf{k}')t_{l'm'lm}^{s's}(W)Y_l^{m*}(\mathbf{k}) \tag{11.49b}$$

is a scalar, and we have defined

$$t^{s's}_{l'm'lm}(W) =$$
$$\frac{2m}{\hbar^2} \int \int j_{l'}(k'r') Y^{m'*}_{l'}(\hat{\mathbf{r}}) U^{s'\dagger}(\mathbf{k}') t(\mathbf{r}',\mathbf{r}'',W) U^s(\mathbf{k}) Y^m_l(\hat{\mathbf{r}''}) j_l(kr'') d\mathbf{r}' d\mathbf{r}''$$

$$(11.49c)$$

which becomes

$$t^{s's}_{l'm'lm}(E) =$$
$$\frac{2m}{\hbar^2} \int \int j_{l'}(k'r') Y^{m'*}_{l'}(\hat{\mathbf{r}}) \chi^{m's\dagger}_s t(\mathbf{r}',\mathbf{r}'',E) \chi^{ms}_s Y^m_l(\hat{\mathbf{r}''}) j_l(kr'') d\mathbf{r}' d\mathbf{r}'' \quad (11.49d)$$
$$= \frac{2m}{\hbar^2} \int \int j_{l'}(k'r') Y^{m'*}_{l'}(\hat{\mathbf{r}}) t(\mathbf{r}',\mathbf{r}'',E) Y^m_l(\hat{\mathbf{r}''}) j_l(kr'') d\mathbf{r}' d\mathbf{r}'' \delta_{m'_s m_s}$$

in the non-relativistic limit. Equation (11.49d) is the usual expression for $t_{l'm'lm}(W)$ in non-relativistic scattering theory. The *t*-matrix $t(\mathbf{r},\mathbf{r}',W)$ is the *T*-operator for single-site scattering. The quantity $f^{s's}(\mathbf{k}',\mathbf{k})$ is called the scattering amplitude and has units of distance. In general, $f^{s's}(\mathbf{k}',\mathbf{k})$ is defined for all values of $\mathbf{k}$ and $\mathbf{k}'$, but if we are describing elastic scattering so that $\mathbf{k}^2 = \mathbf{k}'^2$ we usually say the scattering amplitude (or *t*-matrix) is *on the energy shell*.

Equation (11.48) is similar to its non-relativistic counterpart

$$\psi(\mathbf{k},\mathbf{r}) = e^{i\mathbf{k}\cdot\mathbf{r}} + f(\theta,\phi)\frac{e^{ikr}}{r} \qquad (11.50)$$

The only differences between (11.48) and (11.50) are the four-component nature of U^s and $f^{s's}(\mathbf{k}',\mathbf{k})$, and the fact that spin flips are allowed. We can change the parameters of the scattering amplitude in (11.48) to angles $(\theta,\phi) = \Omega$ representing the angle between the incident and outgoing wavevectors. Clearly (11.48) is what we expect the wavefunction of the scattered relativistic particle to look like a long way from the scattering site. The incident particle is a plane wave. After scattering the wavefunction is a spherical wave multiplied by some function of Ω.

11.6 The Scattering Experiment

Let us consider further the scattering experiment of figure 11.1. The particles generated by the source are either scattered by, or transmitted through, the target. The incident beam is assumed to consist of well-collimated, homogeneous, monoenergetic particles directed towards the centre of the scattering target which is defined by a potential, or force field. The beam is characterized by its intensity or flux density I_0 which is the number of particles crossing a unit area perpendicular to the beam direction per unit time. In what follows we consider a single particle interacting with the target, so the beam intensity must be low enough that the incident particles do not interfere with one another, and high enough to give adequate statistics for interpretation of the results. In particular a certain fraction of the particles will be scattered into a solid angle $d\Omega$ and detected well beyond where the potential has decayed to zero. Thus, we are assuming that the potential is of finite range, i.e.

$$\lim_{r\to\infty} rV(r) = 0 \qquad (11.51)$$

The Coulomb potential is a notorious example of a potential that does not obey this condition. In this section we discuss what is actually measured in such an experiment and relate it to the wavefunction discussed in the previous section. In figure 11.1 we introduced polar coordinates centred on the target, and we insisted that the beam be along the z-direction. We can define the differential cross section $\sigma(\theta, \phi)$ as the ratio of the flux $I(\theta, \phi)$ scattered at an angle (θ, ϕ) (the number of particles deflected into a solid angle $d\Omega$ per unit time) to the incident flux I_0.

$$\sigma(\theta, \phi) = \frac{d\sigma}{d\Omega} = \frac{I(\theta, \phi)}{I_0} = \frac{\text{scattered flux/solid angle}}{\text{incident flux/area}} \qquad (11.52)$$

So the units of $\sigma(\theta, \phi)$ are area. The total cross section can be defined by integrating the differential cross section over all possible angles

$$\sigma = \int \frac{d\sigma}{d\Omega} d\Omega \qquad (11.53)$$

This tells us the number of particles scattered out of the beam per unit time divided by the incident flux. Clearly the cross section is based on the observation of large numbers of particles and is statistical in nature.

If the de Broglie wavelength of the moving particles is of the same order as, or smaller than, the potential region their wave-like nature dictates a quantum mechanical treatment of scattering. If the spin of the particles is also involved, or if they are travelling at an appreciable fraction of the speed of light, a relativistic quantum mechanical treatment becomes necessary, and that is what we are presenting in this chapter.

The scattered particles in figure 11.1 must obey the continuity equation (4.99) with the current defined by (4.100). Consider the volume enclosed by the sphere S. Integrating the continuity equation in this volume gives

$$\int \nabla \cdot \mathbf{j}(\mathbf{r}) d\mathbf{r} = \frac{\partial}{\partial t} \int \rho(\mathbf{r}) d\mathbf{r} \qquad (11.54)$$

where we have interchanged the order of the integration and differentiation on the right side. The integral on the right side is just the charge contained within the sphere which does not change with time in a stationary state, so the right hand side is zero. The left side can be reduced to a surface integral using Gauss's divergence theorem. So

$$\int \mathbf{j(r)} \cdot d\mathbf{S} = 0 \tag{11.55}$$

and this is just the flux of particles through the surface S. Let us look at the contribution of the scattered part of the beam to this integral. In this context *Zitterbewegung* is a nuisance, so let us replace the velocity operator in (4.100) by the average velocity operator of equation (7.41) (and we are restricting ourselves to positive energy states only). Then we find

$$\mathbf{j_{out}(r)} = \frac{1}{V} \frac{\hbar k c^2}{W} \frac{1}{r^2} |f^{s's}(\Omega)|^2 \hat{\mathbf{r}} \tag{11.56}$$

but $\hbar k c^2 / V W$ is the current of the incident particles, so comparison with equation (11.52) yields

$$\sigma = \int |f^{s's}(\Omega)|^2 d\Omega \tag{11.57}$$

This equation links the differential cross section, which is an observable, to the scattering amplitude which we calculate from microscopic quantum mechanics. Much of the rest of this chapter will deal with methods of calculating the scattering amplitude, or equivalently, the t-matrix.

11.7 Single-Site Scattering in Zero Field

In this section we describe the theory of scattering of a particle from a single site. In this model we have a single scattering site which is entirely described by a potential $V(r)$ which we assume is spherically symmetric and of finite range. The potential is continuous and goes smoothly to zero at distance r_m from its centre. It might be argued that this precludes Coulomb scattering which is surely the most important type of scattering in microscopic physics as it is long range. However, in a solid, for example, the nuclear charge is coulombic, but it is strongly screened by surrounding electrons and tends to fall off much faster than $1/r$, so the theory presented below finds much application in condensed matter physics. The non-relativistic version of this theory has been written down many times (Newton 1966, Faulkner 1979, Györffy and Stocks 1979, Stocks and Winter 1984, Weinberger 1990 for example), and the relativistic theory not so many times (Weinberger 1990, Arola 1991). The model of scattering we are describing is that already shown in figure 11.1.

Inside the region of the potential ($r < r_m$) the wavefunction of the scattered electron is a linear combination of solutions of the Dirac equation in a central field (equation (8.10)).

$$\psi(\mathbf{r}) = \sum_{\kappa m_j} \begin{pmatrix} g_\kappa(r)\chi_\kappa^{m_j} \\ if_\kappa(r)\chi_{-\kappa}^{m_j} \end{pmatrix} \tag{11.58}$$

where the $g_\kappa(r)$ and $f_\kappa(r)$ are solutions of the radial Dirac equation (equations (8.12)). In the region of zero potential ($r > r_m$) we have already solved the radial Dirac equation (see section 8.2) and have seen that the solutions are given in terms of Bessel functions. The solution regular at the origin is given by equation (8.19). In this scattering problem there is no reason to assume the irregular solution can be ignored. Therefore we must include it in our analysis. The wavefunction for $r \geq r_m$ must be a linear combination of the regular and irregular zero-potential solutions

$$\psi(\mathbf{r}) = \sum_{\kappa m_j} \begin{pmatrix} (\cos \delta_\kappa(W)j_l(kr) - \sin \delta_\kappa(W)n_l(kr))\chi_\kappa^{m_j} \\ \frac{ik c\hbar S_\kappa}{W+mc^2}(\cos \delta_\kappa(W)j_{\bar{l}}(kr) - \sin \delta_\kappa(W)n_{\bar{l}}(kr))\chi_{-\kappa}^{m_j} \end{pmatrix} \tag{11.59}$$

Here $j_l(kr)$ is a spherical Bessel function and $n_l(kr)$ is a spherical Neumann function, with k being the magnitude of the wavevector of the scattered particle. Equation (11.59) is the most general solution to the Dirac equation in the region where the potential vanishes. At this stage $\sin \delta_\kappa(W)$ and $\cos \delta_\kappa(W)$ are just coefficients of the Bessel functions which will be defined by the boundary condition that this wavefunction must be equal to the solution inside the potential region at the boundary of the potential.

The next step is to make the physically reasonable assumption that the wavefunction of the scattered electron is continuous and differentiable across the boundary of the potential. Consider equation (8.12a) and divide it through by $g(r)$. The left hand side is then the derivative of $g(r)$ over $g(r)$. We require that this quantity, known as the logarithmic derivative, be smoothly varying. This will be so if the right hand side is also a smooth function. It has already been stipulated that the potential be continuous so we simply require that $f(r)/g(r)$ be a continuous function. The condition for matching wavefunctions at the boundary is

$$g_\kappa(r_m) = \cos \delta_\kappa(W)j_l(kr_m) - \sin \delta_\kappa(W)n_l(kr_m) \tag{11.60a}$$

$$f_\kappa(r_m) = \frac{kc\hbar S_\kappa}{W + mc^2}(\cos \delta_\kappa(W)j_{\bar{l}}(kr_m) - \sin \delta_\kappa(W)n_{\bar{l}}(kr_m)) \tag{11.60b}$$

This can easily be seen by comparison of equations (11.58) and (11.59). Dividing (11.60b) by (11.60a) and defining $L_\kappa(W, r_m) = f_\kappa(r_m)/g_\kappa(r_m)$ we can find an expression for $\delta_\kappa(W)$, which is known as a phase shift

$$\tan \delta_\kappa(W) = \frac{L_\kappa(W, r_m)j_l(kr_m) - \frac{\hbar kcS_\kappa}{W+mc^2}j_{\bar{l}}(kr_m)}{L_\kappa(W, r_m)n_l(kr_m) - \frac{\hbar kcS_\kappa}{W+mc^2}n_{\bar{l}}(kr_m)} \tag{11.61}$$

The expansion of the scattered electron wavefunctions in (11.59) is not unique. Another formulation is

$$\psi(\mathbf{r}) = \sum_{\kappa m_j} \begin{pmatrix} (j_l(kr) - ikt_\kappa(W)h_l^+(kr))\chi_\kappa^{m_j} \\ \frac{ikc\hbar S_\kappa}{W+mc^2}(j_{\bar{l}}(kr) - ikt_\kappa(W)h_{\bar{l}}^+(kr))\chi_{-\kappa}^{m_j} \end{pmatrix} \tag{11.62}$$

Here we have defined the t-matrix $t_\kappa(W)$. Another useful expansion is

$$\psi(\mathbf{r}) = \sum_{\kappa m_j} \begin{pmatrix} (h_l^-(kr) - S_\kappa(W)h_l^+(kr))\chi_\kappa^{m_j} \\ \frac{ikc\hbar S_\kappa}{W+mc^2}(h_{\bar{l}}^-(kr) - S_\kappa(W)h_{\bar{l}}^+(kr))\chi_{-\kappa}^{m_j} \end{pmatrix} \tag{11.63}$$

This time our expansion coefficient is known as the S-matrix. Obviously the phase shift, t-matrix and S-matrix are all connected to one another and the equations joining them are summarized here

$$t_\kappa(W) = -\frac{1}{k}\sin\delta_\kappa(W)e^{i\delta_\kappa(W)} \tag{11.64a}$$

$$S_\kappa(W) = e^{2i\delta_\kappa(W)} \tag{11.64b}$$

$$S_\kappa(W) = 1 - 2ikt_\kappa(W) \tag{11.64c}$$

It also turns out to be very useful to define a scattering amplitude

$$f_\kappa(W) = -kt_\kappa(W) \tag{11.65}$$

From (11.65) and (11.64a) we see that the scattering amplitude can take on only a very restricted set of values. The real part of the scattering amplitude can only take on values such that $-0.5 \le \mathrm{Re}f_\kappa(W) \le +0.5$, and the imaginary part is such that $0.0 \le \mathrm{Im}f_\kappa(W) \le 1.0$. Indeed, if we plot the imaginary part of $f_\kappa(W)$ against the real part of $f_\kappa(W)$ as a function of energy W we find the plot is a circle centred on $(0.0, 0.5i)$. When the phase shift passes through $\pi/2$ the scattering amplitude is imaginary.

So far these quantities, the phase shifts, t-matrix, S-matrix and scattering amplitude, look like mathematical entities without much physical significance. The rest of this section is devoted to interpreting what happens in a single-site scattering event in terms of these quantities and hence will (I hope) put some physical flesh on these mathematical bones.

Figure 11.2 displays all the phase shifts for scattering of an electron from the potential due to a single platinum site. The potential was actually calculated self-consistently for crystalline platinum, so this does not represent scattering from a platinum atom. Clearly the s and p phase shifts are small and have fairly uninteresting behaviour. The d phase shifts contain a resonance and become large. From equation (11.64) we see that a phase

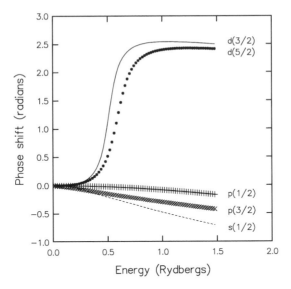

Fig. 11.2. Phase shifts for a platinum potential. Values of the l and j quantum numbers for each phase shift are indicated. The d-electron phase shifts ($l = 2$) contain a resonance and hence are responsible for the bulk of the scattering.

shift around $\pi/2$ means a large t-matrix, and hence a large scattering amplitude, leading to strong scattering, whereas a phase shift around 0 or π implies a small scattering amplitude and hence weak scattering. Clearly in this figure it is the platinum d-electrons that are responsible for most of the scattering.

From (11.64b) it is immediately clear that $S_\kappa^*(W)S_\kappa(W) = 1$. If we use the definition (11.64c) in here we find

$$\mathrm{Im}\, t_\kappa(W) = -k t_\kappa^*(W) t_\kappa(W) \tag{11.66a}$$

or alternatively, using (11.64a)

$$\mathrm{Im}\, t_\kappa(W) = -k^{-1} \sin^2 \delta_\kappa(W) \tag{11.66b}$$

Equations (11.66) are two alternative versions of the optical theorem which is simply a statement of the conservation of particles. We will have more to say about this later.

Now, the scattering amplitude has been introduced in two apparently completely separate ways. In (11.65) the scattering amplitude for the κ channel scattering is defined in terms of the t-matrix for that channel. In (11.48) and (11.49) it is defined as part of the asymptotic behaviour of the wavefunction of the scattered particle. Of course the t-matrix also comes from the asymptotic behaviour of the wavefunction so one should not be surprised that the scattering amplitude appears in these two guises.

However, the theory would be incomplete if we did not show how these two scattering amplitudes are related to one another. To do this we need to expand the plane wave part of (11.48) in spherical waves. Luckily, we have already done this and the expansion coefficients are given by equations (2.146) and (2.147). In the four-component form in which we require the expansion it is

$$U^s e^{i\mathbf{p}\cdot\mathbf{r}/\hbar} = 4\pi \left(\frac{\hbar^2 c^2}{2W}\right)^{1/2} \sum_{\kappa,m_j} C(l\tfrac{1}{2}j;m_j - ss)i^l Y_l^{m_j-s*}(\hat{\mathbf{k}})J_\kappa^{m_j}(\mathbf{r}) \quad (11.67)$$

where $J_\kappa^{m_j}(\mathbf{r})$ is given by equation (11.39a). Now, equation (11.48) is valid in the limit $r \to \infty$, so the Bessel functions take their asymptotic values

$$\lim_{r\to\infty} j_l(kr) = \frac{\sin(kr - \frac{l\pi}{2})}{kr}, \qquad \lim_{r\to\infty} n_l(kr) = -\frac{\cos(kr - \frac{l\pi}{2})}{kr} \quad (11.68)$$

Next we substitute (11.67), in this limit, into our expansion of the scattered particle wavefunction (11.48) to give

$$\psi(\mathbf{r}) = f^{s's}(\mathbf{k}',\mathbf{k})\frac{e^{ikr}}{r} + \frac{4\pi}{kr}\left(\frac{W+mc^2}{2W}\right)^{\frac{1}{2}} \times$$

$$\sum_{\kappa m_j} i^l C(l\tfrac{1}{2}j;m_j - ss)Y_l^{m_j-s*}(\hat{\mathbf{k}}) \begin{pmatrix} \sin(kr - \frac{l\pi}{2})\chi_\kappa^{m_j}(\hat{\mathbf{r}}) \\ \frac{ipcS_\kappa \sin(kr-\frac{l\pi}{2})}{(W+mc^2)}\chi_{-\kappa}^{m_j}(\hat{\mathbf{r}}) \end{pmatrix} \quad (11.69)$$

as the wavefunction of the scattered particle. The asymptotic form of the wavefunction is also given by (11.59). Let us make the $|r| \to \infty$ expansion of the Bessel functions in that equation. With this and simple trigonometry (11.59) can be written

$$\psi(\mathbf{r}) = \frac{1}{kr}\sum_{\kappa m_j} A_{\kappa m_j}^s \begin{pmatrix} \sin(kr - \frac{l\pi}{2} + \delta_\kappa(W))\chi_\kappa^{m_j}(\hat{\mathbf{r}}) \\ \frac{ikchS_\kappa}{W+mc^2}\sin(kr - \frac{\bar{l}\pi}{2} + \delta_\kappa(W))\chi_{-\kappa}^{m_j}(\hat{\mathbf{r}}) \end{pmatrix} \quad (11.70)$$

where the $A_{\kappa m_j}^s$ are constants to be determined. We write the trigonometric functions in equations (11.69) and (11.70) in exponential form and set these two equations equal. Luckily $1/r$ cancels in each term and the only dependence on r is in the e^{ikr} and e^{-ikr}. These are linearly independent so their coefficients can be set equal. Firstly we equate coefficients of e^{-ikr} in the upper part of the wavefunction. Premultiplying the resulting equation by $\chi_\kappa^{m_j\dagger}(\mathbf{r})$ and integrating over real space angles gives

$$A_{\kappa m_j}^s = 4\pi i^l \left(\frac{W+mc^2}{2W}\right)^{1/2} C(l\tfrac{1}{2}j;m_j - ss)Y_l^{m_j-s*}(\hat{\mathbf{k}})e^{i\delta_\kappa(W)} \quad (11.71)$$

Next we equate coefficients of e^{ikr} in the upper part of the wavefunction in (11.69) and (11.70). Doing this and substituting in the resulting equation for $A^s_{\kappa m_j}$ leads to

$$f^{s's}(\Omega) = f^{s's}(\mathbf{k}',\mathbf{k}) = \frac{4\pi}{k}\left(\frac{W+mc^2}{2W}\right)^{\frac{1}{2}} \times$$

$$\sum_{\kappa m_j} C(l\tfrac{1}{2}j;m_j-s,s)Y_l^{m_j-s*}(\hat{\mathbf{k}})f_\kappa(W)\begin{pmatrix} \chi_\kappa^{m_j}(\hat{\mathbf{k}}') \\ \frac{ipcS_\kappa}{W+mc^2}\chi_{-\kappa}^{m_j}(\hat{\mathbf{k}}') \end{pmatrix} \qquad (11.72)$$

where $f_\kappa(W)$ is given in terms of the phase shift by equations (11.64a) and (11.65). The lower part of the wavefunction yields identical results. We see that the scattering amplitude defined in equation (11.65) is essentially one element in a partial wave expansion of the scattering amplitude defined in (11.49). We have also made use of the definition of the vector $\mathbf{k}'$ as being in the same direction as $\mathbf{r}$ in the asymptotic limit. Therefore we have been able to replace $\hat{\mathbf{r}}$ by $\hat{\mathbf{k}}'$ in (11.72). This completes what we set out to do, we have explicitly demonstrated the relationship between the two definitions of the scattering amplitude. However, it is profitable to develop equation (11.72) further. From equations (11.57) and (11.72) we can find an expression for the total scattering cross section in relativistic theory. To do this we need to evaluate $f^{s's\dagger}(\Omega)f^{s's}(\Omega)$ and integrate over angles $\hat{\mathbf{k}}'$. The spin-angular functions give the usual Kronecker δ-functions and we have

$$\int |f^{s's}(\Omega)|^2 d\hat{\mathbf{k}}' = \frac{16\pi^2}{k^2}\sum_{\kappa,m_j} C^2(l\tfrac{1}{2}j;m_j-s,s)\sin^2\delta_\kappa Y_l^{m_j-s*}(\hat{\mathbf{k}})Y_l^{m_j-s}(\hat{\mathbf{k}})$$

$$(11.73)$$

This can be reduced further if (as in figure 11.1) the incident beam is directed along the z-axis. Then we can use equation (C.22) to simplify the spherical harmonics in this expression. From equation (11.57) the left hand side of (11.73) is the total scattering cross section, so we have

$$\sigma = \int |f^{s's}(\mathbf{k}',\mathbf{k})|^2 d\hat{\mathbf{k}}' = \frac{4\pi}{k^2}\sum_\kappa (2l+1)C^2(l\tfrac{1}{2}j;0,s)\sin^2\delta_\kappa$$

$$= \frac{4\pi}{k^2}\sum_\kappa (l+1)\sin^2\delta_\kappa \qquad (j=l+1/2) \qquad (11.74)$$

$$+ \frac{4\pi}{k^2}\sum_\kappa l\sin^2\delta_\kappa \qquad (j=l-1/2)$$

where we have used the definition of the Clebsch–Gordan coefficients from table 2.2. Finally, these two cases can be combined into one as

$$\sigma = \frac{4\pi}{k^2}\sum_\kappa (j+1/2)\sin^2\delta_\kappa \qquad (11.75)$$

This equation shows that a phase shift around 0 or π means weak scattering, whereas one close to $\pi/2$ means strong scattering. It is an alternative form of the optical theorem, as is seen by combining equations (11.66a) and (11.66b).

11.8 Radial Dirac Equation in a Magnetic Field

Interesting effects arise in the solid state as a result of the interaction of spin–orbit coupling and magnetism. To understand these one has to solve the Dirac equation with a magnetic field. Here we will derive the t-matrices for this case. The simplest form of the Dirac equation we can solve is given by equation (10.254) with the field along the z-direction. To find the radial form of this equation we can follow the procedure in section 8.1 and we get to

$$
\left(ic\hbar\tilde{\gamma}_5\tilde{\sigma}_r \left(\frac{\partial}{\partial r} + \frac{1}{r} - \frac{\tilde{\beta}\hat{K}}{\hbar r} \right) + V(r) + \tilde{\beta}mc^2 - \tilde{\beta}\frac{e}{m}\hat{S}_z B(r) - W \right) \psi(\mathbf{r}) = 0
$$

$$(11.76)$$

where we assume that both the potential $V(r)$ and the magnetic field $B(r)$ are spherically symmetric. The wavefunctions can then be separated into radial- and angular-dependent forms as

$$
\psi(\mathbf{r}) = \sum_{\kappa'm'_j} a_{\kappa'} \left(\begin{array}{c} g_{\kappa',\kappa}^{m'_jm_j}(r)\chi_{\kappa'}^{m'_j}(\hat{\mathbf{r}}) \\ if_{\kappa',\kappa}^{m'_jm_j}(r)\chi_{-\kappa'}^{m'_j}(\hat{\mathbf{r}}) \end{array} \right)
$$

$$(11.77)$$

Now we can substitute (11.77) into (11.76) and write everything out in matrix form. This is quite cumbersome, but if we do it we see that the four-component equation (11.76) separates into two two-component equations. Writing these out separately

$$
c\left(-\hbar\frac{\partial}{\partial r} - \frac{\hbar}{r} - \frac{\hat{K}}{r} \right) \sum_{\kappa'm'_j} f_{\kappa',\kappa}^{m'_jm_j}(r)\chi_{\kappa'}^{m'_j}(\hat{\mathbf{r}})
$$

$$
+ \left(V(r) - \frac{e\hbar}{2m}\tilde{\sigma}_z B(r) + mc^2 - W \right) \sum_{\kappa'm'_j} g_{\kappa',\kappa}^{m'_jm_j}(r)\chi_{\kappa'}^{m'_j}(\hat{\mathbf{r}}) = 0 \qquad (11.78a)
$$

$$
c\left(\hbar\frac{\partial}{\partial r} + \frac{\hbar}{r} - \frac{\hat{K}}{r} \right) \sum_{\kappa'm'_j} g_{\kappa',\kappa}^{m'_jm_j}(r)\chi_{-\kappa'}^{m'_j}(\hat{\mathbf{r}})
$$

$$
+ \left(V(r) + \frac{e\hbar}{2m}\tilde{\sigma}_z B(r) - mc^2 - W \right) \sum_{\kappa'm'_j} f_{\kappa',\kappa}^{m'_jm_j}(r)\chi_{-\kappa'}^{m'_j}(\hat{\mathbf{r}}) = 0 \qquad (11.78b)
$$

where we have made use of equation (2.136). We can premultiply (11.78a) by $\chi_{\kappa_1}^{m_{j1}\dagger}(\hat{\mathbf{r}})$ and integrate the equation over angles $\hat{\mathbf{r}}$. Similarly we premultiply (11.78b) by $\chi_{-\kappa_1}^{m_{j1}\dagger}(\hat{r})$ and integrate over angles. Making use of the orthonormality of the functions $\chi_\kappa^{m_j}$ we can reduce this to

$$c\left(\hbar\frac{d}{dr} + \frac{\hbar}{r} - \frac{\kappa_1}{r}\right) f_{\kappa_1,\kappa}^{m_{j1}m_j}(r) - (V(r) + mc^2 - W)g_{\kappa_1,\kappa}^{m_{j1}m_j}(r)$$

$$+ \frac{e\hbar}{2m}B(r)\sum_{\kappa'm'_j} G(\kappa_1, m_{j1}, \kappa', m'_j)g_{\kappa',\kappa}^{m'_j m_j}(r) = 0 \qquad (11.79a)$$

$$c\left(\hbar\frac{d}{dr} + \frac{\hbar}{r} + \frac{\kappa_1}{r}\right) g_{\kappa_1,\kappa}^{m_{j1}m_j}(r) + (V(r) - mc^2 - W)f_{\kappa_1,\kappa}^{m_{j1}m_j}(r)$$

$$- \frac{e\hbar}{2m}B(r)\sum_{\kappa'm'_j} G(-\kappa_1, m_{j1}, -\kappa', m'_j)f_{\kappa',\kappa}^{m'_j m_j}(r) = 0 \qquad (11.79b)$$

where we have defined the coefficients

$$G(\kappa_1, m_{j1}, \kappa_2, m_{j2}) = \int \chi_{\kappa_1}^{m_{j1}\dagger}(\hat{r})\tilde{\sigma}_z\chi_{\kappa_2}^{m_{j2}}(\hat{r})d\hat{r}$$

$$= \left(C(l_1\tfrac{1}{2}j_1; m_{j1} - \tfrac{1}{2}, \tfrac{1}{2})C(l_2\tfrac{1}{2}j_2; m_{j2} - \tfrac{1}{2}, \tfrac{1}{2})\right.$$

$$\left. - C(l_1\tfrac{1}{2}j_1; m_{j1} + \tfrac{1}{2}, -\tfrac{1}{2})C(l_2\tfrac{1}{2}j_2; m_{j2} + \tfrac{1}{2}, -\tfrac{1}{2})\right)$$

$$(11.80)$$

These coefficients can be evaluated from table 2.2. They are non-zero only if $\kappa_1 = \kappa_2$ or $-\kappa_2 - 1$ and if $m_{j1} = m_{j2}$, so the first m_j in the arguments of G is redundant and we shall drop it from now on.

Even with these restrictions, equations (11.79) form two infinite sets of coupled partial differential equations. If we cut out coupling between states with quantum numbers (l, j) and $(l \pm 2, j \pm 1)$ the equations are simplified considerably. This is consistent with our treatment of the Zeeman effect in chapter 8. Our final form for the radial Dirac equation in a spin-only magnetic field is

$$\frac{dg_{\kappa,\kappa}^{m_j}(r)}{dr} = -\frac{\kappa+1}{r}g_{\kappa,\kappa}^{m_j}(r) + \frac{1}{c\hbar}(W + mc^2 - V(r))f_{\kappa,\kappa}^{m_j}(r)$$

$$- \frac{e}{2mc}B(r)G(-\kappa, -\kappa, m_j)f_{\kappa,\kappa}^{m_j}(r) \qquad (11.81a)$$

$$\frac{df_{\kappa,\kappa}^{m_j}(r)}{dr} = \frac{\kappa-1}{r}f_{\kappa,\kappa}^{m_j}(r) + \frac{1}{c\hbar}(V(r) + mc^2 - W)g_{\kappa\kappa}^{m_j}(r)$$

$$- \frac{e}{2mc}B(r)\left(G(\kappa, \kappa, m_j)g_{\kappa,\kappa}^{m_j}(r) + G(\kappa, \kappa', m_j)g_{\kappa',\kappa}^{m_j}(r)\right)$$

$$(11.81b)$$

$$\frac{dg_{\kappa',\kappa}^{m_j}(r)}{dr} = -\frac{\kappa+1}{r}g_{\kappa',\kappa}^{m_j}(r) + \frac{1}{c\hbar}(W + mc^2 - V(r))f_{\kappa',\kappa}^{m_j}(r)$$

$$- \frac{e}{2mc}B(r)G(-\kappa', -\kappa', m_j)f_{\kappa',\kappa}^{m_j}(r) \qquad (11.81c)$$

$$\frac{df_{\kappa',\kappa}^{m_j}(r)}{dr} = \frac{\kappa-1}{r}f_{\kappa',\kappa}^{m_j}(r) + \frac{1}{c\hbar}(V(r) + mc^2 - W)g_{\kappa',\kappa}^{m_j}(r)$$

$$- \frac{e}{2mc}B(r)\left(G(\kappa', \kappa', m_j)g_{\kappa'\kappa}^{m_j}(r) + G(\kappa', \kappa, m_j)g_{\kappa\kappa}^{m_j}(r)\right)$$

$$(11.81d)$$

where $\kappa' = -\kappa - 1$ in equations (11.81). Similar equations follow imme-
diately for $g_{\kappa,\kappa'}^{m_j}(r)$, $g_{\kappa',\kappa'}^{m_j}(r)$, $f_{\kappa,\kappa'}^{m_j}(r)$ and $f_{\kappa',\kappa'}^{m_j}(r)$ by interchanging κ and κ'
in (11.81). So, with this approximation, the radial Dirac equation reduces
from an infinite number of coupled partial differential equations down to
just two sets of four which can be solved using a generalization of standard methods (Loucks 1967). For more rigorous derivations of equations
(11.81) the reader is referred to the original literature (Feder et $al.$ 1983,
Strange et $al.$ 1984). The assumption that coupling between l and $l \pm 2$
can be neglected has been explored by Jenkins and Strange (1994) who
retained this coupling, but cut off the summations in κ at $l = 5, 6$, and
found the effect on the scattering amplitudes to be small, but not necessarily negligible for all applications. They surmised that it may be important
to include this coupling when calculating quantities that are very small
on the scale of electronic energies, such as magnetocrystalline anisotropy
energy. Clearly equations (11.81) become the radial Dirac equation (8.12)
when the magnetic field is zero, and a little trying will convince you that
there is no way they can be solved analytically.

11.9 Single-Site Scattering in a Magnetic Field

Now that the wavefunctions for a Dirac particle in a spherically symmetric
potential and magnetic field can be found (at least in principle), they can
be used to determine scattering t-matrices analogous to those of equation
(11.64a). In this section we show how to achieve this following the work
of Strange et $al.$ (1984). Again we assume that the potential (and field) are
zero outside the muffin tin sphere of radius r_m around the scattering centre,
and that the wavefunctions inside and outside the inscribed sphere match
each other smoothly at the sphere boundary. Then to define the t-matrices
we require the generalization to the magnetic case of equation (11.62).
To this end, it is convenient to define an eight-component 'wavefunction'.

One that evidently contains just the required coupling is

$$\psi_{\kappa_1}^{m_j}(\mathbf{r}) = \begin{pmatrix} (j_l(kr)\delta_{\kappa\kappa_1} - ikt_{\kappa,\kappa_1}^{m_j}(W)h_l^+(kr))\chi_{\kappa_1}^{m_j}(\hat{\mathbf{r}}) \\ \frac{ikc\hbar S_{\kappa_1}}{W+mc^2}(j_{\bar{l}}(kr)\delta_{\kappa\kappa_1} - ikt_{\kappa,\kappa_1}^{m_j}(W)h_{\bar{l}}^+(kr))\chi_{-\kappa_1}^{m_j}(\hat{\mathbf{r}}) \\ (j_l(kr)\delta_{\kappa-\kappa_1-1} - ikt_{\kappa,-\kappa_1-1}^{m_j}(W)h_l^+(kr))\chi_{-\kappa_1-1}^{m_j}(\hat{\mathbf{r}}) \\ \frac{ikc\hbar S_{\kappa_2}}{W+mc^2}(j_{\bar{l}}(kr)\delta_{\kappa-\kappa_1-1} - ikt_{\kappa,-\kappa_1-1}^{m_j}(W)h_{\bar{l}}^+(kr))\chi_{\kappa_1+1}^{m_j}(\hat{\mathbf{r}}) \end{pmatrix} \tag{11.82}$$

These wavefunctions have to be matched onto the solutions of (11.81) inside the sphere written in the form

$$\psi_{\kappa_1}^{m_j}(\mathbf{r}) = \begin{pmatrix} g_{\kappa_1}^{m_j}(r)\chi_{\kappa_1}^{m_j}(\hat{\mathbf{r}}) \\ if_{\kappa_1}^{m_j}(r)\chi_{-\kappa_1}^{m_j}(\hat{\mathbf{r}}) \\ g_{-\kappa_1-1}^{m_j}(r)\chi_{-\kappa_1-1}^{m_j}(\hat{\mathbf{r}}) \\ if_{-\kappa_1-1}^{m_j}(r)\chi_{\kappa_1+1}^{m_j}(\hat{\mathbf{r}}) \end{pmatrix} = \sum_\kappa a_\kappa \begin{pmatrix} g_{\kappa,\kappa_1}^{m_j}(r)\chi_{\kappa_1}^{m_j}(\hat{\mathbf{r}}) \\ if_{\kappa,\kappa_1}^{m_j}(r)\chi_{-\kappa_1}^{m_j}(\hat{\mathbf{r}}) \\ g_{\kappa,-\kappa_1-1}^{m_j}(r)\chi_{-\kappa_1-1}^{m_j}(\hat{\mathbf{r}}) \\ if_{\kappa,-\kappa_1-1}^{m_j}(r)\chi_{\kappa_1+1}^{m_j}(\hat{\mathbf{r}}) \end{pmatrix} \tag{11.83}$$

where the sum extends over $\kappa = \kappa_1, -\kappa_1 - 1$. Now it is simply a case of setting the equivalent elements of these two column vectors equal at r_m. Doing this yields four equations and four unknowns a_{κ_1}, $a_{-\kappa_1-1}$, $t_{\kappa_1,\kappa_1}^{m_j}(W)$ and $t_{\kappa_1,-\kappa_1-1}^{m_j}(W)$ for each set of the quantum numbers. These can be solved with tons of algebra. We find

$$t_{\kappa_1,\kappa_1}^{m_j}(W) = \frac{\Delta^{(1)}j_{l_1}(kr_m) - \frac{ikc\hbar S_{\kappa_1}}{W+mc^2}j_{\bar{l}_1}(kr_m)}{ik\Delta^{(1)}h_{l_1}^+(kr_m) - ik\left[\frac{ikc\hbar S_{\kappa_1}}{W+mc^2}\right]h_{\bar{l}_1}^+(kr_m)} \tag{11.84a}$$

$$t_{\kappa_1,-\kappa_1-1}^{m_j}(W) = \frac{j_l(kr_m) - ikt_{\kappa_1,\kappa_1}^{m_j}(W)h_{l_1}^+(kr_m)}{F_1^{(1)}g_{\kappa_1,\kappa_1}^{m_j}(r_m) + F_2^{(1)}g_{\kappa_1,-\kappa_1-1}^{m_j}(r_m)} \tag{11.84b}$$

where

$$\Delta^{(1)} = i\frac{F_1^{(1)}f_{\kappa_1,\kappa_1}^{m_j}(r_m) + F_2^{(1)}f_{\kappa_1,-\kappa-1}^{m_j}(r_m)}{F_1^{(1)}g_{\kappa_1,\kappa_1}^{m_j}(r_m) + F_2^{(1)}g_{\kappa_1,-\kappa-1}^{m_j}(r_m)} \tag{11.85a}$$

$$F_1^{(1)} = -ik \begin{pmatrix} \frac{ikc\hbar S_{\kappa_2}}{W+mc^2}h_{\bar{l}_2}^+(kr_m) - L_{-\kappa-1,-\kappa-1}^{m_j}h_{\bar{l}_2}^+(kr_m) \\ if_{-\kappa-1,\kappa}^{m_j}(r_m) - L_{-\kappa-1,-\kappa-1}^{m_j}g_{-\kappa-1,\kappa}^{m_j}(r_m) \end{pmatrix} \tag{11.85b}$$

and

$$F_2^{(1)} = -ik \left(\frac{\frac{ikc\hbar S_{\kappa_2}}{W+mc^2} h_{\bar{l}_2}^+(kr_m) - L_{-\kappa-1,\kappa}^{m_j} h_{l_2}^+(kr_m)}{if_{-\kappa-1,-\kappa-1}^{m_j}(r_m) - L_{-\kappa-1,\kappa}^{m_j} g_{-\kappa-1,-\kappa-1}^{m_j}(r_m)} \right) \qquad (11.85c)$$

where the imaginary logarithmic derivative is defined by

$$L_{\kappa,\kappa'}^{m_j} = if_{\kappa,\kappa'}^{m_j}/g_{\kappa,\kappa'}^{m_j} \qquad (11.85d)$$

Similarly $t_{-\kappa_1-1,\kappa_1}^{m_j}(W)$ and $t_{-\kappa-1,-\kappa-1}^{m_j}(W)$ can be found by interchanging the indices in these equations. These t-matrices only depend upon ratios of wavefunctions and so there is no need to normalize the wavefunctions after they have been evaluated from (11.81). The t-matrices obey a generalized version of the optical theorem

$$t_{\kappa,\kappa'}^{m_j}(W) - t_{\kappa',\kappa}^{m_j*}(W) = -2ik \sum_{\kappa''} t_{\kappa,\kappa''}^{m_j}(W) t_{\kappa',\kappa''}^{m_j*}(W) \qquad (11.86)$$

and we can define scattering amplitudes analogously to (11.65) as

$$f_{\kappa,\kappa'}^{m_j}(W) = -kt_{\kappa,\kappa'}^{m_j}(W) \qquad (11.87)$$

An important property of the t-matrices for scattering in the presence of a spin-only magnetic field is that they are no longer diagonal in the quantum numbers. It turns out that $t_{\kappa,\kappa'}^{m_j}(W) = t_{\kappa',\kappa}^{m_j}(W)$ so the t-matrix is still symmetric. In figure 11.3 we plot a phasor diagram displaying the imaginary part of the scattering amplitude against the real part. In zero field figures 11.3a and 11.3c would be unitarity circles centred at $(0.0, 0.5)$ and of radius 0.5. This is easy to check from (11.64a) and (11.65) in the zero field case. However, the field introduces κ character into the $-\kappa-1$ wavefunction, and *vice versa*, so as field increases, these amplitudes leave the unitarity circle and decrease in area, eventually curling back on themselves. The scattering amplitude representing the cross term is shown in figure 11.3b and grows at a corresponding rate, so that particles continue to be conserved during scattering. The off-diagonal components of the t-matrices mean that it is no longer so easy to define an S-matrix or phase shifts. To find the S-matrices we generalize equation (11.64c)

$$S_{\kappa,\kappa'}^{m_j}(W) = \delta_{\kappa\kappa'} - 2ikt_{\kappa\kappa'}^{m_j}(W) \qquad (11.88)$$

A unitarity transform on the S-matrix puts it into diagonal form

$$\begin{pmatrix} S_{11} & 0 \\ 0 & S_{22} \end{pmatrix} = \tilde{U}_\kappa^\dagger \begin{pmatrix} S_{\kappa,\kappa}^{m_j} & S_{\kappa,\kappa'}^{m_j} \\ S_{\kappa',\kappa}^{m_j} & S_{\kappa',\kappa'}^{m_j} \end{pmatrix} \tilde{U}_\kappa \qquad (11.89)$$

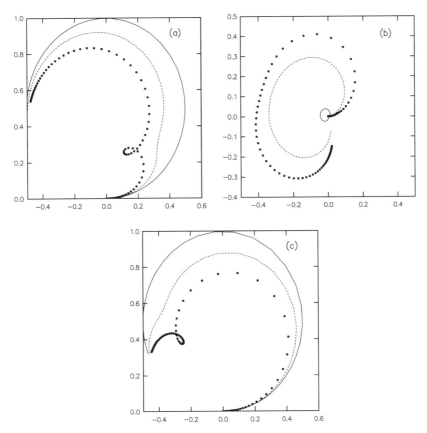

Fig. 11.3. The $l = 2$ scattering amplitudes in the (κ, m_j) representation for platinum as a function of energy and magnetic field. The vertical axis is the imaginary part of the scattering amplitude and the horizontal axis is the real part. Full lines are for low field, dashed lines are for an intermediate field, and dotted lines are for a large field. The zero of energy is at the origin and energy increases as you proceed round the scattering amplitude anticlockwise. All figures are for $m_j = 1/2$. (a) $\kappa = \kappa' = -3$, (b) $\kappa = -3$, $\kappa' = 2$, (c) $\kappa = \kappa' = 2$.

The matrix $\tilde{U}_\kappa$ can be written in the form

$$\tilde{U}_\kappa = \begin{pmatrix} \cos(\theta_\kappa/2) & \sin(\theta_\kappa/2) \\ -\sin(\theta_\kappa/2) & \cos(\theta_\kappa/2) \end{pmatrix} \tag{11.90}$$

where θ is a mixing angle which is a function of both energy and magnetic field. The disadvantage of transforming like this to a representation where the S-matrix is diagonal is that we do not know the good quantum numbers in this representation. However, the diagonal representation is useful because it enables us to define a generalized phase shift using (11.64b). The generalized phase shifts are shown in figure 11.4 at a field

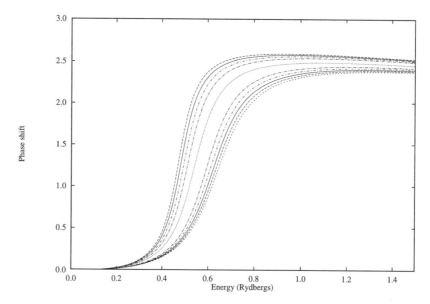

Phase shift

Energy (Rydbergs)

Fig. 11.4.   The $l = 2$ phase shifts for platinum. These are the two $l = 2$ phase shifts of figure 11.2 split by the magnetic field. This is the intermediate field regime; note that one phase shift has split off from the set with higher resonance energy and is midway between the two sets of phase shifts.

of intermediate strength. These are the two resonant phase shifts from figure 11.2 split into all components by a magnetic field, so all degeneracies are now removed. At low values of the field there are six phase shifts in the set at higher energy, and four in the set at lower energy, owing to spin–orbit splitting. At this intermediate field one of the higher set of phase shifts has split off and is moving towards the lower set as a function of spin-only field. Note that we are in the diagonal representation here, so no quantum numbers can be assigned to the individual phase shifts other than $l = 2$.

In figure 11.5 we plot the resonance energy (the energy at which the generalized phase shift is equal to $\pi/2$) against spin-only magnetic field. Magnetic field is in units of Rydbergs per μ_B. At low fields the splitting of the resonance energies is linear and the spin–orbit coupling dominates the magnetic splitting due to the spin-only field. At this end of the figure there are four $j = 3/2$ levels and six $j = 5/2$ levels slightly split by the field. Then we go through an intermediate regime (field equal to one on this figure is the field used as an intermediate field in figures 11.3, 11.4, and 11.6). In this region the splitting does not proceed linearly. At the high field end of this figure we have five spin-up levels and five spin-down levels each slightly split by spin–orbit coupling. Clearly at the low field end

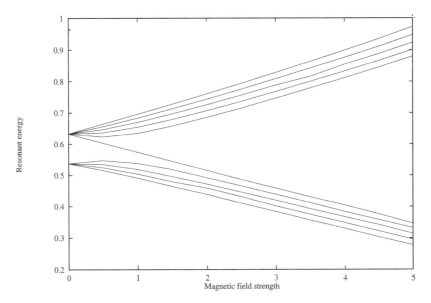

Fig. 11.5. The resonance energy of each of the phase shifts in figure 11.4 plotted as a function of field in atomic units. This is another method of calculating the Zeeman effect.

it is sensible to work in a representation where j and m_j are the quantum numbers, whereas at the high field end the appropriate quantum numbers should include spin. It is also clear that this is a scattering theory method of calculating the Zeeman effect. Furthermore, it is a method that does not involve any perturbation or expansion in small parameters, and so is valid in all field regimes, unlike the more traditional approaches to this topic (Eisberg and Resnick 1985 for example).

11.10 The Single-Site Scattering Green's Function

In the first sections of this chapter we discussed Green's functions, and yet when we discussed scattering theory the Green's functions were conspicuous by their absence. Therefore in this section we derive the single-site scattering Green's function. It is not just a sadistic desire on the part of the author to make you read this. The Green's function will be found to involve some of the scattering theory parameters we have been discussing. This section follows Strange *et al.* (1989*a*) which is a relativistic generalization of the derivation of a computationally convenient form of the Green's function by Faulkner and Stocks (1980). In section 11.17 we will show how to use the Green's function to calculate observables.

The single-site scattering Green's function $G_n(\mathbf{r}, \mathbf{r}', W)$ is written in terms of the free-particle Green's function in equation (11.29). The free-particle

Green's function is given by equation (11.43), which can be substituted into (11.29) to give

$$
G_n(\mathbf{r}, \mathbf{r}', W) = - ik \sum_{\kappa m_j} J_\kappa^{m_j}(\mathbf{r}) H_\kappa^{m_j\dagger}(\mathbf{r}')
$$

$$
- k^2 \sum_{\kappa' m_j'} \sum_{\kappa'' m_j''} H_{\kappa'}^{m_j'}(\mathbf{r}) t_{\kappa'\kappa''}^{m_j',m_j''}(W) H_{\kappa''}^{m_j''\dagger}(\mathbf{r}') \tag{11.91}
$$

under suitable conditions, and

$$
t_{\kappa\kappa'}^{m_j,m_j'}(W) = \int\int J_\kappa^{m_j\dagger}(\mathbf{r}'') t(\mathbf{r}'', \mathbf{r}_2, W) J_{\kappa'}^{m_j'}(\mathbf{r}_2) d\mathbf{r}'' d\mathbf{r}_2 \tag{11.92a}
$$

which is consistent with (11.49) and (11.67) provided

$$
t^{s's}(\mathbf{k}', \mathbf{k}) = 4\pi^2 \left(\frac{mc^2}{W}\right)^{1/2} \times
$$

$$
\sum_{\kappa' m_j'} \sum_{\kappa m_j} i^{l-l'} C(l'\tfrac{1}{2}j'; m', s') Y_{l'}^{m'}(\hat{\mathbf{k}}') t_{\kappa',\kappa}^{m_j' m_j}(W) Y_l^{m*}(\hat{\mathbf{k}}) C(l\tfrac{1}{2}j; m, s) \tag{11.92b}
$$

where $m' = m_j' - s'$ and $m = m_j - s$. We can define the inverse of this *t*-matrix such that

$$
\sum_{\kappa'' m_j''} \left(t_{\kappa,\kappa''}^{m_j,m_j''}\right)^{-1} t_{\kappa''\kappa'}^{m_j'' m_j'} = \delta_{\kappa\kappa'} \delta_{m_j' m_j} \tag{11.93}
$$

Substituting this into (11.91) yields

$$
G_n(\mathbf{r}, \mathbf{r}', W) = \frac{k}{i} \sum_{\kappa m_j} \sum_{\kappa_3 m_{j3}} \sum_{\kappa_4 m_{j4}} J_\kappa^{m_j}(\mathbf{r}) \left(t_{\kappa,\kappa_3}^{m_j,m_{j3}}\right)^{-1} t_{\kappa_3,\kappa_4}^{m_{j3} m_{j4}} H_{\kappa_4}^{m_{j4}\dagger}(\mathbf{r}')
$$

$$
- k^2 \sum_{\kappa' m_j'} \sum_{\kappa'' m_j''} H_{\kappa'}^{m_j'}(\mathbf{r}) t_{\kappa',\kappa''}^{m_j' m_j''} H_{\kappa''}^{m_j''\dagger}(\mathbf{r}') \tag{11.94}
$$

Now we are in a position to bring both terms in the Green's function together, a little rearrangement gives

$$
G_n(\mathbf{r}, \mathbf{r}', W) = -ik \sum_{\kappa' m_j'} \sum_{\kappa'' m_j''} Z_{\kappa'}^{m_j'}(\mathbf{r}) t_{\kappa',\kappa''}^{m_j' m_j''} H_{\kappa''}^{m_j''\dagger}(\mathbf{r}') \tag{11.95}
$$

with

$$
Z_{\kappa'}^{m_j'}(\mathbf{r}) = \sum_{\kappa m_j} J_\kappa^{m_j}(\mathbf{r}) \left(t_{\kappa,\kappa'}^{m_j,m_j'}\right)^{-1} - ik H_{\kappa'}^{m_j'}(\mathbf{r}) \tag{11.96}
$$

Compare equation (11.96) with equation (11.62): they are identical except for a few constants and a multiplicative factor. So $Z_{\kappa'}^{m_j'}(\mathbf{r})$ is clearly a solution of the free-particle Dirac equation outside the potential which matches smoothly onto the solutions inside the potential. Indeed, inside the spherically symmetric potential $Z_\kappa^{m_j}(\mathbf{r})$ is defined as the solution of the Kohn–Sham–Dirac equation. So it is then defined over all space. The expression (11.95) for the Green's function is valid only for $r' > r$. We can rearrange (11.96) and substitute back into equation (11.95) for $H_{\kappa_1}^{m_{j1}\dagger}(\mathbf{r}')$. A little further algebra and use of equation (11.93) leads us to the result we want

$$G_n(\mathbf{r}, \mathbf{r}', W) = \sum_{\kappa'm_j' \, \kappa''m_j''} Z_{\kappa'}^{m_j'}(\mathbf{r}) t_{\kappa',\kappa''}^{m_j'm_j''}(W) Z_{\kappa''}^{m_j''\dagger}(\mathbf{r}')$$

$$- \sum_{\kappa m_j} Z_\kappa^{m_j}(\mathbf{r}) J_\kappa^{m_j\dagger}(\mathbf{r}') \tag{11.97}$$

One can interpret $Z_\kappa^{m_j}(\mathbf{r})$ as being a superposition of an outgoing scattered wave with amplitude normalized to $H_\kappa^{m_j}(\mathbf{r})$ plus a sum of contributions from all possible incident waves $J_\kappa^{m_j}(\mathbf{r})$ with an amplitude determined by the appropriate element of the t-matrix.

All the quantities in equation (11.97) can easily be determined, but it is only one of many possible expressions for the Green's function. This formulation has some convenient features. In particular we write it in this form because it has a natural generalization to an arbitrary array of scatterers, as we shall see later. Also this form is computationally tractable and convenient for going further and calculating observable quantities.

11.11 Transforming Between Representations

Following up our earlier comments about representations, we have now written the t-matrices in two different representations. The (l, m, s) and (κ, m_j) representations are used in equations (11.49) and (11.92) respectively. The t-matrices in the two representations are connected by the following two transformations

$$t_{\kappa',\kappa}^{m_j}(W) = \frac{W}{mc^2} \sum_{m_s, m_s'} C(l'\tfrac{1}{2}j'; m_j - m_s', m_s') \times$$

$$t_{l',m_j-m_s',lm_j-m_s}^{m_s'm_s}(W) C(l\tfrac{1}{2}j; m_j - m_s, m_s) \tag{11.98a}$$

$$t_{l,m_j-m_s,l',m_j'-m_s'}^{m_s'm_s}(W) =$$

$$\frac{mc^2}{W} \sum_{\kappa\kappa'} C(l\tfrac{1}{2}jm_j - m_s, m_s) t_{\kappa,\kappa'}^{m_j}(W) C(l'\tfrac{1}{2}j'; m_j' - m_s', m_s') \tag{11.98b}$$

The energy-dependent factor outside the summation comes because of the different numerical factors outside the definitions of the t-matrices in the

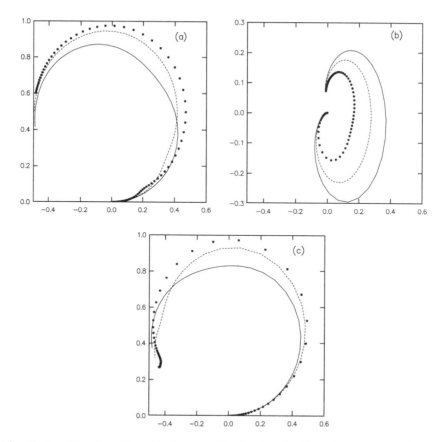

Fig. 11.6. The $l = 2$ scattering amplitudes in the (l, m, s) representation for platinum as a function of energy and magnetic field. In each case the vertical axis is the imaginary part of the scattering amplitude and the horizontal axis is the real part. Magnetic fields are the same as in figure 11.3. The zero of energy is at the origin and energy increases as you proceed round the scattering amplitude anticlockwise. All figures are for $m_j = 1/2$. (a) $l = 2, m'_s = 1/2$, (b) $l = 2$, $m_s = 1/2, m'_s = -1/2$, (c) $l = 2, m'_s = -1/2$.

different representations, and these arise because we normalize spherical free particle wavefunctions differently in the four-component relativistic theory and the single-component non-relativistic theory. This numerical factor is unity in the non-relativistic limit and it is a good approximation to treat it as unity for typical energies of electrons in solids.

As an illustration of the transformations (11.98) we show the scattering amplitudes in the (l, m, s) representation in figure 11.6 as a function of increasing field and energy. These can be compared with figure 11.3. Field increases in the opposite direction in this representation. Because of spin–orbit coupling there is spin-flip scattering even in zero field. In the non-magnetic limit $t_{\kappa\kappa'}^{m_j}(W)$ becomes independent of m_j and diagonal in κ, i.e. it reduces to the t-matrix of the previous section. On the other hand $t_{l,m,l',m'}^{m_s m_s'}(W)$ has off-diagonal elements which take on a maximum value in zero field, and gradually decrease as the field is increased. Figures 11.6a and 11.6c approach the unitarity circle as the field tends to infinity. In the non-relativistic limit the off-diagonal terms become zero and the scattering amplitudes become independent of m_l but not of m_s, i.e.

$$\lim_{c\to\infty} t_{\kappa,\kappa'}^{m_j}(W) = t_\kappa(W)\delta_{\kappa,\kappa'} \qquad (11.99a)$$

$$\lim_{c\to\infty} t_{l,m,l',m'}^{m_s m_s'}(W) = t_{l,m_s,m_s'}\delta_{l,l'}\delta_{m,m'} \qquad (11.99b)$$

We can get to the diagonal representation from the (l,m,s) representation by defining a matrix like equation (11.90) but with a different mixing angle. Of course, the diagonal representation is independent of which representation we start from to get to it (there wouldn't be much point in defining it, if it wasn't). We have now covered all the essential topics associated with scattering of Dirac particles from a spin-polarized target. For further details the reader is referred to the original literature (Strange *et al.* 1984, 1989a).

11.12 The Scattering Path Operator

So far in this chapter we have worked in the coordinate $(\mathbf{r}, \mathbf{r}')$, momentum $(\mathbf{k}, \mathbf{k}')$ and angular momentum (l, m, s) or (κ, m_j) representations. In this section, though, we will consider the operator equivalents of some of the foregoing equations and derive from them an expression for the scattering path operator, a quantity whose role in multiple scattering is similar to that played by the t-matrix in single-site scattering. We are using the operator equations for two reasons. Firstly, they are more general than a specific representation, so are valid in both direct and reciprocal space for example. Secondly, this approach reduces equations to their most basic form and makes equations appear as simple as possible. To get back to a particular representation, one just inserts the relevant bras and kets, e.g. $t(\mathbf{r}, \mathbf{r}', E) = \langle \mathbf{r}|t(E)|\mathbf{r}'\rangle$, and integrates over repeated variables.

We are now considering scattering from more than one site, i.e. multiple scattering. The potential seen by the incident particle is now a superposition of non-overlapping spherical potentials known as muffin tin potentials. This is the origin of the subscript on r_m in the discussion of the range of the potentials at the beginning of section 11.7. They get this name because a two-dimensional square lattice of these potentials looks like a muffin tin as seen in figure 9.9b – how imaginative! Anyway we write our total potential as

$$V(\mathbf{r}) = \sum_i V_i(\mathbf{r} - \mathbf{R}_i) \tag{11.100}$$

where $\mathbf{R}_i$ represents the positions of the scattering centres. Now we represent the T-matrix for multiple scattering with a capital T to distinguish it from the single-site t-matrix. Let us write the operator equivalent of equation (11.30) for T (Weinberger 1990)

$$T(E) = V + VG_0T(E) = \sum_i (V_i + V_iG_0T(E)) = \sum_i P_i(E) \tag{11.101}$$

where

$$P_i(E) = (V_i + V_iG_0(E)T(E)) = V_i + V_iG_0(E)\sum_j P_j(E)$$

$$= V_i + V_iG_0(E)P_i(E) + \sum_{j \neq i} V_iG_0(E)P_j(E) \tag{11.102}$$

A little rearrangement gives

$$P_i(E) = (1 - V_iG_0(E))^{-1}V_i(1 + \sum_{j \neq i} G_0(E)P_j(E)) \tag{11.103}$$

Next consider the single-site version of (11.30) with a single potential centred at $\mathbf{R}_i$. In operator form we can write $t^i(E)$ in two ways

$$t^i(E) = V_i + V_iG_0(E)t^i(E) \tag{11.104a}$$

$$t^i(E) = (1 - V_iG_0(E))^{-1}V_i \tag{11.104b}$$

This can be substituted into (11.103) to give

$$P^i(E) = t^i(E) + \sum_{j \neq i} t^i(E)G_0(E)P^j(E) \tag{11.105}$$

Let us introduce a new set of quantities

$$\tau^{ij}(E) = t^i(E)\delta_{ij} + \sum_{k \neq i} t^i(E)G_0(E)\tau^{kj}(E) \tag{11.106}$$

If we let $Q^i \equiv \sum_j \tau^{ij}$ and then expand (11.105) and Q^i as a Born series, the sums in both cases are identical, which tells us that $Q^i = P^i$ (convince yourself of this). This leads to

$$T(E) = \sum_i P^i(E) = \sum_{ij} \tau^{ij}(E) \tag{11.107}$$

The $\tau^{ij}(E)$ are known as scattering path operators and were first introduced by Györffy and Stott (1972). Although they have been derived in a fairly

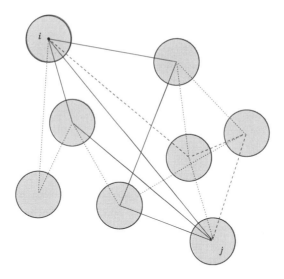

Fig. 11.7. Figure illustrating the operation of the scattering path operator $\tau^{ij}(\mathbf{r}_i, \mathbf{r}_j, W)$. The figure shows several possible routes for scattering from site i to site j, which are among those included in the work of the scattering path operator.

pure mathematical way they do have an interpretation within scattering theory. They are useful because they give us a way of writing down the solution to a multiple scattering problem in terms of the solution to the single-site scattering problem. We have already seen that the single-site t-matrix generates the scattered wave due to a single potential. In multiple scattering $T(E)$ gives us the scattered wave due to an array of potentials. The scattering path operator itself, τ^{ij} (figure 11.7), gives us the scattered wave from site j due to a wave incident upon site i with all possible scatterings in between (Györffy and Stocks 1979). Equation (11.106) can be expanded in a Born series:

$$\tau^{ij}(E) = t^i(E)\delta_{ij} + \sum_{k \neq i} t^i(E)G_0(E)t^k(E)\delta_{kj}$$

$$+ \sum_{k \neq i}\sum_{l \neq k} t^i(E)G_0(E)t^k(E)G_0(E)t^l(E)\delta_{lj} + \cdots \qquad (11.108)$$

This equation shows explicitly the meaning of τ^{ij}. In each term in equation (11.108) the rightmost site operator $(t^j(E))$ defines where the scattering starts and the leftmost operator $(t^i(E))$ defines the site from which the outgoing wave emanates. For example, the third term represents a wave incident at site j which scatters to site k via the propagator $G_0(E)$ and is scattered from there by $t^k(E)$ to site i where it is scattered again.

This term represents scattering involving three particular sites, and this is summed over all possible sets of three sites where the scattering begins at site j and ends at site i, so all three-site scattering is described by this term. All possible scattering processes are represented by $\sum_{i,j} \tau^{ij}$. This quantity takes the incoming waves to all scattering sites and turns them into the outgoing waves from all scattering sites. By definition this is what $T(E)$ does and hence we have the equality (11.107).

11.13 The Non-Relativistic Free-Particle Green's Function

In this section we consider the non-relativistic free-particle Green's function. The reason for this is that relativistic multiple scattering theory can be written most easily in terms of the scattering path operators of the previous section and a suitably transformed version of the partial wave expansion of the non-relativistic free particle Green's function.

Consider equation (11.11), which was derived with the assumption that the system under investigation has a discrete energy spectrum. If the energy spectrum is continuous the summation becomes an integral, and we will work with the retarded Green's function in the limit $\eta \to 0$. Substituting in the non-relativistic free particle wavefunctions

$$G_0^{NR}(\mathbf{r}_i + \mathbf{R}_i, \mathbf{r}_j + \mathbf{R}_j, E) = \frac{1}{8\pi^3} \int \frac{e^{i\mathbf{k}\cdot(\mathbf{r}_i - \mathbf{r}_j + \mathbf{R}_i - \mathbf{R}_j)}}{E - E(\mathbf{k}) + i\eta} d\mathbf{k} \quad i \neq j \quad (11.109)$$

Here $\mathbf{r}_i$ is within the muffin tin radius of the scatterer at $\mathbf{R}_i$, and $\mathbf{r}_j$ is within the muffin tin radius of the scatterer at $\mathbf{R}_j$. For free particles

$$E(\mathbf{k}) = \frac{\hbar^2 k^2}{2m} \quad (11.110)$$

Henceforth in this section only we will drop the $\mathbf{r}_i + \mathbf{R}_i, \mathbf{r}_j + \mathbf{R}_j, E$ in the brackets and just call the Green's function G_0. This is simply to accommodate the following mathematics on the page. We can substitute the usual partial wave expansion of a complex exponential (equation (C.21)) in equation (11.109) to give some absolutely horrendous equations – don't panic, they simplify soon:

$$G_0^{NR} = 8 \int_0^\infty k^2 dk \sum_{lm} \sum_{l'm'} \sum_{l''m''} i^{l-l'+l''} C_{lml'm'}^{l''m''}$$

$$\frac{j_l(kr_i) j_{l'}(kr_j) j_{l''}(k|\mathbf{R}_i - \mathbf{R}_j|)}{E - E(\mathbf{k}) + i\eta} Y_l^m(\hat{\mathbf{r}}_i) Y_{l'}^{m*}(\hat{\mathbf{r}}_j) Y_{l''}^{m''}(\widehat{\mathbf{R}_i - \mathbf{R}_j})$$

$$(11.111)$$

where the $C_{lml'm'}^{l''m''}$ are known as the Gaunt coefficients, which are integrals over spherical harmonics and are given explicitly by equation (C.19). They have the property that they are zero unless $l + l' + l''$ is even. This is useful because if we combine it with the well-known property of the spherical

Bessel functions $j_l(-x) = (-1)^l j_l(x)$ we see that the integrand in (11.111) is even in k. Therefore we have

$$G_0^{NR} = -\sum_{lm} \sum_{l'm'} \sum_{l''m''} i^{l-l'+l''} C_{lml'm'}^{l''m''} Y_l^m(\hat{\mathbf{r}}_i) Y_{l'}^{m'*}(\hat{\mathbf{r}}_j) Y_{l''}^{m''}(\widehat{\mathbf{R}_i - \mathbf{R}_j})$$
$$\times \frac{8m}{\hbar^2} \int_{-\infty}^{\infty} k^2 dk \frac{j_l(kr_i) j_{l'}(kr_j) j_{l''}(k|\mathbf{R}_i - \mathbf{R}_j|)}{(k-q-i\eta')(k+q+i\eta')}$$

(11.112)

where $q^2 = 2mE/\hbar^2$. We can make this step because η is infinitesimal and is going to be set equal to zero at the end of the calculation. Clearly the spherical harmonics have no k dependence and can be taken outside the integral. Furthermore recall that $\mathbf{r}_i$ and $\mathbf{r}_j$ are defined as being inside their respective muffin tin spheres. Therefore the asymptotic behaviour of (11.112) is entirely dependent upon $j_{l''}(k|\mathbf{R}_i - \mathbf{R}_j|)$. Recall also the definition of the spherical Hankel functions

$$h_l^{(1)}(x) = j_l(x) + in_l(x)$$

(11.113a)

$$h_l^{(2)}(x) = j_l(x) - in_l(x)$$

(11.113b)

So, in equation (11.112), let us substitute

$$j_{l''}(k|\mathbf{R}_i - \mathbf{R}_j|) = \frac{1}{2}(h_{l''}^{(1)}(k|\mathbf{R}_i - \mathbf{R}_j|) + h_{l''}^{(2)}(k|\mathbf{R}_i - \mathbf{R}_j|))$$

(11.114)

So

$$G_0^{NR} = -\sum_{lm} \sum_{l'm'} \sum_{l''m''} i^{l-l'+l''} C_{lml'm'}^{l''m''} Y_l^m(\hat{\mathbf{r}}_i) Y_{l'}^{m'*}(\hat{\mathbf{r}}_j) Y_{l''}^{m''}(\widehat{\mathbf{R}_i - \mathbf{R}_j}) \times$$
$$\frac{4m}{\hbar^2} \int_{-\infty}^{\infty} k^2 dk \frac{j_l(kr_i) j_{l'}(kr_j)(h_{l''}^{(1)}(k|\mathbf{R}_i - \mathbf{R}_j|) + h_{l''}^{(2)}(k|\mathbf{R}_i - \mathbf{R}_j|))}{(k-q-i\eta')(k+q+i\eta')}$$

(11.115)

Now the asymptotic behaviour of the Hankel functions is

$$\lim_{x\to\infty} h^{(1)}(x) = (-i)^{l+1}\frac{e^{ix}}{x}, \qquad \lim_{x\to\infty} h^{(2)}(x) = (i)^{l+1}\frac{e^{-ix}}{x}$$

(11.116)

Substituting these into the integral in (11.115) allows us to evaluate it as a contour integral with the help of Jordan's lemma. The integral with $h^{(1)}$ is done in the upper half plane and the integral with $h^{(2)}$ is done in the lower half plane.

$$G_0^{NR} = -\sum_{lm} \sum_{l'm'} \sum_{l''m''} i^{l-l'+l''} C_{lml'm'}^{l''m''} Y_l^m(\hat{\mathbf{r}}_i) Y_{l'}^{m'*}(\hat{\mathbf{r}}_j) Y_{l''}^{m''}(\widehat{\mathbf{R}_i - \mathbf{R}_j})$$
$$\times \frac{4m}{\hbar^2} 2\pi i q j_l(qr_i) j_{l'}(qr_j) h_{l''}^{(1)}(q|\mathbf{R}_i - \mathbf{R}_j|)$$

(11.117)

This enables us to write

$$G_0^{NR}(\mathbf{r}_i + \mathbf{R}_i, \mathbf{r}_j + \mathbf{R}_j, E) =$$

$$\frac{2m}{\hbar^2} \sum_{lm} \sum_{l'm'} Y_l^m(\hat{\mathbf{r}}_i) j_l(qr_i) G_{0\ lml'm'}(\mathbf{R}_i - \mathbf{R}_j, E) j_{l'}(qr_j) Y_{l'}^{m'*}(\hat{\mathbf{r}}_j) \quad (11.118)$$

in analogy with (11.49c), where

$$G_{0\ lml'm'}(\mathbf{R}_i - \mathbf{R}_j, E) =$$

$$- (1 - \delta_{\mathbf{R}_i,\mathbf{R}_j}) 4\pi i q \sum_{l''m''} i^{l-l'+l''} C_{lml'm'}^{l''m''} h_{l''}^{(1)}(q|\mathbf{R}_i - \mathbf{R}_j|) Y_{l''}^{m''}(\widehat{\mathbf{R}_i - \mathbf{R}_j}) \quad (11.119)$$

Note that $G_{0\ lml'm'}(\mathbf{R}_i - \mathbf{R}_j, E)$ can be regarded as one element of a matrix whose elements are defined by their values of i, j, l, m, l', m'. It is important to bear in mind that (11.118) is valid only for $\mathbf{R}_i \neq \mathbf{R}_j$ as it is, by definition, a two-site Green's function. That is why the δ-function term in brackets is included in (11.119)

Finally in this section we show how the non-relativistic free particle Green's function is related to its relativistic counterpart. Consider equation (11.33). This is the equation for the Schrödinger Green's function give or take a few constants. If we put these constants in we have

$$G_0^{NR}(\mathbf{r}, \mathbf{r}', E) = -\frac{2m}{\hbar^2} \frac{e^{iqR}}{4\pi R} \quad (11.120)$$

We can substitute this into (11.32) to see that the relativistic free particle Green's function is given by

$$G_0(\mathbf{r}, \mathbf{r}', W) = \frac{1}{2mc^2}(c\tilde{\boldsymbol{\alpha}} \cdot \hat{\mathbf{p}} + \tilde{\beta}mc^2 + W) G_0^{NR}(\mathbf{r}, \mathbf{r}', E) \quad (11.121a)$$

Here we see very transparently the relation between the relativistic and non-relativistic Green's functions. In the non-relativistic limit the $c\tilde{\boldsymbol{\alpha}} \cdot \hat{\mathbf{p}}$ term is negligible compared to the rest mass term. Furthermore, in this limit $W \approx E + mc^2 \approx mc^2$ so the consistency of this equation can be seen. Also note that for the negative energy solutions in the non-relativistic limit $\tilde{\beta}mc^2$ cancels W and hence these solutions are not allowed. Equation (11.121a) can be written in matrix form as

$$G_0(\mathbf{r}, \mathbf{r}', W) = \frac{1}{2mc^2} \begin{pmatrix} W + mc^2 & c\tilde{\boldsymbol{\sigma}} \cdot \hat{\mathbf{p}} \\ c\tilde{\boldsymbol{\sigma}} \cdot \hat{\mathbf{p}} & W - mc^2 \end{pmatrix} G_0^{NR}(\mathbf{r}, \mathbf{r}', E)$$

$$= \frac{1}{2mc^2} \begin{pmatrix} W + mc^2 & -c\tilde{\boldsymbol{\sigma}} \cdot \hat{\mathbf{p}}' \\ c\tilde{\boldsymbol{\sigma}} \cdot \hat{\mathbf{p}} & -c^2 \frac{\tilde{\boldsymbol{\sigma}} \cdot \hat{\mathbf{p}} \tilde{\boldsymbol{\sigma}} \cdot \hat{\mathbf{p}}'}{(W + mc^2)} \end{pmatrix} G_0^{NR}(\mathbf{r}, \mathbf{r}', E) \quad (11.121b)$$

where each element of the matrix is itself a 2×2 matrix. In the final step here $\hat{\mathbf{p}}'$ acts only on the $\mathbf{r}'$ coordinate of the Green's function and we have used equations (1.19) and (4.19).

11.14 Multiple Scattering Theory

Now that we have seen how to solve the Dirac–Kohn–Sham equation for scattering of an electron from a single site, we will move on to a solution for the case when there are many scattering centres present. Multiple scattering is illustrated in figure 11.7. The scattering path operator allows us to write the solution to the Dirac–Kohn–Sham equations for this case in terms of the single-site scattering t-matrix. In this section we derive an explicit method of carrying out this solution.

Firstly let us put the operator equation (11.106) in the coordinate representation. It is easy to do this, we insert normalized (undefined) wavefunctions between each operator and integrate over the positional dependence

$$\tau^{ij}(\mathbf{r}_i, \mathbf{r}_j, W) = t^i(\mathbf{r}_i, \mathbf{r}_j, W)\delta_{ij} +$$
$$\sum_{k \neq i} \int \int t^i(\mathbf{r}_i, \mathbf{r}'_i, W)G_0(\mathbf{r}'_i + \mathbf{R}'_i, \mathbf{r}_k + \mathbf{R}_k, W)\tau^{kj}(\mathbf{r}_k, \mathbf{r}_j, W)d\mathbf{r}_k d\mathbf{r}'_i \quad (11.122)$$

The first term here takes care of the case when the ith and j^{th} sites are the same, and the second term represents the case when they are different with all possible intermediate states.

Next we need some definitions. In equation (11.49c) we made an angular momentum decomposition of the t-matrix. Now we define an angular momentum decomposition of the scattering path operator analogously

$$\tau^{ij}_{l'm's'lms}(W) = \frac{2m}{\hbar^2} \times$$
$$\int \int j_{l'}(k'r_i)Y_{l'}^{m'*}(\hat{\mathbf{r}}_i)U^{s'\dagger}(\mathbf{k}')\tau^{ij}(\mathbf{r}_i, \mathbf{r}_j, W)U^s(\mathbf{k})Y_l^m(\hat{\mathbf{r}}_j)j_l(kr_j)d\mathbf{r}_i d\mathbf{r}_j \quad (11.123)$$

In analogy with (11.92) we also define

$$\tau^{ij}_{\Lambda\Lambda'}(W) = \int \int J_\Lambda^\dagger(\mathbf{r}_i)\tau^{ij}(\mathbf{r}_i, \mathbf{r}_j, W)J_{\Lambda'}(\mathbf{r}_j)d\mathbf{r}_i d\mathbf{r}_j \quad (11.124)$$

where Λ is shorthand for κ, m_j. Now we premultiply equation (11.122) by $J_\Lambda^\dagger(\mathbf{r}_i)$ and postmultiply it by $J_{\Lambda'}(\mathbf{r}_j)$. Then we integrate over $\mathbf{r}_i$ and $\mathbf{r}_j$, which finally gives

$$\tau^{ij}_{\Lambda\Lambda'}(W)$$
$$= t^i_{\Lambda\Lambda'}(W)\delta_{ij} + \sum_{k \neq i}\sum_{\Lambda_1\Lambda_2} t^i_{\Lambda\Lambda_1}(W)G_{0\,\Lambda_1\Lambda_2}(\mathbf{R}_i - \mathbf{R}_k, W)\tau^{kj}_{\Lambda_2\Lambda'}(W)$$
$$= t^i_{\Lambda\Lambda'}(W)\delta_{ij} + \sum_{k \neq j}\sum_{\Lambda_1\Lambda_2} \tau^{ik}_{\Lambda\Lambda_1}(W)G_{0\,\Lambda_1\Lambda_2}(\mathbf{R}_k - \mathbf{R}_j, W)t^j_{\Lambda_2\Lambda'}(W)$$

$$(11.125)$$

Note that this procedure is independent of the normalization of the $J_\Lambda(\mathbf{r})$s. The last equality in (11.125) illustrates the fact that the whole theory can be written with the operators defined in reverse order. We have defined

$$G_0(\mathbf{r}_i + \mathbf{R}_i, \mathbf{r}_j + \mathbf{R}_j, W) = \sum_{\Lambda\Lambda'} J_\Lambda(\mathbf{r}_i) G_{0\,\Lambda\Lambda'}(\mathbf{R}_i - \mathbf{R}_j, W) J_{\Lambda'}^\dagger(\mathbf{r}_j) \quad (11.126)$$

In exactly the same way as (11.92) is equation (11.49c) written in the (κ, m_j) representation, so equation (11.126) is the (κ, m_j) version of (11.118). Equation (11.125) is the fundamental equation of multiple scattering. The fact that we have been able to decouple the *on-the-energy-shell* part of the problem from the *off-the-energy-shell* part has enabled us to write this equation. Note that we have a further decoupling also. In equation (11.125) $G_{0\,\kappa m_j\kappa'm_j'}(\mathbf{R}_k - \mathbf{R}_j, W)$ depends only upon the arrangement of scatterers, it has no dependence on the potential within the muffin tin sphere. The other quantity in (11.125) is the single-site t-matrix which depends only on the potential at a single site through the phase shift and the logarithmic derivatives of equations (11.61) and (11.64a) and has no dependence on the positions of the scatterers. When there is no spin polarization the single-site t-matrices in (11.125) are diagonal in the quantum numbers, so the summation over Λ_1 disappears. The derivation of (11.125) has not depended on relativity in any fundamental way. It can be written in any representation and is equally valid in non-relativistic scattering theory. Indeed non-relativistic derivations of this equation are available (Györffy and Stocks 1979, Stocks and Winter 1984 for example). Equation (11.125) is valid for an arbitrary arrangement of scattering centres, so it is a good starting point for discussion of scattering in pure metals, random alloys, atomic clusters, amorphous materials, liquids, or indeed anything else. The only limitation is that the potentials around the scattering centres must not overlap. The reason for this is obvious. We have matched the spherical solutions within r_m onto free-particle spherical waves outside r_m. If outside one muffin tin sphere is already inside another, the plane wave solutions would be invalid.

Next we show that the separation (11.126) of $G_0(\mathbf{r}_i + \mathbf{R}_i, \mathbf{r}_j + \mathbf{R}_j, W)$ is valid. We do this by evaluating explicit expressions for the Green's function from equation (11.121b) and showing that (11.126) gives the same result. To facilitate this procedure we need to push a couple of earlier equations a little bit further. Firstly, we write equation (11.118) in the (κ, m_j) representation

$$G_0^{NR}(\mathbf{r}_i + \mathbf{R}_i, \mathbf{r}_j + \mathbf{R}_j, E) =$$
$$\frac{2m}{\hbar^2} \sum_{\kappa, m_j} \sum_{\kappa', m_j'} j_l(qr_i) \chi_\kappa^{m_j}(\hat{\mathbf{r}}_i) G_{0\,\kappa m_j\kappa'm_j'}(\mathbf{R}_i - \mathbf{R}_j, E) \chi_{\kappa'}^{m_j'}(\hat{\mathbf{r}}_j) j_{l'}(qr_j) \quad (11.127a)$$

with

$$G_{0\ \kappa m_j \kappa' m_j'}(\mathbf{R}_i - \mathbf{R}_j, W) =$$

$$\sum_{s=\pm 1/2} C(l\tfrac{1}{2}j; m_j - ss) G_{0\ lm_j-sl'm_j'-s}(\mathbf{R}_i - \mathbf{R}_j, W) C(l'\tfrac{1}{2}j'; m_j' - ss)$$

$$(11.127b)$$

$G_{0\ lml'm'}(\mathbf{R}_i - \mathbf{R}_j, W)$ are the structure constants of (11.119) and equation (11.127b) tells us how to transform them into the relativistic representation. Secondly, we can write equation (11.67) in four-component form. Then comparison of the lower two components with the upper two yields

$$\tilde{\boldsymbol{\sigma}} \cdot \hat{\mathbf{p}} j_l(qr) \chi_\kappa^{m_j}(\hat{\mathbf{r}}) = i p S_\kappa j_{\bar{l}}(qr) \chi_{-\kappa}^{m_j}(\hat{\mathbf{r}}) \qquad (11.128)$$

The Green's function is a 4×4 quantity, and we label its quadrants as shown in equation (11.34). We will compare (11.121b) with (11.126) quadrant by quadrant. For the upper left quadrant, putting (11.127a) into (11.121b) gives

$$[G_0(\mathbf{r}_i + \mathbf{R}_i, \mathbf{r}_j + \mathbf{R}_j, W)]_{11} = \frac{(W + mc^2)}{\hbar^2 c^2} \times$$

$$\sum_{\kappa m_j} \sum_{\kappa' m_j'} \chi_\kappa^{m_j}(\hat{\mathbf{r}}_i) j_l(qr_i) G_{0\kappa m_j \kappa' m_j'}(\mathbf{R}_i - \mathbf{R}_j, W) j_{l'}(qr_j) \chi_{\kappa'}^{m_j'\dagger}(\hat{\mathbf{r}}_j) \qquad (11.129a)$$

Now, multiplying out equation (11.126) using the definition (11.39) also gives this formula, so the upper left quadrant of (11.126) works out OK. The lower left and upper right quadrants give us $\tilde{\boldsymbol{\sigma}} \cdot \hat{\mathbf{p}}$ or $\tilde{\boldsymbol{\sigma}} \cdot \hat{\mathbf{p}}'$ operating on (11.127a). This can be simplified using equation (11.128) to give

$$[G_0(\mathbf{r}_i + \mathbf{R}_i, \mathbf{r}_j + \mathbf{R}_j, W)]_{21} = -[G_0(\mathbf{r}_i + \mathbf{R}_i, \mathbf{r}_j + \mathbf{R}_j, W)]_{12} =$$

$$\frac{iq}{\hbar c} \sum_{\kappa m_j} \sum_{\kappa' m_j'} S_{\kappa'} \chi_\kappa^{m_j}(\hat{\mathbf{r}}_i) j_l(qr_i) G_{0\kappa m_j \kappa' m_j'}(\mathbf{R}_i - \mathbf{R}_j, W) j_{\bar{l}'}(qr_j) \chi_{-\kappa'}^{m_j'\dagger}(\hat{\mathbf{r}}_j) \qquad (11.129b)$$

Again, this is what we get when we multiply out equation (11.126). Finally, if we insert (11.127a) into equation (11.121b) and consider the lower right quadrant we have both $\tilde{\boldsymbol{\sigma}} \cdot \hat{\mathbf{p}}$ and $\tilde{\boldsymbol{\sigma}} \cdot \hat{\mathbf{p}}'$ acting on the non-relativistic Green's function and this yields

$$[G_0(\mathbf{r}_i + \mathbf{R}_i, \mathbf{r}_j + \mathbf{R}_j, W)]_{22} = \frac{(W - mc^2)}{\hbar^2 c^2} \times$$

$$\sum_{\kappa m_j} \sum_{\kappa' m_j'} S_\kappa \chi_{-\kappa}^{m_j}(\hat{\mathbf{r}}_i) j_{\bar{l}}(qr_i) G_{0\kappa m_j \kappa' m_j'}(\mathbf{R}_i - \mathbf{R}_j, W) S_{\kappa'} j_{\bar{l}'}(qr_j) \chi_{-\kappa'}^{m_j'\dagger}(\hat{\mathbf{r}}_j) \qquad (11.129c)$$

This, too, is in agreement with equation (11.126). We have now confirmed that all quadrants of the two definitions of the Green's function (11.121b)

and (11.126) correspond with each other, and this proves that the definition (11.126) is a correct one.

It is clear that equation (11.125) can be written more succinctly as a matrix equation where each element is defined by its $\kappa, m_j, \kappa', m'_j$ values. Let us write (11.125) this way and premultiply it by $\tilde{t}^{i\,-1}(W)$. This gives us

$$\tilde{t}^{i\,-1}(W)\tilde{\tau}^{ij}(W) = \tilde{I}\delta_{ij} + \sum_{k\neq i}\tilde{G}^{ik}(W)\tilde{\tau}^{kj}(W) \tag{11.130a}$$

or, since $\tilde{t}^{i\,-1}(W)\tilde{\tau}^{ij}(W) = \sum_k \tilde{t}^{k\,-1}(W)\tilde{\tau}^{kj}(W)\delta_{ik}$ and $\tilde{G}^{ii}(W) = 0$, we can write

$$\sum_k (\tilde{t}^{k\,-1}\delta_{ik} - \tilde{G}^{ik})\tau^{kj}(W) = \tilde{I}\delta_{ij} \tag{11.130b}$$

This notation clearly suggests creating a matrix out of these matrices where each element of this supermatrix is defined by its site indices, using the obvious notation

$$\tilde{\tau} = (\tilde{t}^{-1} - \tilde{G})^{-1} \tag{11.131}$$

where each element of the supermatrix is

$$\tilde{\tau}^{ij} = (\tilde{t}^{-1} - \tilde{G})^{-1}_{ij} \tag{11.132}$$

If we have N scattering centres and the angular momentum expansions go up to $l^{\max}$, the order of the matrix $\tilde{\tau}$ is $2N\sum_{l=0}^{l^{\max}}(2l+1)$.

Now that we have written down the fundamental equations of multiple scattering theory, the next thing to do is to show how to solve them.

Luckily, solving these equations is fairly easy for scattering in an infinite regular periodic lattice. We define the lattice Fourier transforms

$$\tau_{\Lambda\Lambda'}(\mathbf{q}, W) = \frac{1}{N}\sum_{i,j}e^{-i\mathbf{q}\cdot(\mathbf{R}_i - \mathbf{R}_j)}\tau^{ij}_{\Lambda\Lambda'}(W) \tag{11.133a}$$

$$G_{\Lambda\Lambda'}(\mathbf{q}, W) = \frac{1}{N}\sum_{i,j}e^{-i\mathbf{q}\cdot(\mathbf{R}_i - \mathbf{R}_j)}G_{0\,\Lambda\Lambda'}(\mathbf{R}_i - \mathbf{R}_j, W) \tag{11.133b}$$

If we assume the t-matrix is the same at every lattice site (i.e. we are treating an element) we find that

$$[\tilde{t}^{-1}(W) - \tilde{G}(\mathbf{q}, W)]\tau(\mathbf{q}, W) = \tilde{I} \tag{11.134a}$$

using the obvious matrix notation. The solution to (11.134a) is

$$\tau(\mathbf{q}, W) = [\tilde{t}^{-1}(W) - \tilde{G}(\mathbf{q}, W)]^{-1} \tag{11.134b}$$

Clearly the scattering path operator in (11.134b) will be singular when

$$|\tilde{t}^{-1}(W) - \tilde{G}(\mathbf{q}, W)| = 0 \tag{11.135}$$

This is known as the KKR determinant, after its discoverers Korringa (1947) and Kohn and Rostoker (1954). The relativistic KKR version of KKR theory was developed by Onodera and Okazaki (1966), and a particularly useful discussion of relativistic multiple scattering theory is given by Shen (1974). In practice it is usual to subtract mc^2 from all energies to facilitate comparison with non-relativistic KKR theory. This is allowed because the origin of energy is arbitrary (provided we don't overlap negative energy states). When equation (11.135) is satisfied, the scattering path operator has a singularity, which corresponds to a pole in the Green's function, and poles in the Green's function occur at the same energies as the eigenvalues of the Hamiltonian as shown by equation (11.11). It follows that, if the Hamiltonian is that of relativistic density functional theory, we now have a way of finding its eigenvalues for a crystalline material. Hence, finding the zeros of the KKR determinant is a method of finding the electronic energy levels or band structure of the element in question. A different, but equivalent, interpretation of these equations is to regard the scattering path operator as a response function. If we feed it the incident wave it tells us the scattered wave. Hence where $\tilde{\tau}$ diverges there may be a scattered wave even in the absence of an incident wave.

It is a practically viable project to solve equation (11.135). To do it we require the *t*-matrices which can be evaluated for any value of the quantum numbers using (11.64*a*). The lattice part of the calculation is given by equations (11.119), (11.127*b*) and (11.133*b*). The summation over lattice vectors in (11.133*b*) does not converge easily and it turns out to be most efficient to do this sum partially in real space and partially in reciprocal space using the Ewald technique (Davis 1971). Once the matrices of (11.135) have been set up the zeros of the determinant can be found either at constant $\mathbf{q}$ and searching all energies, or at constant E and searching all values of $\mathbf{q}$ in the Brillouin zone. The optimum method depends on the applications for which the eigenvalues are required. Later in the chapter we will show a band structure calculated using (11.135).

11.15 The Multiple Scattering Green's Function

Equations (11.95) and (11.97) are convenient and succinct expressions for the single-site scattering Green's function. They can be used as the starting point in our calculation of the multiple scattering Green's function, if we adopt the following strategy. We consider a system of a single scatterer at position *n*, surrounded by vacuum. This is the reference system. Our perturbation will be all of the rest of the scattering centres. In a crystal, for example, that would be the whole crystal, with one atom at site *n* missing. Having the perturbation as a much bigger system than the reference system is not the usual way round, but nonetheless it is remarkably successful. This approach was first suggested by Faulkner and Stocks (1980) (see also Stocks and Winter 1984), and this section is a relativistic generalization of their derivation. With this model we can write the Green's function for the whole system in terms of the Green's function for the single-site scattering as

$$G(\mathbf{r},\mathbf{r}',W) = G_n(\mathbf{r},\mathbf{r}',W)$$

$$+ \int\int G_n(\mathbf{r},\mathbf{r}_n,W)T_{nn}(\mathbf{r}_n,\mathbf{r}'_n,W)G_n(\mathbf{r}'_n,\mathbf{r}',W)d\mathbf{r}_n d\mathbf{r}'_n$$

$$(11.136)$$

where $\mathbf{r}_n > \mathbf{r}$ and $\mathbf{r}'_n > \mathbf{r}'$. The quantity $T_{nn}(\mathbf{r}_n,\mathbf{r}'_n,W)$ is the scattering matrix for the system of all scatterers except the nth. It can be written in terms of the scattering path operator as (convince yourself of this)

$$T_{nn}(W) = \sum_{i\neq n}\sum_{j\neq n}\tau^{ij}(W) \qquad (11.137)$$

Our task is to find a computationally convenient expression for the Green's function of equation (11.136). For brevity, in this section we again use the subscript Λ to mean a set of quantum numbers κ and m_j. As you are painfully aware we have already calculated the single-site scattering Green's function. The most useful form for going on to find the multiple scattering Green's function is equation (11.95). Let us substitute this into (11.137). A little manipulation gives

$$G(\mathbf{r},\mathbf{r}',W) = G_n(\mathbf{r},\mathbf{r}',W) + \sum_{\Lambda_1}\sum_{\Lambda_2}Z_{\Lambda_1}(\mathbf{r})X_{\Lambda_1\Lambda_2}(W)Z^\dagger_{\Lambda_2}(\mathbf{r}') \qquad (11.138a)$$

where

$$X_{\Lambda_1\Lambda_2}(W) = -k^2\times$$

$$\sum_{\Lambda_3,\Lambda_4}\int\int t^n_{\Lambda_1\Lambda_3}(W)H^\dagger_{\Lambda_3}(\mathbf{r}_n)T_{nn}(\mathbf{r}_n,\mathbf{r}'_n,W)H_{\Lambda_4}(\mathbf{r}'_n)t^n_{\Lambda_4\Lambda_2}(W)d\mathbf{r}_n d\mathbf{r}'_n$$

$$(11.138b)$$

The four-component spherical Hankel functions in this expression can be put in terms of four-component spherical Bessel functions. This is a nasty derivation, the bulk of which is done in appendix C. Writing equation (C.31) in the Λ representation, with the help of equation (11.127b) we obtain

$$-ikH_\Lambda(\mathbf{r}_p) = \sum_{\Lambda'}G_{0\,\Lambda\Lambda'}(\mathbf{R}_p-\mathbf{R}_q,W)J_{\Lambda'}(\mathbf{r}_q) \qquad (11.139)$$

where the $G_{0\,\Lambda\Lambda'}(\mathbf{R}_n-\mathbf{R}_m,W)$ are given by equations (11.119) and (11.127b). Actually, a different element of G_0 multiplies each element of (11.139), but writing that out in full is very messy, and makes no difference to what follows. We can only make this expansion because the spherical Hankel function diverges at $\mathbf{R}_n$, but is regular at all other points in space. Hence it can be expanded in spherical Bessel functions when it is in the vicinity of $\mathbf{R}_m, m \neq n$. Now we can substitute (11.139) into (11.138). This gives a very long and complicated expression which can be simplified a little using (11.124) to give

$$X_{\Lambda_1\Lambda_2}(W) =$$
$$\sum_{m\neq n}\sum_{p\neq n}\sum_{\Lambda_3}\sum_{\Lambda_4}\sum_{\Lambda'}\sum_{\Lambda''} t^n_{\Lambda_1\Lambda_3}(W)G^{nm}_{0\Lambda_3\Lambda'}\tau^{mp}_{\Lambda'\Lambda''}(W)G^{pn}_{0\Lambda''\Lambda_4}t^n_{\Lambda_4\Lambda_2}(W)$$

(11.140)

where the integral in (11.124) can be done over all space as required because the expansion (11.139) means we expand around sites $m \neq n$ and $p \neq n$, so the space restrictions below equation (11.136) are not applicable. What a lot of summations! Never mind, it starts getting better now. Consider the sum over p and the last three quantities in (11.140). We can use equation (11.125) to replace them:

$$X_{\Lambda_1\Lambda_2}(W) = \sum_{m\neq n}\sum_{\Lambda_3}\sum_{\Lambda'} t^n_{\Lambda_1\Lambda_3}(W)G^{nm}_{0\Lambda_3\Lambda'}\left(\tau^{mn}_{\Lambda'\Lambda_2}(W) - t^m_{\Lambda'\Lambda_2}(W)\delta_{mn}\right)$$

(11.141)

The single-site t-matrix in the brackets here contributes nothing because the δ-function requires $m = n$ for a non-zero contribution, and here the summation is over $m \neq n$, so that can be removed. Then we can use (11.125) again to simplify the resulting equation still further

$$X_{\Lambda_1\Lambda_2}(W) = \tau^{nn}_{\Lambda_1\Lambda_2}(W) - t^n_{\Lambda_1\Lambda_2}(W)$$

(11.142)

This can easily be substituted back into (11.138a) and if we use (11.97) for the single-site Green's function in (11.138a) we see that the terms in the single-site t-matrix cancel and we are left with the following expression for the multiple scattering Green's function

$$G(\mathbf{r}, \mathbf{r}', W) = \sum_{\Lambda_1}\sum_{\Lambda_2} Z_{\Lambda_1}(\mathbf{r})\tau^{nn}_{\Lambda_1\Lambda_2}(W)Z^\dagger_{\Lambda_2}(\mathbf{r}') - \sum_{\Lambda} Z_\Lambda(\mathbf{r})J^\dagger_\Lambda(\mathbf{r}')$$ (11.143)

Obviously this has a very similar structure to the single-site Green's function of (11.97) (Strange *et al.* 1989a), the only difference being the replacement of the t-matrix by the τ-matrix. This equation has a number of nice features. Firstly, in deriving it we have not made any statement about the array of scatterers so it is valid for any array of non-overlapping potentials. Secondly, the multiple scattering information (in τ) is completely separated from the wavefunction information (in Z and J). Thirdly, as we shall see in the following section, many observables depend only upon the imaginary part of the Green's function, and taking the imaginary part is trivial. Finally, calculating observables from the Green's function in this form usually turns out to involve a few numerical integrations which are computationally trivial.

11.16 The Average T-Matrix Approximation

Let us extend our multiple scattering description of electronic structure to another class of materials. We consider a simple lattice, but instead of placing the atoms periodically, we put them down on the lattice randomly, i.e. there is no translational periodicity. If the concentration of atoms A is c, then that of atoms B is $1 - c$ and we say we have a random substitutional alloy $A_c B_{1-c}$. The question that then arises is whether we can find a method of solving the multiple scattering equations for this case. There are, in fact, a whole host of approximations to describe such a material, but here we will only discuss one, known as the average T-matrix approximation (ATA). This was first introduced by Korringa (1958) and was exhaustively explored by Beeby (1964). The first application to a real material ($Cu_{1-c}Zn_c$), using a Lloyd formalism, was by Bansil *et al.* (1974). A discussion is given by Rickayzen (1991).

In such a material as this, it is not even completely apparent what one should calculate. Two experimenters preparing the same random substitutional alloy will not prepare samples with every atom in the same position, there will be some statistical distribution of properties among samples, even if they are perfectly prepared. The best we can do is calculate some average value of the scattering properties. Mathematically we define a configurational average. For an alloy containing N atoms a quantity is calculated for every possible configuration of the atoms on the lattice, and the average value is taken. The symbol for such an averaged quantity will be to enclose it in triangular brackets.

We start by considering the Born series, equation (11.108). This can be summed over all i, j to give the total scattering matrix, as prescribed by equation (11.107). The resulting equation is

$$T(E) = \sum_i t^i(E) + \sum_i t^i(E)G_0(E)\sum_{k\neq i} t^k(E) + \cdots \qquad (11.144)$$

Now, let us take the configurational average of this equation. The free particle Green's function is known and fixed, of course, so we don't average that, we have to average all quantities containing t-matrices

$$\langle T(E)\rangle = \sum_i \langle t^i(E)\rangle + \sum_i \langle t^i(E)G_0(E)\sum_{k\neq i} t^k(E)\rangle + \cdots \qquad (11.145)$$

This is very difficult to evaluate correctly, but we can make approximations for it, and the simplest approximation is to say that the average of the product of the t-matrices is equal to the product of the averages, i.e.

$$\langle t_1 t_2 t_3 \cdots t_n \rangle \approx \langle t_1 \rangle \langle t_2 \rangle \langle t_3 \rangle \cdots \langle t_n \rangle \tag{11.146}$$

This approximation is exact if all the t-matrices are the same, but not otherwise. It means that we write the total scattering matrix as

$$\langle T(E) \rangle_{\text{ATA}} = \sum_i \langle t^i(E) \rangle$$
$$+ \sum_i \langle t^i(E) \rangle G_0(E) \sum_{k \neq i} \langle t^k(E) \rangle + \cdots \tag{11.147}$$

and the corresponding Green's function is

$$\langle G(E) \rangle_{\text{ATA}} = G_0(E) + G_0(E) \langle T(E) \rangle_{\text{ATA}} G_0(E) \tag{11.148}$$

Now we only require an expression for the average value of the single site t-matrix $\langle t^i(E) \rangle = t^{\text{ATA}}$. This is easy: labelling the t-matrix of element $A(B)$ as $t^{A(B)}$ we have

$$t^{\text{ATA}} = c t^A + (1 - c) t^B \tag{11.149}$$

So in the ATA we simply replace the element-specific t-matrices by a concentration-weighted average. This gives us back the crystal symmetry and the multiple scattering problem can be solved using lattice Fourier transforms as was done in section 11.14.

The important thing about this discussion from our point of view is that it has no dependence upon the representation, or the internal structure, of the t-matrix. So it holds whether we are discussing a non-magnetic or a magnetic alloy and whether we are treating relativistically or non-relativistically. The above equations represent a first-principles theory of random substitutional alloys, and can describe alloys throughout the entire concentration range. Furthermore, the ATA scattering amplitudes correspond to inelastic scattering, i.e. they do not sit on the unitarity circle. This tells us that the single-site potential to which t^{ATA} corresponds is complex and energy dependent. Bloch states then have a finite lifetime, and energy bands are broadened. What this tells us is that the ATA does describe disorder in a very fundamental way.

As a matter of fact the ATA is not widely used in research into alloy electronic structure for a number of reasons. If the alloy has a tendency for like atoms to cluster together, or at the opposite extreme, for ordering of the different atoms, the ATA misses that effect completely. Furthermore there is a much more sophisticated approximation known as the coherent potential approximation (CPA) (Soven 1967). The advantage of the CPA is that it treats the disorder self-consistently, whereas the ATA does not. In the language of multiple scattering theory this means that the Green's function of an atom embedded in the CPA medium, averaged over the possible occupations of the single site, should be equal to the Green's function of the medium itself. The scattering theory version of the CPA, known as the KKR-CPA, has been set up by Stocks *et al.* (1971, 1977), Temmerman *et al.* (1978), and Bansil *et al.* (1978). A relativistic version of it (RKKR-CPA) was derived by Staunton *et al.* (1980) (See also Ginatempo and Staunton 1988, Arola 1991). A relativistic spin-polarized version (SP-RKKR-CPA) has also been implemented (Ebert *et al.* 1992, Gotsis *et al.* 1994, Jenkins and Strange 1995). However, it is a very involved and difficult theory and cannot be discussed at the level of this book in a few pages. Therefore for further discussion of this topic the reader is referred to the original literature.

11.17 The Calculation of Observables

In equations (11.97) and (11.143) we now have expressions for the scattering Green's functions in a unified form. So far we have not really shown why the Green's function is a good thing to calculate. Here we will correct that. The main utility of the Green's functions in scattering theory is that knowledge of it enables us to calculate many observable quantities straightforwardly and directly. In this section we will show how to calculate a few simple quantities from the relativistic Green's function.

You may ask what sort of observables we would want to calculate. If our scattering centres represent ions and the electrons surrounding them, we may want to calculate the charge density of the electrons, or the density of states. If the scattering site has a magnetic moment associated with it we may wish to calculate that. A rather different type of observable is the difference in total energy between two states of a system. In this section we will illustrate the method with these examples. We will need to make repeated use of the result

$$\lim_{\epsilon \to 0} \frac{1}{x - a + i\epsilon} = \frac{1}{x - a} - i\pi\delta(x - a) \qquad (11.150)$$

which is easy to derive from the theory of complex variables.

Much of the interpretation of experiments is done in terms of the band structure and Fermi surface, and it is these that we examine first.

The Band Structure

The band structure is not strictly an observable, although angle resolved photoemission experiments do see something that very closely approximates it. If we are using multiple scattering theory to solve the density functional Hamiltonian, the band structure is a plot of the density functional eigenvalues as a function of position in the Brillouin zone. It is not really correct to call it the electron dispersion relation because, rigorously,

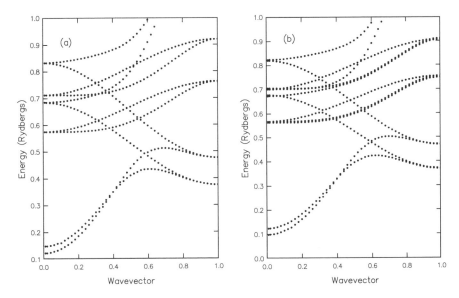

Fig. 11.8. The energy bands of ferromagnetic iron along $\Gamma(0.0)$ to $H(1.0)$ in the BCC Brillouin zone. (a) Calculated using the non-relativistic KKR method. (b) Calculated using the relativistic KKR method.

the eigenvalues do not correspond to single-electron eigenvalues. However, many experiments are interpreted in terms of the band structure, so it makes sense to discuss relativistic effects on band structure here. The best way to do this is by example.

In figure 11.8 we show the energy bands of iron, calculated using the potential of Moruzzi *et al.* (1978) along a particular direction in the body centred cubic (BCC) Brillouin zone. Both figures were calculated using multiple scattering theory, i.e. the bands were found by finding the zeros of the KKR determinant as given by equation (11.135), and both the calculations were performed using the same computer program, with only the speed of light varied. Therefore all numerical methods in the two calculations are the same, so the figures are truly comparable.

Figure 11.8a is a typical non-relativistic BCC metal band structure. The lowest two bands have predominantly $l = 0$ character. The upper bands are d-bands with some small amount of s character mixed in. In figure 11.8b we have the same bands calculated in a fully relativistic theory, with the t-matrices calculated as shown in section 11.9. Clearly, on this scale, the bands are very similar in both cases. There are some changes due to relativity. The principal one is the lifting of degeneracies, as can be seen in some of the d-bands in this figure (e.g. the lower of the two bands that have $E \approx 0.55$ Rydbergs at Γ in figure 11.8a is split into two in figure 11.8*b*). Further examination of these two figures shows that band widths are also affected by relativity. For example, the widths of the two *s*-bands are increased by about 6 mRyd in the relativistic case. The Fermi energy in the non-relativistic calculation is at 0.756 Ryd, whereas in the relativistic calculation it is at 0.743 Ryd. For heavier elements all these effects become more pronounced, but even in an element as light as iron, significant relativistic effects can be seen. Deeper examination of the figures reveals some rather more subtle effects of relativity. Look at the bands around $E = 0.7$ Ryd and $k = 0.3 - 0.4$. Close inspection shows that relativity has a profound effect on the connectivity of the bands. To illustrate the effects of relativity on band structure further, the bands of ferromagnetic cobalt are shown in figure 11.9. These were calculated from a self-consistent potential from a non-relativistic density functional theory calculation. Figure 11.9 shows the energy bands along $\Gamma - X$ in the face centred cubic (FCC) zone for three different cases. However, this time we have magnified them and only show a small energy range compared with that in figures 11.8. In figure 11.9*a* we have the energy bands calculated using non-relativistic scattering theory. In figure 11.9*b* we show the energy bands along $\Gamma - X$ where X is in the direction parallel to the magnetic moment. Here we can clearly see the shifting of the bands and the lifting of degeneracies, particularly in the band that goes from $E \approx 0.615$ Ryd at Γ to $E \approx 0.78$ Ryd at X. It is clear from comparison of these two diagrams that our conclusions as to the effect of relativity in metals are much the same as in the case of iron.

Now consider figure 11.9*c*. Again, this is a relativistic calculation of the energy bands in cobalt. This time, though, the bands are calculated along $\Gamma - X$ where X is in a direction which is crystallographically equivalent to the direction shown in figure 11.9*b*, but perpendicular to the direction of the magnetic moment. Because the magnetic moment picks out a direction in the unit cell, and hence in the Brillouin zone, these two directions become distinct. In a non-relativistic calculation, or in a non-magnetic case of a relativistic calculation, the bands along these two directions would be identical. Clearly in the relativistic spin-polarized case the band structure in these two directions is rather different. The lifting of the degeneracies is much more pronounced in figure 11.9*b* than in figure 11.9*c*. On the other hand, there are no band crossings in figure 11.9*c*, all the bands curve away abruptly, to avoid each other. This can have a profound effect on the Fermi surface and hence on the electronic properties of the material if the Fermi energy falls in a judicious position. As an example, suppose the Fermi energy in figure 11.9*c* fell just below 0.70 Ryd. There is no band crossing the Fermi energy in 11.9*c* around $k = 0.2$ whereas there is in the equivalent position in the non-relativistic

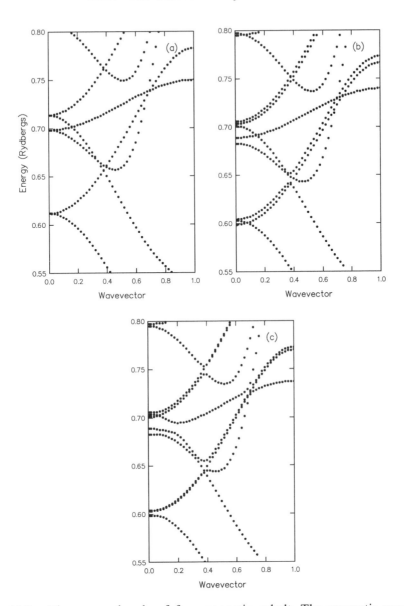

Fig. 11.9. The energy bands of ferromagnetic cobalt. The magnetic moment points in the z-direction. (a) Calculated along $\Gamma - X$ using the non-relativistic KKR method. Identical results occur regardless of whether X is parallel or perpendicular to the moment. (b) Calculated along $\Gamma - X$, with X parallel to the magnetic moment, using the relativistic KKR method. (c) Calculated along $\Gamma - X$, with X perpendicular to the moment using the relativistic KKR method. In (a) the Fermi energy is at 0.809 Rydbergs, while in (b) and (c) it is at 0.794 Rydbergs.

theory as seen in figure 11.9*a*. So non-relativistically there would be a piece of Fermi surface there, whereas relativistically there would not.

The Fermi Surface

Let us recall what is meant by the Fermi surface. It is a defining property of metals and determines many of their solid state properties. It is useful in condensed matter physics because it can be both measured and calculated in several different ways without much difficulty. Therefore it forms a natural meeting point for experiment and theory. Positron annihilation and de Haas–van Alphen measurements are among the experimental methods. The latter, in particular, is a very sensitive and accurate probe, which uses the properties of Landau levels discussed in chapter 9. It is also very easy to calculate the Fermi surface from band theory. One simply looks at the band structure, and anywhere a band crosses the Fermi energy defines a point. If we do this over the whole Brillouin zone, the points map out a surface in reciprocal space, known as the Fermi surface.

There is some subtlety about the Fermi surface in relativistic quantum theory which occurs as a consequence of the avoided band crossings mentioned earlier, and leads us to re-evaluate the behaviour of electrons at the Fermi surface. Suppose we calculate the band structure of a magnetic material in non-relativistic theory. We solve separate Schrödinger equations for the spin-up and the spin-down electrons. These do not 'see' each other except to the extent that we fill states to a Fermi level which contains the right number of electrons. There are, in fact, two Fermi surfaces, one for spin-up, and one for spin-down, electrons. A cross section through the Fermi surface of cobalt calculated in this way is shown in figure 11.10*a*. Cobalt is a fairly light element, so if we calculate the Fermi surface within relativistic quantum theory, we must get much the same answer as in the non-relativistic case. However, this time, electrons with different spins do 'see' one another, through spin–orbit coupling, so there is only one Fermi surface. For cobalt this is shown in figure 11.10*b*. However, with relativity, avoided band crossings occur like those shown in figure 11.9. This leads to the relativistic Fermi surface having different connectivity, which means that a particular piece of Fermi surface may be predominantly spin up in one region of the Brillouin zone and, owing to an avoided crossing with a spin-down band, may be mainly spin-down in another.

Figure 11.10 illustrates a case where this has occurred. In figure 11.10*a* the piece of Fermi surface on the *A–L* line nearer *A* comes from a majority spin band. However the doubly degenerate piece on the *Γ–A* line is from a minority spin band. These are completely separate in the non-relativistic

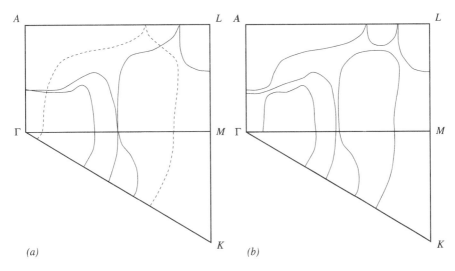

Fig. 11.10. Sketches of some cross sections through the Fermi surface of cobalt. Γ is the centre of the zone and A, L, M, and K represent other points of high symmetry in the zone. (a) Non-relativistic calculation with separate majority spin (dashed line) and minority spin (full line) Fermi surfaces. (b) Relativistic calculation, which yields a Fermi surface with different connectivity to the non-relativistic one.

case. When calculated relativistically these two pieces of Fermi surface have joined up as shown in 11.10b. However, the eigenfunctions at the Fermi energy on the A–L line are around 95% majority spin, whereas on the Γ–A line they are around 95% minority spin at this energy. Hence our picture of electrons in a solid maintaining the same spin as they drift around the Fermi surface becomes less valid. The electrons may continually change their spin character as they move round the Fermi surface (A.R. Mackintosh, private communication).

How about that? Two consecutive (sub) sections with no equations!

The Density of States

The retarded Green's function for a system may be written

$$G^{R}(\mathbf{r}, \mathbf{r}', E) = \lim_{\epsilon \to 0} \sum_n \frac{\psi_n(\mathbf{r})\psi_n^{\dagger}(\mathbf{r}')}{E - E_n + i\epsilon} \tag{11.151}$$

where ψ_n and E_n are the eigenfunctions and eigenvalues of the Hamiltonian. We can let $\mathbf{r}$ and $\mathbf{r}'$ in (11.151) be in the same muffin tin to obtain the site-diagonal Green's function. If we take the trace of both sides and integrate over $\mathbf{r}$ the right hand side simplifies because of the normalization

$$\int \psi_n^\dagger(\mathbf{r})\psi_n(\mathbf{r})d\mathbf{r} = 1 \tag{11.152}$$

and we have

$$\int \mathrm{Tr}\, G^R(\mathbf{r}, \mathbf{r}, E)d\mathbf{r} = \lim_{\epsilon \to 0} \sum_n \frac{1}{E - E_n + i\epsilon} \tag{11.153}$$

The trace on the left hand side of (11.153) is simply the trace of the 4×4 matrix (11.143). In the first term of (11.143) it picks out the leading diagonal of $Z_\kappa^{m_j}(\mathbf{r})Z_\kappa^{m_j\dagger}(\mathbf{r})$. It is easy to see from equation (11.150) that

$$-\frac{1}{\pi}\mathrm{Im}\left[\lim_{\epsilon \to 0}\frac{1}{E - E_n + i\epsilon}\right] = \delta(E - E_n) \tag{11.154}$$

Combining the above equations gives

$$-\frac{1}{\pi}\mathrm{Im}\int \mathrm{Tr}\, G^R(\mathbf{r}, \mathbf{r}, E)d\mathbf{r} = \sum_n \delta(E - E_n) \tag{11.155}$$

The right side of (11.155) is clearly just the density of states because

$$\int_E^{E+\Delta E} \sum_n \delta(E - E_n)dE = N \tag{11.156}$$

where N is the number of states between E and $E + \Delta E$. Therefore we can conclude that

$$n(E) = -\frac{1}{\pi}\mathrm{Im}\int \mathrm{Tr}\, G^R(\mathbf{r}, \mathbf{r}, E)d\mathbf{r} \tag{11.157}$$

Equation (11.157) is a simple expression for the density of states at energy E in terms of the Green's function. With the Green's function of equation (11.143) we can decompose the density of states into its κ- and m_j-dependent parts, if required. Clearly the integrated density of states or the number of states below energy E_{max} is given by

$$N(E_{max}) = -\frac{1}{\pi}\mathrm{Im}\int_{-\infty}^{E_{max}}\int \mathrm{Tr}\, G^R(\mathbf{r}, \mathbf{r}, E)d\mathbf{r}dE \tag{11.158}$$

Calculations of the density of states (DOS) using this method are shown in figures 11.11 and 11.12. In figure 11.11 we have the density of states for ferromagnetic iron calculated according to equation (11.157) and decomposed into the s, p and d contributions. Clearly the density of states is dominated by the contribution from the d-electrons. The large peaks correspond to crystal field and magnetic splittings. In figure 11.12 we decompose the d density of states in figure 11.11 into its m_j components,

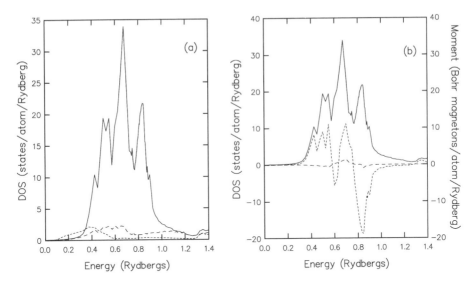

Fig. 11.11. The density of states for iron (Fermi energy 0.743 Rydbergs). (*a*) The short dashed line is the *s*-electron contribution, the long dashed line is due to the *p*-electrons and the full line is the *d*-electron contribution. (*b*) The full line is the *d*-electron contribution to the density of states again, the short dashed line is the spin contribution to the magnetic moment of iron and the long dashed line is the orbital contribution to the magnetic moment of iron. Magnetic moment is shown in units of the Bohr magneton.

to facilitate a deeper analysis of its structure. This shows, for example, that the main peak above the Fermi energy in figure 11.11 is predominantly composed of contributions from $m_j = -5/2$, $-1/2$ and $+3/2$, and the large central peak mainly has $m_j = +1/2$ and $m_j = -3/2$ character. We have not decomposed by κ because this is a relativistic and spin-polarized density of states and so κ and $-\kappa-1$ are mixed as discussed in section 11.9, although m_j remains a good quantum number. One other feature to note is that the curves for $m_j = \pm 5/2$ only contain one electron, as they only occur for $j = l + 1/2$. The other curves contain two electrons each (which would be one electron for each of $j = l \pm 1/2$ if there were no mixing due to the magnetism). It is easy to see that the area under the $j = \pm 5/2$ curves is around half of the area under the other curves. Furthermore the $j = \pm 5/2$ curves have similar shapes and clearly demonstrate the splitting of these states due to the magnetism.

There is an alternative method of determining the integrated density of states which we will introduce here. The best way to discuss this is to write down the formula and then justify it. A full derivation is given by Lloyd and Smith (1972) and is discussed in detail by Faulkner (1977). The

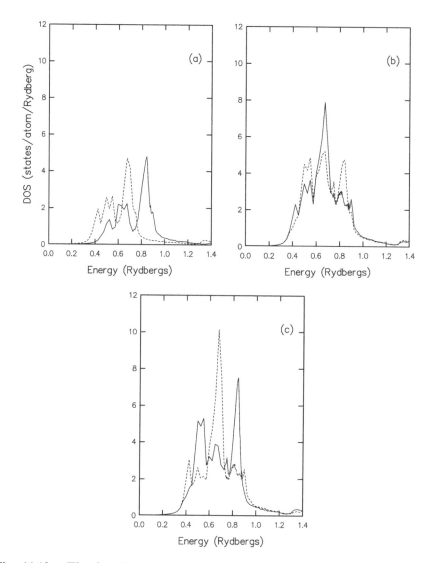

Fig. 11.12.   The $l = 2$ density of states for iron (Fermi energy 0.743 Ryd) decomposed by its m_j values. (a) $m_j = -5/2$ (full line), $m_j = 5/2$ (dashed line), (b) $m_j = -3/2$ (full line), $m_j = 3/2$ (dashed line), (c) $m_j = -1/2$ (full line), $m_j = 1/2$ (dashed line).

relativistic case is specifically dealt with by Arola (1991). In this approach the integrated density of states is given by the Lloyd formula

$$N(E) = N_0(E) - \frac{1}{\pi\Omega_{BZ}} \mathrm{Im} \int_{\Omega_{BZ}} \ln |t^{m_j}_{\kappa\kappa'}(E) - G_{\kappa m_j \kappa' m'_j}(\mathbf{k}, E)| d\mathbf{k} \qquad (11.159)$$

where Ω_{BZ} is the volume of the Brillouin zone. Now let us understand how this comes about. The determinant of the KKR matrix is always real. So if we regard the determinant as a complex number it will have a magnitude and phase, and the phase must be some integral multiple of π, i.e.

$$|t_{\kappa\kappa'}^{m_j}(E) - G_{\kappa m_j \kappa' m_j'}(\mathbf{k}, E)| = A e^{in\pi} \tag{11.160}$$

Taking the imaginary part of the logarithm of this equation yields

$$\text{Im} \ln |t_{\kappa\kappa'}^{m_j}(E) - G_{\kappa m_j \kappa' m_j'}(\mathbf{k}, E)| = n\pi \tag{11.161}$$

But this determinant passes through zero, as that is the definition of the eigenvalues. This phase must change discontinuously through π every time the determinant passes through zero. Thus, if we take a particular value of $\mathbf{k}$ and look at the change in the phase of the KKR determinant as a function of energy E we will find it has changed by an integer multiple of π. That integer will be the number of times the KKR determinant has passed through zero, i.e. the number of states below E. Hence if we then integrate this quantity over the Brillouin zone we have π times the integrated density of states for the material. Dividing by π and normalizing to the volume of the Brillouin zone yields equation (11.159) for the integrated density of states. In figure 11.13 we show a schematic band structure and the corresponding phase of the KKR determinant at the Γ and X points. As you can see the phase changes discontinuously every time we pass through a band.

The Charge Density

The potential that is seen by the scattered particle will often be due to a charge density acting upon the charge of the scattered particle. From equation (10.255) the charge density is given by

$$\rho(\mathbf{r}) = e \sum_n^{\text{occ}} \psi_n^\dagger(\mathbf{r})\psi_n(\mathbf{r}) \tag{11.162}$$

where the summation extends over all occupied states. Again, putting $\mathbf{r} = \mathbf{r}'$ in equation (11.151) and then integrating over energy up to the highest occupied state at the Fermi energy E_F gives

$$\int_{-\infty}^{E_F} \text{Tr}\, G^R(\mathbf{r}, \mathbf{r}, E) dE = \lim_{\epsilon \to 0} \sum_n \int_{-\infty}^{E_F} \text{Tr}\, \frac{\psi_n(\mathbf{r})\psi_n^\dagger(\mathbf{r})}{E - E_n + i\epsilon} dE$$

$$= \lim_{\epsilon \to 0} \sum_n \text{Tr}\, \left\{ \psi_n(\mathbf{r})\psi_n^\dagger(\mathbf{r}) \right\} \int_{-\infty}^{E_F} \frac{1}{E - E_n + i\epsilon} dE$$

$$= \sum_n \psi_n^\dagger(\mathbf{r})\psi_n(\mathbf{r}) \int_{-\infty}^{E_F} \left(\frac{1}{E - E_n} - i\pi\delta(E - E_n) \right) dE \tag{11.163}$$

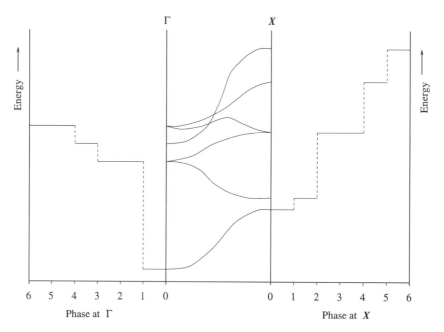

Fig. 11.13.　A schematic band structure and the changes in phase of the Lloyd determinant (in units of π) at the Γ and X points as a function of energy.

where we have used (11.150) again, and taking the trace has allowed us to reverse the order of ψ and $\psi^\dagger$. The wavefunction part of the final line of (11.163) is now definitely real. Therefore, if we take the imaginary part of each side of (11.163), we only have to deal with the δ-function in the integral. Integration of the delta function over the specified range changes the range of the summation from all states to all occupied states. Combining (11.162) and (11.163) gives

$$\rho(\mathbf{r}) = -\frac{e}{\pi}\mathrm{Im}\int_{-\infty}^{E_F} \mathrm{Tr}\, G^R(\mathbf{r}, \mathbf{r}, E)dE \qquad (11.164)$$

This is the expression we use to calculate the charge density for use in the next iteration of density functional theory, or if we want to map out the charge density to investigate bonding in metals for example.

Magnetic Moments

From equation (10.256) the spin magnetization density associated with our potential is

$$\mathbf{m}(\mathbf{r}) = -\mu_B \sum_n^{\mathrm{occ}} \psi_n(\mathbf{r})\tilde{\beta}\tilde{\sigma}\psi_n^*(\mathbf{r}) \qquad (11.165)$$

Clearly this is similar to the charge density and a similar derivation gives

$$\mathbf{m}(\mathbf{r}) = \frac{\mu_B}{\pi} Im \int^{E_F} \mathrm{Tr}\, \tilde{\beta}\tilde{\sigma} G(\mathbf{r},\mathbf{r},E)dE \qquad (11.166)$$

In (11.166) the trace includes the spin. The spin magnetic moment associated with the site is then the integral of (11.166) over $\mathbf{r}$. We have been slightly lax in our derivation here. The integral over energy does not include antiparticle states, and so does not extend down to $-\infty$; it is to be understood that the integral is only over positive energy states.

The spin contribution to the magnetic moment may be obtained directly from the density of states. If we write it in the (l, m, s) representation, the total spin contribution to the magnetic moment is

$$M_{\text{spin}} = \frac{e\hbar}{2m} \int^{E_F} \sum_{lm_l} \left(n_{lm_l}^{\uparrow}(W) - n_{lm_l}^{\downarrow}(W) \right) dW$$

$$= \frac{e\hbar}{m} \int^{E_F} \sum_{lm_l} m_s n_{lm_l m_s}(W) dW \qquad (11.167)$$

This can, of course, be decomposed into individual l and m_l components as well. In figure 11.11b we show the spin contribution to the magnetic moment of iron from the d-electrons as a function of energy, i.e. the integrand of equation (11.167). Compare it with the density of states calculated from the same potential. Evidently, the peaks in each plot coincide. The peak above the Fermi energy is very much a spin-down peak, and below the Fermi energy it is predominantly of spin-up character.

As we discussed in chapter 2 there may also be a contribution to the magnetic moment at a site from the orbital motion of the electrons if they retain some of their atomic character in the solid. This can be calculated in an analogous way to the spin moment:

$$M_{\text{orb}}(\mathbf{r}) = -\frac{1}{\pi}\frac{e\hbar}{2m} Im \int^{E_F} \mathrm{Tr}\, \tilde{\beta}\hat{l}_z G(\mathbf{r},\mathbf{r},W)dW \qquad (11.168a)$$

or, assuming the eigenstates are normalized to unity,

$$M_{\text{orb}} = \frac{e\hbar}{2m} \int^{E_F} \sum_{lm_l} \sum_{m_s} m_l n_{lm_l m_s}(W)dW \qquad (11.168b)$$

The orbital contribution M_{orb} is non-zero because relativistically states with quantum number m_l and $-m_l$ are not equally populated. The value of $M_{\text{orb}}(W)$, essentially the integrand of (11.168b), is also plotted for the d-electrons in iron in figure 11.11. The only impression we can gain from this figure is how small the orbital moment of iron is compared with the spin moment. A comparison of calculated total spin and orbital magnetic moments with experimental magnetic moments for the ferromagnetic transition elements is shown in table 10.12 in the previous chapter. The calculated values are essentially integrals of the curves in figure 11.11*b*.

Table 10.12 indicates very good, although not perfect, agreement between experiment and theory, and you might think that we could explain the ground state magnetic moments of materials with what we have learnt up to now. Life does not turn out to be as simple as that. If we perform a self-consistent density functional theory calculation for a heavy magnetic material, e.g. an actinide compound, there is often little agreement with the measured magnetic moments. The reason for this is a rather subtle failing of the local approximation to density functional theory which we will discuss in a qualitative fashion.

In rare earths and actinides the *f*-electrons retain a lot of their atomic character, and the distribution of electrons among the states is determined, at least to some extent, by Hund's rules which are based on a non-relativistic approach to the quantum mechanics of many-electron atoms. Hund's first rule decrees that we should fill up states with spins parallel before we fill up the antiparallel states, hence maximizing the spin contribution to the magnetic moment. This is taken care of in density functional theory by making a local approximation for the exchange–correlation potential which depends on the electron spin, e.g. the Von Barth–Hedin approximation of equations (10.206 and 207) which fills up one spin direction preferentially. Hund's third rule, which states which value of the total angular momentum *J* has minimum energy, is taken care of by the fact that we are doing relativistic, rather than non-relativistic, calculations. However, Hund's second rule, that we should fill up states to maximize the total orbital angular momentum of an atom, is not included in any local approximation to DFT. Furthermore, it is difficult to see how to incorporate such an atomic effect into an exchange-correlation potential based on that of the free electron gas.

Brooks (1985) has proposed an approximate method for correcting this problem and including some orbital polarization. This idea has been elaborated since this first suggestion (Eriksson *et al.* 1990*a*). Their reasoning is as follows. Hund's first rule can be written in mathematical form as being due to a Hamiltonian of the form $-I \sum_{i \neq j} \mathbf{s}^i \cdot \mathbf{s}^j$. This is difficult to solve, but a mean field approximation to it yields an energy $E = -I S_z^2$ where I is a material dependent constant known as the Stoner exchange parameter. Now let us follow a similar route for the orbital motion. Hund's second rule corresponds to a Hamiltonian $-\frac{1}{2} \sum_{i \neq j} \mathbf{l}^i \cdot \mathbf{l}^j$ and a mean field approximation to this will lead to a term in the energy proportional to $\mathbf{L}^2$ where $\mathbf{L}$ is the total orbital angular momentum of the partially filled electron shell. This energy leads to a corresponding shift in the density functional eigenvalues of $-E^3 |\mathbf{L}| m_l$ for the state with quantum number

Table 11.1. The magnetic moment on the actinide site (in Bohr magnetons) for three actinide–iron intermetallic compounds. The first column is the magnetic moment on the actinide site as measured by neutron scattering. The next three columns are the spin, orbital and total magnetic moment on the actinide site calculated using standard band theory. The final three columns are the spin, orbital and total moment calculated with the self-consistent inclusion of an orbital polarization. The spin contribution to the magnetic moment is only that due to the f-electrons as that is what is measured by neutron scattering, the s-, p- and d-electrons being itinerant. The theoretical values were found by Eriksson *et al.* (1990*a*) and the experiments are from Lander *et al.* (1977).

	μ_{exp}	μ_s^B	μ_L^B	μ_T^B	μ_s	μ_L	μ_T
UFe$_2$	0.06	−0.58	0.47	−0.11	−0.83	0.88	0.05
NpFe$_2$	1.09	−1.17	0.70	−0.47	−2.29	3.49	1.20
PuFe$_2$	0.45	−3.48	1.45	−2.03	−3.19	3.52	0.33

m_l. The quantity E^3 is another constant known as the Racah parameter. This splits the seven spin-up and seven spin-down f-levels into equally spaced levels when the total orbital angular momentum is non-zero.

This eigenvalue shift can be included self-consistently into an electronic structure calculation based on density functional theory, and a series of results is shown in table 11.1. Clearly, when orbital polarization is included, much better agreement between theoretical and experimental values of the total magnetic moment is obtained. In all these cases the orbital and spin moments are in opposite directions and partially cancel.

Orbital polarization is not restricted to the calculation of magnetic moments, it has also been found to improve the agreement between experiment and theory for other observable quantities, e.g. the volume collapse of the light rare earths under pressure (Eriksson *et al.* 1990*b*).

Energetic Quantities

It is rare that we want to know an absolute value of an energy. Usually physicists are more interested in energy differences. Suppose we want to know which is the more stable of two states in an electronic system at zero temperature. The correct procedure is to perform two fully self-consistent density functional calculations, and calculate the total energy in each case from equation (10.249). The state with the lower total energy is the more stable one. This procedure is computationally intensive and involves taking differences of large numbers to get small numbers. Therefore numerical algorithms have to be robust and reliable. Any simplification of this procedure would be welcome, particularly if it could be used to cancel some of the large numbers involved when evaluating energy differences. Here we will illustrate such a simplification by example.

Suppose we want to know whether the ground state crystal structure of a material is BCC or FCC with the same volume per atom in each structure. The self-consistent 4-current density for a BCC lattice is $J_{\mathrm{BCC}}^{\mu}(\mathbf{r})$ which yields an energy $W_{\mathrm{BCC}}[J_{\mathrm{BCC}}^{\mu}(\mathbf{r})]$ using equation (10.249). Similarly the self-consistent 4-current density and energy for an FCC lattice are $J_{\mathrm{FCC}}^{\mu}(\mathbf{r})$ and $W_{\mathrm{FCC}}[J_{\mathrm{FCC}}^{\mu}(\mathbf{r})]$ respectively.

Let us suppose the difference in 4-current density between the two structures is small. Instead of generating the two 4-current densities via a large number of RDFT self-consistency iterations, let us simply generate an approximate 4-current density $J_0^{\mu}(\mathbf{r})$ which we assume is not very different from either $J_{\mathrm{FCC}}^{\mu}(\mathbf{r})$ or $J_{\mathrm{BCC}}^{\mu}(\mathbf{r})$, i.e.

$$J_0^{\mu}(\mathbf{r}) = J_{\mathrm{BCC}}^{\mu}(\mathbf{r}) + \delta J_{\mathrm{BCC}}^{\mu}(\mathbf{r}) \tag{11.169a}$$

$$J_0^{\mu}(\mathbf{r}) = J_{\mathrm{FCC}}^{\mu}(\mathbf{r}) + \delta J_{\mathrm{FCC}}^{\mu}(\mathbf{r}) \tag{11.169b}$$

Now we can determine the total energy in a BCC geometry using $J_0^{\mu}(\mathbf{r})$ and use (11.169a) to perform a Taylor expansion as follows

$$
\begin{aligned}
W_{\mathrm{BCC}}[J_0^{\mu}(\mathbf{r})] &= W_{\mathrm{BCC}}[J_{\mathrm{BCC}}^{\mu}(\mathbf{r}) + \delta J_{\mathrm{BCC}}^{\mu}(\mathbf{r})] \\
&\approx W_{\mathrm{BCC}}[J_{\mathrm{BCC}}^{\mu}(\mathbf{r})] + \delta J_{\mathrm{BCC}}^{\mu}(\mathbf{r}) \left(\frac{\delta W_{\mathrm{BCC}}[J^{\mu}(\mathbf{r})]}{\delta J^{\mu}(\mathbf{r})} \right)_{J^{\mu}(\mathbf{r})=J_{\mathrm{BCC}}^{\mu}(\mathbf{r})} \\
&\quad + (\delta J_{\mathrm{BCC}}^{\mu}(\mathbf{r}))^2 \left(\frac{\delta^2 W_{\mathrm{BCC}}[J^{\mu}(\mathbf{r})]}{\delta (J^{\mu}(\mathbf{r}))^2} \right)_{J^{\mu}(\mathbf{r})=J_{\mathrm{BCC}}^{\mu}(\mathbf{r})} + \cdots
\end{aligned}
$$
$$\tag{11.170a}$$

and we can determine the total energy in a FCC geometry using $J_0^{\mu}(\mathbf{r})$ and (11.169b) to perform a Taylor expansion

$$
\begin{aligned}
W_{\mathrm{FCC}}[J_0^{\mu}(\mathbf{r})] &= W_{\mathrm{FCC}}[J_{\mathrm{FCC}}^{\mu}(\mathbf{r}) + \delta J_{\mathrm{FCC}}^{\mu}(\mathbf{r})] \\
&\approx W_{\mathrm{FCC}}[J_{\mathrm{FCC}}^{\mu}(\mathbf{r})] + \delta J_{\mathrm{FCC}}^{\mu}(\mathbf{r}) \left(\frac{\delta W_{\mathrm{FCC}}[J^{\mu}(\mathbf{r})]}{\delta J^{\mu}(\mathbf{r})} \right)_{J^{\mu}(\mathbf{r})=J_{\mathrm{FCC}}^{\mu}(\mathbf{r})} \\
&\quad + (\delta J_{\mathrm{FCC}}^{\mu}(\mathbf{r}))^2 \left(\frac{\delta^2 W_{\mathrm{FCC}}[J^{\mu}(\mathbf{r})]}{\delta (J^{\mu}(\mathbf{r}))^2} \right)_{J^{\mu}(\mathbf{r})=J_{\mathrm{FCC}}^{\mu}(\mathbf{r})} + \cdots
\end{aligned}
$$
$$\tag{11.170b}$$

The left hand sides of these two equations contain the same current density $J_0^{\mu}(\mathbf{r})$. Equation (10.249) shows that the only terms in the expressions for $W_{\mathrm{BCC}}[J_0^{\mu}(\mathbf{r})]$ and $W_{\mathrm{FCC}}[J_0^{\mu}(\mathbf{r})]$ that are not identical are those in the density functional Lagrange multipliers. So if we subtract (11.170b) from (11.170a) all that remains on the left hand side is $\sum_i w_i^{\mathrm{BCC}} - \sum_j w_j^{\mathrm{FCC}}$. Now consider the right hand side. The term in the first order functional derivative in each of (11.170a and b) is zero. This is because the second postulate of density functional theory says that the total energy is a minimum at the correct 4-current density, i.e. the derivative of the total energy with respect to the 4-current density is zero. The final terms on the right of (11.170) are both of second order in the small parameter, and so we assume that they can be ignored. On the right hand side of (11.170), then, only the first terms remain when we subtract. Hence we have

$$W_{BCC}[J^\mu_{BCC}(\mathbf{r})] - W_{FCC}[J^\mu_{FCC}(\mathbf{r})] \approx \sum_i w_i^{BCC} - \sum_j w_j^{FCC} \qquad (11.171)$$

But this is just the total energy difference, and we have equated it to the difference in the sum of the single-electron-like energies. So, instead of having to evaluate all the terms in the total energy (10.249) we only have to calculate the simplest, the single-electron-like Lagrange multipliers. Furthermore, instead of having to evaluate two 4-current densities self-consistently and accurately within RDFT, we have only had to find one, and that only approximately. Hence this shortcut is capable of providing a great saving in computer time.

11.18 Magnetic Anisotropy

In this section we are going to look at the scattering of a single electron from two sites, both of which have a spin associated with them. This will lead us to a description of magnetic anisotropy.

Magnetic anisotropy in a crystal is the difference in energy between two spin arrangements when the relative direction of the spins with respect to one another remains constant. For example, in ferromagnetic iron, cobalt and nickel the total energy of the solid calculated within relativistic density functional theory depends upon the direction of the magnetic moment. The difference in the total energy between two directions of the moment is known as the magnetocrystalline anisotropy energy. Several attempts have been made to calculate this quantity in real materials, using the methods of the previous section, with variable success. Fritsche *et al.* (1987), Strange *et al.* (1989b, 1989c, 1991), and Daalderop *et al.* (1990, 1992) were among the first to make such calculations. Nowadays it is possible to calculate anisotropy energies and gain reasonable agreement with experiment for many materials. However, cases where the magnetocrystalline anisotropy energy is very low, such as transition metal elements, still present a challenge to theory. Also in cases where the orbital contribution to the magnetic moment is large, such as some actinide materials, this energy difference cannot be well explained from first principles.

In this section we set up as simple a model as possible to illustrate the fact that magnetic anisotropy originates from the interplay of relativity and magnetism. We have two fixed ions, each of which has a magnetic

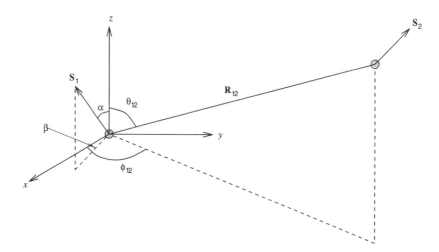

Fig. 11.14.   Indirect exchange scattering. The electron is scattered at one site
where the spin is S_1, and this influences the later scattering at a second site where
the spin is S_2. The axes represent the laboratory frame where the components of
the unit length vector $\hat{S}_1$ are $S_x = \sin\beta\cos\alpha$, $S_y = \sin\beta\sin\alpha$ and $S_z = \cos\beta$. The
vector joining the two spins is R_{12} and that is in the direction (θ_{12}, ϕ_{12}) relative
to the laboratory frame.

moment, and ask how the indirect exchange mechanism determines the
direction of the magnetic moments. By indirect exchange we mean that
an electron sees one of the magnetic ions and is scattered by it. Later on
it sees the second ion, and its interaction with that is affected by how it
interacted with the first ion, so we are specifically interested in scattering
from two separate sites, such as those illustrated in figure 11.14.

We assume that the chemical potential, density of states and magnitude
of the magnetization do not change as we rotate the ionic moments. We
also assume that the magnetic anisotropy energy can be found from the
single-particle-like energies as in equation (11.171). So we only need

$$E_{sp} = \int_0^{E_F} wn(w)dw \tag{11.172}$$

where $n(w)$ is the density of single-particle-like states at energy w and E_F
is the Fermi energy. Integrating (11.172) by parts enables us to write it in
terms of the integrated density of states $N(w)$

$$E_{sp} = [wN(w)]_0^{E_F} - \int_0^{E_F} N(w)dw = ZE_F - \int_0^{E_F} N(w)dw \tag{11.173}$$

The magnetic anisotropy is contained in the integral here, so henceforth
we will ignore the first term. We can use the Lloyd formula (for two sites)

to replace the integrated density of states as

$$E_{sp}^{(1)} = \frac{1}{\pi} \mathrm{Im} \int_0^{E_F} dw \, \ln \left| \begin{matrix} t_1^{-1} & -G(1,2) \\ -G(2,1) & t_2^{-1} \end{matrix} \right| \tag{11.174}$$

where G is given by (11.119). We can evaluate the determinant and factor out the t-matrices to give

$$E_{sp}^{(1)} = \frac{1}{\pi} \mathrm{Im} \int_0^{E_F} dw \left(\ln |t_1^{-1}| + \ln |t_2^{-1}| + \ln |1 - t_1 G(1,2) t_2 G(2,1)| \right) \tag{11.175}$$

where we have dropped the explicit dependence on energy for neatness. The first two terms in the brackets here are contributions to the single particle energy from single-site scattering (they must be, they only contain information about one site). The indirect exchange contribution to the energy is contained in the final term.

It is a useful property of matrices that the logarithm of the determinant of the matrix is equal to the trace of its logarithm. Using this and dropping everything except the final term in (11.175), we obtain

$$E_{sp}^{(2)} = \frac{1}{\pi} \mathrm{Im} \int_0^{E_F} dw \, \mathrm{Tr} \, \ln(1 - t_1 G(1,2) t_2 G(2,1)) \tag{11.176}$$

Now, we are going to assume our spins are well separated. In this case the $G(1,2)$s are small so we can make an expansion of the logarithm

$$\ln(1 + x) = x - \tfrac{1}{2}x^2 + \cdots \tag{11.177}$$

Keeping only the first term in this expansion of (11.177) yields

$$E_{sp}^{(2)} \approx -\frac{1}{\pi} \mathrm{Im} \int_0^{E_F} dw \, \mathrm{Tr} \, t_1 G(1,2) t_2 G(2,1) \tag{11.178}$$

It is this quantity that we will evaluate to discover the origin of magnetic anisotropy. As we have seen, the t-matrices and structure constants should have a lot of subscripts indicating their angular momentum character. Here we are only interested in spin, so where possible we will label them with only the spin subscript, i.e.

$$E_{sp}^{(2)} \approx -\frac{1}{\pi} \mathrm{Im} \int_0^{E_F} dw \, \mathrm{Tr} \, t_{1 s_1 s_1'} G(1,2) t_{2 s_2 s_2'} G(2,1) \tag{11.179}$$

The t-matrices here are in the (l, m, s) representation of equation (11.98b). However, this form is derived on the assumption that the magnetic moment is in the z-direction. Clearly, this can be true in some local frame associated with each ion, but in general the magnetic moments point in different directions, so in the laboratory frame (the frame of figure 11.14) it will not be sufficient to write both moments pointing in the z-direction. We can get to the laboratory frame by evaluating the t-matrices in the local frame and rotating them through the Euler angles α, β and γ, using standard rotation matrices (Messiah 1965), to the laboratory frame. A spinor is rotated via the following operation

$$\chi_s^{m_s} = \sum_{m_s'=\pm\frac{1}{2}} R_{s,s'}^{1/2}(\alpha,\beta,\gamma)\chi_{s'}^{m_s'} \tag{11.180}$$

with

$$R_{s,s'}^{1/2}(\alpha,\beta,\gamma) = \begin{pmatrix} e^{-i\alpha/2}\cos(\beta/2)e^{-i\gamma/2} & -e^{-i\alpha/2}\sin(\beta/2)e^{i\gamma/2} \\ e^{i\alpha/2}\sin(\beta/2)e^{-i\gamma/2} & e^{i\alpha/2}\cos(\beta/2)e^{i\gamma/2} \end{pmatrix} \tag{11.181}$$

We transform the t-matrices from one frame to another using

$$t_{m_s,m_s'} = \sum_{m_{s1},m_{s2}} R_{m_s,m_{s1}}^{1/2}(\alpha,\beta,\gamma)t_{m_{s1}m_{s2}}R_{m_{s2},m_s'}^{1/2\dagger}(\alpha,\beta,\gamma) \tag{11.182}$$

It is easy to become confused about the meaning of the subscripts in what follows, so let us state explicitly what they mean from here on in. Subscript $+(-)$ means spin parallel (antiparallel) to the z-axis in the local frame. Subscript $\uparrow$ ($\downarrow$) means parallel (antiparallel) to the z-axis in the laboratory frame. So in matrix form the t-matrix is

$$\tilde{t} = \begin{pmatrix} t_{++} & t_{+-} \\ t_{-+} & t_{--} \end{pmatrix} \tag{11.183a}$$

in the local frame and

$$\tilde{t} = \begin{pmatrix} t_{\uparrow\uparrow} & t_{\uparrow\downarrow} \\ t_{\downarrow\uparrow} & t_{\downarrow\downarrow} \end{pmatrix} \tag{11.183b}$$

in the laboratory frame. In our discussion the Euler angle γ is set equal to zero. Then if we put these matrices into (11.182) we find

$$t_{\uparrow\uparrow} = t_{++}\cos^2(\beta/2) + t_{--}\sin^2(\beta/2) - (t_{-+}+t_{+-})\cos(\beta/2)\sin(\beta/2) \tag{11.184a}$$

$$t_{\downarrow\downarrow} = t_{++}\sin^2(\beta/2) + t_{--}\cos^2(\beta/2) + (t_{-+}+t_{+-})\sin(\beta/2)\cos(\beta/2) \tag{11.184b}$$

$$t_{\uparrow\downarrow} = e^{-i\alpha}(\tfrac{1}{2}(t_{++}-t_{--})\sin(\beta) + t_{+-}\cos^2(\beta/2) - t_{-+}\sin^2(\beta/2)) \tag{11.184c}$$

$$t_{\downarrow\uparrow} = e^{i\alpha}(\tfrac{1}{2}(t_{++}-t_{--})\sin(\beta) - t_{+-}\sin^2(\beta/2) + t_{-+}\cos^2(\beta/2)) \tag{11.184d}$$

This is the t-matrix in the laboratory frame. As in the local frame it is a 2×2 matrix with the rest of the quantum numbers being equal. We have seen that the Pauli matrices plus the identity matrix form a complete set of 2×2 matrices, so let us write (11.183b) in vector form

$$\tilde{t} = t_0\tilde{I} + \mathbf{t}\cdot\tilde{\sigma} \tag{11.185}$$

It is not hard to see that

$$t_0 = \frac{1}{2}(t_{\uparrow\uparrow} + t_{\downarrow\downarrow}), \qquad t_x = \frac{1}{2}(t_{\uparrow\downarrow} + t_{\downarrow\uparrow})$$
$$t_y = \frac{i}{2}(t_{\uparrow\downarrow} - t_{\downarrow\uparrow}), \qquad t_z = \frac{1}{2}(t_{\uparrow\uparrow} - t_{\downarrow\downarrow}) \tag{11.186}$$

Now we can use equations (11.184) to write these Cartesian components of the t-matrices in the laboratory frame in terms of the t-matrices calculated relative to the local z-axis. This gives

$$t_0 = \frac{1}{2}(t_{++} + t_{--}) \tag{11.187a}$$

$$t_x = \frac{1}{2}(t_{++} - t_{--}) \sin\beta \cos\alpha + \frac{1}{2}\left(e^{-i\alpha} \cos^2(\beta/2) - e^{i\alpha} \sin^2(\beta/2)\right) t_{+-}$$
$$- \frac{1}{2}\left(e^{-i\alpha} \sin^2(\beta/2) - e^{i\alpha} \cos^2(\beta/2)\right) t_{-+} \tag{11.187b}$$

$$t_y = \frac{1}{2}(t_{++} - t_{--}) \sin\beta \sin\alpha + \frac{i}{2}\left(e^{-i\alpha} \cos^2(\beta/2) + e^{i\alpha} \sin^2(\beta/2)\right) t_{+-}$$
$$- \frac{1}{2}\left(e^{-i\alpha} \sin^2(\beta/2) + e^{i\alpha} \cos^2(\beta/2)\right) t_{-+} \tag{11.187c}$$

$$t_z = \frac{1}{2}(t_{++} - t_{--}) \cos\beta - \frac{1}{2}(t_{+-} + t_{-+}) \sin\beta \tag{11.187d}$$

These equations are still complicated and difficult to interpret. They can be simplified further using elementary trigonometric identities and

$$t_{+-} = \tfrac{1}{2}(t_{+-} + t_{-+}) + \tfrac{1}{2}(t_{+-} - t_{-+}) \tag{11.188a}$$

$$t_{-+} = \tfrac{1}{2}(t_{+-} + t_{-+}) - \tfrac{1}{2}(t_{+-} - t_{-+}) \tag{11.188b}$$

Putting these into (11.187) and substituting the results into (11.185) gives

$$\tilde{t} = \tfrac{1}{2}(t_{++} + t_{--})\tilde{I} + \tfrac{1}{2}(t_{++} - t_{--})\mathbf{S} \cdot \tilde{\sigma}$$
$$+ \left(\tfrac{1}{2}(t_{+-} + t_{-+}) \cos\beta \cos\alpha - \tfrac{i}{2} \sin\alpha(t_{+-} - t_{-+})\right) \tilde{\sigma}_x$$
$$+ \left(\tfrac{1}{2}(t_{+-} + t_{-+}) \cos\beta \sin\alpha + \tfrac{i}{2} \cos\alpha(t_{+-} - t_{-+})\right) \tilde{\sigma}_y$$
$$- \left(\tfrac{1}{2}(t_{+-} + t_{-+}) \sin\beta\right) \tilde{\sigma}_z \tag{11.189}$$

where we have used the components of $\mathbf{S}$ written in the form shown in the caption of figure 11.14. Although (11.189) is somewhat simpler than (11.187) it doesn't get us much further forward in terms of our physical understanding. We will remedy this shortly.

Firstly we have to think a bit more about rotation matrices. Equation
(11.183) is the matrix used to rotate a spinor. If we want to rotate a vector
through these angles we have (in the simple case when $\gamma = 0$)

$$R(\alpha\beta) = \begin{pmatrix} \cos\beta\cos\alpha & -\sin\alpha & \sin\beta\cos\alpha \\ \cos\beta\sin\alpha & \cos\alpha & \sin\beta\sin\alpha \\ -\sin\beta & 0 & \cos\beta \end{pmatrix} \qquad (11.190)$$

Now let us act with this matrix on a unit vector in the x-direction in a
local frame

$$R(\alpha\beta)\begin{pmatrix} 1 \\ 0 \\ 0 \end{pmatrix} = \begin{pmatrix} \cos\beta\cos\alpha \\ \cos\beta\sin\alpha \\ -\sin\beta \end{pmatrix} \qquad (11.191a)$$

and on a unit vector in the y-direction

$$R(\alpha\beta)\begin{pmatrix} 0 \\ 1 \\ 0 \end{pmatrix} = \begin{pmatrix} -\sin\alpha \\ \cos\alpha \\ o \end{pmatrix} \qquad (11.191b)$$

The right hand side of equation (11.191a) is the unit vector in the x-
direction in the local frame written in the laboratory frame. Similarly
(11.191b) is the unit vector in the y-direction in the local frame written
in terms of the laboratory frame coordinates. Compare (11.191) with
(11.189). If we call the unit vectors in the x,- y- and z-directions in the
local frame $\mathbf{i}_1$, $\mathbf{j}_1$ and $\mathbf{k}_1$, respectively, and write them as vectors in the
laboratory frame, (11.189) can be written

$$\tilde{t} = \tfrac{1}{2}(t_{++} + t_{--})\tilde{I} + \tfrac{1}{2}(t_{++} - t_{--})\mathbf{S}\cdot\tilde{\sigma}$$
$$+ \tfrac{1}{2}(t_{+-} + t_{-+})\mathbf{i}_1\cdot\tilde{\sigma} + \tfrac{i}{2}(t_{+-} - t_{-+})\mathbf{j}_1\cdot\tilde{\sigma} \qquad (11.192)$$

This is the t-matrix in the laboratory frame written entirely in terms of
the quantities defined in the local frame where the spin is pointing in the
z-direction, so we can use the formalism of section 11.9 to calculate them.
The only other thing we need to know to evaluate equation (11.179) is the
structure factors from equation (11.119).

The Non-Relativistic Limit, the RKKY Interaction

Firstly, let us look at the non-relativistic limit of the anisotropy energy of
(11.179). Then $t_{+-} = t_{-+} = 0$, there is no spin flip scattering. Therefore
the complicated terms in (11.192) disappear and we are left with

$$\tilde{t} = \frac{1}{2}(t_{++} + t_{--})\tilde{I} + \frac{1}{2}(t_{++} - t_{--})\mathbf{S}\cdot\tilde{\sigma} \qquad (11.193)$$

The first term here is nothing to do with the spin, it is the charge scattering
part of the t-matrix. The second term is the one we are interested in. It is

the magnetic part of the scattering t-matrix. Henceforth in this section we will concentrate solely on this term.

We are going to make some drastic approximations now. The reason for this is to make contact with familiar non-relativistic equations which also make such approximations. So in the structure factors of (11.119) we will assume that only $l'' = 0$ contributions are non-zero. The Gaunt coefficient becomes $1/\sqrt{4\pi}$, and we are left with

$$G_{0\,0000}(\mathbf{R}_1 - \mathbf{R}_2, E) = -\sqrt{4\pi}iqh_0^{(1)}(q|\mathbf{R}_1 - \mathbf{R}_2|)Y_0^0(\widehat{\mathbf{R}_1 - \mathbf{R}_2}) = \frac{e^{iqR_{12}}}{R_{12}}$$
(11.194)

where R_{12} is the magnitude of $\mathbf{R}_1 - \mathbf{R}_2$ and we have used the explicit form of the $l = 0$ spherical Hankel function to write it as a complex exponential. Inserting equation (11.194) and the second term on the right hand side of equation (11.193) (to examine magnetic scattering only) into (11.179) we have

$$E_{\mathrm{sp}}^{(2)m} =$$

$$\frac{-1}{4\pi R_{12}^2}\mathrm{Im}\int_0^{E_F} dw e^{2iqR_{12}}(t_{1++} - t_{1--})(t_{2++} - t_{2--})\mathrm{Tr}\,\mathbf{S}_1\cdot\tilde{\sigma}\mathbf{S}_2\cdot\tilde{\sigma}$$
(11.195)

This can be simplified further using our old friend equation (4.19). Then because the trace of the Pauli matrices is zero, the trace of the term in $\mathbf{S}_1 \times \mathbf{S}_2$ is zero and we only have $\mathbf{S}_1 \cdot \mathbf{S}_2$ left. For $l = 0$ we can assume that the t-matrices are constant as a function of energy, then we are left with a simple integral. In the non-relativistic limit we write the energy of the scattered electron as

$$w = \frac{\hbar^2 q^2}{2m} + mc^2$$
(11.196)

Our integral then becomes

$$\int_0^{E_F} e^{2iq|R_{12}|}dw = \frac{\hbar^2}{4mR_{12}^2}\left(2ik_f R_{12}e^{2ik_f|R_{12}|} - \left(e^{2ik_f|R_{12}|} - 1\right)\right)$$
(11.197)

This integral is most easily proved by changing the variable from w to q using equation (11.196). Finally we put this back into equation (11.195). We assume that the imaginary parts of the t-matrices cancel in the subtraction (I said there were some big assumptions) and then we just have to take the imaginary part of (11.197). After a bit of messing around this yields

$$E_{\mathrm{sp}}^{(2)m} = -\frac{\hbar^2}{16\pi m}(t_{++}^1 - t_{--}^1)(t_{++}^2 - t_{--}^2)\times$$

$$\left(\frac{2k_f R_{12}\cos 2k_f R_{12} - \sin 2k_f R_{12}}{R_{12}^4}\right)\mathbf{S}_1\cdot\mathbf{S}_2 = J\mathbf{S}_1\cdot\mathbf{S}_2 \qquad (11.198)$$

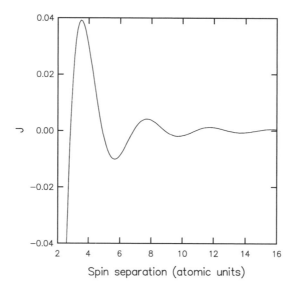

Fig. 11.15.   Plot of J from equation (11.198) as a function of separation of S_1 and S_2. A value of $k_f = 1.5 \times 10^{10}$ m^{-1} was chosen arbitrarily. The prefactors outside the brackets in (11.198) were also arbitrarily set at 1.22.

This is what we were looking for. It is the non-relativistic form of the indirect exchange interaction known as the RKKY interaction after its discoverers (Ruderman and Kittel 1954, Kasuya 1956 and Yosida 1957). Its dependence on the separation of the spins is shown in figure 11.15. The oscillatory nature of J tells us that, as a function of separation, the ground state magnetic ordering of the two ions in figure 11.14 is alternately ferromagnetic and antiferromagnetic. Equation (11.198) is a very rudimentary form of the indirect exchange energy, as we have made some wild approximations to get here. Nonetheless it is this type of interaction which governs the magnetic ordering in localized magnets such as those containing rare earth elements. Note that it only depends upon the magnitude of the vector joining the two scattering sites. There is no dependence upon direction of $\mathbf{R}_{12}$ relative to the spin directions, i.e. this is an isotropic interaction. Several attempts have been made to make (11.198) more realistic (Watson and Freeman 1966, Narita and Kasuya 1984, Fairbairn *et al.* 1979). These have tended to show that the singularity at $R_{12} = 0$ can easily be made to disappear. In fact this singularity is not really serious because we have already assumed that the sites are far apart, so $R_{12} \to 0$ is not included in (11.198).

The Origin of Anisotropy

It turns out that it is not so sensible to ask why magnetic anisotropy occurs in relativistic quantum theory, but rather to ask why it does not occur in non-relativistic theory. We will answer this shortly.

Our t-matrix is given by (11.192), but that is written in terms of the unit vectors in the local frames. It is easy to set up satisfactory local frames. For the frame of spin 1 we use

$$\hat{\mathbf{k}}_1 = \hat{\mathbf{S}}_1, \quad \hat{\mathbf{j}}_1 = \widehat{\mathbf{S}_1 \times \mathbf{S}_2}, \quad \hat{\mathbf{i}}_1 = \widehat{\mathbf{S}_1 \times \mathbf{S}_2} \times \mathbf{S}_1 \qquad (11.199a)$$

$$\hat{\mathbf{k}}_2 = \hat{\mathbf{S}}_2, \quad \hat{\mathbf{j}}_2 = \widehat{\mathbf{S}_2 \times \mathbf{S}_1}, \quad \hat{\mathbf{i}}_2 = \widehat{\mathbf{S}_2 \times \mathbf{S}_1} \times \mathbf{S}_2 \qquad (11.199b)$$

Note that these definitions do produce a set of orthogonal axes, but they are not unique. There is a direction in the problem we are discussing, the vector joining the two spins. We could have set up reference frames in terms of that. However, doing this introduces magnetic anisotropies which are dependent on the choice of frame, and hence are not real. It is necessary to set up the local frames only in terms of the spins. We can use (11.199) and multiply out $\tilde{t}_1$ and $\tilde{t}_2$ as given by (11.192) and we will find that Tr $\tilde{t}_1\tilde{t}_2$ yields t-matrix components multiplied by complicated functions of $\mathbf{S}_1$ and $\mathbf{S}_2$. These do not have any $\mathbf{R}_{12}$ dependence and so cannot result in any anisotropy.

The directional dependence of the magnetization comes from the geometrical part of (11.179). Let us write out (11.179) explicitly including the expansion (11.119) and all the quantum numbers.

$$E_{sp}^{(2)} = -\frac{1}{\pi} \int_0^{E_F} dw \, \mathrm{Tr} \, (4\pi)^2 q^2 \sum_{l_1,l_2} \sum_{m_1,m_2} \sum_{m_3,m_4} \sum_{m_{s1},m_{s2}} t^{(1)}_{l_1 m_1 m_{s1} m_2 m_{s2}}$$

$$\times \left(\sum_{l'm'} i^{l'} C^{l'm'}_{l_1 m_2, l_2 m_3} h^+_{l'}(q|R_{12}|) Y^{m'}_{l'}(\hat{\mathbf{R}}_{12}) \right) \times t^{(2)}_{l_2 m_3 m_{s2} m_4 m_{s1}}$$

$$\times \left(\sum_{l''m''} i^{l''} C^{l''m''}_{l_2 m_4, l_1 m_1} h^+_{l''}(q|R_{12}|) Y^{m''}_{l''}(\hat{\mathbf{R}}_{21}) \right) \qquad (11.200)$$

This is a very complicated expression, and it is difficult to evaluate under even the simplest conditions. It has to be calculated numerically using a computer. However, we will not stop here. We can analyse (11.200) to see if it leads to any anisotropy and, if so, what form it takes. The only place a direction comes into this expression is in the arguments of the spherical harmonics. If we consider just the summations over the m_i

quantum numbers we can extract the important part of (11.200)

$$A = \sum_{m_1,m_2} \sum_{m_3,m_4} \sum_{l'm'} \sum_{l''m''} t^{(1)}_{l_1 m_1 m_{s1} m_2 m_{s2}} t^{(2)}_{l_2 m_3 m_{s2} m_4 m_{s1}} \times$$

$$C^{l'm'}_{l_1 m_2, l_2 m_3} C^{l''m''}_{l_2 m_4, l_1 m_1} Y^{m'}_{l'}(\hat{\mathbf{R}}_{12}) Y^{m''}_{l''}(\hat{\mathbf{R}}_{21})$$

(11.201)

Now, in the non-relativistic limit the t-matrices (in a magnetic field) become independent of the m_i quantum number, i.e. they take the form of equation (11.99b). This means that they can come outside the summations in equation (11.201). The remaining sum over products of Gaunt numbers and spherical harmonics can be shown to be independent of $\hat{\mathbf{R}}_{12}$. This is straightforward and tedious to verify, but just entails inserting the values of the Gaunt numbers, writing the spherical harmonics explicitly, and using some simple trigonometric identities.

If we do not take the non-relativistic limit above, the t-matrices stay inside the summation over m_l and the symmetry that cancelled the dependence of A on $\hat{\mathbf{R}}_{12}$ is broken. For every different possible set of t-matrices there will be a different total anisotropy. But we have not reached the end of this discussion, we can analyse the anisotropy further. Recall that $\hat{\mathbf{R}}_{12}$ stands for two angles θ_{12} and ϕ_{12}, as shown in figure 11.14 for the local frame of spin 1. Now, the spherical harmonics in (11.201) are simple trigonometric functions of these angles. But, with our local reference frames defined by (11.199), we can easily see that

$$\cos\theta_{12} \propto \mathbf{S}_1 \cdot \mathbf{R}_{12}, \qquad \sin\phi_{12} \propto \mathbf{R}_{12} \cdot (\mathbf{S}_1 \times \mathbf{S}_2)$$

(11.202)

and from these $\sin\theta_{12}$ and $\cos\phi_{12}$ can also be found. If the relationships in (11.202) are not obvious, spend a few minutes studying figure 11.14 to convince yourself. In fact, the way we have set up the axes determines the way we express the anisotropies. So the product of spherical harmonics in (11.201) will lead to anisotropies like

$$A_1 \propto (\hat{\mathbf{S}}_1 \cdot \hat{\mathbf{R}}_{12})(\hat{\mathbf{S}}_2 \cdot \hat{\mathbf{R}}_{12})$$

(11.203)

which is a pseudo-dipolar interaction, and

$$A_2 \propto \left(\hat{\mathbf{R}}_{12} \cdot \left(\mathbf{S}_1 \widehat{\times \mathbf{S}}_2\right)\right)^2$$

(11.204)

which is known as the Dzyaloshinsky–Moriya interaction (except for the fact that it is to the power two here) (Dzyaloshinsky 1958, Moriya 1960). Both these anisotropies are thought to be responsible for the anisotropy fields found in spin glasses. Together with the isotropic interaction $\hat{\mathbf{S}}_1 \cdot \hat{\mathbf{S}}_2$ they may also determine the magnetic structure of magnetic impurities in a non-magnetic host, and influence (along with effects from larger numbers of sites) the magnetic structures of rare earth materials, and any other

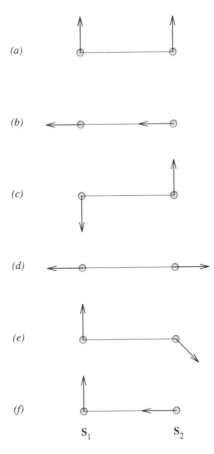

Fig. 11.16. Magnetic anisotropy introduced by relativity. For an explanation of parts (a) to (f) see text.

materials containing localized spins. Staunton *et al.* (1988) have written down a general form for the anisotropy energy of (11.200):

$$E_{sp}^{(2)} = \sum_{i=0}^{2l_{max}} \sum_{j=0}^{2l_{max}} a_{ij} (\hat{\mathbf{S}}_1 \cdot \hat{\mathbf{S}}_2) \left[(\hat{\mathbf{R}}_{12} \cdot \hat{\mathbf{S}}_1)(\hat{\mathbf{R}}_{12} \cdot \hat{\mathbf{S}}_2) \right]^i \left[\hat{\mathbf{R}}_{12} \cdot (\mathbf{S}_1 \widehat{\times} \mathbf{S}_2) \right]^{2j}$$

(11.205)

where a_{ij} contains the t-matrices, Hankel functions and sundry numerical factors. The bracketed $\hat{\mathbf{S}}_1 \cdot \hat{\mathbf{S}}_2$ in (11.205) indicates that it is also a function of the angle between the spins. So, for example, the two spin arrangements at the top of figure 11.16 are both ferromagnetic, both have parallel spins and so both have the same value of a_{ij}.

We will now evaluate (11.205) for some simple cases shown in figure 11.16. In this figure we show six simple spin arrangements, two

ferromagnetic, two antiferromagnetic and two with the spins at right angles to one another, labelled (a) to (f) respectively. Equation (11.205) is
evaluated for each of these in turn. We will assume, in the examples of
figure 11.16, that angles of 90° and 180° are infinitesimally different from
these values so that the scalar and vector products in (11.205) are not
zero, but raising such tiny factors to any power other than zero gives a
negligible contribution to the anisotropy energy. We will also set $l_{max} = 1$
to make the sums as simple as possible.

(a) Here we have two parallel spins, so $\mathbf{S}_1 \times \mathbf{S}_2$ is zero. Both spins are
perpendicular to the vector joining them and so $(\hat{\mathbf{R}}_{12} \cdot \hat{\mathbf{S}}_1)(\hat{\mathbf{R}}_{12} \cdot \hat{\mathbf{S}}_2)$
is also zero. Therefore the only non-zero term in the summations in
(11.205) is $i = j = 0$ and so we have

$$E_{sp}^{(2)} = a_{0,0}(\uparrow\uparrow) \qquad\qquad (11.206a)$$

(b) Here we have two parallel spins again, so $\mathbf{S}_1 \times \mathbf{S}_2$ is zero. This time
both spins are parallel to the vector joining them. Therefore $\hat{\mathbf{R}}_{12} \cdot \hat{\mathbf{S}}_1 =
\hat{\mathbf{R}}_{12} \cdot \hat{\mathbf{S}}_2 = 1$, so we only require the $j = 0$ term. However, we need all
terms in the sum over i and we have

$$E_{sp}^{(2)} = a_{0,0}(\uparrow\uparrow) + a_{10}(\uparrow\uparrow) + a_{20}(\uparrow\uparrow) \qquad\qquad (11.206b)$$

(c) Here we have an antiparallel arrangement of spins. Therefore $\mathbf{S}_1 \times \mathbf{S}_2$
is zero again. Both spins are also perpendicular to the vector joining
them and so $\hat{\mathbf{R}}_{12} \cdot \hat{\mathbf{S}}_1 = \hat{\mathbf{R}}_{12} \cdot \hat{\mathbf{S}}_2 = 0$ Again we only have to consider
the $i = j = 0$ term in the summations in (11.205). But this does not
give the same energy as example (a) because the coefficients a_{ij} vary
with relative spin direction, so

$$E_{sp}^{(2)} = a_{00}(\uparrow\downarrow) \qquad\qquad (11.206c)$$

(d) This is another antiparallel arrangement, but this time along the line
joining the spins. Again $\hat{\mathbf{S}}_1 \times \mathbf{S}_2$ is zero. However $\hat{\mathbf{R}}_{12} \cdot \hat{\mathbf{S}}_1 = 1$ and
$\hat{\mathbf{R}}_{12} \cdot \hat{\mathbf{S}}_2 = -1$ so the pseudo-dipolar term is -1. In the summations
this time $j = 0$, and we only require the sum over i. This gives

$$E_{sp}^{(2)} = a_{00}(\uparrow\downarrow) - a_{10}(\uparrow\downarrow) + a_{20}(\uparrow\downarrow) \qquad\qquad (11.206d)$$

(e) Here the spins are at right angles to one another and $\hat{\mathbf{S}}_2$ points out
of the page. For this arrangement $\hat{\mathbf{R}}_{12} \cdot (\hat{\mathbf{S}}_1 \times \mathbf{S}_2) = 1$, but both dot
products in the pseudo-dipolar term are zero. Therefore in this case
we only have $i = 0$ and we have to do the sum over j

$$E_{sp}^{(2)} = a_{00}(\uparrow\rightarrow) + a_{01}(\uparrow\rightarrow) + a_{02}(\uparrow\rightarrow) \qquad\qquad (11.206e)$$

(*f*) Here we have another perpendicular arrangement of the spins. In this case both the Dzyaloshinsky–Moriya and the pseudo-dipolar terms are zero and so only the $i = j = 0$ terms in the summation contribute

$$E_{sp}^{(2)} = a_{00}(\uparrow\rightarrow) \tag{11.206f}$$

In examples (*a*) to (*f*) we see some examples of anisotropy. Take (*e*) and (*f*) for example: if the energy associated with these two spin arrangements were evaluated using equation (11.198) both arrangements would have the same energy. Here we have seen that relativistic quantum theory predicts different values for the energy yielding an anisotropy. A similar thing has occurred for the pair (*a*) and (*b*) and the pair (*c*) and (*d*). If we were to take spin arrangements for which the angles were not all multiples of 90° we would start getting numbers in the scalar and vector products of equation (11.205) that would have to be evaluated explicitly. Which of the spin arrangements (*a*) to (*f*) (or some other) is most stable will depend on the detailed values of the a_{ij} coefficients, and whether they are positive or negative.

From the examples above we can also infer some facts about the role played by the various terms in (11.205). It is clear that $\hat{\mathbf{S}}_1 \cdot \hat{\mathbf{S}}_2$ and the pseudo-dipolar term are dominant if the spins are arranged parallel or antiparallel to one another. If, on the other hand, the spins are perpendicular to one another the interaction is dominated by the Dzyaloshinsky–Moriya terms.

In summary then we have seen that magnetic isotropy does not occur in non-relativistic quantum mechanics because the *t*-matrices are independent of the m_l quantum number and hence the summations in equation (11.201) become independent of $\hat{\mathbf{R}}_{12}$. This symmetry is broken in relativistic theory, and anisotropy occurs. Although it is difficult to evaluate the anisotropy energy of (11.200), even for a simple model we have seen that the basic building blocks of the anisotropy will be of the Dzyaloshinsky–Moriya and pseudo-dipolar types, and we have shown some examples of the effect of such terms.

Finally, Staunton *et al.* (1989) have shown that three-site interaction energies can be of the same order as the two-site energies discussed here. Therefore, although the above discussion clearly shows that the origin of magnetic anisotropy is in the subtle interplay between relativistic and magnetic effects, it cannot be said to describe anisotropy in real materials in any realistic way.

12

Electrons and Photons

One of the principal applications of quantum theory in physics is in the interpretation of spectroscopies. Spectroscopy is one of the key tools for learning about condensed matter on the microscopic scale. In this chapter we are going to consider the theory of spectroscopy on a quantum mechanical level with particular emphasis on effects that are intrinsically relativistic. This boils down to a study of the interaction of the electrons in the material with incident photons.[*]

A bit of quantum field theory is unavoidable in this chapter. The field theory here is as elementary as it gets. Furthermore, it is conceptually easy and is introduced as a natural extension of the relativistic quantum theory already covered in this book, so it should present no difficulty.

If you scan through this chapter the mathematics appears pretty daunting, (so what's new). Don't worry, it could be worse, at least we make plenty of use of the Dirac notation introduced in chapter 2. Many of the equations in this chapter would be horrendous without it. In the first sections we discuss some properties of the photon and quantization of the electromagnetic field. Then we come to the backbone of the chapter, time-dependent perturbation theory, from which the Golden Rule for transition rates is derived to second order. The Golden Rule contains all the physics we want to discuss, and the rest of the chapter is a description of some of its applications. In the third section we discuss the first order Golden Rule and its use to describe photon absorption and emission processes. We see that the dichroism discussed for the one-electron atom earlier also occurs in condensed matter systems that exhibit a magnetic moment. Equating the power absorbed as calculated from the Golden Rule with a macroscopic expression for the power absorbed enables us to

---

[*] Both chapters 11 and 12 have benefited substantially from numerous discussions with Dr E. Arola.

derive equations to describe Faraday and Kerr rotations. These are shown to be purely relativistic phenomena. In the next part of the chapter we consider some general points concerning the second order Golden Rule, and finally we describe its application to four different types of photon scattering, Thomson scattering, Rayleigh scattering, Compton scattering and magnetic scattering. We discuss the physics illustrated by each of these types of scattering in turn.

12.1 Photon Polarization and Angular Momentum

Consider an electromagnetic plane wave in a volume V containing a large number of wavelengths and assume the wave is propagating in the z-direction with angular frequency ω. Classically the energy associated with this wave is given approximately by

$$E_t = \frac{1}{2}\epsilon_0 \int |\mathbf{E}(\mathbf{r})|^2 d\mathbf{r} = \frac{1}{2}\epsilon_0 |\mathbf{E}(\mathbf{r})|^2 V \tag{12.1}$$

where $\mathbf{E}(\mathbf{r})$ is the electric vector of the wave and $\epsilon_0 |\mathbf{E}(\mathbf{r})|^2/2$ is the average energy per unit volume of the wave. From the quantum mechanical point of view this must be equal to an integer times $\hbar\omega$. If we have a single photon we can write

$$\frac{\epsilon_0 |\mathbf{E}(\mathbf{r})|^2 V}{2} = \hbar\omega \tag{12.2}$$

As the photon is travelling in the z-direction its electric vector can only have x- and y-components. Let us write a state vector to describe it

$$|\Psi\rangle = \begin{pmatrix} \psi_x \\ \psi_y \end{pmatrix} = \sqrt{\frac{V\epsilon_0}{2\hbar\omega}} \begin{pmatrix} E_x \\ E_y \end{pmatrix} \tag{12.3}$$

Comparison with (12.2) shows that this state vector is normalized

$$|\psi_x|^2 + |\psi_y|^2 = 1 \tag{12.4}$$

For propagation in the z-direction, possible photon state vectors are

$$|\Psi\rangle = |\epsilon_x\rangle = \left| \begin{pmatrix} 1 \\ 0 \\ 0 \end{pmatrix} \right\rangle, \qquad |\Psi\rangle = |\epsilon_y\rangle = \left| \begin{pmatrix} 0 \\ 1 \\ 0 \end{pmatrix} \right\rangle \tag{12.5a}$$

which represent linear polarization in the x- and y-directions, and

$$|\Psi\rangle = |\epsilon_+\rangle = \left| \frac{1}{\sqrt{2}} \begin{pmatrix} 1 \\ i \\ 0 \end{pmatrix} \right\rangle, \qquad |\Psi\rangle = |\epsilon_-\rangle = \left| \frac{1}{\sqrt{2}} \begin{pmatrix} 1 \\ -i \\ 0 \end{pmatrix} \right\rangle \tag{12.5b}$$

which represent right and left circular polarized waves respectively. It is easy to see from these that circular polarized radiation may be regarded as a linear combination of linear polarized radiation and *vice versa*.

The photon, like any other particle, can possess an angular momentum. However, with the photon there are some added difficulties of interpretation, not found in other particles. In chapter 5 we defined the spin of the electron in the electron rest frame. As we know the photon has no rest frame, so defining the spin is not so straightforward. Furthermore the photon has no rest mass, so any discussion of orbital angular momentum as $L = mv \times r$ is also doomed to failure. It is really only possible to discuss the total angular momentum of a photon. The photon angular momentum can be discussed by considering the angular momentum of classical fields and quantizing that. This is rather mathematically complicated so we will not adopt this approach (since when has complication put us off?). Instead we will appeal to our understanding of the relationship between angular momentum and rotations, and to experiment.

Consider the circular polarization vectors above. Let us rotate them through an infinitesimal angle $\delta\theta$ about the propagation direction using rotation matrices (equation (11.181) with $\alpha = \gamma = 0$). This gives

$$|\epsilon_\pm\rangle \rightarrow |\epsilon_\pm\rangle \mp i\delta\theta|\epsilon_\pm\rangle \tag{12.6}$$

However, it would certainly be possible to adopt the method of section 2.2 and use the angular momentum operators to perform the rotation. According to equation (2.21) this would yield

$$|\epsilon_\pm\rangle \rightarrow |\epsilon_\pm\rangle - \frac{i}{\hbar}\delta\theta\hat{L}_z|\epsilon_\pm\rangle \tag{12.7}$$

where we have taken the z-component of angular momentum only, because by construction of this problem the rotation is about the z-axis. The only way for (12.6) and (12.7) to be consistent with each other is if

$$\hat{L}_z|\epsilon_\pm\rangle = \pm\hbar|\epsilon_\pm\rangle \tag{12.8}$$

So it looks as though the z-component of angular momentum of the photon takes on the value $+\hbar$ for left circularly polarized light and $-\hbar$ for right circularly polarized light. This is verified by experiment. Whenever a photon incident in the z-direction is absorbed by an atom (or anything else) it transfers angular momentum $\pm\hbar$ in the z-direction to the atom (never zero). The correct formula for the angular momentum of a general photon of state vector $|\Psi\rangle$ is

$$L_z = \hbar\left(|\langle\epsilon_+|\Psi\rangle|^2 - |\langle\epsilon_-|\Psi\rangle|^2\right) \tag{12.9}$$

where $|\langle\epsilon_+|\Psi\rangle|^2$ is the probability of the photon's being right circularly polarized and $|\langle\epsilon_-|\Psi\rangle|^2$ is the probability of its being left circularly polarized. A general photon which is not in either of these polarization states can be regarded as having a probability of being in each state, and hence only has a probability of having z-component of angular momentum $\hbar$

or $-\hbar$. With these possible z-components we can say that the total intrinsic angular momentum quantum number for a photon is $s = 1$, and hence the matrix operators of equations (2.40) determine the angular momentum. Furthermore (2.41) shows us what the eigenvectors with $m_s = 1$ should be (compare equations (2.41) and (12.5)).

There are a couple of final points to discuss to make this section complete. Firstly we have defined the polarization of light, but we have not defined unpolarized light. Unpolarized light is light that has equal probability of being in any polarization state. Secondly we have only discussed the intrinsic angular momentum of the photon here. In nuclear physics multipolar transitions occur with the emission of photons which can carry off angular momentum in units of an integer greater than one times $\hbar$. These may also occur (usually very weakly) in condensed matter physics. Such a photon carries off its intrinsic spin plus some units of orbital angular momentum. However, if such a photon were then to take part in some absorption process then the angular momentum it imparted parallel to its direction of motion would still be only $\pm\hbar$.

12.2 Quantizing the Electromagnetic Field

In the first chapter, we saw in equations (1.45) and (1.46) that electromagnetism is intimately related to the wave theory of light. Of course this is a purely classical theory, and for a complete theory of electromagnetism we must unify electromagnetism and the quantum theory.

What follows is as simple a discussion of field quantization as possible. This book is not about quantum field theory, so we do the minimum for the rest of this chapter to make sense. We assume the electromagnetic field can be quantized and build up an expression for it (actually for the vector potential) step by step. More advanced treatments are given by Berestetskii *et al.* (1982) and by Mandl and Shaw (1984).

Let us define a couple of operators: $c^\dagger(\mathbf{k}, \lambda)$ is a creation operator for a photon of wavevector $\mathbf{k}$ and polarization λ, and $c(\mathbf{k}, \lambda)$ is an annihilation operator for a photon of wavevector $\mathbf{k}$ and polarization λ. The photon is a boson, so these operators obey equations (6.85), (6.89) and (6.90).

Now we write down several familiar and useful operators in terms of these operators. Consider a system that contains a number N of free photons. From (6.93) the number of photons is

$$\hat{N} = \sum_{\mathbf{k},\lambda} c^\dagger(\mathbf{k}, \lambda) c(\mathbf{k}, \lambda) \tag{12.10}$$

In analogy with (6.113) the Hamiltonian is

$$\hat{H} = \sum_{\mathbf{k},\lambda} c^\dagger(\mathbf{k}, \lambda) c(\mathbf{k}, \lambda) \hbar c k \tag{12.11}$$

and the momentum is

$$\hat{\mathbf{p}} = \sum_{\mathbf{k},\lambda} c^{\dagger}(\mathbf{k}, \lambda) c(\mathbf{k}, \lambda) \hbar \mathbf{k} \qquad (12.12)$$

where we have summed over the continuous variable $\mathbf{k}$ and polarization
state λ. We have defined these operators in terms of $\mathbf{k}$, i.e. in momentum
space. They can be Fourier transformed into coordinate space

$$c^{\dagger}(\mathbf{r}, \lambda) = \frac{1}{\sqrt{V}} \sum_{\mathbf{k}} c^{\dagger}(\mathbf{k}, \lambda) e^{-i\mathbf{k}\cdot\mathbf{r}} \qquad (12.13a)$$

$$c(\mathbf{r}, \lambda) = \frac{1}{\sqrt{V}} \sum_{\mathbf{k}} c(\mathbf{k}, \lambda) e^{i\mathbf{k}\cdot\mathbf{r}} \qquad (12.13b)$$

where V is the normalization volume. There is nothing particularly unex-
pected so far. The subtlety comes because we make the assumption that
the vector potential associated with the photon can be written in terms of
the operators in equation (12.13) as

$$A(\mathbf{r}) = \sum_{\mathbf{k},\lambda} \left(\frac{\hbar \mu_0 c^2}{2V \omega_{\mathbf{k}}} \right)^{1/2} (c(\mathbf{k}, \lambda) \epsilon_{\lambda} e^{i\mathbf{k}\cdot\mathbf{r}} + c^{\dagger}(\mathbf{k}, \lambda) \epsilon_{\lambda}^* e^{-i\mathbf{k}\cdot\mathbf{r}}) \qquad (12.14)$$

where μ_0 is the permeability of free space and ϵ_{λ} is the polarization vector
which depends on the propagation direction $\mathbf{k}$. However, we have not
written this $\mathbf{k}$ dependence explicitly in (12.14). In this theory, it is the
polarization vector that gives the vector potential its vector nature. The
coefficient can be understood on the basis of equation (12.3) and the
relation (1.50) for the electric field in terms of the vector potential. We
have almost finished the quantum field theory now. There is just one more
point to make. It may well be that the vector potential is a function of
time. In that case we have to multiply by a phase factor

$$c_t(\mathbf{k}, \lambda) \rightarrow c(\mathbf{k}, \lambda) e^{-i\omega_{\mathbf{k}} t} \qquad (12.15a)$$

$$c_t^{\dagger}(\mathbf{k}, \lambda) \rightarrow c^{\dagger}(\mathbf{k}, \lambda) e^{i\omega_{\mathbf{k}} t} \qquad (12.15b)$$

Now that we have completed the mathematics of this section, let us think
about what it means. We have come to a rather weird interpretation of the
electromagnetic field, in which we regard the magnetic and electric fields
as being mathematically related to this vector potential, but the vector
potential is now a radiation field made up of harmonic oscillators, one
for every value of $\mathbf{k}$ and λ. The fundamental quanta of these oscillators
are the photons. This is the description we will use of the photon in our
discussion of the interaction of electrons and photons.

12.3 Time-Dependent Perturbation Theory

Perturbation theory is at the heart of the calculation of virtually all spectroscopies. In this section we are going to determine the equations of time-dependent perturbation theory consistent with the 4×4 nature of the Hamiltonian and the multi-component wavefunction.

Consider a system describable by a Hamiltonian $\hat{H}_0$. The Hamiltonian contains external potentials $V(\mathbf{r})$ and $\mathbf{A}(\mathbf{r})$ and has eigenstates such that

$$\hat{H}_0 \psi_n(\mathbf{r}) = W_n \psi_n(\mathbf{r}) \tag{12.16}$$

The time-dependent solutions of this equation are

$$\Psi_n(\mathbf{r}, t) = e^{-iW_n t/\hbar} \psi_n(\mathbf{r}) \tag{12.17}$$

where $\Psi_n(\mathbf{r}, t)$ satisfy

$$i\hbar \frac{\partial \Psi_n(\mathbf{r}, t)}{\partial t} = (c\tilde{\boldsymbol{\alpha}} \cdot (\hat{\mathbf{p}} - e\mathbf{A}(\mathbf{r})) + \tilde{\beta} mc^2 + V(\mathbf{r}))\Psi_n(\mathbf{r}, t) \tag{12.18}$$

Now let us introduce a small time-dependent perturbing potential so that

$$V'(\mathbf{r}, t) = V(\mathbf{r}) + v(\mathbf{r}, t) \tag{12.19}$$

where $v(\mathbf{r}, t)$ may include a perturbation in the vector, as well as scalar, potential (as in equation (8.82)). Now we assume the solution to the Dirac equation for the potential of equation (12.19) can be written as a linear combination of the solutions to the unperturbed Hamiltonian, i.e.

$$\Psi'(\mathbf{r}, t) = \sum_n a_n(t) \Psi_n(\mathbf{r}, t) \tag{12.20}$$

where the coefficients $a_n(t)$ are, in general, functions of time. Putting $\Psi'(\mathbf{r}, t)$ into the time-dependent Dirac equation gives

$$i\hbar \frac{\partial \Psi'(\mathbf{r}, t)}{\partial t} = (c\tilde{\boldsymbol{\alpha}} \cdot (\hat{\mathbf{p}} - e\mathbf{A}(\mathbf{r})) + \tilde{\beta} mc^2 + V'(\mathbf{r}, t))\Psi'(\mathbf{r}, t) \tag{12.21}$$

If we substitute into here the expansion of $\Psi'(\mathbf{r}, t)$ in terms of the unperturbed eigenstates, equation (12.20), and rearrange slightly we find

$$\sum_n a_n(t) \left[i\hbar \frac{\partial}{\partial t} - c\tilde{\boldsymbol{\alpha}} \cdot (\hat{\mathbf{p}} - e\mathbf{A}(\mathbf{r})) - \tilde{\beta} mc^2 - V(\mathbf{r}) \right] \Psi_n(\mathbf{r}, t)$$

$$- \sum_n a_n(t) v(\mathbf{r}, t) \Psi_n(\mathbf{r}, t) + \sum_n i\hbar \Psi_n(\mathbf{r}, t) \frac{da_n(t)}{dt} = 0 \tag{12.22}$$

Now the term in square brackets acting on the unperturbed eigenfunctions is related to the unperturbed Dirac Hamiltonian, so that term is identically equal to zero, and we are left with

$$i\hbar \sum_n \Psi_n(\mathbf{r}, t) \frac{da_n(t)}{dt} = \sum_n a_n(t) v(\mathbf{r}, t) \Psi_n(\mathbf{r}, t) \tag{12.23}$$

To proceed further we premultiply each side of (12.23) by $\Psi_m^\dagger(\mathbf{r}, t)$, i.e. by a particular eigenfunction of the unperturbed Hamiltonian. If we then integrate over all spatial coordinates we get

$$i\hbar \sum_n \frac{da_n(t)}{dt} e^{-i(W_n - W_m)t/\hbar} \int \psi_m^\dagger(\mathbf{r})\psi_n(\mathbf{r})d\mathbf{r} =$$

$$\sum_n a_n(t) e^{-i(W_n - W_m)t/\hbar} \int \psi_m^\dagger(\mathbf{r})v(\mathbf{r}, t)\psi_n(\mathbf{r})d\mathbf{r} \qquad (12.24)$$

As the $\psi_n(\mathbf{r})$ form an orthonormal set this becomes

$$\frac{da_m(t)}{dt} = -\frac{i}{\hbar} \sum_n \langle m|v(t)|n\rangle e^{i(W_m - W_n)t/\hbar} a_n(t) \qquad (12.25)$$

where $\langle m|v(t)|n\rangle$ is the matrix element of the perturbation defined using Dirac notation. Comparison of equations (12.25) and (12.20) tells us that (12.25) describes the time development of the perturbed eigenfunction, provided we know the values of $a_n(t)$ at the time the perturbation is switched on. A special case of the above, which is actually of very wide applicability, is that before the perturbation is switched on only one state is occupied. In that case $a_n(0) = \delta_{nk}$ where k represents the quantum numbers defining the initial state. Then the first order approximation (represented by superscript 1) to (12.25) reads

$$\frac{da_m^{(1)}(t)}{dt} = -\frac{i}{\hbar} e^{i(W_m - W_k)t/\hbar} \langle m|v|k\rangle \qquad (12.26)$$

and this is easy to solve. We just integrate both sides with respect to t to get

$$a_m^{(1)}(t) = \delta_{mk} - \frac{i}{\hbar} \int_0^t dt' \langle m|v(t')|k\rangle e^{i(W_m - W_k)t'/\hbar} \qquad (12.27)$$

This may look like a trivial integral, but, of course, the matrix element $\langle m|v|k\rangle$ depends on t' as it contains the time-dependent part of the perturbation, so it cannot be done analytically. Equation (12.27) can be used to find $a_m^{(1)}(t)$ for all values of m.

It may be that $a_m^{(1)}(t)$ is zero or that we require a more accurate estimation of $a_m(t)$. If so, we can obtain a better value by substituting $a_m(t) = a_m^{(1)}(t) + a_m^{(2)}(t)$ into the left side of (12.25), and $a_n^{(1)}(t)$ into the right side, and we have

$$\frac{da_m^{(1)}(t)}{dt} + \frac{da_m^{(2)}(t)}{dt} = -\frac{i}{\hbar} \sum_n \langle m|v(t)|n\rangle e^{i(W_m - W_n)t/\hbar} \times$$

$$\left[\delta_{nk} - \frac{i}{\hbar} \int_0^t \langle n|v(t')|k\rangle e^{i(W_n - W_k)t'/\hbar} dt' \right] \qquad (12.28)$$

We can integrate over time, and recall the initial conditions $a_m^{(1)}(0) = 1$ and $a_m^{(2)}(0) = 0$, then with the help of (12.26) we find that

$$a_m^{(2)}(t) = -\frac{1}{\hbar^2} \sum_n \int_0^t dt'' \int_0^{t''} dt'$$

$$\langle m|v(t'')|n\rangle e^{i(W_m - W_n)t''/\hbar} \langle n|v(t')|k\rangle e^{i(W_n - W_k)t'/\hbar} \tag{12.29}$$

Equation (12.27) is the amplitude for a direct transition from state $|k\rangle$ to state $|m\rangle$ whereas (12.29) is the amplitude for the transition to go via some intermediate virtual states $|n\rangle$, i.e. they are the first and second order amplitudes for this transition. The $a_m^{(1)}(t)$ and $a_m^{(2)}(t)$ (and all higher order terms) have to be added to give the total amplitude for the transition. To go further we could use $a_m^{(3)}(t)$ in (12.25) to get the third order term and so on. This is just a perturbation series reminiscent of the Born series we considered in chapter 11 on electron scattering; see equation (11.144) for example. Henceforth we shall restrict ourselves to going no higher than second order in our expansion of $a_m(t)$. This sum can then be squared to give the probability of the transition taking place. The total probability of the transition's occurring at time t is

$$P_{mk} = |a_m(t)|^2 = |a_m^{(1)}(t) + a_m^{(2)}(t)|^2 \tag{12.30}$$

and the average transition rate can be defined as

$$T_{mk} = \lim_{t\to\infty} P_{mk}/t = \lim_{t\to\infty} |a_m^{(1)}(t) + a_m^{(2)}(t)|^2/t \tag{12.31}$$

As seen earlier (equation (8.82)), the perturbation due to a photon is

$$\hat{H}'(\mathbf{r}, t) = v(\mathbf{r}, t) = -ec\tilde{\alpha} \cdot \mathbf{A}(\mathbf{r}, t) \tag{12.32}$$

To use this perturbation we have to substitute the quantized field form of the vector potential, (12.14), into (12.32). If we do that the perturbation contains two terms, i.e.

$$\hat{H}'(\mathbf{r}, t) = -ec \sum_{\mathbf{k},\lambda} \left(\frac{\hbar\mu_0 c^2}{2V\omega_\mathbf{k}}\right)^{\frac{1}{2}} c_t(\mathbf{k}, \lambda)\tilde{\alpha} \cdot \epsilon_\lambda e^{i\mathbf{k}\cdot\mathbf{r}}$$

$$- ec \sum_{\mathbf{k},\lambda} \left(\frac{\hbar\mu_0 c^2}{2V\omega_\mathbf{k}}\right)^{\frac{1}{2}} c_t^\dagger(\mathbf{k}, \lambda)\tilde{\alpha} \cdot \epsilon_\lambda^* e^{-i\mathbf{k}\cdot\mathbf{r}} = \hat{H}_1' + \hat{H}_2' \tag{12.33}$$

Each term in this operator is linear in the photon creation or annihilation operators. Equation (12.33) is going to be used in the evaluation of matrix elements of the perturbation like those in (12.27) and (12.29). Let us see how this works. As we shall see, in relativistic quantum theory, $a_m^{(1)}(t)$ and $a_m^{(2)}(t)$ describe different transitions. Hence, only one of them ever

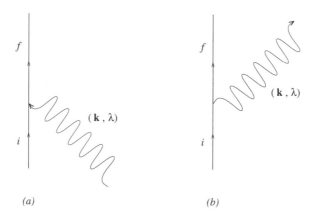

Fig. 12.1. (*a*) Feynman diagram illustrating absorption, (*b*) Feynman diagram illustrating emission.

contributes to a given transition. Both terms are important and we will consider both in turn.

Firstly let's think about what happens when $a_m^{(2)}(t)$ is zero. Look at what happens to the photons when we put (12.32) and (12.33) into (12.27). You can see immediately that the photon creation and annihilation operators can act only once; we can either create or destroy a single photon with a corresponding change of state of the electron system. This means we are describing either absorption or emission of a single photon. These processes are shown schematically in figure 12.1. Note that in all these diagrams time proceeds forwards as we go up the page. In figure 12.1*a* we have electrons represented by the straight line in state *i* and a photon $(\mathbf{k}, \lambda)$. The photon then interacts with the electrons, disappearing itself, and leaving the electron system in a state *f*. This process represents absorption of a photon. In figure 12.1*b* we have an electron system in state *i*. At some time a change of state of the electron system occurs and a single photon $(\mathbf{k}, \lambda)$ is given off. After this the electron system is in its final state *f* and there exists a photon which has become separate from the electron system and which can, in principle, be detected. This process is known as emission. Figures 12.1*a* and *b* are examples of Feynman diagrams. They are a pictorial representation of the perturbation theory we are in the process of deriving.

In describing emission or absorption only one of $\hat{H}_1'$ or $\hat{H}_2'$ need to be considered. This is best illustrated by example and next, almost as an aside, we show how this is true for an absorption process.

At this point the Dirac notation really comes into its own. We can write down the eigenstates of the system without knowing exactly what they are. Recall that $|m\rangle$ represents an eigenstate of the whole system, and that means photons as well as electrons and other particles. Although the expressions for the scattering amplitudes $a_m^{(1)}(t)$ and $a_m^{(2)}(t)$ have been derived using standard quantum mechanics, they are also correct in the context of quantum field theory. Let us specify the initial, intermediate and final states a bit more precisely. We will consider a general state consisting of $n_t + 1$ electrons, and one or zero photons. The process we will use for illustration involves the excitation of one electron from state 1 to state 2, with the absorption of a photon. If n_1 is the number of electrons in state 1, n_2 is the number of electrons in state 2, n_t represents the electrons in all other states, and n_{ph} is the number of photons, then the initial state and final states are respectively

$$|i\rangle = |n_t, n_1 = 1, n_2 = 0, n_{ph}(\mathbf{k}, \lambda) = 1\rangle = |n_t, 1, 0, 1\rangle \tag{12.34a}$$

$$|f\rangle = |n_t, n_1 = 0, n_2 = 1, n_{ph}(\mathbf{k}, \lambda) = 0\rangle = |n_t, 0, 1, 0\rangle \tag{12.34b}$$

We may separate the electron and photon parts of these states

$$|i\rangle = |n_t, 1, 0\rangle \otimes |1\rangle, \qquad\qquad |f\rangle = |n_t, 0, 1\rangle \otimes |0\rangle \tag{12.35}$$

The symbol $\otimes$ is simply the operator defining the operation of separating the electron and photon parts of these states. Consider the photon part of $\hat{H}_2'$ operating between the photon parts of the states $|f\rangle$ and $|i\rangle$

$$\langle 0|c^\dagger(\mathbf{k}, \lambda)|1\rangle = \sqrt{2}\langle 0|2\rangle = 0 \tag{12.36a}$$

where we have used equation (6.89) and we know that all photon states must be orthonormal to one another. So in the absorption process $\hat{H}_2'$ plays no role. Let us look at the photon part of $\hat{H}_1'$, using (12.13b)

$$\langle 0|c(\mathbf{k}, \lambda)|1\rangle = \langle 0|0\rangle = 1 \tag{12.36b}$$

where we have again used the orthonormality. Hence only $\hat{H}_1'$ is required to describe an absorption process. If we were thinking about emission the reverse would be true: only $\hat{H}_2'$ is necessary to describe emission.

Now, let us return to the calculation of $a_n^{(1)}(t)$. Equation (12.27) is

$$a_f^{(1)}(t)$$

$$= \frac{i}{\hbar} \int_0^t dt' \langle f|ec \left(\frac{\hbar\mu_0 c^2}{2V\omega_k}\right)^{\frac{1}{2}} \tilde{\alpha} \cdot \epsilon_\lambda^{\{*\}} c^{\{\dagger\}} e^{\pm i\mathbf{k}\cdot\mathbf{r}}|i\rangle e^{i(W_f - W_i \mp \hbar\omega_k)t'/\hbar}$$

$$= -\frac{i}{\hbar} \langle f|\hat{H}_{a(e)}'|i\rangle \int_0^t dt' e^{i(W_f - W_i \mp \hbar\omega_k)t'/\hbar} \tag{12.37}$$

where the time-independent part of the perturbation is now

$$\hat{H}'_{a(e)} = -ec \left(\frac{\hbar \mu_0 c^2}{2V \omega_k} \right)^{\frac{1}{2}} \tilde{\alpha} \cdot \epsilon_\lambda^{\{*\}} c^{\{\dagger\}} e^{\pm i\mathbf{k} \cdot \mathbf{r}} \tag{12.38a}$$

and the time-dependent part is

$$\hat{H}'_{a(e)}(t) = \hat{H}'_{a(e)} e^{\mp i\omega t} \tag{12.38b}$$

Here the summation has gone because we are considering only one photon. Absorption is described by the subscript a, the upper signs, the curly bracketed symbols absent and the annihilation operator. Emission is described by the subscript e, lower signs, the curly bracketed symbols present and the creation operator. Equation (12.37) can be substituted into equation (12.31)

$$T_{fi} = |\langle f|\hat{H}'_{a(e)}|i\rangle|^2 \lim_{t \to \infty} \frac{1}{t} \left| \frac{i}{\hbar} \int_0^t e^{i(W_f - W_i \mp \hbar\omega)t'/\hbar} dt' \right|^2 \tag{12.39}$$

Now one of the definitions of the Dirac δ-function is

$$\lim_{t \to \infty} \frac{1}{t} \left| \frac{i}{\hbar} \int_0^t e^{iEt'/\hbar} dt' \right|^2 = \frac{2\pi}{\hbar} \delta(E) \tag{12.40}$$

and clearly we can use this in (12.39) to give

$$T_{fi} = \frac{2\pi}{\hbar} |\langle f|\hat{H}'_{a(e)}|i\rangle|^2 \delta(W_f - W_i \mp \hbar\omega) \tag{12.41}$$

This equation is known as Fermi's Golden Rule. An important point to note about this derivation is that at no time have we assumed anything about the form of the electronic wavefunctions, so the rule is applicable in relativistic and non-relativistic quantum theory, for perturbations linear in the vector potential. It is important to note that, although the states $\langle i|$ and $\langle f|$ contain both photons and electrons, the energies W_i and W_f refer to the electron states only. Equation (12.41) describes processes involving a single photon, and so is applicable to absorption and emission only. This is in contrast to non-relativistic theory where the Hamiltonian is quadratic in the vector potential. Then two-photon processes can occur at the first order level.

Real experiments often involve scattering, where photons exist in the initial and final states. To describe scattering we need to consider $a_m^{(2)}(t)$ only, because $a_m^{(1)}(t) = 0$, as can easily be shown. In the following we evaluate this and find the transition rate for scattering. Scattering to this order requires two Feynman diagrams, shown in figures 12.2a and b.

In figure 12.2a there is an electron system and a photon $(\mathbf{k}, \lambda)$ in the initial state. At some time t_1 the photon interacts with the electrons and is absorbed. Later at time t such that $t_1 < t < t_2$ the electron system is

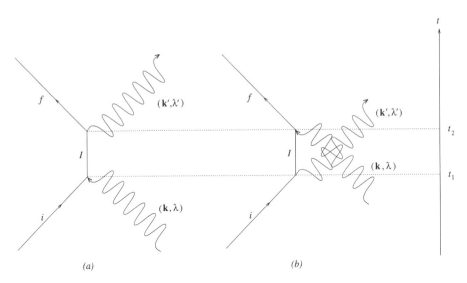

Fig. 12.2. The Feynman diagrams necessary to describe photon scattering. (*a*) With no photons in the intermediate state, (*b*) with two photons in the intermediate state.

in some intermediate state I and there are no photons. At t_2 the electron system emits a photon $(\mathbf{k}', \lambda')$ and for $t > t_2$ the electrons are in state f and the second photon has been emitted and can be detected.

Figure 12.2*b* is surprising, but it is essential to include it for the theory to be complete and in order to calculate observables in agreement with experiment. We start in the same initial state as in figure 12.2*a* with the electrons in state i and the free photon in state $(\mathbf{k}, \lambda)$. However, this time at $t_1 < t_2$ the electron system emits a photon $(\mathbf{k}', \lambda')$. The intermediate state this time has electrons in a state I' and two photons. At t_2 the electron system absorbs the photon $(\mathbf{k}, \lambda)$ and we are left with the electrons in state f and the single photon $(\mathbf{k}', \lambda')$. This diagram may seem to violate conservation of energy, but it describes a virtual process which only takes place within a time allowed by the uncertainty principle. That it is necessary to include this diagram is an illustration of the power of the uncertainty principle. As the initial and final states in both diagrams are the same the processes involved cannot be distinguished.

Now that we have described in words and pictures the mathematics required to calculate scattering rates we actually have to work it out. After so many chapters containing so much mathematics it would have been more of a surprise if we weren't going to work it out! The amplitude $a^{(2)}(t)$ is given by equation (12.29) and we have the two intermediate states discussed above. This time we will label our states by two symbols, a letter which tells us whether the electron system is in its initial i, intermediate I or final f state, and a number of photons which will be $0, 1$ or 2. With the aid of (12.38) equation (12.29) becomes

$$a_f^{(2)}(t) = -\frac{1}{\hbar^2} \sum_I \langle f, 1 | \hat{H}_e' | I, 0 \rangle \langle I, 0 | \hat{H}_a' | i, 1 \rangle \times$$

$$\int_0^t dt'' \int_0^{t''} dt' e^{i(W_f - W_I + \hbar\omega')t''/\hbar} e^{i(W_I - W_i - \hbar\omega)t'/\hbar}$$

$$-\frac{1}{\hbar^2} \sum_I \langle f, 1 | \hat{H}_a' | I', 2 \rangle \langle I', 2 | \hat{H}_e' | i, 1 \rangle$$

$$\int_0^t dt'' \int_0^{t''} dt' e^{i(W_f - W_I - \hbar\omega)t''/\hbar} e^{i(W_I - W_i + \hbar\omega')t'/\hbar} \qquad (12.42)$$

where it is understood that $\hat{H}_a$ and $\hat{H}_e$ are characterized by $(\mathbf{k}, \lambda)$ for the incident photon and $(\mathbf{k}', \lambda')$ for the outgoing photon respectively. The time integrals can be done straightforwardly. It is easy to show that

$$\int_0^t dt_2 \int_0^{t_2} dt_1 e^{iAt_2/\hbar} e^{iBt_1/\hbar} = \frac{\hbar}{iB} \int_0^t dt_2 (e^{i(A+B)t_2/\hbar} - e^{iAt_2/\hbar}) \qquad (12.43)$$

Applying this result to the integrals in (12.42) gives

$$\int_0^t dt'' \int_0^{t''} dt' e^{i(W_f - W_I + \hbar\omega')t''/\hbar} e^{i(W_I - W_i - \hbar\omega)t'/\hbar} =$$

$$\frac{\hbar}{i(W_I - W_i - \hbar\omega)} \int_0^t dt'' \left[e^{i(W_f - W_i + \hbar\omega' - \hbar\omega)t''/\hbar} - e^{i(W_f - W_I + \hbar\omega')t''/\hbar} \right]$$

$$(12.44a)$$

$$\int_0^t dt'' \int_0^{t''} dt' e^{i(W_f - W_I - \hbar\omega)t''/\hbar} e^{i(W_I - W_i + \hbar\omega')t'/\hbar} =$$

$$\frac{\hbar}{i(W_I - W_i + \hbar\omega')} \int_0^t dt'' \left[e^{i(W_f - W_i + \hbar\omega' - \hbar\omega)t''/\hbar} - e^{i(W_f - W_I - \hbar\omega)t''/\hbar} \right]$$

$$(12.44b)$$

In (12.44a and b) the final term on the right side does not conserve energy and gives zero when integrated. The exponent in the first term on the right side conserves energy in both cases and when put into (12.42) gives a non-vanishing contribution

$$a_f^{(2)}(t) = \frac{i}{\hbar} \int_0^t dt'' e^{i(W_f - W_i + \hbar\omega' - \hbar\omega)t''/\hbar}$$

$$\sum_I \left[\frac{\langle f, 1 | \hat{H}_e' | I, 0 \rangle \langle I, 0 | \hat{H}_a' | i, 1 \rangle}{W_I - W_i - \hbar\omega} + \frac{\langle f, 1 | \hat{H}_a' | I, 2 \rangle \langle I, 2 | \hat{H}_e' | i, 1 \rangle}{W_I - W_i + \hbar\omega'} \right]$$

$$(12.45)$$

We can write this more succinctly by redefining our energies. In the above equations W is the energy of the electron system. If we let E denote the energy of the electrons and photons together and unite the intermediate states with zero and two photons into a single symbol we have

$$E_i = W_i + \hbar\omega \qquad E_f = W_f + \hbar\omega' \qquad (12.46a)$$

If we have no photons in the intermediate state

$$E_I = W_I \qquad (12.46b)$$

and if we have two photons in the intermediate state

$$E_I = W_I + \hbar\omega + \hbar\omega' \qquad (12.46c)$$

From these it is easy to see that the denominator in the first fraction in (12.45) is $E_I - E_i$, and the denominator in the second term is also $E_I - E_i$. We can also recombine the perturbations

$$\hat{H}' = \hat{H}'_a + \hat{H}'_e \qquad (12.47)$$

The perturbation $\hat{H}'$ defined in (12.47) has no time dependence. In defining $\hat{H}'$ here it is understood that it only acts between states that have the right number of photons. If any part of the operator leads to the wrong number of photons then it is set equal to zero because states with different numbers of photons are assumed to be orthogonal. The way this works will be seen explicitly when we come to look at some examples of scattering. The sum over I now includes photon, as well as electron, intermediate states, and the state vectors themselves also contain the photons. We can write (12.45) as

$$a_f^{(2)}(t) = \frac{i}{\hbar} \sum_I \frac{\langle f|\hat{H}'|I\rangle\langle I|\hat{H}'|i\rangle}{E_I - E_i} \int_0^t dt'' \, e^{i(E_f - E_i)t''/\hbar} \qquad (12.48)$$

The time integral gives us a δ-function again from (12.40). Combining (12.37), (12.40) and (12.48) according to equation (12.31) gives us Fermi's Golden Rule correct to second order.

$$T_{fi} = \frac{2\pi}{\hbar} \left| \langle f|\hat{H}'|i\rangle - \sum_I \frac{\langle f|\hat{H}'|I\rangle\langle I|\hat{H}'|i\rangle}{E_I - E_i} \right|^2 \delta(E_f - E_i) \qquad (12.49)$$

It may be possible for a photon to induce several different transitions, in which case it may be necessary to sum over many possible final states. If we are only interested in scattering in relativistic quantum theory (but not in the non-relativistic theory) we can write

$$T_{fi} = \frac{2\pi}{\hbar} \left| \sum_I \frac{\langle f|\hat{H}'|I\rangle\langle I|\hat{H}'|i\rangle}{E_I - E_i} \right|^2 \delta(E_f - E_i) \qquad (12.50)$$

It took a lot of complicated mathematics to get the relatively simple, but exceedingly powerful formula (12.49). Let us just take a moment to think about its units. Each matrix element has units of energy, so the modulus squared gives us units of energy^2. The units of $\hbar$ are energy × time. The δ-function units are somewhat deceptive. We must be able to integrate it over energy to get unity which is dimensionless, therefore its units must be energy^{-1}. Putting all these together confirms that T_{fi} has units of time^{-1} which are the units we would expect for a transition rate. One final point is worth mentioning in this section. An approximation often used in evaluating transition rates using (12.49) is the following. The operator $\hat{H}'$ contains an exponential and for small wavevectors this can be approximated using

$$e^{i\mathbf{k}\cdot\mathbf{r}} \approx 1 + i\mathbf{k}\cdot\mathbf{r} - \frac{1}{2}(\mathbf{k}\cdot\mathbf{r})^2 \cdots \qquad (12.51)$$

Cutting off this series at the first term is called the electric dipole approximation, cutting it off at the second term is adding an electric quadrupole and magnetic dipole correction etc. It is often the case that the dipole approximation is sufficient, and higher order terms are negligible. In the following sections we will apply (12.49) in various circumstances and we will make some use of the dipole approximation.

12.4 Photon Absorption and Emission in Condensed Matter

In this first application of the Golden Rule, we will consider how to calculate photon absorption rates in condensed matter. Let's think about a photon exciting an electron from a core level up to a state above the Fermi energy in a metal.

Here we will not go through the detailed mathematics of evaluating transition rates for absorption or emission, rather we will just indicate how it is done. The mathematical details here would be analogous to those found in section 8.6 for the one-electron atom anyway.

We will work in the single-particle approximation in this section, so the states $|i\rangle$ and $|f\rangle$ are single-particle wavefunctions and we assume that all other electrons in the material are unaffected by the absorption. If a photon enters a material it may be able to excite many different electrons. Furthermore, even if we can identify which electron has been excited, there may be many possible final states, so to describe the interaction between photons and electrons in condensed matter properly it is necessary to sum over all possible initial and final states. We can decouple the photon and electron states as described in equations (12.34) to (12.36). Then the first order Golden Rule is

$$\sum_{f,i} T_{fi} = \frac{2\pi}{\hbar} \sum_i \sum_f |\langle f|\hat{H}'|i\rangle|^2 \delta(E_f - E_i)$$

$$= \frac{2\pi}{\hbar} \sum_i \sum_f \langle i|\hat{H}'^\dagger|f\rangle\langle f|\hat{H}'|i\rangle \delta(W_f - W_i - \hbar\omega)$$

$$= \frac{2\pi}{\hbar} \sum_i \int dE \delta(E - W_i - \hbar\omega)\langle i|\hat{H}'^\dagger\left[\sum_f |f\rangle\delta(W_f - E)\langle f|\right]\hat{H}'|i\rangle$$

$$(12.52)$$

Now, the part of this in square brackets is proportional to the imaginary part of the scattering Green's function, which can be defined in this way using equations (11.24) and (11.150). So our final expression for the transition rate is

$$T_{fi} = -\frac{2}{\hbar} \sum_i \langle i|\hat{H}'^\dagger \mathrm{Im}[G(E)]\hat{H}'|i\rangle \delta(E - W_i - \hbar\omega) \qquad (12.53)$$

This expression happens to represent absorption, but a very similar expression can be worked out for emission. Now it is clear how to calculate photon absorption and emission rates for real materials. One does a standard electronic structure calculation for the material using the methods of chapter 11 and evaluates the Green's function of equation (11.143). This can be substituted into (12.53) to give

$$T_{fi} = -\frac{2}{\hbar} \sum_i M_i^\Lambda(E) \mathrm{Im}\tau_{\Lambda\Lambda'}(E) M_i^{\Lambda'\dagger}(E) \qquad (12.54a)$$

where $M_i^\Lambda(E)$ is a matrix element

$$M_i^{\Lambda'\dagger}(E) = \int \psi_i^\dagger(\mathbf{r})\hat{H}' Z_\Lambda(\mathbf{r}, E)d\mathbf{r} \qquad (12.54b)$$

where the integral is over the Wigner–Seitz cell, $\psi_i(\mathbf{r})$ is the wavefunction of the core level being excited, $Z_\Lambda(\mathbf{r})$ is the solution of the Dirac–Kohn–Sham equation and $\tau_{\Lambda\Lambda'}$ is a matrix element of the scattering path operator. These latter two quantities describe the electronic state to which the electron is excited and are evaluated at that energy. So, the transition rate can be calculated from quantities easily determined from a standard band theory calculation. The perturbation $\hat{H}'$ is given by equation (12.47) which contains the polarization of the incident (emitted) photon. If the calculations are relativistic, and the material is magnetic, right and left circularly polarized photons lead to different absorption (emission) rates. Equation (12.54b) can be evaluated in the same way as the matrix elements (8.80) are evaluated in section 8.6, and yields identical selection rules, (8.89),

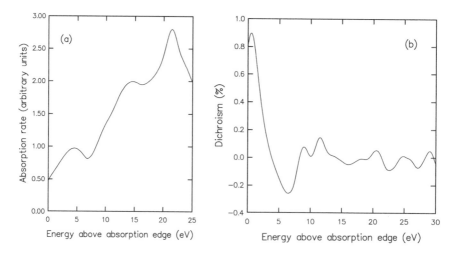

Fig. 12.3. (a) The absorption rate for X-rays at the K-edge of iron. (b) Dichroism at the K-edge of iron. Both curves have been convoluted with a Lorentzian of width 1.8 eV to simulate the various broadening mechanisms. These theoretical curves are in good agreement with the experimental results of Schütz et al. (1987).

for left and right circularly polarized radiation. Many materials exhibit spontaneous magnetism and for a dichroic effect the internal magnetism is sufficient, with no need for an external field. Indeed dichroism in absorption is a very powerful probe of magnetism in condensed matter. In figure 12.3a we show the absorption rate for circularly polarized X-rays in iron. The X-ray energy is chosen at the K-edge, i.e. so that it is sufficient to raise an electron from the $1s$ level to just above the Fermi level. This curve is convoluted with a Lorentzian of width 1.8 eV to simulate lifetime broadening and instrumental resolution effects. In figure 12.3b we show the dichroism defined by equation (8.94) in the same energy range.

Dichroism in metals was first discovered by Schütz et al. (1987), and the interpretation of the effect in terms of electron band theory was provided by Ebert et al. (1988b). Since then several authors have shown that dichroism can be interpreted as a direct measure of the various components of the magnetic moment of a material. Defining T^+, T^- and T^0 as the transition rates for left, right, and linearly polarized light respectively, Thole et al. (1992) have constructed the following expression relating these absorption rates directly to the expectation value of the orbital magnetic moment $\langle \hat{L}_z \rangle$

$$\frac{\int_{j\pm}(T^+ - T^-)dE}{\int_{j\pm}(T^+ + T^- + T^0)dE} = \frac{1}{2}\frac{l(l+1) - c(c+1) + 2}{l(l+1)(4l+2-n)}\langle \hat{L}_z \rangle \qquad (12.55)$$

where the integral over energy starts at the Fermi energy and extends up to the top of the band. In this equation c is the l quantum number of the initial state and l is the l quantum number of the state to which a dipole transition from state c is allowed. The $j\pm$ indicates that the integral is over both the $j = c - 1/2$ and $j = c + 1/2$ initial core states. The number of electrons in the conduction band with quantum number l is n, so $4l + 2 - n$ is the number of holes in that band. Carra *et al.* (1993) have derived a similar formula for the spin magnetic moment

$$\frac{\int_{j_+}(T^+ - T^-)dE - \frac{c+1}{c}\int_{j_-}(T^+ - T^-)dE}{\int_{j\pm}(T^+ + T^- + T^0)dE} = \frac{l(l+1) - 2 - c(c+1)}{3c(4l+2-n)}\langle\hat{S}_z\rangle$$

$$+ \frac{l(l+1)[l(l+1) + 2c(c+1) + 4] - 3(c-1)^2(c+2)^2}{6lc(l+1(l+1)(4l+2-n)}\langle\hat{T}_z\rangle \qquad (12.56)$$

Here $\langle\hat{S}_z\rangle$ is the expectation value of the spin magnetic moment, and $\langle\hat{T}_z\rangle$ is the expectation value of a dipolar operator which is negligible in cubic systems. Equations (12.55) and (12.56) give us the integrated magnetic moments. It has been shown (Strange 1994, Wu *et al.* 1993, Wu and Freeman 1994) that these rules are often obeyed energy by energy, not just in integrated form. An elementary derivation of these sum rules is given by Altarelli (1993) and by Ankudinov and Rehr (1995). Strange and Györffy (1995) have derived the following expression

$$\frac{T_{j_-}^+ - T_{j_-}^-}{T_{j_-}^+ + T_{j_-}^- + T_{j_-}^0} = -\frac{1}{j_-}\frac{\langle\hat{j}_{z_-}\rangle}{\langle n_{j_-}\rangle} \qquad (12.57)$$

where all absorption rates are measured at the $j_- = c - 1/2$ edge. Here $\langle\hat{j}_{z_-}\rangle$ is the total magnetic moment of the $j_- = l - 1/2$ conduction band states only, and n_{j_-} is the density of states of those electrons, both evaluated at the same energy. This equation is derived to be valid energy by energy and an example of its use is shown in figure 12.4. Equation (12.57) was derived assuming L–S coupling, and so its accuracy decreases when applied to materials with a high atomic number.

The derivation of these rules is very complicated, even by the standards of this book, and uses mathematics that is unfamiliar to many physics students (and to more senior physicists). Therefore the formulae (12.55) to (12.57) can only be quoted. For the derivations the reader is referred to the original literature.

12.5 Magneto-Optical Effects

This section follows on directly from the previous one, and is simply an application of the absorption rate calculations discussed above. Here they

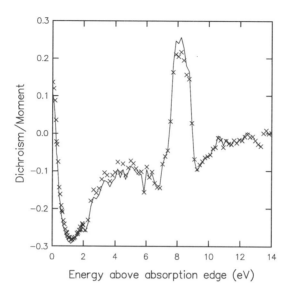

Fig. 12.4.   The left side of equation (12.57) calculated using equation (12.54) (full line) for the L_2 edge (exciting the $2p_{1/2}$ electrons) in iron, and the right side of equation (12.54) calculated directly from band theory (crosses) for the $d_{3/2}$ states in the conduction band region of iron. These curves are not broadened.

are combined with macroscopic electromagnetic theory to describe some magneto-optical effects (Bennett and Stern 1965, Pershan 1967).

In equation (12.54) we have found a method of calculating the absorption rate for photons of any polarization passing through materials, i.e. the number of photons absorbed per unit time. If we multiply this by the energy of each photon we have an expression for the power absorbed

$$P = \hbar\omega \sum_f T_{fi} \tag{12.58}$$

But in macroscopic electromagnetic theory the power absorbed by a solid interacting with an electric field $\mathbf{E}$ is

$$P = \frac{1}{2} \int \text{Re}(\mathbf{J}^* \cdot \mathbf{E}) dV \tag{12.59}$$

where $\mathbf{J}$ is the induced current and V is the volume of the solid. If we assume that the spatial variation of $\mathbf{E}$ can be neglected (in our case this means that the photon wavelength must be much greater than the electron mean free path), this integral can be done with no difficulty and with the help of Ohm's law (equation (1.48)) we find that

$$P = \frac{V}{2} \mathrm{Re} \left(\sum_{i,j=x,y,z} \sigma_{ij}^*(\omega) E_i^* E_j \right) \tag{12.60}$$

Here $\sigma_{ij}(\omega)$ is one element of the optical conductivity tensor which depends on the frequency of the incident radiation. In general it is complex

$$\sigma_{ij}(\omega) = \sigma_{1ij}(\omega) + i\sigma_{2ij}(\omega) \tag{12.61}$$

The electric vector describing the photons in a general polarization state can be written

$$\mathbf{E} = |E|\epsilon_\lambda e^{i\omega t} \tag{12.62}$$

If the photon wavevector points along the z-axis, the polarization vector is given by equation (12.5). For plane polarized light (12.60) gives

$$P_x = \frac{V|E|^2}{2} \sigma_{1xx}(\omega) \tag{12.63}$$

and for right (r) or left (l) circularly polarized light we find

$$P_{\mathrm{r(l)}} = \frac{V|E|^2}{2} (\sigma_{1xx}(\omega) \mp \sigma_{2xy}(\omega)) \tag{12.64}$$

where the symmetry of the optical conductivity tensor is

$$\sigma = \begin{pmatrix} \sigma_{xx} & \sigma_{xy} & 0 \\ -\sigma_{xy} & \sigma_{xx} & 0 \\ 0 & 0 & \sigma_{zz} \end{pmatrix} \tag{12.65}$$

which is the usual form used to discuss magneto-optical effects. Equation (12.64) gives us expressions for elements of the optical conductivity tensor in terms of the power absorbed, i.e.

$$\sigma_{1xx}(\omega) = \frac{P_1 + P_{\mathrm{r}}}{V|E|^2} \tag{12.66a}$$

$$\sigma_{2xy}(\omega) = \frac{P_1 - P_{\mathrm{r}}}{V|E|^2} \tag{12.66b}$$

These quantities are response functions. The imaginary part of σ_{xx} and the real part of σ_{xy} can be found using the Kramers–Kronig relations

$$\sigma_{2xx}(\omega) = -\frac{2\omega}{\pi} \mathbf{P} \int_0^\infty \frac{\sigma_{1xx}(\omega')}{\omega'^2 - \omega^2} d\omega' \tag{12.67a}$$

$$\sigma_{1xy}(\omega) = \frac{2}{\pi} \mathbf{P} \int_0^\infty \frac{\omega' \sigma_{2xy}(\omega')}{\omega'^2 - \omega^2} d\omega' \tag{12.67b}$$

where **P** indicates that the principal part of the integral should be taken. With these we have two of the three independent elements of the optical conductivity tensor and this is sufficient to describe many interesting physical effects. In equation (12.65) we have assumed that the photon propagates along the z-axis. Using Ohm's law together with equations (12.5) and (12.62) it is easy to show that the conductivities for left and right circularly polarized radiation are

$$\sigma^{r(l)} = \sigma_{xx} \mp i\sigma_{xy} \tag{12.68}$$

Without proof we state that if the calculations above had been done in a non-relativistic framework there would be no difference between the absorption rates for left and right circularly polarized radiation and so σ_{xy} would be identically equal to zero. Hence any observables we find that are dependent upon σ_{xy} are of relativistic origin.

To proceed further we need Maxwell's equations which were written down in chapter 1, together with the vector identity

$$\nabla \times \nabla \times \mathbf{E} = \nabla\nabla \cdot \mathbf{E} - \nabla^2 \mathbf{E} \tag{12.69}$$

Now we are considering an electromagnetic wave, so the charge density is zero. Hence the first term on the right hand side of (12.69) is zero as can be seen from equation (1.42a). We can substitute Maxwell's equations into the left hand side of (12.69) and assume our fields are of the form $E = E_0 e^{i\omega t}$, $B = B_0 e^{-i\omega t}$. This leads to

$$-\nabla^2 \mathbf{E} = -i\omega \nabla \times \mathbf{B} = -i\omega(\mu_0 \mathbf{J} + \mu_0 \epsilon_0 \frac{\partial \mathbf{E}}{\partial t}) = \left(\frac{-i\omega}{\epsilon_0 c^2} \sigma \mathbf{E} + \frac{\omega^2}{c^2} \mathbf{E} \right) \tag{12.70}$$

where we have used Ohm's law from equation (1.48) and equation (1.46) for c^2. We end up with

$$-\nabla^2 \mathbf{E} = \frac{\omega^2}{c^2} \epsilon(\omega) \mathbf{E} \tag{12.71}$$

where the frequency-dependent dielectric constant is

$$\epsilon(\omega) = \left(1 - \frac{i\sigma}{\epsilon_0 \omega} \right) \tag{12.72a}$$

Equation (12.72a) is written in the usual isotropic form. It can be immediately generalized to describe anisotropic propagation by putting it in tensor form

$$\epsilon_{ij}(\omega) = \left(\delta_{ij} - \frac{i\sigma_{ij}}{\epsilon_0 \omega} \right) \tag{12.72b}$$

and in analogy to (12.68) we have

$$\epsilon^{r(l)}(\omega) = \epsilon_{xx}(\omega) \mp i\epsilon_{xy}(\omega) \tag{12.73}$$

So, only a single function of ω is necessary to describe the response of the medium to an electric field. Now we will take up another thread of the argument (Dillon 1977) and return to (12.72) shortly. We will consider the second Maxwell equation (1.42b) and discuss a solution to this representing an electromagnetic wave propagating through some medium

$$\mathbf{E} = \mathbf{E}_0 e^{i\omega(t-nz/c)} \qquad\qquad \mathbf{B} = \mathbf{B}_0 e^{i\omega(t-nz/c)} \qquad (12.74)$$

whose phase velocity is $v_p = c/n$. We have chosen the z-axis as the direction of propagation and n represents the refractive index of the material. Solutions (12.74) have x- and y-components that depend on z, but as the wave propagates along the z-direction it has no z-component. These solutions can be substituted into the Maxwell equation to give

$$\nabla \times \mathbf{E} = \left(-\frac{\partial E_y}{\partial z}, \frac{\partial E_x}{\partial z}, 0\right) = \left(\frac{i\omega n E_y}{c}, -\frac{i\omega n E_x}{c}, 0\right) = -i\omega\mathbf{B} \qquad (12.75)$$

Now let us define $\hat{\mathbf{k}}$ as a unit length vector in the direction of propagation of the wave. In our case this is just a unit vector in the z-direction, of course. It is easy to see that

$$\hat{\mathbf{k}} \times \mathbf{E} = (-E_y, E_x, 0) \qquad (12.76)$$

so we can substitute this into (12.75) to give us

$$n(\hat{\mathbf{k}} \times \mathbf{E}) = c\mathbf{B} \qquad (12.77)$$

Next, let's do a similar analysis on the fourth Maxwell equation (1.42d). Using Ohm's law and again substituting in the solutions (12.74) leads directly to

$$\frac{n}{c}(B_y, -B_x, 0) = \left(\frac{\mu_0\sigma}{i\omega} + \mu_0\epsilon_0\right)\mathbf{E} \qquad (12.78)$$

and introducing the unit wavevector again gives

$$nc(\hat{\mathbf{k}} \times \mathbf{B}) = \left(1 - \frac{i\sigma}{\epsilon_0\omega}\right)\mathbf{E} = -\epsilon(\omega)\mathbf{E} \qquad (12.79)$$

where we have made use of equation (12.72). We can substitute from (12.77) for $\mathbf{B}$ in (12.79) to get

$$n^2(\hat{\mathbf{k}} \times \hat{\mathbf{k}} \times \mathbf{E}) = -\epsilon(\omega)\mathbf{E} \qquad (12.80)$$

Now we know $\hat{\mathbf{k}} \times \mathbf{E}$ from (12.76), and taking the vector product of this with $\hat{\mathbf{k}}$ gives

$$n^2(E_x, E_y, 0) = \epsilon(\omega)\mathbf{E} \qquad (12.81)$$

Comparison of (12.72*b*) and (12.65) shows that, for our high symmetry case, the dielectric tensor must take on the form

$$\epsilon = \begin{pmatrix} \epsilon_{xx} & -i\epsilon_{xy} & 0 \\ i\epsilon_{xy} & \epsilon_{xx} & 0 \\ 0 & 0 & \epsilon_{zz} \end{pmatrix} \tag{12.82}$$

where the *i* and minus sign have been chosen conventionally, the requirement being only that this has the same symmetry as (12.65). With this we can write out equation (12.81) in component form

$$(n^2 - \epsilon_{xx})E_x + i\epsilon_{xy}E_y = 0 \tag{12.83a}$$

$$(n^2 - \epsilon_{xx})E_y - i\epsilon_{xy}E_x = 0 \tag{12.83b}$$

We will drop the *z*-component henceforth as nothing further can be learnt from it. These are just a pair of simultaneous equations which have a non-trivial solution if

$$\begin{vmatrix} n^2 - \epsilon_{xx} & i\epsilon_{xy} \\ -i\epsilon_{xy} & n^2 - \epsilon_{xx} \end{vmatrix} = 0 \tag{12.84}$$

which is only satisfied if the refractive index is

$$n = \sqrt{\epsilon_{xx} \pm \epsilon_{xy}} \tag{12.85}$$

Here the positive sign corresponds to left circular polarization and the negative sign to right circular polarization (Pershan 1967).

This is a good point at which to summarize where we have got to so far. We have seen in equation (12.54) that photon absorption rates can be found using the relativistic quantum mechanical scattering theory description of the electronic structure of condensed matter. By analogy with the dichroism calculations in chapter 8 we have asserted that different photon polarizations result in different absorption rates. In equation (12.58) we have seen how to calculate the power absorbed by a medium when photons are absorbed. This will also be different for different photon polarizations. Using macroscopic electromagnetic theory we have then related the powers absorbed to elements of the optical conductivity tensor as explicitly shown in equations (12.66) and (12.67). These can be related directly to the dielectric tensor as in equation (12.72). Finally in equation (12.85) we have shown that the dielectric tensor determines the complex refractive index of our material, and we can see that left and right circularly polarized light have differing refractive indices. To take the discussion full circle, the difference in the refractive indices may, in principle, be used to probe the electronic structure of magnetic materials.

We can ask if the theory above leads to any new physical phenomena. Indeed it does and we discuss them here. Consider the experiment shown

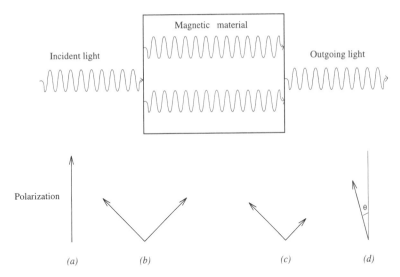

Fig. 12.5. An experiment illustrating Faraday effects. In region (*a*) linear polarized light travels towards a magnetic material. At the boundary of the material (*b*) it splits into equal amounts of left and right circularly polarized light. These travel through the material, but with different refractive indices. At the other end of the material (*c*) more of one polarization has been absorbed than the other. So after the waves recombine (*d*) the apparent polarization direction has been rotated through θ. In this figure we are dealing with absorption which is the complex part of the refractive index, so the rotation is into the complex plane and therefore the light has become elliptically polarized.

in figure 12.5. Linearly polarized light enters an anisotropic medium from the left as shown. Now, linearly polarized light can be regarded as a combination of equal amounts of left and right circularly polarized light, so let us assume that on entry into the material the light splits into two beams, one of left, and one of right, circular polarization, and that these travel through the material without interference. We know that left and right polarized beams are absorbed differently, so when the beams reach the other side of the material there is more of one polarization remaining than the other. Hence when the beams recombine on leaving the material they are no longer linearly polarized along the same direction. Absorption is described by the imaginary part of the refractive index, and the effect of the preferential absorption is to rotate the incoming linear (real) polarization vector slightly into the complex plane, i.e. to introduce a small degree of elliptical polarization to the outgoing wave.

In fact the situation is even more complicated. Because the real parts of the refractive indices are also different, left and right circularly polarized waves travel through the material at different velocities and hence emerge

out of phase. This means that when the beams recombine there is also
a rotation of the polarization direction. These two effects together, the
rotation of the major axis of polarization and the induced ellipticity, are
known as Faraday effects if the experiment is done in transmission and
as Kerr effects if the experiment is done in reflection (Argyres 1955).

It is easy to find an expression for the phase difference between the two
emerging waves. Assume one polarization passes through the material in
time t_1 and the other in time t_2. The phase difference is then

$$\theta = \omega(t_2 - t_1) \tag{12.86}$$

but the time taken to travel through the material is just the thickness d
of material divided by the velocity of light in the material. Of course the
velocity of light in the material is just the velocity of light in vacuum
divided by the real part of the refractive index, so

$$\theta = \frac{\omega d}{c} \mathrm{Re}(n_r - n_l) \tag{12.87}$$

and the Faraday rotation is just half of this

$$\theta_F = \frac{\omega d}{2c} \mathrm{Re}(n_r - n_l) \tag{12.88a}$$

The Faraday ellipticity, i.e. the ratio of the minor axis to the major axis of
the ellipse, arises from the difference in the imaginary part of the refractive
indices, i.e. from the difference in the absorption coefficients

$$\eta_F = -\tanh\left(\frac{\omega d}{2c} \mathrm{Im}(n_r - n_l)\right) \tag{12.88b}$$

Equations (12.88) constitute the Faraday effect. If the experiments are
done in reflection rather than in transmission the underlying physics is the
same, but the expressions for the angle of rotation and induced ellipticity
are different

$$\theta_K = -\mathrm{Im}\frac{n_r - n_l}{n_r n_l - 1} \tag{12.89a}$$

$$\eta_K = -\mathrm{Re}\frac{n_r - n_l}{n_r n_l - 1} \tag{12.89b}$$

Equations (12.89) describe the Kerr effect. The angles θ_F and θ_K are the
angles through which the linear polarization is rotated about the direction
of incidence, and η_F and η_K are the amount of elliptical polarization that
has been induced in the linearly polarized beam. After all this mathematics
we have finally got to something of technological importance. Faraday and
Kerr rotations are the underlying physical processes involved in magneto-
optic storage which is so important in the computer industry. These effects
can also be used to measure the electronic structure of magnetic materials

(Wang and Callaway 1974, Ebert *et al.* 1988c). They have been measured
in the X-ray (Siddons *et al.* 1990), as well as the optical, region of the
electromagnetic spectrum. Although we have at least implied that the
difference in absorption rates is due to the symmetry breaking introduced
by magnetism, there are alternative symmetry breaking mechanisms which
result in preferential absorption rates of one polarization over another.
One example is that such effects also occur in non-magnetic materials
where the crystal lattice does not have a centre of inversion symmetry.

This completes our discussion of the consequences of the first order
part of the Golden Rule (12.50). In this and the previous section we
have focussed on applications of the Golden Rule which are intrinsically
relativistic. This has led us to see that a relativistic quantum theory
description really becomes essential when the polarization of the photon
plays an important role in the interaction.

12.6 Photon Scattering Theory

We are now going to look at some of the consequences of the second
order term in the Golden Rule. In this section we will consider photon
scattering in general and in the following sections we will describe several
specific examples. The examples are characterized either by the energy of
the photon or by the initial and final states involved. Throughout we will
use the convention that we have an incident photon with wavevector and
polarization $(\mathbf{k}, \lambda)$ and after scattering the wavevector and polarization of
the emitted photon are $(\mathbf{k}', \lambda')$.

In relativistic theory, scattering involves evaluating the second term in
equation (12.49), and this means we have to evaluate scattering amplitudes,
a typical one being

$$M_{fi} = \sum_{I} \frac{\langle f|\hat{H}'|I\rangle\langle I|\hat{H}'|i\rangle}{E_i - E_I} \tag{12.90}$$

We know from figures 12.2 that the intermediate state may contain zero
or two photons, so (12.90) can be separated into two contributions

$$M_{fi} = \sum_{I} \left(\frac{\langle f,1|\hat{H}'|I,0\rangle\langle I,0|\hat{H}'|i,1\rangle}{E_i - E_{I,0}} \right.$$
$$\left. + \frac{\langle f,1|\hat{H}'|I,2\rangle\langle I,2|\hat{H}'|i,1\rangle}{E_i - E_{I,2}} \right) \tag{12.91}$$

Next we substitute the perturbation Hamiltonian into this expression. This
is given by equation (12.38). We also know whether it is the absorptive or
emissive part of $\hat{H}'$ that contributes to each of these terms from (12.45),

so

$$M_{fi} = -\sum_I \left(\frac{\langle f,1|\hat{H}'_e|I,0\rangle\langle I,0|\hat{H}'_a|i,1\rangle}{W_I - W_i - \hbar\omega_k} + \frac{\langle f,1|\hat{H}'_a|I,2\rangle\langle I,2|\hat{H}'_e|i,1\rangle}{W_I - W_i + \hbar\omega_{k'}} \right)$$

$$= -\sum_{k'\lambda'\,k\lambda,I} K \frac{\langle f,1|\tilde{\alpha}\cdot\epsilon^*_{\lambda'}c^\dagger_{k'}e^{-ik'\cdot r}|I,0\rangle\langle I,0|\tilde{\alpha}\cdot\epsilon_\lambda c_k e^{ik\cdot r}|i,1\rangle}{W_I - W_i - \hbar\omega_k}$$

$$- \sum_{k'\lambda'\,k\lambda,I} K \frac{\langle f,1|\tilde{\alpha}\cdot\epsilon_\lambda c_k e^{ik\cdot r}|I,2\rangle\langle I,2|\tilde{\alpha}\cdot\epsilon^*_{\lambda'}c^\dagger_{k'}e^{-ik'\cdot r}|i,1\rangle}{W_I - W_i + \hbar\omega_{k'}}$$

(12.92)

where we have shortened the $\mathbf{k}$ and λ arguments of c and $c^\dagger$ so as to fit the equation on the page. Furthermore we have collected together all the constants into a single quantity with dimensions of energy squared:

$$K = e^2 c^2 \left(\frac{\hbar\mu_0 c^2}{V\sqrt{\omega_{k_1}\omega_{k_2}}} \right)$$

(12.93)

Consider the first term in equation (12.92): the initial state of the photon is $(\mathbf{k},\lambda)$. That is the only photon state that the annihilation operator can annihilate because there are no other photons. Also the creation operator has to create a photon in the state $(\mathbf{k}',\lambda')$, as that is the only photon in the final state. A similar argument holds for the second term, so the summations over wavevector and polarization in equation (12.92) disappear

$$M_{fi} = -K\sum_I \left(\frac{\langle f,1|\tilde{\alpha}\cdot\epsilon^*_{\lambda'}c^\dagger_{k'}e^{-ik'\cdot r}|I,0\rangle\langle I,0|\tilde{\alpha}\cdot\epsilon_\lambda c_k e^{ik\cdot r}|i,1\rangle}{W_I - W_i - \hbar\omega_k} \right)$$

$$- K\sum_I \left(\frac{\langle f,1|\tilde{\alpha}\cdot\epsilon_\lambda c_k e^{ik\cdot r}|I,2\rangle\langle I,2|\tilde{\alpha}\cdot\epsilon^*_{\lambda'}c^\dagger_{k'}e^{-ik'\cdot r}|i,1\rangle}{W_I - W_i + \hbar\omega_{k'}} \right)$$

$$= -K\sum_I \left(\frac{\langle f|\tilde{\alpha}\cdot\epsilon^*_{\lambda'}e^{-ik'\cdot r}|I\rangle\langle I|\tilde{\alpha}\cdot\epsilon_\lambda e^{ik\cdot r}|i\rangle}{W_I - W_i - \hbar\omega_k} \right)$$

$$- K\sum_I \left(\frac{\langle f|\tilde{\alpha}\cdot\epsilon_\lambda e^{ik\cdot r}|I\rangle\langle I|\tilde{\alpha}\cdot\epsilon^*_{\lambda'}e^{-ik'\cdot r}|i\rangle}{W_I - W_i + \hbar\omega_{k'}} \right)$$

(12.94)

This is an expression for the amplitude for photon scattering which is written only in terms of the electronic part of the problem. But it is not in a very convenient form. The intermediate states I include states involving excitations from the negative energy sea. A negative energy electron can absorb the incident photon and become a positive energy electron. Later, the initial electron can fall into the negative energy state vacated by the excited electron. The excited electron then becomes the electron in the final state. Although this is a virtual transition, the matrix element for it is non-zero, in general. The reverse process is also possible, where we first have photon emission and excitation of an electron from the negative energy sea followed by absorption of the initial photon as the initial electron falls into the negative energy hole. We must include such intermediate states when calculating scattering amplitudes, and it is convenient to separate them from the intermediate states involving positive energy excited states. There is a final subtlety. Terms involving negative energy intermediate states are of opposite sign, and in opposite order, to those involving positive energy states, because the former involve exchange of electrons. Our final expression for the scattering amplitude is

$$
M_{fi} = -K \sum_I \frac{\langle f | \tilde{\alpha} \cdot \epsilon^*_{\lambda'} e^{-i\mathbf{k}' \cdot \mathbf{r}} | I \rangle \langle I | \tilde{\alpha} \cdot \epsilon_\lambda e^{i\mathbf{k} \cdot \mathbf{r}} | i \rangle}{W_I - W_i - \hbar\omega_{\mathbf{k}}}
$$

$$
- K \sum_I \frac{\langle f | \tilde{\alpha} \cdot \epsilon_\lambda e^{i\mathbf{k} \cdot \mathbf{r}} | I \rangle \langle I | \tilde{\alpha} \cdot \epsilon^*_{\lambda'} e^{-i\mathbf{k}' \cdot \mathbf{r}} | i \rangle}{W_I - W_i + \hbar\omega_{\mathbf{k}'}}
$$

$$
+ K \sum_I \frac{\langle f | \tilde{\alpha} \cdot \epsilon_\lambda e^{i\mathbf{k} \cdot \mathbf{r}} | I \rangle \langle I | \tilde{\alpha} \cdot \epsilon^*_{\lambda'} e^{-i\mathbf{k}' \cdot \mathbf{r}} | i \rangle}{W_f + |W_I| - \hbar\omega_{\mathbf{k}}}
$$

$$
+ K \sum_I \frac{\langle f | \tilde{\alpha} \cdot \epsilon^*_{\lambda'} e^{-i\mathbf{k}' \cdot \mathbf{r}} | I \rangle \langle I | \tilde{\alpha} \cdot \epsilon_\lambda e^{i\mathbf{k} \cdot \mathbf{r}} | i \rangle}{W_f + |W_I| + \hbar\omega_{\mathbf{k}'}} \tag{12.95}
$$

Here I represents different intermediate states in each term, in general, although we have used the same label for all of them. The energy denominators for the terms in (12.95) involving negative energy states require some explanation. The denominators are given by the Golden Rule (12.49) as $E_I - E_i$. To write down these quantities we have to define a zero of energy which we take as the vacuum energy E_{vac}. This is the energy when all negative energy states are filled and all positive energy states are empty. The initial state contains one electron and one photon and so has energy $E_i = E_{vac} + W_i + \hbar\omega_{\mathbf{k}}$. If we excite a negative energy electron, initially with energy W_I, up to the positive energy states the total energy of the electron system in the intermediate state is the sum of the energies of the initial electron and the excited electron plus the energy due to the absence of a negative energy state, i.e. $W'' = E_{vac} + W_i + W_f - (-|W_I|)$. So, if we have no photons in the intermediate state,

$$
E_I - E_i = (E_{vac} + W_i + W_f + |W_I|) - (E_{vac} + W_i + \hbar\omega_{\mathbf{k}}) = W_f + |W_I| - \hbar\omega_{\mathbf{k}} \tag{12.96a}
$$

and similarly, if we have two photons in the intermediate state

$$
E_I - E_i = W_f + |W_I| + \hbar\omega_{\mathbf{k}'} \tag{12.96b}
$$

and these are the denominators in equation (12.95).

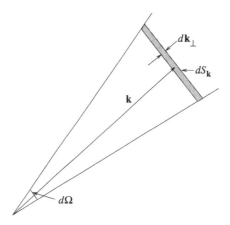

Fig. 12.6. Photons are scattered into a small solid angle $d\Omega$, which corresponds to an area $dS_{\mathbf{k}} = k^2 d\Omega$. Therefore photons of wavevector $\mathbf{k}$ occupy a volume of reciprocal space $k^2 d\Omega dk_\perp$.

In scattering experiments it is usually the cross section that is measured rather than the scattering rate of equation (12.50). However, the two quantities are simply related. The cross section per unit solid angle, per unit energy is (Blume 1985)

$$\frac{\partial^2 \sigma}{\partial \Omega \partial E_f} = \frac{T_{fi} n(E_f)}{I_0} = \frac{2\pi}{\hbar} \frac{|M_{fi}|^2 \delta(E_f - E_i) n(E_f)}{I_0} \tag{12.97}$$

where $n(E_f)$ is the density of available states at the energy of the final state, and I_0 is the flux of the incident photons

$$I_0 = \frac{c}{V} \tag{12.98}$$

If the target is stationary, $n(E_f)$ refers to the states of the outgoing photon. Suitable expressions for the density of final photon states are easy to derive (Merzbacher 1970). If we consider a volume L^3 in real space, one allowed photon state occupies a volume $\Delta^3 \mathbf{k} = (2\pi/L)^3$ in k-space. As seen in figure 12.6 a thin shell of width $dk_\perp$ has volume $dS_{\mathbf{k}} dk_\perp$. If $n(\hbar\omega)$ is the number of allowed photon states per unit solid angle per unit energy, the total number of allowed states in the shell is

$$n(\hbar\omega)d\Omega d(\hbar\omega) = \frac{dS_{\mathbf{k}} dk_\perp}{\Delta^3 \mathbf{k}} = \frac{k^2 d\Omega dk_\perp}{8\pi^3} V = \frac{\omega^2 V d\Omega d\omega}{8\pi^3 c^3} = \frac{\omega^2 V d\Omega d(\hbar\omega)}{8\pi^3 \hbar c^3} \tag{12.99a}$$

and so

$$n(\hbar\omega) = \frac{V\omega^2}{8\pi^3 \hbar c^3} \tag{12.99b}$$

Equation (12.97) is our final general expression for the scattering cross section, which can be evaluated and compared directly with experiment. In the rest of this chapter we are going to examine some examples of the application of equations (12.95) and (12.97).

12.7 Thomson Scattering

The first application of (12.95) will be to the scattering of a low energy photon by a free electron, known as Thomson scattering (Sakurai 1967). This is a particularly convenient example as we already know the wave-function describing a free electron from chapter 5. Let's assume that the electron is initially at rest. If we use equations (5.23) for the electron wavefunctions (with $\mathbf{p} = 0$ for the initial state) a single matrix element from the first two terms of (12.95) becomes

$$\langle I | \tilde{\alpha} \cdot \epsilon_\lambda e^{i\mathbf{k}\cdot\mathbf{r}} | i \rangle = U^{\pm\dagger}(\mathbf{p}'') \tilde{\alpha} \cdot \epsilon_\lambda U^{\pm}(0) \int e^{-i\mathbf{p}''\cdot\mathbf{r}/\hbar} d\mathbf{r}$$

$$= U^{\pm\dagger}(\mathbf{p}'') \tilde{\alpha} \cdot \epsilon_\lambda U^{\pm}(0) \delta_{\mathbf{p}'',0} = (\chi_I^{m_s''\dagger}, 0) \begin{pmatrix} 0 & \tilde{\sigma} \cdot \epsilon_\lambda \\ \tilde{\sigma} \cdot \epsilon_\lambda & 0 \end{pmatrix} \begin{pmatrix} \chi^{m_s} \\ 0 \end{pmatrix} = 0$$

$$(12.100)$$

In the final line χ^{m_s}, $\chi_I^{m_s''}$ and 0 represent two-component spinors in the wavefunctions. In the matrix 0 is a 2×2 zero matrix and $\tilde{\sigma}$ are the 2×2 Pauli spin matrices. Thomson scattering is defined as the scattering of a low energy photon and in (12.100) we have assumed this by letting $\mathbf{k} \to 0$ (the dipole approximation). So we have the surprising result that the terms with intermediate states that involve excitations of positive energy states contribute nothing to the scattering. This arises from the fact that the perturbation contains a matrix $\tilde{\alpha}$ which is odd, so it connects the upper two components of one four-component wavefunction with the lower two components of another.

Next we consider the contribution to the scattering rate from the last two terms in (12.95). As before, we assume the electron is initially at rest. Because of the space integration in the term connecting the initial and intermediate states we again get that the electron in the intermediate state is at rest. In turn the space integration between the intermediate and final states means that the final state is also at rest. If the intermediate state is at rest the summation over intermediate states reduces to a sum over possible spin states. So a typical matrix element becomes

$$\langle s'' | \tilde{\alpha} \cdot \epsilon_\lambda | s \rangle = (0, \chi_I^{m_{s''}\dagger}) \begin{pmatrix} 0 & \tilde{\sigma} \cdot \epsilon_\lambda \\ \tilde{\sigma} \cdot \epsilon_\lambda & 0 \end{pmatrix} \begin{pmatrix} \chi^{m_s} \\ 0 \end{pmatrix} = \chi_I^{m_{s''}\dagger} \tilde{\sigma} \cdot \epsilon_\lambda \chi^{m_s} \quad (12.101)$$

Furthermore, we still assume that $\hbar\omega << mc^2$ so $\hbar\omega$ in the denominators can be neglected. Then the second two terms in (12.95) give us

$$M_{fi} = \frac{K}{2mc^2} \sum_{m_{s''}} (\chi^{m_{s'}\dagger}\tilde{\boldsymbol{\sigma}} \cdot \boldsymbol{\epsilon}_\lambda \chi_I^{m_{s''}} \chi_I^{m_{s''}\dagger}\tilde{\boldsymbol{\sigma}} \cdot \boldsymbol{\epsilon}_{\lambda'}^* \chi^{m_s}$$

$$+ \chi^{m_{s'}\dagger}\tilde{\boldsymbol{\sigma}} \cdot \boldsymbol{\epsilon}_{\lambda'}^* \chi_I^{m_{s''}} \chi_I^{m_{s''}\dagger}\tilde{\boldsymbol{\sigma}} \cdot \boldsymbol{\epsilon}_\lambda \chi^{m_s}) \tag{12.102}$$

Now this can be simplified considerably by noting that

$$\sum_{m_{s''}} \chi^{m_{s''}} \chi^{m_{s''}\dagger} = \begin{pmatrix} 1 \\ 0 \end{pmatrix}(1,0) + \begin{pmatrix} 0 \\ 1 \end{pmatrix}(0,1) = \begin{pmatrix} 1 & 0 \\ 0 & 1 \end{pmatrix} \tag{12.103}$$

i.e. the summation over the intermediate spin states just gives us the identity matrix. So

$$M_{fi} = \frac{K}{2mc^2}\left(\chi^{m_{s'}\dagger}\tilde{\boldsymbol{\sigma}} \cdot \boldsymbol{\epsilon}_\lambda\tilde{\boldsymbol{\sigma}} \cdot \boldsymbol{\epsilon}_{\lambda'}^* \chi^{m_s} + \chi^{m_{s'}\dagger}\tilde{\boldsymbol{\sigma}} \cdot \boldsymbol{\epsilon}_{\lambda'}^*\tilde{\boldsymbol{\sigma}} \cdot \boldsymbol{\epsilon}_\lambda \chi^{m_s}\right)$$

$$= \frac{K}{2mc^2}\chi^{m_{s'}\dagger}(\tilde{\boldsymbol{\sigma}} \cdot \boldsymbol{\epsilon}_\lambda\tilde{\boldsymbol{\sigma}} \cdot \boldsymbol{\epsilon}_{\lambda'}^* + \tilde{\boldsymbol{\sigma}} \cdot \boldsymbol{\epsilon}_{\lambda'}^*\tilde{\boldsymbol{\sigma}} \cdot \boldsymbol{\epsilon}_\lambda)\chi^{m_s}$$

$$= \frac{K}{mc^2}\boldsymbol{\epsilon}_{\lambda'}^* \cdot \boldsymbol{\epsilon}_\lambda\delta_{m_s,m_{s'}} \tag{12.104}$$

In the last line we have used the all too familiar identity (4.19). If we replace K using (12.93), and assume linear polarization of the photons, our final expression for the scattering rate is

$$T_{fi} = \left(\frac{e^2}{m}\right)^2\left(\frac{h}{\epsilon_0^2 V^2 \omega_k \omega_{k'}}\right)\cos^2\theta\delta_{m_s,m_{s'}}\delta(W_f + \hbar\omega' - W_i - \hbar\omega) \tag{12.105}$$

where θ is the angle between $\boldsymbol{\epsilon}_\lambda$ and $\boldsymbol{\epsilon}_{\lambda'}^*$. Assuming elastic scattering, and using (12.97) and (12.99b), we find that the Thomson scattering cross section is

$$\frac{\partial^2\sigma_T}{\partial\Omega\partial E_f} = 4\left(\frac{e^2}{4\pi\epsilon_0 mc^2}\right)^2 |\boldsymbol{\epsilon}_{\lambda'}^* \cdot \boldsymbol{\epsilon}_\lambda|^2\delta_{m_{s'},m_s}\delta(W_f + \hbar\omega' - W_i - \hbar\omega')$$

$$= 4\alpha^2\lambda_C^2\cos^2\theta\delta_{m_{s'},m_s}\delta(W_f + \hbar\omega' - W_i - \hbar\omega) \tag{12.106}$$

where α is the fine structure constant and λ_C is the Compton wavelength.

Equation (12.106) is our final expression for the cross section for Thomson scattering. Notice that it is independent of frequency. Equation (12.106) is identical to the cross section found in a non-relativistic treatment. However, it illustrates an astonishing feature of relativistic quantum theory. It is absolutely imperative to include the excitations from negative energy states in the intermediate states in perturbation theory to get results that give the correct non-relativistic limit. This is true even though the negative energy states are totally alien to non-relativistic quantum theory. It should be noted that this theory is approximate. By letting $\mathbf{k} \to 0$ we have ignored the photon momentum and so momentum is not really conserved in this derivation. Therefore (12.106) should really be regarded as a limiting case only. Furthermore, we have ignored $\hbar\omega$ in comparison with mc^2, although this approximation will be lifted later in this chapter.

12.8 Rayleigh Scattering

Rayleigh scattering is another special case of the general formulae (12.95) and (12.97). Here we are thinking of the case when the initial and final states are the same and we will label them both $|a\rangle$. Therefore we have elastic scattering $\hbar\omega = \hbar\omega'$. We are not scattering from free electrons this time, but rather we are considering electrons bound to atoms, so the sort of physical problem we are trying to understand is how light scatters in gases. Actually, this is a type of scattering that can be described adequately within classical theory (Jackson 1962), and non-relativistic quantum theory (Sakurai 1967). Nonetheless, it is instructive to see it derived within a relativistic quantum mechanical framework. In this section we will work entirely in the dipole approximation, so all exponentials in (12.95) are set equal to one. In fact this is a difficult problem to solve, so rather than evaluating the cross section fully, we will simply extract the dominant frequency dependence.

Before we start evaluating (12.95) for this case, let us mention a couple of mathematical tricks we will use during the derivation below.

Let us start by writing down what is probably the most obvious equation in the whole book

$$\hat{\mathbf{r}}\hat{\mathbf{r}} - \hat{\mathbf{r}}\hat{\mathbf{r}} = 0 \qquad (12.107a)$$

We can take a scalar product of these operators with a pair of polarization vectors. It doesn't matter which way round we do this so let us write

$$\epsilon \cdot \hat{\mathbf{r}}\hat{\mathbf{r}} \cdot \epsilon' - \epsilon' \cdot \hat{\mathbf{r}}\hat{\mathbf{r}} \cdot \epsilon = 0 \qquad (12.107b)$$

and we can insert some intermediate states between these operators, and sum over them

$$\sum_I (\epsilon \cdot \hat{\mathbf{r}}|I\rangle\langle I|\hat{\mathbf{r}} \cdot \epsilon' \ - \ \epsilon' \cdot \hat{\mathbf{r}}|I\rangle\langle I|\hat{\mathbf{r}} \cdot \epsilon) = 0 \qquad (12.107c)$$

and finally we can put state vectors round this expression

$$\sum_I (\langle a|\epsilon \cdot \hat{\mathbf{r}}|I\rangle\langle I|\hat{\mathbf{r}} \cdot \epsilon'|a\rangle - \langle a|\epsilon' \cdot \hat{\mathbf{r}}|I\rangle\langle I|\hat{\mathbf{r}} \cdot \epsilon|a\rangle) = 0 \qquad (12.107d)$$

Secondly, recalling equation (4.113a), we know that the velocity operator is $c\tilde{\alpha}$. Using equation (3.80) we can write

$$\langle a|\tilde{\alpha}\cdot\epsilon|b\rangle = \frac{i}{c\hbar}\langle a|[\hat{H},\hat{r}]\cdot\epsilon|b\rangle = \frac{i}{c\hbar}(W_a - W_b)\langle a|\hat{r}\cdot\epsilon|b\rangle \qquad (12.108)$$

assuming $|a\rangle$ and $|b\rangle$ are eigenfunctions of $\hat{H}$, so

$$\langle a|\hat{r}\cdot\epsilon|b\rangle = \frac{c\hbar}{i(W_a - W_b)}\langle a|\tilde{\alpha}\cdot\epsilon|b\rangle \qquad (12.109)$$

Now, we can substitute (12.109) into (12.107d)

$$\sum_I \left(\frac{\langle a|\tilde{\alpha}\cdot\epsilon_{\lambda'}^*|I\rangle\langle I|\tilde{\alpha}\cdot\epsilon_\lambda|a\rangle}{(W_I - W_a)^2} - \frac{\langle a|\tilde{\alpha}\cdot\epsilon_\lambda|I\rangle\langle I|\tilde{\alpha}\cdot\epsilon_{\lambda'}^*|a\rangle}{(W_I - W_a)^2} \right) = 0 \qquad (12.110)$$

Now we move on to consider the scattering amplitudes themselves. For our purposes the most convenient form for them is the last equality of equation (12.94). So, if we use the dipole approximation for the case we are considering, this becomes

$$M_{fi} = -K\sum_I \left(\frac{\langle a|\tilde{\alpha}\cdot\epsilon_{\lambda'}^*|I\rangle\langle I|\tilde{\alpha}\cdot\epsilon_\lambda|a\rangle}{W_I - W_a - \hbar\omega} \right)$$

$$-K\sum_I \left(\frac{\langle a|\tilde{\alpha}\cdot\epsilon_\lambda|I\rangle\langle I|\tilde{\alpha}\cdot\epsilon_{\lambda'}^*|a\rangle}{W_I - W_a + \hbar\omega} \right) \qquad (12.111)$$

We can put this over a common denominator

$$M_{fi} = -K\sum_I \left(\frac{(W_I - W_a + \hbar\omega)\langle a|\tilde{\alpha}\cdot\epsilon_{\lambda'}^*|I\rangle\langle I|\tilde{\alpha}\cdot\epsilon_\lambda|a\rangle}{(W_I - W_a)^2 - \hbar^2\omega^2} \right)$$

$$-K\sum_I \left(\frac{(W_I - W_a - \hbar\omega)\langle a|\tilde{\alpha}\cdot\epsilon_\lambda|I\rangle\langle I|\tilde{\alpha}\cdot\epsilon_{\lambda'}^*|a\rangle}{(W_I - W_a)^2 - \hbar^2\omega^2} \right) \qquad (12.112)$$

Now, we assume $\hbar\omega \ll W_I - W_a$ so we can make the approximation

$$\frac{1}{(W_I - W_a)^2 - \hbar^2\omega^2} = \frac{1}{(W_I - W_a)^2 \left(1 - \frac{\hbar^2\omega^2}{(W_I - W_a)^2}\right)}$$

$$\approx \frac{1}{(W_I - W_a)^2}\left(1 + \frac{\hbar^2\omega^2}{(W_I - W_a)^2}\right) \qquad (12.113)$$

Now we are going to replace the denominator in (12.112) using equation (12.113). We will start with the leading order term, and split the resulting equation into two parts

$$M_{fi}^{(1)} = -K \sum_I \left(\frac{\langle a|\tilde{\alpha} \cdot \epsilon_{\lambda'}^*|I\rangle\langle I|\tilde{\alpha} \cdot \epsilon_\lambda|a\rangle}{(W_I - W_a)} + \frac{\langle a|\tilde{\alpha} \cdot \epsilon_\lambda|I\rangle\langle I|\tilde{\alpha} \cdot \epsilon_{\lambda'}^*|a\rangle}{(W_I - W_a)} \right)$$

$$- \hbar\omega K \sum_I \left(\frac{\langle a|\tilde{\alpha} \cdot \epsilon_{\lambda'}^*|I\rangle\langle I|\tilde{\alpha} \cdot \epsilon_{\lambda'}|a\rangle}{(W_I - W_a)^2} - \frac{\langle a|\tilde{\alpha} \cdot \epsilon_\lambda|I\rangle\langle I|\tilde{\alpha} \cdot \epsilon_{\lambda'}^*|a\rangle}{(W_I - W_a)^2} \right)$$

$$(12.114)$$

The second line here is zero, as can be seen immediately on comparison
with (12.110). In the first line we can replace the matrix element in $\tilde{\alpha} \cdot \epsilon_{\lambda'}^*$
in each term using equation (12.109), and then we use the completeness
of the intermediate states to give

$$M_{fi}^{(1)} = -\frac{iK}{c\hbar} \left(\langle a|\tilde{\alpha} \cdot \epsilon_\lambda \hat{\mathbf{r}} \cdot \epsilon_{\lambda'}^*|a\rangle - \langle a|\hat{\mathbf{r}} \cdot \epsilon_{\lambda'}^* \tilde{\alpha} \cdot \epsilon_\lambda|a\rangle \right) = 0 \qquad (12.115)$$

So the leading order term in the expansion (12.113) contributes nothing
to the scattering. We have to substitute the second order term from that
equation into (12.112). The $\hbar^2\omega^2$ will come outside the summation over
intermediate states and we have

$$M_{fi}^{(2)} = -K\hbar^2\omega^2 \sum_I \left(\frac{(W_I - W_a + \hbar\omega)\langle a|\tilde{\alpha} \cdot \epsilon_{\lambda'}^*|I\rangle\langle I|\tilde{\alpha} \cdot \epsilon_\lambda|a\rangle}{(W_I - W_a)^4} \right)$$

$$- K\hbar^2\omega^2 \sum_I \left(\frac{(W_I - W_a - \hbar\omega)\langle a|\tilde{\alpha} \cdot \epsilon_\lambda|I\rangle\langle I|\tilde{\alpha} \cdot \epsilon_{\lambda'}^*|a\rangle}{(W_I - W_a)^4} \right)$$

$$(12.116)$$

As $\hbar\omega \ll W_I - W_a$ we can ignore the $\hbar\omega$ in the numerator of the terms
in large brackets in (12.116), and write

$$M_{fi}^{(2)} = -K\hbar^2\omega^2 \sum_I \left(\frac{\langle a|\tilde{\alpha} \cdot \epsilon_{\lambda'}^*|I\rangle\langle I|\tilde{\alpha} \cdot \epsilon_\lambda|a\rangle}{(W_I - W_a)^3} \right.$$

$$\left. + \frac{\langle a|\tilde{\alpha} \cdot \epsilon_\lambda|I\rangle\langle I|\tilde{\alpha} \cdot \epsilon_{\lambda'}^*|a\rangle}{(W_I - W_a)^3} \right) \qquad (12.117)$$

This is the scattering amplitude. We can put this into the Golden Rule
and use equations (12.93), and (12.97) to (12.99) to get the final form for
the scattering cross section per unit solid angle per unit energy

$$\frac{\partial^2\sigma_R}{\partial\Omega\partial E_f} = 4\alpha^2\hbar^2c^2\hbar^4\omega^4\delta(W_f - W_i)\times$$

$$\left| \sum_I \frac{1}{(W_I - W_a)^3} \left(\langle a|\tilde{\alpha} \cdot \epsilon_{\lambda'}^*|I\rangle\langle I|\tilde{\alpha} \cdot \epsilon_\lambda|a\rangle + \langle a|\tilde{\alpha} \cdot \epsilon_\lambda|I\rangle\langle I|\tilde{\alpha} \cdot \epsilon_{\lambda'}^*|a\rangle \right) \right|^2$$

$$(12.118)$$

where α is the fine structure constant. This is the Rayleigh cross section. It
has the same form as the Rayleigh cross section derived non-relativistically.
The important point to note about this expression is that it depends on ω^4,

so the scattering is strongly dependent on frequency. For the atmosphere $W_I - W_a$ is in the far ultraviolet region of the spectrum so the condition $\hbar\omega << W_I - W_a$ is fulfilled for visible light. Rayleigh scattering is what happens to sunlight as it passes through the atmosphere. If you look towards the sun at sunset or sunrise it appears a reddish colour. This is because more of the higher frequencies have been scattered away leaving us with sunlight dominated by the red end of the spectrum. On the other hand if you look up at the sky away from the sun during the day, the only sunlight you see is that which has been scattered and this, as we all know, is blue. This example shows that higher frequency light is scattered much more strongly than lower frequency light. Obviously, I don't recommend anybody to look directly at the sun. It is not worth ruining your eyesight just to verify a relativistic quantum mechanical scattering formula. For completeness we point out that other mechanisms (aerosol scattering for example) also scatter light in the atmosphere, so the colours of the sky and sun, as viewed from the surface of the earth, are not solely determined by Rayleigh scattering.

12.9 Compton Scattering

It is appropriate to have a section on Compton scattering near the end of this book as we had a section on the elementary theory close to the beginning (see section 1.6). We are going to derive the Klein–Nishina* formula for Compton scattering from a free electron. This formula is Thomson scattering without the restrictions, i.e. we do not assume that $\hbar\omega << mc^2$, or that the electron in the intermediate state is at rest. However, if you try to derive it you will see why detailed derivations of this formula are rarely reproduced in textbooks. I pieced together the full derivation from clues from several different sources (actually I bribed a graduate student to do it for me).[†] The derivation is difficult and the equations become exceedingly long (Bjorken and Drell 1964, Rose 1961, Sakurai 1967). Writing some of them would take a whole page of this book! Therefore we do not reproduce them here. Instead, we show enough of the derivation, together with a detailed description of the steps between the lines of mathematics, to enable readers to reproduce it themselves (if they have a lot of time to spare).

You may think we have already understood the Compton effect, as it is described by equation (1.63). All that equation tells us is a wavelength shift. It says nothing about the probability of Compton scattering occurring, or the intensity of the Compton peaks. Equation (1.63) is simply a statement

---

[*] Yes! It's the ubiquitous O. Klein again.
[†] I would like to thank H.J. Gotsis for collaborating on this derivation.

of conservation of energy and momentum during Compton scattering. It can be rewritten in terms of photon wavevectors as

$$\mathbf{k}_i \cdot \mathbf{k}_f = k_i k_f - \frac{mc}{\hbar}(k_i - k_f) \tag{12.119}$$

What we will describe here is again Compton scattering from a free electron. The process is the following. In the initial state we have a photon with wavevector $\mathbf{k}_i$ and a stationary electron with energy $W_i = mc^2$. In the final state the photon has been scattered and has wavevector $\mathbf{k}_f$ and the electron has acquired momentum $\mathbf{p}_f = \hbar(\mathbf{k}_i - \mathbf{k}_f)$ and energy

$$W_f = (p_f^2 c^2 + m^2 c^4)^{1/2} \tag{12.120}$$

where p_f is the magnitude of $\mathbf{p}_f$, etc.

In analogy with equation (12.99a) the appropriate expression to use for the density of states per unit solid angle is

$$n(E_f) = \frac{V k_f^2}{8\pi^3} \left(\frac{\partial k_f}{\partial E_f} \right) d\Omega \tag{12.121}$$

We can replace $\hbar\omega$ by E_f in this formula because the final state of the photon uniquely determines the final state of the electron. Equation (12.121) is actually more difficult to work out than you might imagine. The total energy of the electron and photon after scattering is

$$E_f = W_f + \hbar c k_f \tag{12.122}$$

and differentiating both sides of this with respect to k_f leads to

$$\frac{\partial k_f}{\partial E_f} = \frac{1}{\hbar c + \partial W_f / \partial k_f} \tag{12.123}$$

Now we can work out $\partial W_f / \partial k_f$ explicitly. Writing (12.120) in terms of wavevectors gives

$$W_f = (\hbar^2 c^2 (\mathbf{k}_f - \mathbf{k}_i)^2 + m^2 c^4)^{1/2}$$
$$= (\hbar^2 c^2 (k_f^2 + k_i^2 - 2\mathbf{k}_f \cdot \mathbf{k}_i) + m^2 c^4)^{1/2} \tag{12.124}$$

and differentiation gives us

$$\frac{\partial W_f}{\partial k_f} = \hbar^2 c^2 \frac{k_f^2 - \mathbf{k}_f \cdot \mathbf{k}_i}{k_f W_f} \tag{12.125}$$

Now we substitute (12.125) into (12.123) and find that

$$\frac{\partial k_f}{\partial E_f} = \frac{k_f W_f}{\hbar c k_f W_f + \hbar^2 c^2 k_f^2 - \hbar^2 c^2 \mathbf{k}_i \cdot \mathbf{k}_f} \tag{12.126}$$

Equation (12.124) is an expression for W_f but it has an awkward scalar product in it. However, we can get rid of this by substituting for it from equation (12.119)

$$W_f = hc \left((k_f^2 + k_i^2 - 2k_ik_f + \frac{2mc}{\hbar}(k_i - k_f) + \frac{m^2c^2}{\hbar^2} \right)^{1/2}$$

$$= \hbar ck_i - \hbar ck_f + mc^2 \qquad (12.127)$$

We can replace W_f in (12.126) with this and use (12.119) again to get

$$\frac{\partial k_f}{\partial E_f} = \frac{k_f W_f}{\hbar mc^3 k_i} \qquad (12.128)$$

Finally we can substitute this into (12.121) to get our final expression for the density of final states per final energy

$$n(E_f) = \frac{V}{8\pi^3 mc^3 \hbar} \frac{k_f^3 W_f}{k_i} \qquad (12.129)$$

We are trying to find the Compton cross section using equation (12.97). The flux I_0 is given by equation (12.98) and we now have $n(E_f)$ in equation (12.129). Let us write down our expression for the differential scattering cross section

$$\frac{\partial \sigma_C}{\partial \Omega} = \frac{2\pi}{\hbar} \frac{V}{c} \frac{V}{8\pi^3 mc^3 \hbar} \frac{k_f^3 W_f}{k_i} K^2 |M'_{fi}|^2 = \frac{e^4 \mu_0^2 c^2}{4\pi^2 m} \frac{k_f^2 W_f}{k_i^2} |M'_{fi}|^2 \qquad (12.130)$$

where M'_{fi} is M_{fi} of equation (12.94) with the constant K omitted. Our next task is to evaluate the scattering amplitude M'_{fi} for this case.

It is convenient to use the expression (12.94) for the scattering amplitude and we will do this. To make life a little bit easier we will deal with the case of linearly polarized X-rays only. That means that both $\epsilon_\lambda = \epsilon$ and $\epsilon_{\lambda'} = \epsilon'$ are real. For our case

$$M'_{fi} = \left(\sum_{I'} \frac{\langle f | \tilde{\alpha} \cdot \epsilon' e^{-i\mathbf{k}_f \cdot \mathbf{r}} | I' \rangle \langle I' | \tilde{\alpha} \cdot \epsilon e^{i\mathbf{k}_i \cdot \mathbf{r}} | i \rangle}{W_i - W' + \hbar \omega_{\mathbf{k}_i}} \right)$$

$$+ \sum_{I''} \left(\frac{\langle f | \tilde{\alpha} \cdot \epsilon e^{i\mathbf{k}_i \cdot \mathbf{r}} | I'' \rangle \langle I'' | \tilde{\alpha} \cdot \epsilon' e^{-i\mathbf{k}_f \cdot \mathbf{r}} | i \rangle}{W_i - W'' - \hbar \omega_{\mathbf{k}_f}} \right) \qquad (12.131)$$

The next step is to multiply the top and bottom of the first term in (12.131) by $W' + mc^2 + \hbar \omega_{\mathbf{k}_i}$. The denominator becomes

$$W'^2 - (mc^2 + \hbar \omega_{\mathbf{k}_i})^2 = -2\hbar \omega_{\mathbf{k}_i} mc^2 \qquad (12.132a)$$

In the second term in (12.131) we multiply the top and bottom by $W'' + mc^2 - \hbar \omega_{\mathbf{k}_f}$ and the denominator becomes

$$W''^2 - (mc^2 - \hbar \omega_{\mathbf{k}_f})^2 = 2\hbar \omega_{\mathbf{k}_f} mc^2 \qquad (12.132b)$$

This is a useful thing to do because it means that the denominators in (12.131) are independent of the intermediate state. However, we still have to worry about the numerators. For the first term we have

$$\sum_{I'} (W' + mc^2 + \hbar\omega_{\mathbf{k}_i})\langle f|\tilde{\alpha} \cdot \epsilon' e^{-i\mathbf{k}_f \cdot \mathbf{r}}|I'\rangle\langle I'|\tilde{\alpha} \cdot \epsilon e^{i\mathbf{k}_i \cdot \mathbf{r}}|i\rangle$$

$$= \sum_{I'} \langle f|\tilde{\alpha} \cdot \epsilon' e^{-i\mathbf{k}_f \cdot \mathbf{r}}(mc^2 + \hbar\omega_{\mathbf{k}_i} + \hat{H}')|I'\rangle\langle I'|\tilde{\alpha} \cdot \epsilon e^{i\mathbf{k}_i \cdot \mathbf{r}}|i\rangle$$

(12.133a)

and for the second term

$$\sum_{I''} (W'' + mc^2 - \hbar\omega_{\mathbf{k}_f})\langle f|\tilde{\alpha} \cdot \epsilon e^{i\mathbf{k}_i \cdot \mathbf{r}}|I''\rangle\langle I''|\tilde{\alpha} \cdot \epsilon' e^{-i\mathbf{k}_f \cdot \mathbf{r}}|i\rangle$$

$$= \sum_{I''} \langle f|\tilde{\alpha} \cdot \epsilon e^{i\mathbf{k}_i \cdot \mathbf{r}}(mc^2 - \hbar\omega_{\mathbf{k}_f} + \hat{H}'')|I''\rangle\langle I''|\tilde{\alpha} \cdot \epsilon' e^{-i\mathbf{k}_f \cdot \mathbf{r}}|i\rangle$$

(12.133b)

The sums over I' and I'' here are over all possible momenta, energies and spins of the intermediate states. If we put the two terms in (12.133) into a position representation all the position dependence appears in exponents. Integrating over repeated coordinates soon yields that only $\mathbf{p}' = \hbar\mathbf{k}_i$ of the intermediate state contributes to the sum in (12.133a) and only $\mathbf{p}'' = -\hbar\mathbf{k}_f$ contributes in (12.133b). This procedure also produces a δ-function giving us conservation of momentum, and a volume which cancels the volume in the plane wave normalization of equation (5.23). However, there are still four possible intermediate states of the electron with each of these momenta. It can have positive or negative energy and can be spin up or spin down. The summations in equation (12.131) thus reduce to sums over these four states. So, equations (12.133) simplify and the scattering amplitude becomes

$$M'_{fi} = \sum_{I'} \frac{\langle U_f|\tilde{\alpha} \cdot \epsilon'(mc^2 + \hbar\omega_{\mathbf{k}_i} + \hat{H}')|U_{I'}\rangle\langle U_{I'}|\tilde{\alpha} \cdot \epsilon|U_i\rangle}{-2\hbar\omega_{\mathbf{k}_i}mc^2}$$

$$+ \sum_{I''} \frac{\langle U_f|\tilde{\alpha} \cdot \epsilon(mc^2 - \hbar\omega_{\mathbf{k}_f} + \hat{H}'')|U_{I''}\rangle\langle U_{I''}|\tilde{\alpha} \cdot \epsilon'|U_i\rangle}{2\hbar\omega_{\mathbf{k}_f}mc^2}$$

(12.134)

where only values of U_f and U_i that conserve momentum are acceptable. The values of U are given by equation (5.22). The Hamiltonians $\hat{H}'$ and $\hat{H}''$ can be written entirely in terms of quantities associated with the initial and/or final states. This is because the momenta of the intermediate states can be written in terms of initial or final state quantities, as discussed below equation (12.133). Therefore the only place the intermediate states appear is in the state vectors, and we can invoke equation (2.102) to help us do the summation. Thus equation (12.134) becomes

$$M'_{fi} = \frac{\langle U_f | \tilde{\alpha} \cdot \epsilon'(mc^2 + \hbar\omega_{k_i} + \hat{H}')\tilde{\alpha} \cdot \epsilon | U_i \rangle}{-2\hbar\omega_{k_i}mc^2}$$

$$+ \frac{\langle U_f | \tilde{\alpha} \cdot \epsilon(mc^2 - \hbar\omega_{k_f} + \hat{H}'')\tilde{\alpha} \cdot \epsilon' | U_i \rangle}{2\hbar\omega_{k_f}mc^2} \quad (12.135)$$

Now we are working with plane waves, so we can use our knowledge of their Hamiltonian to simplify the numerators of equation (12.135)

$$mc^2 + \hbar\omega_{k_i} + \hat{H}' = mc^2 + \hbar\omega_{k_i} + c\tilde{\alpha} \cdot \mathbf{p}' + \tilde{\beta}mc^2$$

$$= (\tilde{I} + \tilde{\beta})mc^2 + \hbar c(\tilde{I} + \tilde{\alpha} \cdot \hat{\mathbf{k}}_i)k_i \quad (12.136a)$$

and similarly

$$mc^2 - \hbar\omega_{k_f} + \hat{H}'' = mc^2 - \hbar\omega_{k_f} + c\tilde{\alpha} \cdot \mathbf{p}'' + \tilde{\beta}mc^2$$

$$= (\tilde{I} + \tilde{\beta})mc^2 - \hbar c(\tilde{I} + \tilde{\alpha} \cdot \hat{\mathbf{k}}_f)k_f \quad (12.136b)$$

Here $\hat{\mathbf{k}}_i$ ($\hat{\mathbf{k}}_f$) is a unit wavevector in the direction of $\mathbf{k}_i$ ($\mathbf{k}_f$). We can use these expressions and the fact that the electron is initially at rest to simplify the scattering amplitudes further.

The electron is initially at rest ($\mathbf{p} = 0$ obviously), therefore

$$\lim_{\mathbf{p} \to 0}(c\tilde{\alpha} \cdot \mathbf{p} + \tilde{\beta}mc^2)U_i = \tilde{\beta}mc^2U_i = W_iU_i = mc^2U_i$$

$$\longrightarrow mc^2(\tilde{I} - \tilde{\beta})U_i = 0 \quad (12.137)$$

Now look what happens if we put the first term on the right hand side of (12.136) into (12.135). We get terms like

$$(\tilde{I} + \tilde{\beta})mc^2\tilde{\alpha} \cdot \epsilon U_i = \tilde{\alpha} \cdot \tilde{\epsilon}(\tilde{I} - \tilde{\beta})mc^2U_i = 0 \quad (12.138)$$

where we have used (12.137) and equation (4.8). So when we substitute the right hand side of (12.136) into (12.135) only the second terms contribute and we are left with

$$M'_{fi} = \frac{1}{2mc^2} \times$$

$$\left(\frac{\langle U_f | \tilde{\alpha} \cdot \epsilon' c[\tilde{I} + \tilde{\alpha} \cdot \hat{\mathbf{k}}_i]k_i\tilde{\alpha} \cdot \epsilon | U_i \rangle}{-\omega_{k_i}} - \frac{\langle U_f | \tilde{\alpha} \cdot \epsilon c[\tilde{I} + \tilde{\alpha} \cdot \hat{\mathbf{k}}_f]k_f\tilde{\alpha} \cdot \epsilon' | U_i \rangle}{\omega_{k_f}} \right)$$

$$= -\frac{1}{2mc^2} \left(\langle U_f | 2\epsilon \cdot \epsilon' + \frac{1}{k_i}\tilde{\alpha} \cdot \epsilon'\tilde{\alpha} \cdot \mathbf{k}_i\tilde{\alpha} \cdot \epsilon + \frac{1}{k_f}\tilde{\alpha} \cdot \epsilon\tilde{\alpha} \cdot \mathbf{k}_f\tilde{\alpha} \cdot \epsilon' | U_i \rangle \right)$$

$$= -\frac{1}{2mc^2}\langle U_f | \hat{Q} | U_i \rangle \quad (12.139)$$

This would be the form of the scattering amplitude that we would substitute into (12.130) if the spin states of the initial and final electrons were specified. Unfortunately they cannot be known. The electron in the initial state could be spin up or spin down, so we have to take an average value. We do this by summing over the two possible states and halving the result. Compton scattering may occur with or without a flip in the spin of the electron, so we also have to sum over the possible spin states of the final electron. So

$$|\langle M'_{fi}\rangle|^2 = \frac{1}{2}\sum_{S_i}\sum_{S_f}\left|\frac{1}{2mc^2}\langle U_f|\hat{Q}|U_i\rangle\right|^2 \qquad (12.140)$$

The summations here are over spin directions only, but it turns out to be convenient to change them to sums over spin and energy (with the same value of the momentum). We can do this by introducing the energy projection operator discussed in section 6.2.

$$\hat{\Gamma}^+ = \frac{1}{2}(\tilde{I} + \hat{\Gamma}) = \frac{1}{2}\left(\tilde{I} + \frac{\tilde{\alpha}\cdot\mathbf{p} + \tilde{\beta}mc}{(p^2 + m^2c^2)^{1/2}}\right) = \frac{1}{2}\left(\tilde{I} + \frac{\hat{H}}{W}\right) \qquad (12.141)$$

as this will force any contribution from the negative energy states to the summation to be zero.

$$\begin{aligned}|\langle M'_{fi}\rangle|^2 &= \frac{1}{8m^2c^4}\sum_{S_i}\sum_{S_f}(\langle U_f|\hat{Q}|U_i\rangle)^\dagger\langle U_f|\hat{Q}|U_i\rangle\\ &= \frac{1}{8m^2c^4}\sum_i\sum_f\langle U_i|\hat{Q}^\dagger\tfrac{1}{2}(\tilde{I} + \tfrac{\hat{H}_f}{W_f})|U_f\rangle\langle U_f|\hat{Q}\tfrac{1}{2}(\tilde{I} + \tfrac{\hat{H}_i}{W_i})|U_i\rangle\\ &= \frac{1}{32m^2c^4W_iW_f}\mathrm{Tr}\,\hat{Q}^\dagger(\hat{H}_f + W_f)\hat{Q}(\hat{H}_i + W_i) \qquad (12.142)\end{aligned}$$

where Tr means trace, $\hat{Q}^\dagger$ is the hermitian conjugate of $\hat{Q}$ and

$$\hat{H}_i + W_i = c\tilde{\alpha}\cdot\mathbf{p}_i + \tilde{\beta}mc^2 + mc^2 = mc^2(\tilde{I} + \tilde{\beta}) \qquad (12.143a)$$

and, from (12.127)

$$\begin{aligned}\hat{H}_f + W_f &= c\tilde{\alpha}\cdot\mathbf{p}_f + \tilde{\beta}mc^2 + \hbar\omega_{\mathbf{k}_i} - \hbar\omega_{\mathbf{k}_f} + mc^2\\ &= (\tilde{I} + \tilde{\beta})mc^2 + \hbar(\omega_{\mathbf{k}_i} - \omega_{\mathbf{k}_f}) + c\hbar\tilde{\alpha}\cdot(\mathbf{k}_i - \mathbf{k}_f) \qquad (12.143b)\end{aligned}$$

Recall that taking the hermitian conjugate of a matrix operator means taking the complex conjugate and transpose of the matrices and reversing their order in the expressions. However, we are aided slightly by the fact that the α-matrices are their own hermitian conjugates. So we have to evaluate the trace

$$|\langle M'_{fi}\rangle|^2 = \frac{1}{32m^3c^6W_f}\text{Tr}\bigg\{\bigg(2\epsilon\cdot\epsilon' + \frac{\tilde{\alpha}\cdot\epsilon\tilde{\alpha}\cdot\mathbf{k}_i\tilde{\alpha}\cdot\epsilon'}{k_i} + \frac{\tilde{\alpha}\cdot\epsilon'\tilde{\alpha}\cdot\mathbf{k}_f\tilde{\alpha}\cdot\epsilon}{k_f}\bigg)$$
$$\times\, (\hbar\omega_{\mathbf{k}_i} - \hbar\omega_{\mathbf{k}_f} + mc^2 + \tilde{\beta}mc^2 + c\hbar\tilde{\alpha}\cdot\mathbf{k}_i - c\hbar\tilde{\alpha}\cdot\mathbf{k}_f)$$
$$\times\, (2\epsilon\cdot\epsilon' + \frac{1}{k_i}\tilde{\alpha}\cdot\epsilon'\tilde{\alpha}\cdot\mathbf{k}_i\tilde{\alpha}.\epsilon + \frac{1}{k_f}\tilde{\alpha}\cdot\epsilon\tilde{\alpha}\cdot\mathbf{k}_f\tilde{\alpha}\cdot\epsilon')mc^2(\tilde{I} + \tilde{\beta})\bigg\}$$

$$(12.144)$$

This is a long trace! It can be simplified considerably by recalling several properties of the matrices and of traces in general.

- The α-matrices are odd, and when squared give the identity matrix. Therefore any term containing a product of an odd number of them will contribute zero to the trace and so can be ignored.
- The lower two components of the leading diagonal of $\tilde{\beta}$ are the negative of the upper two components and the square of $\tilde{\beta}$ is also the identity matrix. Therefore any term containing an odd number of powers of $\tilde{\beta}$ can be ignored as it contributes nothing to the trace.
- In a trace we can alter the order of the matrices cyclically without changing the value of the trace.
- The polarization vector and the wavevector for a particular photon are always at right angles to each other, so $\mathbf{k}_i\cdot\epsilon = \mathbf{k}_f\cdot\epsilon' = 0$.
- The polarization vectors ϵ and ϵ' are real and have unit length.
- From equation (4.19) for any two vectors $2\mathbf{A}\cdot\mathbf{B} = \tilde{\alpha}\cdot\mathbf{A}\tilde{\alpha}\cdot\mathbf{B} + \tilde{\alpha}\cdot\mathbf{B}\tilde{\alpha}\cdot\mathbf{A}$.

After pages of tiresome mathematics we eventually arrive at

$$|\langle M'_{fi}\rangle|^2 = \frac{1}{m^2c^3W_f}\bigg(mc(\epsilon\cdot\epsilon')^2 + \hbar\frac{k_i - k_f}{4k_ik_f}(k_ik_f - \mathbf{k}_i\cdot\mathbf{k}_f)\bigg) \quad (12.145)$$

and use of equation (12.119) to remove the scalar product leads to

$$|\langle M'_{fi}\rangle|^2 = \frac{1}{4mc^2W_f}\bigg(4(\epsilon\cdot\epsilon')^2 + \frac{k_f}{k_i} + \frac{k_i}{k_f} - 2\bigg) \quad (12.146)$$

The final step is to incorporate this expression into equation (12.130) to find the differential cross section for Compton scattering.

$$\frac{\partial\sigma_C}{\partial\Omega} = \frac{e^4\mu_0^2}{16\pi^2m^2}\frac{k_f^2}{k_i^2}\bigg(4(\epsilon\cdot\epsilon')^2 + \frac{k_f}{k_i} + \frac{k_i}{k_f} - 2\bigg)$$
$$= \alpha^2\lambda_C^2\frac{k_f^2}{k_i^2}\bigg(4(\epsilon\cdot\epsilon')^2 + \frac{k_f}{k_i} + \frac{k_i}{k_f} - 2\bigg) \quad (12.147)$$

Equation (12.147) is known as the Klein–Nishina formula, named after its discoverers (Klein and Nishina 1929). The first thing to note about it is that if we have elastic scattering $k_f = k_i$ and then (12.147) reduces to the Thomson scattering cross section of equation (12.106).

Let us look at the limits of (12.147). Firstly we'll take the non-relativistic limit. It is useful to rewrite equation (1.63) here

$$\lambda_f - \lambda_i = \frac{h}{mc}(1 - \cos\theta) = 2.43 \times 10^{-12}(1 - \cos\theta) \text{ metres} \qquad (12.148)$$

In the non-relativistic limit $\hbar\omega_{k_i} \ll mc^2$, i.e. $\lambda_i \gg \lambda_C$. Then the change in wavelength during a Compton scattering event is very small. Hence $k_f \approx k_i$ and equation (12.147) again tends to the Thomson limit, i.e. the cross section becomes independent of energy of the incident photon. Another way of looking at this is to state the obvious and point out that quantum effects disappear in the long wavelength limit.

For the ultra-relativistic limit it is more useful to write the Klein–Nishina formula in terms of energies. Recalling that we have linearly polarized light, we can see that

$$\epsilon \cdot \epsilon' = \frac{1}{k_i k_f}\mathbf{k}_i \cdot \mathbf{k}_f = \cos\theta = 1 - \frac{mc}{\hbar}\frac{(k_i - k_f)}{k_i k_f} \qquad (12.149)$$

Substituting this into (12.147) easily gives

$$\frac{\partial\sigma_C}{\partial\Omega} =$$

$$\alpha^2\lambda_C^2\frac{W_f^2}{W_i^2}\left(2 - 8mc^2\frac{W_i - W_f}{W_i W_f} + 4m^2c^4\frac{(W_i - W_f)^2}{W_i^2 W_f^2} + \frac{W_f^2 + W_i^2}{W_i W_f}\right)$$

$$(12.150)$$

In the ultra-relativistic limit $\lambda_i \ll h/mc$ and in that case equation (12.148) tells us that for moderate scattering angles $\lambda_f \approx h/mc$. This illustrates the fact that the Compton wavelength shift is much more pronounced for high energy photons and when the energy of the scattered photon is about mc^2. This information can be substituted into (12.150) to see that the cross section for incident photons of very high energy is proportional to W_i^{-1}. However, it should be noted that this is only true in the rest frame of the initial electron. It may not be true in any other frame. At energies of a few MeV Compton scattering is dominated by pair production and so the Klein–Nishina formula becomes inapplicable.

Compton scattering is often used as a probe of condensed matter. Of course, in condensed matter the electrons are not free, so the above analysis is at best an approximation. The full analysis of Compton scattering when the initial electron is in a Bloch state or an atomic orbital is very complicated. Nonetheless, the above analysis does give insight into the physics of Compton scattering and serves as a firm foundation and guide when trying to understand Compton scattering in solids.

12.10 Magnetic Scattering of X-Rays

Here we are going to consider the scattering of X-rays by a gas of electrons, a problem first treated relativistically by Low (1954) and by Gell-mann and Goldberger (1954). Of course, X-rays scatter from the charge density in such a system, and we will see this, but the photon is also sensitive to the spin of the electrons and, as we shall see later, can be used as a probe of magnetism in materials (Platzman and Tzoar 1970). In a series of key experiments de Bergevin and Brunel (1981) (see also Brunel and de Bergevin 1981) demonstrated the feasibility of magnetic X-ray scattering as a probe of magnetic structures in solids.

In the present analysis we derive an expression for the cross section for scattering of X-rays by magnetic materials and show how the magnetic terms can be used in a measurement of magnetic structure. By taking the non-relativistic limit of our expression we find the cross section of Blume (1985) which is written in terms of quantities familiar from the traditional theory of magnetism. We do not use the second quantized formulation of quantum mechanics, as we wish to make the interpretation of the equations more straightforward. Our starting point is the Hamiltonian

$$\hat{H} = \sum_j c\tilde{\boldsymbol{\alpha}} \cdot (\hat{\mathbf{p}}_j - e\mathbf{A}(\mathbf{r}_j)) + \sum_j \tilde{\beta}mc^2 + \sum_{ij} V(\mathbf{r}_i - \mathbf{r}_j) \qquad (12.151)$$

The summation in the first two terms is a sum over all electrons in the system, representing the kinetic energy of the electrons in an electromagnetic field and the rest mass energy. The final term is a sum over all pairs of electrons and represents their mutual Coulomb repulsion, Breit interaction, etc. Now we are treating the photons as quantized harmonic oscillators, so there should also be a term to represent their energy in the Hamiltonian. However, we assume the photon field does not change the underlying electron states of the unperturbed Hamiltonian, and ignore it in the crystal Hamiltonian. Let us rewrite (12.151) as

$$\hat{H} = \sum_j \left(c\tilde{\boldsymbol{\alpha}} \cdot \hat{\mathbf{p}}_j + \tilde{\beta}mc^2 + \sum_i V(\mathbf{r}_i - \mathbf{r}_j) \right) - \sum_j ec\tilde{\boldsymbol{\alpha}} \cdot \mathbf{A}(\mathbf{r}_j) = \hat{H}_0 + \hat{H}'$$

$$(12.152)$$

Because we have many electrons there is a summation in the perturbation.

Therefore the cross section of equation (12.97) becomes

$$
\frac{\partial^2 \sigma_M}{\partial W' \partial \Omega} = \frac{e^4 c^4 \mu_0^2}{4\pi^2} \delta(W_f + \hbar\omega_{k'} - W_i - \hbar\omega_k) \times
$$

$$
\left| -\sum_I \frac{\langle f | \sum_j \tilde{\alpha} \cdot \epsilon^*_{\lambda'} e^{-i\mathbf{k}'\cdot\mathbf{r}_j} | I \rangle \langle I | \sum_l \tilde{\alpha} \cdot \epsilon_\lambda e^{i\mathbf{k}\cdot\mathbf{r}_l} | i \rangle}{W_I - W_i - \hbar\omega_k} \right.
$$

$$
- \sum_I \frac{\langle f | \sum_j \tilde{\alpha} \cdot \epsilon_\lambda e^{i\mathbf{k}\cdot\mathbf{r}_j} | I \rangle \langle I | \sum_l \tilde{\alpha} \cdot \epsilon^*_{\lambda'} e^{-i\mathbf{k}'\cdot\mathbf{r}_l} | i \rangle}{W_I - W_i + \hbar\omega_{k'}}
$$

$$
+ \sum_I \frac{\langle f | \sum_j \tilde{\alpha} \cdot \epsilon_\lambda e^{i\mathbf{k}\cdot\mathbf{r}_j} | I \rangle \langle I | \sum_l \tilde{\alpha} \cdot \epsilon^*_{\lambda'} e^{-i\mathbf{k}'\cdot\mathbf{r}_l} | i \rangle}{W_f + |W_I| - \hbar\omega_k}
$$

$$
+ \sum_I \frac{\langle f | \sum_j \tilde{\alpha} \cdot \epsilon^*_{\lambda'} e^{-i\mathbf{k}'\cdot\mathbf{r}_j} | I \rangle \langle I | \sum_l \tilde{\alpha} \cdot \epsilon_\lambda e^{i\mathbf{k}\cdot\mathbf{r}_l} | i \rangle}{W_f + |W_I| + \hbar\omega_{k'}} \left. \right|^2 \qquad (12.153)
$$

and we have used equation (12.99*b*) for the photon density of states. A cautionary note should be sounded here. The denominators in the latter two terms in (12.95) were derived in a single-particle picture. We have carried this through to a many-particle system, i.e. we assume the negative energy excitations in an interacting electron gas are the same as those in a non-interacting electron gas. There is no *a priori* reason to assume that this is valid. Let us do it anyway. Equation (12.153) is our expression for the scattering cross section, and in a full relativistic calculation it is this expression that must be evaluated. It contains all scattering up to second order in the Golden Rule. However, it is not very transparent and one can ask whether it can be written in more familiar terms. The answer is that it can in certain limits, which we examine now.

We start by examining the last two terms in equation (12.153), which contain intermediate states involving excitations from negative energies

$$
M_{fi}^- = \sum_I \frac{\langle f | \sum_j \tilde{\alpha} \cdot \epsilon_\lambda e^{i\mathbf{k}\cdot\mathbf{r}_j} | I \rangle \langle I | \sum_l \tilde{\alpha} \cdot \epsilon^*_{\lambda'} e^{-i\mathbf{k}'\cdot\mathbf{r}_l} | i \rangle}{W_f + |W_I| - \hbar\omega_k}
$$

$$
+ \sum_I \frac{\langle f | \sum_j \tilde{\alpha} \cdot \epsilon^*_{\lambda'} e^{-i\mathbf{k}'\cdot\mathbf{r}_j} | I \rangle \langle I | \sum_l \tilde{\alpha} \cdot \epsilon_\lambda e^{i\mathbf{k}\cdot\mathbf{r}_l} | i \rangle}{W_f + |W_I| + \hbar\omega_{k'}} \qquad (12.154)
$$

Now, we will consider elastic scattering so $\omega_k = \omega_{k'} = \omega$ and we will also assume that all energies are dwarfed by mc^2, as is usual in condensed matter experiments. So $W_f \approx |W_I| \approx mc^2$ (at least this is consistent with our earlier statement that the intermediate states are one-particle-like). In this limit the denominators of (12.154) can be written

$$
(2mc^2 \mp \hbar\omega_k)^{-1} = (2mc^2)^{-1} \left(1 \mp \frac{\hbar\omega}{2mc^2} \right)^{-1} \approx \frac{1}{2mc^2} \pm \frac{\hbar\omega}{4m^2 c^4} \qquad (12.155)
$$

So in both terms in (12.154) the dominant term is of order $1/2mc^2$. Let us consider this first.

$$M_{fi}^{(1)-} = \sum_I \frac{1}{2mc^2} \langle f | \sum_j \tilde{\alpha} \cdot \epsilon_\lambda e^{i\mathbf{k}\cdot\mathbf{r}_j} | I \rangle \langle I | \sum_l \tilde{\alpha} \cdot \epsilon_{\lambda'}^* e^{-i\mathbf{k}'\cdot\mathbf{r}_l} | i \rangle$$

$$+ \sum_I \frac{1}{2mc^2} \langle f | \sum_j \tilde{\alpha} \cdot \epsilon_{\lambda'}^* e^{-i\mathbf{k}'\cdot\mathbf{r}_j} | I \rangle \langle I | \sum_l \tilde{\alpha} \cdot \epsilon_\lambda e^{i\mathbf{k}\cdot\mathbf{r}_l} | i \rangle$$

$$(12.156)$$

The approximation $W_I \approx mc^2$ is, essentially, taking the non-relativistic limit. Then the intermediate states involving negative energy states separate from those involving positive energy states only, and both form complete sets. The only dependence on the intermediate states in (12.156) is in $|I\rangle\langle I|$ and we use closure to do the summation, then we have

$$M_{fi}^{(1)-} = \frac{1}{2mc^2} \langle f | \sum_{j,l} \tilde{\alpha} \cdot \epsilon_\lambda e^{i\mathbf{k}\cdot\mathbf{r}_j} e^{-i\mathbf{k}'\cdot\mathbf{r}_l} \tilde{\alpha} \cdot \epsilon_{\lambda'}^* | i \rangle$$

$$+ \frac{1}{2mc^2} \langle f | \sum_{j,l} \tilde{\alpha} \cdot \epsilon_{\lambda'}^* e^{-i\mathbf{k}'\cdot\mathbf{r}_j} e^{i\mathbf{k}\cdot\mathbf{r}_l} \tilde{\alpha} \cdot \epsilon_\lambda | i \rangle \qquad (12.157)$$

We now use the irritatingly familiar identity (4.19) to write this as

$$M_{fi}^{(1)-} = \frac{1}{mc^2} \langle f | \sum_{j,l} e^{-i\mathbf{k}'\cdot\mathbf{r}_j} e^{i\mathbf{k}\cdot\mathbf{r}_l} \epsilon_{\lambda'}^* \cdot \epsilon_\lambda | i \rangle \qquad (12.158)$$

where we have used the fact that j and l can be interchanged in the summations. The summation can be simplified

$$\sum_{j,l} e^{-i\mathbf{k}'\cdot\mathbf{r}_j} e^{i\mathbf{k}\cdot\mathbf{r}_l} = \sum_{j,l} e^{-i\mathbf{k}'\cdot\mathbf{r}_j} e^{i\mathbf{k}\cdot(\mathbf{r}_j + \Delta\mathbf{r}_{jl})} = \sum_{j,l} e^{i(\mathbf{k}-\mathbf{k}')\cdot\mathbf{r}_j} e^{-i\mathbf{k}\cdot\Delta\mathbf{r}_{jl}} \qquad (12.159)$$

We can make the so called 'incoherence approximation' which means that the sum over l of the second exponential on the right side here is set equal to unity. Then

$$M_{fi}^{(1)-} = \frac{1}{mc^2} \langle f | \sum_j e^{i\mathbf{K}\cdot\mathbf{r}_j} | i \rangle \epsilon_{\lambda'}^* \cdot \epsilon_\lambda \qquad (12.160)$$

with $\mathbf{K} = \mathbf{k} - \mathbf{k}'$. This is the first order contribution to the scattering amplitude from the negative energy states. We will discuss its physical implications later. Next let us consider the second order term, i.e. taking the second terms in (12.155). Using the closure of the intermediate states in the same way as before yields

$$M_{fi}^{(2)-} = \frac{\hbar\omega}{4m^2c^4} \langle f| \sum_{j,l} \tilde{\alpha} \cdot \epsilon_\lambda e^{i\mathbf{k}\cdot\mathbf{r}_j} e^{-i\mathbf{k}'\cdot\mathbf{r}_l} \tilde{\alpha} \cdot \epsilon_{\lambda'}^* |i\rangle$$

$$- \frac{\hbar\omega}{4m^2c^4} \langle f| \sum_{j,l} \tilde{\alpha} \cdot \epsilon_{\lambda'}^* e^{-i\mathbf{k}'\cdot\mathbf{r}_l} e^{i\mathbf{k}\cdot\mathbf{r}_l} \tilde{\alpha} \cdot \epsilon_\lambda |i\rangle \qquad (12.161)$$

Again we can use equation (12.159) to reduce the number of summations from two to one. The two terms in equation (12.161) have opposite signs now, so when we use (4.19) it is the term in the cross product that yields a non-zero result. This means that this term in the scattering amplitude is imaginary. We are left with

$$M_{fi}^{(2)-} = -\frac{i\hbar\omega}{2m^2c^4} \langle f| \sum_j e^{i\mathbf{K}\cdot\mathbf{r}_j} \sigma |i\rangle \cdot \epsilon_{\lambda'}^* \times \epsilon_\lambda$$

$$= -\frac{i\omega}{m^2c^4} \langle f| \sum_j e^{i\mathbf{K}\cdot\mathbf{r}_j} \hat{\mathbf{S}}_j |i\rangle \cdot \epsilon_{\lambda'}^* \times \epsilon_\lambda \qquad (12.162)$$

This is the first magnetic scattering term. Note that it is of order $\hbar\omega/mc^2$ of the first order scattering (which is small for the X-rays usually used in condensed matter experiments). Equation (12.162) is complex. We will have more to say about this later when we have derived the full expression for the cross section. We are still not quite at the non-relativistic limit. The spin operator in equation (12.162) is still a 4×4 matrix. Furthermore, the operator in equation (12.160) is also implicitly multiplied by the identity matrix. To get the non-relativistic limit we have to take the upper components of these expressions only. That is a trivial operation because all the matrices are diagonal so our final expression for $M_{\bar{f}i}$ is

$$M_{\bar{f}i} = \frac{1}{mc^2} \langle f| \sum_j e^{i\mathbf{K}\cdot\mathbf{r}_j} |i\rangle \epsilon_{\lambda'}^* \cdot \epsilon_\lambda$$

$$- \frac{i\hbar\omega}{m^2c^4} \langle f| \frac{1}{\hbar} \sum_j e^{i\mathbf{K}\cdot\mathbf{r}_j} \hat{\mathbf{S}}_j |i\rangle \cdot \epsilon_{\lambda'}^* \times \epsilon_\lambda \qquad (12.163)$$

where the spin operator is now the 2×2 operator of Pauli theory given by equation (2.30).

The next step is to take the non-relativistic limit of the first two terms in (12.153).

$$M_{fi}^+ = -\sum_I \frac{\langle f| \sum_j \tilde{\alpha} \cdot \epsilon_{\lambda'}^* e^{-i\mathbf{k}'\cdot\mathbf{r}_j} |I\rangle \langle I| \sum_l \tilde{\alpha} \cdot \epsilon_\lambda e^{i\mathbf{k}\cdot\mathbf{r}_l} |i\rangle}{W_I - W_i - \hbar\omega_\mathbf{k}}$$

$$- \sum_I \frac{\langle f| \sum_j \tilde{\alpha} \cdot \epsilon_\lambda e^{i\mathbf{k}\cdot\mathbf{r}_j} |I\rangle \langle I| \sum_l \tilde{\alpha} \cdot \epsilon_{\lambda'}^* e^{-i\mathbf{k}'\cdot\mathbf{r}_l} |i\rangle}{W_I - W_i + \hbar\omega_{\mathbf{k}'}} \qquad (12.164)$$

We already know the procedure for taking the non-relativistic limit of a relativistic four-component equation. It is to decouple the upper and lower two components of the equations. Before this let us simplify our notation slightly. The matrix elements in (12.164) can be written as

$$\langle f | \sum_j \tilde{\boldsymbol{\alpha}} \cdot \boldsymbol{\epsilon} e^{i\mathbf{k}\cdot\mathbf{r}_j} | i \rangle = \langle f | \sum_j \tilde{\boldsymbol{\alpha}} \cdot \mathbf{A}_{\mathbf{k}}^j(\mathbf{r}) | i \rangle \qquad (12.165a)$$

$$\langle f | \sum_j \tilde{\boldsymbol{\alpha}} \cdot \boldsymbol{\epsilon}' e^{-i\mathbf{k}'\cdot\mathbf{r}_j} | i \rangle = \langle f | \sum_j \tilde{\boldsymbol{\alpha}} \cdot \mathbf{A}_{\mathbf{k}'}^j(\mathbf{r}_j) | i \rangle \qquad (12.165b)$$

where

$$\mathbf{A}_{\mathbf{k}}^j(\mathbf{r}) = \boldsymbol{\epsilon} e^{i\mathbf{k}\cdot\mathbf{r}_j} = \epsilon_\lambda e^{i\mathbf{k}\cdot\mathbf{r}_j}, \qquad \mathbf{A}_{\mathbf{k}'}^j(\mathbf{r}) = \boldsymbol{\epsilon}' e^{-i\mathbf{k}'\cdot\mathbf{r}_j} = \epsilon_{\lambda'}^* e^{-i\mathbf{k}'\cdot\mathbf{r}_j} \qquad (12.166)$$

Now, whatever form the wavefunctions take here, they will always be able to be written as a large and small component, so there is no loss of generality if we write

$$|f\rangle = \begin{pmatrix} |f_1\rangle \\ |f_2\rangle \end{pmatrix}, \qquad |i\rangle = \begin{pmatrix} |i_1\rangle \\ |i_2\rangle \end{pmatrix} \qquad (12.167)$$

Using the definition of the α-matrices in terms of the Pauli matrices of equation (4.13) we can write a typical matrix element as

$$\langle f | \sum_j \tilde{\boldsymbol{\alpha}} \cdot \mathbf{A}_{\mathbf{k}}^j(\mathbf{r}) | i \rangle = \langle f_1 | \sum_j \tilde{\boldsymbol{\sigma}} \cdot \mathbf{A}_{\mathbf{k}}^j(\mathbf{r}) | i_2 \rangle + \langle f_2 | \sum_j \tilde{\boldsymbol{\sigma}} \cdot \mathbf{A}_{\mathbf{k}}^j(\mathbf{r}) | i_1 \rangle \qquad (12.168)$$

Following closely the discussion in chapter 4, where we take the non-relativistic limit of the single-particle Dirac equation, we can take the non-relativistic limit of the Dirac equation with the Hamiltonian given by equation (12.151) and find the analogue of equation (4.55). If we also assume locality, i.e. that the small part of the wavefunction at position $\mathbf{r}_j$ is determined by the large part of the wavefunction at this point, then we can remove the summations and the analogue of (4.55) reduces to

$$|f_2\rangle = \frac{1}{2mc} (\hat{\mathbf{p}}_j - e\mathbf{A}(\mathbf{r}_j)) \cdot \tilde{\boldsymbol{\sigma}} |f_1\rangle \qquad (12.169a)$$

$$|i_2\rangle = \frac{1}{2mc} (\hat{\mathbf{p}}_j - e\mathbf{A}(\mathbf{r}_j)) \cdot \tilde{\boldsymbol{\sigma}} |i_1\rangle \qquad (12.169b)$$

These can be inserted into (12.168) to give

$$\langle f | \sum_j \tilde{\boldsymbol{\alpha}} \cdot \mathbf{A}_{\mathbf{k}}^j(\mathbf{r}) | i \rangle = \frac{1}{2mc} \langle f_1 | \sum_j \tilde{\boldsymbol{\sigma}} \cdot \mathbf{A}_{\mathbf{k}}^j(\mathbf{r}) \tilde{\boldsymbol{\sigma}} \cdot \hat{\mathbf{p}}_j + \tilde{\boldsymbol{\sigma}} \cdot \hat{\mathbf{p}}_j \tilde{\boldsymbol{\sigma}} \cdot \mathbf{A}_{\mathbf{k}}^j(\mathbf{r}) | i_1 \rangle$$

$$- \frac{e}{2mc} \langle f_1 | \sum_j \tilde{\boldsymbol{\sigma}} \cdot \mathbf{A}_{\mathbf{k}}^j(\mathbf{r}) \tilde{\boldsymbol{\sigma}} \cdot \mathbf{A}(\mathbf{r}) + \tilde{\boldsymbol{\sigma}} \cdot \mathbf{A}(\mathbf{r}) \tilde{\boldsymbol{\sigma}} \cdot \mathbf{A}_{\mathbf{k}}^j(\mathbf{r}) | i_1 \rangle$$

$$(12.170)$$

In the weak field limit the second term here can be neglected, so we are left with the first term only. This can be simplified straightforwardly

$$\sum_j \left(\tilde{\sigma} \cdot \mathbf{A}_{\mathbf{k}}^j(\mathbf{r}) \tilde{\sigma} \cdot \hat{\mathbf{p}}_j + \tilde{\sigma} \cdot \hat{\mathbf{p}}_j \tilde{\sigma} \cdot \mathbf{A}_{\mathbf{k}}^j(\mathbf{r}) \right)$$

$$= \sum_j \left(\tilde{\sigma} \cdot \epsilon e^{i\mathbf{k}\cdot\mathbf{r}_j} \tilde{\sigma} \cdot \hat{\mathbf{p}}_j + \tilde{\sigma} \cdot \hat{\mathbf{p}}_j \tilde{\sigma} \cdot \epsilon e^{i\mathbf{k}\cdot\mathbf{r}_j} \right)$$

$$= \sum_j \left(\tilde{\sigma} \cdot \epsilon \tilde{\sigma} \cdot \hat{\mathbf{p}}_j e^{i\mathbf{k}\cdot\mathbf{r}_j} - \hbar \tilde{\sigma} \cdot \epsilon \tilde{\sigma} \cdot \mathbf{k} e^{i\mathbf{k}\cdot\mathbf{r}_j} + \tilde{\sigma} \cdot \hat{\mathbf{p}}_j \tilde{\sigma} \cdot \epsilon e^{i\mathbf{k}\cdot\mathbf{r}_j} \right)$$

$$= \sum_j \left(2\epsilon \cdot \hat{\mathbf{p}}_j + i\hbar \tilde{\sigma} \cdot (\mathbf{k} \times \epsilon) \right) e^{i\mathbf{k}\cdot\mathbf{r}_j} \tag{12.171}$$

So our final expressions for the non-relativistic limits of the individual matrix elements in a weak field can be found by substituting (12.170) and (12.171) into (12.165). We end up with

$$\langle f | \sum_j \tilde{\alpha} \cdot \epsilon e^{i\mathbf{k}\cdot\mathbf{r}_j} | i \rangle \rightarrow \langle f_1 | \sum_j \left(\frac{\hat{\mathbf{p}}_j \cdot \epsilon + \frac{i\hbar}{2} \tilde{\sigma} \cdot (\mathbf{k} \times \epsilon)}{mc} \right) e^{i\mathbf{k}\cdot\mathbf{r}_j} | i_1 \rangle \tag{12.172a}$$

$$\langle f | \sum_j \tilde{\alpha} \cdot \epsilon' e^{-i\mathbf{k}'\cdot\mathbf{r}_j} | i \rangle \rightarrow \langle f_1 | \sum_j \left(\frac{\hat{\mathbf{p}}_j \cdot \epsilon' - \frac{i\hbar}{2} \tilde{\sigma} \cdot (\mathbf{k}' \times \epsilon')}{mc} \right) e^{-i\mathbf{k}'\cdot\mathbf{r}_j} | i_1 \rangle$$

$$\tag{12.172b}$$

On the left of (12.172) $\langle f |$ and $\langle i |$ are written in the Dirac representation, whereas on the right $\langle f_1 |$ and $\langle i_1 |$ are in the Pauli representation.

Our next task is to substitute (12.172) into equation (12.164) to give the scattering amplitude M_{fi}^+ in the Pauli representation. This is straightforward and leads to another of those gigantic expressions that we now know arise with surprising frequency in relativistic quantum mechanics

$$\frac{1}{\hbar^2} M_{fi}^+ = \sum_I$$

$$- \frac{\langle f | \sum_j \left(\frac{\hat{\mathbf{p}}_j \cdot \epsilon'}{\hbar} - \frac{i}{2} \tilde{\sigma} \cdot \mathbf{k}'_{\epsilon'} \right) e^{-i\mathbf{k}'\cdot\mathbf{r}_j} | I \rangle \langle I | \sum_l \left(\frac{\hat{\mathbf{p}}_l \cdot \epsilon}{\hbar} + \frac{i}{2} \tilde{\sigma} \cdot \mathbf{k}_{\epsilon} \right) e^{i\mathbf{k}\cdot\mathbf{r}_l} | i \rangle}{m^2 c^2 (W_I - W_i - \hbar\omega_{\mathbf{k}})}$$

$$- \frac{\langle f | \sum_j \left(\frac{\hat{\mathbf{p}}_j \cdot \epsilon}{\hbar} + \frac{i}{2} \tilde{\sigma} \cdot \mathbf{k}_{\epsilon} \right) e^{i\mathbf{k}\cdot\mathbf{r}_j} | I \rangle \langle I | \sum_l \left(\frac{\hat{\mathbf{p}}_l \cdot \epsilon'}{\hbar} - \frac{i}{2} \tilde{\sigma} \cdot \mathbf{k}'_{\epsilon'} \right) e^{-i\mathbf{k}'\cdot\mathbf{r}_l} | i \rangle}{m^2 c^2 (W_I - W_i + \hbar\omega_{\mathbf{k}'})}$$

$$\tag{12.173}$$

where we have defined $\mathbf{k}_\epsilon = \mathbf{k} \times \epsilon$ and $\mathbf{k}'_{\epsilon'} = \mathbf{k}' \times \epsilon'$ for the purposes of getting the equation on the page. Now, for these to be combined sensibly with the terms we have already found whose intermediate states arise from excitations from the negative energy sea of electrons, we have to make the same level of approximation. So, we will again assume elastic scattering and have that $W_I \approx W_i \approx mc^2$. This leads us to a position where we can use the closure of the intermediate states to obtain

$$
M_{fi}^+ = \frac{\hbar \langle f | \sum_j \left(\frac{\hat{\mathbf{p}}_j \cdot \boldsymbol{\epsilon}'}{\hbar} - \frac{i}{2} \tilde{\boldsymbol{\sigma}} \cdot \mathbf{k}'_{\epsilon'} \right) e^{-i\mathbf{k}' \cdot \mathbf{r}_j} \sum_l \left(\frac{\hat{\mathbf{p}}_l \cdot \boldsymbol{\epsilon}}{\hbar} + \frac{i}{2} \tilde{\boldsymbol{\sigma}} \cdot \mathbf{k}_\epsilon \right) e^{i\mathbf{k} \cdot \mathbf{r}_l} | i \rangle}{m^2 c^2 \omega}
$$

$$
- \frac{\hbar \langle f | \sum_j \left(\frac{\hat{\mathbf{p}}_j \cdot \boldsymbol{\epsilon}}{\hbar} + \frac{i}{2} \tilde{\boldsymbol{\sigma}} \cdot \mathbf{k}_\epsilon \right) e^{i\mathbf{k} \cdot \mathbf{r}_j} \sum_l \left(\frac{\hat{\mathbf{p}}_l \cdot \boldsymbol{\epsilon}'}{\hbar} - \frac{i}{2} \tilde{\boldsymbol{\sigma}} \cdot \mathbf{k}'_{\epsilon'} \right) e^{-i\mathbf{k}' \cdot \mathbf{r}_l} | i \rangle}{m^2 c^2 \omega}
$$

$$
= \frac{\hbar \langle f | \left[\sum_j \left(\frac{\hat{\mathbf{p}}_j \cdot \boldsymbol{\epsilon}'}{\hbar} - \frac{i}{2} \tilde{\boldsymbol{\sigma}} \cdot \mathbf{k}'_{\epsilon'} \right) e^{-i\mathbf{k}' \cdot \mathbf{r}_j}, \sum_l \left(\frac{\hat{\mathbf{p}}_l \cdot \boldsymbol{\epsilon}}{\hbar} + \frac{i}{2} \tilde{\boldsymbol{\sigma}} \cdot \mathbf{k}_\epsilon \right) e^{i\mathbf{k} \cdot \mathbf{r}_l} \right] | i \rangle}{m^2 c^2 \omega}
$$

$$\tag{12.174}$$

In the last line of (12.174) we have written the operators in commutator form, which we now have to evaluate. This is not an enthralling task, but we have never let that stop us before.

The commutator in equation (12.174) is actually four separate commutators and we will outline how to evaluate each of them in turn. Firstly it is useful to note that we can take the summations in (12.174) outside the matrix element.

Our first commutator is

$$
C_1 = \left[\frac{-i}{2} \tilde{\boldsymbol{\sigma}}_j \cdot (\mathbf{k}' \times \boldsymbol{\epsilon}') e^{-i\mathbf{k}' \cdot \mathbf{r}_j}, \frac{i}{2} \tilde{\boldsymbol{\sigma}}_l \cdot (\mathbf{k} \times \boldsymbol{\epsilon}) e^{i\mathbf{k} \cdot \mathbf{r}_l} \right]
$$

$$
= \frac{1}{4} e^{i(\mathbf{k}-\mathbf{k}') \cdot \mathbf{r}_j} \left[\tilde{\boldsymbol{\sigma}}_j \cdot (\mathbf{k}' \times \boldsymbol{\epsilon}'), \tilde{\boldsymbol{\sigma}}_j \cdot (\mathbf{k} \times \boldsymbol{\epsilon}) \right] \tag{12.175}
$$

where we have used the fact that this commutator is clearly zero unless $j = l$. Multiplying it out and using the identity (4.19) we can show that

$$
C_1 = \frac{i}{2} e^{i\mathbf{K} \cdot \mathbf{r}_j} \tilde{\boldsymbol{\sigma}}_j \cdot (\mathbf{k}' \times \boldsymbol{\epsilon}') \times (\mathbf{k} \times \boldsymbol{\epsilon}) \delta_{jl} = \frac{i}{\hbar} e^{i\mathbf{K} \cdot \mathbf{r}_j} \hat{\mathbf{S}}_j \cdot (\mathbf{k}' \times \boldsymbol{\epsilon}') \times (\mathbf{k} \times \boldsymbol{\epsilon}) \delta_{jl} \tag{12.176}
$$

The second commutator is

$$
C_2 = \left[\frac{\hat{\mathbf{p}}_j \cdot \boldsymbol{\epsilon}'}{\hbar} e^{-i\mathbf{k}' \cdot \mathbf{r}_j}, \frac{i}{2} (\mathbf{k} \times \boldsymbol{\epsilon}) \cdot \tilde{\boldsymbol{\sigma}}_l e^{i\mathbf{k} \cdot \mathbf{r}_l} \right] \tag{12.177}
$$

This one is also straightforward. Again it is zero unless $j = l$. Writing out (12.177) and operating with the momentum operator explicitly gives

$$
C_2 = e^{i\mathbf{K} \cdot \mathbf{r}_j} \frac{i}{2} (\mathbf{k} \times \boldsymbol{\epsilon}) \cdot \tilde{\boldsymbol{\sigma}}_j \mathbf{K} \cdot \boldsymbol{\epsilon}' + e^{i\mathbf{K} \cdot \mathbf{r}_j} \frac{i}{2} (\mathbf{k} \times \boldsymbol{\epsilon}) \cdot \tilde{\boldsymbol{\sigma}}_j \frac{\hat{\mathbf{p}}_j \cdot \boldsymbol{\epsilon}'}{\hbar}
$$

$$
- \frac{i}{2} (\mathbf{k} \times \boldsymbol{\epsilon}) \cdot \tilde{\boldsymbol{\sigma}}_j e^{i\mathbf{K} \cdot \mathbf{r}_j} \frac{\hat{\mathbf{p}}_j \cdot \boldsymbol{\epsilon}'}{\hbar} + \frac{i}{2} (\mathbf{k} \times \boldsymbol{\epsilon}) \cdot \tilde{\boldsymbol{\sigma}}_j e^{i\mathbf{K} \cdot \mathbf{r}_j} (\boldsymbol{\epsilon}' \cdot \mathbf{k}')
$$

$$
= \frac{i}{2} e^{i\mathbf{K} \cdot \mathbf{r}_j} (\mathbf{k} \times \boldsymbol{\epsilon}) \cdot \tilde{\boldsymbol{\sigma}}_j (\boldsymbol{\epsilon}' \cdot \mathbf{k}) \delta_{jl} \tag{12.178}
$$

where we have used $\mathbf{K} = \mathbf{k} - \mathbf{k}'$.

The third commutator can be evaluated in a similar way to the second

$$C_3 = \left[-\frac{i}{2} \tilde{\sigma} \cdot (\mathbf{k}' \times \boldsymbol{\epsilon}') e^{-i\mathbf{k}' \cdot \mathbf{r}_j}, \frac{\hat{\mathbf{p}}_j \cdot \boldsymbol{\epsilon}}{\hbar} e^{i\mathbf{k} \cdot \mathbf{r}_j} \right]$$

$$= -\frac{i}{2} e^{i\mathbf{K} \cdot \mathbf{r}_j} (\mathbf{k}' \times \boldsymbol{\epsilon}') \cdot \tilde{\sigma}_j (\boldsymbol{\epsilon} \cdot \mathbf{k}') \delta_{jl} \qquad (12.179)$$

The final commutator C_4 is somewhat more obscure to evaluate, but is straightforward once you know how.

$$C_4 = \left[\frac{\hat{\mathbf{p}}_j \cdot \boldsymbol{\epsilon}'}{\hbar} e^{-i\mathbf{k}' \cdot \mathbf{r}_j}, \frac{\hat{\mathbf{p}}_l \cdot \boldsymbol{\epsilon}}{\hbar} e^{i\mathbf{k} \cdot \mathbf{r}_l} \right]$$

$$= \frac{\hat{\mathbf{p}}_j \cdot \boldsymbol{\epsilon}'}{\hbar} e^{-i\mathbf{k}' \cdot \mathbf{r}_j} (\boldsymbol{\epsilon} \cdot \mathbf{k}) e^{i\mathbf{k} \cdot \mathbf{r}_j} + \frac{\hat{\mathbf{p}}_j \cdot \boldsymbol{\epsilon}'}{\hbar} e^{-i\mathbf{k}' \cdot \mathbf{r}_j} e^{i\mathbf{k} \cdot \mathbf{r}_l} \frac{\hat{\mathbf{p}}_l \cdot \boldsymbol{\epsilon}}{\hbar}$$

$$- \frac{\hat{\mathbf{p}}_l \cdot \boldsymbol{\epsilon}}{\hbar} e^{i\mathbf{k} \cdot \mathbf{r}_j} (-\boldsymbol{\epsilon}' \cdot \mathbf{k}') e^{-i\mathbf{k}' \cdot \mathbf{r}_j} - \frac{\hat{\mathbf{p}}_l \cdot \boldsymbol{\epsilon}}{\hbar} e^{i\mathbf{k} \cdot \mathbf{r}_j} e^{-i\mathbf{k}' \cdot \mathbf{r}_j} \frac{\hat{\mathbf{p}}_j \cdot \boldsymbol{\epsilon}'}{\hbar} \qquad (12.180)$$

where we have written out the commutator in full and done the differentiations associated with some of the momentum operators. In (12.180) the first and third terms are zero because the polarization vector and wavevector of the photon are, by definition, perpendicular. So

$$C_4 = \frac{\hat{\mathbf{p}}_j \cdot \boldsymbol{\epsilon}'}{\hbar} e^{i\mathbf{K} \cdot \mathbf{r}_j} \frac{\hat{\mathbf{p}}_j \cdot \boldsymbol{\epsilon}}{\hbar} - \frac{\hat{\mathbf{p}}_j \cdot \boldsymbol{\epsilon}}{\hbar} e^{i\mathbf{K} \cdot \mathbf{r}_j} \frac{\hat{\mathbf{p}}_j \cdot \boldsymbol{\epsilon}'}{\hbar} \qquad (12.181)$$

because C_4 certainly vanishes if $j \neq l$. Next we operate again with sufficient of the momentum operators to enable us to bring the exponentials in equation (12.181) to the front of the expression. This leads to

$$C_4 = e^{i\mathbf{K} \cdot \mathbf{r}_j} \left(\boldsymbol{\epsilon}' \cdot \mathbf{K} \frac{\hat{\mathbf{p}}_j \cdot \boldsymbol{\epsilon}}{\hbar} - \boldsymbol{\epsilon} \cdot \mathbf{K} \frac{\hat{\mathbf{p}}_j \cdot \boldsymbol{\epsilon}'}{\hbar} + \frac{1}{\hbar^2} [\boldsymbol{\epsilon}' \cdot \hat{\mathbf{p}}_j, \boldsymbol{\epsilon} \cdot \hat{\mathbf{p}}_j] \right) \qquad (12.182)$$

A horribly messy little piece of algebra shows us that the first two terms in (12.182) can be written more succinctly as

$$\boldsymbol{\epsilon}' \cdot \mathbf{K} \frac{\hat{\mathbf{p}}_j \cdot \boldsymbol{\epsilon}}{\hbar} - \boldsymbol{\epsilon} \cdot \mathbf{K} \frac{\hat{\mathbf{p}}_j \cdot \boldsymbol{\epsilon}'}{\hbar} = \frac{1}{\hbar} (\boldsymbol{\epsilon}' \times \boldsymbol{\epsilon}) \cdot (\mathbf{K} \times \hat{\mathbf{p}}_j) \qquad (12.183)$$

The commutator in (12.182) vanishes because any component of the momentum operator commutes with any other, so we are left with

$$C_4 = e^{i\mathbf{K} \cdot \mathbf{r}_j} \frac{1}{\hbar} (\boldsymbol{\epsilon}' \times \boldsymbol{\epsilon}) \cdot (\mathbf{K} \times \hat{\mathbf{p}}_j) \qquad (12.184)$$

This completes our evaluation of the commutators in (12.174). We can now bring them all together to give us the full expression for M_{fi}^+

$$M_{fi}^+ = -\frac{\hbar}{m^2 c^2 \omega} \langle f | \sum_j C_1 + C_2 + C_3 + C_4 | i \rangle$$

$$= -\frac{\hbar}{m^2 c^2 \omega} \left(\langle f | \sum_j e^{i\mathbf{K} \cdot \mathbf{r}_j} \frac{1}{\hbar} (\mathbf{K} \times \hat{\mathbf{p}}_j) \cdot (\boldsymbol{\epsilon}' \times \boldsymbol{\epsilon}) | i \rangle + \right.$$

$$\left. \frac{i}{\hbar} \langle f | \sum_j e^{i\mathbf{K} \cdot \mathbf{r}_j} \hat{\mathbf{S}}_j | i \rangle \cdot (\mathbf{k}'_{\epsilon'} \times \mathbf{k}_\epsilon + \mathbf{k}_\epsilon (\boldsymbol{\epsilon}' \cdot \mathbf{k}) - \mathbf{k}'_{\epsilon'} (\boldsymbol{\epsilon} \cdot \mathbf{k}')) \right)$$

$$= -\frac{\hbar \omega}{m^2 c^4} \langle f | \sum_j e^{i\mathbf{K} \cdot \mathbf{r}_j} \frac{(\mathbf{K} \times \hat{\mathbf{p}}_j)}{\hbar k^2} (\boldsymbol{\epsilon}' \times \boldsymbol{\epsilon}) | i \rangle$$

$$- \frac{i \hbar \omega}{m^2 c^4} \langle f | \frac{1}{\hbar} \sum_j e^{i\mathbf{K} \cdot \mathbf{r}_j} \hat{\mathbf{S}}_j | i \rangle \cdot (\hat{\mathbf{k}}'_{\epsilon'} \times \hat{\mathbf{k}}_\epsilon + \hat{\mathbf{k}}_\epsilon (\boldsymbol{\epsilon}' \cdot \hat{\mathbf{k}}) - \hat{\mathbf{k}}'_{\epsilon'} (\boldsymbol{\epsilon} \cdot \hat{\mathbf{k}}'))$$

$$\tag{12.185}$$

where in the last line we have taken k^2 outside the matrix element and replaced the wavevectors by unit vectors (as indicated by the hats).

We are now in a position to write down a useful expression for the scattering cross section. Equation (12.163) can be substituted into (12.154) and from there into (12.153). Equation (12.185) can be substituted into (12.164) and then also into (12.153) and we have

$$\frac{\partial^2 \sigma_M}{\partial W' \partial \Omega} = 4\alpha^2 \lambda_C^2 \delta(W_f - W_i) \Big| \langle f | \sum_j e^{i\mathbf{K} \cdot \mathbf{r}_j} | i \rangle \boldsymbol{\epsilon}' \cdot \boldsymbol{\epsilon}$$

$$- \frac{i \hbar \omega}{mc^2} \langle f | \sum_j e^{i\mathbf{K} \cdot \mathbf{r}_j} \left(i \frac{(\mathbf{K} \times \hat{\mathbf{p}}_j)}{\hbar k^2} \cdot \mathscr{A} + \frac{\hat{\mathbf{S}}_j \cdot \mathscr{B}}{\hbar} \right) | i \rangle \Big|^2 \tag{12.186}$$

where

$$\mathscr{A} = \boldsymbol{\epsilon}' \times \boldsymbol{\epsilon} \tag{12.187a}$$

$$\mathscr{B} = \boldsymbol{\epsilon}' \times \boldsymbol{\epsilon} - (\hat{\mathbf{k}}' \times \boldsymbol{\epsilon}') \times (\hat{\mathbf{k}} \times \boldsymbol{\epsilon}) - (\hat{\mathbf{k}} \times \boldsymbol{\epsilon}) \cdot (\boldsymbol{\epsilon}' \cdot \hat{\mathbf{k}}) + (\hat{\mathbf{k}}' \times \boldsymbol{\epsilon}')(\boldsymbol{\epsilon} \cdot \hat{\mathbf{k}}') \tag{12.187b}$$

Equation (12.186) is our final expression for the scattering cross section for a system of electrons. It is derived as the non-relativistic limit of equation (12.153) with relativistic corrections to order $1/c^2$, and was first written down in this way by Blume (1985).

Let us be completely clear about the status of (12.186). A full relativistic calculation of the scattering cross section from a system of electrons requires the use of equation (12.153). However, it is difficult to relate this equation directly to the familiar physical properties of materials. Equation (12.186) is much more physically transparent and we will see now why this is.

It is clear from the derivation that the leading order term in this scattering cross section is the first one, indeed it is the only term in a fully non-relativistic derivation. Note that it has the same polarization dependence as equation (12.106), and it is indeed Thomson scattering. In a solid crystal where we have a lattice, taking account of the periodicity in this term leads us to the Laue condition for Bragg scattering. It is this term that is used in standard X-ray diffraction experiments to determine crystal structures. This is easy to see, so let us show it explicitly.

The Thomson term contains a summation over all electrons. Let us replace this by a summation over all unit cells in the crystal and a summation over electrons in each unit cell – we are explicitly invoking the periodicity of the lattice. Then

$$\sum_j \langle f|e^{i\mathbf{K}\cdot\mathbf{r}_j}|i\rangle \boldsymbol{\epsilon}' \cdot \boldsymbol{\epsilon} = \sum_i \langle f| \sum_n e^{i\mathbf{K}\cdot(\mathbf{r}_i - \mathbf{R}_n)}|i\rangle \boldsymbol{\epsilon}' \cdot \boldsymbol{\epsilon}$$

$$= \sum_i \sum_n \rho_{in}(\mathbf{K})e^{-i\mathbf{K}\cdot\mathbf{R}_n} \boldsymbol{\epsilon}' \cdot \boldsymbol{\epsilon} \tag{12.188}$$

Here $\rho_{in}(\mathbf{K})$ is the Fourier transform of the charge density associated with the ith electron in the nth unit cell. The summation over i just gives us the scattering from all the electrons in the unit cell, so we will drop the summation over electrons and say we are scattering from unit cells rather than electrons henceforth. Now we have to take the square modulus of this to find the cross section due to the Thomson scattering

$$\frac{\partial^2 \sigma_{\mathrm{M_T}}}{\partial E_f \partial \Omega} \propto \sum_n \sum_m \rho_m^*(\mathbf{K}) \rho_n(\mathbf{K}) e^{-i\mathbf{K}\cdot(\mathbf{R}_n - \mathbf{R}_m)} \tag{12.189}$$

where we have ignored the constants as they are irrelevant to this discussion. Now if all our unit cells are the same, as in a perfect crystal, and $\mathbf{K}$ is a reciprocal lattice vector, constructive interference will occur in this expression and the scattering will be large. If $\mathbf{K}$ takes on any other value, the interference will be destructive and no scattering will be observed (at least in theory). It is useful to note that if $\mathbf{K} = 0$ in (12.189) the scattering cross section is a direct measure of the charge density.

Now it is time to consider the other terms in equation (12.186). They are a factor of $\hbar\omega/mc^2$ smaller than the charge scattering. In a real material there is a further reason why the contribution to the cross section from these terms is small. Taking iron as an example, it has 26 electrons, but a net magnetic moment equivalent to only about 2 electrons, so there is a further factor of 2/26 which reduces the cross section further. In a typical condensed matter experiment photons of order 5 keV are used and this means that the scattering cross section due to these terms is a factor of 10^{-6} times that due to the Thomson term. The scattering due to these terms is known as magnetic X-ray scattering. This scattering can be used in a diffraction experiment in much the same way as Thomson scattering is used in standard diffraction experiments. The spin and orbital magnetic moments appear in equation (12.186) and, in conjunction with the structure factor, they lead to diffraction from the magnetic structure, rather than the crystal structure. A review of magnetic X-ray scattering has been published by Lovesey (1993).

Consider first the term containing $\mathscr{B}$ in (12.186). In much the same way as for the Thomson term, this term is dependent upon the spin density of the electrons through $\hat{\mathbf{S}}_j$. The term in $\mathscr{A}$ in (12.186) depends upon the angular momentum of the electrons and is a measure of their orbital magnetic moments. Furthermore, the polarization dependence of these two terms is different, so from a polarization analysis of a diffraction experiment, equation (12.186) tells us that it is possible to separate the orbital and magnetic components of the total magnetic moment. No other experiment can make such a direct separation.

Sadly, a magnetic diffraction experiment is very difficult for ferromagnetic materials because the magnetic scattering peaks occur in the same place in reciprocal space as the charge scattering peaks and are completely hidden by them. However, for materials where the magnetic and the crystal periodicity are different some (at least) of the magnetic peaks are separate from the charge peaks and can be investigated using these methods. Many rare earth and actinide materials have very exotic magnetic structures and are fertile ground for application of magnetic X-ray scattering (Gibbs *et al.* 1991, Gibbs 1992, 1993, Langridge *et al.* 1994a, 1994b).

Finally one should note the factor of i in front of the magnetic scattering terms in (12.186). One can imagine interference between the charge and magnetic scattering terms occurring. However, because of this factor, interference can only be observed if the polarization of the radiation contains a complex factor, i.e. if it is circularly or elliptically polarized.

An interesting point to note in the comparison of the discussion here with the non-relativistic derivation of Blume (1985) is the following. The terms in the scattering amplitude that we derived from the positive energy intermediate states also come from the second order part of the Golden Rule in the non-relativistic derivation. However, the terms we derived from the intermediate states involving excitations from the negative energy sea correspond with the scattering that comes from the first order Golden Rule in the non-relativistic theory! (Remember, the perturbation is quadratic in the vector potential in non-relativistic quantum theory, so scattering occurs at first, as well as second, order.)

Clearly, all the physics described here can be seen immediately (by an experienced physicist) from equation (12.186). However, none of it is apparent from (12.153). Ultimately, this is because our human experience is non-relativistic rather than relativistic. The full theory described by equation (12.153), and its implications for solids, have been developed by Arola *et al.* (1997).

Several approximations went into the derivation of (12.186). These will not always hold. In particular if in (12.153) we choose an input frequency such that $\hbar\omega \approx W_I - W_i$ there will be a resonance in the first term and this will lead to large scattering not expected on the basis of (12.186).

12.11 Resonant Scattering of X-Rays

This is really a continuation of the previous section. We are going to discuss briefly a method of amplifying the magnetic scattering of X-rays relative to the charge scattering. Let us start by rewriting equation (12.153), without assuming elastic scattering, and with the substitutions (12.163) and (12.173). We will also make the dipole approximation so that all exponentials in the positive energy terms are set equal to one

$$
\frac{\partial^2 \sigma_M}{\partial E_f \partial \Omega} = \frac{e^4 \mu_0^2}{4\pi^2 m^2} \delta(W_f - W_i + \hbar\omega_{\mathbf{k}'} - \hbar\omega_{\mathbf{k}}) \times
$$

$$
\left| \langle f | \sum_j e^{i\mathbf{K}\cdot\mathbf{r}_j} | i \rangle \boldsymbol{\epsilon}' \cdot \boldsymbol{\epsilon} - \frac{i\hbar\omega}{mc^2} \langle f | \frac{1}{\hbar} \sum_j e^{i\mathbf{K}\cdot\mathbf{r}_j} \hat{\mathbf{S}}_j | i \rangle \cdot \boldsymbol{\epsilon}' \times \boldsymbol{\epsilon} \right.
$$

$$
- \frac{\hbar^2}{m} \sum_I \frac{\langle f | \sum_j \left(\frac{\hat{\mathbf{p}}_j \cdot \boldsymbol{\epsilon}'}{\hbar} - \frac{i}{\hbar}\hat{\mathbf{S}}_j \cdot \mathbf{k}'_{\epsilon'} \right) | I \rangle \langle I | \sum_l \left(\frac{\hat{\mathbf{p}}_l \cdot \boldsymbol{\epsilon}}{\hbar} + \frac{i}{\hbar}\hat{\mathbf{S}}_l \cdot \mathbf{k}_\epsilon \right) | i \rangle}{W_I - W_i - \hbar\omega_{\mathbf{k}} + i\Lambda/2}
$$

$$
\left. - \frac{\hbar^2}{m} \sum_I \frac{\langle f | \sum_j \left(\frac{\hat{\mathbf{p}}_j \cdot \boldsymbol{\epsilon}}{\hbar} + \frac{i}{\hbar}\hat{\mathbf{S}}_j \cdot \mathbf{k}_\epsilon \right) | I \rangle \langle I | \sum_l \left(\frac{\hat{\mathbf{p}}_l \cdot \boldsymbol{\epsilon}'}{\hbar} - \frac{i}{\hbar}\hat{\mathbf{S}}_l \cdot \mathbf{k}'_{\epsilon'} \right) | i \rangle}{W_I - W_i + \hbar\omega_{\mathbf{k}'} + i\Lambda/2} \right|^2
$$

$$
\tag{12.190}
$$

where the complex term in the denominators in equation (12.190) is included to take account of broadening of the cross section due to the breadth Λ of the energy levels.

Let us consider this expression when $W_I = W_i + \hbar\omega_{\mathbf{k}}$. Under these circumstances the first of the two terms arising from the positive energy intermediate states completely dominates the second, because the denominator is almost equal to zero, and then the latter term can be neglected. It is convenient to use

$$
\frac{1}{W_i - W_I + \hbar\omega_{\mathbf{k}} - i\Lambda/2} = \frac{W_i - W_I + \hbar\omega_{\mathbf{k}} + i\Lambda/2}{(W_i - W_I + \hbar\omega_{\mathbf{k}})^2 + \Lambda^2/4} = R_{il} \tag{12.191}
$$

for the denominator. Bringing together like terms in equation (12.190) leads to the following expression

$$\frac{\partial^2 \sigma_M}{\partial E_f \partial \Omega} = \frac{e^4 \mu_0^2}{4\pi^2 m^2} \delta(W_f - W_i + \hbar\omega_{\mathbf{k}'} - \hbar\omega_{\mathbf{k}}) \times$$

$$\left| \langle f| \sum_j e^{i\mathbf{K}\cdot\mathbf{r}_j}|i\rangle \boldsymbol{\epsilon}'\cdot\boldsymbol{\epsilon} - \frac{i\hbar\omega}{mc^2} \langle f| \sum_j e^{i\mathbf{K}\cdot\mathbf{r}_j}\hat{\mathbf{S}}_j|i\rangle \cdot \boldsymbol{\epsilon}' \times \boldsymbol{\epsilon} \right.$$

$$+ \frac{\hbar^2}{m} \sum_I \langle f| \sum_j \frac{\hat{\mathbf{p}}_j \cdot \boldsymbol{\epsilon}'}{\hbar}|I\rangle \langle I| \sum_l \frac{\hat{\mathbf{p}}_l \cdot \boldsymbol{\epsilon}}{\hbar}|i\rangle R_{iI}$$

$$+ \frac{\hbar^2}{m} \sum_I \langle f| \sum_j \frac{\hat{\mathbf{p}}_j \cdot \boldsymbol{\epsilon}'}{\hbar}|I\rangle \langle I| \sum_l i\hat{\mathbf{S}}_l \cdot (\mathbf{k} \times \boldsymbol{\epsilon})|i\rangle R_{iI}$$

$$- \frac{\hbar^2}{m} \sum_I \langle f| \sum_j i\hat{\mathbf{S}}_j \cdot (\mathbf{k}' \times \boldsymbol{\epsilon}')|I\rangle \langle I| \sum_l \frac{\hat{\mathbf{p}}_l \cdot \boldsymbol{\epsilon}}{\hbar}|i\rangle R_{iI}$$

$$\left. - \frac{\hbar^2}{m} \sum_I \langle f| \sum_j i\hat{\mathbf{S}}_j \cdot (\mathbf{k}' \times \boldsymbol{\epsilon}')|I\rangle \langle I| \sum_l i\hat{\mathbf{S}}_l \cdot (\mathbf{k} \times \boldsymbol{\epsilon})|i\rangle R_{iI} R_{iI} \right|^2$$

$$(12.192)$$

Now we can reduce this expression somewhat by writing the matrix elements involving momentum in terms of matrix elements of position, using the following substitution

$$\langle a| \frac{\hat{\mathbf{p}}_j \cdot \boldsymbol{\epsilon}}{\hbar}|b\rangle = \frac{\boldsymbol{\epsilon}}{\hbar} \cdot \langle a|\hat{\mathbf{p}}_j|b\rangle = \frac{m\boldsymbol{\epsilon}}{\hbar} \cdot \langle a|\frac{d\hat{\mathbf{r}}}{dt}|b\rangle$$

$$= \frac{m\boldsymbol{\epsilon}}{\hbar} \cdot \langle a|\frac{i}{\hbar}[\hat{H}, \hat{\mathbf{r}}]|b\rangle = \frac{im\boldsymbol{\epsilon}}{\hbar^2} \cdot \left(\langle a|\hat{H}\hat{\mathbf{r}}|b\rangle - \langle a|\hat{\mathbf{r}}\hat{H}|b\rangle \right)$$

$$= \frac{im\boldsymbol{\epsilon}}{\hbar^2} \cdot \left(\langle\hat{H}^\dagger a|\hat{\mathbf{r}}|b\rangle - \langle a|\hat{\mathbf{r}}W_b|b\rangle \right) = \frac{im}{\hbar^2}(W_a - W_b)\langle a|\hat{\mathbf{r}}|b\rangle \cdot \boldsymbol{\epsilon}$$

$$(12.193)$$

Clearly we have invoked a classical expression for the momentum–velocity relationship here. Also we have made use of the hermiticity of the Hamiltonian, and assumed $|a\rangle$ and $|b\rangle$ are eigenfunctions of the Hamiltonian. Substituting this into (12.192) gives

$$\frac{\partial^2 \sigma_M}{\partial E_f \partial \Omega} = \frac{e^4 \mu_0^2}{4\pi^2 m^2} \delta(W_f - W_i + \hbar\omega_{\mathbf{k}'} - \hbar\omega_{\mathbf{k}})$$

$$= \left| \langle f| \sum_j e^{i\mathbf{K}\cdot\mathbf{r}_j}|i\rangle \boldsymbol{\epsilon}' \cdot \boldsymbol{\epsilon} - \frac{i\hbar\omega}{mc^2} \langle f| \sum_j e^{i\mathbf{K}\cdot\mathbf{r}_j}\hat{\mathbf{S}}_j|i\rangle \cdot \boldsymbol{\epsilon}' \times \boldsymbol{\epsilon} \right.$$

$$\left. + \frac{m}{\hbar^2} \sum_I \sum_{j,l} (W_f - W_I)(W_i - W_I)\boldsymbol{\epsilon}' \cdot \langle f|\hat{\mathbf{r}}_j|I\rangle \langle I|\hat{\mathbf{r}}_l|i\rangle \cdot \boldsymbol{\epsilon} R_{iI} \right.$$

$$- \sum_{I} \sum_{j,l} (W_f - W_I) \boldsymbol{\epsilon}' \cdot \langle f | \hat{\mathbf{r}}_j | I \rangle \langle I | \hat{\mathbf{S}}_l \cdot (\mathbf{k} \times \boldsymbol{\epsilon}) | i \rangle R_{il}$$

$$- \sum_{I} \sum_{j,l} (W_i - W_I) \langle f | \hat{\mathbf{S}}_j \cdot (\mathbf{k}' \times \boldsymbol{\epsilon}') | I \rangle \langle I | \hat{\mathbf{r}}_l | i \rangle \cdot \boldsymbol{\epsilon} R_{il}$$

$$+ \frac{\hbar^2}{m} \sum_{I} \sum_{j,l} \langle f | \hat{\mathbf{S}}_j \cdot (\mathbf{k}' \times \boldsymbol{\epsilon}') | I \rangle \langle I | \hat{\mathbf{S}}_l \cdot (\mathbf{k} \times \boldsymbol{\epsilon}) | i \rangle \bigg|^2 \qquad (12.194)$$

This expression can be simplified further if we assume elastic scattering and so $W_I - W_i = W_I - W_f = \hbar \omega$. Now let us examine each term in (12.194). The first two are the same as previously, they have no resonance and describe Thomson scattering and non-resonant magnetic scattering respectively. The third term is the expression used to describe resonant Raman scattering (Baym 1967). The fourth and fifth terms are resonant magnetic scattering. The sixth term is pure spin scattering. Resonances occur when the energy of the incident photon corresponds to the energy difference between available intermediate states and the initial state of the system. This occurs at absorption edges when there are a large number of empty states which become available to the system just above the Fermi energy. Intensities at such energies may be raised by several orders of magnitude.

Working close to the resonance energies is a case of gaining on the roundabouts what you lose on the swings. Although the cross section becomes larger and hence the experiments are easier to do, it is less clear what is being measured and the interpretation of the experiments is more risky. Nonetheless several authors have been able to use the results of such experiments to elucidate details of the magnetic structure of the materials. (Gibbs *et al.* 1991, Gibbs 1992, Langridge *et al.* 1994a, 1994b), This concludes all we want to say about resonant X-ray scattering.

13

Superconductivity

Superconductivity was discovered by Kammerlingh Onnes (1911) (see Gorter 1964). It turned out to be one of the most difficult problems in condensed matter physics of the twentieth century. There were over 40 years between the discovery of the effect and the development of a satisfactory theory (Cooper 1956, Bardeen, Cooper and Schrieffer 1957). The theory was based on the insightful suggestion by Frohlich (1950) that under some circumstances electrons in a lattice could actually attract one another. The theory of superconductivity divides neatly into two parts. Firstly, there is the theory required to describe the mutual attraction of electrons to form Cooper pairs. Secondly, there is the theory that accepts pairing as a fact and then goes on to calculate observables and properties of superconductors. In this chapter we will be principally concerned with the latter aspects of superconductivity theory. We will start from the many-body theory developed in chapter 6 together with a pairing interaction to get to observables such as the superconducting energy gap, critical fields and temperatures, and to describe the electrodynamics of superconductors. On the whole, though, we will not reproduce superconductivity theory that appears in other textbooks. There are several good books on the non-relativistic theory of superconductivity. Tilley and Tilley (1990) give an excellent introduction to the subject. More advanced treatments of the microscopic theory are given by de Gennes (1966), Saint-James *et al.* (1969) and Tinkham (1980) among others.

In the first section we discuss pairing very briefly, as any discussion of the microscopic theory of superconductivity would be incomplete without it. Then we go on to set up the Hamiltonian for an interacting electron gas including the pairing interaction. This is diagonalized, leading to a relativistic generalization of the Bogolubov–de Gennes equations. We can use the self-consistency condition on these equations to give expressions for the critical temperatures and fields, and other observables. Then we use

536

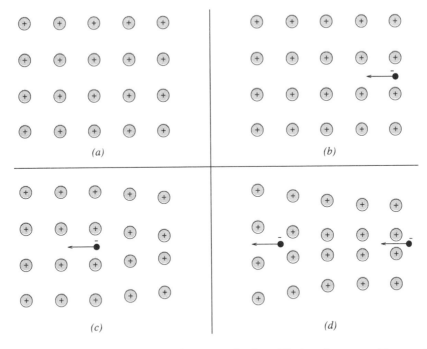

Fig. 13.1. (*a*) A region of simple square lattice. (*b*) An electron with negative charge enters the region of lattice and attracts the positively charged ions. (*c*) The electron moves on quickly, but the ions are still responding to the attraction and move together. (*d*) The first electron continues through the lattice and leaves a *tunnel* of ions brought close together. A second electron, seeing the excess positive charge, is attracted and follows the first through the lattice.

the relativistic current density operator to determine the London equation which governs the electrodynamics of superconductors. Our strategy will be to set up the relativistic theory from scratch, but to use the non-relativistic theory of superconductivity as a guide in doing so.

13.1 Do Electrons Find Each Other Attractive?

The isotope effect (the fact that the superconducting transition temperature varies for different isotopes of the same element) tells us that the lattice, and not just the electrons, must play a crucial role in the creation of the superconducting state. The basic mechanism for superconductivity in the BCS theory is the attraction of the pair of electrons via the electron–phonon interaction. How this may occur in a perfect two-dimensional lattice is shown in figure 13.1. It depends upon two criteria being met. Firstly, the lattice must be *floppy*, so lattice vibrations are easy to set up. Secondly, the motion of the ions in the solid must be sluggish in

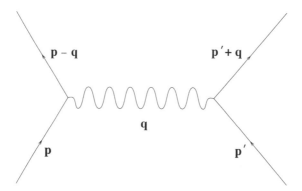

Fig. 13.2. The interaction described by equation (13.1) in diagrammatic form.

comparison with electron velocities. In figure 13.1*a* we see a perfect lattice of positive ions. Let us assume it also contains a uniform negatively charged cloud of electrons. Now suppose that an extra electron enters this region of the lattice from the right as shown in figure 13.1*b*. It will proceed through the lattice as shown in figure 13.1*c*. However, in the meantime, the slow ions have reacted to seeing the excess negative charge between them and have been attracted towards each other, as also seen in figure 13.1*c*. The extra electron will leave a narrow passage of excess positive charge behind it and, as shown in figure 13.1*d*, this tunnel of positive charge will attract a second electron which will follow the trail of positive charge left by the first. Hence, we have the electrons attracted to one another via a lattice vibration. However, we emphasize that this picture is schematic only, showing how pairing can occur. In no way can it be taken to be a realistic description of pairing.

In figure 13.1 we have effective pairing of two electrons. A schematic diagram showing pairing of electrons via exchange of a phonon of wavevector **q** is shown in figure 13.2, which also indicates how momentum is conserved in such an interaction. It should be noted that electron pairing is possible via other low-lying excitations of the electron–lattice system. The isotope effect is absent or much reduced even in some elements, for example ruthenium and osmium, and particularly in the cuprate superconductors. A different pairing mechanism is very possible in those cases. Now we make a big jump: we assume that the phonon (or other excitation) is irrelevant to superconductivity beyond the fact that it pairs up electrons. Then we can represent the interaction in figure 13.2 by

$$\hat{H}'' = -\sum_q \sum_{pp'} V_q a_{p'+q}^\dagger a_{p'} a_{p-q}^\dagger a_p \qquad (13.1)$$

The $a_{\mathbf{p}}$ and $a_{\mathbf{p}}^{\dagger}$ are the fermion annihilation and creation operators of equation (6.86). The quantity $V_{\mathbf{q}}$ is positive, so the whole interaction is attractive. In fact $V_{\mathbf{q}}$ is, in principle, a complicated function of $\mathbf{q}$, but we will usually take a simplified form for it in what follows. Equation (13.1) by-passes the phonon in this problem, and just rewrites the interaction as the annihilation of a pair of electrons with momenta $\mathbf{p}$ and $\mathbf{p}'$ and the creation of a pair with momenta $\mathbf{p} - \mathbf{q}$ and $\mathbf{p}' + \mathbf{q}$.

13.2 Superconductivity, the Hamiltonian

Early attempts to set up a relativistic theory of superconductivity (Bailin and Love 1982) met with limited success. In this section we follow the more recent work of Capelle and Gross (1995) who set up the relativistic superconducting Hamiltonian for the simplified model of a homogeneous electron gas in an external potential, including all two-particle interactions. The Hamiltonian is

$$\hat{H} = \hat{H}_0 + \hat{V} \tag{13.2}$$

where $\hat{H}_0$ is the single-particle term, given by

$$\hat{H}_0 = \int \varphi^{\dagger}(\mathbf{r}) \left(c\tilde{\boldsymbol{\alpha}} \cdot (\hat{\mathbf{p}} - e\mathbf{A}(\mathbf{r})) + \tilde{\beta}mc^2 + U_0(\mathbf{r}) \right) \varphi(\mathbf{r}) d\mathbf{r} \tag{13.3}$$

Here the φs are the relativistic field operators defined in equation (6.145). The terms in brackets are the usual ones, the kinetic energy, the rest mass energy and the interaction of the electrons with an external potential. The second term in (13.2) represents the electron–electron interactions. It is given by

$$\hat{V} = \int \int \varphi^{\dagger}(\mathbf{r})\varphi^{\dagger}(\mathbf{r}')U(\mathbf{r},\mathbf{r}')\varphi(\mathbf{r}')\varphi(\mathbf{r})d\mathbf{r}'d\mathbf{r} \tag{13.4}$$

In (13.4) $U(\mathbf{r},\mathbf{r}')$ represents the full two-electron interaction, i.e. the sum of the Coulomb and Breit interactions and the attraction of (13.1).

This Hamiltonian is hard to solve, there is no doubt about that. It is here that we must let ourselves be guided by the non-relativistic theory of superconductivity (de Gennes 1966). There, an effective one-electron Hamiltonian is set up as an approximation to the full Hamiltonian in terms of an effective one-electron potential and a pair potential. The total energy is then minimized with respect to changes in the wavefunctions and occupation numbers. We will not reproduce this standard non-relativistic derivation, rather we will just outline the important results and then make a plausibility argument for the relativistic generalization of this theory.

The effective non-relativistic Hamiltonian is chosen as

$$\hat{H}_{\text{eff}} = \int \sum_s \varphi_s^\dagger(\mathbf{r}) \left(\frac{1}{2m} (\hat{\mathbf{p}} - e\mathbf{A}(\mathbf{r}))^2 + U_1(\mathbf{r}) + U_0(\mathbf{r}) - \mu \right) \varphi_s(\mathbf{r}) d\mathbf{r}$$
$$- \int \Delta^*(\mathbf{r}) \varphi_\uparrow(\mathbf{r}) \varphi_\downarrow(\mathbf{r}) d\mathbf{r} - \int \Delta(\mathbf{r}) \varphi_\downarrow^\dagger(\mathbf{r}) \varphi_\uparrow^\dagger(\mathbf{r}) d\mathbf{r} \qquad (13.5)$$

where $\Delta(\mathbf{r})$ and $U_1(\mathbf{r})$ are to be determined. The second term here destroys two particles and the third term creates two particles. Any eigenfunction of this Hamiltonian will not be a simultaneous eigenfunction of the number operator. If we are not working with a fixed number of particles we have a grand canonical ensemble and that is why we include μ times the number operator. This should not upset your physical intuition too much. Recall that the superconductors on which experiments are carried out have current flowing into and out of them. They are not eigenfunctions of the number operator either. The factor μ turns out to be the chemical potential. If we make a local approximation for the interaction potential, minimizing the expectation value of this Hamiltonian gives us the conditions

$$U_1(\mathbf{r}) = \int \sum_s U(\mathbf{r}, \mathbf{r}') \langle \varphi_s^\dagger(\mathbf{r}') \varphi_s(\mathbf{r}') \rangle d\mathbf{r}' = \int U(\mathbf{r}, \mathbf{r}') n(\mathbf{r}') d\mathbf{r}' \qquad (13.6a)$$

$$\Delta(\mathbf{r}) = \int U(\mathbf{r}, \mathbf{r}') \langle \varphi_\downarrow(\mathbf{r}') \varphi_\uparrow(\mathbf{r}') \rangle d\mathbf{r}' = \int U(\mathbf{r}, \mathbf{r}') \langle \varphi_\uparrow(\mathbf{r}') \varphi_\downarrow(\mathbf{r}') \rangle d\mathbf{r}' \qquad (13.6b)$$

The first equation here represents a standard Hartree–Fock potential. The second is known as a pair potential. In BCS theory it is usual to treat $U(\mathbf{r}, \mathbf{r}')$ as a point interaction characterized by a single coefficient as suggested by equation (13.1).

Now we want to find the relativistic generalization of the Hamiltonian (13.5) and the effective potentials (13.6). The one-electron-like terms are easy, they are already in the first term of the full relativistic Hamiltonian (13.3). We can also include in the first term the standard Hartree–Fock potential (like (10.113)) which is felt by an individual electron. This leaves us with the terms in the pair potential to deal with. In non-relativistic BCS theory pairing occurs between electrons of opposite spin, and with opposite momenta, and we assume this is still true in relativistic theory. The question is how we can build this into the pair potential terms in (13.5). Well, it is quite simple really (when you know how – it is impossibly difficult if you don't).* We write the attractive pair potential terms in (13.5)

---

* I would like to thank K. Capelle for some very useful discussions on this topic.

in a less than familiar way as matrix equations

$$V_a = \int \Delta^*(\mathbf{r})\varphi_\uparrow(\mathbf{r})\varphi_\downarrow(\mathbf{r})d\mathbf{r} + \int \Delta(\mathbf{r})\varphi_\downarrow^*(\mathbf{r})\varphi_\uparrow^*(\mathbf{r})d\mathbf{r}$$

$$= \frac{1}{2}\int \Delta^*(\mathbf{r})(\varphi_\uparrow(\mathbf{r}), \varphi_\downarrow(\mathbf{r}))\begin{pmatrix} 0 & 1 \\ -1 & 0 \end{pmatrix}\begin{pmatrix} \varphi_\uparrow(\mathbf{r}) \\ \varphi_\downarrow(\mathbf{r}) \end{pmatrix}d\mathbf{r}$$

$$+ \frac{1}{2}\int \Delta(\mathbf{r})(\varphi_\uparrow^*(\mathbf{r}), \varphi_\downarrow^*(\mathbf{r}))\begin{pmatrix} 0 & 1 \\ -1 & 0 \end{pmatrix}\begin{pmatrix} \varphi_\uparrow^*(\mathbf{r}) \\ \varphi_\downarrow^*(\mathbf{r}) \end{pmatrix}d\mathbf{r} \qquad (13.7)$$

where in the last line we have written the wavefunction in two-component form and have invoked the anticommutation relations for the field operators. This may seem an odd way to simplify the problem, but consider the matrix in equation (13.7). It is $i\tilde{\sigma}_y$ and $\tilde{\sigma}_y$ is the non-relativistic time-reversal matrix. This way of writing the attractive interaction demonstrates explicitly that Cooper pairs consist of time-reversed single-particle states. Recall the action of the time-reversal operator, it transforms an electron with spin m_s and momentum $\mathbf{p}$ into an electron with opposite spin and momentum $-\mathbf{p}$. That is exactly what we want, so let us assume that the electrons that attract one another are a Kramers degenerate pair. We can incorporate these ideas into equations (13.3) and (13.4) by inserting the time-reversal matrix between each pair of Dirac field operators in the pair potential terms in the relativistic analogue of equation (13.5). Then the relativistic Hamiltonian is

$$\hat{H} = \int \varphi^\dagger(\mathbf{r})\left(c\tilde{\alpha}\cdot(\hat{\mathbf{p}} - e\mathbf{A}(\mathbf{r})) + \tilde{\beta}mc^2 + U_{1R}(\mathbf{r}) + U_0(\mathbf{r})\right)\varphi(\mathbf{r})d\mathbf{r}$$

$$- \int \left(\varphi^T(\mathbf{r})\tilde{\eta}\varphi(\mathbf{r})D^*(\mathbf{r}) + \varphi^\dagger(\mathbf{r})\tilde{\eta}\varphi^*(\mathbf{r})D(\mathbf{r})\right)d\mathbf{r} \qquad (13.8)$$

where $U_{1R}(\mathbf{r})$ is a relativistic Hartree–Fock potential felt by a single electron. The matrix in this equation plays a central role in the following theory, so let us write it out explicitly

$$\tilde{\eta} = -\frac{1}{2}\tilde{\alpha}_x\tilde{\alpha}_z = \begin{pmatrix} 0 & 1 & 0 & 0 \\ -1 & 0 & 0 & 0 \\ 0 & 0 & 0 & 1 \\ 0 & 0 & -1 & 0 \end{pmatrix} \qquad (13.9)$$

and we have also defined the relativistic pair potential

$$D(\mathbf{r}) = \frac{1}{2}\langle U(\mathbf{r}, \mathbf{r}')\varphi^T(\mathbf{r}')\tilde{\alpha}_x\tilde{\alpha}_z\varphi(\mathbf{r}')\rangle \qquad (13.10)$$

The Hamiltonian (13.8) has not really been derived, rather it has been postulated to have this form on the basis of time-conjugate pairing and having the correct non-relativistic limit. However, these constraints do uniquely determine the structure of the Hamiltonian. Furthermore, the relativistic order parameter is $\varphi^T(\mathbf{r})\tilde{\eta}\varphi(\mathbf{r})$ and it can be shown that this transforms as a scalar quantity under a Lorentz transformation.

13.3 The Dirac–Bogolubov–de Gennes Equation

We have found the Hamiltonian (13.8) and, of course, we want to solve it as accurately as possible. To do this we use a relativistic generalization of the Bogolubov–Valatin transformation (Bogolubov 1958, Valatin 1958) to write the field operators in terms of quasiparticle creation and annihilation operators

$$\varphi_i(\mathbf{r}) = \sum_{k,j} \left(u_{ikj}(\mathbf{r})\gamma_{kj} + v_{ikj}^*(\mathbf{r})\gamma_{kj}^\dagger \right) \tag{13.11}$$

In (13.11), subscripts i and j label individual elements of a 4-vector, so $\varphi_i(\mathbf{r})$ is only one element of the full four-component field operator. The factors $u_{ikj}(\mathbf{r})$ and $v_{ikj}(\mathbf{r})$ are amplitudes which are to be determined, and $\gamma_{kj}^\dagger$ and γ_{kj} are creation and annihilation operators for quasiparticles with quantum numbers k (in a crystalline solid k may represent wavevector and band index for example). The operators $\gamma_{kj}^\dagger$ and γ_{kj} obey fermion anticommutation rules:

$$\gamma_{lj}^\dagger\gamma_{mi} + \gamma_{mi}\gamma_{lj}^\dagger = \delta_{lm}\delta_{ij}$$
$$\gamma_{lj}\gamma_{mi} + \gamma_{mi}\gamma_{lj} = 0 \tag{13.12}$$
$$\gamma_{lj}^\dagger\gamma_{mi}^\dagger + \gamma_{mi}^\dagger\gamma_{lj}^\dagger = 0$$

The transformation (13.11) enables us to write the Hamiltonian in the following form

$$\hat{H} = W_g + \sum_k \epsilon_k \gamma_k^\dagger \gamma_k \tag{13.13}$$

Here W_g is the ground state energy and the ϵ_k describe the excitation spectrum. Using equations (13.12) and (13.13) it is completely straightforward to prove the following commutation relations

$$[\gamma_{kj}, \hat{H}] = \epsilon_{kj}\gamma_{kj}, \qquad [\gamma_{kj}^\dagger, \hat{H}] = -\epsilon_{kj}\gamma_{kj}^\dagger \tag{13.14}$$

In fact equation (13.13) is the only form of the Hamiltonian that yields the commutation relations (13.14), so equations (13.14) can be taken as an equivalent mathematical statement to (13.13).

The next step in this theory is to work out the commutators $[\varphi_i, \hat{H}]$ for all four possible values of the subscript i. Dirac field operators obey the equal time anticommutation relations (equation (6.146)). Using them and equation (13.8) written in component form enables us to work out this commutator. It is a long and tedious business, but there are no shortcuts that ensure the maths is right. Little extra insight is gained by writing out the intermediate expressions; the result is

$$[\varphi(\mathbf{r}'), \hat{H}] = (c\tilde{\alpha} \cdot (\hat{\mathbf{p}} - e\mathbf{A}(\mathbf{r})) + \tilde{\beta}mc^2 + U(\mathbf{r}) - \mu)\varphi(\mathbf{r}) + D(\mathbf{r})\tilde{\eta}\varphi^*(\mathbf{r}) \quad (13.15)$$

where $U(\mathbf{r}) = U_{1R}(\mathbf{r}) + U_0(\mathbf{r})$. Again we have to decompose (13.15) into component form. This is easy

$$[\varphi_1(\mathbf{r}'), \hat{H}] = (U(\mathbf{r}) - \mu + mc^2)\varphi_1(\mathbf{r}) + c\hat{\pi}_z\varphi_3(\mathbf{r}) + c\hat{\pi}_-\varphi_4(\mathbf{r}) - D(\mathbf{r})\varphi_2^*(\mathbf{r})$$
$$(13.16a)$$

$$[\varphi_2(\mathbf{r}'), \hat{H}] = (U(\mathbf{r}) - \mu + mc^2)\varphi_2(\mathbf{r}) - c\hat{\pi}_z\varphi_4(\mathbf{r}) + c\hat{\pi}_+\varphi_3(\mathbf{r}) + D(\mathbf{r})\varphi_1^*(\mathbf{r})$$
$$(13.16b)$$

$$[\varphi_3(\mathbf{r}'), \hat{H}] = (U(\mathbf{r}) - \mu - mc^2)\varphi_3(\mathbf{r}) + c\hat{\pi}_z\varphi_1(\mathbf{r}) + c\hat{\pi}_-\varphi_2(\mathbf{r}) - D(\mathbf{r})\varphi_4^*(\mathbf{r})$$
$$(13.16c)$$

$$[\varphi_4(\mathbf{r}'), \hat{H}] = (U(\mathbf{r}) - \mu - mc^2)\varphi_4(\mathbf{r}) - c\hat{\pi}_z\varphi_2(\mathbf{r}) + c\hat{\pi}_+\varphi_1(\mathbf{r}) - D(\mathbf{r})\varphi_3^*(\mathbf{r})$$
$$(13.16d)$$

where we have used the shorthand $\hat{\pi} = \hat{\mathbf{p}} - e\mathbf{A}(\mathbf{r})$ and $\hat{\pi}_\pm = \pi_x \pm i\pi_y$ etc. Now we substitute the transformations (13.11) into these expressions. Again we get some very long equations. On the left hand side of (13.16) we can use the commutators (13.14) and as γ_{kj} and $\gamma_{kj}^\dagger$ are linearly independent we can equate their coefficients. If we do this we end up with a set of equations for $u_{ikj}(\mathbf{r})$ and $v_{ikj}(\mathbf{r})$. In matrix form these are

$$(c\tilde{\alpha} \cdot (\hat{\mathbf{p}} - e\mathbf{A}(\mathbf{r})) + \tilde{\beta}mc^2 + U(\mathbf{r}) - \mu)u_{kj}(\mathbf{r}) = \epsilon_{kj}u_{kj}(\mathbf{r}) - 2D(\mathbf{r})\tilde{\eta}v_{kj}(\mathbf{r}) \quad (13.17a)$$

$$(c\tilde{\alpha} \cdot (\hat{\mathbf{p}} - e\mathbf{A}(\mathbf{r})) + \tilde{\beta}mc^2 + U(\mathbf{r}) - \mu)^*v_{kj}(\mathbf{r}) = -\epsilon_{kj}v_{kj}(\mathbf{r}) - 2D^*(\mathbf{r})\tilde{\eta}u_{kj}(\mathbf{r})$$
$$(13.17b)$$

where

$$u_{kj}(\mathbf{r}) = \begin{pmatrix} u_{1kj}(\mathbf{r}) \\ u_{2kj}(\mathbf{r}) \\ u_{3kj}(\mathbf{r}) \\ u_{4kj}(\mathbf{r}) \end{pmatrix}, \qquad v_{kj}(\mathbf{r}) = \begin{pmatrix} v_{1kj}(\mathbf{r}) \\ v_{2kj}(\mathbf{r}) \\ v_{3kj}(\mathbf{r}) \\ v_{4kj}(\mathbf{r}) \end{pmatrix} \qquad (13.18)$$

Equations (13.17) are the important equations in this section. They are the relativistic generalization of the Bogolubov–de Gennes equations. Henceforth we will refer to these equations as the Dirac–Bogolubov–de Gennes equations and many of the properties of superconductors can be derived from them. We will illustrate this in subsequent sections. They embody much of the second half of the microscopic theory of superconductivity discussed in the introduction. Given some form of the pairing interaction, they determine the properties of the superconductor. However, they have nothing to say about the mechanism of electron pairing.

Let us consider $U(\mathbf{r})$ and $D(\mathbf{r})$ for a moment. The potential $U(\mathbf{r})$ is the sum of the external potential which is assumed to be constant, and a

Hartree–Fock potential defined by equation (13.8). At zero temperature there is a contribution to the Hartree–Fock potential from all electrons below some Fermi level, and it contains terms like $\langle \gamma_{kj}^{\dagger} \gamma_{kj} \rangle$ which are zero if the state k is empty and unity if it is occupied. At elevated temperatures this is modified to the mean value rules

$$\langle \gamma_{kj}^{\dagger} \gamma_{k'j'} \rangle = \delta_{kk'} \delta_{jj'} \frac{1}{e^{\epsilon_{kj}/k_{\mathrm{B}}T} + 1} = \delta_{kk'} \delta_{jj'} f(\epsilon_{kj}) \qquad (13.19a)$$

and

$$\langle \gamma_{kj}^{\dagger} \gamma_{k'j'}^{\dagger} \rangle = \langle \gamma_{kj} \gamma_{k'j'} \rangle = 0 \qquad (13.19b)$$

These smooth out the abrupt cut off in occupation number at the Fermi energy, but their effect on the occupancy of electron states at room temperature and below is small. Furthermore, we know from the non-relativistic theory that it is electrons within $\hbar\omega_{\mathrm{D}}$ (where ω_{D} is the Debye frequency) of the Fermi energy that do the superconducting. However, their contribution to the Hartree–Fock energy is very small. So, to a very good approximation, the Hartree–Fock potential for the material in its normal state, calculated at zero temperature, can be used in $U(\mathbf{r})$.

The pair potential $D(\mathbf{r})$ is given by equation (13.10). Inserting (13.11) in (13.10) and using the mean value rules (13.19) we find that

$$D(\mathbf{r}) = \sum_{k,j} \int V(\mathbf{r}, \mathbf{r}') u_{kj}^{T}(\mathbf{r}') \tilde{\alpha}_x \tilde{\alpha}_z v_{kj}^{*}(\mathbf{r}')(1 - 2f(\epsilon_{kj})) d\mathbf{r}' \qquad (13.20)$$

The non-relativistic BCS theory tells us that this pair potential is large only close to the Fermi energy, so this function will depend strongly on temperature. An approximation that is frequently made is to assume that $D(\mathbf{r})$ is a local function, i.e

$$D(\mathbf{r}) = \sum_{k,j} V(\mathbf{r}) u_{kj}^{T}(\mathbf{r}) \tilde{\alpha}_x \tilde{\alpha}_z v_{kj}^{*}(\mathbf{r})(1 - 2f(\epsilon_{kj})) \qquad (13.21)$$

and this is the form of the pair potential that we will use henceforth.

13.4 Solution of the Dirac–Bogolubov–de Gennes Equations

There is not much point in setting up the Dirac–Bogolubov–de Gennes equations above if we are not going to solve them. The simplest case to look at is when the pair potential is zero. Then the terms in $D(\mathbf{r})$ in equation (13.8) become zero and the problem reduces to the relativistic Hartree–Fock theory which can be solved straightforwardly (in principle). Although nothing to do with superconductivity, it is useful that this limit exists and of course, it is a property we would expect a sensible theory to have.

In this section we solve the Dirac–Bogolubov–de Gennes equations under very restricted conditions (Capelle and Gross 1995). We will assume that the electrons are non-interacting (except for the pairing) and that there is no external potential or field. Furthermore we will assume that both $D(\mathbf{r})$ and $V(\mathbf{r})$ in equation (13.21) are real and constant over all space. With these restrictions the Dirac–Bogolubov–de Gennes equations become

$$(c\tilde{\boldsymbol{\alpha}} \cdot \hat{\mathbf{p}} + \tilde{\beta}mc^2 - \mu)u_k(\mathbf{r}) = \epsilon_k u_k(\mathbf{r}) - 2D\tilde{\eta}v_k(\mathbf{r}) \qquad (13.22a)$$

$$(c\tilde{\boldsymbol{\alpha}} \cdot \hat{\mathbf{p}} + \tilde{\beta}mc^2 - \mu)v_k(\mathbf{r}) = -\epsilon_k v_k(\mathbf{r}) - 2D\tilde{\eta}u_k(\mathbf{r}) \qquad (13.22b)$$

where we have dropped the subscript j. It is irrelevant here, as it only describes how the quasiparticle operators are obtained from the electron operators. Let us consider the left side of (13.22). We are describing a system of free independent particles in this model (except for the small pairing interaction). For this model we know from chapter 5 that

$$(c\tilde{\boldsymbol{\alpha}} \cdot \hat{\mathbf{p}} + \tilde{\beta}mc^2)\psi_k(\mathbf{r}) = W_k\psi_k(\mathbf{r}) \qquad (13.23)$$

where $\psi_k(\mathbf{r})$ are free-particle eigenfunctions. Let us assume we can use this in (13.22). They become

$$(\epsilon_k - (W_k - \mu))u_k(\mathbf{r}) - 2D\tilde{\eta}v_k(\mathbf{r}) = 0 \qquad (13.24a)$$

$$(\epsilon_k + (W_k - \mu))v_k(\mathbf{r}) + 2D\tilde{\eta}u_k(\mathbf{r}) = 0 \qquad (13.24b)$$

These can easily be written out in component form. They form four pairs of simultaneous equations, one pair being

$$(\epsilon_k - (W_k - \mu))u_{1k}(\mathbf{r}) + Dv_{2k}(\mathbf{r}) = 0 \qquad (13.25a)$$

$$(\epsilon_k + (W_k - \mu))v_{2k}(\mathbf{r}) + Du_{1k}(\mathbf{r}) = 0 \qquad (13.25b)$$

The other three pairs of equations look very similar and they all have non-trivial solutions provided the determinant of the coefficients is set equal to zero, i.e.

$$\begin{vmatrix} \epsilon_k - (W_k - \mu) & D \\ D & \epsilon_k + (W_k - \mu) \end{vmatrix} = 0 \qquad (13.26)$$

and the solution to this is

$$\epsilon_k = \pm\sqrt{(W_k - \mu)^2 + D^2} \qquad (13.27)$$

This equation requires some discussion. It is the excitation spectrum for our model system. The energy $W_k = +\sqrt{(\hbar^2 k^2 c^2 + m^2 c^4)}$ is that of a free Dirac particle. In a system without pairing the energy spectrum would be a continuous function of k. Here, however, there is no value of k that gives a value of ϵ_k in the region $\mu - |D|$ to $\mu + |D|$, so the pairing interaction

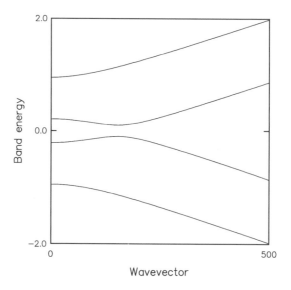

Fig. 13.3.   Schematic diagram of the energy band spectrum of equations (13.27) and (13.28). Unrealistically large values of D have been used to enable us to see both the superconducting energy gap (the minimum distance between the two central bands) of width $2|D|$ and both the positive and negative energy solutions on the same diagram. Zero represents the Fermi level, and the units are arbitrary.

has introduced an energy gap. Note that we have ignored the positrons in this discussion; if they had been included we would have ended up with an energy spectrum

$$\epsilon_k = \pm\sqrt{(W_k + \mu)^2 + D^2} \tag{13.28}$$

The spectra in equations (13.27) and (13.28) are shown in figure 13.3 as a function of wavevector k. The energy gap introduced in this theory is $2|D|$ which is the same as that found in the non-relativistic theory. In figure 13.4 we show a schematic picture of the density of states for this system for the electrons only. On the left hand side we have the free-particle density of states without pairing, on the right hand side is the density of states with pairing. The energy gap is clearly visible, and the consequent increase in the density of states just above and just below the Fermi level μ is also illustrated.

The Fermi wavevector is defined by $E_F = \hbar^2 k_F^2 / 2m$. Clearly equation (13.27) takes on its minimum value when

$$W_k = \mu = E_F \tag{13.29}$$

We can substitute for E_F and W_k in here and perform a binomial expansion on the expression for W_k. Retaining terms to order $1/c^2$ we find that the

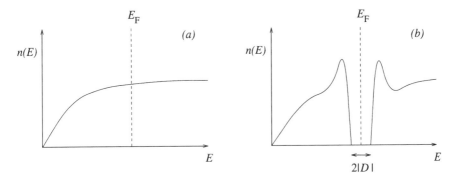

Fig. 13.4.   Schematic diagram illustrating the density of states of a superconductor, (a) for a free-electron gas, (b) for a free-electron gas with an attractive
interaction between electrons. This figure shows the energy gap of width $2|D|$,
and the build up of states just above and below the Fermi energy.

energy gap is at

$$k_{\text{gap}}^2 = k_F^2 \left(1 + \frac{v_F^2}{4c^2} \right)$$
(13.30)

so the position of the minimum value of the energy gap is shifted from
its non-relativistic value of k_F by a factor $v_F^2/4c^2$ where v_F is the Fermi
velocity. This shift is the same order of magnitude as other relativistic
effects such as the relativistic increase in the mass of a Cooper pair
(Cabrera and Peskin 1989). Of course we have derived (13.27) for a model
system which is in no way a true representation of reality. For real systems
with interacting electrons a numerical solution of the Dirac–Bogolubov–de
Gennes equations is necessary.

13.5 Observable Properties of Superconductors

Let us continue to explore the model we have developed in this section,
and see if we can find some simple expressions for observable properties of
superconductors. Of course we can, otherwise I wouldn't be saying that.
Consider equations (13.25) and the equivalent equations for the other
components of $u_k(\mathbf{r})$ and $v_k(\mathbf{r})$. These can be rearranged to give

$$u_{1k}(\mathbf{r}) = -\frac{D}{\epsilon_k - (W_k - \mu)} v_{2k}(\mathbf{r})$$
(13.31a)

$$v_{2k}(\mathbf{r}) = -\frac{D}{\epsilon_k + (W_k - \mu)} u_{1k}(\mathbf{r})$$
(13.31b)

What this tells us is that $u_{1k}(\mathbf{r})$ and $v_{2k}(\mathbf{r})$ differ from one another only by a
multiplicative constant. The same is true of the other pairs of components.

If this is so we can write

$$u_{1k}(\mathbf{r}) = u_{1k}w_k(\mathbf{r}), \qquad v_{2k}(\mathbf{r}) = v_{2k}w_k(\mathbf{r}) \qquad (13.32)$$

where we have separated out the common dependence on $\mathbf{r}$ from both functions. The same thing can be done for the other pairs of components of $u_k(\mathbf{r})$ and $v_k(\mathbf{r})$ although there is no mathematical reason to suppose that the part of the wavefunction that depends on $\mathbf{r}$ is the same for the different pairs. However, in our model we have free particles (except for the small attractive interaction) and we have used the free-particle Dirac equation to substitute W_k into the Dirac–Bogolubov–de Gennes equations. As we saw in chapter 5, for free particles the part of the eigenfunction that depends on $\mathbf{r}$ can be taken outside the 4-vector part of the eigenfunction. So let us assume that $w_k(\mathbf{r})$ in (13.32) is the same function for all pairs of components. Then we can write

$$u_k(\mathbf{r}) = \begin{pmatrix} u_{1k} \\ u_{2k} \\ u_{3k} \\ u_{4k} \end{pmatrix} w_k(\mathbf{r}), \qquad v_k(\mathbf{r}) = \begin{pmatrix} v_{1k} \\ v_{2k} \\ v_{3k} \\ v_{4k} \end{pmatrix} w_k(\mathbf{r}) \qquad (13.33)$$

There is still some arbitrariness in the separation (13.33). We will assume that the $w_k(\mathbf{r})$ are normalized to unity. Then for the quasiparticle wavefunctions to be normalized we must insist that

$$\sum_{n=1}^{4}(|u_{nk}|^2 + |v_{nk}|^2) = 1 \qquad (13.34)$$

We know that D, ϵ_k, W_k and μ are real quantities, so we can square (13.31) and the equivalent relations for the other components and substitute them into (13.34) to get

$$\sum_{n=1}^{4}\left(|u_{nk}|^2 + \frac{D^2}{(\epsilon_k + (W_k - \mu))^2}|u_{nk}|^2\right) = 1 \qquad (13.35)$$

Rearrangement of this and use of equation (13.27) yields

$$\sum_{n=1}^{4}|u_{nk}|^2 = \frac{1}{2}\left(1 + \frac{W_k - \mu}{\epsilon_k}\right) \qquad (13.36a)$$

and similarly

$$\sum_{n=1}^{4}|v_{nk}|^2 = \frac{1}{2}\left(1 - \frac{W_k - \mu}{\epsilon_k}\right) \qquad (13.36b)$$

Now, let us go back to our local expression for the pair potential in equation (13.21). If we multiply out the matrices and substitute in equation

(13.32) the pair potential is given by

$$D(\mathbf{r}) = V \sum_k (1 - 2f(\epsilon_k))|w_k(\mathbf{r})|^2 (u_{2k}v_{1k}^* - u_{1k}v_{2k}^* + u_{4k}v_{3k}^* - u_{3k}v_{4k}^*) \quad (13.37)$$

but from equations (13.31) and (13.32) and the equivalents for other components we can see that

$$v_{1k}^* = \frac{D}{\epsilon_k + (W_k - \mu)} u_{2k}^*, \qquad v_{2k}^* = -\frac{D}{\epsilon_k + (W_k - \mu)} u_{1k}^* \qquad (13.38a)$$

$$v_{3k}^* = \frac{D}{\epsilon_k + (W_k - \mu)} u_{4k}^*, \qquad v_{4k}^* = -\frac{D}{\epsilon_k + (W_k - \mu)} u_{3k}^* \qquad (13.38b)$$

Substituting these into (13.37) gives

$$D(\mathbf{r}) = V \sum_k (1 - 2f(\epsilon_k))|w_k(\mathbf{r})|^2 \frac{D}{\epsilon_k + (W_k - \mu)} \sum_{n=1}^{4} |u_{nk}|^2 \qquad (13.39)$$

But the summation over the components of $|u_k|^2$ is just given by the condition (13.36a). Substituting this into (13.39) and using (13.27) gives

$$D(\mathbf{r}) = V \sum_k (1 - 2f(\epsilon_k))|w_k(\mathbf{r})|^2 \frac{1}{2} \frac{D}{((W_k - \mu)^2 + D^2)^{1/2}} \qquad (13.40)$$

This is not self-consistent because we have assumed D on the right hand side has no dependence on $\mathbf{r}$. However, it has emerged that it has the same $\mathbf{r}$ dependence as $|w_k(\mathbf{r})|^2$. Let us force the self-consistency by assuming that $|w_k(\mathbf{r})|^2$ has only weak or zero dependence on $\mathbf{r}$. This probably isn't true in general and is a weak point in the argument. However, we are not pretending to model real materials, we are only trying to get a feel for the relativistic theory of superconductivity. We can then cancel D on both sides of (13.40) and we have

$$1 = \frac{V}{2} \sum_k (1 - 2f(\epsilon_k))|w_k|^2 \frac{1}{((W_k - \mu)^2 + D^2)^{1/2}} \qquad (13.41)$$

We will solve this in the simplest case, i.e. at zero temperature. Then the Fermi function is unity. It is also useful to convert the summation over k to an integral and to change the variables from k to energy. This brings in a factor of dk/dE which when multiplied by $|w_k|^2$ gives us the density of states per unit energy range. The resulting equation is

$$1 = -\frac{V}{2} \int_{-\hbar\omega_D}^{\hbar\omega_D} n(\xi_k) \frac{1}{((\xi_k)^2 + D^2)^{1/2}} d\xi_k \qquad (13.42)$$

where we have set the Fermi energy $E_F = \mu$ as the zero of energy. The reason for this is that it makes the integral symmetric. We have assumed that the electron attraction is constant in magnitude when

$$E_F - \hbar\omega_D < W_k < E_F + \hbar\omega_D \tag{13.43}$$

and zero otherwise, as is done in the non-relativistic theory. Now, in the integral in (13.42) the energy range $2\hbar\omega_D$ is very small on the scale of electronic energies, and to a good approximation the density of free-particle states is constant over this region. Therefore it can be taken outside the integral and set equal to its value at the Fermi energy. We are then left with a standard integral. Evaluating it yields

$$1 = Vn(E_F)\sinh^{-1}\frac{\hbar\omega_D}{D} \quad \text{or} \quad D = \hbar\omega_D\frac{1}{\sinh\left((Vn(E_F))^{-1}\right)} \tag{13.44}$$

Because V is assumed to be very small we can write sinh in its exponential form in (13.44) and ignore $\exp(-1/Vn(E_F))$ relative to $\exp(1/Vn(E_F))$ and we obtain the familiar expression

$$D \approx 2\hbar\omega_D e^{-1/Vn(E_F)} \tag{13.45}$$

Evaluating the full expression (13.41) at finite temperature is very difficult. However, if we keep the temperature finite, but set $D = 0$ in (13.41), careful integration yields

$$k_B T_c = 1.14\hbar\omega_D e^{-1/Vn(E_F)} \tag{13.46}$$

This is the temperature at which $D \to 0$, i.e. it is an equation for the critical temperature T_c for the onset of superconductivity. In the harmonic lattice approximation the Debye frequency is proportional to $M^{-1/2}$ where M is the isotopic mass. Therefore this equation also explains the isotope effect. Approximately, $k_B T_c$ is the energy which the electron gas gains when it is heated to T_c. We can increase the energy of the system in other ways. For example, application of a magnetic field B increases the energy by $\mu_B B$ and forces the electron spins in a Cooper pair to align with one another. Therefore the critical field is given by

$$\mu_B B_c = 1.14\hbar\omega_D e^{-1/Vn(E_F)} \tag{13.47}$$

We have predicted $T_c \propto B_c$. Although this is approximately true, it is not a particularly good approximation. Furthermore, as stated earlier, the isotope effect is reduced or almost zero in some materials. This can be attributed to the many approximations that have gone into the above analysis. In particular the assumed functional form of V as a function of $\mathbf{r}$ and energy is a gross simplification. It is likely to be a complicated function and to differ markedly for different materials. Here we have assumed that the electron attraction is mediated by the electron–phonon interaction, but it is also possible that it comes about because of the exchange of some other low energy excitation in the solid such as a polaron or magnon. Such a mechanism would also be a possible explanation for the absence of an isotope effect.

One final observable worthy of mention is the dichroic response found in superconductors. Capelle *et al.* (1997) have shown that a superconductor can be expected to exhibit dichroism in absorption, as is seen in magnetic materials. They have shown that there are two distinct mechanisms for spin–orbit coupling to cause dichroism which can be distinguished through their temperature dependence. There are also two further mechanisms. Orbital currents caused by external fields may lead to dichroism even in the absence of the spin–orbit interaction and, most interestingly, the pair potentials themselves may yield a dichroic signal if they are complex or if they break inversion symmetry. In fact, the quality all these mechanisms have in common is that they break symmetries of time-reversal or parity. Dichroism has been observed in superconductors (Lyons *et al.* 1990, 1991, Karrai *et al.* 1992, Weber *et al.* 1990) and the results can be understood, at least qualitatively, on the basis of this theory.

13.6 Electrodynamics of Superconductors

In this section we examine the electrodynamics of a superconductor in a magnetic field which is treated perturbatively. The results will be used to find the current density induced by that field. Our vector potential $\mathbf{A}(\mathbf{r})$ is, in some sense, small, so we work consistently to first order in $\mathbf{A}(\mathbf{r})$ (Kittel 1963).

The superconductor many-body wavefunction will be written as

$$\Phi = \Phi_0 + \Phi_A + \Phi_{A^2} + \cdots \tag{13.48}$$

where Φ_0 is the wavefunction of the unperturbed system with energy W_0, Φ_A is linear in $\mathbf{A}(\mathbf{r})$, Φ_{A^2} is quadratic in $\mathbf{A}(\mathbf{r})$ etc. To first order in perturbation theory we can say that

$$\Phi \approx \Phi_0 + \Phi_A = \Phi_0 + \sum_{\mathbf{k}}{}' |\mathbf{k}\rangle \frac{\langle \mathbf{k}|\hat{H}'|\Phi_0\rangle}{W_0 - W_{\mathbf{k}}} \tag{13.49}$$

where $W_{\mathbf{k}}$ is the energy of the excited state $|\mathbf{k}\rangle$ which may include excitations from the negative energy states. The prime on the summation indicates that $\mathbf{k} = \Phi_0$ is excluded from the sum. We will assume the current density is zero in the absence of the field, i.e.

$$\langle \Phi_0 | \hat{\mathbf{J}}(\mathbf{r}) | \Phi_0 \rangle = 0 \tag{13.50}$$

The perturbation Hamiltonian is

$$\hat{H}' = \int \varphi^\dagger(\mathbf{r}, t) ec\tilde{\boldsymbol{\alpha}} \cdot \mathbf{A}(\mathbf{r}) \varphi(\mathbf{r}, t) d\mathbf{r} \tag{13.51}$$

where $\varphi^\dagger(\mathbf{r}, t)$ and $\varphi(\mathbf{r}, t)$ are the Dirac field operators in the form of equation (6.145). It is the expectation value of the current density operator given by equation (6.139) that we are aiming to calculate.

Before going on to calculate the expectation value of $\hat{\mathbf{J}}(\mathbf{r})$, a little thought can be used to simplify the problem considerably. The superconductor is defined as a system that consists of a large number of electrons and no positrons or negative energy particles. In the current density operator the field operators occur as an ordered product $\varphi^\dagger(\mathbf{r}, t)\varphi(\mathbf{r}, t)$ so we will be multiplying (6.145b) by (6.145a) (with the appropriate operators in between, of course). Matrix elements between states with an inappropriate number of positrons will automatically give zero if the operator does not create or destroy the right number of positrons.

The current operator does not contain $\mathbf{A}(\mathbf{r})$. We want the expectation value $\langle \hat{\mathbf{J}}(\mathbf{r}) \rangle = \langle \Phi | \hat{\mathbf{J}}(\mathbf{r}) | \Phi \rangle$ to first order in $\mathbf{A}(\mathbf{r})$, i.e.

$$\langle \hat{\mathbf{J}}(\mathbf{r}) \rangle = \langle \Phi_0 | \hat{\mathbf{J}}(\mathbf{r}) | \Phi_A \rangle + \langle \Phi_A | \hat{\mathbf{J}}(\mathbf{r}) | \Phi_0 \rangle = \langle \hat{\mathbf{J}}(\mathbf{r}) \rangle_1 + \langle \hat{\mathbf{J}}(\mathbf{r}) \rangle_2 \qquad (13.52)$$

Here we will evaluate $\langle \hat{\mathbf{J}}(\mathbf{r}) \rangle_1$ and we will be able to determine the value of $\langle \hat{\mathbf{J}}(\mathbf{r}) \rangle_2$ by inspection of that result.

Substituting for Φ_A from equation (13.49) in $\langle \hat{\mathbf{J}}(\mathbf{r}) \rangle_1$ gives

$$\langle \hat{\mathbf{J}}(\mathbf{r}) \rangle_1 = {\sum_{\mathbf{k}}}' \frac{1}{W_0 - W_{\mathbf{k}}} \langle \Phi_0 | \hat{\mathbf{J}}(\mathbf{r}) | \mathbf{k} \rangle \langle \mathbf{k} | \hat{H}' | \Phi_0 \rangle \qquad (13.53)$$

Now, as we have already stated, $|\mathbf{k}\rangle$ includes states that arise from the creation of electron–positron pairs. It turns out to be convenient to split the summation in equation (13.53) into two. Firstly we have states $|\mathbf{k}_+\rangle$ which involve electrons only, and secondly states $|\mathbf{k}_-\rangle$ which involve pair creation. So (13.53) becomes

$$\langle \hat{\mathbf{J}}(\mathbf{r}) \rangle_1 = {\sum_{\mathbf{k}_+}}' \frac{1}{W_0 - W_{\mathbf{k}_+}} \langle \Phi_0 | \hat{\mathbf{J}}(\mathbf{r}) | \mathbf{k}_+ \rangle \langle \mathbf{k}_+ | \hat{H}' | \Phi_0 \rangle$$

$$+ \sum_{\mathbf{k}_-} \frac{1}{W_0 - W_{\mathbf{k}_-}} \langle \Phi_0 | \hat{\mathbf{J}}(\mathbf{r}) | \mathbf{k}_- \rangle \langle \mathbf{k}_- | \hat{H}' | \Phi_0 \rangle \qquad (13.54)$$

Let us calculate the contribution to the current density from the first term in equation (13.54). We will start by evaluating the matrix element $\langle \Phi_0 | \hat{\mathbf{J}}(\mathbf{r}) | \mathbf{k}_+ \rangle$.

Now, both $|\mathbf{k}_+\rangle$ and $|\Phi_0\rangle$ contain only positive energy particles so from the arguments given above we only need the terms in $a^\dagger a$ and $bb^\dagger$ when we insert the Dirac field operators into this matrix element. The exponentials cancel immediately and we have the following two terms

$$\langle\Phi_0|\hat{\mathbf{J}}(\mathbf{r})|k_+\rangle = \frac{ec}{V}\sum_{\mathbf{p}\mathbf{p}'}\sum_{j,l=1}^{2}\langle\Phi_0|a_{\mathbf{p}}^{(j)\dagger}a_{\mathbf{p}'}^{(l)}U^{(j)\dagger}(\mathbf{p})\tilde{\alpha}U^{(l)}(\mathbf{p}')|k_+\rangle$$

$$+\frac{ec}{V}\sum_{\mathbf{p}\mathbf{p}'}\sum_{j,l=1}^{2}\langle\Phi_0|b_{\mathbf{p}}^{(j)}b_{\mathbf{p}'}^{(l)\dagger}V^{(j)\dagger}(\mathbf{p})\tilde{\alpha}V^{(l)}(\mathbf{p}')|k_+\rangle \qquad (13.55)$$

Without loss of generality we can let $\mathbf{p}' = \mathbf{p}+\mathbf{q}$ and then the summations over $\mathbf{p}'$ and $\mathbf{p}$ in (13.55) become sums over $\mathbf{p}$ and $\mathbf{q}$. Now, the electron interaction in figure 13.2 is mediated by a phonon. The phonon momentum is very small compared with that of the electrons so we can take the $\mathbf{q}\rightarrow 0$ limit. We also know from chapter 5 that

$$U^{(j)\dagger}(\mathbf{p})\tilde{\alpha}U^{(l)}(\mathbf{p}) = \frac{\mathbf{p}c}{W} \quad j=l, \qquad V^{(j)\dagger}(\mathbf{p})\tilde{\alpha}V^{(l)}(\mathbf{p}) = \frac{\mathbf{p}c}{W} \quad j=l$$

$$= 0 \quad j\neq l \qquad\qquad\qquad = 0 \quad j\neq l$$
$$(13.56)$$

Then (13.55) becomes

$$\langle\Phi_0|\hat{\mathbf{J}}(\mathbf{r})|k_+\rangle = \frac{ec}{V}\sum_{\mathbf{p}}\sum_{j=1}^{2}\frac{\mathbf{p}c}{W}\langle\Phi_0|a_{\mathbf{p}}^{(j)\dagger}a_{\mathbf{p}}^{(j)}|k_+\rangle$$

$$+\frac{ec}{V}\sum_{\mathbf{p}}\sum_{j=1}^{2}\frac{\mathbf{p}c}{W}\langle\Phi_0|(1-b_{\mathbf{p}}^{(j)\dagger}b_{\mathbf{p}}^{(j)})|k_+\rangle \qquad (13.57)$$

where we have used the fermion anticommutation rules again. In the first term in (13.57) the operator is just the number operator. It acts on $|k_+\rangle$ and yields a number which can be taken outside the matrix element. Then we just have a matrix element between two different eigenstates, $\langle\Phi_0|k_+\rangle$, which must be zero because both states are different eigenfunctions of the unperturbed Hamiltonian, and therefore are orthogonal. For the second term in (13.57) the term in the positron creation and annihilation operators is a number operator. It counts the number of positrons in $|k_+\rangle$ which is zero by definition so that disappears. We are then left with $\langle\Phi_0|k_+\rangle$ again which is zero for the reasons stated above. Finally we have

$$\langle\Phi_0|\hat{\mathbf{J}}(\mathbf{r})|k_+\rangle = 0 \qquad (13.58)$$

and substituting this back into equation (13.54) shows that the contribution to $\langle\hat{\mathbf{J}}(\mathbf{r})\rangle_1$ from the states $|k\rangle$ not involving the creation of particle–antiparticle pairs is zero. We are left with the term containing excitations from negative energy states only

$$\langle\hat{\mathbf{J}}(\mathbf{r})\rangle_1 = \sum_{k_-}\frac{1}{W_0 - W_{k_-}}\langle\Phi_0|\hat{\mathbf{J}}(\mathbf{r})|k_-\rangle\langle k_-|\hat{H}'|\Phi_0\rangle \qquad (13.59)$$

Here, the energy denominator $W_0 - W_{k_-}$ is dominated by the rest mass energy of the created electron–positron pair. Therefore, to a very good approximation, we can replace it by $2mc^2$. Also we can use the fact that the eigenfunctions of the unperturbed Hamiltonian form a complete set

$$\sum_k |k\rangle\langle k| = \sum_{k_-} |k_-\rangle\langle k_-| + \sum_{k_+} |k_+\rangle\langle k_+| = 1 \qquad (13.60)$$

to write (13.59) as

$$\langle \hat{\mathbf{J}}(\mathbf{r})\rangle_1 = \frac{1}{2mc^2}\langle\Phi_0|\hat{\mathbf{J}}(\mathbf{r})\hat{H}'|\Phi_0\rangle$$

$$- \frac{1}{2mc^2}\sum_{k_+}\langle\Phi_0|\hat{\mathbf{J}}(\mathbf{r})|k_+\rangle\langle k_+|\hat{H}'|\Phi_0\rangle \qquad (13.61)$$

We have shown that $\langle\Phi_{rm0}|\hat{\mathbf{J}}(\mathbf{r})|k_+\rangle = 0$ for all $|k_+\rangle$ except $|k_+\rangle = |\Phi_0\rangle$, and that $\langle\Phi_0|\hat{\mathbf{J}}(\mathbf{r})|\Phi_0\rangle = 0$ from equation (13.50). So the second term on the right of (13.61) is identically equal to zero. Substituting the operators from (13.51) and (6.139) into the first term of (13.61) gives

$$\langle \hat{\mathbf{J}}(\mathbf{r})\rangle_1 = -\frac{e^2c^2}{2mc^2}\langle\Phi_0|\varphi^\dagger(\mathbf{r})\tilde{\boldsymbol{\alpha}}\varphi(\mathbf{r})\int\varphi^\dagger(\mathbf{r}')\tilde{\boldsymbol{\alpha}}\cdot\mathbf{A}(\mathbf{r}')\varphi(\mathbf{r}')d\mathbf{r}'|\Phi_0\rangle$$

$$= -\frac{e^2}{2m}\langle\Phi_0|\int\varphi^\dagger(\mathbf{r})\tilde{\boldsymbol{\alpha}}\delta(\mathbf{r}-\mathbf{r}')\tilde{I}_4\tilde{\boldsymbol{\alpha}}\cdot\mathbf{A}(\mathbf{r}')\varphi(\mathbf{r}')d\mathbf{r}'|\Phi_0\rangle \qquad (13.62)$$

Here we have used the equal time anticommutators (6.146), and kept only the δ-function part of them, i.e. we have imposed locality on the expectation value of the charge density. There is no good mathematical case for this although physical intuition may lead us to suspect that non-local effects are small. Anyway, the integral then goes and we get

$$\langle \hat{\mathbf{J}}(\mathbf{r})\rangle_1 = -\frac{e^2}{2m}\langle\Phi_0|\varphi^\dagger(\mathbf{r})\tilde{\boldsymbol{\alpha}}\tilde{\boldsymbol{\alpha}}\cdot\mathbf{A}(\mathbf{r})\varphi(\mathbf{r})d\mathbf{r}|\Phi_0\rangle \qquad (13.63)$$

The current density is a vector, let's calculate its x-component by taking the scalar product of both sides of (13.63) with a unit vector in the x-direction $\hat{\mathbf{i}}$. On the right side of (13.63) we take the scalar product of the $\hat{\mathbf{i}}$ with the $\tilde{\boldsymbol{\alpha}}$. Then we can use equation (4.19a) to write it as

$$\langle \hat{\mathbf{J}}(\mathbf{r})\rangle_{1x} = -\frac{e^2}{2m}\langle\Phi_0|\varphi^\dagger(\mathbf{r})A_x(\mathbf{r})\varphi(\mathbf{r})|\Phi_0\rangle$$

$$- \frac{ie^2}{2m}\langle\Phi_0|\varphi^\dagger(\mathbf{r})(\tilde{\boldsymbol{\sigma}}\cdot\hat{\mathbf{i}}\times\mathbf{A}(\mathbf{r}))\varphi(\mathbf{r})|\Phi_0\rangle \qquad (13.64)$$

The other components work out similarly. To find the full current density we also have to work out $\langle\hat{\mathbf{J}}(\mathbf{r})\rangle_2$. The argument goes through very like the derivation of $\langle \hat{\mathbf{J}}(\mathbf{r}) \rangle_1$ with no contribution from positive energy excitations and we find

$$\langle \hat{\mathbf{J}}(\mathbf{r}) \rangle_{2x} = -\frac{e^2}{2m} \langle \Phi_0 | \varphi^\dagger(\mathbf{r}) A_x(\mathbf{r}) \varphi(\mathbf{r}) | \Phi_0 \rangle$$
$$-\frac{ie^2}{2m} \langle \Phi_0 | \varphi^\dagger(\mathbf{r})(\tilde{\boldsymbol{\sigma}} \cdot \mathbf{A}(\mathbf{r}) \times \hat{\mathbf{i}}) \varphi(\mathbf{r}) | \Phi_0 \rangle \quad (13.65)$$

So, to get the total current density we have to add (13.64) and (13.65) and the equivalent expressions for the y- and z-components, according to (13.52). As $\mathbf{A} \times \mathbf{B} = -\mathbf{B} \times \mathbf{A}$ the terms containing the cross product cancel and we are left with

$$\langle \hat{\mathbf{J}}(\mathbf{r}) \rangle = -\frac{e^2}{m} \mathbf{A}(\mathbf{r}) \langle \Phi_0 | \varphi^\dagger(\mathbf{r}) \varphi(\mathbf{r}) | \Phi_0 \rangle \quad (13.66a)$$

Now the matrix element on the right hand side of this equation is just the expectation value of the number density operator at point $\mathbf{r}$. Let's call that $\langle n(\mathbf{r}) \rangle$. Then equation (13.66a) becomes

$$\langle \hat{\mathbf{J}}(\mathbf{r}) \rangle = -\frac{\langle n(\mathbf{r}) \rangle e^2}{m} \mathbf{A}(\mathbf{r}) \quad (13.66b)$$

If $\langle n(\mathbf{r}) \rangle$ is fairly constant we can take the curl of both sides to give

$$\nabla \times \langle \hat{\mathbf{J}}(\mathbf{r}) \rangle = -\frac{\langle n \rangle e^2}{m} \nabla \times \mathbf{A}(\mathbf{r}) = \frac{\langle n \rangle e^2}{m} \mathbf{B}(\mathbf{r}) \quad (13.67)$$

This is the famous London equation that describes the electrodynamics of superconductors (London and London 1935). It is difficult to derive macroscopically, but plausibility arguments for it are given in elementary discussions of superconductivity (Ashcroft and Mermin 1976, Kittel 1986). Here we have derived it from the relativistic microscopic theory of superconductivity. The remarkable thing about the derivation is that there is no contribution to it from the positive energy excitations. It all comes from excitations involving pair creation.

When used in conjunction with Maxwell's equations (13.67) describes many of the electrodynamic properties of superconductors. In particular it can be used to explain the Meissner effect (the exclusion of magnetic flux from a superconductor), and one can also derive an expression for the penetration depth (one of the fundamental lengths in a superconductor) from it. It is not appropriate to go through the derivation of such quantities here as that would just be repeating topics covered in texts on the non-relativistic theory of superconductivity. Therefore we will end our discussion of superconductivity here.

Appendix A
The Uncertainty Principle

There are two familiar forms of the uncertainty principle in quantum mechanics, the position–momentum relation

$$\Delta p \Delta x \geq \hbar \qquad (A.1)$$

and the energy–time relation

$$\Delta E \Delta t \geq \hbar \qquad (A.2)$$

These two relations actually have rather different meanings (Landau and Lifschitz 1982). The former says that, although it is possible to make an arbitrarily accurate measurement of the momentum or position of a particle, these two observables cannot be measured *simultaneously* such that the accuracy of both measurements disobeys equation $(A.1)$. The latter uncertainty relation states that if we make two measurements of the energy of a system where the measurements are separated by a time interval Δt the measured energies will differ by an amount ΔE which is at least large enough for $(A.2)$ to be satisfied. The larger the time interval between the two measurements the smaller the necessary energy difference. This comes about because in order to measure quantities like the momentum and energy of a particle we have to perturb the particle and hence change its momentum and/or energy. One can ask whether there are any changes to the interpretation of the uncertainty principle when we move from a non-relativistic to a relativistic theory, and I wouldn't have written this appendix if the answer to that was negative.

Consider equation $(A.1)$ first in the low velocity limit. Then the momentum can be written as $\mathbf{p} \approx m v$ and it becomes

$$m \Delta v \Delta x \geq \hbar \qquad (A.3)$$

Now velocities are bounded by $\pm c$ in relativity, so the maximum possible uncertainty Δv is $2c$ and that defines the minimum uncertainty in position Δx as

$$\Delta x \geq \frac{\hbar}{2mc} \qquad (A.4)$$

It is impossible to localize a particle to better than this amount, half of its Compton wavelength. The uncertainty in momentum ($2mc$) in ($A.4$), if the $\geq$ is set to $=$, just corresponds to the threshold energy for the production of particle–antiparticle pairs. We have already seen in section 7.3 that the *Zitterbewegung* also predicts $\hbar/2mc$ for the *size* of a particle. However, there we attributed this to the existence of antiparticle states. Now we are attributing it to the uncertainty principle. In this forced single-particle picture, it seems that one could almost say that the uncertainty principle exists because of the existence of the negative energy states!

Next consider the uncertainty relation ($A.2$). Let us apply it to some idealized thought experiment (Landau and Lifschitz 1982). Suppose we have a particle and we want to know its momentum. We can only measure this by interacting with it in some way. Let us perform the measurement by making it interact with a photon. The photon initially has energy E_i and momentum k_i. After the interaction it has energy and momentum E_f and k_f. The particle whose energy and momentum we are measuring has initial momentum W_i and p_i and final energy and momentum W_f and p_f. The conservation of momentum gives

$$p_i + k_i = p_f + k_f \qquad (A.5)$$

If we assume we can know the momentum of the photon to arbitrary accuracy (there is no measurement of position for a photon, so that is feasible), this implies that the uncertainty in the momentum of the particle prior to the interaction is equal to the uncertainty in its momentum after the collision.

Similarly we can write down the conservation of energy. However, from ($A.2$) there is an uncertainty in the energy depending upon the time between the measurements of the photon energy. Therefore the conservation of energy only holds to the level

$$W_f + E_f - W_i - E_i \approx \frac{\hbar}{\Delta t} \qquad (A.6)$$

Again, let us assume we can know the photon energy before and after the interaction exactly. Then ($A.6$) tells us that

$$\Delta W_f - \Delta W_i \geq \frac{\hbar}{\Delta t} \qquad (A.7)$$

where ΔW is the uncertainty in W etc. The energy of a particle depends only on its momentum, so

$$\Delta W = \frac{dW}{dp}\Delta p = v\Delta p \tag{A.8}$$

as can easily be verified by differentiating equation (1.19). Using this in (A.7), we obtain

$$v_f\Delta p_f - v_i\Delta p_i = (v_f - v_i)\Delta p \geq \frac{\hbar}{\Delta t} \tag{A.9}$$

where we have made use of the earlier statement that the uncertainty in p_i is equal to the uncertainty in p_f. This shows that a measurement of a particle's momentum to accuracy Δp requires a change in the velocity of the particle, and hence in the momentum itself. The required change in velocity increases as the time for the measurements becomes small. Equation (A.9) tells us that the measurement of a particle's momentum is unrepeatable in a fundamental way. An exact measurement of momentum could only occur if an infinite time elapsed between measuring the energy and momentum of the initial state of the photon, and the energy and momentum of the final state.

This is all very well, but in relativity the change in the velocity of the particle before and after interaction with the photon cannot be greater than $2c$. In this case (A.9) can be rewritten

$$\Delta p \Delta t \geq \frac{\hbar}{2c} \tag{A.10}$$

Equation (A.1) implies we can make an arbitrarily accurate measurement of momentum provided the uncertainty in position is correspondingly large in a reasonable time. However, equation (A.10) tells us that we cannot make such an arbitrarily accurate measurement of momentum in a reasonable time. We can conclude that the uncertainty relations in relativistic theory are more restrictive of our ability to know observable quantities than in non-relativistic theory.

Appendix B

The Confluent Hypergeometric Function

It tends to be rather late in one's physics education that one comes across the hypergeometric function and the confluent hypergeometric function. Even then it takes some time for its miraculous nature to sink into one's consciousness. I exaggerate a little, but these functions revolutionize one's understanding of the theory of functions and differential equations because it turns out that virtually every standard function one meets as an undergraduate is simply a limit of the confluent hypergeometric function. A good discussion of the properties of this function is given by Abramowitz and Stegun (1972). We follow their notation in this appendix, and will only cover those properties of the confluent hypergeometric function necessary for an understanding of its applications in this book.

The confluent hypergeometric function is defined as the solution of the differential equation

$$z\frac{d^2 M}{dz^2} + (b - z)\frac{dM}{dz} - aM = 0 \qquad (B.1)$$

The solution of this is

$$M(a, b, z) = 1 + \frac{az}{b} + \frac{(a)_2 z^2}{2!(b)_2} + \cdots \frac{(a)_n z^n}{n!(b)_n} \cdots \qquad (B.2)$$

where

$$(a)_0 = 1, \qquad (a)_n = a(a + 1)(a + 2)\cdots(a + n - 1) \qquad (B.3)$$

A very important point to note about equation (B.2) is that it is an infinite series in general. However, it can be terminated if a is equal to a negative integer $-m$ and b is not equal to a negative integer. In that case the mth term in the series for $(a)_n$ and all subsequent terms are zero. Useful

559

asymptotic forms of the confluent hypergeometric function are

$$\lim_{z \to \infty} M(a, b, z) = \frac{\Gamma(b)}{\Gamma(a)} e^z z^{a-b}(1 + O(|z|^{-1})) \qquad \text{Re } z > 0$$

$$\lim_{z \to \infty} M(a, b, z) = \frac{\Gamma(b)}{\Gamma(b-a)}(-z)^{-a}(1 + O(|z|^{-1})) \quad \text{Re } z < 0$$

(B.4)

where $\Gamma(b)$ is the Gamma function (Abramowitz and Stegun 1972) and $O(x)$ means terms of order x. Also note that

$$M(0, b, z) = 1, \qquad M(a, 0, z) = \infty \qquad\qquad (B.5)$$

$M(a, b, z)$ is sometimes known as Kummer's function. The functions

$$y_1 = M(a, b, z), \qquad\qquad y_2 = z^{1-b} M(1 + a - b, 2 - b, z)$$

$$y_3 = e^z M(b - a, b, -z), \qquad y_4 = z^{1-b} e^z M(1 - a, 2 - b, -z)$$

(B.6)

are all independent of one another. There are further solutions which are linear combinations of these basic functions.

B.1 Relations to Other Functions

A very simple limit of the confluent hypergeometric function comes when $a = b$. Then a lot of the complicated fractions in (B.2) cancel and the function becomes independent of a and b. In fact, it is easy to see that

$$M(a, a, z) = e^z \qquad\qquad (B.7)$$

So if z is real $M(a, a, z)$ is the exponential function. If z is complex the real part of $M(a, a, z)$ is $\cos z$ and the imaginary part is $\sin z$.

Another set of functions that are very easily obtained from the confluent hypergeometric function are the Laguerre polynomials. Laguerre polynomials obey the differential equation

$$x \frac{d^2 L_n^\alpha(x)}{dx^2} + (\alpha + 1 - x) \frac{d L_n^\alpha(x)}{dx} + n L_n^\alpha(x) = 0 \qquad (B.8)$$

This is obviously a special case of (B.1) and we immediately have

$$L_n^\alpha(x) = M(-n, \alpha + 1, x) \qquad\qquad (B.9)$$

The Laguerre polynomials appear frequently in physics. They are the solutions of the radial Schrödinger equation in a coulombic potential. So the one-electron atom in non-relativistic, as well as relativistic, quantum mechanics is described by a confluent hypergeometric function.

Next, let's take equation (B.1) and substitute $x = z^{1/2}$. This yields

$$\frac{1}{4} \frac{d^2 M}{dx^2} - \left(\frac{1}{4x} - \frac{b}{2x} + \frac{x}{2} \right) \frac{dM}{dx} - aM = 0 \qquad (B.10)$$

If we let $b = 1/2$ this simplifies further to

$$\frac{d^2 M}{dx^2} - 2x\frac{dM}{dx} - 4aM = 0 \qquad (B.11)$$

But the defining equation for Hermite polynomials is (Merzbacher 1970)

$$\frac{d^2 H_n(x)}{dx^2} - 2x\frac{dH_n(x)}{dx} + 2nH_n(x) = 0 \qquad (B.12)$$

Comparison of $(B.11)$ and $(B.12)$ shows that the Hermite polynomials can be written in terms of the confluent hypergeometric function as

$$H_n(x) = M\left(-\frac{n}{2}, \frac{1}{2}, x^2\right) \qquad (B.13)$$

The Hermite polynomials are the eigenfunctions of the quantum mechanical harmonic oscillator. So, here we have another familiar model which is just a particular limit of the confluent hypergeometric function.

As a final example let us consider Bessel functions. These arise in many areas of physics. Their defining differential equation is

$$\frac{d^2 J_v(z)}{dz^2} + \frac{1}{z}\frac{dJ_v(z)}{dz} + \left(1 - \frac{v^2}{z^2}\right) J_v(z) = 0 \qquad (B.14)$$

In this equation we will make the substitution

$$J_v(z) = z^v e^{-iz} K(z) \qquad (B.15)$$

The mechanics of this substitution is involved, but doing it gives

$$z\frac{d^2 K(z)}{dz^2} + (2v + 1 - 2iz)\frac{dK(z)}{dz} - (2v + 1)iK(z) = 0 \qquad (B.16)$$

and changing the variable to $y = 2iz$ leads to

$$y\frac{d^2 K(y)}{dy^2} + (2v + 1 - y)\frac{dK(y)}{dy} - \frac{1}{2}(2v + 1)K(y) = 0 \qquad (B.17)$$

Comparison with $(B.1)$ shows that $K(y)$ is a confluent hypergeometric function, so

$$J_v(z) = z^v e^{-iz} M(\tfrac{1}{2}(2v + 1), 2v + 1, 2iz) \qquad (B.18)$$

So all the physics that is done using Bessel functions in many areas of physics could be described as an application of a particular limit of the confluent hypergeometric function.

We will stop our explicit examples here. Suffice it to say that with suitable manipulation the confluent hypergeometric function can be related to all the well-known standard functions found in physics, including the Airy function which we saw in chapter 9.

Appendix C
Spherical Harmonics

It is difficult to overestimate the importance of the spherical harmonics Y_l^m for our understanding of the behaviour of atoms, molecules and solids. Their properties are discussed in most quantum mechanics text books. Throughout this one we have made copious use of them. Therefore, in this appendix we have brought together a list of them for low values of the quantum numbers and a brief description of some of their more useful mathematical properties.

The spherical harmonics form a complete set of functions. The first few of them for low values of the quantum numbers are

$$Y_0^0(\theta, \phi) = \frac{1}{\sqrt{4\pi}} \tag{C.1}$$

$$Y_1^0(\theta, \phi) = \sqrt{\frac{3}{4\pi}} \cos\theta \tag{C.2}$$

$$Y_1^1(\theta, \phi) = -\sqrt{\frac{3}{8\pi}} e^{i\phi} \sin\theta \tag{C.3}$$

$$Y_2^0(\theta, \phi) = \sqrt{\frac{5}{16\pi}} (3\cos^2\theta - 1) \tag{C.4}$$

$$Y_2^1(\theta, \phi) = -\sqrt{\frac{15}{8\pi}} e^{i\phi} \sin\theta \cos\theta \tag{C.5}$$

$$Y_2^2(\theta, \phi) = \sqrt{\frac{15}{32\pi}} e^{2i\phi} \sin^2\theta \tag{C.6}$$

$$Y_3^0(\theta, \phi) = \sqrt{\frac{7}{16\pi}} (5\cos^3\theta - \cos\theta) \tag{C.7}$$

$$Y_3^1(\theta, \phi) = -\frac{1}{4}\sqrt{\frac{21}{4\pi}} e^{i\phi} \sin\theta(5\cos^3\theta - 1) \tag{C.8}$$

$$Y_3^2(\theta, \phi) = \frac{1}{4}\sqrt{\frac{105}{2\pi}} e^{2i\phi} \sin^2\theta \cos\theta \tag{C.9}$$

$$Y_3^3(\theta, \phi) = -\frac{1}{4}\sqrt{\frac{35}{4\pi}} e^{3i\phi} \sin^3\theta \tag{C.10}$$

$$Y_4^0(\theta, \phi) = \frac{1}{8}\sqrt{\frac{9}{4\pi}} (35\cos\theta - 30\cos^2\theta + 3) \tag{C.11}$$

$$Y_4^1(\theta, \phi) = -\frac{3}{4}\sqrt{\frac{5}{4\pi}} e^{i\phi} \sin\theta(7\cos^3\theta - 3\cos\theta)e^{i\phi} \tag{C.12}$$

$$Y_4^2(\theta, \phi) = \frac{3}{4}\sqrt{\frac{5}{8\pi}} e^{2i\phi} \sin^2\theta(7\cos^2\theta - 1) \tag{C.13}$$

$$Y_4^3(\theta, \phi) = -\frac{3}{4}\sqrt{\frac{35}{4\pi}} e^{3i\phi} \sin^3\theta \cos\theta \tag{C.14}$$

$$Y_4^4(\theta, \phi) = \frac{3}{8}\sqrt{\frac{35}{8\pi}} e^{4i\phi} \sin^4\theta \tag{C.15}$$

Spherical harmonics with negative values of m are related to these by

$$Y_l^{-m}(\theta, \phi) = (-1)^m Y_l^{m*}(\theta, \phi) \tag{C.16}$$

The spherical harmonics are written in terms of Legendre functions as

$$Y_l^m(\theta, \phi) = (-1)^m \left(\frac{2l+1}{4\pi}\frac{(l-m)!}{(l+m)!}\right)^{1/2} P_l^m(\cos\theta)e^{im\phi} \tag{C.17}$$

The normalization is such that

$$\int_0^{2\pi} d\phi \int_0^\pi \sin\theta\, Y_l^{m*}(\theta, \phi)Y_{l'}^{m'}(\theta, \phi)d\theta = \delta_{ll'}\delta_{mm'} \tag{C.18}$$

and it is this integral that often determines selection rules in physical processes involving electronic transitions. The integral of three spherical harmonics is also a very useful quantity and it defines the Gaunt numbers which appeared so frequently in our discussion of scattering theory and magnetic anisotropy in terms of the Clebsch–Gordan coefficients that were discussed in chapter 2

$$
C_{lml'm'}^{l''m''} = \int Y_{l''}^{m''*}(\Omega) Y_{l'}^{m'}(\Omega) Y_l^m(\Omega) d\Omega
$$

$$
= \left(\frac{(2l+1)(2l'+1)}{4\pi(2l''+1)} \right)^{1/2} C(ll'l''; m, m') C(ll'l''; 0, 0) \delta_{m''m'+m}
$$ (C.19)

This equation determines the selection rules for dipolar radiation from particles with zero spin, i.e. in non-relativistic quantum theory. The spherical harmonic with argument $\hat{\mathbf{r}}$ is related to the spherical harmonic with argument $-\hat{\mathbf{r}}$ via the equation

$$
Y_l^m(-\hat{\mathbf{r}}) = (-1)^{l+m} Y_l^{-m*}(\hat{\mathbf{r}})
$$ (C.20)

Finally, it is well known and extremely useful in scattering theory that the exponential function can be written as a sum of terms involving Bessel functions and spherical harmonics

$$
e^{i\mathbf{k}\cdot\mathbf{r}} = 4\pi \sum_{l,m} i^l j_l(kr) Y_l^{m*}(\hat{\mathbf{k}}) Y_l^m(\hat{\mathbf{r}})
$$ (C.21)

This enables us to write a plane wave in terms of spherical waves, exactly what is required in scattering theory. Two further properties of the spherical harmonics prove themselves useful, mainly because they give us ways of removing the m-dependence in mathematical expressions. If we restrict the angles so that the angles (θ, ϕ) correspond to the z-axis, i.e. $\theta = 0$, we have

$$
Y_l^m(\hat{z}) = \left(\frac{2l+1}{4\pi} \right)^{1/2} \delta_{m,0}
$$ (C.22)

The spherical harmonics are related to the Legendre polynomials $P_l(\cos\theta)$ by the addition formula

$$
P_l(\cos\theta) = \frac{4\pi}{2l+1} \sum_{m=-l}^{+l} Y_l^{m*}(\hat{\mathbf{k}}) Y_l^m(\hat{\mathbf{r}})
$$ (C.23)

where θ is the angle between the directions $\hat{\mathbf{k}}$ and $\hat{\mathbf{r}}$.

Finally in this appendix we prove our assertion that the expansion coefficients in equation (11.139) are the same as those given by equations (11.119) and (11.127b). The proof rests heavily on the expansion (C.21), and essentially shows how to expand spherical harmonics centred on one site around another site.

We start by defining two vectors $\mathbf{r}_p$ and $\mathbf{r}_q$ and the difference between them $\mathbf{R}_{pq} = \mathbf{r}_p - \mathbf{r}_q$. Now

$$e^{-i\mathbf{k}\cdot(\mathbf{R}_{pq}-\mathbf{r}_p)} = e^{-i\mathbf{k}\cdot\mathbf{R}_{pq}}e^{i\mathbf{k}\cdot\mathbf{r}_p} = e^{-i\mathbf{k}\cdot\mathbf{R}_{pq}}4\pi \sum_{l,m} i^l j_l(kr_p)Y_l^{m*}(\hat{\mathbf{k}})Y_l^m(\hat{\mathbf{r}}_p)$$

$$= 4\pi \sum_{l',m'} i^{-l'} j_{l'}(k|\mathbf{R}_{pq}-\mathbf{r}_p|)Y_{l'}^{m'}(\hat{\mathbf{k}})Y_{l'}^{m'*}(\widehat{\mathbf{R}_{pq}-\mathbf{r}_p}) \tag{C.24}$$

Taking only the last two equalities in $(C.24)$, we can multiply each side by $Y_{l''}^{m''*}(\hat{\mathbf{k}})$ and integrate over angles $d\hat{\mathbf{k}}$. From the orthonormality of the spherical harmonics, $(C.18)$, the right hand side simplifies considerably to give

$$j_{l''}(k|\mathbf{R}_{pq}-\mathbf{r}_p|)Y_{l''}^{m''*}(\widehat{\mathbf{R}_{pq}-\mathbf{r}_p})$$

$$= \sum_{l,m} i^{l+l''} j_l(kr_p)Y_l^m(\hat{\mathbf{r}}_p) \int e^{-i\mathbf{k}\cdot\mathbf{R}_{pq}} Y_l^{m*}(\hat{\mathbf{k}})Y_{l''}^{m''*}(\hat{\mathbf{k}})d\hat{\mathbf{k}} \tag{C.25}$$

Now if we expand the $e^{-i\mathbf{k}\cdot\mathbf{R}_{pq}}$ as a series using $(C.21)$, the angles only appear in three spherical harmonics which can be integrated and give a Gaunt coefficient according to equation $(C.19)$. Then we have

$$j_{l''}(k|\mathbf{R}_{pq}-\mathbf{r}_p|)Y_{l''}^{m''*}(\widehat{\mathbf{R}_{pq}-\mathbf{r}_p})$$

$$= 4\pi \sum_{lm}\sum_{l'm'} i^{l-l'+l''} C_{lml'm'}^{l''m''} j_l(kr_p)Y_l^m(\hat{\mathbf{r}}_p)j_{l'}(kR_{pq})Y_{l'}^{m'*}(\hat{\mathbf{R}}_{pq}) \tag{C.26}$$

Now consider a special case of the expansion (11.35)

$$\frac{e^{ik|\mathbf{r}_p-\mathbf{a}|}}{|\mathbf{r}_p-\mathbf{a}|} = 4\pi ik \sum_{l,m} h_l(kr_p)j_l(ka)Y_l^m(\hat{\mathbf{r}}_p)Y_l^{m*}(\hat{\mathbf{a}}) \tag{C.27}$$

But

$$\mathbf{r}_p - \mathbf{a} = \mathbf{r}_p - \mathbf{r}_q - \mathbf{a} + \mathbf{r}_q = \mathbf{R}_{pq} - (\mathbf{a} - \mathbf{r}_q) \tag{C.28}$$

So we can write equation $(C.27)$ as

$$\frac{e^{ik|\mathbf{R}_{pq}-(\mathbf{a}-\mathbf{r}_q)|}}{|\mathbf{R}_{pq}-(\mathbf{a}-\mathbf{r}_q)|}$$

$$= 4\pi ik \sum_{l'',m''} h_{l''}(kR_{pq})Y_{l''}^{m''}(\hat{\mathbf{R}}_{pq})j_{l''}(k|\mathbf{a}-\mathbf{r}_q|)Y_{l''}^{m''*}(\widehat{\mathbf{a}-\mathbf{r}_q}) \tag{C.29}$$

Now, the left hand sides of $(C.27)$ and $(C.29)$ are the same by definition, so we can equate the right hand sides. If we do this, and replace the last two terms on the right of equation (C.29) using (C.26), we have

$$\sum_{l,m} h_l(kr_p)j_l(ka)Y_l^m(\hat{\mathbf{r}}_p)Y_l^{m*}(\hat{\mathbf{a}}) = 4\pi \sum_{l'',m''} h_{l''}(kR_{pq})Y_{l''}^{m''}(\hat{\mathbf{R}}_{pq}) \times$$

$$\sum_{lm}\sum_{l'm'} i^{l-l'+l''} C_{lml'm'}^{l''m''} j_l(kr_q)Y_l^m(\hat{\mathbf{r}}_q)j_{l'}(ka)Y_{l'}^{m'*}(\hat{\mathbf{a}})$$

$$(C.30)$$

Now the products $j_l(ka)Y_l^{m*}(\hat{\mathbf{a}})$ form a set of linearly independent functions, so we can compare their coefficients, and that gives

$$-ikh_l(kr_p)Y_l^m(\hat{\mathbf{r}}_p) = \sum_{l',m'} G_{lml'm'}(\mathbf{R}_{pq})j_{l'}(kr_q)Y_{l'}^{m'}(\hat{\mathbf{r}}_q) \qquad (C.31)$$

with

$$G_{lml'm'}(\mathbf{R}_{pq}) = -4\pi ik \sum_{l''m''} i^{l-l'+l''} C_{lml'm'}^{l''m''} h_{l''}(kR_{pq})Y_{l''}^{m''}(\hat{\mathbf{R}}_{pq}) \qquad (C.32)$$

and this is identical to equation (11.119). The $-ik$ on each side of (C.31) is unnecessary, but convenient for when we calculate the multiple scattering Green's function in section 11.15. Equations (C.31) and (C.32) verify that we were correct to state that the expansion coefficient in equation (11.139) can be identified with the components of the partial wave expansion of the free-particle Green's function in equation (11.119). Clearly we could combine these expansions with equations (11.39) to give us the four-component version of equation (C.31). Note that we have a spherical harmonic centred on site $\mathbf{r}_p$ on the left and one centred on $\mathbf{r}_q$ in equation (C.31).

Appendix D
Unit Systems

Throughout this book we have worked almost exclusively within the SI unit system. The only exceptions have been where it would be clearly ludicrous, as in the local density approximations, or where relativistic effects would not show up because they are just too small, as in the dichroism in the one-electron atom.

There are many unit systems in existence in physics of course, and their proliferation can lead to a lot of unnecessary confusion (as if there wasn't enough confusion caused by the subject itself). In this appendix we will discuss two unit systems which commonly occur in the research literature and their relation to the SI system.

The central quantity in any system of units on the quantum and relativistic level is the fine structure constant, which we repeat here

$$\alpha = \frac{e^2}{4\pi\epsilon_0 \hbar c} \approx \frac{1}{137} \qquad (D.1)$$

α is dimensionless and so takes on the same value in any unit system.

The first unit system we discuss is atomic units (Loucks 1967). These are designed so that atomic (quantum) phenomena are emphasized. In atomic units we put all the fundamental constants up to manageable numbers. In particular we set

$$\frac{e^2}{4\pi\epsilon_0} = 2, \quad \hbar = 1, \quad m = \tfrac{1}{2} \qquad (D.2)$$

The fundamental constants themselves have not changed, we are simply working in a unit system in which they take on these values. With these definitions, and equation $(D.1)$, the speed of light in atomic units is

$$c = \frac{2}{\alpha} \approx 2 \times 137 \qquad (D.3)$$

Now let us think about what the units of simple quantities are in this system. Firstly consider energy. We can substitute the values $(D.2)$ into the

Rydberg energy (the first term on the right of equation (8.49) with $Z = 1$ and $n = 1$). This comes out as -1, so the unit energy in atomic units is the magnitude of the Rydberg energy (the negative of the energy of the $1s$ electron in hydrogen). Next consider the expression for the first Bohr radius (equation (8.60)). Again if we insert the definitions ($D.2$) we find that $a_0 = 1$, so the unit of distance is the Bohr radius. With $\hbar = 1$, energy and frequency take on the same numerical values and once we have a frequency we can invert it to find the unit of time. So, the unit of time in atomic units is the inverse of the frequency of the ionization energy of hydrogen in its ground state.

The second unit system is relativistic units (Loucks 1967). These are designed to emphasize relativistic effects by decreasing the value of the speed of light so that it is much the same as the other fundamental constants. In fact we choose them as simply as we possibly can:

$$c = 1, \quad m = 1, \quad \hbar = 1 \qquad (D.4)$$

and from ($D.1$) this leaves us with

$$\frac{e^2}{4\pi\epsilon_0} \approx \frac{1}{137} \qquad (D.5)$$

Again we can write the units of time, space and energy in terms of quantities we know and love. The unit distance is $\hbar/mc = \lambda_C$, the Compton wavelength, and the unit of energy is mc^2, the rest energy of the electron. The unit of time is $\hbar/mc^2$. It is this system we have used in plotting figures for the Dirac oscillator.

Finally, we have also occasionally used the electronvolt in this book, a unit that finds much favour in spectroscopy and doesn't belong to any particular unit system. It is related to other systems by 1 Rydberg $= 13.6057$ eV, 1 eV $= 1.602177 \times 10^{-19}$ joules and $mc^2 = 510.9991$ keV.

Any calculation must yield the same answer regardless of the units, and the above systems can simplify calculations. However, there are also drawbacks in such systems of units. For example it becomes impossible to check the dimensions of equations if the fundamental constants have all been set equal to simple numbers. Also they can be a barrier to a full understanding; for instance, if c has been set equal to a number it becomes difficult to take a non-relativistic limit. Furthermore, such unit systems are only an aid to understanding if you feel fully capable and confident in their use. Transforming from relativistic to atomic units, for example, can be a nightmare if you are not certain of what you are doing. So, unless you are fully confident in your ability to transform between different sets of units, I recommend you stick to the SI system.

Appendix E
Fundamental Constants

Planck's Constant h	6.626076×10^{-34} J s
Planck's Constant $\hbar = h/2\pi$	1.054572×10^{-34} J s
Speed of Light in Vacuum c	2.997924×10^{8} m s^{-1}
Charge of Electron e	$-1.602177 \times 10^{-19}$ C
Permeability of a Vacuum μ_0	$4\pi \times 10^{-7}$ H m^{-1}
Permittivity of a Vacuum $\epsilon_0 = 1/\mu_0 c^2$	8.854188×10^{-12} F m^{-1}
Fine Structure Constant α	7.297353×10^{-3}
Fine Structure Constant α^{-1}	137.03599
Electronvolt eV	1.602177×10^{-19} J
Rydberg Energy Ryd	2.179874×10^{-18} J
	$= 13.605701$ eV
Electron Rest Mass m	9.109390×10^{-31} kg
	$= 510.9991$ keV
Proton Rest Mass m_P	1.672623×10^{-27} kg
Neutron Rest Mass m_N	1.674929×10^{-27} kg
Bohr Radius a_0	5.291772×10^{-11} m
Bohr Magneton μ_B	9.274015×10^{-24} J T^{-1}
Electron Compton Wavelength λ_C	2.426311×10^{-12} m
Boltzmann's Constant k_B	1.380658×10^{-23} J K^{-1}
Avogadro's Number N	6.022137×10^{23} mol^{-1}

References

Abramowitz. M and Stegun I.A. (1972). *Handbook of Mathematical Functions* (Dover, New York).

Altarelli M. (1993). Orbital Magnetization Sum Rule for X-Ray Circular Dichroism: A Simple Proof, *Phys. Rev. B* **47,** 597.

Altmann S.L. (1970). *Band Theory of Metals, The Elements* (Pergamon Press, Oxford, UK).

Anderson C.D. (1933). The Positive Electron, *Phys. Rev.* **43,** 491.

Ankudinov A. and Rehr J.J. (1995). Sum Rules for Polarization Dependent X-Ray Absorption, *Phys. Rev. B* **51,** 1282.

Argyres P.N. (1955). Theory of Faraday and Kerr Effects in Ferromagnetics, *Phys. Rev.* **97,** 334.

Arola E. (1991). The Relativistic KKR–CPA Method: A Study of Electronic Structures of $Cu_{75}Au_{25}$, $Au_{70}Pd_{30}$ and $Cu_{75}Pt_{25}$ Disordered Alloys, *Acta Polytechnica Scandinavica* **Ph174,** 1.

Arola E. Strange P. and Györffy B.L. (1997). Relativistic Theory of Magnetic Scattering of X-rays: Application to Ferromagnetic Iron, *Phys. Rev. B* **55,** 472.

Ashcroft N.W. and Mermin N.D. (1976). *Solid State Physics* (Holt, Rinehart and Winston, New York).

Bailin D. and Love A. (1982). Superconductivity for Relativistic Electrons, *J. Phys. A: Math. Gen.* **15,** 3001.

Bansil A. (1978). Application of Coherent Potential Approximation to Disordered Muffin-Tin Alloys, *Phys. Rev. Lett.* **41,** 1670.

Bansil A. Ehrenreich H. Schwartz L. and Watson R.E. (1974). Complex Energy Bands in α-Brass, *Phys. Rev. B* **9,** 445.

Bardeen J. Cooper L.N. and Schreiffer J.R. (1957). Theory of Superconductivity, *Phys. Rev.* **108,** 1175.

Barker W.A. and Glover F.N. (1955). Reduction of Relativistic Two-Particle Wave Equations to Approximate Forms III, *Phys. Rev.* **99,** 317.

Baur G. Boero G. Brauksiepe S. Buzzo A. Eyrich W. Geyer R. Grzonka D. Hauffe J. Killian K. LoVetere M. Macri M. Moosburger M. Nellen R. Oelert W. Passaggio S. Pozzo A. Roehrich K. Sachs K. Schepers G. Sefzick T. Simon R.S. Stratmann R. Stinzing F. Wolke M. (1996). Production of Antihydrogen, *Phys. Lett. B* **368,** 251.

Baym, G. (1967). *Lectures on Quantum Mechanics* (Addison-Wesley, California).

Bearden J.A. and Burr A.F. (1967). Re-evaluation of X-Ray Atomic Energy Levels, *Rev. Mod. Phys.* **39,** 125.

Beeby J.L. (1964). Electronic Structure of Alloys, *Phys. Rev.* **135,** A130.

Benitez J. Martinez y Romero. R.P. Nunez-Yepez. H.N. and Salas-Brito A.L. (1990). Solution and Hidden Supersymmetry of the Dirac Oscillator, *Phys. Rev. Lett.* **64,** 1643, 2085.

Bennett H.S. and Stern E.A. (1965). Faraday Effects in Solids, *Phys. Rev.* **137,** A448.

Berestetskii V.B. Lifshitz E.M. and Pitaevskii L.P. (1982). *Landau and Lifshitz Course on Theoretical Physics Vol. 4: Quantum Electrodynamics* (Pergamon Press, Oxford, UK).

Bethe H.A. and Salpeter E.E. (1957). *Quantum Mechanics of One- and Two-Electron Systems* (Springer-Verlag, Berlin).

Bjorken J.D. and Drell S.D. (1964). *Relativistic Quantum Mechanics* (McGraw-Hill Book Company, New York).

Blume M. (1985). Magnetic Scattering of X-Rays, *J. Appl. Phys.* **57,** 3615.

Bogolubov N.N. (1958) *Nuovo Cimento* **7,** 794.

Bose S.K. Gamba G.A. and Sudarshan E.C.G. (1959). Representations of the Dirac Equation, *Phys. Rev.* **113,** 1661.

Breit G. (1929). The Effect of Retardation on the Interaction of Two Electrons, *Phys. Rev.* **34,** 553.

Breit G. (1930). The Fine Structure of He as a Test of the Spin Interactions of Two Electrons, *Phys. Rev.* **36,** 383.

Breit G. (1932). Dirac's equation for the Spin–Spin Interaction of Two Electrons, *Phys. Rev.* **39,** 616.

Brooks M.S.S. (1985). Calculated Ground State Properties of Light Actinide Metals and their Compounds, *Physica* **130B,** 6.

Brouder Ch. Alouani M. and Bennemann K.H. (1996). Multiple Scattering Theory of X-Ray magnetic Circular Dichroism: Implementation and Results for the Iron K-Edge, *Phys. Rev. B* **54,** 7334.

Brunel M. and de Bergevin F. (1981). Diffraction of X-Rays by Magnetic Materials 2: Measurements on Antiferromagnetic Fe_2O_3, *Acta Cryst.* **A37,** 325.

Cabrera B. and Peskin M.E. (1989). Cooper-Pair Mass, *Phys. Rev. B* **39,** 6425.

Callaway J. (1964). *Energy Band Theory* (Academic Press, New York).

Capelle K. and Gross E.K.U. (1995). Relativistic Theory of Superconductivity, *Phys. Lett. A* **198,** 261.

Capelle K. Griss E.K.U. and Györffy B.L. (1997). Theory of Dichroism in the Electromagnetic Response of Superconductors, *Phys. Rev. Lett.* **78,** 3753.

Carra P. Thole B.T. Altarelli M. and Wang X.-D. (1993). X-ray Circular Dichroism and Local Magnetic Fields, *Phys. Rev. Lett.* **70,** 694.

Chraplyvy Z.V. (1953*a*). Reduction of Relativistic Two-Particle Wave Equations to Approximate Forms I, *Phys. Rev.* **91,** 388.

Chraplyvy Z.V. (1953*b*). Reduction of Relativistic Two-Particle Wave Equations to Approximate Forms II, *Phys. Rev.* **92,** 1310.

Cini M. and Touschek B. (1958) *Nuovo Cimento* **7**, 422.

Compton A.H. (1923*a*). The Spectrum of Scattered X-Rays, *Phys. Rev.* **22,** 409.

Compton A.H. (1923*b*). A Quantum Theory of the Scattering of X-Rays by Light Elements, *Phys. Rev.* **21,** 483.

Cooper L.N. (1956). Bound Electron Pairs in a Degenerate Fermi Gas, *Phys. Rev.* **104,** 1189.

Cooper J.R.A. (1965). Electron Interaction Coefficients in Relativistic Self-Consistent Field Theory, *Proc. Phys. Soc.* **86,** 529.

Corinaldesi E. and Strocchi F. (1963). *Relativistic Wave Mechanics* (North Holland Publishing, Amsterdam).

Cortona P. Doniach S. and Sommers C. (1985). Relativistic Extension of the Spin-Polarised Local Density Functional Theory: Study of the Electronic and Magnetic Properties of the Rare Earth Ions, *Phys. Rev. A* **31,** 2842.

Costella J.P. and McKellar B.H.J. (1995). The Foldy–Wouthuysen Transformation, *Am. J. Phys.* **63,** 1119.

Coulthard M.A. (1967). A Relativistic Hartree–Fock Atomic Field Calculation, *Proc. Phys. Soc.* **91,** 44.

Daalderop G.H.O. Kelly P.J. and Schuurmans M.F.H. (1990). First Principles Calculation of the Magnetocrystalline Anisotropy Energy of Iron, Cobalt and Nickel, *Phys. Rev. B* **41,** 11919.

Daalderop G.H.O. Kelly P.J. and Schuurmans M.F.H. (1992). Magnetocrystalline Anisotropy and Orbital Moments in Transition Metal Compounds, *Phys. Rev. B* **44,** 12054.

Das M.P. Ramana M.V. and Rajagopal A.K. (1980). Self-Consistent Relativistic Density Functional Theory: Application to Neutral Uranium Atom and some Ions of Lithium Isoelectronic Sequence, *Phys. Rev. A* **22,** 9.

Davis H.L. (1971). Efficient Numerical Techniques for the Calculation of KKR Structure Constants, in *Computational Methods in Band Theory,* ed. P.M. Marcus, J.F. Janak and A.R. Williams (Plenum Press, New York).

de Bergevin F. and Brunel M. (1981). Diffraction of X-Rays by Magnetic Materials I: General Formulae and Measurements on Ferro and Ferrimagnetic Compounds, *Acta Cryst. A* **37,** 314.

de Gennes P. (1966). *Superconductivity of Metals and Alloys* (W.A.Benjamin, New York).

de Lange O. (1991). Algebraic Properties of the Dirac Oscillator, *J. Phys. A: Math. Gen.* **24,** 667.

Desclaux J.P. (1973) *Atom. Data. Nucl. Data Tables* **12**, 311.

Desclaux J.P. and Kim Y-K. (1975). Relativistic Effects in Outer Shells of Heavy Atoms, *J. Phys. B: Atom. Molec. Phys.* **8,** 1177.

Desclaux J.P. Mayers D.F. and O'Brien F (1971). Relativistic Atomic Wavefunctions, *J. Phys. B: Atom. Molec. Phys.* **4**, 631.

Desclaux J.P. and Pyykkö P (1974). Relativistic and Non-Relativistic Hartree–Fock One Centre Expansion Calculations for the Series CH_4 to PbH_4 within the Spherical Approximation, *Chem. Phys. Lett.* **29**, 534.

Desiderio A.M. and Johnson W.R. (1971). Lamb Shift and Binding Energies of K Electrons in Heavy Atoms, *Phys. Rev. A* **3**, 1267.

Dicke R.H. and Wittke J.P. (1974). *Introduction to Quantum Mechanics* (Addison-Wesley, Massachusetts).

Dillon J.F. (1977). Magneto-Optical Properties of Magnetic Crystals, in *Magnetic Properties of Materials (Inter University Electronics Series)*, ed. J. Smit (McGraw-Hill, New York).

Dirac P.A.M. (1928*a*) *Proc. Roy. Soc. (London) Series A*, **117**, 610.

Dirac P.A.M. (1928*b*) *Proc. Roy. Soc. (London) Series A*, **118**, 351.

Dirac P.A.M. (1958). *The Principles of Quantum Mechanics* (Clarendon Press, Oxford).

Dominguez-Adame F. (1989). A Generalized Dirac Kronig–Penney Model, *J. Phys.: Condens. Matter.* **1**, 109.

Dominguez-Adame F. and Gonzalez M.A. (1990). Solvable Linear Potentials in the Dirac Equation, *Europhysics Letters* **13**, 193.

Dominguez-Adame F. and Méndez B. (1991). A Solvable Two-Body Dirac Equation in One Space Dimension, *Can. J. Phys.* **69**, 780.

Drazin P.G. and Johnson R.S. (1989). *Solitons: An Introduction* (Cambridge University Press, Cambridge).

Duffin W.J. (1973). *Electricity and Magnetism, 2nd Edition* (McGraw-Hill, London).

Dzyaloshinsky I. (1958). A Thermodynamic Theory of "Weak" Ferromagnetism in Antiferromagnetics, *J. Phys. Chem. Solids* **4**, 241.

Ebert H. Strange P. and Györffy B.L. (1988*a*). The Influence of Relativistic Effects on the Magnetic Moments and Hyperfine Fields of Fe, Co and Ni, *J. Phys. F: Metal Phys.* **18**, L135.

Ebert H. Strange P. and Györffy B.L. (1988*b*). A Relativistic Theory of X-Ray Absorption by Spin-Polarized Targets, *Z. Phys. B* **73**, 77.

Ebert H. Strange P. and Györffy B.L. (1988*c*). Present Status of Magneto-Optical Effects, *Journal de Physique, Colloque C8* **49**, C8-31.

Ebert H. Drittler B. and Akai H. (1992). Spin-Polarized Relativistic Electronic Structure Calculations for Disordered Alloys using the CPA: Application to Fe_xCo_{1-x} and Co_xPt_{1-x}, *J. Magn. Magn. Mater.* **104-107**, 733.

Edmonds A.R. (1957). *Angular Momentum in Quantum Mechanics* (Princeton University Press, Princeton, New Jersey).

Einstein A. (1905). *Ann. Physik* **17**, 891; translated by W. Perrett and G.B. Jeffrey (1923), in *The Principle of Relativity* (Dover, New York).

Eisberg R. and Resnick R. (1985). *Quantum Physics of Atoms, Molecules, Solids, Nuclei and Particles, 2nd Edition* (John Wiley and Sons, New York).

Englisch H. and Englisch R. (1983). Hohenberg–Kohn Theorem and Non-V-Representable Densities, *Physica* **121A,** 253.

Erhard S. and Gross E.K.U. (1996). Scaling and Virial Theorems in Current Density Functional Theory, *Phys. Rev. A* **53,** R5.

Eriksson O. Brooks M.S.S. and Johansson B. (1990*a*). Orbital Polarization in Narrow Band Systems: Application to Volume Collapses in Light Lanthanides, *Phys. Rev. B* **41,** 7311.

Eriksson O. Brooks M.S.S. and Johansson B. (1990*b*). Theoretical Aspects of the Magnetism in the Ferromagnetic AFe$_2$ Systems (A = U, Np, Pu, and Am), *Phys. Rev. B* **41,** 9087.

Eschrig H. Seifert G. and Zeische P. (1985). Current Density Functional Theory of Quantum Electrodynamics, *Solid State Communications* **56,** 777.

Fairbairn W.M. Glasser M.L. and Steslicka M. (1973). Relativistic Theory of Surface States, *Surf. Sci.* **36,** 462.

Fairbairn W.M. Sharland A.J. and Strange P. (1979). Anisotropic Effects from Spin Split Bands, *Journal de Physique, Colloque C5* **40,** C5.81.

Faulkner J.S. (1977). Scattering Theory and Cluster Calculations, *J. Phys. C: Solid State* **10,** 4661.

Faulkner J.S. (1979). Multiple Scattering Approach to Band Theory, *Phys. Rev. B* **19,** 6186.

Faulkner J.S. and Stocks G.M. (1980). Calculating Properties with the Coherent Potential Approximation, *Phys. Rev. B* **21,** 3222.

Feder R. Rosicky F. and Ackerman B. (1983). *Z. Phys. B* **52** 31.

Feshbach H. and Villars F. (1958). Elementary Relativistic Wave Mechanics of Spin 0 and Spin 1/2 Particles, *Rev. Mod. Phys* **30,** 24.

Feynman R.P. (1949) *Phys. Rev.* **76,** 749.

Feynman R.P. (1962). *Quantum Electrodynamics* (W.A. Benjamin, New York).

Feynman R.P. and Gell-mann M. (1958). Theory of the Fermi Interaction, *Phys. Rev.* **109,** 193.

Foldy L. and Wouthuysen S. (1950). On the Dirac Theory of Spin 1/2 Particles and its Non-Relativistic Limit, *Phys. Rev.* **78,** 29.

Freeman A.J. Fu C.L. Ohnishi S. and Weinert M. (1985). Electronic and Magnetic Structure of Solid Surfaces, in *Polarized Electrons in Surface Physics,* ed. R. Feder (World Scientific, Singapore).

French A.P. (1968). *Special Relativity* (Thomas Nelson and Sons, London).

Fritsche L. Noffke J and Eckardt H. (1987). A Relativistic Treatment of Interacting Spin-Aligned Electron Systems – Application to Ferromagnetic Iron, Nickel and Palladium Metal, *J. Phys. F: Met. Phys.* **17,** 943.

Froese-Fischer C. (1973) *Atom. Data Nucl. Data Tables* **12,** 87.

Fröhlich H. (1950). Theory of the Superconducting State 1. The Ground State at the Absolute Zero of Temperature, *Phys. Rev.* **79,** 845.

Gasiorowicz S. (1974). *Quantum Physics* (John Wiley and Sons, New York).

Geller M.R. and Vignale G. (1994). Currents in the Compressible and Incompressible Region of the Two-Dimensional Electron Gas, *Phys. Rev. B.* **50,** 11714.

Gell-mann M. and Goldberger M.L. (1954). Scattering of Low Energy Photons by Particles of Spin 1/2, *Phys. Rev.* **96**, 1433.

Gibbs D. (1992). Resonant X-Ray Magnetic Scattering in Holmium, *J. Magn. Magn. Mater.* **104**, 1489.

Gibbs D. (1993). X-Ray Magnetic Scattering–New Developments, *J. Appl. Phys.* **73**, 6883.

Gibbs D. Grubel G. Harshman D.R. Isaacs E.D. McWhan D.B. Mills D. and Vettier C. (1991). Polarization and Resonance Studies of X-Ray Magnetic Scattering in Holmium, *Phys. Rev. B* **43**, 5663.

Ginatempo B. and Staunton J.B. (1988). The Electronic Structure of Disordered Alloys Containing Heavy Elements – An Improved Calculational Method Illustrated by a Study of Copper-Gold Alloy, *J. Phys. F: Metal. Phys.* **18**, 1827.

Glasser M.L. and Davison S.G. (1970). Analytic Solution of the Dirac Equation for the Kronig-Penney Potential, *Int. J. Quant. Chem.* **35**, 867.

Gordon W. (1926), *Z. Physik* **40**, 117.

Gordon W. (1928), *Z. Physik* **50**, 630.

Gorter C.J. (1964). Superconductivity until 1940, in Leiden, and as Seen from There, *Rev. Mod. Phys.* **36**, 1.

Gotsis H.J. Strange P. and Staunton J.B. (1994). Relativistic Spin-Polarized Electronic Structure Calculations for Random Substitutional Alloys, *Solid State Commun.* **92**, 449.

Gradshteyn I.S. and Ryzhik I.M (1980). *Table of Integrals, Series and Products, 4th Edition* (Academic Press, London).

Grant I.P. (1961). Relativistic Self-Consistent Fields, *Proc. Roy. Soc. (London), Series A* **262**, 555.

Grant I.P. (1965). Relativistic Self-Consistent Fields, *Proc. Phys. Soc.* **86**, 523.

Grant I.P. (1970). Relativistic Calculation of Atomic Structures, *Advances in Physics* **19**, 747.

Grant I.P. (1986). Variational Methods for Dirac Wave Equations, *J. Phys. B: Atom. Molec. Phys.* **19**, 3187.

Greiner W. (1990). *Relativistic Quantum Mechanics Wave Equations* (Springer-Verlag, Berlin).

Gunnarsson O. (1979). The Density Functional Theory of Metallic Surfaces, in *NATO ASI Series B Vol. 42, Electrons in Disordered Metals and at Metallic Surfaces,* ed. P. Phariseau, B.L. Györffy and L. Scheire (Plenum Press, New York).

Györffy B.L. and Stott M.J. (1972), in *Band Structure Spectroscopy of Metals and Alloys,* ed. D.J. Fabian and L.M. Watson (Academic Press, New York).

Györffy B.L. and Stocks G.M. (1979). First Principles Band Theory for Random Metallic Alloys, in *Electrons in Disordered Metals and at Metallic Surfaces, NATO ASI Series B, Vol. 42,* ed. P. Phariseau, B.L. Györffy and L. Scheire (Plenum Press, New York).

Györffy B.L. Staunton J.B. Ebert H. Strange P. and Ginatempo B. (1991). Relativistic Density Functional Theory for Electrons in Solids, in *The Effects of Relativity in Atoms, Molecules and the Solid State,* ed. S. Wilson, I.P. Grant and B.L. Györffy (Plenum Press, New York).

Hall G.G. (1951). The Molecular Orbital Theory of Chemical Valency VIII, A Method of Calculating Ionization Potentials, *Proc. Roy. Soc. (London) Series A* **205**, 541.

Harris J and Jones R.O. (1974). The Surface Energy of a Bounded Electron gas, *J. Phys. F: Metal Phys.* **4**, 1170.

Hartley A.C. and Sandars P.G.H. (1991). Relativistic Calculations of Parity Non-Conserving Effects in Atoms, in *The Effects of Relativity in Atoms, Molecules and the Solid State,* ed. S. Wilson, I.P. Grant and B.L. Györffy (Plenum Press, New York).

Hedin L. and Lundqvist B.I. (1971). Explicit Local Exchange–Correlation Potentials, *J. Phys. C: Solid State Phys* **4**, 2064.

Herman F. and Skillman S. (1963). *Atomic Structure Calculations* (Prentice-Hall, Englewood Cliffs, New Jersey).

Hohenberg P. and Kohn W. (1964). Inhomogeneous Electron Gas, *Phys. Rev.* **136B**, 864.

Holm P. (1988). Relativistic Compton Cross Section for General Central-Field Hartree–Fock Wavefunctions, *Phys. Rev. A* **37**, 3706.

Inkson J.C. (1983). *Many-Body Theory of Solids* (Plenum Press, New York).

Ito D. Mori K. and Carrieri E. (1967). *Nuovo Cimento* **51A**, 1119.

Itzykson C. and Zuber J.B. (1980). *Quantum Field Theory* (McGraw-Hill, New York).

Jackson J.D. (1962). *Classical Electrodynamics* (John Wiley, New York).

Jansen H.J.F. (1988). Magnetic Anisotropy in Density Functional Theory, *Phys. Rev. B* **38**, 8022.

Jenkins A.C. and Strange P. (1993). Magnetic Dichroism in the One-Electron Atom, *European Journal of Physics* **14**, 80.

Jenkins A.C. and Strange P. (1994). Relativistic Spin-Polarized Single-Site Scattering Theory, *J. Phys.: Condens. Matter* **6**, 3499.

Jenkins A.C. and Strange P. (1995). Electronic Structure and X-Ray Magnetic Dichroism in Random Substitutional Alloys of f-Electron Elements, *Phys. Rev. B* **51**, 7279.

Kagawa T. (1975). Relativistic Hartree–Fock–Roothan Theory for Open Shell Atoms, *Phys. Rev. A* **12**, 2245.

Kagawa T. (1980). Multiconfigurational Relativistic Hartree–Fock–Roothan Theory for Atomic Systems, *Phys. Rev. A* **22**, 2340.

Kammerlingh Onnes H. (1911). *Leiden. Comm.* **122b, 124c**.

Karrai K. Choi E. Dunmore F. Liu S. Ying X. Qi.Li. Venkatesan T. Drew H.D. and Fenner D.B. (1992). Far-Infrared Magneto-Optical Activity in Type *II* Superconductors, *Phys. Rev. Lett.* **69**, 355.

Kasuya T. (1956). A Theory of Metallic Ferro- and Antiferromagnetism on Zener's Model, *Prog. Theor. Phys.* **16**, 45.

Kenny S.D. Rajagopal G. Needs R.J. Leung W.-K. Godfrey M.J. Williamson A.J. and Foulkes W.M.C. (1996). Quantum Monte Carlo Calculations of the Energy of the Relativistic Homogeneous Electron Gas, *Phys. Rev. Lett.* **77,** 1099.

Kim Y.K. (1967). Relativistic Self-Consistent Field Theory for Closed Shell Atoms, *Phys. Rev.* **154,** 17.

Kittel C. (1963). *Quantum Theory of Solids* (John Wiley and Sons, New York).

Kittel C. (1986). *Introduction to Solid State Physics* (John Wiley and Sons, New York).

Kittel C. Knight W.D. Ruderman M.A. Helmholz A.C. and Moyer B.J. (1973). *Berkeley Physics Course Vol. 1, Mechanics* (McGraw-Hill, New York).

Klein O. (1927). *Z. Physik,* **41,** 407.

Klein O. (1929). *Z. Physik,* **53,** 157.

Klein O. and Nishina Y. (1929). *Z. Physik,* **52,** 853.

Kohn W. and Rostoker N. (1954). Solution of the Schrödinger Equation in Periodic Lattices with an Application to Metallic Lithium, *Phys. Rev.* **94,** 1111.

Kohn W. and Sham L.J. (1965). Self-Consistent Equations Including Exchange and Correlation Effects, *Phys. Rev.* **140,** A1133.

Kohn W. and Vashishta P. (1982). General Density Functional Theory, in *Theory of the Inhomogeneous Electron Gas,* ed. S. Lundqvist and N.H. March (Plenum Press, New York).

Koopmans T.A. (1933). *Physica* **1,** 104.

Korringa J. (1947). On the Calculation of the Energy of a Bloch Wave in a Metal, *Physica* **13,** 392.

Korringa J. (1958). Dispersion Theory for Electrons in a Random Lattice with Applications to the Electronic Structure of Alloys, *J. Phys. Chem. Solids* **7,** 252.

Kursunoglu B. (1956). Transformation of Relativistic Wave Equations, *Phys. Rev.* **101,** 149.

Kutzelnigg W. (1984). Basis Set Expansion of the Dirac Operator without Variational Collapse, *Int. J. Quant. Chem.* **25,** 107.

Lam L. (1970*a*). New Exact Solutions of the Dirac Equation, *Can. J. Phys.* **48,** 1935.

Lam L. (1970*b*). Dirac Electron in Parallel Electric and Magnetic Fields, *Phys. Lett.* **31A,** 406.

Lamb W.E. and Retherford R.C. (1947). *Phys. Rev.* **72,** 241.

Landau L.D. and Lifschitz E.M (1982). *Quantum Electrodynamics, 2nd Edition* (Pergamon Press, Oxford).

Landau L.D. and Lifschitz E.M (1977). *Quantum Mechanics (Non-Relativistic Theory), 3rd Edition* (Pergamon Press, Oxford).

Lander G.H. Aldred A.T. Dunlap B.D. and Shenoy G.K. (1977) *Physica* **86-88B,** 152.

Langridge S. Stirling W.G. Lander G.H. and Rebizant J. (1994*a*). Resonant Magnetic X-Ray Scattering Studies of NpAs. 1: Magnetic and Lattice Structure, *Phys. Rev. B* **49,** 12010.

Langridge S. Stirling W.G. Lander G.H. and Rebizant J. (1994*b*). Resonant Magnetic X-Ray Scattering Studies of NpAs. 2: The Critical Regime, *Phys. Rev. B* **49,** 12022.

Layzer D. and Bachall J.N. (1962). *Ann. Phys. N.Y.* **8,** 271.

Lee T.D. and Yang C.N. (1957). *Phys. Rev.* **105,** 167.

Levy M. (1982). Electron Densities in Search of Hamiltonians, *Phys. Rev. A* **26,** 1200.

Liberman D. Waber J.T. and Cromer D.T. (1965). Self-Consistent Field Dirac–Slater Wavefunctions for Atoms and Ions I: Comparison with Previous Calculations, *Phys. Rev.* **137,** A27.

Lloyd P. and Smith P.V. (1972). Multiple Scattering Theory in Condensed Materials, *Adv. in Phys.* **21,** 69.

London F. and London H. (1935). The Electrodynamic Equations of the Supraconductor, *Proc. Roy. Soc. (London) Series A* **149,** 71.

Loucks T.L. (1967). *The Augmented Plane Wave Method* (Benjamin, New York).

Lovesey S.W. (1993). Photon Scattering by Magnetic Solids, *Reports on Progress in Physics* **56,** 257.

Low F.E. (1954). Scattering of Light of Very Low Frequency by Systems of Spin 1/2, *Phys. Rev.* **96,** 1428.

Löwdin P.-O. (1955). *Phys. Rev.* **97,** 1490.

Ludders R. (1954). *Kgl. Dansk. Vid. Sels. Mat.-Fys. Medd* **28**, no 5.

Lyons K.B. Kwo J. Dillon J.F. Espinosa G.P. McGlashen-Powell M. Ramirez A.P. and Schneemeyer L.F. (1990). Search for Circular Dichroism in High-T_c Superconductors, *Phys. Rev. Lett* **64,** 2949.

Lyons K.B. Dillon J.F. Hellman E.S. Hartford E.H. and McGlashen-Powell M. (1991). Circular Dichroism Observed in Bismuthate Sperconductors, *Phys. Rev. B* **43,** 11408.

MacDonald A.H. (1983). Spin-Polarised Relativistic Exchange Energies and Potentials, *J. Phys. C: Solid State Phys.* **16,** 3869.

MacDonald A.H. and Vosko S.H. (1979). A Relativistic Density Functional Formalism, *J. Phys. C: Solid State Phys.* **12,** 2977.

MacDonald A.H. Daams J.M. Vosko S.H. and Koelling D.D. (1981). Influence of Relativistic Contributions to the Effective Potential on the Electronic Structure of Pd and Pt, *Phys. Rev. B* **23,** 6377.

MacDonald A.H. Daams J.M. Vosko S.H. and Koelling D.D. (1982). Non-Muffin-Tin and Relativistic Interaction Effects on the Electronic Structure of Noble Metals, *Phys. Rev. B* **25,** 713.

Malli G. and Oreg J. (1975). Relativistic Self-Consistent Field (RSCF) Theory for Closed Shell Molecules, *J. Chem. Phys.* **63,** 830.

Mandl F. and Shaw G. (1984). *Quantum Field Theory* (John Wiley and Sons, Chichester).

Mann J.B. and Waber J.T. (1970). Self-Consistent Field Relativistic Hartree–Fock Calculations on the Superheavy Elements 118, *J. Chem. Phys.* **53,** 2397.

Mayers D.F. and O'Brien F. (1968). The Calculation of Atomic Wavefunctions, *J. Phys. B* **1,** 145.

McKellar B.H.J. and Stephenson G.J. (1987). Klein Paradox and the Dirac–Kronig–Penney Model, *Phys. Rev. A* **36**, 2566.

McMurry S.M. (1993). *Quantum Mechanics* (Addison-Wesley, Wokingham).

Mermin N.D. (1965). Thermal Properties of the Inhomogeneous Electron Gas, *Phys. Rev.* **137**, A1441.

Merzbacher E. (1970). *Quantum Mechanics* (John Wiley and Sons, New York).

Messiah A. (1965). *Quantum Mechanics Vol II* (Interscience, New York).

Michelson A.A. and Morley E.W. (1887). *Am. J. Sci.* **134**, 33.

Moller C. (1931). *Z. Physik* **70**, 786.

Moller C. (1932). *Ann. Physik* **14**, 531.

Moriya T. (1960). New Mechanisms of Anisotropic Superexchange Interaction, *Phys. Rev. Lett.* **4**, 228.

Moruzzi V.L. Janak J.F. and Williams A.R. (1978). *Calculated Electronic Properties of Metals* (Pergamon Press, New York).

Moshinsky. M and Szczepaniak A. (1989). The Dirac Oscillator, *J. Phys. A* **22**, L817.

Nagle J.K. Balch A.L. and Olmstead M.M. (1988). $Tl_2Pt(CN)_4$: A Non-Columnar, Luminescent Form of $Pt(CN)_4$ Containing $Pt - Tl$ Bonds, *J. Am. Chem. Soc.* **110**, 319.

Narita A. and Kasuya T. (1984). Effects of Band Structure and Matrix Element on RKKY Interaction, *J. Magn. Magn. Materials* **43**, 21.

Newton R.G. (1966). *Scattering Theory of Waves and Particles* (McGraw-Hill, New York).

Newton T.D. and Wigner E.P. (1949). Localized States for Elementary Systems, *Rev. Mod. Phys.* **21**, 400.

Nogami Y. and Toyama F.M. (1992). Transparent Potential for the One-Dimensional Dirac Equation, *Phys. Rev. A* **45**, 5258.

Nogami Y. Toyama F.M. and Zhao Z. (1995). Nonlinear Dirac Soliton in an External Field, *J. Phys. A* **28**, 1413.

Oliveira I.N. Gross E.K.U. and Kohn W. (1988). Density Functional Theory for Superconductors, *Phys. Rev. Lett.* **60**, 2430.

Onodera Y. and Okazaki M. (1966). Relativistic Theory for Energy Band Calculation, *J. Phys. Soc. Japan* **21**, 1273.

Pac P.Y. (1959*a*). On the Transformation Properties of the Dirac Equation, *Progr. Theor. Phys.* **21**, 640.

Pac P.Y. (1959*b*). Remark on the Transformation Properties of the Dirac Equation, *Progr. Theor. Phys.* **22**, 857.

Pauli W. (1927). *Z. Physik* **43**, 601.

Pershan P.S. (1967). Magneto-Optical Effects, *J. Appl. Phys.* **38**, 1482.

Pettifor D. (1995). *Bonding and Structure of Molecules and Solids* (Clarendon Press, Oxford).

Platzman P.M. and Tzoar N. (1970). Magnetic Scattering of X-Rays from Electrons in Molecules and Solids, *Phys. Rev. B* **2**, 3556.

Pyykkö P. (1978). Relativistic Quantum Chemistry, *Adv. in Quant. Chem.* **11**, 353.

Pyykkö P. (1991). Relativistic Effects on Periodic Trends, in *The Effects of Relativity in Atoms, Molecules and the Solid State,* ed. S. Wilson, I.P. Grant and B.L. Györffy (Plenum Press, New York).

Racah G. (1942). Theory of Complex Spectra: II, *Phys. Rev.* **62,** 438.

Rajagopal A.K. (1978). Inhomogeneous Relativistic Electron Gas, *J. Phys. C: Solid State Phys.* **11,** L943.

Rajagopal A.K. (1980). Theory of Inhomogeneous Electron Systems: Spin Density Functional Formalism, in *Advances in Chemical Physics,* ed. I. Prigogine and S.A. Rice (John Wiley and Sons, New York).

Rajagopal A.K. (1994). Time Dependent Functional Theory of Coupled Electron and Electromagnetic Fields in Condensed Matter Systems, *Phys. Rev. A* **50,** 3759.

Rajagopal A.K. and Callaway J. (1973). Inhomogeneous Electron Gas, *Phys. Rev. B* **7,** 1912.

Ramana M.V. and Rajagopal A.K. (1979). Relativistic Spin-Polarised Electron Gas, *J. Phys. C: Solid State Phys.* **12,** L845.

Ramana M.V. and Rajagopal A.K. (1981). Inhomogeneous Relativistic Electron Gas: Correlation Potential, *Phys. Rev. A* **24,** 1689.

Ramana M.V. and Rajagopal A.K. (1983). Inhomogeneous Relativistic Electron Systems: A Density Functional Formalism, *Adv. Chem. Phys.* **54,** 231.

Ramana M.V. Rajagopal A.K. and Johnson W.R. (1982). Effects of Correlation and Breit and Transverse Interactions in the Relativistic Local Density Theory for Atoms, *Phys. Rev. A* **25,** 96.

Reitz J.R. and Milford F.J. (1972). *Foundations of Electromagnetic Theory* (Addison-Wesley, Massachusetts).

Rickayzen G. (1980). *Green's Functions and Condensed Matter* (Academic Press, London).

Roothan C.J. (1951). New Developments in Molecular Orbital Theory, *Rev. Mod. Phys.* **23,** 69.

Rose M.E. (1957). *Elementary Theory of Angular Momentum* (John Wiley, London).

Rose M.E. (1961). *Relativistic Electron Theory* (John Wiley, New York).

Ruderman M.A. and Kittel C. (1954). Indirect Exchange Coupling of Nuclear Magnetic Moments by Conduction Electrons, *Phys. Rev.* **96,** 99.

Saint-James D. Thomas E.J. and Sarma G. (1969). *Type II Superconductivity* (Pergamon Press, Oxford).

Sakurai J.J. (1967). *Advanced Quantum Mechanics* (Addison-Wesley, Massachusetts).

Salpeter E.E. and Bethe H.A. (1951). A Relativistic Equation for Bound State Problems, *Phys. Rev.* **84,** 1232.

Saxon D.S. and Hutner R.A. (1949). Some Electronic Properties of a One-Dimensional Crystal Model, *Phillips Research Reports* **4,** 81.

Schrödinger E. (1926). *Ann. Physik,* **81,** 109.

Schütz G. Wagner W. Wilhelm W. Kienle P. Zeller R. Frahm H. and Materlik C. (1987). Absorption of Circularly Polarized X-Rays in Iron, *Phys. Rev. Lett.* **58**, 737.

Schwinger J. (1948). Quantum Electrodynamics. I. A Covariant Formulation, *Phys. Rev.* **74**, 1439.

Schwinger J. (1949). Quantum Electrodynamics. II. Vacuum Polarization and Self-Energy, *Phys. Rev.* **75**, 651.

Sen Gupta N.D. (1974). On the Relativistic Electronic States of a Monatomic and a Diatomic One-Dimensional Lattice, *Phys. Stat. Sol.* **65**, 351.

Sewell M.J. (1987). *Maximum and Minimum Principles* (Cambridge University Press, Cambridge, UK).

Sham L.J. and Kohn W. (1966). One-Particle Properties of an Inhomogeneous Interacting Electron Gas, *Phys. Rev.* **145**, 561.

Shankland R.S. McCuskey S.W. Leone F.C. and Kuerti G. (1955). New Analysis of the Interferometer Observations of Dayton C. Miller, *Rev. Mod. Phys.* **27**, 167.

Shen A.P. (1974). Relativistic Green's Function Electron Diffraction Theory, *Phys. Rev. B* **9**, 1328.

Siddons D.P. Hart M. Amemiya Y. and Hastings J.B. (1990). X-Ray Optical Activity and the Faraday Effect in Cobalt and its Compounds, *Phys. Rev. Lett.* **64**, 1967.

Slater J.C. (1960). *Quantum Theory of Atomic Structure Vols I and II* (McGraw-Hill, New York).

Slater J.C. (1963). *Quantum Theory of Molecules and Solids* (McGraw-Hill, New York).

Smith F.C. and Johnson W.R. (1967). *Phys. Rev.* **160**, 136.

Soven P. (1967). Coherent Potential Model of Substitutional Disordered Alloys, *Phys. Rev.* **156**, 809.

Stahlhofen A.A. (1994). Supertransparent Potentials for the Dirac Equation , *J. Phys. A* **27**, 8279.

Staunton J.B. Györffy B.L. and Weinberger P. (1980). On the Electronic Structure of Random Metallic Alloys Containing Heavy Elements: A Relativistic Theory, *J. Phys. F: Metal Phys.* **10**, 2665.

Staunton J.B. Poulter J. Györffy B.L. and Strange P. (1988). A Relativistic RKKY Interaction Between Two Magnetic Impurities: The Origin of a Magnetic Anisotropic Effect, *J. Phys. C: Solid State Phys.* **21**, 1595.

Staunton J.B. Györffy B.L. Poulter J. and Strange P. (1989). The Relativistic RKKY Interaction, Uniaxial and Unidirectional Magnetic Anisotropies and Spin Glasses, *J. Phys. Condens. Matter* **1**, 5157.

Stocks G.M. Williams R.W. and Faulkner J.S. (1971). Densities of States of Paramagnetic Cu–Ni Alloys, *Phys. Rev. B* **4**, 4390.

Stocks G.M. Györffy B.L. Guilliano E.S. and Ruggeri R. (1977). The Coherent Potential Approximation for Nonoverlapping Muffin Tin Potentials: Paramagnetic Ni_xCu_{1-x}, *J. Phys. F: Metal Phys.* **7**, 1859.

Stocks G.M. Temmerman W.M. and Györffy B.L. (1978). Complete Solution of the Korringa–Kohn–Rostoker Coherent Potential Approximation Equations: Cu–Ni Alloys, *Phys. Rev. Lett.* **41**, 339.

Stocks G.M. and Winter H. (1984). A First Principles Approach to the Band Theory of Random Metallic Alloys, in *The Electronic Structure of Complex Systems, NATO ASI Series B, Vol. 113,* ed. P. Phariseau and W.M. Temmerman (Plenum Press, New York).

Strange P. (1994). Magnetic Absorption Dichroism and Sum Rules in Itinerant Magnets, *J. Phys.: Condens. Matter* **6**, L491.

Strange P. Ebert H. Staunton J.B. and Györffy B.L. (1989*a*). A Relativistic Spin-Polarized Multiple Scattering Theory, with Applications to the Calculation of the Electronic Structure of Condensed Matter , *J. Phys.: Condens. Matter* **1**, 2959.

Strange P. Ebert H. Staunton J.B. and Györffy B.L. (1989*b*). A First Principles Theory of Magnetocrystalline Anisotropy in Metals, *J. Phys.: Condens. Matter* **1**, 3947.

Strange P. and Györffy B.L. (1995). Interpretation of X-Ray Absorption Dichroism Experiments, *Phys. Rev. B* **52**, R13091.

Strange P. Staunton J.B. and Ebert H. (1989*c*). A First Principles Theory of Magnetocrystalline Anisotropy in Metals – Prediction of a Flip in the Easy Axis of Magnetization, *Europhys. Lett.* **9**, 169.

Strange P. Staunton J.B. and Györffy B.L. (1984). Relativistic Spin-Polarized Scattering Theory – Solution of the Single Site Problem, *J. Phys. C: Solid State Phys.* **17**, 3355.

Strange P. Staunton J.B. Györffy B.L. and Ebert H. (1991). First Principles Theory of Magnetocrystalline Anisotropy, *Physica B* **172**, 51.

Stückelberg E.C.G. (1941). Creation of Particle Pairs on the Relativity Theory, *Helv. Phys. Acta* **14**, 588.

Su R. and Zhang Y. (1984). Exact Solutions of the Dirac Equation with a Linear Scaling Confining Potential in a Uniform Electric Field, *J. Phys. A: Math. Gen.* **17**, 851.

Subramanian R. and Bhagwat K.V. (1971). Relativistic Generalization of the Saxon–Hutner Theorem, *Phys. Stat. Sol.* **48**, 399.

Sutherland B and Mattis D.C. (1981). Ambiguities with the Relativistic δ-Function Potential, *Phys. Rev. A* **24**, 1194.

Swamy N.V.V.J. (1969). *Phys. Rev.* **180**, 1225.

Swirles B. (1935). The Relativistic Self-Consistent Field, *Proc. Roy. Soc (London)* **152**, 625.

Swirles B. (1936). The Relativistic Interaction of Two Electrons in the Self-Consistent Field Method, *Proc. Roy. Soc (London)* **157**, 680.

Temmerman W.M. Györffy B.L. and Stocks G.M. (1978). The Atomic Sphere Approximation to the KKR–CPA: Electronic Structure of $Cu_c Ni_{1-c}$ Alloys, *J. Phys. F: Metal Phys* **8**, 2461.

Thaller B. (1992). *The Dirac Equation* (Springer, Berlin).

Thole B.T. Carra P. Sette F. and Van der Laan G. (1992). X-Ray Circular Dichroism as a Probe of Orbital Magnetization, *Phys. Rev. Lett.* **68**, 1943.

Tilley D.R. and Tilley J. (1990). *Superfluidity and Superconductivity* (Adam Hilger, IOP Publishing Ltd, Bristol, UK).

Tinkham M. (1964). *Group Theory and Quantum Mechanics* (McGraw-Hill, New York).

Tinkham M. (1980). *Introduction to Superconductivity* (Kreiger, New York).

Tomonaga S. (1946). *Progr. Theor. Phys. (Kyoto)* **1**, 27.

Valatin J.G. (1958). *Nuovo Cimento* **7**, 843.

Vignale G. (1993). Current Density Functional Theory of the Two-Dimensional Wigner Crystal in a Strong Magnetic Field, *Phys. Rev. B* **47**, 10105.

Vignale G. and Rasolt M. (1987). Density Functional theory in Strong Magnetic Fields, *Phys. Rev. Lett.* **59**, 2360.

Vignale G. and Rasolt M. (1988). Current and Spin Density Functional Theory for Inhomogeneous Electronic Systems in Strong Magnetic Fields, *Phys. Rev. B* **37**, 10685.

Vignale G. Skudlarski P. and Rasolt M. (1992). Current Density Functional Theory of the Surface Properties of Electron–Hole Droplets in a Strong Magnetic Field, *Phys. Rev. B* **45**, 8494.

Von Barth U. (1983). Density Functional Theory for Solids, in *NATO ASI Series B Vol. 113, The Electronic Structure of Complex Systems,* ed. P. Phariseau and W.M. Temmerman (Plenum Press, New York).

Von Barth U. and Hedin L. (1972). A Local Exchange–Correlation Potential for the Spin-Polarized Case, *J. Phys. C: Solid State Phys.* **5**, 1629.

Wan F.Y.M. (1993). *Introduction to the Calculus of Variations and its Applications* (Chapman and Hall, New York).

Wang C.S. and Callaway J. (1974). Band Structure of Nickel: Spin–Orbit Coupling, The Fermi Surface and Optical Conductivity, *Phys. Rev. B* **9**, 4897.

Watson R.E. and Freeman A.J. (1966). Exchange Coupling and Conduction Electron Polarization in Metals, *Phys. Rev.* **152**, 566.

Weber H.J. Weitbrecht D. Brach D. Shelankov A.L. Keiter H. Weber W. Wolf Th. Geerk J. Linker G. Roth G. Splittgerber-Hünnekes P.C. and Guntherodt G. (1990). Evidence for Broken Time-Reversal Symmetry in Cuprate Superconductors, *Solid State Communications* **76**, 511.

Weinberger P. (1990). *Electron Scattering Theory for Ordered and Disordered Matter* (Oxford University Press, Oxford).

Welton T. (1948). *Phys. Rev.* **74**, 1157.

West D. (1958). Mesonic Atoms, *Rep. Progr. Phys* **21**, 271.

Wichman E.H. (1971). *Berkeley Physics Course Volume 4, Quantum Physics* (McGraw-Hill, New York).

Wigner E.P. (1931). *Gruppentheorie* (Friedrich Vieweg und Sohn, Braunschweig).

Wigner E.P. (1932). *Göttinger Nachrichten* **31**, 546.

Wilson M.T. and Györffy B.L. (1995). Auxiliary Field Quantum Monte Carlo Calculations for the Relativistic Electron Gas, *J. Phys: Condens. Matter* **7**, 1565.

Wilson S. Grant I.P. and Györffy B.L. eds. (1991). *The Effect of Relativity in Atoms, Molecules and the Solid State* (Plenum Press, New York).

Wu R. and Freeman A.J. (1994). Limitation of the Magnetic Circular Dichroism Spin Sum Rule for Transition Metals and the Importance of the Magnetic Dipole Term, *Phys. Rev. Lett.* **73**, 1994.

Wu R. Wang D. and Freeman A.J. (1993). First Principles Investigation of the Validity and Range of Applicability of the X-Ray Magnetic Circular Dichroism Sum Rule, *Phys. Rev. Lett.* **71**, 3581.

Xing Xu B. Rajagopal A.K. and Ramana M.V. (1984). Theory of Spin-Polarised Inhomogeneous Relativistic Electron Systems: II, *J. Phys. C.: Solid State Phys.* **17**, 1339.

Yosida K. (1957). Magnetic Properties of Cu–Mn Alloys, *Phys. Rev.* **106**, 893.

Index

585